SCHAUM'S SOLVED PROBLEMS SERIES

2000 SOLVED PROBLEMS IN

NUMERICAL ANALYSIS

by

Francis Scheid, Ph.D

Boston University

McGRAW-HILL PUBLISHING COMPANY

New York   St. Louis   San Francisco   Auckland   Bogotá   Caracas
Hamburg   Lisbon   London   Madrid   Mexico   Milan   Montreal
New Delhi   Oklahoma City   Paris   San Juan   São Paulo
Singapore   Sydney   Tokyo   Toronto

FRANCIS SCHEID is Emeritus Professor of Mathematics at Boston University where he has been a faculty member since receiving his doctorate from MIT in 1948, serving as department chairman from 1956 to 1968. From 1961 to 1962 he was Fulbright Professor at Rangoon University in Burma. Professor Scheid has lectured extensively on educational television and the videotapes are used by the U.S. Navy. His research now centers on computer simulations in golf. Among his publications are the Schaum's Outlines of *Numerical Analysis*, *Computer Science* and *Computers and Programming*.

Project supervision by Science Typographers, Inc.

Library of Congress Cataloging-in-Publication Data

Scheid, Francis J.
    2000 solved problems in numerical analysis / by Francis Scheid.
        p.   cm. --(Schaum's solved problems series)
    Includes index.
    ISBN 0-07-055233-9
    1. Numerical analysis—Problems, exercises, etc.   I. Title.
II. Series.
QA297.3.S34   1990
519.4′076—dc20
ISBN 0-07-055233-9

1 2 3 4 5 6 7 8 9 0 SHP/SHP 8 9 2 1 0 9

CONTENTS

TO THE STUDENT

One of the great attractions of mathematics, to its devotees, may be that a well posed problem does in fact have a unique and exact solution. This idea of absolute precision has a sort of beauty that in its own way is unsurpassable. It may also be what frightens away many who feel more comfortable where there are shades of gray. In numerical analysis we enjoy a little bit of both worlds but are much closer to the former than the latter. The perfect solution is definitely out there somewhere in thought space, and knowledge of this fact has sustained many a wearisome effort to get close to it. In what direction does it lie? How close to it are we now?

In the year 1225 Leonardo of Pisa studied the equation

$$x^3 + 2x^2 + 10x - 20 = 0$$

seeking its one real root, and produced $x = 1.368808107$. Nobody knows his method, but it is a remarkable achievement for his time. Leonardo surely knew it was not perfection and may have wondered at the true identity of the target number, but must have derived great pleasure from realizing how close he had come.

Numerical analysis problems do have exact solutions, but the thrill of victory does not wait for their discovery. It is enough to come close. Error is expected. Without it we would be out of business. There is no need of approximation where the real thing is within grasp. The following problems illustrate mathematics with some controlled shades of gray. It has been a pleasure to work them through. I was almost sorry to come to the 2000th. Though error is the substance of our subject, mistakes and blunders are not. I hope there are none, but experience suggests otherwise. I will be grateful to anyone who takes the time to point them out, in the kinder and gentler way that my now advanced years can withstand.

Francis Scheid

May, 1989

CHAPTER 1

The Representation of Numbers

1.1   To what extent is the issue of number representation of interest in numerical analysis?

▌ Since the subject is after all numerical, there is bound to be a connection, but it is one that becomes apparent in a limited number of ways. The ideas of floating-point, overflow, underflow, double precision and such are sure to surface occasionally, and the problem of conversions between the decimal input and output that we prefer and the more or less binary internal representations preferred by computers also makes itself known. Anyone who has done even a little computing has seen an input 1 returned in a form such as .9999999. The important question of roundoff is also related to the number representations used. So in brief, the issue is usually submerged but worthy of a moment's thought.

1.2   The general binary representation of a positive integer is

$$d_n 2^n + d_{n-1} 2^{n-1} + \cdots + d_1 2^1 + d_0 2^0$$

with all the binary digits (bits) d_i either 0 or 1. Review the well-known division method for finding these bits, using the decimal 43.

▌ This method involves continued division by 2, noting all remainders, until a quotient of zero is reached. The remainders are then retrieved in reverse order.

$$43/2 = 21 + 1 \qquad 5/2 = 2 + 1$$
$$21/2 = 10 + 1 \qquad 2/2 = 1 + 0$$
$$10/2 = 5 + 0 \qquad 1/2 = 0 + 1$$

The resulting 101011 is the binary equivalent of 43, as the sum $32 + 8 + 2 + 1$ assures. For binary to decimal conversion this summing of place values will be adequate for our limited needs.

1.3   The general binary representation of a positive number less than 1 is

$$d_{-1} 2^{-1} + d_{-2} 2^{-2} + d_{-3} 2^{-3} + \cdots$$

with each binary digit d_i either 0 or 1. Review the well-known multiplication method for finding these digits, using the decimal .625.

▌ Here we continually multiply by 2 and extract the digits that appear to the left of the point, stopping when zero is reached. Thus

$$.625 \times 2 = 1.250 \quad \text{(extract the 1)}$$
$$.250 \times 2 = 0.500 \quad \text{(extract the 0)}$$
$$.500 \times 2 = 1.000 \quad \text{(extract the 1)}$$

The digits extracted are taken in the order extracted, not the reverse. In this case .101 is the result and is confirmed by reading off the place values of the 1 bits:

$$1/2 + 1/8 = 5/8 = .625$$

For this number both the decimal and binary representations terminate. This is, of course, not the usual case.

1.4   Apply the method of the preceding problem to convert the fraction 1/3 to binary.

▌ This can be brought off while retaining fractional representations:

$$1/3 \times 2 = 2/3 \quad (<1, \text{ extract a } 0)$$
$$2/3 \times 2 = 4/3 \quad (>1, \text{ extract a } 1)$$
$$1/3 \times 2 = 2/3 \quad (<1, \text{ extract a } 0)$$
$$2/3 \times 2 = 4/3 \quad (>1, \text{ etc.})$$

The infinite, repeating representation

$$1/3 = .01010101\ldots$$

develops and its validity is attested by the geometric series

$$2^{-2} + 2^{-4} + 2^{-6} + \cdots = 1/3$$

1.5   One source of error in computations is the need to convert back and forth between decimal symbols, convenient for input and output, and binary symbols, the preferred medium for internal machine activity. Illustrate this using the sum $.6 + .6 = 1.2$.

▌ First we convert .6 to binary using the multiplicative method, extracting each left side digit before the next step (except from the original .6).

$$.6 \quad 1.2 \quad 0.4 \quad 0.8 \quad 1.6 \quad 1.2 \quad 0.4 \quad 0.8 \quad 1.6 \quad \text{etc.}$$

The indicated binary symbol is $.1001100110011\ldots$ and does not terminate. Suppose we are limited to eight bits in our number representations. This is unrealistically small but will serve to make the point. The sum $.6 + .6$ is then done:

$$
\begin{aligned}
&\ .10011001\\
+\ &\ .10011001\\
=\ &1.0011001
\end{aligned}
$$

Converting back to decimal,

$$1 + 1/8 + 1/16 + 1/128 = 1.195$$

to three places. The error of .005 is due to the roundoffs made during the conversions. In a real machine it would be possible to carry many more binary digits than eight, and the error would be less, but present.

1.6   What is the ASCII code?

▌ It is one of the major codes used to transfer information. A depressed key, for example, transmits a one character message. In an ASCII system this is translated into a seven bit code. The digits 0 to 9 are represented by

$$0110000 \quad 0110001 \quad 0110010 \quad \cdots \quad 0111001$$

in which the corresponding binary forms can be recognized as the final four bit groups. The leading groups 011 indicate that the information is numerical. Other leading triples mark alphabetic characters or punctuation. To solve numerical problems we must transmit such information to the computer, but the nature of the code used in the process is not part of our subject.

1.7   What are hexadecimal symbols?

▌ These are number representations of the form

$$d_n16^n + d_{n-1}16^{n-1} + \cdots + d_1 16^1 + d_0 16^0$$

for integers, with negative powers of 16 appearing for numbers less than 1. A purported use of hexadecimal symbols is to abbreviate binary symbols for the convenience of human users, but in fact most numerical analysts prefer to get along without them. However, to illustrate, the binary symbol 1011101 is converted to hexadecimal by grouping its digits in 4s, starting from the right: 101 1101. Interpreted in binary, such four bit groups are equivalent to decimal 0 to 15, but our purpose here is to equate them with hexadecimal digits

$$0, 1, 2, 3, 4, 5, 6, 7, 8, 9, A, B, C, D, E, F.$$

The last six are to be thought of as aliases for decimal symbols 10 to 15. The given binary is thus $5D$ in hexadecimal. Looking at things in reverse,

$$5D_{hex} = 5 \times 16^1 + 13 \times 16^0 = 93_{dec}$$

1.8   Convert $F.F4$ from hexadecimal to binary and then to decimal.

▮ Replacing F by 1111 and 4 by 0100 we have

$$1111.11110100$$

which in decimal means

$$8 + 4 + 2 + 1 + 1/2 + 1/4 + 1/8 + 1/16 + 1/64$$

and comes eventually to 15.953125.

As a further illustration of the multiplication method, suppose we convert .953125 directly from decimal to hexadecimal. The multiplier will now be 16 instead of 2:

$$.953125 \times 16 = 15.25 \quad \text{(remove the leading 15)}$$
$$.25 \times 16 = 4.00 \quad \text{(remove the leading 4)}$$

The removed digits form $.F4$ in hexadecimal.

1.9   Find a 16 bit representation of the approximation 1.414 to the square root of 2.

▮ The digit 1 to the left of the point is the same in both symbolisms. To the right we use the multiplicative method

0.828	1.656	1.312	0.624	1.248	0.496	0.992	1.984
1.968	1.936	1.872	1.744	1.488	0.976	1.952	

and finish with 1.0110 1001 1111 101 as the required 16 bits. In hexadecimal this becomes (right-side digits already grouped in 4s) 1.69FA which may be checked by converting back to decimal as

$$1 + 6/16 + 9/256 + 15/4096 + 10/65536$$

to find 1.41397 more or less.

1.10   Find a 16 bit representation of 3.1416.

▮ The 3 will become 11 in binary and the rest will again be treated by the multiplicative method.

0.2832	0.5664	1.1328	0.2656	0.5312	1.0624	0.1248
0.2496	0.4992	0.9984	1.9968	1.9936	1.9872	1.9744

The representation is 11.0010 0100 0011 11 and translates to 3.243C in hexadecimal. As a check,

$$3 + 2/16 + 4/256 + 3/4096 + 12/65536$$

comes to 3.1415 to four decimal places. Sixteen bits are not quite enough to manage the last place.

In comparing decimal and hexadecimal representations it is clear that for integers the hexadecimal will usually be shorter. Thus 15 becomes F and 1492 turns into 5D4. The powers of 16 are larger than those of 10 and each place can absorb more numbers. To the right of the point, however, the powers of 1/16 are smaller than those of 1/10, and we have found 1.69FA replacing 1.414 as an approximate square root of 2, while 3.243C is the hexadecimal substitute for 3.1416.

1.11   Interpret the floating-point decimal $+.1066 \times 10^4$.

▌ Clearly the exponent shifts the decimal point four places to the right to make 1066. Similarly, $+.1066 \times 10^{-2}$ is .001066. The term floating is used to indicate that the point can be shifted at will if the exponent is adjusted accordingly.

Similarly, in the floating-point binary symbol

$$.1011010011111101 \times 2^1$$

the exponent shifts the point one place to the right and we have once again the 16 bit representation of the square root of 2.

1.12   Express the 16 bit approximation 11.00100100001111 to pi in normalized floating-point form.

▌ The form $.1100100100001111 \times 2^2$ is called the normalized form. In normalized floating-point forms the leading 1 digit appears immediately to the right of the point. The preceding problem also offered a normalized form.

1.13   A floating-point symbol has three parts, the sign, the mantissa and the exponent. Identify these in $+.1492 \times 10^0$.

▌ There is certainly no mystery here. The sign is plus, the mantissa .1492 and the exponent 0. The symbol is normalized.

1.14   The decimal 14.92 converts to binary as

$$1110.111010111\ldots$$

Round this to 11 digits (for a reason to be described) and express the result as a normalized binary symbol.

▌ The trailing 11 pair must be removed. Since this amounts to more than half a unit in the final place to be retained (it comes to $1/2 + 1/4$ units) we round up and have 1110.1110110, the sixth and seventh places to the right of the point changing from 01 to 10. The normalized binary symbol is

$$+.11101110110 \times 2^4$$

The sign, mantissa and exponent are readily identified.

1.15   Interpret the floating-point representation

$$0111011101100100$$

▌ The number is that of the preceding problem, the coding only slightly changed. If we group the bits in this way,

$$0\ 11101110110\ 0\ 100$$

the parts may be identified. The conventions used in this example are: The leading zero is the sign of the mantissa, 0 meaning plus; the 11 bit group is the mantissa itself, a binary point being assumed at its left end; the next 0 is the sign of the exponent, again plus; the final group is the exponent 4 in binary. With these conventions we have once again $+1110.1110110$ as above. The floating-point form of this example is a miniaturized version of those in use, the point being that the number is now represented by nothing but 0s and 1s.

1.16   Interpret the string 0110111000000100 with the same conventions used in the preceding problem. What is the decimal equivalent?

▌ The groups are $0\ 11011100000\ 0\ 100$ and we see that both the mantissa and the exponent are again positive. (A 1 bit would mean minus.) Shifting the implied point four places to the right gives us 1101.11 which responds to a quick mental conversion as 13.75 decimal.

In practice the mantissa and exponent would have more bits, the sign conventions and positioning of groups might differ from those just described, but the floating point concept is pivotal to modern computing, and so to numerical analysis.

1.17   Add these numbers, represented in the floating-point form just described:

$$0101101110000010$$

$$0100011000001100$$

▌ One way or another it will be necessary to line up the binary points. Interpretations lead to the sum

$$\begin{aligned}
&10.110111000\\
+\,&.000010001100000\\
=\,&10.1110010011
\end{aligned}$$

if all digits except some trailing 0s are retained. However, it is reasonable to expect an output result in the same form as the inputs, and this means reducing the mantissa to 11 bits. If the final 1 bit is assumed to trigger a rounding up, then we finish with this floating-point representation of the sum:

$$0101110010100010$$

1.18   Illustrate overflow.

▌ The largest number that can be represented using floating-point symbols in the 16 bit form just described is

$$0111111111110111$$

having mantissa and exponent of solid 1s. This is equivalent to $127 + 15/16$. Any number larger than this cannot be represented in the given form and is called an overflow.

1.19   Illustrate underflow.

▌ The smallest positive number that can be represented in the form being used is

$$0000000000011111$$

or 2^{-18}. If, however, we make the usual assumption that symbols must be normalized, thereby fixing the exponent, the smallest becomes

$$0100000000001111$$

or 2^{-8}. Any positive number smaller than this cannot be represented and is called an underflow. Any floating-point system of number representation will have such limitations and the ideas of overflow and underflow will be relevant.

1.20   Illustrate the uneven distribution of numbers represented in floating-point form.

▌ For simplicity we imagine a floating-point system in which mantissas have just three bits and the exponent is limited to -1, 0 or 1. Assuming normalization, these numbers have the form $.1xx$ apart from exponent. The entire set, therefore, consists of three subsets of four numbers each, as follows:

.0100	.0101	.0110	.0111	(exponent -1)
.100	.101	.110	.111	(exponent 0)
1.00	1.01	1.10	1.11	(exponent 1)

These are plotted in Fig. 1-1. Notice the denser packing of the smaller numbers, the separation increasing from 1/16 to 1/4 as we pass from group to group. This is due, of course, to the fact that we have only three significant digits (with the leader fixed at 1) with the exponent supplying progressive magnification as it increases. The number 1.005, for example, is not available here. The set is not dense enough in this part of its range. A fourth digit would be needed. Realistic floating-point systems have this same feature, uneven distribution of the available resources.

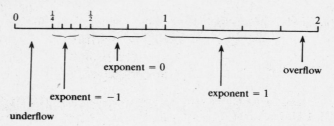

Fig. 1-1

1.21 Imitating the preceding problem, imagine a floating-point system in which normalized mantissas have four bits and the exponents are −1, 0 and 1. Discuss the density of such numbers and the ideas of overflow and underflow.

▌ There will be three groups of eight numbers.

.01000	.01001	.01010	.01011	.01100	.01101	.01110	.01111
.1000	.1001	.1010	.1011	.1100	.1101	.1110	.1111
1.000	1.001	1.010	1.011	1.100	1.101	1.110	1.111

The top group, for exponent −1, falls in the interval 1/4 to 1/2 and has a spacing of 1/32. The middle group, for exponent 0, falls in the interval 1/2 to 1 and has spacing 1/16. The last group, exponent 1, falls between 1 and 2 with spacing 1/8. Any positive number less than 1/4 will cause underflow. Any number greater than 15/8 will cause overflow.

1.22 What is the difference between rounding and chopping?

▌ An abbreviated example will serve. The binary .10101 can be rounded to four places by added a 1 bit to the fifth place and then saving only four places. This is the usual procedure. The result is .1011. Similarly .10100 would round to .1010.

To round to n places a 1 bit is added in the $(n+1)$th place, which amounts to rounding up when there is a 1 in the $(n+1)$th place and rounding down otherwise.

In chopping, all unwanted places are simply removed. The binary symbols .10101 and .10100 both become .1010 to four places.

1.23 What is double precision?

▌ Computers ordinarily deal with number representations having a fixed length, called the *word length* for that machine. The numbers so represented are said to have single precision. For some applications the accuracy available by single precision arithmetic proves to be inadequate and the device of using two separate word lengths for each number is substituted. The details vary, but the numbers are said to have double precision and the resulting procedures are double-precision arithmetic.

1.24 Illustrate the operation of multiplication in double-precision arithmetic.

▌ Sometimes the needed procedures are built-in; in other instances they have to be programmed. Either way something of the following nature must take place. Let one double-precision number have the parts 1001 and 1011, it being understood that these four bit single-precision words really form the double precision .10011011. Similarly let the parts 1100 and 0111 represent .11000111. If only single-precision devices are built-in, then the product must be developed by the familiar algebraic theorem

$$(a+b)(c+d) = ac + ad + bc + bd$$

with the four right-side products requiring only single-precision arithmetic. Thus $ac = .1001 \times .1100 = .01101100$. The partial products will be available to double length. Even in single-precision work this is true, for appropriate rounding if for nothing else. The product $ad = .1001 \times .00000111$ will have to be done as $.1001 \times .0111 = .00111111$ and then shifted (and perhaps chopped) to .00000011 or else rounded. The bc term is treated in the same way and is .00001000. The bd product will be zero to eight places. Presumably these various products will now have to be split into single-precision parts, perhaps as

$$ac = (0110)(1100)$$
$$ad = (0000)(0011)$$
$$bc = (0000)(1000)$$

and single-precision sums found for the principal and secondary parts. Taking the secondary parts first, we find 10111, the leading 1 bit interpreted as a carry to the principal part, which all by itself manages 0110. The combination is then

.01110111

which may be split into the two single-precision words 0111 and 0111. It will be evident that double-precision arithmetic may take at least four times as long as single precision for certain operations.

1.25 It has been suggested that the following message be broadcast to outer space as a sign that this planet supports intelligent life. The idea is that any form of intelligent life elsewhere will surely comprehend its intellectual content and so deduce our own intelligent presence here. What is the meaning of the message?

11.001001000011111101110

I It is an approximation to pi in binary form, only slightly longer than the representation found in Problem 1.16.

1.26 Interpret this 32 bit representation given in the code known as *EBCDIC*:

11110001111101001111100111110010

I Surely Sherlock Holmes would have been able to break the code, which in fact is much like ASCII. Each block of eight bits represents a character and the leading 1111 group of each block signals a numerical digit. With these clues, everyone will discover the binary nature of the remaining four bits per block. The number is 1492. When it is known in advance that only numerical characters are being transmitted, the leading 1111 parts can be dropped. This produces *binary coded decimal* (BCD) representations. The number 1775 is then represented by

0001011101110101

and 0001000001100110 will be of further interest to students of history.

1.27 Not so very long ago the punched card was the axis around which the world of computation revolved. Figure 1-2 shows (at the left) how various characters were represented by hole patterns and offers (at the right) a four digit number. What is this number?

I It is, of course, 1492. Numerical analysts of only a decade or so past had many opportunities to squint through such holes, with the hope that the message would be good news.

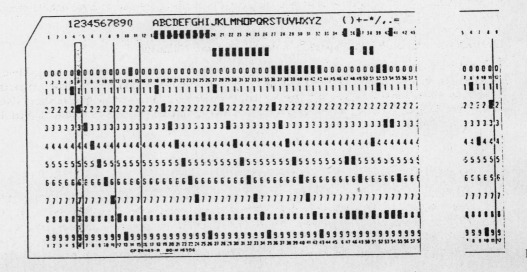

Fig. 1-2

1.28   Figure 1-3 simulates a short strip of magnetic tape. The 1 bits stand for magnetized spots and blanks for unmagnetized. Looking at the bottom four rows of spots and reading from the bottom upward, binary patterns will be recognized. The strip at the left shows patterns for the digits 0 to 9. What number is represented by the strip at the right?

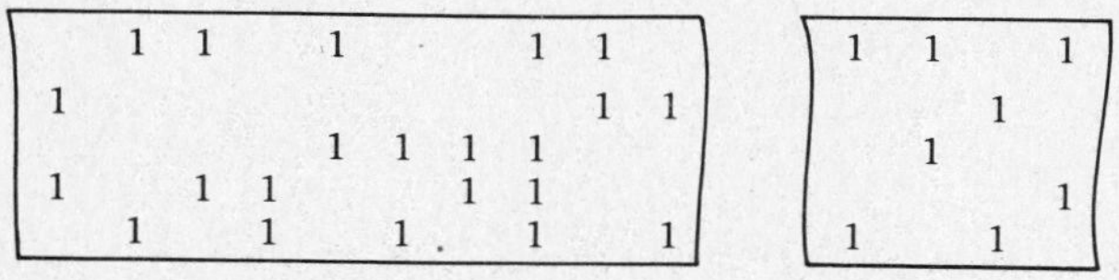

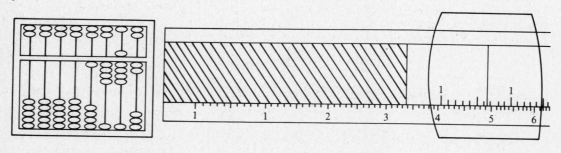

Fig. 1-3

❚ The quincentennial will be celebrated in 1992. The top row on the tape creates what is called an even parity code, arranging an even number of 1 bits per column. This serves as a partial test of data validity.

1.29   Decode both the old and the ancient number representation offered in Fig. 1-4.

Fig. 1-4

❚ MCCCCXCII and MLXVI.

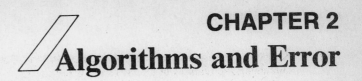

CHAPTER 2
Algorithms and Error

2.1 ALGORITHMS

2.1 What is an algorithm?

▌ An algorithm is a way of reducing a more or less difficult task to a sequence of simpler steps. In numerical analysis the simpler steps are the four operations of arithmetic and some truly difficult tasks have been reduced to them. As a first example, however, consider the problem of dividing 24 by 17 using the method of successive subtractions. To the mathematically sophisticated this may sound tongue-in-cheek, but the idea is to perform the "difficult" operation of division by doing a sequence of simpler steps, so the example is fully in the algorithmic spirit. We will subtract 17s until less than 17 remains and then count the number of such subtractions.

To begin, *one* such subtraction leaves a remainder of 7. Attaching a zero, for reasons which are not entirely obscure, we have 70. *Four* subtractions then reduce this to 2. Attaching another zero, making 20, *one* subtraction is again enough, leaving a new remainder of 3. Clearly the next step will again find *one* subtraction sufficient, and putting the various counts together we have

$$24/17 = 1.411\ldots$$

with more digits available through persistence. Needless to say, the algorithm involved is usually presented in this layout.

$$
\begin{array}{r}
1.411 \\
17 \,\overline{\smash{\big)}\, 24.000} \\
\underline{17} \\
7\,0 \\
\underline{6\,8} \\
20 \\
\underline{17} \\
30 \\
\underline{17} \\
13
\end{array}
$$

2.2 Find the square root of 2 to four decimal places.

▌ Many algorithms, using only the four basic operations of arithmetic, exist. The best known and most used is

$$x_1 = 1 \qquad x_{n+1} = \frac{1}{2}\left(x_n + \frac{2}{x_n}\right)$$

from which a few mental calculations quickly manage

$$x_2 = \tfrac{3}{2} \qquad x_3 = \tfrac{17}{12} \qquad x_4 = \tfrac{1}{2}\left(\tfrac{17}{12} + \tfrac{24}{17}\right)$$

or rounded to four places

$$x_2 = 1.5000 \qquad x_3 = 1.4167 \qquad x_4 = 1.4142$$

the last being correct to all four places. This numerical algorithm has a long history.

2.3 Generalizing the preceding problem, the algorithm

$$x_{n+1} = \tfrac{1}{2}(x_n + Q/x_n)$$

produces a square root of Q. Apply it with $Q = 63$.

❚ With $Q = 63$ and $x_1 = 8$ we obtain the successive approximations

$$7.9375 \quad 7.9372535 \quad 7.9372535$$

the last two of which are correct as far as they go. The fast convergence to this root is due to the excellent initial approximation.

2.4 There are sometimes several good algorithms for a given task. If x_0 is an approximate square root of Q, we may let

$$s = (Q/x_0^2) - 1$$

and use the binomial theorem to estimate $\sqrt{Q}$ in the form

$$x_0(1+s)^{1/2} = x_0(1 + s/2 - s^2/8 + s^3/16 - 5s^4/128 + 7s^5/256)$$

using six terms. Apply this with $Q = 2$ and $x_0 = 1$.

❚ First we find $s = 1$ and then the series terms

$$1 + 1/2 - 1/8 + 1/16 - 5/128 + 7/256$$

which surely will not provide a good approximation to the square root of 2. The convergence is slow and several more terms would be required for a good result. Suppose, however, that we try $x_0 = 1.4$. Now $s = 2/1.96 - 1 = .0204081$ and

$$\sqrt{2} = 1.4(1 + .0102040 - .000052 + .0000005)$$
$$= 1.414214$$

correct to six places with only four terms being used. The importance of a good initial approximation is clear.

2.5 Which of the two square root algorithms seems to be the more convenient?

❚ Public opinion stands squarely on the side of the algorithm of Problem 2.3.

2.6 Use the algorithm

$$x_{n+1} = 2x_n/3 + 1/x_n^2$$

to find the cube root of 3.

❚ Taking the uninspired initial approximation of 1, we find successively

$$1.3333333$$
$$1.4513888$$
$$1.4423069$$
$$1.4422495$$
$$1.4422494$$

which is correct as far as it goes.

2.7 If x_0 is an approximate cube root of Q and $s = (Q/x_0^3) - 1$, then the exact root is

$$x_0(1 + s/3 - s^2/9 + 5s^3/81 - 10s^4/243 + 22s^5/729 - \cdots)$$

according to the binomial theorem. Apply this algorithm to the case $Q = 3$.

❚ Choosing $x_0 = 1.5$ we have $s = 3/3.375 - 1 = -.1111112$ so the algorithm offers

$$1.5(1 - .0370370 - .0013717 - .0000846 - .0000062)$$

which proves to be 1.442250.

2.8   The quotient $1/.985$ can be found by long division, but it can also be found by the algorithm

$$1/(1 - x) = 1 + x + x^2 + x^3 + \cdots$$

which is another case of the binomial theorem. How fast is the convergence?

▌ With $x = .015$ we have

$$1/.985 = 1 + .015 + .000225 + .0000033 + \cdots = 1.0152283$$

and the algorithm may be faster than long division in this case.

2.9   Consider algorithms for computing $1 - \cos 5°$.

▌ The example illustrates a phenomenon familiar to numerical analysts, the disappearance of significant digits when nearly equal numbers are subtracted. The value of $\cos 5^0$ is .99619 to five places. Subtracting from 1 we have .00381, which is correct to five places but has only three significant digits. To emphasize the loss we can rewrite our result in floating-point form as $.381 \times 10^{-2}$.

But now rewrite $1 - \cos 5^0$ as $\sin^2 5^0/(1 + \cos 5^0)$. The computation runs a different course (proceeds by a different algorithm) and arrives at $.0075957/1.9962 = .0038050$. Five figures are now in hand, instead of three.

These computations ignore the question of how the values of the cosine function itself are to be found. The familiar series can be converted for present purposes to

$$1 - \cos x = x^2/2! - x^4/4! + x^6/6! - \cdots$$

and with $x = 5^0 = .087266$ rad produces

$$.0038076 - .0000024 + .0000004 = .0038056$$

carrying five figures (apart from leading zeros) all the way.

The point of the example is that five digit accuracy from start to finish is possible if a suitable algorithm is selected, but when near equals are subtracted, not all of the available information may penetrate through.

2.10   Solve the quadratic equation $x^2 - 16x + 1 = 0$ for its smaller root.

▌ The quadratic formula determines this root in the form $8 - \sqrt{63}$. For emphasis, suppose we are limited to three digit computation. From Problem 2.3 we have $\sqrt{63} = 7.94$ so $x = 8.00 - 7.94 = .06$ and only a single digit emerges from the algorithm. This is another example of loss caused by the subtraction of near equals, highlighted by the restrictions imposed on our computing power.

Now, by elementary algebra we have

$$8 - \sqrt{63} = 1/(8 + \sqrt{63})$$

and an alternative algorithm involves computing the right side instead of the left:

$$1/(8 + \sqrt{63}) = 1/(8.00 + 7.94) = 1/15.9 = .0629$$

Only three digits have been carried, but now we have a three digit result. The correct value is close to .0627 so the new algorithm makes less than a tenth the error of the original.

2.11   Infinite series make excellent algorithms for computing function values. Obtain $\sin(1)$ from the familiar

$$\sin x = x - x^3/3! + x^5/5! - \cdots$$

getting five place accuracy. How many terms are needed?

▌ Substituting into the series we have these terms,

$$1 - .166667 + .008333 - .000198 + .000003 = .84147$$

which is correct to five places. Five terms were needed for this relatively large argument (roughly 60°).

2.12   Rework the computation of the preceding problem for the argument 30°. How many terms are needed for five place accuracy?

▮ The equivalent in radians is .523599 and substituting this value brings the terms

$$.523599 - .023925 + .000328 - .000002 = .500000$$

Carrying six places, in order to safeguard five, has in this case left us with a bonus sixth place. Frequently the final digit, or more in very lengthy computations, is corrupted by unavoidable roundoffs.

2.13 Compute the product 45×17 by the *Russian peasant algorithm*.

▮ This ancient method involves continually doubling one factor while halving the other, noting where the halving leaves a remainder. One then sums those multiples for which remainders did occur (starred in this computation):

$$
\begin{array}{cccccc}
45 & 22 & 11 & 5 & 2 & 1 \\
17^* & 34 & 68^* & 136^* & 272 & 544^*
\end{array}
$$

Just why the algorithm works is not too hard to discover. The idea does have the algorithmic flavor if one concedes that doubling and halving are simpler than the usual routine for multiplication.

Exploiting the reasonable assumption that one factor deserves doubling as well as the other, the algorithm can be run the other way:

$$
\begin{array}{ccccc}
17 & 8 & 4 & 2 & 1 \\
45^* & 90 & 180 & 360 & 720^*
\end{array}
$$

2.14 Using the familiar algorithm for solving the set of simultaneous equations

$$ax + by = e$$
$$cx + dy = f$$

how many steps are required?

▮ The algorithm in mind multiplies the top equation by c, the bottom one by a and subtracts to eliminate x:

$$(cb - ad)y = ce - af$$

We may view this last equation as determining y via four multiplications, two subtractions and a division. The number x is then obtained by substituting y into one of the given equations, three more steps being needed, for a final total of 10. For larger scale problems it is useful to estimate the number of steps that an algorithm will take, as a way of judging the time and cost of the job.

2.15 How many arithmetic steps are needed to solve the quadratic equation $ax^2 + bx + c = 0$?

▮ Suppose the quadratic formula is used:

$$x = \left(-b \pm \sqrt{b^2 - 4ac}\right)\big/2a$$

Four multiplications, a division and two additions or subtractions are evident, once the square root is in hand. Finding this root is really the principal chore. As Problem 2.2 shows, three or four steps may have to be repeated a few times in order to do this. Overall, perhaps 20 steps may be enough to get the first root of the equation. If the ingredients are saved, two steps will get the second.

2.16 Compute the value of the polynomial

$$p(x) = 2x^3 - 4x^2 + 6x - 8$$

for the argument $x = 5$.

▮ Following the natural course we have $x^2 = 25$ and $x^3 = 125$, and assembling the pieces

$$p(5) = 250 - 100 + 30 - 8 = 172$$

a count of the operations shows that five multiplications and three additions or subtractions have been performed.

Now rearrange the polynomial as

$$p(x) = ((2x - 4)x + 6)x - 8$$

and try again. From $x = 5$ we have successively 10, 6, 30, 36, 180 and 172. This time three multiplications have been made instead of five. The reduction is not dramatic, but it is suggestive. For a polynomial of degree n the first algorithm uses $2n - 1$ multiplications, the second only n. In a large operation, with many polynomial evaluations, the saving in time and roundoff errors could be significant.

2.2 ERROR

2.17 Define the error of an approximation.

▌ The traditional definition is

$$\text{true value} = \text{approximation} + \text{error}$$

so that, for example,

$$\sqrt{2} = 1.41421 + \text{error (of about .000004)}$$

$$\pi = 3.141592654 + \text{error (of about } - .0000000004)$$

2.18 What is relative error?

▌ This is error measured relative to true value:

$$\text{relative error} = \text{error/true value}$$

In the common event that the true value is unknown, or unwieldy, the approximation is sometimes substituted for it and the result still called, somewhat loosely, the relative error. Thus the familiar approximation 1.414 to $\sqrt{2}$ has a relative error of about

$$.0002/1.414 = .00014$$

while the cruder approximation 1.41 has relative error near

$$.004/1.414 = .003$$

The idea of relative error has application throughout numerical analysis.

2.19 What are significant digits?

▌ Suppose the number .1492 is correct to four decimal places, as given. In other words, it is an approximation to a true value that lies somewhere in the interval between .14915 and .14925. The error is then at most five units in the fifth place or half a unit in the fourth. In such cases the approximation is said to have four significant digits. Similarly, 14.92 has two correct decimal places and four significant digits, provided its error does not exceed .005. Significant digits are counted from the leading nonzero digit to the last correct one.

2.20 What are rounding down and rounding up?

▌ The number .16204 is said to be rounded to four decimal places when abbreviated to .1620, while .16206 would be rounded to .1621. In both cases the error made by rounding is no more than .00005, assuming the given figures correct. The first is an example of rounding down, the second of rounding up. A borderline case such as .16205 is usually rounded to the nearest even digit, here to .1620, to avoid long range prejudice between ups and downs.

Roundoffs to n places can be effected by adding a 5 digit in the $(n + 1)$th place and then chopping all places beyond the nth. This applies to decimal representations, of course. With binary representations, rounding can be done by adding a 1 bit in the $(n + 1)$th place and then chopping.

2.21 Illustrate the occurrence of roundoff error in multiplication and addition.

▌ Suppose 1.492 is multiplied by 1.066, the product being 1.590472. Assume for simplicity that a four digit computer is doing the job. Then for future use in a single-precision computation, rounding to 1.590 is necessary. The roundoff error of .000472 has to be accepted because of the machine limitation. All computers have some such limitation and make roundoff errors by the unavoidable millions.

As for addition, suppose $x = 1.492$ and $y = 1.066$ to three places, that is, the errors are no greater than .0005 units one way or the other. Again suppose a four digit computer and ask it to compute $x + y$ and $x + .01y$. Here are the computations:

$$
\begin{array}{llcll}
x & 1.492 & \qquad & x & 1.492 \\
y & 1.066 & & .01y & .011 \\
x+y & 2.558 & & x+.01y & 1.503
\end{array}
$$

In the first column the errors in x and y are carried along invisibly but no further errors are introduced. In the second column, however, it has been necessary to align the decimal points and round .01066 to .011.

2.22 What are some of the error sources with which numerical analysis must contend?

❚ First there is input error. The data provided as input to an algorithm are seldom exact, perhaps measured values or the results of prior computations. Then there is algorithm error, made during the course of the algorithm's run. Two kinds of algorithm error that will get frequent mention are truncation error and roundoff error. Finally there is output error, which may be viewed as the bottom line, the result of processing all the individual errors that have entered the computation from its beginning. There is no way to follow the details of error propagation through a nontrivial computation. Error bounds or rough estimates must usually be accepted, but the error question is important and should never be far out of mind.

2.23 Analyze error development in the computation of $x + .01y$ just completed.

❚ Let $X = 1.492 + e_1$ and $Y = 1.066 + e_2$, with X and Y the true values and the two errors e_1 and e_2 assumed to be no larger than .0005. Then

$$
X + .01Y - 1.503 = 1.492 + e_1 + .01066 + .01e_2 - 1.503
$$
$$
= e_1 + .01e_2 - .00034
$$
$$
|X + .01Y - 1.503| \le .0005 + .000005 + .00034 = .000845
$$

The third decimal place of our result appears to be open to suspicion, but the blame falls mostly on the input error in x.

Tracing error takes patience and gives little pleasure or thrill of victory, but the credibility of output is at stake.

2.24 Suppose the numbers $x_1, \ldots, x_n$ are approximations to $X_1, \ldots, X_n$ and that in each case the maximum possible error is e. Prove that the maximum possible error in the sum of the x_i is ne.

❚ Since

$$
x_i - e \le X_i \le x_i + e
$$

it follows by addition that

$$
\sum x_i - ne \le \sum X_i \le \sum x_i + ne
$$

so that

$$
-ne \le \sum X_i - \sum x_i \le ne
$$

which is what was to be proved. The content of this theorem is quite intuitive. When n numbers are added it is at least possible that the errors they contain are all of maximum size e and of the same sign, making the output error of size ne.

2.25 Compute the sum $\sqrt{1} + \sqrt{2} + \sqrt{3} + \cdots + \sqrt{100}$ with all the roots given to two decimal places. What is the maximum possible error in the result?

❚ Whether by a few well chosen lines of programming or by an old-fashioned appeal to tables, the roots in question can be found and summed. The result is

$$
1.00 + 1.41 + 1.73 + \cdots + 10.00 = 671.38
$$

Since each root has a maximum error of $e = .005$, the maximum possible error in the sum is $ne = 100 \times .005 = .5$, suggesting that the sum as found may not be correct to even one decimal place.

2.26   What is meant by the probable error of a computed result?

▮ This is an estimate such that the actual error will exceed the estimate with probability $1/2$. In other words, the actual error is as likely to be greater than the estimate as less. It does depend on the way errors are distributed and so is not an easy target, but the assumption of normally distributed errors leads to

$$\text{probable error} = \sqrt{n}\, e$$

with e the maximum error in the x_i of Problem 2.24.

2.27   What is the actual error of the sum computed in Problem 2.25 and how does it compare with the maximum possible and the probable errors?

▮ A new computation, using square roots to five decimal places, managed the sum

$$1.00000 + 1.41421 + 1.73205 + \cdots + 10.00000 = 671.46288$$

with maximum error $100 \times .000005$ or $.0005$ of its own. This means that we now have the sum correct to three places as 671.463. Comparing with our computed 671.38, we find the actual error to have been about .08, compared to the maximum possible error of .50 and the probable error of $\sqrt{100} \times .005 = .05$. So one of our estimates was much too pessimistic and the other slightly too optimistic. Maximum error estimates generally prove to be pessimistic.

2.28   Suppose 1000 square roots are to be summed instead of a mere 100. If three place accuracy is to be guaranteed, how accurately should the individual roots be computed?

▮ For a solid guarantee we must assume the worst, that the maximum possible error might be reached. The formula ne of Problem 2.24 becomes $1000e$ and shows that three decimal places may be lost in a computation of this length. Since three are wanted in the output we must have six correct places in the input. In long jobs there is time for even very small errors to make a collective contribution.

2.29   What is truncation error?

▮ The classical use of this term involves the use of a partial sum to approximate the value of an infinite series. Consider, for example, the series for $\log 2$

$$1 - 1/2 + 1/3 - 1/4 + \cdots$$

the value of which is .693 to three places. How many terms of the series are needed to obtain this value? A well-known theorem of analysis states that when terms alternate in sign and steadily decrease, the partial sums of the series will dodge back and forth across the series value, making the error at any point less than the first term omitted. To get the specified accuracy we therefore need n terms, with $1/n \le .0005$ or $n \ge 2000$. Two thousand terms will have to be summed. Working to eight decimal places, the 2000 roundoff errors might accumulate to

$$ne = 2000 \times .000000005 = .00001$$

which seems negligible. So we allow the algorithm to proceed, round the result to three places and have .693. Eight decimal place input is needed to guarantee three decimal output.

Note that in this problem there are no input errors, only algorithm errors. First we truncate the series to a partial sum. Then we make 2000 roundoffs in trying to evaluate this sum. The first of these is an example of truncation error and appears to be the larger of the two error sources in this case. In summary,

$$\text{actual error} = \text{truncation error} + \text{roundoff error}$$
$$= .0005 + .00001$$

more or less.

2.30   Prove that if the series

$$a_1 - a_2 + a_3 - a_4 + \cdots$$

is convergent, all the a_i being positive, then

$$(1/2)\, a_1 + (1/2)(a_1 - a_2) - (1/2)(a_2 - a_3) + (1/2)(a_3 - a_4) - \cdots$$

is also convergent and represents the same number.

❚ With A_n and B_n representing the nth partial sums of the two series, it is easy to see that $A_n - B_n = \pm(1/2)a_n$. Since the first series is convergent, $\lim a_n = 0$ and the result follows.

2.31 Apply the theorem of the preceding problem to evaluate the series of Problem 2.29, again to three places.

❚ A little algebra finds $B_1 = 1/2$ and for $n > 1$

$$B_n = 1/2 + \sum_2^n (-1)^k / 2k(k-1)$$

so we have another alternating series with monotone decreasing terms and the theorem of Problem 2.29 is again available. For three place accuracy we need $1/2n(n+1) \le .0005$ or $n \ge 32$. This is far fewer than 2000 terms and roundoff error will hardly be an issue. The new algorithm is much quicker than the other and manages the same .693 with far less effort.

2.32 Given that the numbers .1492 and .1498 are correct as far as they go, that is, they contain errors no larger than five units in the fifth place, find the relative error in the quotient

$$1/(.1498 - .1492)$$

❚ The problem gives us a chance to watch how relative error can develop in a continuing computation. The given numbers have relative errors of about 5/15000, which is near .03%. For their sums and differences a maximum error of one unit in the fourth place is possible. In the case of the sum this again means a relative error of about .03%, but with the .0006 difference we have an error of one part in six, which comes to 17%. Coming to the quotient it may be just as well to take the pessimistic view. As given, a quotient of 1667 would be calculated, to the nearest integer. But conceivably it is $1/(.14985 - .14915)$, which ought to have been found instead, and this would have brought us 1429. At the other extreme is $1/(.14975 - .14925) = 2000$. Clearly a large relative error generated at some interior stage of a computation can lead to large absolute errors down the line.

2.3 CONDITION AND STABILITY

2.33 Define the condition of a function $f(x)$.

❚ Condition measures the error in the function caused by an error in the argument. If the argument x is altered to x', which is a relative change of $(x - x')/x$, then the relative change in the function is

$$(f(x) - f(x'))/f(x)$$

and the ratio of relative changes is

$$\frac{f(x) - f(x')}{f(x)} \bigg/ \frac{x - x'}{x} = \frac{xf'(x)}{f(x)}$$

by the mean value theorem. This is the condition number of the function. When it is large the function is ill-conditioned at argument x.

2.34 What is the condition of $f(x) = \sqrt{x}$?

❚ We find

$$xf'(x)/f(x) = x(1/2\sqrt{x})/\sqrt{x} = 1/2$$

so that computation of square roots is well-conditioned. It actually cuts the relative error in half. For example, $\sqrt{100}$ is 10; a 10% change in the argument would have us calculating $\sqrt{110}$, which is 10.5. The change in output is only 5%.

2.35 What is the condition of $f(x) = 1/(1 - x)$?

❚ We find

$$xf'(x)/f(x) = x/(1 - x)$$

which is large near $x = 1$. The function is ill-conditioned in this neighborhood, which is, of course, no surprise. With $x = .99$ we have $f(x) = 100$, but shifting to $x = .98$ brings $f(x) = 50$ instead. A 1% change in x causes a 50% change in the function.

2.36   Determine the condition of $f(x) = x^2$.

   ▮ We find

$$xf'(x)/f(x) = 2$$

which is interpreted to mean that the relative, or percent, error in the function will be about twice that of the argument x. Suppose the true argument is $x = 2$ but we have the poor approximation 2.5. The input error is thus 25%. The output error will be $6.25 - 4 = 2.25$, corresponding to a relative error of $2.25/4$ or just over 50%.

2.37   Find the condition number of $f(x) = x^n$.

   ▮ An easy mental computation manages

$$xf'(x)/f(x) = n$$

and we learn that high powers of x are not well-conditioned. For example, if $f(x) = x^8$, then $f(1) = 1$ while $f(1.05)$ proves to be close to 1.48. A 5% change in argument leads to a 48% change in the function, more or less as predicted by the condition number, and for the relatively unchallenging argument of 1.

2.38   Find the condition number of $f(x) = \sin x$.

   ▮ It is, of course,

$$xf'(x)/f(x) = (x \cos x)/\sin x$$

At $x = \pi/2$ it will be zero, which is reasonable since the sine function is flat in this vicinity. Comparing $f(1.57) = 1.0000$ with $f(1.50) = .9975$ we have a relative change in argument of, say, 5% and a relative change in the function of about 1/4%, roughly about a twentieth of the input error.

2.39   What is meant by the condition of a numerical problem?

   ▮ A problem is well-conditioned if small changes in the input information cause small changes in the output. This qualitative definition extends the idea of condition beyond its use for functions as just described. For instance, the system

$$x + y = 1$$
$$1.1x + y = 2$$

presents obvious troubles, representing the intersection of nearly parallel lines. It has the unique solution $x = 10$, $y = -9$. Now change the 1.1 to 1.05:

$$x + y = 1$$
$$1.05x + y = 2$$

The new solution is $x = 20$, $y = -19$. A 5% change in one coefficient has caused a 100% change in the solution.

2.40   What is a stable algorithm?

   ▮ In extended calculations it is likely that many small errors will be introduced, perhaps most of them roundoffs. Each of them plays the role of an input error for the remainder of the computation and has an impact on the eventual output. Algorithms for which the cumulative effect of all such errors is limited, so that a useful result is generated, are called stable algorithms. Unfortunately there are algorithms for which error accumulation is devastating and the solution is overwhelmed by error. Such algorithms are unstable. Clearly, the ideas of instability and ill-conditioning are closely related.

2.4 NORMS OF VECTORS AND MATRICES

2.41   What is a norm of a vector and in what way is it used?

▌ The most familiar example of a vector norm is the Euclidean length

$$\left(v_1^2 + v_2^2 + \cdots + v_n^2 \right)^{1/2}$$

for the vector V with components v_i. It is given the symbol $\|V\|$ and has these three important properties.

(1) $\|V\| \geq 0$ and equals 0 if and only if $V = 0$.
(2) $\|cV\| = c \cdot \|V\|$ for any number c.
(3) $\|V + W\| \leq \|V\| + \|W\|$.

The last property is called the triangle inequality.

Several other real functions also have these properties and are also called norms. All are used as measures of vector size, often in error studies. Of particular interest are the L_p norms

$$\|V\|_p = \left(\sum_{i=1}^{n} |v_i|^p \right)^{1/p}$$

for $p \geq 1$. With $p = 1$ it is the L_1 norm, the sum of the component magnitudes. With $p = 2$ it is the Euclidean length or L_2 norm. As p tends to infinity the dominant v_i takes over and we have the maximum norm

$$\lim \|V\|_p = \max_i |v_i| = \|V\|_\infty$$

2.42   Using the three norms just described, for $p = 1$, $p = 2$ and $p = \infty$, describe the corresponding sets of unit vectors having their initial points at $0, 0$.

▌ With L_1, vectors such as $(1,0)$, $(1/2, 1/2)$ and $(0,1)$ among others have norm 1 and are called unit vectors. It is easy to see that the set of all such vectors will have terminal points forming the square of Fig. 2-1(a). The more familiar unit vectors of the Euclidean norm terminate on the unit circle of Fig. 2-1(b). Using the L_∞ norm it is vectors such as $(1,0)$, $(1,1)$, $(0,1)$ that have norm 1. Their plot is Fig. 2-1(c). If norms are taken to be measures of distance, as they are, then even these two squares must be considered unit circles.

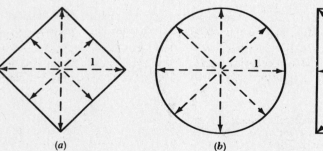 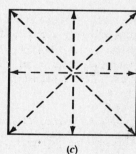

(a)   (b)   (c)   **Fig. 2-1**

2.43   Let P be a point on the real line between 0 and 1. If its coordinate is x and we define the vector V with components x and $1 - x$, for what value of x does V have minimum norm?

$$\begin{array}{ccc} | & | & | \\ 0 & x & 1 \end{array}$$

▌ Our vector V is $(x, 1 - x)$. Taking the L_1 norm first we have

$$\|V\| = x + (1 - x) = 1$$

for any x, so any x provides a minimum and V is always a unit vector. With the L_2 norm we must minimize

$$\|V\|_2 = x^2 + (1 - x)^2$$

and find easily $x = 1/2$. Finally

$$\|V\|_\infty = \text{the larger of } x \text{ and } 1 - x$$

which will be minimized if we choose $x = 1/2$.

2.44 Given the three points $(0,0)$, $(1,1)$ and $(2,0)$ as shown in Fig. 2-2, let L be a horizontal line at altitude a. Let V be the vector of vertical deviations of the line from the given points. For what value of a will V have minimum L_1 norm?

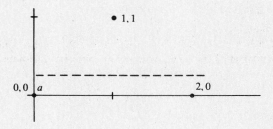

Fig. 2-2

❚ Clearly $V = (a, 1 - a, a)$. In the L_1 norm,

$$\|V\| = a + (1 - a) + a = 1 + a$$

which is minimal if $a = 0$. That the line through the two low points should be the "best" line may come as a slight surprise, but look at it this way. Any effort to raise this line, say by an amount r, will increase the two end deviations by a total of $2r$ and diminish the center deviation by r. So there would be a net gain in total deviation of amount r.

2.45 Rework the preceding problem using the L_2 norm.

❚ Now we find

$$\|V\|^2 = a^2 + (1 - a)^2 + a^2 = 3a^2 - 2a + 1$$

which is least for $a = 1/3$. Why are we now allowed to raise the line to this level? It is because the deviations are now being squared. The function $f(x) = x^2$ grows very slowly near $x = 0$ but comes down fairly rapidly near $x = 1$. For our problem this means that as the line is raised slightly from the zero level the two end deviations (squared) do not grow as fast as the center one (squared) diminishes. At $a = 1/3$ this ceases to be true and we have the line of minimum norm.

2.46 Again rework Problem 2.44, this time using the maximum norm.

❚ We now want the larger of a and $1 - a$ to be as small as we can make it. Clearly this happens when they are equal, that is, when $a = 1/2$.

2.47 Given the three points $(0,0)$, $(1,0)$ and $(2,1)$ and a line L that threads its way between them, more or less as shown in Fig. 2-3, find the "best" position for this line.

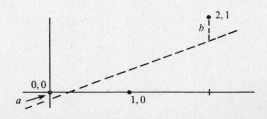

Fig. 2-3

▮ Presumably this means we are to find the line that misses the three points by as little as possible. The vector V whose components are the three deviations will be

$$V = \left(a, \frac{1-a-b}{2}, b \right)$$

and we come to a choice of norm. Taking L_2 first,

$$\|V\|^2 = a^2 + (1-a-b)^2/4 + b^2$$

and a bit of algebra and calculus shows the minimum to occur for $a = b = 1/6$. The center deviation is then $1/3$. This line is called the least-squares line for the point set.

2.48   Continue the preceding problem using the maximum norm.

▮ Intuition, perhaps supported by the experiences of Problems 2.43 and 2.46, with extensive support to follow in our analysis of min-max lines, recommends that the three deviations be of equal size. Following this clue we are led quickly to $a = b = 1/4$.

2.49   Also apply the L_1 norm to Problem 2.47.

▮ We now have

$$\|V\| = a + (1-a-b)/2 + b = (1+a+b)/2$$

and must consider the limitations on a and b. For the line L to be as described, these two deviations may be restricted to keep the point (a, b) within the shaded triangle of an a, b plane shown as Fig. 2-4. This implies that $a + b$ is least at $a = b = 0$, putting the "best" line through the two extreme points. The norm of V is then $1/2$. The solution method just used is called linear programming, though it is, of course, only a minor example of this powerful tool.

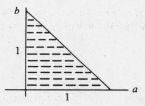

Fig. 2-4

Which line L is really the best one? The answer depends on the use to which the line will be put, as will be discussed more fully in later chapters.

2.50   Three lines of position cross as shown in Fig. 2-5. They were supposed to meet in a point, but measurement errors have intervened. What is the best estimate of the desired point?

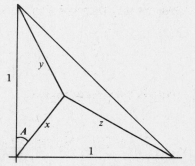

Fig. 2-5

I The problem is a simplified version of the navigator's problem at sea, but Gauss first encountered it in his almost lifelong work as a surveyor. His solution uses the L_2 norm. From trigonometry we borrow

$$y^2 = 1 + x^2 - 2x \cos A$$
$$z^2 = 1 + x^2 - 2x \sin A$$

so that with $V = (x, y, z)$,

$$\|V\|_2^2 = x^2 + y^2 + z^2 = 2 + 3x^2 - 2x(\sin A + \cos A)$$

Partial derivatives relative to x and A are

$$6x - 2(\sin A + \cos A) = 0$$
$$-2x(\cos A - \sin A) = 0$$

from which come $\cos A = \sin A = \sqrt{2}/2$ and $x = (2/3)(1/\sqrt{2})$. The required point is soon discovered to be at the center of gravity of the triangle.

2.51 Why did Gauss choose the L_2 norm?

I His formula for error distribution made the probability of an error of size x proportionate to $\exp(-x^2/c)$. Considering the three corners of the triangle as erroneous "fixes" and the required point as true, the probability of this trio of errors is proportionate to $\exp(-(x^2 + y^2 + z^2)/c)$. This is maximized by minimizing the sum $x^2 + y^2 + z^2$, the L_2 norm of V.

2.52 Continuing Problem 2.50, find the L_1 estimate of the best fix.

I We now have $\|V\| = x + 2y = x + 2\sqrt{1 + x^2 - \sqrt{2}\,x}$ and setting the derivative to zero find $x = .3$ approximately.

2.53 Also apply the maximum norm to Problem 2.50.

I The midpoint of the hypotenuse is equidistant from all three vertices of the triangle. Any departure from this point will increase one or more components of V, so this is the best fix according to the maximum norm.

2.54 Define the norm of a matrix A.

I The usual definition is

$$\|A\| = \max\|AV\|$$

The maximum is taken over all unit vectors V and the meaning of unit depends on the type of vector norm being used. There is, therefore, a matrix norm corresponding to each type of vector norm.

2.55 List some basic properties of matrix norms.

I The following are parallel to those listed earlier for vector norms.

(1) $\|A\| \geq 0$ and equals 0 if and only if $A = 0$.
(2) $\|cA\| = c\|A\|$ for any number c.
(3) $\|A + B\| \leq \|A\| + \|B\|$.

In addition, for matrices A, B and vector V, the properties

(4) $\|AV\| \leq \|A\| \cdot \|V\|$,
(5) $\|AB\| \leq \|A\| \cdot \|B\|$

will be useful.

2.56 What are the L_1, L_2 and L_∞ norms of the identity matrix?

I They are all 1. We have

$$\|I\| = \max\|IV\| = \max\|V\| = 1$$

since V is a unit vector.

2.57 Find the usual three norms for this matrix $\begin{pmatrix} 1 & 1 \\ 1 & 1 \end{pmatrix}$.

■ With $V = (v_1, v_2)$ we have

$$AV = \begin{pmatrix} v_1 + v_2 \\ v_1 + v_2 \end{pmatrix}$$

Assume for simplicity that v_1, v_2 are nonnegative. Then for the L_1 norm $\|AV\| = 2(v_1 + v_2) = 2$ since V is a unit vector in the L_1 norm. Thus $\|A\|_1 = 2$. For the L_2 norm we must square and add the two components, obtaining $2(v_1^2 + 2v_1v_2 + v_2^2)$. In this norm $v_1^2 + v_2^2 = 1$ so we maximize v_1v_2. Elementary calculus then manages $v_1 = v_2 = 1/\sqrt{2}$, leading quickly to $\|A\|_2 = 2$. Finally $\|AV\|_\infty = v_1 + v_2$ since with this norm we seek the maximum component. But here again the maximum is 2, because with this norm neither v_i can exceed 1. The L_1 and L_∞ norms could have been read instantly using the results of the following two problems.

2.58 Show that

$$\|A\|_\infty = \max_i \sum_{j=1}^n |a_{ij}|$$

■ Choose a vector V with all components of size 1 and signs matching the a_{ij} such that $\Sigma |a_{ij}| v_j$ is maximal. Then $\Sigma a_{ij} v_i$ is an element of AV equaling this maximal value and clearly cannot be exceeded. Since this V has norm 1, the norm of A also takes this value. This norm is thus the maximum row sum of absolute elements of A.

2.59 Show that

$$\|A\|_1 = \max_j \sum_{i=1}^n |a_{ij}|$$

making this norm the maximum column sum of absolute elements of A.

■ Choose a vector V with a single component of 1, the others being 0. Then the product AV is a column of A. Position the 1 so that the column with maximal absolute sum is obtained. This maximal sum must be the L_1 norm of A.

2.60 Find the L_1 and L_∞ norms of $\begin{pmatrix} 1 & 1 \\ 1 & 0 \end{pmatrix}$.

■ The maximum row and column sums are both 2 which is thus the common value of these two norms.

2.61 Also find the L_2 norm of the matrix of Problem 2.60.

■ Unfortunately there is no shortcut to the L_2 norm and this innocent appearing matrix does not yield its value without some resistance. By definition, the L_2 norm of the matrix is the maximum L_2 norm of the vector

$$\begin{pmatrix} 1 & 1 \\ 1 & 0 \end{pmatrix} \begin{pmatrix} x \\ y \end{pmatrix} = \begin{pmatrix} x + y \\ x \end{pmatrix}$$

for $x^2 + y^2 = 1$, that is, for (x, y) on the unit circle of Fig. 2-1(b). The square of this norm is

$$(x + y)^2 + x^2 = 1 + 2xy + x^2 = 1 + 2x\sqrt{1 - x^2} + x^2$$

which can be maximized by elementary calculus. The assumption that y is positive here is not restrictive since the norm takes the same value for (x, y) and $(-x, -y)$. Eventually one discovers that a maximum occurs for $x^2 = (1/2) + 5/10$ and that

$$\|A\|_2^2 = (3 + \sqrt{5})/2$$

2.62 Prove the triangle property of vector length, the L_2 norm, by first proving the Cauchy–Schwarz inequality

$$\left(\sum a_i b_i \right)^2 \le \left(\sum a_i^2 \right) \left(\sum b_i^2 \right)$$

■ One interesting proof begins by noting that $\Sigma(a_i - b_i x)^2$ is nonnegative, so that the quadratic equation

$$\left(\sum b_i^2 \right) x^2 - 2 \left(\sum a_i b_i \right) x + \sum a_i^2 = 0$$

cannot have distinct real roots. This requires

$$4 \left(\sum a_i b_i \right)^2 - 4 \sum a_i^2 \cdot \sum b_i^2 \le 0$$

and cancelling the 4s, we have the Cauchy–Schwarz inequality.

The triangle inequality now follows quite directly, but with a bit of algebra. Written in component form it states

$$\left[(v_1 + w_1)^2 + \cdots + (v_n + w_n)^2\right]^{1/2} \le \left(v_1^2 + \cdots + v_n^2\right)^{1/2} + \left(w_1^2 + \cdots + w_n^2\right)^{1/2}$$

Squaring, removing common terms, squaring again and using the Cauchy–Schwarz inequality will bring the desired result.

2.63 Show that the vector L_p norm approaches $\max|v_i|$ for p tending to infinity.

$\blacksquare$ Suppose v_m is the absolutely largest component and rewrite the sum as

$$|v_m|\left(1 + \sum_{i \ne m} |v_i/v_m|^p\right)^{1/p}$$

Within the parentheses all terms except the first approach zero as limit and the required result follows.

2.64 Show that the definition of $\|A\|$ satisfies properties (1) to (3) as listed in Problem 2.55.

$\blacksquare$ These follow fairly easily from the corresponding properties of the companion vector norm. Since AV is a vector, $\|AV\| \ge 0$ and so $\|A\| \ge 0$. If $\|A\| = 0$ and even one element of A were not zero, then V could be chosen to make a component of AV positive, a contradiction of $\max\|AV\| = 0$. This proves the first.

Next we find

$$\|cA\| = \max\|cAV\| = \max|c| \cdot \|AV\| = |c| \cdot \|A\|$$

proving the second. The third is handled similarly.

2.65 Prove that $\|AV\| \le \|A\| \cdot \|V\|$.

$\blacksquare$ For a unit vector U we have by definition of $\|A\|$

$$\|AU\| \le \max_u \|AU\| = \|A\|$$

so choosing $U = V/\|V\|$ and applying property (2),

$$\|A(V/\|V\|)\| \le \|A\| \qquad \|AV\| \le \|A\| \cdot \|V\|$$

2.66 Prove $\|AB\| \le \|A\| \cdot \|B\|$.

$\blacksquare$ We make repeated use of the result of Problem 2.65:

$$\|AB\| = \max\|ABU\| \le \max\|A\| \cdot \|BU\| \le \max\|A\| \cdot \|B\| \cdot \|U\| = \|A\| \cdot \|B\|$$

2.67 In a city of square blocks, which is the suitable norm for taxicab travel?

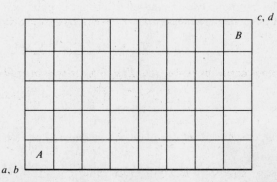

Fig. 2-6

I Even without Fig. 2-6 one decides that the taxicab will travel 13 blocks to reach an objective 8 blocks east and 5 blocks north. So the appropriate norm adds the two components and is the L_1 norm. More generally, to travel from a corner at a, b to one at c, d will involve

$$|c - a| + |d - b|$$

blocks. If the skyscrapers are tall enough even the crows must abide by this norm.

2.68 Again using Fig. 2-6, but now thinking of it as part of a chessboard, what is the appropriate norm for the travel of a chess king?

I Measuring distance by the minimum number of moves the king would need and recalling that diagonal moves are permitted, we see that seven moves are enough for travel between squares A and B. Diagonal moves allow the four upward steps to be covered while concentrating on the seven sideways steps. More generally,

$$\text{distance} = \max(|c - a|, |d - b|) = L_\infty \text{ norm}$$

2.5 ERROR ANALYSIS

2.69 Assume a number x represented by a floating-point binary symbol, rounded to a mantissa of n bits. Also assume normalization. What are the bounds for the absolute and relative errors caused by the rounding?

I Rounding will cause an error of at most a unit in the $(n + 1)$th binary place or half a unit in the nth place. So

$$\text{absolute error} \le 2^{-n-1}$$

while for the relative error we must take into account the true value x. Normalization means a mantissa no smaller than $\frac{1}{2}$ and this leads to the bound

$$|\text{relative error}| \le \frac{2^{-n-1}}{2^{-1}} = 2^{-n}$$

It is useful to rewrite this letting $\text{fl}(x)$ represent the floating-point symbol for x. Then

$$\text{relative error} = \frac{\text{fl}(x) - x}{x} = E$$

or

$$\text{fl}(x) = x(1 + E) = x + xE$$

with $|E| \le 2^{-n}$. The operation of rounding off can thus be viewed as the replacement of x by a perturbed value $x + xE$, the perturbation being relatively small.

2.70 Find a bound for the relative error made by the addition of two floating-point numbers.

I Let the numbers be $x = m_1 * 2^e$ and $y = m_2 * 2^f$ with y the smaller. Then m_2 must be shifted $e - f$ places to the right (lining up the binary points). The mantissas are then added, the result normalized and rounded. There are two possibilities. Either overflow occurs to the left of the binary point (not overflow in the sense of Problem 1.18) or it does not. The first possibility is characterized by

$$1 \le |m_1 + m_2 * 2^{f-e}| < 2$$

and the second by

$$\tfrac{1}{2} \le |m_1 + m_2 * 2^{f-e}| < 1$$

If overflow does occur, a right shift of one place will be required and we have

$$\text{fl}(x + y) = \left[(m_1 + m_2 * 2^{f-e})2^{-1} + \epsilon \right] * 2^{e+1}$$

where ϵ is the roundoff error. This can be rewritten

$$\text{fl}(x+y) = (x+y)\left(1 + \frac{2\epsilon}{m_1 + m_2 * 2^{f-e}}\right)$$

$$= (x+y)(1+E)$$

with $|E| \leq 2\epsilon \leq 2^{-n}$.

If there is no overflow, then

$$\text{fl}(x+y) = \left[(m_1 + m_2 * 2^{f-e}) + \epsilon\right] * 2^e$$

$$= (x+y)\left(1 + \frac{\epsilon}{m_1 + m_2 * 2^{f-e}}\right)$$

$$= (x+y)(1+E)$$

with E bounded as before.

2.71 Find a bound for the relative error made by multiplying two floating-point numbers.

$\blacksquare$ Again let the two numbers be $x = m_1 * 2^e$ and $y = m_2 * 2^f$. Then $xy = m_1 m_2 * 2^{e+f}$ with $\frac{1}{4} \leq |m_1 m_2| < 1$ because of normalization. This means that to normalize the product there will be a left shift of at most one place. Rounding will, therefore, produce either $m_1 m_2 + \epsilon$ or $2m_1 m_2 + \epsilon$, with $|\epsilon| \leq 2^{-n-1}$. This can be summarized as

$$\text{fl}(xy) = \begin{cases} (m_1 m_2 + \epsilon) * 2^{e+f} & \text{if } |m_1 m_2| \geq \frac{1}{2} \\ (2m_1 m_2 + \epsilon) * 2^{e+f-1} & \text{if } \frac{1}{2} > |m_1 m_2| \geq \frac{1}{4} \end{cases}$$

$$= m_1 m_2 * 2^{e+f} \begin{cases} 1 + \dfrac{\epsilon}{m_1 m_2} & \text{if } |m_1 m_2| \geq \frac{1}{2} \\ 1 + \dfrac{\epsilon}{2m_1 m_2} & \text{if } \frac{1}{2} > |m_1 m_2| \geq \frac{1}{4} \end{cases}$$

$$= xy(1+E)$$

with $|E| \leq 2|\epsilon| \leq 2^{-n}$.

This means that in all four arithmetic operations, using floating-point numbers, the relative error introduced does not exceed 1 in the least significant place of the mantissa.

2.72 Estimate the error generated in computing the sum

$$x_1 + x_2 + \cdots + x_k$$

using floating-point operations.

$\blacksquare$ We consider the partial sums s_i. Let $s_1 = x_1$. Then

$$s_2 = \text{fl}(s_1 + x_2) = (s_1 + x_2)(1 + E_1)$$

with E_1 bounded by 2^{-n} as shown in Problem 2.69. Rewriting,

$$s_2 = x_1(1 + E_1) + x_2(1 + E_1)$$

Continuing

$$s_3 = \text{fl}(s_2 + x_3) = (s_2 + x_3)(1 + E_2)$$

$$= x_1(1 + E_1)(1 + E_2) + x_2(1 + E_1)(1 + E_2) + x_3(1 + E_2)$$

and eventually

$$s_k = \text{fl}(s_{k-1} + x_k) = (s_{k-1} + x_k)(1 + E_{k-1})$$

$$= x_1(1 + c_1) + x_2(1 + c_2) + \cdots + x_k(1 + c_k)$$

where, for $i = 2, \ldots, k$,

$$1 + c_i = (1 + E_{i-1})(1 + E_i) \cdots (1 + E_{k-1})$$

and $1 + c_1 = 1 + c_2$. In view of the uniform bound on the E_j, we now have this estimate for the $1 + c_i$:

$$(1 - 2^{-n})^{k-i+1} \leq 1 + c_i \leq (1 + 2^{-n})^{k-i+1}$$

Summarizing

$$\text{fl}\left(\sum_{j=1}^{k} x_j\right) = \left(\sum_{j=1}^{k} x_j\right)(1 + E)$$

where

$$E = \sum_{j=1}^{k} x_j c_j \bigg/ \sum_{j=1}^{k} x_j$$

Note that if the true sum Σx_j is small compared with the x_j, then the relative error E can be large. This is the cancellation effect caused by subtractions, observed earlier in Problem 2.9.

2.73 Illustrate a forward error analysis.

▮ Suppose the value of $A(B + C)$ is to be computed, using approximations a, b, c which are in error by amounts e_1, e_2, e_3. Then the true value is

$$A(B + C) = (a + e_1)(b + e_2 + c + e_3) = ab + ac + \text{error}$$

where

$$\text{error} = a(e_2 + e_3) + be_1 + ce_1 + e_1e_2 + e_1e_3$$

Assuming the uniform bound $|e_i| \leq e$ and that error products can be neglected, we find

$$|\text{error}| \leq (2|a| + |b| + |c|)e$$

This type of procedure is called forward error analysis. In principle it could be carried out for any algorithm. Usually, however, the analysis is tedious if not overwhelming. Besides, the resulting bounds are usually very conservative, suitable if what is needed is an idea of the worst that might happen. In the present example one point of minor interest does surface. The value of a seems to be twice as sensitive as the values of b and c.

2.74 What is backward error analysis?

▮ The underlying idea of backward error analysis is to take the result of a computation and try to determine the range of input data that could have produced it. It is important not to misunderstand the motive here. There is no intention of modifying the data to accommodate the answer. If a backward error analysis is completed and shows that the result found is consistent with the input data, within the range of observational or roundoff error, then one may have some confidence in the result. If this does not happen, then a major source of error exists elsewhere, presumably within the algorithm itself.

2.75 Show that the error analysis in Problem 2.70 was a backward error analysis.

▮ The result obtained was

$$\text{fl}(x + y) = (x + y)(1 + E)$$

with $|E| \leq 2^{-n}$, where n is the number of binary places in the mantissas. Rewriting this as

$$\text{fl}(x + y) = x(1 + E) + y(1 + E)$$

and recalling Problem 2.69, we see that the sum as computed, namely $\text{fl}(x + y)$, is also the true sum of numbers differing from the original x and y by no more than the roundoff error bound E. That is, the output can be explained by input data well within the recognized error limit.

2.76 Show that the analysis done in Problem 2.71 was a backward error analysis.

▮ We found

$$\text{fl}(xy) = xy(1 + E)$$

which we can think of as the product of x by $y(1 + E)$. This means that the computed $\mathrm{fl}(xy)$ is also the true product of numbers differing from the original x and y by no more than the roundoff error. It is consistent with input data well within our recognized error limit.

2.77 What does the backward error analysis made in Problem 2.72 indicate?

▮ First, the equation

$$\mathrm{fl}\left(\sum_{j=1}^{k} x_j\right) = x_1(1 + c_1) + \cdots + x_k(1 + c_k)$$

shows that the floating-point sum of k numbers x_1 to x_k is also the true sum of k numbers which differ from the x_j by relative errors of size c_j. Unfortunately the estimates then obtained in Problem 2.72 also show that these errors can be much greater than simple roundoffs.

2.6 ADDITIONAL PROBLEMS

2.78 Use the method of Archimedes to estimate the value of pi.

▮ Archimedes imagined regular polygons of $4, 8, 16, \ldots, 2^k$ sides placed inside a circle of radius 1, as shown in Fig. 2-7. The triangle corresponding to each side is easily found to have area $\frac{1}{2} \sin A$, while the area of the circle is, of course, pi. The polygon area of $\frac{1}{2}n \sin A$, where $n = 2^k$ is the number of sides, is an approximation to pi which improves with increasing n.

For $n = 4$ we have $A = 90°$ making $\sin A = 1$ and $\cos A = 0$. Each time the value of n is doubled the angle A will be halved, so the new sine and cosine can be found economically using a half angle formula:

$$\sin \tfrac{1}{2}A = \sqrt{\frac{1 - \cos A}{2}} \qquad \cos \tfrac{1}{2}A = \sqrt{1 - \sin^2 A}$$

Some computational results appear in the figure, and it will be noted that they do not improve indefinitely. The best result seems to be for $n = 2^9$, and beyond 2^{15} we have disaster. There is a numerical instability in the algorithm. Archimedes himself found the approximation $3\frac{1137}{8069} = 3.1409$ by hand using $n = 96$. He also found the upper bound 3.1428. Midway between these is the very respectable estimate of 3.14185.

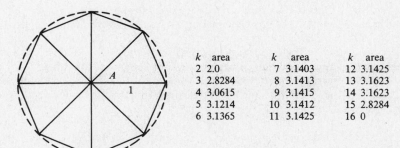

k	area	k	area	k	area
2	2.0	7	3.1403	12	3.1425
3	2.8284	8	3.1413	13	3.1623
4	3.0615	9	3.1415	14	3.1623
5	3.1214	10	3.1412	15	2.8284
6	3.1365	11	3.1425	16	0

Fig. 2-7

2.79 A sequence $J_0, J_1, J_2, \ldots$ is defined by

$$J_{n+1} = 2nJ_n - J_{n-1}$$

with $J_0 = .765198$ and $J_1 = .440051$ correct to six places. Compute $J_2, \ldots, J_7$ and compare with the correct values that follow. (These correct values were obtained by an altogether different process. See the next problem for explanation of errors.)

n	2	3	4	5	6	7
correct J_n	.114903	.019563	.002477	.000250	.000021	.000002

❚ One computation found the values

$$.114904 \quad .019565 \quad .002486 \quad .000323 \quad .000744 \quad .008605$$

The results may differ slightly from computer to computer but the numerical instability should be apparent in any case. The error has already grown from one unit in the sixth decimal place to more than eight units in the third.

2.80 Show that for the sequence of the preceding problem,

$$J_7 = 36767 J_1 - 21144 J_0$$

exactly. Compute this from the given values of J_0 and J_1. The same erroneous value will be obtained. The large coefficients multiply the roundoff errors in the given J_0 and J_1 values and the combined results then contain a large error.

❚ Exact calculations give us

$$
\begin{aligned}
J_2 &= 2J_1 - J_0 \\
J_3 &= 4J_2 - J_1 = 7J_1 - 4J_0 \\
J_4 &= 6J_3 - J_2 = 40J_1 - 23J_0 \\
J_5 &= 8J_4 - J_3 = 313J_1 - 180J_0 \\
J_6 &= 10J_5 - J_4 = 3090J_1 - 1777J_0 \\
J_7 &= 12J_6 - J_5 = 36767J_1 - 21144J_0
\end{aligned}
$$

as suggested. A double-precision calculation is now in order and will output the same .008605 just found.

To six places the number J_8 should be all zeros. The formula of Problem 2.79 manages .119726.

2.81 If only an 8 qt pail and a 5 qt pail are available, plus an infinite supply of water, how can exactly 4 qt be measured?

❚ This problem is an old classic and is solved by a historic algorithm. Call the two pails E and F. The algorithm follows.

    Repeat as necessary:
        fill E from the infinite source,
        fill F from E, empty F, repeat if possible,
        transfer remainder from E to F.

Any desired whole number of quarts from 1 to 8 will be achieved at some point. The first fill of E manages one fill of F, then transfers 3 qt to F. But 3 was not the target number so we repeat. The second fill of E is used to top up F, leaving 6 qt in E. But this was not the target number either, so we empty F, refill it at once from E, empty F again. The remainder of 1 qt is then transferred from E to F. Since 1 was not the target number we begin a third pass through the algorithm. Filling E from the source, 4 qt are needed to top up F, leaving the target number of 4 in E. A continuation would find the missing measures of 2 and 7 qt following.

There is a companion algorithm in which F is taken to the source and used to fill and refill E. The problem is numerical, and easily analyzed by modular arithmetic. No error study will be attempted since the accuracy of pouring is unknown.

2.82 One coin in a set of nine is heavier than the rest, though all of them look alike. Develop an algorithm for identifying the odd coin using only a balance scale.

❚ This is another old classic and one of the easier of its type since the odd coin is known in advance to be heavier than the rest, not just different. Limiting the procedure to use of a balance scale only requires that the identification be made by comparing the coins with one another, not by weighing each one separately. Two weighings are enough. Here is the algorithm.

    Place three coins in each tray of the scale.
    If the scale is in balance,
        then the odd coin is one of the unused three;
        otherwise it is in the lower tray.
    (The field has been narrowed to three coins, say, X, Y, Z.)
    Weigh X against Y.
    If they balance,
        then Z is the odd coin;
        otherwise it is in the lower tray.

CHAPTER 3
Classical Numerical Analysis to Newton's Formula

3.1 THE COLLOCATION POLYNOMIAL

3.1 Prove that any polynomial $p(x)$ may be expressed as

$$p(x) = (x - r)q(x) + R$$

where r is any number, $q(x)$ is a polynomial of degree $n - 1$ and R is a constant.

▌ This is an example of the *division algorithm*. Let $p(x)$ be of degree n:

$$p(x) = a_n x^n + a_{n-1} x^{n-1} + \cdots + a_0$$

Then

$$p(x) - (x - r)a_n x^{n-1} = q_1(x) = b_{n-1} x^{n-1} + \cdots$$

will be of degree $n - 1$ or less. Similarly,

$$q_1(x) - (x - r)b_{n-1} x^{n-2} = q_2(x) = c_{n-2} x^{n-2} + \cdots$$

will be of degree $n - 2$ or less. Continuing in this way, we eventually reach a polynomial $q_n(x)$ of degree zero, a constant. Renaming this constant R, we have

$$p(x) = (x - r)(a_n x^{n-1} + b_{n-1} x^{n-2} + \cdots) + R = (x - r)q(x) + R$$

3.2 Prove $p(r) = R$. This is called the *remainder theorem*.

▌ Let $x = r$ in Problem 3.1. At once, $p(r) = 0 \cdot q(r) + R$.

3.3 Illustrate the *synthetic division method* for performing the division described in Problem 3.1, using $r = 2$ and $p(x) = x^3 - 3x^2 + 5x + 7$.

▌ Synthetic division is merely an abbreviated version of the same operations described in Problem 3.1. Only the various coefficients appear. For the $p(x)$ and r above, the starting layout is

$$r = 2 \underline{|\quad 1 \;\; -3 \;\; 5 \;\; 7} \leftarrow \text{coefficients of } p(x)$$
$$1$$

Three times we "multiply by r and add" to complete the layout.

$$r = 2 \underline{\begin{array}{rrrr} 1 & -3 & 5 & 7 \\ & 2 & -2 & 6 \end{array}}$$
$$\underbrace{1 \;\; -1 \;\; 3}_{\substack{\text{coefficients} \\ \text{of } q(x)}} \;\; 13 \leftarrow \text{the number } R$$

Thus, $q(x) = x^2 - x + 3$ and $R = f(2) = 13$. This may be verified by computing $(x - r)q(x) + R$, which will be $p(x)$. It is also useful to find $q(x)$ by the long division method, starting from this familiar layout:

$$(x - 2) \overline{)\; x^3 - 3x^2 + 5x + 7}$$

Comparing the resulting computation with the synthetic algorithm just completed, one easily sees the equivalence of the two.

3.4 Apply synthetic division to divide $p(x) = x^3 - x^2 + x - 1$ by $x - 1$. What is the quotient? Also compute $p(7)$:

$$
\begin{array}{r|rrrr}
r=1 & 1 & -1 & 1 & -1 \\
 & & 1 & 0 & 1 \\
\hline
 & 1 & 0 & 1 & 0
\end{array}
$$

∎ It is no surprise that $R = p(1) = 0$, making $x - 1$ a factor and 1 a zero of $p(x)$. The quotient is $q(x) = x^2 + 1$, so the other roots of this cubic are $\pm i$. The computation of $p(7) = 300$ is routine

$$
\begin{array}{r|rrrr}
r=7 & 1 & -1 & 1 & -1 \\
 & & 7 & 42 & 301 \\
\hline
 & 1 & 6 & 43 & 300
\end{array}
$$

3.5 For the case of a cubic polynomial $p(x) = a_0 x^3 + a_1 x^2 + a_2 x + a_3$ run the synthetic division process through using divisor $x - r$, to note the development of the remainder theorem:

$$
\begin{array}{r|cccc}
r=r & a_0 & a_1 & a_2 & a_3 \\
 & & a_0 r & a_0 r^2 + a_1 r & a_0 r^3 + a_1 r^2 + a_2 r \\
\hline
 & a_0 & a_0 r + a_1 & a_0 r^2 + a_1 r + a_2 & a_0 r^3 + a_1 r^2 + a_2 r + a_3 = R
\end{array}
$$

∎ The appearance of $p(r)$ in the R position at lower right is the content of the remainder theorem. Suppose r happens to be a root of $p(x) = 0$. The sum of all three roots being $-a_1/a_0$, it may also be worth noting that the second coefficient in the quotient $q(x)$, when divided by $-a_0$, is the sum of the remaining two roots, that is, $-r - a_1/a_0$. The third coefficient also plays its role involving root products.

3.6 Prove that if $p(r) = 0$, then $x - r$ is a factor of $p(x)$. This is the *factor theorem*. The other factor has degree $n - 1$.

∎ If $p(r) = 0$, then $0 = 0q(x) + R$ making $R = 0$. Thus, $p(x) = (x - r)q(x)$.

3.7 Prove that a polynomial of degree n can have at most n zeroes, meaning that $p(x) = 0$ can have at most n roots.

∎ Suppose n roots exist. Call them $r_1, r_2, \ldots, r_n$. Then by n applications of the factor theorem,

$$p(x) = A(x - r_1)(x - r_2) \cdots (x - r_n)$$

where A has degree 0, a constant. This makes it clear that there can be no other roots. (Note also that $A = a_n$.)

3.8 Prove that at most one polynomial of degree n can take the specified values y_k at given arguments x_k, where $k = 0, 1, \ldots, n$.

∎ Suppose there were two such polynomials, $p_1(x)$ and $p_2(x)$. Then the difference $p(x) = p_1(x) - p_2(x)$ would be of degree n or less, and would have zeros at all the arguments x_k: $p(x_k) = 0$. Since there are $n + 1$ such arguments this contradicts the result of the previous problem. Thus, at most one polynomial can take the specified values. It is called the *collocation polynomial*.

3.9 Suppose a polynomial $p(x)$ of degree n takes the same values as a function $y(x)$ for $x = x_0, x_1, \ldots, x_n$. [This is called collocation of the two functions and $p(x)$ is the collocation polynomial.] Obtain a formula for the difference between $p(x)$ and $y(x)$.

∎ Since the difference is zero at the points of collocation, we anticipate a result of the form

$$y(x) - p(x) = C(x - x_0)(x - x_1) \cdots (x - x_n) = C\pi(x)$$

which may be taken as the definition of C. Now consider the function $F(x)$:

$$F(x) = y(x) - p(x) - C\pi(x)$$

This $F(x)$ is zero for $x = x_0, x_1, \ldots, x_n$ and if we choose a new argument x_{n+1} and

$$C = \frac{y(x_{n+1}) - p(x_{n+1})}{\pi(x_{n+1})}$$

then $F(x_{n+1})$ will also be zero. Now $F(x)$ has $n + 2$ zeros at least. By Rolle's theorem $F'(x)$ then is guaranteed $n + 1$ zeros between those of $F(x)$, while $F''(x)$ is guaranteed n zeros between those of $F'(x)$. Continuing to apply Rolle's theorem in this way eventually shows that $F^{(n+1)}(x)$ has at least one zero in the interval from x_0 to x_n, say at $x = \xi$. Now calculate this derivative, recalling that the $(n + 1)$th derivative of $p(x)$ will be zero, and put $x = \xi$:

$$0 = y^{(n+1)}(\xi) - C(n + 1)!$$

This determines C, which may now be substituted back:

$$y(x_{n+1}) - p(x_{n+1}) = \frac{y^{(n+1)}(\xi)}{(n+1)!} \pi(x_{n+1})$$

Since x_{n+1} can be any argument between x_0 and x_n except for $x_0, \ldots, x_n$ and since our result is clearly true for $x_0, \ldots, x_n$ also, we replace x_{n+1} by the simpler x:

$$y(x) - p(x) = \frac{y^{(n+1)}(\xi)}{(n+1)!} \pi(x)$$

This result is often quite useful in spite of the fact that the number ξ is usually undeterminable, because we can estimate $y^{(n+1)}(\xi)$ independently of ξ.

3.10 The line $p(x) = 1 - x$ includes the two points $(0,1)$ and $(1,0)$. So does the function $y(x) = \cos \frac{1}{2} \pi x$. Use Problem 3.9 to produce a formula for the difference.

▮ With $n = 1$, all that is needed is the second difference of cosine function. Inserting it we have

$$y(x) - p(x) = -\frac{\pi^2 \cos \frac{1}{2} \pi \xi}{8} x(x - 1)$$

Even without determining ξ we can estimate this difference by $|y(x) - p(x)| \le (\pi^2/8)x(x - 1)$. Viewing $p(x)$ as a linear approximation to $y(x)$, this error estimate is simple, though generous. At $x = \frac{1}{2}$ it suggests an error of size roughly .3, while the actual error is approximately $\cos \frac{1}{4}\pi - (1 - \frac{1}{2}) = .2$.

3.11 The line $p(x) = x$ passes through $(0,0)$ and $(1,1)$. So does the function $y(x) = \sqrt{x}$. Why is the result in Problem 3.9 of little use in estimating the difference?

▮ The derivatives of the square root function are infinite at $x = 0$, which is within the active interval and so a viable candidate for the role of ξ. The difference between the two functions could be represented by a formula including the factor $1/(4\xi^{-3/2})$, but what is to be done with ξ?

There is no difficulty finding the maximum difference over the interval. The derivative of $y - p$ is $1/2\sqrt{x} - 1$, and set to zero manages $x = \frac{1}{4}$. This is the ideal value for ξ, giving the upper bound $\frac{1}{4}$ for the difference. But to be of any use, a value for ξ must lead the way, not follow.

3.12 The function $y(x) = x^p$ also passes through the points $(0,0)$ and $(1,1)$ of the preceding problem. Use Problem 3.9 to estimate the difference between it and the line $p(x) = x$.

▮ Again $n = 1$ and it is the second derivative that is wanted. We find $y''(x) = p(p - 1)x^{p-2}$ and, substituting,

$$y(x) - p(x) = [p(p - 1)/2]\xi^{p-2}x(x - 1)$$

with ξ somewhere in $(0,1)$. To be on the safe side we choose $\xi = 1$, since the slope of $y(x)$ increases across this interval. The difference between the two functions is then bounded by

$$[p(p - 1)/2]x(1 - x)$$

This time we do at least have a bound. For $p = 2$ it is $x(1 - x)$ with a correct maximum of $\frac{1}{4}$ at the center point.

For $p = 3$ it is three times as large, estimating the maximum difference as $\frac{3}{4}$, whereas the true maximum is only $2\sqrt{3}/9$. With increasing p the functions are a mismatch and the bound useless.

3.13 To find a second-degree polynomial that takes values

x_k	0	1	2
y_k	0	1	0

we could write $p(x) = A + Bx + Cx^2$ and substitute to find the conditions

$$0 = A \qquad 1 = A + B + C \qquad 0 = A + 2B + 4C$$

Solve for A, B and C and so determine this collocation polynomial. Theoretically the same procedure applies for higher-degree polynomials, but more efficient algorithms will be developed.

▮ One easily discovers $B = 2$ and $C = -1$, making $p(x) = 2x - x^2$.

The function $y(x) = \sin(\pi x/2)$ also takes these same values so the error estimate in Problem 3.9 applies. Since the degree of our polynomial is $n = 2$ we need the third derivative of $y(x)$, which is $-(\pi/2)^3 \cos(\pi x/2)$. The error of $p(x)$ as an approximation to this trigonometric function is then $x(x - 1)(x - 2)/6$ times this derivative, or

$$y(x) - p(x) = -\frac{\pi^3 \cos\frac{1}{2}\pi\xi}{48} x(x - 1)(x - 2)$$

where ξ depends upon x.

Moreover, since $\cos x$ is bounded by 1, we also have

$$|y(x) - p(x)| \leq \left| \frac{\pi^3}{48} x(x - 1)(x - 2) \right|$$

At $x = \frac{1}{2}$ this is $3\pi^3/384 = .24$ approximately. The actual error is $2x - x^2 - \sin(\pi/4) = .04$, so the estimate is somewhat pessimistic.

3.14 For the $y(x)$ and $p(x)$ of Problem 3.13, compare values of the first and second derivatives at $x = \frac{1}{2}$:

▮
$$y'(x) = (\pi/2)\cos(\pi x/2) = 1.11$$
$$p'(x) = 2 - 2x = 1$$
$$y''(x) = -(\pi^2/4)\sin(\pi x/2) = -1.75$$
$$p''(x) = -2$$

The approximation of derivatives is not as accurate as that of $y(x)$ itself.

3.15 Using the same $y(x)$ and $p(x)$, compare their integrals over the interval $(0, 2)$.

▮ We find

$$\int_0^2 (2x - x^2)\, dx = 4 - 8/3 = 4/3$$

$$\int_0^2 \sin(\pi x/2)\, dx = -\cos(\pi x/2)(2/\pi) = 4/\pi$$

3.16 Find the unique cubic polynomial that takes the values

x	0	1	2	3
y	0	1	16	81

▮ As an alternate procedure, take the cubic in the form

$$p = a + bx + cx(x-1) + dx(x-1)(x-2)$$

Successive substitution of $x = 0, 1, 2$ and 3 finds a convenient triangular system determining the coefficients

$$
\begin{aligned}
a &= 0 \\
a + b &= 1 & b &= 1 \\
a + 2b + 2c &= 16 & c &= 7 \\
a + 3b + 6c + 6d &= 81 & d &= 6
\end{aligned}
$$

The polynomial is thus available as

$$p = x + 7x(x-1) + 6x(x-1)(x-2)$$

The function $y(x) = x^4$ also takes the preceding values, so we can use Problem 3.9 to express the difference between it and the cubic polynomial. Since $y^{(4)}(x)$ is the constant 24, we have

$$y(x) - p(x) = \pi(x) = x(x-1)(x-2)(x-3)$$

3.17 For the preceding $y(x)$ and $p(x)$, what is the maximum absolute value of $y - p$ on the interval $(0,3)$?

▮ The exact maximum can be found by the following device. Let $u = x - \frac{3}{2}$ to obtain the symmetric form

$$
\begin{aligned}
y - p &= \left(u + \tfrac{3}{2}\right)\left(u + \tfrac{1}{2}\right)\left(u - \tfrac{1}{2}\right)\left(u - \tfrac{3}{2}\right) \\
&= \left(u^2 - \tfrac{9}{4}\right)\left(u^2 - \tfrac{1}{4}\right) = \left(t - \tfrac{9}{4}\right)\left(t - \tfrac{1}{4}\right)
\end{aligned}
$$

where $t = u^2$. The derivative of $y - p$ relative to the original argument x is now $(2t - 5/2)(2u)$ and vanishes for $u = 0$ or $t = \frac{5}{4}$. The first of these leads to a relative maximum of size $\frac{9}{16}$ at the point $x = \frac{3}{2}$. The second leads to $y - p = -1$ at $x = (3 \pm \sqrt{5})/2$. The required absolute maximum of $y - p$ is thus 1.

3.18 The preceding problem may be viewed as the approximation of $y(x) = x^4$ by the cubic polynomial

$$p(x) = x + 7x(x-1) + 6x(x-1)(x-2) = 6x^3 - 11x^2 + 6x$$

The error of the approximation has been found by Problem 3.9 to be exactly $\pi(x) = x(x-1)(x-2)(x-3)$; the fact that $y^{(4)}(x)$ is constant makes this simple result possible. Verify this error result directly.

▮ We have

$$
\begin{aligned}
\text{error} = y(x) - p(x) &= x^4 - 6x^3 + 11x^2 - 6x \\
&= x(x-1)(x-2)(x-3) = \pi(x)
\end{aligned}
$$

3.19 Having approximated x^4 by a cubic polynomial, now apply the same procedure to the simpler problem represented by the data

x	0	1	2
y	0	1	8

▮ Obviously the y values are cubes, but first let us find the unique second-degree polynomial fitting the data

$$p(x) = a + bx + cx(x-1)$$

Mental computations will be adequate to manage $a = 0$, $b = 1$, $c = 3$ and yield $p(x) = x + 3x(x-1) = 3x^2 - 2x$.

The data are, as noted, also compatible with the function $y(x) = x^3$. The difference between the two functions $y(x)$ and $p(x)$ is thus accessible by Problem 3.9:

$$y - p = y'''(\xi)x(x-1)(x-2)/6 = x(x-1)(x-2) = \pi(x)$$

since the derivative is a constant. More directly

$$y - p = x^3 - 3x^2 + 2x = \pi(x)$$

The message of this and the preceding problem can also be verified for the even simpler case of $y(x) = x^2$ and its collocation polynomial of degree 1 for points $(0,0)$ and $(1,1)$:

$$y - p = x^2 - x = x(x - 1) = \pi(x)$$

The formula of Problem 3.9 will confirm this result.

3.20 Suppose a function $y(x)$, defined on the interval (a, b), is to be approximated by a polynomial. Choosing $n + 1$ points we find the unique collocation polynomial of degree n or less. If n is then increased and a new polynomial of higher degree found, is it likely to be a better approximation? Will convergence occur if n is increased indefinitely?

▮ The answer is slightly complicated. For carefully chosen collocation arguments x_k and reasonable functions $y(x)$, convergence is assured, as will appear later. But for the most popular case of equally spaced arguments x_k, divergence may occur. For some $y(x)$ the sequence of polynomials is convergent for all arguments x. For other functions, convergence is limited to a finite interval, with the error $y(x) - p(x)$ oscillating in the manner shown in Fig. 3-1. Within the interval of convergence the oscillation dies out and $\lim(y - p) = 0$, but outside that interval $y(x) - p(x)$ grows arbitrarily large as n increases. The oscillation is produced by the $\pi(x)$ factor, the size being influenced by the derivatives of $y(x)$. This error behavior is a severe limitation on the use of high-degree collocation polynomials.

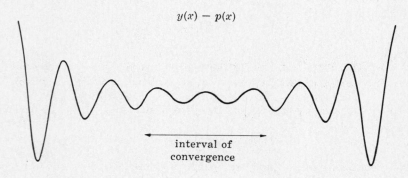

$$y(x) - p(x)$$

interval of
convergence

Fig. 3-1

3.2 FINITE DIFFERENCES

3.21 Define finite differences.

▮ Finite differences have had a strong appeal to mathematicians for centuries. Isaac Newton was an especially heavy user, and much of the subject originated with him. Given a discrete function, that is, a finite set of arguments x_k each having a mate y_k, and supposing the arguments equally spaced, so that $x_{k+1} - x_k = h$, the differences of the y_k values are denoted

$$\Delta y_k = y_{k+1} - y_k$$

and are called first differences. The differences of these first differences are denoted

$$\Delta^2 y_k = \Delta(\Delta y_k) = \Delta y_{k+1} - \Delta y_k = y_{k+2} - 2y_{k+1} + y_k$$

and are called second differences. In general,

$$\Delta^n y_k = \Delta^{n-1} y_{k+1} - \Delta^{n-1} y_k$$

defines the nth differences.

3.22 What is a difference table?

▌ The *difference table* is the standard format for displaying finite differences. Its diagonal pattern makes each entry, except for the x_k and y_k, the difference of its two nearest neighbors to the left:

$$
\begin{array}{llllll}
x_0 & y_0 \\
& & \Delta y_0 \\
x_1 & y_1 & & \Delta^2 y_0 \\
& & \Delta y_1 & & \Delta^3 y_0 \\
x_2 & y_2 & & \Delta^2 y_1 & & \Delta^4 y_0 \\
& & \Delta y_2 & & \Delta^3 y_1 \\
x_3 & y_3 & & \Delta^2 y_2 \\
& & \Delta y_3 \\
x_4 & y_4
\end{array}
$$

3.23 Express $\Delta^2 y_0$ and $\Delta^2 y_1$ in terms of y values.

▌
$$\Delta^2 y_0 = \Delta y_1 - \Delta y_0 = (y_2 - y_1) - (y_1 - y_0)$$
$$= y_2 - 2y_1 + y_0$$

It is now enough to advance all subscripts by 1, converting this into

$$\Delta^2 y_1 = y_3 - 2y_2 + y_1$$

3.24 Express $\Delta^3 y_0$ and $\Delta^4 y_0$ in terms of y values.

▌ Using the preceding problem,

$$\Delta^3 y_0 = \Delta^2 y_1 - \Delta^2 y_0 = y_3 - 3y_2 + 3y_1 - y_0$$

after which

$$\Delta^4 y_0 = \Delta^3 y_1 - \Delta^3 y_0$$
$$= (y_4 - 3y_3 + 3y_2 - y_1) - (y_3 - 3y_2 + 3y_1 - y_0)$$
$$= y_4 - 4y_3 + 6y_2 - 4y_1 + y_0$$

3.25 Compute up through third differences of the discrete function displayed in the x_k and y_k columns of Table 3.1. The integer variable k also appears for convenience.

▌ The required differences appear in the remaining three columns. For example,

$$\Delta y_0 = y_1 - y_0 = 8 - 1 = 7$$
$$\Delta^2 y_0 = \Delta y_1 - \Delta y_0 = 19 - 7 = 12 \quad \text{etc.}$$

Clearly all fourth and higher differences of this tabulated function will be zero. This is a special case of a general result to be obtained shortly.

3.26 Calculate up through fourth differences for this function.

k	0	1	2	3	4	5	6
y	0	1	16	81	256	625	1296

TABLE 3.1

k	x_k	y_k	Δy_k	$\Delta^2 y_k$	$\Delta^3 y_k$
0	1	1			
			7		
1	2	8		12	
			19		6
2	3	27		18	
			37		6
3	4	64		24	
			61		6
4	5	125		30	
			91		6
5	6	216		36	
			127		6
6	7	343		42	
			169		
7	8	512			

▮ The arithmetic is simple enough but there is a point to be made. The fourth differences of k^4 turn out to be constant:

$$
\begin{array}{cccccc}
1 & 15 & 65 & 175 & 369 & 671 \\
\;\;14 & \;\;50 & \;\;110 & \;\;194 & \;\;302 & \\
\;\;\;\;36 & \;\;\;\;60 & \;\;\;\;84 & \;\;\;\;108 & & \\
\;\;\;\;\;\;24 & \;\;\;\;\;\;24 & \;\;\;\;\;\;24 & & &
\end{array}
$$

3.27 Develop an algorithm for the calculation of a difference table.

▮ Suppose values $y(x)$ given for $x_0, \ldots, x_n$. Introducing the variable $k = (x - x_0)/h$, with values 0 to n at the data points, we define

$$D(k, 0) = y(x_k)$$

which initializes the table, the zero-order differences being the function values themselves. The other differences may then be obtained by the loops

$$\text{for } j = 1 \text{ to } n$$
$$\text{for } k = 1 \text{ to } n - j + 1$$
$$D(k, j) = D(k + 1, j - 1) - D(k, j - 1)$$

3.28 Prove that for any positive integer k,

$$\Delta^k y_0 = \sum_{i=0}^{k} (-1)^i \binom{k}{i} y_{k-i}$$

where the familiar symbol for binomial coefficients,

$$\binom{k}{i} = \frac{k!}{i!(k-i)!} = \frac{k(k-1) \cdots (k-i+1)}{i!}$$

has been used.

▮ The proof will be by induction. For $k = 1, 2, 3$ and 4, the result has already been established, by definition when $k = 1$. Assume it true when k is some particular integer p:

$$\Delta^p y_0 = \sum_{i=0}^{p} (-1)^i \binom{p}{i} y_{p-i}$$

By advancing all lower indices we have also

$$\Delta^p y_1 = \sum_{i=0}^{p} (-1)^i \binom{p}{i} y_{p-i+1}$$

and by a change in the summation index, namely, $i = j + 1$,

$$\Delta^p y_1 = y_{p+1} - \sum_{j=0}^{p-1} (-1)^j \binom{p}{j+1} y_{p-j}$$

It is also convenient to make a nominal change of summation index, $i = j$, in our other sum:

$$\Delta^p y_0 = \sum_{j=0}^{p-1} (-1)^j \binom{p}{j} y_{p-j} + (-1)^p y_0$$

Then

$$\Delta^{p+1} y_0 = \Delta^p y_1 - \Delta^p y_0 = y_{p+1} - \sum_{j=0}^{p-1} (-1)^j \left[\binom{p}{j+1} + \binom{p}{j} \right] y_{p-j} - (-1)^p y_0$$

Now using

$$\binom{p}{j+1} + \binom{p}{j} = \binom{p+1}{j+1}$$

(see Problem 3.57) and making a final change of summation index, $j + 1 = i$,

$$\Delta^{p+1} y_0 = y_{p+1} + \sum_{i=1}^{p} (-1)^i \binom{p+1}{i} y_{p+1-i} - (-1)^p y_0 = \sum_{i=0}^{p+1} (-1)^i \binom{p+i}{i} y_{p+1-i}$$

Thus our result is established when k is the integer $p + 1$. This completes the induction.

3.29 Prove that for a constant function all differences are zero.

❚ Let $y_k = C$ for all k. This is a constant function. Then, for all k,

$$\Delta y_k = y_{k+1} - y_k = C - C = 0$$

3.30 Prove $\Delta(Cy_k) = C\Delta y_k$.

❚ This is analogous to a result of calculus. $\Delta(Cy_k) = Cy_{k+1} - Cy_k = C\Delta y_k$.
Essentially this problem involves two functions defined for the same arguments x_k. One function has the values y_k, the other has values $z_k = Cy_k$. We have proved $\Delta z_k = C\Delta y_k$.

3.31 Consider two functions defined for the same set of arguments x_k. Call the values of these functions u_k and v_k. Also consider a third function with values

$$w_k = C_1 u_k + C_2 v_k$$

where C_1 and C_2 are two constants (independent of x_k). Prove

$$\Delta w_k = C_1 \Delta u_k + C_2 \Delta v_k$$

This is the *linearity property* of the difference operation.

❚ The proof is direct from the definitions.

$$\Delta w_k = w_{k+1} - w_k = (C_1 u_{k+1} + C_2 v_{k+1}) - (C_1 u_k + C_2 v_k)$$
$$= C_1(u_{k+1} - u_k) + C_2(v_{k+1} - v_k) = C_1 \Delta u_k + C_2 \Delta v_k$$

Clearly the same proof would apply to sums of any finite length.

3.32 With the same symbolism as in Problem 3.31, consider the function with values $z_k = u_k v_k$ and prove $\Delta z_k = u_k \, \Delta v_k + v_{k+1} \Delta u_k$.

▌ Again starting from the definitions,

$$\Delta z_k = u_{k+1}v_{k+1} - u_k v_k = u_{k+1}v_{k+1} - u_k v_{k+1} + u_k v_{k+1} - u_k v_k$$

$$= v_{k+1}(u_{k+1} - u_k) + u_k(v_{k+1} - v_k) = u_k \, \Delta v_k + v_{k+1} \Delta u_k$$

The result $\Delta z_k = u_{k+1} \, \Delta v_k + v_k \, \Delta u_k$ could also be proved.

3.33 Compute differences of the function displayed in the first two columns of Table 3.2. This may be viewed as a type of "error function," if one supposes that all its values should be zero but the single 1 is a *unit error*. How does this unit error affect the various differences?

▌ Some of the required differences appear in the other columns of the table.

TABLE 3.2

x_0	0				
		0			
x_1	0		0		
		0		0	
x_2	0		0		1
		0		1	
x_3	0		1		-4
		1		-3	
x_4	1		-2		6
		-1		3	
x_5	0		1		-4
		0		-1	
x_6	0		0		1
		0		0	
x_7	0		0		
		0			
x_8	0				

This error influences a triangular portion of the difference table, increasing for higher differences and having a binomial coefficient pattern.

3.34 Compute differences for the function displayed in the first two columns of Table 3.3. This may be viewed as a type of error function, each value being a roundoff error of amount 1 unit. Show that the alternating $\pm$ pattern leads to serious error growth in the higher differences. Hopefully, roundoff errors will seldom alternate in just this way.

TABLE 3.3

x_0	1					
		-2				
x_1	-1		4			
		2		-8		
x_2	1		-4		16	
		-2		8		-32
x_3	-1		4		-16		64
		2		-8		32
x_4	1		-4		16	
		-2		8		
x_5	-1		4			
		2				
x_6	1					

■ Some of the required differences appear in the other columns of the table. The error doubles for each higher difference.

3.35 One number in this list is misprinted. Which one?

$$1 \quad 2 \quad 4 \quad 8 \quad 16 \quad 26 \quad 42 \quad 64 \quad 93$$

■ Calculating the first four differences and displaying them horizontally for a change, we have

1		2		4		8		10		16		22		29
	1		2		4		2		6		6		7	
		1		2		-2		4		0		1		
			1		-4		6		-4		1			

and the impression is inescapable that these binomial coefficients arise from a data error of size 1 in the center entry 16 of the original list. Changing it to 15 brings the new list

$$1 \quad 2 \quad 4 \quad 8 \quad 15 \quad 26 \quad 42 \quad 64 \quad 93$$

from which we find the differences

1		2		4		7		11		16		22		29
	1		2		3		4		5		6		7	

which suggest a job well done. This is a very simple example of data smoothing, which we treat much more fully in a later chapter. There is always the possibility that data such as we have in our original list come from a bumpy process, not from a smooth one, so that the bump (16 instead of 15) is real and not a misprint. The preceding analysis can then be viewed as bump detection, rather than as error correcting.

3.36 Imitating Problem 3.32, prove this formula for the differences of a quotient.

■ We have

$$\Delta \frac{u_k}{v_k} = \frac{u_{k+1}}{v_{k+1}} - \frac{u_k}{v_k} = \frac{v_k u_{k+1} - v_k u_k + v_k u_k - u_k v_{k+1}}{v_{k+1} v_k}$$

$$= \frac{v_k \Delta u_k - u_k \Delta v_k}{v_{k+1} v_k}.$$

3.37 Calculate differences through the fifth order to observe the effect of adjacent "errors" of size 1.

k	0	1	2	3	4	5	6	7
y_k	0	0	0	1	1	0	0	0

■ The differences develop as in Table 3.4. Clearly there is growth and the pattern is not entirely clear.

3.38 Find and correct a single error in these y_k values.

k	0	1	2	3	4	5	6	7
y_k	0	0	1	6	24	60	120	210

■ The differences begin as in Table 3.5 and the suspicious mind will surely pause to consider the bottom row. It consists of binomial coefficients with alternating signs, exactly where they would occur if a unit error were present

TABLE 3.4

0		0		1		0		-1		0		0
	0		1		-1		-1		1		0	
		1		-2		0		2		-1		
			-3		2		2		-3			
				5		0		-5				

TABLE 3.5

0		1		5		18		36		60		90
	1		4		13		18		24		30	
		3		9		5		6		6		
			6		−4		1		0			

(in the original data row) directly above the 6. (See Problem 3.33 or 3.35.) Replacing this 1 by a 0 we have the new data row

$$0 \quad 0 \quad 0 \quad 6 \quad 24 \quad 60 \quad 120 \quad 210$$

and a new computation will find the third differences constant at 6. The orderly mind will prefer to accept this corrected function.

3.39 Use the linearity property to find the first few differences of $y_k = k^p$ for $p = 1, 2, 3, 4$.

▌ Certainly $\Delta k = (k+1) - k = 1$. Then

$$\Delta k^2 = (k+1)^2 - k^2 = 2k + 1$$

and by the linearity property, $\Delta^2 k^2 = 2 \Delta k = 2$. Next

$$\Delta k^3 = (k+1)^3 - k^3 = 3k^2 + 3k + 1$$

The linearity property again comes into action, making

$$\Delta^2 k^3 = 3\Delta k^2 + 3\Delta k = 3(2k + 1) + 3 = 6k + 6$$
$$\Delta^3 k^3 = 6\Delta k = 6$$

Finally,

$$\Delta k^4 = (k+1)^4 - k^4 = 4k^3 + 6k^2 + 4k + 1$$

by the binomial theorem, so using the linearity property

$$\Delta^2 k^4 = 4\Delta k^3 + 6\Delta k^2 + 4\Delta k = 12k^2 + 24k + 14$$
$$\Delta^3 k^4 = 12\Delta k^2 + 24\Delta k = 24k + 36$$
$$\Delta^3 k^4 = 24\Delta k = 24$$

3.40 Show that if $y_k = 2^k$ then $\Delta y_k = y_k$. Also generalize to find the differences of $y_k = C^k$.

▌ First

$$\Delta 2^k = 2^{k+1} - 2^k = 2^k(2 - 1) = 2^k$$

This means that all differences of 2^k are again 2^k, making this function a sort of parallel to $y(x) = e^x$ in calculus. The generalization is not much different.

$$\Delta C^k = C^{k+1} - C^k = C^k(C - 1)$$

It follows that

$$\Delta^2 C^k = C^k(C - 1)^2 \qquad \Delta^3 C^k = C^k(C - 1)^3$$

and so on.

3.41 Compute the missing y_k values.

y_k	0		·		·		·		·		·		·
Δy_k		1		2		4		7		11		16	

▌ The computation is, of course, trivial, yielding these six missing values: 1 3 7 14 25 41. The point is that we have a sort of initial value problem. Given the first differences and a single contiguous value of the function itself, the other function values are determined. The problem may also be viewed as a "finite integration" problem, in which we work our way not from the function to its difference function but backward from the difference function to the function itself.

3.42 Compute the missing y_k and Δy_k values from the data provided.

y_k	·	·	·	6	·	·	·
Δy_k		·	·	5	·	·	·
$\Delta^2 y_k$		1	4	13	18	24	

▌ Clearly the idea is to work from the bottom upward, doing two finite integrations. The Δy_k row becomes 0 1 5 18 36 60 and then the y_k row is 0 0 1 6 24 60 120. This is once again the function of Problem 3.38 with the alleged error restored.

3.43 Compute the missing values in Table 3.6.

▌ Once again the error of Problem 3.38 has been removed, as we discover by working our way upward to the function values

$$0 \quad 0 \quad 0 \quad 6 \quad 24 \quad 60 \quad 120 \quad 210 \quad 336$$

TABLE 3.6

y_k	0		0		0		6		·		·		·		·
Δy_k		0		0		6		·		·		·		·	
$\Delta^2 y_k$			0		6		·		·		·		·		
$\Delta^3 y_k$				6		6		6		6		6		6	

The assumption of solid 6s along the bottom row permits the entire parallelogram at the upper right to be filled in, including the new function value (not in Problem 3.38) of 336.

3.44 Find and correct a misprint in this data:

$$y_k \quad 1 \quad 3 \quad 11 \quad 31 \quad 69 \quad 113 \quad 223 \quad 351 \quad 521 \quad 739 \quad 1011$$

▌ We have some differences to compute. (See Table 3.7.) This looks like a good place to stop, because of the alternation of signs. In fact, the final row is 18 times the revealing

$$0 \quad -1 \quad 4 \quad -6 \quad 4 \quad -1 \quad 0$$

TABLE 3.7

2		8		20		38		44		110		128		170		218		272
	6		12		18		6		66		18		42		48		54	
		6		6		−12		60		−48		24		6		6		
			0		−18		72		−108		72		−18		0			

which accuses the center entry in our y_k function of harboring an error of size -18. We therefore add 18 to this 113 entry and have the corrected 131. Apparently a transposition was made during data entry. The new function then triggers the difference table

y_k	1		3		11		31		69		131		223		351		521		739		1011
Δy_k		2		8		20		38		62		92		128		170		218		272	
$\Delta^2 y_k$			6		12		18		24		30		36		42		48		54		

We need hardly go farther.

3.45 What is the total of all first differences in the final table of the preceding problem?

▌ It comes to 1010, hardly a surprise since these numbers are the steps in the climb from 1 to 1011 in the function row. This illustrates an important process, the summation of differences. Summing differences, like integrating derivatives, can be achieved by subtracting suitable upper and lower limit values. Notice that the limit values here, 1011 and 1, are y_{10} and y_0. The differences summed were Δy_0 to Δy_9.

3.46 What is the sum of the second differences in the final table of Problem 3.44?

▌ Doing it the easy way we have $272 - 2 = 270$. The computation in Table 3.8 is attractive and equivalent. All of the full NW to SE diagonals sum to zero, leaving three entries in each of the remaining corners. Their sum is

$$(y_{n+1} - y_n) - (y_1 - y_0) = \Delta y_n - \Delta y_0$$

TABLE 3.8

$$\Delta y_0 = y_2 - 2y_1 + y_0$$
$$\Delta y_1 = y_3 - 2y_2 + y_1$$
$$\Delta y_2 = y_4 - 2y_3 + y_2$$
$$\vdots$$
$$\Delta y_{n-1} = y_{n+1} - 2y_n + y_{n-1}$$

3.47 Find a function y_k for which $\Delta y_k = 2y_k$.

▌ Suspecting exponential character we recall the result

$$\Delta C^k = C^k(C - 1)$$

which is just what is needed if $C = 3$ and $y_k = 3^k$. Generalizations are apparent.

3.48 Find the missing values in this table:

$$
\begin{array}{ccccccccc}
y_k & 0 & & \cdot & & \cdot & & \cdot & & 0 \\
\Delta y_k & & \cdot & & \cdot & & \cdot & & \cdot & \\
\Delta^2 y_k & & & 0 & & -2 & & 0 &
\end{array}
$$

▌ Once again an elementary problem has an interesting feature. This can be viewed as a second-order boundary value problem, in which we must work from second differences back to the function, with the end values of the function supplied.

Assign the leadoff first difference the temporary alias a. Then working as usual from the bottom upward we have

$$
\begin{array}{ccccccccc}
0 & & a & & 2a & & 3a-2 & & 4a-4 \\
& a & & a & & a-2 & & a-2 & \\
& & 0 & & -2 & & 0 &
\end{array}
$$

from which the upper right element permits the deduction $a = 1$. The function row is thus 0 1 2 1 0.

3.49 Deduce the missing values in Table 3.9a.

▌ Insert aliases a and b for the missing elements on the leading diagonal. The result is Table 3.9b.
The equations $a + 2b = 0$ and $5a + 2b + 8 = 0$ determine $a = -2$ and $b = 1$. The function y_k thus has the values 0 1 0 -2 0.

TABLE 3.9a

0			0			0
	·				·	
		·		·		
			1	5		

TABLE 3.9b

0		b		$a + 2b$		$2a + b + 1$		$5a + 2b + 8$
	b		$a + b$		$2a + b + 1$		$3a + b + 7$	
		a		$a + 1$		$a + 6$		
			1		5			

3.50 Find two functions for which $\Delta^2 y_k = 9 y_k$.

$\blacksquare$ Again recalling $\Delta C^k = C^k(C - 1)$ we see that

$$\Delta^2 C^k = C^k(C - 1)^2$$

and we need only that $(C - 1)^2 = 9$. This means either $C = 4$ or $C = -2$. The functions 4^k and $(-2)^k$ have the required property.

Continuing on, and imitating a procedure of differential equations, we may now form the family of solutions

$$y = A \cdot 4^k + B \cdot (-2)^k$$

If initial conditions $y_0 = 0$ and $y_1 = 1$ are imposed, they can be satisfied by substituting into the family, obtaining the pair of equations

$$0 = A + B$$
$$1 = 4A - 2B$$

with solution $A = 1/6 = -B$.

3.51 Derive formulas for $\Delta \sin(k)$ and $\Delta \cos(k)$.

$\blacksquare$
$$\Delta \sin(k) = \sin(k + 1) - \sin(k) = 2 \sin\left(\tfrac{1}{2}\right) \cos\left(k + \tfrac{1}{2}\right)$$
$$\Delta \cos(k) = \cos(k + 1) - \cos(k) = -2 \sin\left(\tfrac{1}{2}\right) \sin\left(k + \tfrac{1}{2}\right)$$

Clearly these are not as clean as the corresponding formulas for derivatives.

3.52 Derive a formula for $\Delta \log(x_k)$.

$\blacksquare$
$$\Delta \log(x_k) = \log(x_{k+1}) - \log(x_k) = \log(x_k + h) - \log(x_k)$$
$$= \log(1 + h/x_k)$$

where $x_k = x_0 + kh$.

3.3 FACTORIAL POLYNOMIALS

3.53 What are factorial polynomials?

$\blacksquare$ They are defined by

$$k^{(n)} = k(k - 1)(k - 2) \cdots (k - n + 1)$$

where n is a positive integer. For example, $k^{(2)} = k(k - 1) = k^2 - k$. Factorial polynomials play a central role in the theory of finite differences because of their convenient properties.

3.54 Find the first differences of the function

$$y_k = k(k - 1)(k - 2) = k^{(3)}$$

verifying that they are the values of $3k(k - 1) = 3k^{(2)}$.

$\blacksquare$
$$\Delta y_k = y_{k+1} - y_k = (k + 1)k(k - 1) - k(k - 1)(k - 2)$$
$$= [(k + 1) - (k - 2)]k(k - 1) = 3k(k - 1)$$

In tabular form this same result, for the first few integer values of k, looks like

k	0	1	2	3	4	5	6	7
y_k	0	0	0	6	24	60	120	210
y_k		0	0	6	18	36	60	90

3.55 Generalizing the result of the preceding problem, show that the first difference function of

$$y_k = k(k-1) \cdots (k-n+1) = k^{(n)}$$

is $\Delta k^{(n)} = nk^{(n-1)}$.

∎ This is strongly reminiscent of the calculus result on the derivative of x^n:

$$\begin{aligned}
\Delta y_k &= y_{k+1} - y_k \\
&= [(k+1) \cdots (k-n+2)] - [k \cdots (k-n+1)] \\
&= [(k+1) - (k-n+1)] k(k-1) \cdots (k-n+2) \\
&= nk^{(n-1)}
\end{aligned}$$

3.56 Apply the preceding problem successively to find the higher differences of $y_k = k^{(n)}$.

∎ Here again the track will recall earlier experiences. From $\Delta k^{(n)} = nk^{(n-1)}$ it follows that

$$\Delta^2 k^{(n)} = n \, \Delta k^{(n-1)} = n(n-1) k^{(n-2)}$$

$$\Delta^3 k^{(n)} = n(n-1) \, \Delta k^{(n-2)} = n(n-1)(n-2) k^{(n-3)}$$

and so on. Eventually $\Delta^n k^{(n)} = n!$ after which all higher differences will be zero, since $n!$ is constant (independent of k). The symbol $k^{(0)}$ is often used, interpreted as 1. The result in this problem is then valid for integers $n = 1, 2, 3, \ldots$.

3.57 Compute differences through fourth order for $y_k = k^{(4)}$ and $k = 0$ to 7.

∎ The function occupies the top two rows of Table 3.10 and the required differences occupy the rest. The entries can be checked by doing the subtractions or by the result of Problem 3.56. For example,

$$480 = 4 \cdot 6^{(3)} \qquad 144 = 12 \cdot 4^{(2)} \qquad 48 = 24 \cdot 2^{(1)}$$

the appropriate k for any entry being found by following its NW diagonal to the y_k value.

TABLE 3.10

k	0	1	2	3	4	5	6	7
y_k	0	0	0	0	24	120	360	840
		0	0	0	24	96	240	480
			0	0	24	72	144	240
				0	24	48	72	96
					24	24	24	24

3.58 Express the first five differences of $y_k = k^{(5)}$ in terms of factorial polynomials and compare with a tabular version of the same information.

∎ The differences are $5k^{(4)}$, $20k^{(3)}$, $60k^{(2)}$, $120k^{(1)}$ and 120. Table 3.11 contains the same information, a part consisting of solid zeros being omitted at the left. For example,

$$1800 = 5 \cdot 6^{(4)} \qquad 1200 = 20 \cdot 5^{(3)} \qquad 720 = 60 \cdot 4^{(2)} \qquad 360 = 120 \cdot 3^{(1)}$$

and so on. Even $120 = 120 \cdot 2^{(0)}$ is correct.

TABLE 3.11

k	2	3	4	5	6	7
y_k	0	0	0	120	720	2520
		0	0	120	600	1800
		0	120	480	1200	
			120	360	720	
			240	360		
				120		

3.59 The binomial coefficients $\binom{k}{n}$ are related to factorial polynomials in this way:

$$\binom{k}{n} = \frac{k^{(n)}}{n!} = \frac{k!}{n!(k-n)!}$$

For example, $\binom{6}{3} = 6 \cdot 5 \cdot 4/6 = 20$ and $\binom{6}{6} = 6!/6! = 1$. Use the result in Problem 3.55 to prove the famous recursion for these numbers.

$$\binom{k+1}{n+1} = \binom{k}{n+1} + \binom{k}{n}$$

❚ We find

$$\binom{k+1}{n+1} - \binom{k}{n+1} = \frac{(k+1)^{(n+1)}}{(n+1)!} - \frac{k^{(n+1)}}{(n+1)!} = \frac{\Delta k^{(n+1)}}{(n+1)!} = \frac{(n+1)k^{(n)}}{(n+1)!} = \frac{k^{(n)}}{n!} = \binom{k}{n}$$

which transposes at once into the recursion.

3.60 Use the recursion to tabulate the binomial coefficients up to $k = 8$.

❚ The first column of Table 3.12 gives $\binom{k}{0}$ which is defined to be 1. The diagonal, where $k = n$, is 1 by definition. The other entries result from the recursion. The table is easily extended.

TABLE 3.12

					n				
k	0	1	2	3	4	5	6	7	8
1	1	1							
2	1	2	1						
3	1	3	3	1					
4	1	4	6	4	1				
5	1	5	10	10	5	1			
6	1	6	15	20	15	6	1		
7	1	7	21	35	35	21	7	1	
8	1	8	28	56	70	56	28	8	1

3.61 Show that if k is a positive integer, then $k^{(n)}$ and $\binom{k}{n}$ are 0 for $n > k$. [For $n > k$ the symbol $\binom{k}{n}$ is defined as $k^{(n)}/n!$.]

❚ Note that $k^{(k+1)} = k(k-1) \cdots 0$. For $n > k$ the factorial $k^{(n)}$ will contain this 0 factor and so will $\binom{k}{n}$.

3.62 The binomial coefficient symbol and the factorial symbol are often used for nonintegral k. Calculate $k^{(n)}$ and $\binom{k}{n}$ for $k = \frac{1}{2}$ and $n = 2, 3$.

❚

$$k^{(2)} = \tfrac{1}{2}^{(2)} = \tfrac{1}{2}\left(\tfrac{1}{2} - 1\right) = -\tfrac{1}{4} \qquad k^{(3)} = \tfrac{1}{2}^{(3)} = \tfrac{1}{2}\left(\tfrac{1}{2} - 1\right)\left(\tfrac{1}{2} - 2\right) = \tfrac{3}{8}$$

$$\binom{k}{2} = k^{(2)}/2! = \tfrac{1}{2}\left(-\tfrac{1}{4}\right) = -\tfrac{1}{8} \qquad \binom{k}{3} = k^{(3)}/3! = \tfrac{1}{6}\left(\tfrac{3}{8}\right) = \tfrac{1}{16}$$

As further examples,

$$\binom{1/3}{2} = (1/3)^{(2)}/2 = (1/3)(-2/3)/2 = -1/9$$

$$\binom{1/3}{3} = (1/3)^{(3)}/6 = (1/3)(-2/3)(-5/3)/6 = 5/81$$

$$\binom{1/3}{4} = (1/3)^{(4)}/24 = \binom{1/3}{3}(-8/3)/4 = -10/243$$

$$\binom{1/3}{5} = (1/3)^{(5)}/120 = \binom{1/3}{4}(-11/3)/5 = 22/729$$

with a clear track ahead.

3.63 The idea of factorial has also been extended to upper indices that are not positive integers. It follows from the definition that when n is a positive integer, $k^{(n+1)} = (k-n)k^{(n)}$. Rewriting this as

$$k^{(n)} = \frac{1}{k-n}k^{(n+1)}$$

and using it as a *definition* of $k^{(n)}$ for $n = 0, -1, -2, \ldots$, show that $k^{(0)} = 1$ and $k^{(-n)} = 1/(k+n)^{(n)}$.

∎ With $n = 0$ the first result is instant. For the second we find successively

$$k^{(-1)} = \frac{1}{k+1}k^{(0)} = \frac{1}{k+1} = \frac{1}{(k+1)^{(1)}} \qquad k^{(-2)} = \frac{1}{k+2}k^{(-1)} = \frac{1}{(k+2)(k+1)} = \frac{1}{(k+2)^{(2)}}$$

and so on. An inductive proof is indicated but the details will be omitted. For $k = 0$ it is occasionally convenient to define $k^{(0)} = 1$ and to accept the consequences.

As examples,

$$6^{(-1)} = 1/(6+1) = 1/7 \qquad\qquad (1/3)^{(-1)} = 1/(4/3) = 3/4$$

$$6^{(-2)} = 1/(6+2)(6+1) = 1/56 \qquad (1/3)^{(-2)} = 1/(7/3 \cdot 4/3) = 9/28$$

$$6^{(-3)} = 1/9 \cdot 8 \cdot 7 = 1/504 \qquad\qquad (1/3)^{(-3)} = 1/(10/3 \cdot 7/3 \cdot 4/3) = 27/280$$

3.64 Prove that $\Delta k^{(n)} = nk^{(n-1)}$ for all integers n.

∎ For $n > 1$, this has been proved in Problem 3.55. For $n = 1$ and 0, it is immediate. For n negative, say $n = -p$,

$$\Delta k^{(n)} = \Delta k^{(-p)} = \Delta \frac{1}{(k+p)^{(p)}} = \frac{1}{(k+1+p)\cdots(k+2)} - \frac{1}{(k+p)\cdots(k+1)}$$

$$= \frac{1}{(k+p)\cdots(k+2)}\left(\frac{1}{k+1+p} - \frac{1}{k+1}\right) = \frac{-p}{(k+1+p)\cdots(k+1)}$$

$$= \frac{n}{(k+1-n)^{(1-n)}} = nk^{(n-1)}$$

This result is analogous to the fact that the theorem of calculus "if $f(x) = x^n$, then $f'(x) = nx^{n-1}$" is also true for all integers.

3.65 Find $\Delta k^{(-1)}$.

∎ By the previous problems, $\Delta k^{(-1)} = -k^{(-2)} = -1/(k+2)(k+1)$.

3.66 Show that $k^{(2)} = -k + k^2$, $k^{(3)} = 2k - 3k^2 + k^3$ and $k^{(4)} = -6k + 11k^2 - 6k^3 + k^4$.

∎ Directly from the definitions:

$$k^{(2)} = k(k-1) = -k + k^2$$

$$k^{(3)} = k^{(2)}(k-2) = 2k - 3k^2 + k^3$$

$$k^{(4)} = k^{(3)}(k-3) = -6k + 11k^2 - 6k^3 + k^4$$

3.67 Generalizing Problem 3.66, show that in the expansion of a factorial polynomial into standard polynomial form

$$k^{(n)} = S_1^{(n)} k + \cdots + S_n^{(n)} k^n = \sum S_i^{(n)} k^i$$

the coefficients satisfy the recursion

$$S_i^{(n+1)} = S_{i-1}^{(n)} - n S_i^{(n)}$$

These coefficients are called *Stirling's numbers of the first kind*.

▌ Replacing n by $n+1$,

$$k^{(n+1)} = S_1^{(n+1)} k + \cdots + S_{n+1}^{(n+1)} k^{n+1}$$

and using the fact that $k^{(n+1)} = k^{(n)}(k - n)$, we find

$$S_1^{(n+1)} k + \cdots + S_{n+1}^{(n+1)} k^{n+1} = \left[S_1^{(n)} k + \cdots + S_n^{(n)} k^n \right] (k - n)$$

Now compare coefficients of k^i on both sides. They are

$$S_i^{(n+1)} = S_{i-1}^{(n)} - n S_i^{(n)}$$

for $i = 2, \ldots, n$. The special cases $S_1^{(n+1)} = -n S_1^{(n)}$ and $S_{n+1}^{(n+1)} = S_n^{(n)}$ should also be noted, by comparing coefficients of k and k^{n+1}.

3.68 Use the formulas of Problem 3.67 to develop a brief table of Stirling's numbers of the first kind.

▌ The special formula $S_1^{(n+1)} = -n S_1^{(n)}$ leads at once to column one of Table 3.13. For example, since $S_1^{(1)}$ is clearly 1,

$$S_1^{(2)} = -S_1^{(1)} = -1 \qquad S_1^{(3)} = -2 S_1^{(2)} = 2$$

and so on. The other special formula fills the top diagonal of the table with 1s. Our main recursion then completes the table. For example,

$$S_2^{(3)} = S_1^{(2)} - 2 S_2^{(2)} = (-1) - 2(1) = -3$$

$$S_2^{(4)} = S_1^{(3)} - 3 S_2^{(3)} = (2) - 3(-3) = 11$$

$$S_3^{(4)} = S_2^{(3)} - 3 S_3^{(3)} = (-3) - 3(1) = -6$$

and so on. Through $n = 8$ the table reads as given in Table 3.13.

TABLE 3.13

n	1	2	3	i 4	5	6	7	8
1	1							
2	−1	1						
3	2	−3	1					
4	−6	11	−6	1				
5	24	−50	35	−10	1			
6	−120	274	−225	85	−15	1		
7	720	−1,764	1,624	−735	175	−21	1	
8	−5,040	13,068	−13,132	6,769	−1,960	322	−28	1

3.69 Use Table 3.13 to expand $k^{(5)}$. Also develop

$$y_k = 2k^{(3)} - k^{(2)} + 4k^{(1)} - 7$$

as a conventional polynomial.

▮ The first can be read directly from row five of the table:

$$k^{(5)} = 24k - 50k^2 + 35k^3 - 10k^4 + k^5$$

For the other we need a blend of four rows:

$$
\begin{array}{r}
4k - 6k^2 + 2k^3 \\
k - k^2 \\
-7 + 4k \\
\hline
-7 + 9k - 7k^2 + 2k^3
\end{array}
$$

3.70 Use Table 3.13 to express $y_k = k^{(6)} + k^{(3)}$ as a conventional polynomial.

▮
$$
\begin{array}{r}
-120k + 274k^2 - 225k^3 + 85k^4 - 15k^5 + k^6 \\
2k - 3k^2 + k^3 \\
\hline
-118k + 271k^2 - 224k^3 + 85k^4 - 15k^5 + k^6
\end{array}
$$

3.71 Show that $k^2 = k^{(1)} + k^{(2)}$, $k^3 = k^{(1)} + 3k^{(2)} + k^{(3)}$ and $k^4 = k^{(1)} + 7k^{(2)} + 6k^{(3)} + k^{(4)}$.

▮ Using Table 3.13,

$$k^{(1)} + k^{(2)} = k + (-k + k^2) = k^2$$

$$k^{(1)} + 3k^{(2)} + k^{(3)} = k + 3(-k + k^2) + (2k - 3k^2 + k^3) = k^3$$

$$k^{(1)} + 7k^{(2)} + 6k^{(3)} + k^{(4)} = k + 7(-k + k^2) + 6(2k - 3k^2 + k^3) + (-6k + 11k^2 - 6k^3 + k^4) = k^4$$

3.72 As a necessary preliminary to the following problem, prove that a power of k can have only one representation as a combination of factorial polynomials.

▮ Assume that two such representations exist for k^p:

$$k^p = A_1 k^{(1)} + \cdots + A_p k^{(p)} \qquad k^p = B_1 k^{(1)} + \cdots + B_p k^{(p)}$$

Subtracting leads to

$$0 = (A_1 - B_1)k^{(1)} + \cdots + (A_p - B_p)k^{(p)}$$

Since the right side is a polynomial and no polynomial can be zero for all values of k, every power of k on the right side must have coefficient 0. But k^p appears only in the last term; hence A_p must equal B_p. Then k^{p-1} appears only in the last term remaining, which will be $(A_{p-1} - B_{p-1})k^{(p-1)}$; hence $A_{p-1} = B_{p-1}$. This argument prevails right back to $A_1 = B_1$.

This proof is typical of unique representation proofs which are frequently needed in numerical analysis. The analogous theorem, that two polynomials cannot have identical values without also having identical coefficients, is a classic result of algebra and has already been used in Problem 3.67.

3.73 Generalizing Problem 3.71, show that the powers of k can be represented as combinations of factorial polynomials

$$k^n = s_1^{(n)} k^{(1)} + \cdots + s_n^{(n)} k^{(n)} = \sum s_i^{(n)} k^{(i)}$$

and that the coefficients satisfy the recursion $s_i^{(n+1)} = s_{i-1}^{(n)} + i s_i^{(n)}$. These coefficients are called *Stirling's numbers of the second kind*.

▮ We proceed by induction, Problem 3.71 already having established the existence of such representations for small k. Suppose

$$k^n = s_1^{(n)} k^{(1)} + \cdots + s_n^{(n)} k^{(n)}$$

and then multiply by k to obtain

$$k^{n+1} = k s_1^{(n)} k^{(1)} + \cdots + k s_n^{(n)} k^{(n)}$$

Now notice that $k \cdot k^{(i)} = (k - i)k^{(i)} + i k^{(i)} = k^{(i+1)} + i k^{(i)}$ so that

$$k^{n+1} = s_1^{(n)}(k^{(2)} + k^{(1)}) + \cdots + s_n^{(n)}(k^{(n+1)} + n k^{(n)})$$

This is already a representation of k^{n+1}, completing the induction, so that we may write

$$k^{n+1} = s_1^{(n+1)}k^{(1)} + \cdots + s_{n+1}^{(n+1)}k^{(n+1)}$$

By Problem 3.72, coefficients of $k^{(i)}$ in both these last lines must be the same, so that

$$s_i^{(n+1)} = s_{i-1}^{(n)} + is_i^{(n)}$$

for $i = 2, \ldots, n$. The special cases $s_1^{(n+1)} = s_1^{(n)}$ and $s_{n+1}^{(n+1)} = s_n^{(n)}$ should also be noted, by comparing coefficients of $k^{(1)}$ and $k^{(n+1)}$.

3.74 Use the formulas of Problem 3.73 to develop a brief table of Stirling's numbers of the second kind.

The special formula $s_1^{(n+1)} = s_1^{(n)}$ leads at once to column one of Table 3.14, since $s_1^{(1)}$ is clearly 1. The other special formula produces the top diagonal. Our main recursion then completes the table. For example,

$$s_2^{(3)} = s_1^{(2)} + 2s_2^{(2)} = (1) + 2(1) = 3 \qquad s_2^{(4)} = s_1^{(3)} + 2s_2^{(3)} = (1) + 2(3) = 7$$

$$s_3^{(4)} = s_2^{(3)} + 3s_3^{(3)} = (3) + 3(1) = 6$$

and so on. Through $n = 8$, the table is given by Table 3.14.

TABLE 3.14

	i							
n	1	2	3	4	5	6	7	8
1	1							
2	1	1						
3	1	3	1					
4	1	7	6	1				
5	1	15	25	10	1			
6	1	31	90	65	15	1		
7	1	63	301	350	140	21	1	
8	1	127	966	1701	1050	266	28	1

3.75 Use Table 3.14 to express k^5 and then

$$y_k = \tfrac{1}{3}(2k^4 - 8k^2 + 3)$$

as a combination of factorial polynomials.

The first can be read directly from row five of the table:

$$k^5 = k^{(1)} + 15k^{(2)} + 25k^{(3)} + 10k^{(4)} + k^{(5)}$$

For the other, apart from the $\frac{1}{3}$ factor we have

$$\begin{array}{r} 2k^{(1)} + 14k^{(2)} + 12k^{(3)} + 2k^{(4)} \\ 3 - 8k^{(1)} - 8k^{(2)} \phantom{+ 12k^{(3)} + 2k^{(4)}} \end{array}$$

so combining and taking a third yields

$$1 - 2k^{(1)} + 2k^{(2)} + 4k^{(3)} + (2/3)k^{(4)}$$

3.76 Use Table 3.14 to express $y_k = 80k^3 - 30k^4 + 3k^5$ as a combination of factorial polynomials.

$$\begin{array}{rrrrr} 80k^{(1)} & + 240k^{(2)} & + \ 80k^{(3)} & & \\ -30k^{(1)} & - 210k^{(2)} & - 180k^{(3)} & - 30k^{(4)} & \\ 3k^{(1)} & + \ 45k^{(2)} & + \ 75k^{(3)} & + 30k^{(4)} & + 3k^{(5)} \\ \hline 53k^{(1)} & + \ 75k^{(2)} & - \ 25k^{(3)} & & + 3k^{(5)} \end{array}$$

3.76 Use the result of the previous problem to obtain Δy_k in terms of factorial polynomials. Then apply Table 3.13 to convert the result to conventional form.

$$\Delta y_k = 53 + 150k - 75k^{(2)} + 15k^{(4)}$$
$$= 53 + 150k - 75(k^2 - k) + 15(-6k + 11k^2 - 6k^3 + k^4)$$
$$= 53 + 135k + 90k^2 - 90k^3 + 15k^4$$

3.77 Continuing the preceding problem, find $\Delta^2 y_k$ in terms of factorial polynomials. Then convert the result to conventional form.

$$\Delta^2 y_k = 150 - 150k + 60k^{(3)}$$
$$= 150 - 150k + 60(2k - 3k^2 + k^3)$$
$$= 150 - 30k - 180k^2 + 60k^3$$

3.78 Prove that the nth differences of a polynomial of degree n are equal, higher differences than the nth being zero.

▮ Call the polynomial $P(x)$ and consider its values for a discrete set of equally spaced arguments $x_0, x_1, x_2, \ldots$. It is usually convenient to deal with the substitute integer argument k which we have used so frequently, related to x by $x_k - x_0 = kh$ where h is the uniform difference between consecutive x arguments. Denote the value of our polynomial for the argument k by the symbol P_k. Since the change of argument is linear, the polynomial has the same degree in terms of both x and k, and we may write it as

$$P_k = a_0 + a_1 k + a_2 k^2 + \cdots + a_n k^n$$

Problem 3.73 shows that each power of k can be represented as a combination of factorial polynomials, leading to a representation of P_k itself as such a combination.

$$P_k = b_0 + b_1 k^{(1)} + b_2 k^{(2)} + \cdots + b_n k^{(n)}$$

Applying Problem 3.55 and the linearity property

$$\Delta P_k = b_1 + 2b_2 k^{(1)} + \cdots + nb_n k^{(n-1)}$$

and reapplying Problem 3.55 leads eventually to $\Delta^n P_k = n! b_n$. So all the nth differences are this number. They do not vary with k and consequently higher differences are zero.

3.79 Assuming that the following y_k values belong to a polynomial of degree 4, compute the next three values.

k	0	1	2	3	4	5	6	7
y_k	0	1	2	1	0	.	.	.

A fourth-degree polynomial has constant fourth differences, according to Problem 3.78. Calculating from the given data, we obtain the entries to the left of the line in Table 3.15.

Assuming the other fourth differences also to be 4 leads to the entries to the right of the line from which the missing entries may be predicted: $y_5 = 5$, $y_6 = 26$, $y_7 = 77$.

TABLE 3.15

1		1		−1		−1		5		21		51			
	0		−2		0		6		16		30				
		−2		2		6		10		14					
			4		4		4		4						

3.80 Assuming that the following y_k values belong to a polynomial of degree 4, predict the next three values.

k	0	1	2	3	4	5	6	7
y_k	1	−1	1	−1	1	.	.	.

▮ We first compute the differences to the left of the line in Table 3.16. Assuming the polynomial has degree 4 means the bottom row must then be solid 16s, after which the calculation proceeds upward to the right of the line. The missing values prove to be the somewhat surprising 31, 129 and 351.

TABLE 3.16

1		−1		1		−1		1		31		129		351
	−2		2		−2		2		30		98		222	
		4		−4		4		28		68		124		
			−8		8		24		40		56			
				16		16		16		16				

3.81 What is the lowest degree possible for a polynomial that takes these values?

k	0	1	2	3	4	5
y_k	0	3	8	15	24	35

▮ The first and second differences are

$$3 \quad 5 \quad 7 \quad 9 \quad 11$$
$$2 \quad 2 \quad 2 \quad 2$$

and the constant second row shows that degree 2 will serve.

3.82 What is the lowest degree possible for a polynomial that takes these values?

k	0	1	2	3	4	5
y_k	0	1	1	1	1	0

▮ Speculation is permitted, but it is the differences that tell the story:

$$1 \quad 0 \quad 0 \quad 0 \quad -1$$
$$-1 \quad 0 \quad 0 \quad -1$$
$$1 \quad 0 \quad -1$$
$$-1 \quad -1$$

The constant fourth differences give the polynomial degree 4.

3.83 Assuming these y_k values belong to a polynomial of degree 4, predict the next three values:

$$0 \quad 0 \quad 1 \quad 0 \quad 0 \quad \cdot \quad \cdot \quad \cdot$$

▮ Once again speculation is invited, but Table 3.17 has the answer.
Once again we have a polynomial that explodes as it leaves the interval of given values.

TABLE 3.17

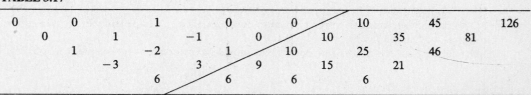

0		0		1		0		0		10		45		126
	0		1		−1		0		10		35		81	
		1		−2		1		10		25		46		
			−3		3		9		15		21			
				6		6		6		6				

3.84 Find a function y_k for which $\Delta y_k = k^{(2)} = k(k-1)$.

▌ Just as x^2 integrates to $x^3/3$ we find that $k^{(2)}$ finite integrates to $k^{(3)}/3$. The proof, if needed, resembles the calculus proof; difference $k^{(3)}/3$ and one is back to $k^{(2)}$. Since any constant will have difference zero, its addition to the result in hand will produce another finite integral, so

$$y_k = k^{(3)}/3 + C$$

is a family of solutions, with C an arbitrary constant.

Similarly, if $\Delta y_k = k(k-1)(k-2)$, it follows that

$$y_k = k^{(4)}/4 + C$$

3.85 Find a function y_k such that $\Delta y_k = k^2$.

▌ Since $k^2 = k^{(2)} + k^{(1)}$ we find $y_k = k^{(3)}/3 + k^{(2)}/2 + C$.

3.86 Solve $\Delta y_k = k^3$.

▌ From Table 3.14 this is equivalent to

$$\Delta y_k = k^{(3)} + 3k^{(2)} + k^{(1)}$$

from which the solution is instant:

$$y_k = k^{(4)}/4 + k^{(3)} + k^{(2)}/2 + C$$

3.87 Solve $\Delta y_k = 1/(k+1)(k+2)$.

▌ Here $\Delta y_k = k^{(-2)}$ and integration brings

$$y_k = k^{(-1)}/-1 = -1/(k+1)$$

3.4 SUMMATION

3.88 In Problems 3.45 and 3.46 examples were given of the process of summing differences. Such sums are also called telescoping sums. Formalize this idea, developing what is known as the fundamental theorem of summation.

▌ Observing that

$$(y_1 - y_0) + (y_2 - y_1) + (y_3 - y_2) + \cdots + (y_n - y_{n-1}) = y_n - y_0$$

or

$$\sum_{k=0}^{n-1} y_k = y_n - y_0$$

we have the result that is often compared to the fundamental theorem of integral calculus, involving the integration of derivatives. The term finite integration, already used in a few examples, is related. It is the process of obtaining y_k from

$$\Delta y_k = z_k$$

where z_k is known. Since it follows that

$$y_n = y_0 + \sum_{k=0}^{n-1} z_k$$

finite integration and summation are companion problems. The z_k can be summed by telescoping, provided the finite integral y_k can be found.

As the earlier examples show, difference tables provide a picturesque view of the theorem. The fragment

$$y_0 \qquad\qquad\qquad\qquad y_n$$
$$\Delta y_0 \quad \Delta y_1 \quad \Delta y_2 \quad \cdots \quad \Delta y_{n-1}$$

makes this clear, the sum of these adjacent differences being $y_n - y_0$. The idea can be applied throughout the table. For example,

$$\Delta y_3 \qquad\qquad\qquad\qquad\qquad\qquad\qquad \Delta y_n$$
$$\Delta^2 y_3 \quad \Delta^2 y_4 \quad \Delta^2 y_5 \quad \cdots \quad \Delta^2 y_{n-1}$$

The sum of the second differences in this row is $\Delta y_n - \Delta y_3$.

3.89   Prove $1^2 + 2^2 + \cdots + n^2 = \sum_{i=1}^{n} i^2 = [n(n+1)(2n+1)]/6$.

▮   We need a function for which $\Delta y_i = i^2$. This is similar to the integration problem of calculus. In this simple example, the y_i could be found almost by intuition, but even so we apply a method that handles harder problems just as well. First replace i^2 by a combination of factorial polynomials, using Stirling's numbers

$$\Delta y_i = i^2 = i^{(2)} + i^{(1)}$$

A function having this difference is

$$y_i = \tfrac{1}{3} i^{(3)} + \tfrac{1}{2} i^{(2)}$$

as may easily be verified by computing Δy_i. Obtaining y_i from Δy_i is called *finite integration*. The resemblance to the integration of derivatives is obvious. Now rewrite the result of Problem 3.88 as $\sum_{i=1}^{n} \Delta y_i = y_{n+1} - y_1$ and substitute to obtain

$$\sum_{i=1}^{n} i^2 = \left[\frac{1}{3}(n+1)^{(3)} + \frac{1}{2}(n+1)^{(2)} \right] - \left[\frac{1}{3}(1)^{(3)} + \frac{1}{2}(1)^{(2)} \right]$$

$$= \frac{(n+1)n(n-1)}{3} + \frac{(n+1)n}{2} = \frac{n(n+1)(2n+1)}{6}$$

3.90   Evaluate the series $\sum_{i=0}^{\infty} 1/[(i+1)(i+2)]$.

▮   By an earlier result $\Delta i^{(-1)} = -1/[(i+1)(i+2)]$. Then, using Problem 3.63 to handle $0^{(-1)}$,

$$S_n = \sum_{i=0}^{n-1} \frac{1}{(i+1)(i+2)} = - \sum_{i=0}^{n-1} \Delta i^{(-1)} = - \left[n^{(-1)} - 0^{(-1)} \right] = 1 - \frac{1}{n+1}$$

The series is defined as $\lim S_n$ and is therefore equal to 1.

3.91   Consider two functions defined for the same set of arguments x_k, having values u_k and v_k. Prove

$$\sum_{i=0}^{n-1} u_i \, \Delta v_i = u_n v_n - u_0 v_0 - \sum_{i=0}^{n-1} v_{i+1} \, \Delta u_i$$

▮   This is called *summation by parts* and is analogous to the result of calculus

$$\int_{x_0}^{x_n} u(x) v'(x) \, dx = u(x_n) v(x_n) - u(x_0) v(x_0) - \int_{x_0}^{x_n} v(x) u'(x) \, dx$$

The proof begins with the result of Problem 3.32, slightly rearranged:

$$u_i \, \Delta v_i = \Delta(u_i v_i) - v_{i+1} \, \Delta u_i$$

Sum from $i = 0$ to $i = n - 1$,

$$\sum_{i=0}^{n-1} u_i \, \Delta v_i = \sum_{i=0}^{n-1} \Delta(u_i v_i) - \sum_{i=0}^{n-1} v_{i+1} \, \Delta u_i$$

and then apply Problem 3.88 to the first sum on the right. The required result follows.

3.92 Evaluate the series $\sum_{i=0}^{\infty} iR^i$ where $-1 < R < 1$.

▌ Since $\Delta R^i = R^{i+1} - R^i = R^i(R-1)$, we may put $u_i = i$ and $v_i = R^i/(R-1)$ and apply summation by parts. Take the finite sum

$$S_n = \sum_{i=0}^{n-1} iR^i = \sum_{i=0}^{n-1} u_i \Delta v_i = n \cdot \frac{R^n}{R-1} - 0 - \sum_{i=0}^{n-1} \frac{R^{i+1}}{R-1}$$

The last sum is *geometric* and responds to an elementary formula, making

$$S_n = \frac{nR^n}{R-1} + \frac{R(1-R^n)}{(1-R)^2}$$

Since nR^n and R^{n+1} both have limit zero, the value of the infinite series is $\lim S_n = R/(1-R)^2$.

3.93 A coin is tossed until heads first shows. A payoff is then made, equal to i dollars if heads first showed on the ith toss ($1 if heads showed at once on the first toss, $2 if the first head showed on the second toss and so on). Probability theory leads to the series.

$$1\left(\tfrac{1}{2}\right) + 2\left(\tfrac{1}{4}\right) + 3\left(\tfrac{1}{8}\right) + \cdots = \sum_{i=0}^{\infty} i\left(\tfrac{1}{2}\right)^i$$

for the average payoff. Use the previous problem to compute this series.

▌ By Problem 3.92 with $R = \tfrac{1}{2}$, $\sum_{i=0}^{\infty} i\left(\tfrac{1}{2}\right)^i = \left(\tfrac{1}{2}\right)/\left(\tfrac{1}{4}\right) = \$2$.

3.94 Apply summation by parts to evaluate the series $\sum_{i=0}^{\infty} i^2 R^i$.

▌ Putting $u_i = i^2$, $v_i = R^i/(R-1)$ we find $\Delta u_i = 2i + 1$ and so

$$S_n = \sum_{i=0}^{n-1} i^2 R^i = \sum_{i=0}^{n-1} u_i \Delta v_i = n^2 \frac{R^n}{R-1} - 0 - \sum_{i=0}^{n-1} \frac{R^{i+1}}{R-1}(2i+1)$$

$$= n^2 \frac{R^n}{R-1} - \frac{2R}{R-1} \sum_{i=0}^{n-1} iR^i - \frac{R}{R-1} \sum_{i=0}^{n-1} R^i$$

The first of the two remaining sums was evaluated in Problem 3.92 and the second is geometric. So we come to

$$S_n = \frac{n^2 R^n}{R-1} - \frac{2R}{R-1}\left[\frac{nR^n}{R-1} + \frac{R(1-R^n)}{(1-R)^2}\right] - \frac{R}{R-1} \cdot \frac{1-R^n}{1-R}$$

and letting $n \to \infty$ finally achieve $\lim S_n = (R + R^2)/(1-R)^3$.

3.95 A coin is tossed until heads first shows. A payoff is then made, equal to i^2 dollars if heads first showed on the ith toss. Probability theory leads to the series $\sum_{i=0}^{\infty} i^2 \left(\tfrac{1}{2}\right)^i$ for the average payoff. Evaluate the series.

▌ By Problem 3.94 with $R = \tfrac{1}{2}$, $\sum_{i=0}^{\infty} i^2 \left(\tfrac{1}{2}\right)^i = \left(\tfrac{1}{2} + \tfrac{1}{4}\right)/\left(\tfrac{1}{8}\right) = \$6$.

3.96 Use finite integration to prove the familiar (arithmetic) sum

$$\sum_{i=1}^{n} i = 1 + 2 + \cdots + n = (n+1)n/2$$

▌ Since $i = i^{(1)}$ it has the finite integral $i^{(2)}/2$. The given sum can then be expressed as

$$(n+1)^{(2)}/2 - 1^{(2)}/2 = (n+1)n/2$$

as anticipated.

3.97   Evaluate $\sum_{i=1}^{n} i^3$ by finite integration.

∎ Using Stirling's numbers we express i^3 as $i^{(1)} + 3i^{(2)} + i^{(3)}$ and the finite integral as $i^{(2)}/2 + i^{(3)} + i^{(4)}/4$. All three of these parts vanish at the lower limit of the summation so we are left with

$$(n+1)n/2 + (n+1)n(n-1) + (n+1)n(n-1)(n-2)/4$$

which reduces eventually to $n^2(n+1)^2/4$.

3.98   Show that $\sum_{i=0}^{n-1} A^i = (A^n - 1)/(A - 1)$ by finite integration.

∎ This is, of course, the geometric sum of elementary algebra. Since $\Delta A^i = A^i(A-1)$ we have the finite integral $A^i/(A-1)$. The sum thus equals $(A^n - A^0)/(A-1)$ as expected.

3.99   Show that

$$\sum_{i=1}^{n-1} \binom{i}{k} = \binom{n}{k+1}$$

∎ Since

$$\binom{i}{k} = i^{(k)}/k! = \Delta i^{(k+1)}/(k+1)!$$

we have a finite integral in hand and can express the sum as

$$\frac{n^{(k+1)} - 1^{(k+1)}}{(k+1)!} = \binom{n}{k+1}$$

since the second term of the numerator is zero. For example, with $k=2$ and $n=6$, we find

$$\binom{1}{2} + \binom{2}{2} + \binom{3}{2} + \binom{4}{2} + \binom{5}{2} = 0 + 1 + 3 + 6 + 10 = 20 = \binom{6}{3}$$

3.100   Evaluate by finite integration: $\sum_{i=0}^{\infty} 1/[(i+1)(i+2)(i+3)]$.

∎ We are summing $i^{(-3)}$ for which a finite integral is $-i^{(-2)}/2$. For the truncated series with limits 0 and $n-1$ this offers us the value

$$\frac{0^{(-2)}}{2} - \frac{n^{(-2)}}{2} = \frac{1}{4} - \frac{1}{2(n+1)(n+2)}$$

which approaches $\frac{1}{4}$ as n tends to infinity.

3.101   Evaluate $\sum_{i=1}^{\infty} 1/[i(i+2)]$.

∎ A simple device is sufficient. Noting that

$$\frac{1}{i(i+2)} = \frac{1}{2}\left(\frac{1}{i} - \frac{1}{i+2}\right)$$

the series can be rewritten as

$$\frac{1}{1 \cdot 3} + \frac{1}{2 \cdot 4} + \frac{1}{3 \cdot 5} \cdots = \frac{1}{2}\left[\left(\frac{1}{1} - \frac{1}{3}\right) + \left(\frac{1}{2} - \frac{1}{4}\right) + \left(\frac{1}{3} - \frac{1}{5}\right) + \left(\frac{1}{4} - \frac{1}{6}\right) + \cdots\right]$$

in which substantial cancellation occurs, only the two terms $\frac{1}{1}$ and $\frac{1}{2}$ perservering. The sum is, therefore $\frac{3}{4}$. The generalization to $\Sigma 1/i(i + n)$ is immediate.

3.102 Evaluate $\sum_{i=0}^{\infty} i^3 R^i$ for $-1 < R < 1$.

‖ There are several ways to approach this sum. Suppose we first follow the lead of Problems 3.92 and 3.94 and apply summation by parts directly. Take u_i to be i^3 and $\Delta v_i = R^i$ as usual. The integrated term will then be $i^3 R^i/(R - 1)$ which is zero at the lower limit and will also vanish at the upper when n tends to infinity. It can be ignored. Since $\Delta u_i = (i + 1)^3 - i^3 = 3i^2 + 3i + 1$ the remaining term becomes

$$\frac{1}{1 - R} R^{i+1}(3i^2 + 3i + 1)$$

all parts of which have been evaluated in the earlier problems. Combining the various results yields

$$\frac{R}{1 - R}\left[\frac{3(R + R^2)}{(1 - R)^3} + \frac{3R}{(1 - R)^2}\frac{1}{1 - R}\right]$$

which reduces gradually to $R(1 + 4R + R^2)/(1 - R)^4$.

3.103 Change Problem 3.93 so that the payoff is i^3. What will now be the average payoff?

‖ Substituting $R = \frac{1}{2}$ in the preceding problem, we have

$$\sum_{i=0}^{\infty} i^3\left(\frac{1}{2}\right)^i = \frac{1}{2}\left(1 + 2 + \frac{1}{4}\right)/\frac{1}{16} = 26$$

The new payoff is quite substantial.

3.104 Change Problem 3.93 so that the payoff is $+1$ when i is even and -1 when i is odd. Again find the average payoff.

‖ It will be $\sum_{i=1}^{\infty}(-\frac{1}{2})^i$. If the $i = 0$ term were included, the result of Problem 3.98 would offer us $1/(1 + \frac{1}{2}) = \frac{2}{3}$. Without it we have only $-\frac{1}{3}$. It is now a losing game.

3.105 Evaluate $\sum_{i=2}^{\infty} \log[(i + 1)/(i - 1)]$.

‖ Since $\log[(i + 1)/(i - 1)] = \log(i + 1) - \log(i - 1)$, we have the telescoping sum

$$(\log 3 - \log 1) + (\log 4 - \log 2) + \cdots + [\log(n + 1) - \log(n - 1)]$$
$$= \log(n + 1) + \log n - \log 2$$

3.106 Evaluate $\sum_{i=1}^{N} i^n$ in terms of Stirling's numbers.

‖ The expansion in Stirling's numbers is $i^n = \sum_{j=1}^{n} s_j^{(n)} i^{(j)}$, so that $\sum_{j=1}^{n} s_j^{(n)} i^{(j+1)}/(j + 1)$ is a finite integral of i^n. At the lower limit $i = 1$ this is zero, leaving us with

$$\sum_{j=1}^{n} s_j^{(n)}(N + 1)^{(j+1)}/(j + 1)$$

3.107 Evaluate $\sum_{i=0}^{\infty} i^n R^i$ for $-1 < R < 1$.

‖ This generalizes several earlier problems. One particularly attractive solution involves differentiation relative to the parameter R. Let $S_n(R) = \Sigma i^n R^i$. Then $S_n'(R) = \Sigma i^{n+1} R^{i-1}$ and

$$RS_n'(R) = \sum i^{n+1} R^i = S_{n+1}(R)$$

This is a recursion which produces S_{n+1} from S_n.

For example, starting from $S_0 = 1/(1 - R)$ we again find

$$S_1 = R/(1 - R)^2$$
$$S_2 = R\left[(1 - R)^2 + 2R(1 - R)\right]\Big/(1 - R)^4 = (R + R^2)/(1 - R)^3$$

and so on.

3.108 Evaluate $T_n(R) = \sum_{i=0}^{\infty} i^{(n)}R^i$ for $-1 < R < 1$.

▮ Using summation by parts as usual we have the integrated term $i^{(n)}R^i/(R - 1)$ which vanishes at both limits. The other term is

$$T_n = -\sum \frac{ni^{(n-1)}R^{i+1}}{R - 1} = \frac{nR}{1 - R}\sum i^{(n-1)}R^i = \frac{nR}{1 - R}T_{n-1}$$

and provides a recursion for the sums T_n. Using it repeatedly,

$$T_n = \frac{nR}{1 - R}T_{n-1} = \frac{nR}{1 - R}\frac{(n - 1)R}{1 - R}T_{n-2} = \cdots = \frac{n!R^n}{(1 - R)^{n+1}}$$

since $T_0 = 1/(1 - R)$. For $n = 1, 2, 3$ this means

$$T_1 = R/(1 - R)^2 \qquad T_2 = 2R^2/(1 - R)^3 \qquad T_3 = 6R^3/(1 - R)^4$$

and so on.

We now have still another way to compute some familiar sums. Since $i^2 = i^{(1)} + i^{(2)}$ and $i^3 = i^{(1)} + 3i^{(2)} + i^{(3)}$,

$$\sum i^2 R^i = T_1 + T_2$$
$$\sum i^3 R^i = T_1 + 3T_2 + T_3$$

and so on.

3.109 Express a finite integral of $\Delta y_k = 1/k$ in the form of a summation, avoiding $k = 0$.

▮
$$\sum_1^{n-1} \Delta y_k = \sum_1^{n-1} 1/k = y_n - y_1$$

from which $y_n = y_1 + 1/1 + 1/2 + \cdots + 1/(n - 1)$, with y_1 arbitrary.

A finite integral of $\Delta y_k = \log k$ can be arranged in the same way:

$$\sum_1^{n-1} \Delta y_k = \sum_1^{n-1} \log k = y_n - y_1$$

so that $y_n = y_1 + \log(n - 1)!$, with y_1 arbitrary.

3.5 NEWTON'S FORMULA

3.110 Prove that

$$y_1 = y_0 + \Delta y_0 \qquad y_2 = y_0 + 2\Delta y_0 + \Delta^2 y_0 \qquad y_3 = y_0 + 3\Delta y_0 + 3\Delta^2 y_0 + \Delta^3 y_0$$

and infer similar results such as

$$\Delta y_2 = \Delta y_0 + 2\Delta^2 y_0 + \Delta^3 y_0 \qquad \Delta^2 y_2 = \Delta^2 y_0 + 2\Delta^3 y_0 + \Delta^4 y_0$$

▮ This is merely a preliminary to a more general result. The first result is obvious. For the second, with one eye on Table 3.18,

$$y_2 = y_1 + \Delta y_1 = (y_0 + \Delta y_0) + (\Delta y_0 + \Delta^2 y_0)$$

leading at once to the required result. Notice that this expresses y_2 in terms of entries in the top diagonal of Table 3.18. Notice also that almost identical computations produce

$$\Delta y_2 = \Delta y_0 + 2 \Delta^2 y_0 + \Delta^3 y_0 \qquad \Delta^2 y_2 = \Delta^2 y_0 + 2 \Delta^3 y_0 + \Delta^4 y_0$$

TABLE 3.18

x_0	y_0				
		Δy_0			
x_1	y_1		$\Delta^2 y_0$		
		Δy_1		$\Delta^3 y_0$	
x_2	y_2		$\Delta^2 y_1$		$\Delta^4 y_0$
		Δy_2		$\Delta^3 y_1$	
x_3	y_3		$\Delta^2 y_2$		
		Δy_3			
x_4	y_4				

etc., expressing the entries on the y_2 diagonal in terms of those on the top diagonal. Finally,

$$y_3 = y_2 + \Delta y_2 = \left(y_0 + 2 \Delta y_0 + \Delta^2 y_0 \right) + \left(\Delta y_0 + 2 \Delta^2 y_0 + \Delta^3 y_0 \right)$$

leading quickly to the third required result. Similar expressions for Δy_3, $\Delta^2 y_3$, etc., can be written by simply raising the upper index on each Δ.

3.111 Prove that for any positive integer k,

$$y_k = \sum_{i=0}^{k} \binom{k}{i} \Delta^i y_0$$

(Here $\Delta^0 y_0$ means simply y_0.)

∎ The proof will be by induction. For $k = 1, 2,$ and 3, see Problem 3.110. Assume the result true when k is some particular integer p:

$$y_p = \sum_{i=0}^{p} \binom{p}{i} \Delta^i y_0$$

Then, as suggested in the previous problem, the definition of our various differences makes

$$\Delta y_p = \sum_{i=0}^{p} \binom{p}{i} \Delta^{i+1} y_0$$

also true. We now find

$$y_{p+1} = y_p + \Delta y_p = \sum_{j=0}^{p} \binom{p}{j} \Delta^j y_0 + \sum_{j=1}^{p+1} \binom{p}{j-1} \Delta^j y_0$$

$$= y_0 + \sum_{j=1}^{p} \left[\binom{p}{j} + \binom{p}{j-1} \right] \Delta^j y_0 + \Delta^{p+1} y_0$$

$$= y_0 + \sum_{j=1}^{p} \binom{p+1}{j} \Delta^j y_0 + \Delta^{p+1} y_0 = \sum_{j=0}^{p+1} \binom{p+1}{j} \Delta^j y_0$$

Problem 3.59 was used in the third step. The summation index may now be changed from j to i if desired. Thus our result is established when k is the integer $p + 1$, completing the induction.

3.112 Prove that the polynomial of degree n,

$$p_k = y_0 + k\,\Delta y_0 + \frac{1}{2!}k^{(2)}\Delta^2 y_0 + \cdots + \frac{1}{n!}k^{(n)}\Delta^n y_0$$

$$= \sum_{i=0}^{n} \frac{1}{i!}k^{(i)}\Delta^i y_0 = \sum_{i=0}^{n}\binom{k}{i}\Delta^i y_0$$

takes the values $p_k = y_k$ for $k = 0, 1, \ldots, n$. This is *Newton's formula*.

$\blacksquare$ Notice first that when k is 0 only the y_0 term on the right contributes, all others being 0. When k is 1 only the first two terms on the right contribute, all others being 0. When $k = 2$ only the first three terms contribute. Thus, using Problem 3.110,

$$p_0 = y_0 \qquad p_1 = y_0 + \Delta y_0 = y_1 \qquad p_2 = y_0 + 2\,\Delta y_0 + \Delta^2 y_0 = y_2$$

and the nature of our proof is indicated. In general, if k is any integer from 0 to n, then $k^{(i)}$ will be 0 for $i > k$. (It will contain the factor $k - k$.) The sum abbreviates to

$$p_k = \sum_{i=0}^{k} \frac{1}{i!}k^{(i)}\Delta^i y_0$$

and by Problem 3.111 this reduces to y_k. The polynomial of this problem therefore takes the same values as our y_k function for the integer arguments $k = 0, \ldots, n$. (The polynomial is, however, defined for any argument k.)

3.113 Express the result of Problem 3.112 in terms of the argument x_k, where $x_k = x_0 + kh$.

$\blacksquare$ Notice first that

$$k = \frac{x_k - x_0}{h} \qquad k - 1 = \frac{x_{k-1} - x_0}{h} = \frac{x_k - x_1}{h} \qquad k - 2 = \frac{x_{k-2} - x_0}{h} = \frac{x_k - x_2}{h}$$

and so on. Using the symbol $p(x_k)$ instead of p_k, we now find

$$p(x_k) = y_0 + \frac{\Delta y_0}{h}(x_k - x_0) + \frac{\Delta^2 y_0}{2!h^2}(x_k - x_0)(x_k - x_1) + \cdots + \frac{\Delta^n y_0}{n!h^n}(x_k - x_0)\cdots(x_k - x_{n-1})$$

which is Newton's formula in is alternative form.

3.114 Find the polynomial of degree 3 that takes the four values listed in the y_k column of Table 3.19 at the corresponding arguments x_k.

$\blacksquare$ The various differences needed appear in the remaining columns of Table 3.19. Substituting the circled numbers in their places in Newton's formula,

$$p(x_k) = 1 + \tfrac{2}{2}(x_k - 4) + \tfrac{3}{8}(x_k - 4)(x_k - 6) + \tfrac{4}{48}(x_k - 4)(x_k - 6)(x_k - 8)$$

TABLE 3.19

k	x_k	y_k	Δy_k	$\Delta^2 y_k$	$\Delta^3 y_k$
0	④	①			
			②		
1	⑥	3		③	
			5		④
2	⑧	8		7	
			12		
3	10	20			

which can be simplified to

$$p(x_k) = \frac{1}{24}(2x_k^3 - 27x_k^2 + 142x_k - 240)$$

though often in applications the first form is preferable.

3.115 Express the polynomial of Problem 3.114 in terms of the argument k.

❚ Directly from Problem 3.112,

$$p_k = 1 + 2k + \frac{3}{2}k^{(2)} + \frac{4}{6}k^{(3)}$$

which is a convenient form for computing p_k values and so could be left as is. It can also be rearranged into

$$p_k = 1 + \frac{11}{6}k - \frac{1}{2}k^2 + \frac{2}{3}k^3$$

3.116 Apply Newton's formula to find a polynomial of degree 4 or less that takes the y_k values of Table 3.20.

TABLE 3.20

k	x_k	y_k	Δ	Δ^2	Δ^3	Δ^4
0	1	①				
			−2			
1	2	−1		④		
			2		−8	
2	3	1		−4		⑯
			−2		8	
3	4	−1		4		
			2			
4	5	1				

❚ The needed differences are circled. Substituting the circled entries into their places in Newton's formula,

$$p_k = 1 - 2k + \frac{4}{2}k^{(2)} - \frac{8}{6}k^{(3)} + \frac{16}{24}k^{(4)}$$

which is also

$$p_k = \frac{1}{3}(2k^4 - 16k^3 + 40k^2 - 32k + 3)$$

Since $k = x_k - 1$, this result can also be written as

$$p(x_k) = \frac{1}{3}(2x_k^4 - 24x_k^3 + 100x_k^2 - 168x_k + 93)$$

3.117 Find a polynomial of degree 4 that takes the values

x	2	4	6	8	10
y	0	0	1	0	0

▮ Keeping the horizontal layout, here is the difference table:

$$
\begin{array}{ccccccccc}
0 & & 0 & & 1 & & 0 & & 0 \\
& 0 & & 1 & & -1 & & 0 & \\
& & 1 & & -2 & & 1 & & \\
& & & -3 & & 3 & & & \\
& & & & 6 & & & &
\end{array}
$$

The required differences are now along the left side. Inserting them into the Newton formula brings

$$p_k = 0 + 0k + \tfrac{1}{2}k(k-1) - \tfrac{3}{6}k(k-1)(k-2) + \tfrac{6}{24}k(k-1)(k-2)(k-3)$$

in which $k = (x - 2)/2$. The equivalent in terms of x is

$$p(x) = [(x-2)(x-4)/64][8 - 4(x-6) + (x-6)(x-8)]$$

This function was encountered earlier in Problems 3.33 and 3.83. Figure 3-2 shows how it prepares for, and recovers from, the leap to the 1 level by well-timed oscillations.

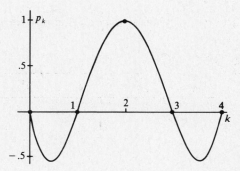

Fig. 3-2

3.118 Find a polynomial that takes the values

$k = x_k$	0	1	2	3	4	5	6	7
y_k	1	2	4	7	11	16	22	29

▮ The second differences prove to be constant,

$$
\begin{array}{ccccccccccccccc}
1 & & 2 & & 4 & & 7 & & 11 & & 16 & & 22 & & 29 \\
& 1 & & 2 & & 3 & & 4 & & 5 & & 6 & & 7 & \\
& & 1 & & 1 & & 1 & & 1 & & 1 & & 1 & &
\end{array}
$$

so Newton's formula produces a quadratic.

$$p(x) = 1 + x + \tfrac{1}{2}x(x-1)$$

3.119 Find a polynomial that takes the values appearing in the top two rows of Table 3.21.

▮ With $k = x - 3$ the polynomial is

$$p = 6 + 18k + \tfrac{18}{2}k(k-1) + \tfrac{6}{6}k(k-1)(k-2)$$

3.120 Find the lowest degree polynomial that takes the values shown in the top two rows of Table 3.22. This is the *adjacent error* function of Problem 3.37.

▮ To accommodate the six data points, a polynomial of degree 5 is expected, but as the table shows, fourth differences are the same, so the fifth-degree term will have a zero coefficient. The fourth-degree polynomial

TABLE 3.21

x	3		4		5		6
y	6		24		60		120
		18		36		60	
			18		24		
				6			

TABLE 3.22

$k = x_k$	0		1		2		3		4		5
y_k	0		0		1		1		0		0
		0		1		0		−1		0	
			1		−1		−1		1		
				−2		0		2			
					2		2				

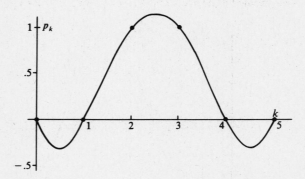

Fig. 3-3

$p(x) = \frac{1}{2}x^{(2)} - \frac{2}{6}x^{(3)} + \frac{2}{24}x^{(4)}$ gets the job done. Figure 3-3 shows how it does it with appropriate preliminary and followup maneuvers.

3.121 What cubic polynomial takes the values

$k = x_k$	0	1	2	3	4	5	6	7	8
y_k	1	2	4	8	15	26	42	64	93

▋ Reference to Problem 3.35, in which a unit error was removed to create the present function, shows that all the differences we need are 1. Newton then provides the polynomial

$$p(x) = 1 + k + \frac{1}{2}k^{(2)} + \frac{1}{6}k^{(3)}$$

3.122 Expressing the general polynomial of degree n in the form

$$p_k = a_0 + a_1 k^{(1)} + a_2 k^{(2)} + \cdots + a_n k^{(n)}$$

calculate $\Delta p_k, \Delta^2 p_k, \ldots, \Delta^n p_k$. Then substitute $k = 0$ to find Newton's formula. This alternate derivation imitates a common approach to the Taylor polynomial.

▋ The differences are

$$\Delta p_k = a_1 + 2a_2 k^{(1)} + 3a_3 k^{(2)} + 4a_4 k^{(3)} + \cdots + na_n k^{(n-1)}$$
$$\Delta^2 p_k = 2a_2 + 6a_3 k^{(1)} + 12a_4 k^{(2)} + \cdots + n(n-1)a_n k^{(n-2)}$$
$$\Delta^3 p_k = 6a_3 + 24a_4 k^{(1)} + \cdots + n(n-1)(n-2)a_n k^{(n-3)}$$

and so on. The requirement that $p_k = y_k$ for $k = 0, \ldots, n$ means that all differences of p agree with those of y, being computed from precisely the same values. So substituting $k = 0$ leads to the equations

$$y_0 = a_0 \qquad \Delta y_0 = a_1 \qquad \Delta^2 y_0 = 2a_2 \qquad \Delta^3 y_0 = 6a_3 \qquad \cdots$$

from which the Newton coefficients are easily recovered.

3.123 Find the quadratic polynomial that collocates with $y(x) = x^4$ at $x = 0, 1, 2$.

▮ The brief table

x	0		1		2
y	0		1		16
		1		15	
			14		

leads quickly to the quadratic

$$p(x) = x + 7x(x-1)$$

3.124 Find the cubic polynomial collocating with $y(x) = \sin(\pi x/2)$ at $x = 0, 1, 2, 3$. Compare the two functions at $x = 4$ and $x = 5$. Is there a polynomial of degree 4 that collocates with this sine function at $x = 0, 1, 2, 3, 4$? (See Fig. 3-4.)

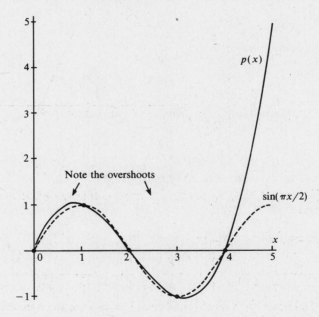

Note the overshoots

Fig. 3-4

▮ The brief table

x	0		1		2		3
y	0		1		0		-1
		1		-1		-1	
			-2		0		
				2			

leads to the cubic

$$p(x) = x - x^{(2)} + \tfrac{1}{3}x^{(3)}$$

At $x = 4$ both y and p take the value 0, so there is a bonus collocation. But at $x = 5$ we find $y = 1$ and $p = 5$. For the five points $x = 0, 1, 2, 3, 4$ the cubic polynomial will do; there is no need for degree 4.

3.125 Is there a polynomial of degree 2 that collocates with $y(x) = x^3$ at $x = -1, 0, 1$?

▮ The points $(-1, -1)$, $(0, 0)$ and $(1, 1)$ fall on the cubic and also on the line $y = x$. Our theorem guarantees a unique polynomial of degree 2 or less. In this case, as in the preceding problem, it is less. The following difference table is relevant though hardly challenging:

x	-1		0		1
y	-1		0		1
		1		1	
			0		

3.126 What polynomial of degree 4 collocates with $y(x) = |x|$ at $x = -2, -1, 0, 1, 2$? Where is it greater than $y(x)$ and where less?

▮ The needed difference table is

x	-2		-1		0		1		2
y	2		1		0		1		2
		-1		-1		1		1	
			0		2		0		
				2		-2			
					-4				

The fourth-order collocation polynomial is

$$p = 2 - k + \tfrac{1}{6}k^{(3)} - \tfrac{1}{6}k^{(4)}$$

with $k = x + 2$. Substitution leads to the equivalent representation

$$p(x) = x^2(7 - x^2)/6$$

The polynomial is above the function in the symmetric intervals $(-2, -1)$ and $(1, 2)$. Figure 3-5 indicates its overall behavior.

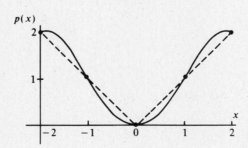

Fig. 3-5

3.127 Why is Newton's formula not available for collocation with $y(x) = \sqrt{x}$ at $x = 0, 1, 4$?

▮ Newton's formula, the one now being illustrated, does not apply because the data points in this problem are not equally spaced. However, if we try a polynomial in our usual form

$$p(x) = a + bx + cx(x - 1)$$

the coefficients can be determined directly. Inserting $x = 0, 1, 4$ together with the accompanying square roots,

$$0 = a \qquad 1 = a + b \qquad 2 = a + 4b + 12c$$

so the three coefficients are 0, 1 and $-\frac{1}{6}$, the quadratic

$$p(x) = x - \tfrac{1}{6}x(x - 1)$$

is indicated. A modification of Newton's formula using divided differences will be developed later, extending the general idea of this preliminary example.

3.128　Solve $\Delta^3 y_k = 1$ subject to $y_0 = \Delta y_0 = \Delta^2 y_0 = 0$.

❚　The constant third difference means a polynomial of third degree, and because the initial conditions reduce Newton's formula to just the final term, the solution must be

$$p = \tfrac{1}{6}k^{(3)}$$

The problem has this equivalent. The vanishing of y_0 and its first two differences requires that y_0, y_1, y_2 all be zero, but then $\Delta^3 y_0 = 1$ forces $y_3 = 1$. This provides four points through which the cubic must pass, enough to determine it. The differences in Table 3.23 then lead to the result already found.

TABLE 3.23

k	0		1		2		3
y_k	0		0		0		1
		0		0		1	
			0		1		
				1			

3.129　Rewrite Newton's formula in algorithmic form, convenient for computer programming.

❚　First observe that the formula can be written as

$$p(x) = y + \frac{x - x_0}{h}\left(\Delta y_0 + \cdots + \frac{x - x_{n-3}}{(n-2)h}\left(\Delta^{n-2} y_0 + \frac{x - x_{n-2}}{(n-1)h}\left(\Delta^{n-1} y_0 + \frac{x - x_{n-1}}{nh}\left(\Delta^n y_0 \right)\right) \cdots \right)\right)$$

which is equivalent to the algorithm

$$p = \Delta^n y_0$$
$$p = \frac{x - x_{n-1}}{nh}p + \Delta^{n-1} y_0$$
$$p = \frac{x - x_{n-2}}{(n-1)h}p + \Delta^{n-2} y_0$$
$$\vdots$$
$$p = \frac{x - x_0}{h}p + y_0$$

The algorithm can also be expressed more compactly as

$$p = \Delta^n y_0$$
$$\text{for } k = n \text{ to } 1$$
$$p = \frac{x - x_{k-1}}{kh}p + \Delta^{k-1} y_0$$

which is almost a computer program.

3.130　Problem 3.20 raises doubts about the usefulness of trying higher and higher degree polynomials as approximations to a given function, if equally spaced points are chosen. Use the algorithm of the preceding problem to test this with $y(x) = \log x$.

❚　Over the interval $(1, 20)$ the following maximum errors were found for polynomials of degree n.

n	6	10	14	18	22
max error	5×10^{-7}	3×10^{-6}	5×10^{-6}	6×10^{-5}	2×10^{-4}

The polynomial of degree 6 has managed six place accuracy, but things definitely go downhill from there.

3.131 Repeat the experiment of the preceding problem using $y(x) = e^{-x^2}$ and trying various intervals.

❚ Here are the numbers:

degree n	6	10	14	18	22
max error on $(0, 20)$	8×10^{-7}	4×10^{-6}	8×10^{-6}	2×10^{-3}	6×10^{-2}
max error on $(-10, 10)$	7×10^{-8}	9×10^{-5}	5×10^{-3}	1×10^{-1}	4

Neither row is particularly encouraging but it is clear that the second is deteriorating faster than the first.

3.132 Repeat the experiment again, with the function $y(x) = 1/(1 + x^2)$.

❚ This apparently simple algebraic function does not respond well to equal interval approximation:

degree n	6	14	22
max error on $(-4, 4)$	1×10^{-7}	6×10^{-4}	6×10^{-2}
max error on $(-5, 5)$	5×10^{-6}	1×10^{-3}	0.5
max error on $(-10, 10)$	1×10^{-6}	2×10^{-3}	5

On $(-4, 4)$ things are not going well and on the wider intervals they are going even worse.

3.133 Try once more, using $y(x) = \sin x$.

❚ Here are the numbers:

degree n	6	14	22
max error on $(0, \pi)$	9×10^{-8}	5×10^{-7}	3×10^{-6}

The smoothness of the sine function is partly responsible for this better behavior, but the outlook is still unfavorable.

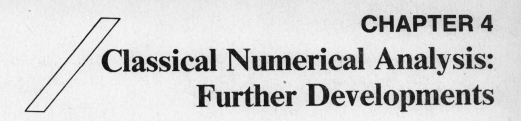

CHAPTER 4
Classical Numerical Analysis: Further Developments

4.1 OPERATORS

4.1 What is the role of operators in numerical analysis?

▮ Operators are used here and there in numerical analysis, in particular for simplifying the development of complicated formulas. Some of the most interesting applications are carried out in a spirit of optimism, without excessive attention to logical precision, the results subject to verification by other methods, or checked experimentally.

A number of the formulas to be derived in this chapter are, in part, of historical interest, providing a view of the numerical priorities of an earlier time. The attached names, such as Newton and Gauss, indicate their importance in those times. Changes in computing hardware have reduced their range of application.

4.2 Define the operators Δ and E.

▮ The *operator* Δ is defined by

$$\Delta y_k = y_{k+1} - y_k$$

We now think of Δ as an operator that when offered y_k as an input produces $y_{k+1} - y_k$ as an output, for all k values under consideration:

$$\boxed{y_k} \rightarrow \boxed{\Delta} \rightarrow \boxed{y_{k+1} - y_k}$$

The analogy between operator and an algorithm (as described in Chapter 2) is apparent. The *operator* E is defined by

$$Ey_k = y_{k+1}$$

Here the input to the operator is again y_k. The output is y_{k+1}:

$$\boxed{y_k} \rightarrow \boxed{E} \rightarrow \boxed{y_{k+1}}$$

Both Δ and E have the linear property, that is,

$$\Delta(C_1 y_k + C_2 z_k) = C_1 \Delta y_k + C_2 \Delta z_k$$

$$E(C_1 y_k + C_2 z_k) = C_1 Ey_k + C_2 Ez_k$$

where C_1 and C_2 are any constants (independent of k). All the operators to be introduced will have this property.

4.3 Define addition of operators and multiplication of an operator by a constant.

▮ Consider two operators L_1 and L_2 that produce the outputs $L_1 y_k$ and $L_2 y_k$ from the input y_k. The sum of these operators is then defined as the operator that outputs $L_1 y_k + L_2 y_k$:

$$\boxed{y_k} \rightarrow \boxed{L_1 + L_2} \rightarrow \boxed{L_1 y_k + L_2 y_k}$$

A similar definition introduces the difference of two operators.

More generally, if C_1 and C_2 are constants (independent of k) the operator $C_1 L_1 + C_2 L_2$ produces the output $C_1 L_1 y_k + C_2 L_2 y_k$:

$$\boxed{y_k} \rightarrow \boxed{C_1 L_1 + C_2 L_2} \rightarrow \boxed{C_1 L_1 y_k + C_2 L_2 y_k}$$

4.4 Define the product of operators.

▮ The *product of operators* L_1 and L_2 is defined as the operator which outputs $L_1 L_2 y_k$. A diagram makes this clearer:

$$\boxed{y_k} \rightarrow \boxed{L_2} \rightarrow \boxed{L_2 y_k} \rightarrow \boxed{L_1} \rightarrow \boxed{L_1 L_2 y_k}$$

The operator L_1 is applied to the output produced by L_2. The center three parts of the diagram together represent the operator $L_1 L_2$:

$$\boxed{y_k} \rightarrow \boxed{L_1 L_2} \rightarrow \boxed{L_1 L_2 y_k}$$

With this definition of product, numbers such as the preceding C_1 and C_2 may also be thought of as operators. For instance, C being any number, the operator C performs a multiplication by the number C.

4.5 Define operator equality.

▮ Two operators L_1 and L_2 are called equal if they produce identical outputs for all inputs under consideration. In symbols,

$$L_1 = L_2 \quad \text{if} \quad L_1 y_k = L_2 y_k$$

for all arguments k under consideration. With this definition a comparison of outputs shows at once that for any operators L_1, L_2 and L_3,

$$L_1 + L_2 = L_2 + L_1$$
$$L_1 + (L_2 + L_3) = (L_1 + L_2) + L_3$$
$$L_1(L_2 L_3) = (L_1 L_2) L_3$$
$$L_1(L_2 + L_3) = L_1 L_2 + L_1 L_3$$

but the commutative law of multiplication is not always true:

$$L_1 L_2 \neq L_2 L_1$$

If either operator is a number C, however, equality is obvious from a comparison of outputs,

$$CL_1 = L_1 C$$

4.6 What are inverse operators?

▮ In a special case of commutativity, the operators L_1 and L_2 are said to be inverse operators if

$$L_1 L_2 = L_2 L_1 = 1$$

In such case we use the symbols

$$L_1 = L_2^{-1} = \frac{1}{L_2} \qquad L_2 = L_1^{-1} = \frac{1}{L_1}$$

The operator 1 is known as the *identity operator* and it is easy to see that it makes $1 \cdot L = L \cdot 1$ for any operator L.

4.7 Define the backward and central difference operators and the averaging operator.

▮ The *backward difference operator* ∇ is defined by

$$\nabla y_k = y_k - y_{k-1}$$

and it is then easy to verify that

$$\nabla E = E \nabla = \Delta$$

The relationship between ∇ and E^{-1} proves to be

$$E^{-1} = 1 - \nabla$$

and leads to the expansion

$$y_k = y_0 + \sum_{i=1}^{-k} \frac{k(k+1)\cdots(k+i-1)}{i!} \nabla^i y_0$$

for negative integers k.

The *central difference operator* is defined by

$$\delta = E^{1/2} - E^{-1/2}$$

It follows that $\delta E^{1/2} = \Delta$. In spite of the fractional arguments this is a heavily used operator. It is closely related to the following operator.

The *averaging operator* is defined by

$$\mu = \tfrac{1}{2}(E^{1/2} + E^{-1/2})$$

and is the principal mechanism by which fractional arguments can be eliminated from central difference operations.

4.8 Prove $E = 1 + \Delta$.

▮ By definition of E, $Ey_k = y_{k+1}$, and by definition of $1 + \Delta$,

$$(1 + \Delta) = 1 \cdot y_k + \Delta y_k = y_k + (y_{k+1} - y_k) = y_{k+1}$$

Having identical outputs for all arguments k, the operators E and $1 + \Delta$ are equal. This result can also be written as $\Delta = E - 1$.

4.9 Prove $E\Delta = \Delta E$.

▮ $\qquad E\Delta y_k = E(y_{k+1} - y_k) = y_{k+2} - y_{k+1} \quad \text{and} \quad \Delta E y_k = \Delta y_{k+1} = y_{k+2} - y_{k+1}$

The equality of outputs makes the operators equal. This is an example in which the commutative law of multiplication is true.

4.10 Prove $\Delta^2 = E^2 - 2E + 1$.

▮ Using various operator properties,

$$\Delta^2 = (E-1)(E-1) = E^2 - 1 \cdot E - E \cdot 1 + 1 = E^2 - 2E + 1$$

4.11 Apply the binomial theorem to prove $\Delta^k y_0 = \sum_{i=0}^{k}(-1)^i \binom{k}{i} y_{k-i}$.

▮ The binomial theorem, $(a+b)^k = \sum_{i=0}^{k} \binom{k}{i} a^{k-i} b^i$, is valid as long as a and b (and therefore $a + b$) commute in multiplication. In the present situation these elements will be E and -1 and these do commute. Thus,

$$\Delta^k = (E-1)^k = \sum_{i=0}^{k} (-1)^i \binom{k}{i} E^{k-i}$$

Noticing that $Ey_0 = y_1$, $E^2 y_0 = y_2$, etc., we have finally

$$\Delta^k y_0 = \sum_{i=0}^{k} (-1)^i \binom{k}{i} y_{k-i}$$

which duplicates the result of Problem 3.35.

4.12 Prove $y_k = \sum_{i=0}^{k} \binom{k}{i} \Delta^i y_0$.

▮ Since $E = 1 + \Delta$, the binomial theorem produces $E^k = (1 + \Delta)^k = \sum_{i=0}^{k} \binom{k}{i} \Delta^i$. Applying this operator to y_0 and using the fact that $E^k y_0 = y_k$, produces the required result at once. Note that this duplicates Problem 3.111.

4.13 Backward differences are normally applied only at the bottom of a table, using negative k arguments as shown in Table 4.1. Using the symbols $\nabla^2 y_k = \nabla \nabla y_k$, $\nabla^3 y_k = \nabla \nabla^2 y_k$, etc., show that $\Delta^n y_k = \nabla^n y_{k+n}$.

TABLE 4.1

k	x	y				
-4	x_{-4}	y_{-4}				
			∇y_{-3}			
-3	x_{-3}	y_{-3}		$\nabla^2 y_{-2}$		
			∇y_{-2}		$\nabla^3 y_{-1}$	
-2	x_{-2}	y_{-2}		$\nabla^2 y_{-1}$		$\nabla^4 y_0$
			∇y_{-1}		$\nabla^3 y_0$	
-1	x_{-1}	y_{-1}		$\nabla^2 y_0$		
			∇y_0			
0	x_0	y_0				

▌ Since $\Delta = E\nabla$, we have $\Delta^n = (E\nabla)^n$. But E and ∇ commute, so the $2n$ factors on the right side may be rearranged to give $\Delta^n = \nabla^n E^n$. Applying this to y_k, $\Delta^n y_k = \nabla^n E^n y_k = \nabla^n y_{k+n}$.

4.14 Prove that

$$y_{-1} = y_0 - \nabla y_0 \qquad y_{-2} = y_0 - 2\nabla y_0 + \nabla^2 y_0 \qquad y_{-3} = y_0 - 3\nabla y_0 + 3\nabla^2 y_0 - \nabla^3 y_0$$

and that in general for k a negative integer,

$$y_k = y_0 + \sum_{i=1}^{-k} \frac{k(k+1)\cdots(k+i-1)}{i!} \nabla^i y_0$$

▌ Take the general case at once: $y_k = E^k y_0 = (E^{-1})^{-k} y_0 = (1 - \nabla)^{-k} y_0$. With k a negative integer the binomial theorem applies, making

$$y_4 = \sum_{i=0}^{-k} (-1)^i \binom{-k}{i} \nabla^i y_0 = y_0 + \sum_{i=1}^{-k} (-1)^i \frac{(-k)(-k-1)\cdots(-k-i+1)}{i!} \nabla^i y_0$$

$$= y_0 + \sum_{i=1}^{-k} \frac{k(k+1)\cdots(k+i-1)}{i!} \nabla^i y_0$$

The special cases now follow for $k = -1, -2, -3$ by writing out the sum.

4.15 Prove that the polynomial of degree n, which has values defined by the following formula, reduces to $p_k = y_k$ when $k = 0, -1, \ldots, -n$ (this is Newton's backward difference formula):

$$p_k = y_0 + k\nabla y_0 + \frac{k(k+1)}{2!} \nabla^2 y_0 + \cdots + \frac{k\cdots(k+n-1)}{n!} \nabla^n y_0$$

$$= y_0 + \sum_{i=1}^{n} \frac{k(k+1)\cdots(k+i-1)}{i!} \nabla^i y_0$$

▌ The proof is very much like the one in Problem 3.112. When k is 0, only the first term on the right side contributes. When k is -1, only the first two terms contribute, all others being zero. In general, if k is any integer from 0 to $-n$, then $k(k+1)\cdots(k+i-1)$ will be 0 for $i > -k$. The sum abbreviates to

$$p_k = y_0 + \sum_{i=1}^{-k} \frac{k(k+1)\cdots(k+i-1)}{i!} \nabla^i y_0$$

and by Problem 4.14 this reduces to y_k. The polynomial of this problem therefore agrees with our y_k function for $k = 0, -1, \ldots, -n$.

4.16 Find the polynomial of degree 3 that takes the four values listed as y_k below at the corresponding x_k arguments.

❚ The differences needed appear in the remaining columns of Table 4.2.
Substituting the circled numbers in their places in Newton's backward difference formula,

$$p_k = 20 + 12k + \tfrac{7}{2}k(k+1) + \tfrac{4}{6}k(k+1)(k+2)$$

TABLE 4.2

k	x_k	y_k	∇y_k	$\nabla^2 y_k$	$\nabla^3 y_k$
-3	4	1			
			2		
-2	6	3		3	
			5		④
-1	8	8		⑦	
			⑫		
0	10	⑳			

Notice that except for the arguments k these data are the same as those of Problem 3.141. Eliminating k by the relation $x_k = 10 + 2k$, the formula found in that problem,

$$p(x_k) = \tfrac{1}{24}\left(2x_k^3 - 27x_k^2 + 142x_k - 240\right)$$

is again obtained. Newton's two formulas are simply rearrangements of the same polynomial. Other rearrangements now follow.

4.17 Apply Newton's backward formula to the following data, to obtain a polynomial of degree 4 in the argument k:

k	-4	-3	-2	-1	0
x_k	1	2	3	4	5
y_k	1	-1	1	-1	1

❚ The needed differences appear in the table

```
  1        -1         1        -1        1
      -2         2        -2         2
           4        -4         4
               -8         8
                     16
```

along the right-side diagonal. The Newton formula yields

$$p_k = 1 + 2k + \tfrac{4}{2}k(k+1) + \tfrac{8}{6}k(k+1)(k+2) + \tfrac{16}{24}k(k+1)(k+2)(k+3)$$

The substitution $x = k + 5$ may now be made if desired and the result compared with that of Problem 3.116.

4.18 Apply Newton's backward formula to find a polynomial of degree 3 that includes the following x_k, y_k pairs:

x_k	3	4	5	6
y_k	6	24	60	120

▌ The needed information appears in the table

$$
\begin{array}{rrr}
k & x & y \\
-3 & 3 & 6 \\
& & & 18 \\
-2 & 4 & 24 & & 18 \\
& & & 36 & & 6 \\
-1 & 5 & 60 & & 24 \\
& & & 60 \\
0 & 6 & 120
\end{array}
$$

Again we use the right-side diagonal, but this time from the bottom up:

$$p_k = 120 + 60k + \tfrac{24}{2}k(k+1) + \tfrac{6}{6}k(k+1)(k+2)$$

The substitution $x = k + 6$ can be made to obtain $p(x)$, which may then be compared with the result of Problem 3.119.

4.19 Apply the change of variable $x_k = x_0 + kh$ to convert Newton's backward formula into a form involving the variable x directly. Note that h is the equal interval size between data points and that x_0 is now at the bottom of the point set.

▌ One need only substitute into the formula of Problem 4.15, using $k = (x - x_0)/h$, $k + 1 = (x - x_{-1})/h$, $k + 2 = (x - x_{-2})/h$ and so on, to obtain

$$p(x_k) = y_0 + \frac{\nabla y_0}{h}(x - x_0) + \frac{\nabla^2 y_0}{2!h^2}(x - x_0)(x - x_{-1}) + \cdots + \frac{\nabla^n y_0}{n!h^n}(x - x_0) \cdots (x - x_{-n+1})$$

in which x_k has been abbreviated to x.

4.20 Rework Problem 4.18 using the formula of the preceding problem to obtain $p(x)$ directly.

▌ The difference table is, of course, reusable and $h = 1$. We find the polynomial

$$p(x) = 120 + 60(x - 6) + \tfrac{24}{2}(x - 6)(x - 5) + \tfrac{6}{6}(x - 6)(x - 5)(x - 4)$$

4.21 The central difference operator δ is defined by $\delta = E^{1/2} - E^{-1/2}$ so that $\delta y_{1/2} = y_1 - y_0 = \Delta y_0 = \nabla y_1$ and so on. Observe that $E^{1/2}$ and $E^{-1/2}$ are inverses and that $(E^{1/2})^2 = E$, $(E^{-1/2})^2 = E^{-1}$. Show that $\Delta^n y_k = \delta^n y_{k+n/2}$.

▌ From the definition of δ, we have $\delta E^{1/2} = E - 1 = \Delta$ and $\Delta^n = \delta^n E^{n/2}$. Applied to y_k, this produces the required result.

4.22 In δ notation, the usual difference table may be rewritten as in Table 4.3. Express $\delta y_{1/2}$, $\delta^2 y_0$ and $\delta^4 y_0$ using the Δ operator.

▌ By Problem 4.21, $\delta y_{1/2} = \Delta y_0$, $\delta^2 y_0 = \Delta^2 y_{-1}$ and $\delta^4 y_0 = \Delta^4 y_{-2}$.

TABLE 4.3

k	y_k	δ	δ^2	δ^3	δ^4
-2	y_{-2}				
		$\delta y_{-3/2}$			
-1	y_{-1}		$\delta^2 y_{-1}$		
		$\delta y_{-1/2}$		$\delta^3 y_{-1/2}$	
0	y_0		$\delta^2 y_0$		$\delta^4 y_0$
		$\delta y_{1/2}$		$\delta^3 y_{1/2}$	
1	y_1		$\delta^2 y_1$		
		$\delta y_{3/2}$			
2	y_2				

4.23 The averaging operator has been defined by $\mu = \frac{1}{2}(E^{1/2} + E^{-1/2})$ so that $\mu y_{1/2} = \frac{1}{2}(y_0 + y_1)$ and so on. Prove $\mu^2 = 1 + \frac{1}{4}\delta^2$.

▌ First we compute $\delta^2 = E - 2 + E^{-1}$. Then $\mu^2 = \frac{1}{4}(E + 2 + E^{-1}) = \frac{1}{4}(\delta^2 + 4) = 1 + \frac{1}{4}\delta^2$.

4.24 Prove $\nabla = \delta E^{-1/2} = 1 - E^{-1} = 1 - (1 + \Delta)^{-1}$.

▌

$$\delta E^{-1/2} y_k = \delta y_{k-1/2} = y_k - y_{k-1} = \nabla y_k$$

$$(1 - E^{-1}) y_k = y_k - y_{k-1} = \nabla y_k$$

Problem 4.8 has already shown that $E = 1 + \Delta$, so the ring is now complete.

4.25 Prove $\sqrt{1 + \delta^2 \mu^2} = 1 + \frac{1}{2}\delta^2$.

▌ From Problem 4.23 we have $\delta^2 = E - 2 + E^{-1}$. This makes

$$1 + \tfrac{1}{2}\delta^2 = \tfrac{1}{2}(E + E^{-1})$$

On the left side observe first that, using $k = 0$ for simplicity,

$$\delta \mu y_0 = \delta \tfrac{1}{2}(y_{1/2} + y_{-1/2}) = \tfrac{1}{2}(y_1 - y_0 + y_0 - y_{-1}) = \tfrac{1}{2}(y_1 - y_{-1})$$

so that $\delta \mu = \frac{1}{2}(E - E^{-1})$. A parallel computation shows that $\mu \delta y_0$ comes out the same. This commutativity allows us to compute

$$1 + \delta^2 \mu^2 = 1 + (\delta \mu)^2 = \tfrac{1}{4}(E^2 + 2 + E^{-2}) = \left(1 + \tfrac{1}{2}\delta^2\right)^2$$

from which the result follows by taking the square root.

4.26 Prove $E^{1/2} = \mu + \frac{1}{2}\delta$ and $E^{-1/2} = \mu - \frac{1}{2}\delta$.

▌ Trying a different tack we note that

$$\mu = \tfrac{1}{2}E^{1/2} + \tfrac{1}{2}E^{-1/2}$$

$$\tfrac{1}{2}\delta = \tfrac{1}{2}E^{1/2} - \tfrac{1}{2}E^{-1/2}$$

so that solving this little system

$$E^{1/2} = \mu + \tfrac{1}{2}\delta \qquad E^{-1/2} = \mu - \tfrac{1}{2}\delta$$

as required.

4.27 Show that μ, δ, E, Δ and ∇ all commute with one another.

▌ Take μ and E as an example. Also choose $k = 0$ for simplicity, the particular data point being irrelevant:

$$\mu E y_0 = \mu y_1 = \tfrac{1}{2}(y_{3/2} + y_{1/2})$$

$$E \mu y_0 = E \tfrac{1}{2}(y_{1/2} + y_{-1/2}) = \tfrac{1}{2}(y_{3/2} + y_{1/2})$$

The other pairs can be treated similarly. The details are not particularly inspiring.

4.28 Prove $\mu \delta = \frac{1}{2} \Delta E^{-1} + \frac{1}{2}\Delta$.

▌ Borrowing our result $\mu \delta = \frac{1}{2}(E - E^{-1})$ for the left side, we also discover

$$\tfrac{1}{2}\Delta(E^{-1} + 1) = \tfrac{1}{2}(E - 1)(E^{-1} + 1) = \tfrac{1}{2}(E - E^{-1})$$

for the right.

4.29 Prove $\Delta = \frac{1}{2}\delta^2 + \delta\sqrt{1 + \frac{1}{4}\delta^2}$.

▮ Trying yet another tack we write the now familiar $\delta = E^{1/2} - E^{-1/2}$ as $E^{1/2} - \delta - E^{-1/2} = 0$ and multiplying by $E^{1/2}$,

$$E - \delta E^{1/2} - 1 = 0$$

which we take as a quadratic equation for $E^{1/2}$. Solving,

$$E^{1/2} = \frac{1}{2}\left(\delta + \sqrt{\delta^2 + 4}\right) = \frac{1}{2}\delta + \sqrt{1 + \frac{1}{4}\delta^2}$$

the other root being of no immediate interest. But then

$$\Delta = E - 1 = \delta E^{1/2} = \frac{1}{2}\delta^2 + \sqrt{1 + \frac{1}{4}\delta^2}$$

which was required.

4.30 Verify the following for the indicated arguments k:

$$k = 0, 1 \qquad y_k = y_0 + \binom{k}{1}\delta y_{1/2}$$

$$k = -1, 0, 1 \qquad y_k = y_0 + \binom{k}{1}\delta y_{1/2} + \binom{k}{2}\delta^2 y_0$$

$$k = -1, 0, 1, 2 \qquad y_k = y_0 + \binom{k}{1}\delta y_{1/2} + \binom{k}{2}\delta^2 y_0 + \binom{k+1}{3}\delta^3 y_{1/2}$$

$$k = -2, -1, 0, 1, 2 \qquad y_k = y_0 + \binom{k}{1}\delta y_{1/2} + \binom{k}{2}\delta^2 y_0 + \binom{k+1}{3}\delta^3 y_{1/2} + \binom{k+1}{4}\delta^4 y_0$$

▮ For $k = 0$ only the y_0 terms on the right contribute. When $k = 1$ all right sides correspond to the operator

$$1 + \delta E^{1/2} = 1 + (E - 1) = E$$

which does produce y_1. For $k = -1$ the last three formulas lead to

$$1 - \delta E^{1/2} + \delta^2 = 1 - (E - 1) + (E - 2 + E^{-1}) = E^{-1}$$

which produces y_{-1}. When $k = 2$ the last two formulas bring

$$1 + 2\delta E^{1/2} + \delta^2 + \delta^3 E^{1/2} = 1 + 2(E - 1) + (E - 2 + E^{-1})(1 + E - 1) = E^2$$

producing y_2. Finally when $k = -2$ the last formula involves

$$1 - 2\delta E^{1/2} + 3\delta^2 - \delta^3 E^{1/2} + \delta^4 = 1 - 2(E - 1) + (E - 2 + E^{-1})\left[3 - (E - 1) + (E - 2 + E^{-1})\right] = E^{-2}$$

leading to y_{-2}.

The formulas of this problem generalize to form the *Gauss forward formula*. It represents a polynomial either of degree $2n$,

$$p_k = y_0 + \sum_{i=1}^{n}\left[\binom{k+i-1}{2i-1}\delta^{2i-1}y_{1/2} + \binom{k+i-1}{2i}\delta^{2i}y_0\right]$$

taking the values $p_k = y_k$ for $k = -n, \ldots, n$, or of degree $2n + 1$,

$$p_k = \sum_{i=0}^{n}\left[\binom{k+i-1}{2i}\delta^{2i}y_0 + \binom{k+i}{2i+1}\delta^{2i+1}y_{1/2}\right]$$

taking the values $p_k = y_k$ for $k = -n, \ldots, n + 1$. (In special cases the degree may be lower.)

4.31 Apply Gauss' formula with $n = 2$ to find a polynomial of degree 4 or less that takes the y_k values in Table 4.4.

▮ The differences needed are listed as usual. This resembles a function used in illustrating the two Newton formulas, with a shift in the argument k and an extra number pair added at the top. Since the fourth difference is

TABLE 4.4

k	x_k	y_k				
−2	2	−2				
			3			
−1	4	1		−1		
			2		4	
0	6	③		③		⓪
			⑤		④	
1	8	8		7		
			12			
2	10	20				

0 in this example, we anticipate a polynomial of degree 3. Substituting the circled entries into their places in Gauss' formula,

$$p_k = 3 + 5k + \tfrac{3}{2}k(k-1) + \tfrac{4}{6}(k+1)k(k-1)$$

If k is eliminated by the relation $x_k = 6 + 2k$, the cubic already found twice before appears once again.

4.32 Apply Gauss' forward Formula to find a polynomial of degree 4 or less that takes the y_k values in Table 4.5.

▮ The needed differences are circled.
  Substituting into their places in the Gauss formula,

$$p_k = 1 - 2k - 4\frac{k(k-1)}{2} + 8\frac{(k+1)k(k-1)}{6} + 16\frac{(k+1)k(k-1)(k-2)}{24}$$

TABLE 4.5

k	x_k	y_k				
−2	1	1				
			−2			
−1	2	−1		4		
			2		−8	
0	3	①		⟨−4⟩		⑯
			⟨−2⟩		⑧	
1	4	−1		4		
			2			
2	5	1				

which simplifies to

$$p_k = \tfrac{1}{3}\left(2k^4 - 8k^2 + 3\right)$$

Since $k = x_k - 3$, this result can also be written as

$$p(x_k) = \tfrac{1}{3}\left(2x_k^4 - 24x_k^3 + 100x_k^2 - 168x_k + 93\right)$$

agreeing, of course, with the polynomial found earlier by Newton's formula.

4.33 Apply the Gauss forward formula to this familiar data.

x	2	4	6	8	10
y	0	0	1	0	0

■ Let $k = 0$ at $x = 6$ and find differences as shown in the table

k	x	y				
-2	2	0				
			0			
-1	4	0		1		
			1		-3	
0	6	1		(-2)		(6)
			(-1)		(3)	
1	8	0		1		
			0			
2	10	0				

Using the circled entries we have

$$p = 1 - k - k(k-1) + \tfrac{1}{2}(k+1)k(k-1) + \tfrac{1}{4}(k+1)k(k-1)(k-2)$$

and by the substitution $k = (x-6)/2$ can recover the result of Problem 3.117 where Newton's formula was applied to the same data.

4.34 Apply the Gauss forward formula to the data of Table 4.6.

■ The $k = 0$ value has been chosen at $x = 2$ and the differences to be used have been circled. We find

$$p = 1 - \tfrac{1}{2}k(k-1) + \tfrac{1}{12}(k+1)k(k-1)(k-2)$$

which may be transformed into the result of Problem 3.120 by using $k = x - 2$.

TABLE 4.6

k_1	x	y						
2	0	0						
			0					
1	1	0		1				
			1		-2			
0	2	①		(-1)		2		
			⓪		⓪		⓪	
1	3	1		-1		2		
			-1		2			
2	4	0		1				
			0					
3	5	0						

4.35 Verify that, for $k = -1, 0, 1$,

$$y_k = y_0 + \binom{k}{1}\delta y_{-1/2} + \binom{k+1}{2}\delta^2 y_0$$

and, for $k = -2, -1, 0, 1, 2$,

$$y_k = y_0 + \binom{k}{1}\delta y_{-1/2} + \binom{k+1}{2}\delta^2 y_0 + \binom{k+1}{3}\delta^3 y_{-1/2} + \binom{k+2}{4}\delta^4 y_0$$

■ For $k = 0$, only the y_0 terms on the right contribute. When $k = 1$ both formulas involve the operator

$$1 + \delta E^{-1/2} + \delta^2 = 1 + (1 - E^{-1}) + (E - 2 + E^{-1}) = E$$

which does produce y_1. For $k = -1$ both formulas involve

$$1 - \delta E^{-1/2} = 1 - (1 - E^{-1}) = E^{-1}$$

which does produce y_{-1}. Continuing with the second formula, we find, for $k = 2$,

$$1 + 2\,\delta E^{-1/2} + 3\delta^2 + \delta^3 E^{-1/2} + \delta^4 = 1 + 2(1 - E^{-1}) + (E - 2 + E^{-1})(3 + 1 - E^{-1} + E - 2 + E^{-1}) = E^2$$

and, for $k = -2$,

$$1 - 2\,\delta E^{-1/2} + \delta^2 - \delta^3 E^{-1/2} = 1 - 2(1 - E^{-1}) + (E - 2 + E^{-1})(1 - 1 + E^{-1}) = E^{-2}$$

as required.

The formulas of this problem can be generalized to form the *Gauss backward formula*. It represents the same polynomial as the Gauss forward formula of even order and can be verified as before:

$$p_k = y_0 + \sum_{i=1}^{n} \left[\binom{k+i-1}{2i-1} \delta^{2i-1} y_{-1/2} + \binom{k+i}{2i} \delta^{2i} y_0 \right]$$

4.36 Apply the Gauss backward formula just developed to the data of Problem 4.33.

▌ The difference table

$$
\begin{array}{ccc}
k & x & y \\
-2 & 2 & 0 \\
& & & 0 \\
-1 & 4 & 0 & & 1 \\
& & & ① & & \boxed{-3} \\
0 & 6 & ① & & \boxed{-2} & & ⑥ \\
& & & -1 & & 3 \\
1 & 8 & 0 & & 1 \\
& & & 0 \\
2 & 10 & 0 \\
\end{array}
$$

is repeated only to show the course taken through it when the Gauss backward formula is used; see the circled entries. The Gauss backward formula as needed here will have the terms

$$p = y_0 + k\,\delta y_{-1/2} + \tfrac{1}{2}(k+1)k\,\delta^2 y_0 + \tfrac{1}{6}(k+1)k(k-1)\,\delta^3 y_{-1/2} + \tfrac{1}{24}(k+2)(k+1)k(k-1)\,\delta^4 y_0$$

Making the substitutions we have

$$p = 1 + k - \tfrac{2}{2}(k+1)k - \tfrac{3}{6}(k+1)k(k-1) + \tfrac{6}{24}(k+2)(k+1)k(k-1)$$

which is equivalent to the result found earlier.

4.37 It is easy to verify that stopping with an odd difference term in the Gauss formulas also leaves a collocation polynomial. For example,

$$y_k = y_0 + \binom{k}{1}\delta y_{-1/2} + \binom{k+1}{2}\delta^2 y_0 + \binom{k+1}{3}\delta^3 y_{-1/2}$$

is a cubic with collocation at $k = -2, -1, 0, 1$. Apply this to the data of Problem 4.18.

▌ The difference table

$$
\begin{array}{ccc}
k & x & y \\
-2 & 3 & 6 \\
& & & 18 \\
-1 & 4 & 24 & & 18 \\
& & & ㊱ & & ⑥ \\
0 & 5 & ㉐ & & ㉔ \\
& & & 60 \\
1 & 6 & 120 \\
\end{array}
$$

is repeated with $k = 0$ now at the $x = 5$ level. Substituting the circled entries we have

$$p = 60 + 36k + \tfrac{24}{2}(k+1)k + \tfrac{6}{6}(k+1)k(k-1)$$

with $k = x - 5$.

4.38 Prove

$$\binom{k+i}{2i} + \binom{k+i-1}{2i} = \frac{k}{i}\binom{k+i-1}{2i-1}$$

$\blacksquare$ From the definitions of binomial coefficients,

$$\binom{k+i}{2i} + \binom{k+i-1}{2i} = \binom{k+i-1}{2i-1}[(k+i)+(k-i)]\frac{1}{2i}$$

as required.

4.39 Deduce *Stirling's formula*, which follows, from the Gauss formulas.

$\blacksquare$ Adding the Gauss formulas for degree $2n$ term by term, dividing by two and using Problem 4.38,

$$p_k = y_0 + \sum_{i=1}^{n}\left[\binom{k+i-1}{2i-1}\delta^{2i-1}\mu\, y_0 + \frac{k}{2i}\binom{k+i-1}{2i-1}\delta^{2i}y_0\right]$$

$$= y_0 + \binom{k}{1}\delta\mu\, y_0 + \frac{k}{2}\binom{k}{1}\delta^2 y_0 + \binom{k+1}{3}\delta^3\mu\, y_0 + \frac{k}{4}\binom{k+1}{3}\delta^4 y_0$$

$$+ \cdots + \binom{k+n-1}{2n-1}\delta^{2n-1}\mu\, y_0 + \frac{k}{2n}\binom{k+n-1}{2n-1}\delta^{2n}y_0$$

This is Stirling's formula.

4.40 Apply Stirling's formula with $n = 2$ to find a polynomial of degree 4 or less that takes the y_k values in Table 4.7.

$\blacksquare$ The differences needed are again listed.
Substituting the circled entries into their places in Stirling's formula,

$$p_k = 3 + \frac{2+5}{2}k + 3\frac{k^2}{2} + \frac{4+4}{2}\frac{(k+1)k(k-1)}{6}$$

which is easily found to be a minor rearrangement of the result found by the Gauss forward formula.

TABLE 4.7

k	x_k	y_k	δ	δ^2	δ^3	δ^4
-2	2	-2				
			3			
-1	4	1		-1		
			②		④	
0	6	③		③		⓪
			⑤		④	
1	8	8		7		
			12			
2	10	20				

4.41 Apply Stirling's formula to find a polynomial of degree 4 or less that takes the y_k values in Table 4.8.

TABLE 4.8

k	x_k	y_k	δ	δ^2	δ^3	δ^4
−2	1	1				
			−2			
−1	2	−1		4		
			②		−8	
0	3	①		−4		16
			−2		8	
1	4	−1		4		
			2			
2	5	1				

▌ The needed differences are circled.
Substituting the circled entries into their places in Stirling's formula,

$$p_k = 1 + \frac{2 + (-2)}{2}k - 4\frac{k^2}{2} + \frac{8 + (-8)}{2}\binom{k+1}{3} + 16\frac{k}{4}\frac{(k+1)k(k-1)}{6}$$

which simplifies to $p_k = \frac{1}{3}(2k^4 - 8k^2 + 3)$ as with Gauss' forward formula.

4.42 Apply Stirling's formula to find a polynomial of degree 4 or less that takes the y_k values in Table 4.9.

▌ The needed differences are circled.
Substituting the circled entries into their places in Stirling's formula,

$$p_k = 1 + \frac{2 + (-2)}{2}k - 4\frac{k^2}{2} + \frac{8 + (-8)}{2}\binom{k+1}{3} + 16\frac{k}{4}\frac{(k+1)k(k-1)}{6}$$

which simplifies to $p_k = \frac{1}{3}(2k^4 - 8k^2 + 3)$ as with Gauss' forward formula.

TABLE 4.9

k	x_k	y_k	δ	δ^2	δ^3	δ^4
−2	1	1				
			−2			
−1	2	−1		4		
			②		−8	
0	3	①		−4		16
			−2		8	
1	4	−1		4		
			2			
2	5	1				

4.43 Applying Stirling's formula to the data of Problem 3.117.

▌ We need only the center strip of the now familiar difference table

$$
\begin{array}{ccccccc}
 & & & 1 & & -3 & \\
0 & 6 & 1 & & -2 & & 6 \\
 & & & -1 & & 3 & \\
\end{array}
$$

with k now 0 at $x = 6$. Substituting appropriately,

$$p = 1 - k^2 + \tfrac{1}{4}k(k+1)k(k-1)$$

4.44   Prove

$$\binom{k+i-1}{2i}\delta^{2i}y_0 + \binom{k+i}{2i+1}\delta^{2i+1}y_{1/2} = \binom{k+i}{2i+1}\delta^{2i}y_1 - \binom{k+i-1}{2i+1}\delta^{2i}y_0$$

▮ The left side becomes

$$\left[\binom{k+i}{2i+1} - \binom{k+i-1}{2i+1}\right]\delta^{2i}y_0 + \binom{k+i}{2i+1}\delta^{2i+1}y_{1/2}$$

$$= \binom{k+i}{2i+1}\left[\delta^{2i}(1+\delta E^{1/2})y_0\right] - \binom{k+i-1}{2i+1}\delta^{2i}y_0$$

$$= \binom{k+i}{2i+1}\delta^{2i}y_1 - \binom{k+i-1}{2i+1}\delta^{2i}y_0$$

where in the last step we used $1 + \delta E^{1/2} = E$.

4.45   Deduce Everett's formula from the Gauss forward formula of odd degree.

▮ Using Problem 4.44, we have at once

$$p_k = \sum_{i=0}^{n}\left[\binom{k+1}{2i+1}\delta^{2i}y_1 - \binom{k+i-1}{2i+1}\delta^{2i}y_0\right]$$

$$= \binom{k}{1}y_1 + \binom{k+1}{3}\delta^2 y_1 + \binom{k+2}{5}\delta^4 y_1 + \cdots + \binom{k+n}{2n+1}\delta^{2n}y_1$$

$$- \binom{k-1}{1}y_0 - \binom{k}{3}\delta^2 y_0 - \binom{k+1}{5}\delta^4 y_0 - \cdots - \binom{k+n-1}{2n+1}\delta^{2n}y_0$$

which is *Everett's formula*. Since it is a rearrangement of the Gauss formula it is the same polynomial of degree $2n + 1$, satisfying $p_k = y_k$ for $k = -n, \ldots, n + 1$. It is a heavily used formula because of its simplicity, only even differences being involved.

4.46   Apply Everett's formula with $n = 1$ to find a polynomial of degree 3 or less that takes the values in Table 4.10.

▮ Simply substitute the circled entries in their proper places in Everett's formula. The result is

$$p = 8k + \tfrac{7}{6}(k+1)k(k-1) - 3(k-1) - \tfrac{3}{6}k(k-1)(k-2)$$

agreeing, of course, with an earlier application of Gauss' forward formula.

TABLE 4.10

k	x	y		
−1	4	1		
			2	
0	6	③		③
			5	
1	8	⑧		⑦
			12	
2	10	20		

4.47   Apply Everett's formula to find a polynomial of degree 5 or less that takes the y_k values of Table 4.11. The needed differences have been circled.

TABLE 4.11

k	x_k	y_k	δ	δ^2	δ^3	δ^4
-2	0	0				
			-1			
-1	1	-1		10		
			9		108	
0	2	⑧		⑪⑱		②⑯
			127		324	
1	3	⑬⑤		④④②		③③⑥
			569		660	
2	4	704		1102		
			1671			
3	5	2375				

▌ Substituting the circled entries into their places in Everett's formula,

$$p_k = 135k + 442\frac{(k+1)k(k-1)}{6} + 336\frac{(k+2)(k+1)k(k-1)(k-2)}{120}$$

$$-8(k-1) - 118\frac{k(k-1)(k-2)}{6} - 216\frac{(k+1)k(k-1)(k-2)(k-3)}{120}$$

which can be simplified, using $x_k = k + 2$, to $p(x_k) = x_k^5 - x_k^4 - x_k^3$.

4.48 Show that

$$\binom{k+i-1}{2i}\mu\,\delta^{2i}y_{1/2} + \frac{k-\frac{1}{2}}{2i+1}\binom{k+i-1}{2i}\delta^{2i+1}y_{1/2} = \binom{k+i}{2i+1}\delta^{2i}y_1 - \binom{k+i-1}{2i+1}\delta^{2i}y_0$$

▌ The left side corresponds to the operator

$$\delta^{2i}\binom{k+i-1}{2i}\frac{1}{2}\left[E+1+\frac{2k-1}{2i+1}(E-1)\right] = \delta^{2i}\binom{k+i-1}{2i}\left(\frac{k+i}{2i+1}E - \frac{k-i-1}{2i+1}\right)$$

The right side corresponds to the operator

$$\delta^{2i}\left[\binom{k+i}{2i+1}E - \binom{k+i-1}{2i+1}\right] = \delta^{2i}\binom{k+i-1}{2i}\left(\frac{k+i}{2i+1}E - \frac{k-i-1}{2i+1}\right)$$

so that both sides are the same.

4.49 Show that *Bessel's formula* is a rearrangement of Everett's formula.

▌ Bessel's formula is

$$p_k = \sum_{i=0}^{n}\left[\binom{k+i-1}{2i}\mu\,\delta^{2i}y_{1/2} + \frac{1}{2i+1}\left(k-\frac{1}{2}\right)\binom{k+i-1}{2i}\delta^{2i+1}y_{1/2}\right]$$

$$= \mu y_{1/2} + \left(k-\frac{1}{2}\right)\delta y_{1/2} + \binom{k}{2}\mu\,\delta^2 y_{1/2} + \frac{1}{3}\left(k-\frac{1}{2}\right)\binom{k}{2}\delta^3 y_{1/2}$$

$$+ \cdots + \binom{k+n-1}{2n}\mu\,\delta^{2n}y_{1/2} + \frac{1}{2n+1}\left(k-\frac{1}{2}\right)\binom{k+n-1}{2n}\delta^{2n+1}y_{1/2}$$

By the previous problem it reduces at once to Everett's.

4.50 Apply Bessel's formula with $n = 1$ to find a polynomial of degree 3 or less that takes the y_k values in Table 4.12.

▌ The needed differences are circled and have been inserted into their places in Bessel's formula. Needless to say, the resulting polynomial is the same one already found by other formulas:

$$p_k = \frac{3+8}{2} + 5\left(k - \frac{1}{2}\right) + \frac{3+7}{2}\frac{k(k-1)}{2} + \frac{1}{3}(4)\left(k - \frac{1}{2}\right)\frac{k(k-1)}{2}$$

This can be verified to be equivalent to earlier results.

TABLE 4.12

k	x_k	y_k				
−1	4	1				
			2			
0	6	③		③		
			⑤		④	
1	8	⑧		⑦		
			12			
2	10	20				

4.51 Apply Bessel's formula with $n = 2$ to find a polynomial of degree 5 or less that takes the y_k values in Table 4.13.

▌ The needed differences are circled.
Inserting the circled entries into their places in the Bessel formula,

$$p_k = \frac{135+8}{2} + 127\left(k - \frac{1}{2}\right) + \frac{442+118}{2}\frac{k(k-1)}{2} + \frac{1}{3}(324)\left(k - \frac{1}{2}\right)\frac{k(k-1)}{2}$$

$$+ \frac{336+216}{2}\frac{(k+1)k(k-1)(k-2)}{24} + \frac{1}{5}(120)\left(k - \frac{1}{2}\right)\frac{(k+1)k(k-1)(k-2)}{24}$$

which can be simplified, using $x_k = k + 2$, to the familiar $p(x_k) = x_k^5 - x_k^4 - x_k^3$.

TABLE 4.13

k	x_k	y_k						
−2	0	0						
			−1					
−1	1	−1		10				
			9		108			
0	2	⑧		⑪⑧		㉒⑯		
			⑫⑦		㉛㉔		⑫⓪	
1	3	⑬⑤		④④②		㉝㉖		
			569		660			
2	4	704		1102				
			1671					
3	5	2375						

4.52 Illustrate the zigzag rule

▌ The zigzag rule states that polynomials that take specified values at given arguments can be constructed in a wide variety of ways, by first drawing a zigzag line from any y_k value to an adjacent first difference, then to an adjacent second difference and so on. At each step there is a choice of two paths. An acceptable line is shown in Table 4.14.

TABLE 4.14

-3	x_{-3}	y_{-3}				
			$\delta y_{-5/2}$	$\delta^2 y_{-2}$		
-2	x_{-2}	y_{-2}			$\delta^3 y_{-3/2}$	
			$\delta y_{-3/2}$	$\delta^2 y_{-1}$		$\delta^4 y_{-1}$
-1	x_{-1}	y_{-1}			$\delta^3 y_{-1/2}$	
			$\delta y_{-1/2}$	$\delta^2 y_0$		$\delta^4 y_0$ $\quad\delta^5 y_{-1/2}$
0	x_0	y_0			$\delta^3 y_{1/2}$	
			$\delta y_{1/2}$	$\delta^2 y_1$		
1	x_1	y_1				
			$\delta y_{3/2}$			
2	x_2	y_2				

Having chosen the line it is only necessary to multiply the differences encountered on that line by suitable factors. In this case, the result would be

$$p_k = y_0 + \binom{k}{1}\delta y_{-1/2} + \binom{k+1}{2}\delta^2 y_0 + \binom{k+1}{3}\delta^3 y_{1/2} + \binom{k+1}{4}\delta^4 y_0 + \binom{k+2}{5}\delta^5 y_{-1/2}$$

the general rule being that after the first two terms the upper index in the binomial coefficient increases after an upward zig but not after a downward zag. With each step the polynomial then matches the data y_k within a triangle determined by the diagonals running back from the highest difference reached. In the preceding example the left (vertical) side of this triangle includes, in successive steps, (y_{-1}, y_0), (y_{-1}, y_0, y_1), (y_{-1}, y_0, y_1, y_2), $(y_{-2}, y_{-1}, y_0, y_1, y_2)$ and finally the full y_k column. The Newton and Gauss formulas are further illustrations of the zigzag rule. Our remaining formulas are obtained as averages of zigzag formulas, often rearranged.

4.53 Diagram the zigzag paths for our various formulas

❚ Where a formula is obtained by averaging over two paths, only the differences that actually appear in the formula are shown. See Fig. 4-1.

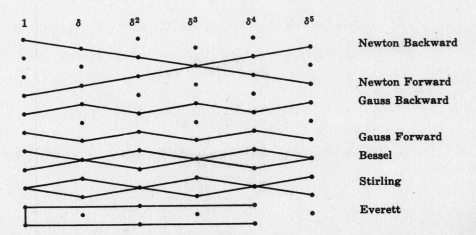

Newton Backward

Newton Forward
Gauss Backward

Gauss Forward
Bessel

Stirling

Everett

Fig. 4-1

4.54 Write a zigzag formula based on the path shown in Fig. 4-2. For which arguments will it take the prescribed y_k values? (The path starts at y_0 and ends at $\delta^6 y_0$.)

▮ Simply add $\binom{k+3}{6}\delta^6 y_0$ to the formula of Problem 4.52.

Fig. 4-2

4.2 LAGRANGE'S FORMULA

4.55 What values does the Lagrange multiplier function

$$L_i(x) = \frac{(x-x_0)(x-x_1)\cdots(x-x_{i-1})(x-x_{i+1})\cdots(x-x_n)}{(x_i-x_0)(x_i-x_1)\cdots(x_i-x_{i-1})(x_i-x_{i+1})\cdots(x_i-x_n)}$$

take at the data points $x = x_0, x_1, \ldots, x_n$?

▮ First notice that the numerator factors guarantee $L_i(x_k) = 0$ for $k \neq i$ and then the denominator factors guarantee that $L_i(x_i) = 1$.

4.56 Verify that the polynomial $p(x) = \sum_{i=0}^n L_i(x) y_i$ takes the value y_k at the argument x_k, for $k = 0, \ldots, n$. This is *Lagrange's formula* for the collocation polynomial.

▮ By Problem 4.55, $p(x_k) = \sum_{i=0}^n L_i(x_k) y_i = L_k(x_k) y_k = y_k$ so that Lagrange's formula does provide the collocation polynomial.

4.57 With $\pi(x)$ defined as the product $\pi(x) = \prod_{i=0}^n (x - x_i)$, show that

$$L_k(x) = \frac{\pi(x)}{(x-x_k)\pi'(x_k)}$$

▮ Since $\pi(x)$ is the product of $n+1$ factors, the usual process of differentiation produces $\pi'(x)$ as the sum of $n+1$ terms, in each of which one factor has been differentiated. If we define

$$F_k(x) = \prod_{i \neq k} (x - x_i)$$

to be the same as $\pi(x)$ except that the factor $x - x_k$ is omitted, then

$$\pi'(x) = F_0(x) + \cdots + F_n(x)$$

But then at $x = x_k$ all terms are zero except $F_k(x_k)$, since this is the only term not containing $x - x_k$. Thus

$$\pi'(x_k) = F_k(x_k) = (x_k - x_0)\cdots(x_k - x_{k-1})(x_k - x_{k+1})\cdots(x_k - x_n)$$

and

$$\frac{\pi(x)}{(x-x_k)\pi'(x_k)} = \frac{F_k(x)}{\pi'(x_k)} = \frac{F_k(x)}{F_k(x_k)} = L_k(x)$$

4.58 Use the Lagrange formula to find the quadratic polynomial that takes these values. Then find $y(2)$ from the quadratic

x	0	1	3
y	0	1	0

▮ Because of the two zero values of the function there is only one active term,

$$p(x) = \frac{(x-0)(x-3)}{(1-0)(1-3)}(1) = -\frac{1}{2}x(x-3)$$

At $x = 2$ it yields $p = 1$.

4.59 Use the Lagrange formula to find the quadratic that takes these values. What values does the quadratic suggest for $x = 2$ and $x = 3$? Check results by computing the difference table for the five point data set

x	0	1	4
y	2	1	4

▮ This time we need the full Lagrange formula:

$$p(x) = \frac{(x-1)(x-4)}{(0-1)(0-4)}(2) + \frac{(x-0)(x-4)}{(1-0)(1-4)}(1) + \frac{(x-0)(x-1)}{(4-0)(4-1)}(4)$$

$$= \frac{1}{2}(x-1)(x-4) - \frac{1}{3}x(x-4) + \frac{1}{3}x(x-1)$$

It has the values $p(2) = 1$ and $p(3) = 2$. The completed function has this table of differences, with those of second order all 1.

$$
\begin{array}{ccccccccc}
y_k & 2 & & 1 & & 1 & & 2 & & 4 \\
& & -1 & & 0 & & 1 & & 2 & \\
& & & 1 & & 1 & & 1 & &
\end{array}
$$

4.60 Find the polynomial of degree 3 that takes the values prescribed by

x_k	0	1	2	4
y_k	1	1	2	5

▮ The polynomial can be written directly:

$$p(x) = \frac{(x-1)(x-2)(x-4)}{(0-1)(0-2)(0-4)}1 + \frac{x(x-2)(x-4)}{1(1-2)(1-4)}1 + \frac{x(x-1)(x-4)}{2(2-1)(2-4)}2 + \frac{x(x-1)(x-2)}{4(4-1)(4-2)}5$$

It can be rearranged into $p(x) = \frac{1}{12}(-x^3 + 9x^2 - 8x + 12)$.

4.61 Use Lagrange's formula to produce a cubic polynomial that includes the following x_k, y_k number pairs. Then evaluate this polynomial for $x = 2, 3, 5$.

x_k	0	1	4	6
y_k	1	-1	1	-1

▮ $$\frac{(x-1)(x-4)(x-6)}{(0-1)(0-4)(0-6)} - \frac{x(x-4)(x-6)}{1(1-4)(1-6)} + \frac{x(x-1)(x-6)}{4(4-1)(4-6)} - \frac{x(x-1)(x-4)}{6(6-1)(6-4)}$$

For this polynomial we find $p(2) = -1$, $p(3) = 0$ and $p(5) = 1$. The difference table for the full data set is

$$
\begin{array}{ccccccccccccc}
1 & & -1 & & -1 & & 0 & & 1 & & 1 & & -1 \\
& -2 & & 0 & & 1 & & 1 & & 0 & & -2 & \\
& & 2 & & 1 & & 0 & & -1 & & -2 & & \\
& & & -1 & & -1 & & -1 & & -1 & & &
\end{array}
$$

4.62 Use Lagrange's formula to produce a fourth-degree polynomial that includes the following x_k, y_k number pairs. Then evaluate the polynomial for $x = 3$.

x_k	0	1	2	4	5
y_k	0	16	48	88	0

▮ The zero end values leave us with three terms:

$$\frac{x(x-2)(x-4)(x-5)}{-12}(16) + \frac{x(x-1)(x-4)(x-5)}{12}(48) + \frac{x(x-1)(x-2)(x-5)}{-24}(88)$$

At $x = 3$ this takes the value 84.

4.63 Deduce Lagrange's formula by determining the coefficients a_i in the partial fractions expansion

$$\frac{p(x)}{\pi(x)} = \sum_{i=0}^{n} \frac{a_i}{x - x_i}$$

▮ Multiply both sides by $x - x_i$ and let x approach x_i as limit. On the right we clearly get a_i. On the left, the numerator has limit $p(x_i) = y_i$ while the denominator tends to the limit $\pi'(x_i)$ except for the factor $x - x_i$ which has been removed by the multiplication. But this has been shown to be $\pi'(x_i)$, so we have

$$a_i = y_i/\pi'(x_i) = y_i/(x - x_0) \cdots (x - x_{i-1})(x - x_{i+1}) \cdots (x - x_n)$$

for the coefficients.

4.64 Apply Problem 4.63 to express

$$\frac{3x^2 + x + 1}{x^3 - 6x^2 + 11x - 6}$$

as a sum of partial fractions

$$\frac{a_0}{x - x_0} + \frac{a_1}{x - x_1} + \frac{a_2}{x - x_2}$$

▮ We think of the denominator as $\pi(x) = (x - x_0)(x - x_1)(x - x_2)$ and discover one way or another that the zero points of this cubic are $x = 1, 2, 3$. The corresponding values of the numerator function are $p = 5, 15, 31$. The values of $\pi'(x_i)$ are

$$(1-2)(1-3) \qquad (2-1)(2-3) \qquad (3-1)(3-2)$$

and the preceding problem puts these pieces together:

$$a_0 = 5/2 \qquad a_1 = 15/-1 \qquad a_2 = 31/2$$

4.65 Express

$$\frac{x^2 + 6x + 1}{(x^2 - 1)(x - 4)(x - 6)}$$

as a sum of partial fractions.

▮ The points x_i are clearly $-1, 1, 4, 6$ and the corresponding y_i are $-4, 8, 41, 73$. The $\pi'(x_i)$ are $(-1-1)(-1-4)(-1-6) = -70$, $(1+1)(1-4)(1-6) = 30$, $(4+1)(4-1)(4-6) = -30$ and $(6+1)(6-1)(6-4) = 70$. The coefficients of the partial fractions representation are then

$$a_0 = \frac{-4}{-70} \qquad a_1 = \frac{8}{30} \qquad a_2 = \frac{41}{-30} \qquad a_3 = \frac{73}{70}$$

giving us the result

$$\frac{2/35}{x+1} + \frac{4/15}{x-1} + \frac{-41/30}{x-4} + \frac{73/70}{x-6}$$

4.66 Show that the determinant equation

$$\begin{vmatrix} p(x) & 1 & x & x^2 & \cdots & x^n \\ y_0 & 1 & x_0 & x_0^2 & \cdots & x_0^n \\ y_1 & 1 & x_1 & x_1^2 & \cdots & x_1^n \\ \vdots & & & & & \vdots \\ y_n & 1 & x_n & x_n^2 & \cdots & x_n^n \end{vmatrix} = 0$$

also provides the collocation polynomial $p(x)$.

❚ Expansion of this determinant using minors of the first row elements would clearly produce a polynomial of degree n. Substituting $x = x_k$ and $p(x) = y_k$ makes two rows identical so that the determinant is zero. Thus $p(x_k) = y_k$ and this polynomial is a collocation polynomial. As attractive as this result is, it is not of much use due to the difficulty of evaluating determinants of large size.

4.67 Write out the determinant of the preceding problem for the function taking the values

x	0	1	2	4
y	1	1	2	5

❚ The determinant takes the form

$$\begin{vmatrix} p(x) & 1 & x & x^2 & x^3 \\ 1 & 1 & 0 & 0 & 0 \\ 1 & 1 & 1 & 1 & 1 \\ 2 & 1 & 2 & 4 & 8 \\ 5 & 1 & 4 & 16 & 64 \end{vmatrix} = 0$$

It is clear that when $x = 4$ and $p(x) = 5$, for example, two rows of the determinant are identical making the determinant zero. A similar conclusion follows for the other three x_i.

4.68 Show that

$$p_{0,k}(x) = \frac{1}{x_k - x_0} \begin{vmatrix} y_0 & x_0 - x \\ y_k & x_k - x \end{vmatrix}$$

represents the collocation polynomial of degree 1 for arguments x_0 and x_k.

❚ Direct evaluation of this determinant produces $p_{0,k}(x_0) = y_0$, $p_{0,k}(x_k) = y_k$.

4.69 Show that

$$p_{0,1,k}(x) = \frac{1}{x_k - x_1} \begin{vmatrix} p_{0,1}(x) & x_1 - x \\ p_{0,k}(x) & x_k - x \end{vmatrix}$$

represents the collocation polynomial of degree 2 for arguments x_0, x_1 and x_k.

❚ Remembering that $p_{0,1}(x)$ and $p_{0,k}(x)$ are collocation polynomials, one finds

$$p_{0,1,k}(x_0) = \frac{y_0(x_k - x_0) - y_0(x_1 - x_0)}{x_k - x_1} = y_0$$

$$p_{0,1,k}(x_1) = \frac{y_1(x_k - x_1)}{x_k - x_1} = y_1$$

$$p_{0,1,k}(x_k) = \frac{-y_k(x_1 - x_k)}{x_k - x_1} = y_k$$

4.70 Show that

$$p_{0,1,\ldots,n}(x) = \frac{1}{x_n - x_{n-1}} \begin{vmatrix} p_{0,1,\ldots,n-2,n-1}(x) & x_{n-1} - x \\ p_{0,1,\ldots,n-2,n}(x) & x_n - x \end{vmatrix}$$

represents the collocation polynomial of degree n for arguments $x_0, x_1, \ldots, x_n$.

▮ For $x = x_k$ and $k = 0, 1, \ldots, n-2$ the first column entries of the determinant agree, with value y_k. Evaluating the determinant produces $p_{0,1,\ldots,n}(x_k) = y_k$ at once. For x_{n-1} we find

$$p_{0,1,\ldots,n}(x_{n-1}) = \frac{1}{x_n - x_{n-1}} y_{n-1}(x_n - x_{n-1}) = y_{n-1}$$

and at x_n the value y_n is found by a similar computation.

4.71 Describe *Aitken's method* for obtaining the collocation polynomial.

▮ This is an iterative method based on the previous three problems. It involves computing the entries in Table 4.15 by successive columns from left to right, given the two columns at the left and the one at the right. The final entry is the collocation polynomial.

Each determinant to be evaluated may be lifted directly from this format. For example, the entry $p_{0,1,3}$ is obtained as the determinant

$$\begin{vmatrix} p_{0,1}(x) & x_1 - x \\ p_{0,3}(x) & x_3 - x \end{vmatrix}$$

divided by $x_3 - x_1$. As an additional convenience, the divisor is the difference between the entries used in the rightmost column.

TABLE 4.15

x_0	y_0				$x_0 - x$
x_1	y_1	$p_{0,1}(x)$			$x_1 - x$
x_2	y_2	$p_{0,2}(x)$	$p_{0,1,2}(x)$		$x_2 - x$
x_3	y_3	$p_{0,3}(x)$	$p_{0,1,3}(x)$	$p_{0,1,2,3}(x)$	$x_3 - x$
⋮	⋮	⋮	⋮	⋮	⋮

4.72 Use Aitken's method to compute $p(3)$ for the collocation polynomial that includes the x_k, y_k pairs in the first two columns of Table 4.16.

▮ The other entries in Table 4.16 are computed as described in Problem 4.71, with $x = 3$. For example, the third column entries are found as follows:

$$p_{0,1}(3) = \frac{(1)(-2) - (1)(-3)}{(-2) - (-3)} = 1$$

$$p_{0,2}(3) = \frac{(1)(-1) - (2)(-3)}{(-1) - (-3)} = \frac{5}{2}$$

$$p_{0,3}(3) = \frac{(1)(1) - (5)(-3)}{1 - (-3)} = 4$$

TABLE 4.16

x_k	y_k				
0	1				$0 - 3 = -3$
1	1	1			$1 - 3 = -2$
2	2	5/2	4		$2 - 3 = -1$
4	5	4	3	7/2	$4 - 3 = 1$

The fourth column entries are then

$$p_{0,1,2}(3) = \frac{(1)(-1) - (5/2)(-2)}{(-1) - (-2)} = 4 \qquad p_{0,1,3}(3) = \frac{(1)(1) - (4)(-2)}{1 - (-2)} = 3$$

and finally

$$p_{0,1,2,3}(3) = \frac{(4)(1) - (3)(-1)}{1 - (-1)} = \frac{7}{2}$$

which is $p(3)$. Here six similar steps lead to the final result. The labor involved may be compared with that of substituting $x = 3$ into the Lagrange formula which, of course, also produces $p(3) = \frac{7}{2}$ since it represents the same cubic collocation polynomial.

4.73 What are divided differences?

I Given the data x_k, y_k the *first divided difference* between 0 and 1 is defined as

$$y(x_0, x_1) = \frac{y_1 - y_0}{x_1 - x_0}$$

with similar formulas applying between the other argument pairs.

Then *higher divided differences* are defined in terms of lower divided differences. For example,

$$y(x_0, x_1, x_2) = \frac{y(x_1, x_2) - y(x_0, x_1)}{x_2 - x_0}$$

is a second difference, while

$$y(x_0, x_1, \ldots, x_n) = \frac{y(x_1, \ldots, x_n) - y(x_0, \ldots, x_{n-1})}{x_n - x_0}$$

is an nth difference. In many ways these differences play roles equivalent to those of the simpler differences used earlier.

A *difference table* is again a convenient device for displaying differences, the standard diagonal form being used:

$$
\begin{array}{llllll}
x_0 & y_0 & & & & \\
 & & y(x_0, x_1) & & & \\
x_1 & y_1 & & y(x_0, x_1, x_2) & & \\
 & & y(x_1, x_2) & & y(x_0, x_1, x_2, x_3) & \\
x_2 & y_2 & & y(x_1, x_2, x_3) & & y(x_0, x_1, x_2, x_3, x_4) \\
 & & y(x_2, x_3) & & y(x_1, x_2, x_3, x_4) & \\
x_3 & y_3 & & y(x_2, x_3, x_4) & & \\
 & & y(x_3, x_4) & & & \\
x_4 & y_4 & & & & \\
\end{array}
$$

4.74 Compute divided differences through the third for the y_k values in Table 4.17.

I The differences are listed in the last three columns. For example,

$$y(2,4) = \frac{5-2}{4-2} = \frac{3}{2} \qquad y(1,2,4) = \frac{3/2 - 1}{4 - 1} = \frac{1}{6}$$

$$y(0,1,2) = \frac{1-0}{2-0} = \frac{1}{2} \qquad y(0,1,2,4) = \frac{1/6 - 1/2}{4 - 0} = \frac{-1}{12}$$

TABLE 4.17

x_k	y_k			
0	1			
		0		
1	1		1/2	
		1		−1/12
2	2		1/6	
		3/2		
4	5			

4.75 Calculate divided differences through third order for both of the data sets

x	0	1	4	6
y	1	−1	1	−1

x	4	1	6	0
y	1	−1	−1	1

▌ The two parts of Table 4.18 show the results. The point is that rearranging the data pairs does not change divided differences. We have two examples of this here. The third difference in each table is −1/6, first coming to us as $y(0,1,4,6)$ and then again as $y(4,1,6,0)$. Also $y(1,4,6)$ and $y(4,1,6)$ both prove to be −1/3. This property of divided differences is called their symmetry. A general proof will be given in a moment.

TABLE 4.18

0	1				4	1			
		−2					2/3		
1	−1		2/3		1	−1		−1/3	
		2/3		−1/6			0		−1/6
4	1		−1/3		6	−1		1/3	
		−1					−1/3		
6	−1				0	1			

4.76 Calculate a fourth divided difference for the data in the first two columns of Table 4.19.

▌ The remaining columns contain the various differences, with the fourth proving to be −1.

TABLE 4.19

0	0			
		16		
1	16		8	
		32		−3
2	48		−4	
		20		−8	−1
4	88		−36	
		−88		
5	0			

4.77 Prove $y(x_0, x_1) = y(x_1, x_0)$. This is called *symmetry* of the first divided difference.

▌ This is obvious from the definition, but can also be seen from the fact that

$$y(x_0, x_1) = \frac{y_0}{x_0 - x_1} + \frac{y_1}{x_1 - x_0}$$

since interchanging x_0 with x_1 and y_0 with y_1 here simply reverses the order of the two terms on the right. This procedure can now be applied to higher differences.

4.78 Prove $y(x_0, x_1, x_2)$ is symmetric.

$\blacksquare$ Rewrite this difference as

$$y(x_0, x_1, x_2) = \frac{y(x_1, x_2) - y(x_0, x_1)}{x_2 - x_0} = \frac{1}{x_2 - x_0}\left(\frac{y_2 - y_1}{x_2 - x_1} - \frac{y_1 - y_0}{x_1 - x_0}\right)$$

$$= \frac{y_0}{(x_0 - x_1)(x_0 - x_2)} + \frac{y_1}{(x_1 - x_0)(x_1 - x_2)} + \frac{y_2}{(x_2 - x_0)(x_2 - x_1)}$$

Interchanging any two arguments x_j and x_k and the corresponding y values now merely interchanges the y_j and y_k terms on the right, leaving the overall result unchanged. Since any permutation of the arguments x_k can be effected by successive interchanges of pairs, the divided difference is invariant under all permutations (of both the x_k and y_k numbers).

4.79 Prove that, for any positive integer n,

$$y(x_0, x_1, \ldots, x_n) = \sum_{i=0}^{n} \frac{y_i}{F_i^n(x_i)}$$

where $F_i^n(x_i) = (x_i - x_0)(x_i - x_1) \cdots (x_i - x_{i-1})(x_i - x_{i+1}) \cdots (x_i - x_n)$. This generalizes the results of the previous two problems.

$\blacksquare$ The proof is by induction. We already have this result for $n = 1$ and 2. Suppose it true for $n = k$. Then by definition,

$$y(x_0, x_1, \ldots, x_{k+1}) = \frac{y(x_1, \ldots, x_{k+1}) - y(x_0, \ldots, x_k)}{x_{k+1} - x_0}$$

Since we have assumed our result true for differences of order k, the coefficient of y_k on the right, for $i = 1, 2, \ldots, k$ will be

$$\frac{1}{x_{k+1} - x_0}\left[\frac{1}{(x_i - x_1) \cdots (x_i - x_{k+1})} - \frac{1}{(x_i - x_0) \cdots (x_i - x_k)}\right]$$

where it is understood that the factor $(x_i - x_i)$ is not involved in the denominator products. But this coefficient reduces to

$$\frac{1}{(x_i - x_0) \cdots (x_i - x_{k+1})} = \frac{1}{F_i^{k+1}(x_i)}$$

as claimed. For $i = 0$ or $i = k + 1$ the coefficient of y_i comes in one piece instead of two, but in both cases is easily seen to be what is claimed in the theorem with $n = k + 1$, that is,

$$\frac{1}{(x_0 - x_1) \cdots (x_0 - x_{k+1})} \qquad \frac{1}{(x_{k+1} - x_0) \cdots (x_{k+1} - x_k)}$$

This completes the induction and proves the theorem.

4.80 Prove that the nth divided difference is symmetric.

$\blacksquare$ This follows at once from the previous problem. If any pair of arguments is interchanged, say x_j and x_k, the terms involving y_j and y_k on the right are interchanged and nothing else changes.

4.81 Evaluate the first few differences of $y(x) = x^2$ and x^3.

$\blacksquare$ Take $y(x) = x^2$ first. Then

$$y(x_0, x_1) = \frac{x_1^2 - x_0^2}{x_1 - x_0} = x_1 + x_0 \qquad y(x_0, x_1, x_2) = \frac{(x_2 + x_1) - (x_1 + x_0)}{x_2 - x_0} = 1$$

Higher differences will clearly be 0. Now take $y(x) = x^3$:

$$y(x_0, x_1) = \frac{x_1^3 - x_0^3}{x_1 - x_0} = x_1^2 + x_1 x_0 + x_0^2$$

$$y(x_0, x_1, x_2) = \frac{(x_2^2 + x_2 x_1 + x_1^2) - (x_1^2 + x_1 x_0 + x_0^2)}{x_2 - x_0} = x_0 + x_1 + x_2$$

$$y(x_0, x_1, x_2, x_3) = \frac{(x_1 + x_2 + x_3) - (x_0 + x_1 + x_2)}{x_3 - x_0} = 1$$

Again higher differences are clearly zero. Notice that in both cases all the differences are symmetric polynomials.

4.82   Prove that the kth divided difference of a polynomial of degree n is a polynomial of degree $n - k$ if $k \leq n$ and is zero if $k > n$.

▮   Call the polynomial $p(x)$. A typical divided difference is

$$p(x_0, x_1) = \frac{p(x_1) - p(x_0)}{x_1 - x_0}$$

Thinking of x_0 as fixed and x_1 as the argument, the various parts of this formula can be viewed as functions of x_1. In particular, the numerator is a polynomial in x_1, of degree n, with a zero at $x_1 = x_0$. By the factor theorem the numerator contains $x_1 - x_0$ as a factor and therefore the quotient, which is $p(x_0, x_1)$, is a polynomial in x_1 of degree $n - 1$. By the symmetry of $p(x_0, x_1)$ it is therefore also a polynomial in x_0 of degree $n - 1$. The same argument may now be repeated. A typical second difference is

$$p(x_0, x_1, x_2) = \frac{p(x_1, x_2) - p(x_0, x_1)}{x_2 - x_0}$$

Thinking of x_0 and x_1 as fixed and x_2 as the argument, the numerator is a polynomial in x_2, of degree $n - 1$, with a zero at $x_2 = x_0$. By the factor theorem $p(x_0, x_1, x_2)$ is therefore a polynomial in x_2 of degree $n - 2$. By the symmetry of $p(x_0, x_1, x_2)$ it is also a polynomial in either x_0 or x_1, again of degree $n - 2$. Continuing in this way the required result is achieved. An induction is called for, but it is an easy one and the details are omitted.

4.83   Prove that *Newton's divided difference formula*

$$p(x) = y_0 + (x - x_0) y(x_0, x_1) + (x - x_0)(x - x_1) y(x_0, x_1, x_2)$$
$$+ \cdots + (x - x_0)(x - x_1) \cdots (x - x_{n-1}) y(x_0, \ldots, x_n)$$

represents the collocation polynomial. That is, it takes the values $p(x_k) = y_k$ for $k = 0, \ldots, n$.

▮   The fact that $p(x_0) = y_0$ is obvious. Next, from the definition of divided differences, and using symmetry,

$$y_k = y_0 + (x_k - x_0) y(x_0, x_k)$$
$$y(x_0, x_k) = y(x_0, x_1) + (x_k - x_1) y(x_0, x_1, x_k)$$
$$y(x_0, x_1, x_k) = y(x_0, x_1, x_2) + (x_k - x_2) y(x_0, x_1, x_2, x_k)$$
$$\vdots$$
$$y(x_0, \ldots, x_{n-2}, x_k) = y(x_0, \ldots, x_{n-1}) + (x_k - x_{n-1}) y(x_0, \ldots, x_{n-1}, x_k)$$

For example, the second line follows from

$$y(x_0, x_1, x_k) = y(x_1, x_0, x_k) = \frac{y(x_0, x_k) - y(x_1, x_0)}{x_k - x_1}$$

For $k = 1$ the first of these proves $p(x_1) = y_1$. Substituting the second into the first brings

$$y_k = y_0 + (x_k - x_0) y(x_0, x_1) + (x_k - x_0)(x_k - x_1) y(x_0, x_1, x_k)$$

which for $k = 2$ proves $p(x_2) = y_2$. Successive substitutions verify $p(x_k) = y_k$ for each x_k in its turn until finally we reach

$$y_n = y_0 + (x_n - x_0) y(x_0, x_1) + (x_n - x_0)(x_n - x_1) y(x_0, x_1, x_2)$$
$$+ \cdots + (x_n - x_0)(x_n - x_1) \cdots (x_n - x_{n-1}) y(x_0, \ldots, x_{n-1}, x_n)$$

which proves $p(x_n) = y_n$.

Since this Newton formula represents the same polynomial as the Lagrange formula, the two are just rearrangements of each other.

4.84 Find the polynomial of degree 3 that takes the values given in Table 4.17.

❚ Using Newton's formula, which involves the differences on the top diagonal,

$$p(x) = 1 + (x - 0)0 + (x - 0)(x - 1)\tfrac{1}{2} + (x - 0)(x - 1)(x - 2)\left(-\tfrac{1}{12}\right)$$

which simplifies to $p(x) = \tfrac{1}{12}(-x^3 + 9x^2 - 8x + 12)$, the same result as found by Lagrange's formula.

4.85 Find the collocation polynomial of degree 3 for the x_k, y_k pairs of Problem 4.75.

❚ We have two choices. From the left half of Table 4.18 we can write

$$p(x) = 1 - 2x + \tfrac{2}{3}x(x - 1) - \tfrac{1}{6}x(x - 1)(x - 4)$$

while from the right half comes

$$p(x) = 1 + \tfrac{2}{3}(x - 4) - \tfrac{1}{3}(x - 4)(x - 1) - \tfrac{1}{6}(x - 4)(x - 1)(x - 6)$$

both of which reduce to $-\tfrac{1}{6}x^3 + \tfrac{3}{2}x^2 - \tfrac{10}{3}x + 1$. After all, there is only one cubic through these four points.

4.86 Apply Newton's formula to find the collocation polynomial for the data of Table 4.19.

❚ $$p(x) = 0 + 16x + 8x(x - 1) - 3x(x - 1)(x - 2) - x(x - 1)(x - 2)(x - 4)$$

At the missing argument $x = 3$ we again get the value 84 offered to us by the Lagrange formula in Problem 4.62.

4.87 Show that for the special case of the function

$$\pi(x) = (x - x_0)(x - x_1) \cdots (x - x_n)$$

the following are true:

$$y(x_0, x_1, \ldots, x_p) = 0 \quad \text{for } p = 0, 1, \ldots, n$$
$$y(x_0, x_1, \ldots, x_n, x) = 1 \quad \text{for all } x$$
$$y(x_0, x_1, \ldots, x_n, x, z) = 0 \quad \text{for all } x, z$$

❚ A proof for the case $n = 2$ provides full insight into the facts involved and appears as Table 4.20. Since $\pi(x) = 0$ for x_0, x_1, x_2, the first part of the theorem is covered by the solid triangle of zeros in the center.

TABLE 4.20

z	$\pi(z)$				
		$\pi(z)/(z - x_0)$			
x_0	0		$\pi(z)/(z - x_0)(z - x_1)$		
		0		$\pi(z)/(z - x_0)(z - x_1)(z - x_2)$	
x_1	0		0		0
		0		$\pi(x)/(x - x_2)(x - x_1)(x - x_0)$	
x_2	0		$\pi(x)/(x - x_2)(x - x_1)$		
		$\pi(x)/(x - x_2)$			
x	$\pi(x)$				

In the column of (two) third differences we have two entries both of which are 1, since the denominators are $\pi(z)$ and $\pi(x)$. This proves part two of the theorem. The final part follows from the zero in the last column.

4.88 The following is a generalization of Stirling's formula and can be verified to have the property $p(x_k) = y_k$ for $k = -2, -1, 0, 1, 2$:

$$p(x) = y_0 + \frac{y(x_1, x_0) + y(x_0, x_{-1})}{2}(x - x_0) + y(x_1, x_0, x_{-1})(x - x_0)\left(x - \frac{x_1 + x_{-1}}{2}\right)$$

$$+ \frac{y(x_2, x_1, x_0, x_{-1}) + y(x_1, x_0, x_{-1}, x_{-2})}{2}(x - x_1)(x - x_0)(x - x_{-1})$$

$$+ y(x_2, x_1, x_0, x_{-1}, x_{-2})(x - x_0)(x - x_1)(x - x_{-1})\left(x - \frac{x_2 + x_{-2}}{2}\right)$$

Apply it to the data of Table 4.19.

❚ The table

is repeated with the entries to be used circled. Inserting them leads to

$$p(x) = 48 + 26(x - 2) - 4(x - 2)\left(x - \tfrac{5}{2}\right) - \tfrac{11}{2}(x - 4)(x - 2)(x - 1) - (x - 2)(x - 4)(x - 1)\left(x - \tfrac{5}{2}\right)$$

which can be verified to assume the five given data values.

4.89 Apply the Stirling formula just presented to find a root of the cubic equation $y(x) = x^3 - 2x - 2 = 0$.

❚ First we note that $y(1)$ is negative while $y(2)$ is positive. Slightly deeper probing then discovers the pairs of values occupying the first two columns of Table 4.21. The remaining columns contain divided differences of x relative to y.

TABLE 4.21

y	x			
-1.104	1.6			
		.1621		
$-.487$	1.7		$-.0245$	
		.1391		.00587
.232	1.8		$-.0118$	.000709
		.1209		.00807
1.059	1.9		.00826	
		.1063		
2.000	2.0			

The idea is to view x as a function of y and use our formula to predict the x value corresponding to $y = 0$. For this purpose we mentally reverse the roles of the two variables in Stirling's formula and write

$$x = 1.8 + .13(0 - .232) - .0118(-.232)(-.286)$$

$$+ .00697(-.232)(.487)(-1.059)$$

$$+ .000709(-.232)(.487)(-1.059)(-.448)$$

which comes to 1.770. More digits could be carried, but the basic idea of using a polynomial approximation to a somewhat less cooperative function, here $x(y)$, is demonstrated.

4.90 Show that for arguments that are equally spaced, so that $x_{k+1} - x_k = h$, we have

$$y(x_0, x_1, \ldots, x_n) = \frac{\Delta^n y_0}{n! h^n}$$

▌ For $n = 1$,

$$y(x_0, x_1) = (y_1 - y_0)/h = \Delta y_0 / h$$

while for $n = 2$,

$$y(x_0, x_1, x_2) = \left(\frac{y_2 - y_1}{h} - \frac{y_1 - y_0}{h} \right) \Big/ 2h = (y_2 - 2y_1 + y_0)/2h^2$$

so we are off to a good start. Proceeding by induction we now assume the theorem true for $n = k$. Then

$$y(x_0, x_1, \ldots, x_{k+1}) = \frac{y(x_1, \ldots, x_{k+1}) - y(x_0, \ldots, x_k)}{(k+1)h}$$

$$= \frac{\Delta^k y_1}{k! h^k} - \frac{\Delta^k y_0}{k! h^k} \cdot \frac{1}{(k+1)h}$$

$$= \frac{\Delta^{k+1} y_0}{(k+1)! h^{k+1}}$$

and the induction is complete.

4.91 Define the *divided difference* $y(x_0, x_0)$ as $\lim y(x, x_0)$ for x approaching x_0 and then discover the simple relation to the derivative of y.

▌ Immediately from the definition

$$y(x_0, x_0) = \lim \frac{y - y_0}{x - x_0} = y'(x_0)$$

In Problem 4.81 we found that if $y(x) = x^2$, then $y(x, x_0) = x + x_0$. The limit of $2x_0$ illustrates the present theorem. In that same problem we found for $y(x) = x^3$ the divided difference

$$y(x, x_0) = x_0^2 + x x_0 + x^2$$

which in the limit becomes $3x^2$ and provides a second example.

4.92 What are the relative merits of our two algorithms for the collocation polynomial over a set of unequally spaced data points, the Lagrange formula and the Newton formula?

▌ The Lagrange formula offers us a direct view of the way in which the polynomial depends on the data values y_k. This is often used, particularly in studying the dependence of output information on these input values. However, to alter the degree of the approximating polynomial requires rebuilding it entirely. This is awkward when one wishes to successively increase the degree until a satisfactory result is found. The Newton formula has the flexibility required for this approach.

4.93 Show that the Newton formula develops the approximation of degree $n + 1$ from that of degree n by a simple additive step.

▌ This is probably clear enough from the structure of the formula as first presented in Problem 4.83 and as applied in several following examples. However, it may be useful to put it in one other way. Suppose $p_n(x)$ is the polynomial taking the values y_k at arguments x_k for $k = 0, 1, \ldots, n$. We now want the polynomial $p_{n+1}(x)$ which also satisfies $p(x_{n+1}) = y_{n+1}$. Try it in the form

$$p_{n+1}(x) = p_n(x) + c(x - x_0) \cdots (x - x_n)$$

which seems suitable since $p_{n+1}(x_k) = p_n(x_k) = y_k$ for $k \leq n$ due to the fact that the added term vanishes at all these points. Now we ask for collocation at x_{n+1}:

$$p_{n+1}(x_{n+1}) = p_n(x_{n+1}) + c(x_{n+1} - x_0) \cdots (x_{n+1} - x_n) = y_{n+1}$$

This determines the coefficient c,

$$c = \frac{y_{n+1} - p_n(x_{n+1})}{(x_{n+1} - x_0) \cdots (x_{n+1} - x_n)}$$

We can now see how this builds up the Newton formula step by step. For example,

$$p_0(x) = y_0$$

$$p_1(x) = p_0(x) + \frac{y_1 - p_0(x_1)}{x_1 - x_0}(x - x_0)$$

$$p_2(x) = p_1(x) + \frac{y_2 - p_1(x_2)}{(x_2 - x_0)(x_2 - x_1)}(x - x_0)(x - x_1)$$

and so on.

4.94 Illustrate the diagonal method of computing a divided difference table.

$\blacksquare$ Figure 4-3 provides the picture. At the right differences are calculated in columns, all the first differences being found first, followed by the second differences and so on. At the left a diagonal ordering is taken. On the first diagonal the single difference $y(x_0, x_1)$ is found. On the next lower diagonal the two entries $y(x_1, x_2)$ and $y(x_0, x_1, x_2)$ are found, and the process continues in an obvious way following the dotted line.

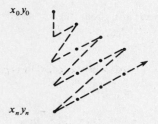 

Fig. 4-3

One advantage of the diagonal ordering is that a new data pair added to the table is easily accommodated. It is only necessary to increase the parameter n in the algorithm, which can be expressed in the form

for $k = 1$ to n
  for $i = 1$ to k

$$y(x_{k-i}, \ldots, x_k) = \frac{y(x_{k-i+1}, \ldots, x) - y(x_{k-i}, \ldots, x_{k-1})}{x_k - x_{k-i}}$$

4.95 For the Newton formula only the elements along the top right diagonal of the difference table are needed. Develop an algorithm for obtaining them.

$\blacksquare$ We take the column by column ordering, with the differences of order j denoted $d(i, j)$. It is convenient to rename the data values y as $d(i, 0)$ and the algorithm begins with this step.

for $i = 0$ to n

$$d(i, 0) = y(i)$$

for $j = 1$ to n
  for $i = 0$ to $n - j$

$$d(i, j) = [d(i+1, j-1) - d(i, j-1)]/(x(i+j) - x(i))$$

The required coefficients are now $d(0, j)$ for $j = 0$ to n.

4.96 Develop the Newton divided difference formula with an error term.

▮ From the definition of divided differences, we have for any x (which we will take to be between x_0 and x_n)

$$y(x) = y_0 + (x - x_0)\,y(x, x_0)$$
$$y(x, x_0) = y(x_0, x_1) + (x - x_1)\,y(x, x_0, x_1)$$
$$y(x, x_0, x_1) = y(x_0, x_1, x_2) + (x - x_2)\,y(x, x_0, x_1, x_2)$$
$$\vdots$$
$$y(x, x_0, \ldots, x_{n-1}) = y(x_0, \ldots, x_n) + (x - x_n)\,y(x, x_0, \ldots, x_n)$$

Multiply the second equation by $(x - x_0)$, the third by $(x - x_0)(x - x_1)$ and so on, the last equation being multiplied by $(x - x_0)(x - x_1)\cdots(x - x_{n-1})$. Then add the resulting equations together. There is extensive cancellation, with the result

$$y(x) = y_0 + (x - x_0)\,y(x_0, x_1) + (x - x_0)(x - x_1)\,y(x_0, x_1, x_2)$$
$$+ \cdots + (x - x_0)(x - x_1)\cdots(x - x_{n-1})\,y(x_0, \ldots, x_n) + R$$

where the *remainder term* R is

$$R = y(x, x_0, \ldots, x_n)(x - x_0)\cdots(x - x_n)$$

When x is one of the arguments x_k, the first factor of R reduces to a derivative and the factor $(x_k - x_k)$ makes $R = 0$. This proves $p(x_k) = y_k$ and verifies Newton's formula.

4.97 We now have two representations of the difference $y - p$ between a given function y and the collocation polynomial p, one being the R of the preceding problem and the other the

$$\frac{y^{(n+1)}(\xi)}{(n+1)!}\pi(x)$$

found in Problem 3.9. By comparing them find another relation between differences and derivatives.

▮ Simply deleting the common factor of $\pi(x)$ we have

$$y(x, x_0, \ldots, x_n) = \frac{y^{(n+1)}(\xi)}{(n+1)!}$$

with ξ somewhere among the x_k.

4.98 Prove that $y(x_0, x_1, \ldots, x_n) = (y^{(n)}(\xi))/n!$

▮ Reducing the count of arguments in Problem 4.97 by 1,

$$y(x, x_0, \ldots, x_{n-1}) = \frac{y^{(n)}(\xi)}{n!}$$

Now let $x = x_n$ and rearrange the arguments (allowable by the symmetry of divided differences) and the result follows.

4.99 Prove that $\Delta^n y_0 = y^{(n)}(\xi)h^n$.

▮ This follows at once by combining the results of the preceding problem and Problem 4.90:

$$\Delta^n y_0 = n!h^n y(x_0, x_1, \ldots, x_n) = h^n y^{(n)}(\xi)$$

4.100 Show that if the derivatives of $y(x)$ have a bound that is independent of n, then for small h the differences of y will have limit zero for $n \to \infty$. This extends the simpler result found earlier for polynomials.

▮ With $y^{(n)}(\xi)$ bounded this follows at once from the result in Problem 4.99, since h^n will have limit zero.

4.101 Develop Leibnitz's rule for the nth difference of a product of two functions $f(x)$ and $g(x)$.

I Let the spacing between arguments be h. Then

$$\Delta f(x)g(x) = f(x+h)g(x+h) - f(x)g(x)$$
$$= f(x+h)g(x+h) - f(x)g(x+h) + f(x)g(x+h) - f(x)g(x)$$
$$= f(x)\,\Delta g(x) + \Delta f(x)\cdot g(x+h)$$

which is equivalent to the result found earlier in Problem 3.32. In differencing a product we difference each factor in its turn but there is one argument advance to be respected. As we now continue on to higher differences we choose the g factor as the recipient of all such advances. Here is the second difference:

$$\Delta^2 f(x)g(x) = f(x)\,\Delta^2 g(x) + \Delta f(x)\,\Delta g(x+h) + \Delta f(x)\,\Delta g(x+h) + \Delta^2 f(x)\cdot g(x+h)$$
$$= f(x)\,\Delta^2 g(x) + 2\,\Delta f(x)\,\Delta g(x+h) + \Delta^2 f(x)\cdot g(x+h)$$

The third follows similarly:

$$\Delta^3 f(x)g(x) = f(x)\,\Delta^3 g(x) + \Delta f(x)\,\Delta^2 g(x+h) + 2\,\Delta f(x)\,\Delta^2 g(x+h)$$
$$+ 2\,\Delta^2 f(x)\,\Delta g(x+2h) + \Delta^2 f(x)\,\Delta g(x+2h) + \Delta^3 f(x)\cdot g(x+3h)$$
$$= f(x)\,\Delta^3 g(x) + 3\,\Delta f(x)\,\Delta^2 g(x+h) + 3\,\Delta^2 f(x)\,\Delta g(x+2h) + \Delta^3 f(x)\cdot g(x+3h)$$

An induction would now establish the general case, which is

$$\Delta^n f(x)\,g(x) = \sum_{k=0}^{n} \binom{n}{k}\Delta^k f(x)\,\Delta^{n-k}g(x+kh)$$

and is called Leibnitz's rule.

4.102 Apply the Leibnitz rule to the function $x\cdot 2^x$.

I Choosing $f = x$ and $g = 2^x$,

$$\Delta^n x\cdot 2^x = x\cdot 2^x + \binom{n}{1}h\cdot 2^{x+h} = x\cdot 2^x + nh\cdot 2^{x+h}$$

all other terms being zero.

4.103 Apply the Leibnitz rule to $k^{(2)}C^k$, where $x_k = x_0 + kh$ making the interval of the argument $k = 1$.

I
$$\Delta^n k^{(2)}\,C^k = k^{(2)}C^k(C-1)^n + \binom{n}{1}2k^{(1)}C^{k+1}(C-1)^{n-1} + \binom{n}{2}2C^{k+2}(C-1)^{n-2}$$

all other terms again being zero.

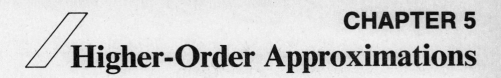

CHAPTER 5

Higher-Order Approximations

5.1 THE HERMITE FORMULA

5.1 What are osculating polynomials?

▌ Osculating polynomials not only agree in value with a given function at specified arguments, which is the collocation idea explored in the preceding two chapters, but their derivatives up to some order also match the derivatives of the function, usually at the same arguments. Thus for the simplest case of osculation we require

$$p(x_k) = y(x_k) \qquad p'(x_k) = y'(x_k)$$

for $k = 0, 1, \ldots, n$. In the language of geometry, this makes the curves representing our two functions tangent to each other at these $n + 1$ points. Higher-order osculation would also require $p''(x_k) = y''(x_k)$ and so on. The corresponding curves then have what is called contact of higher order.

5.2 Verify that $p(x) = \sum_{i=0}^{n} U_i(x) y_i + \sum_{i=0}^{n} V_i(x) y_i'$ will be a polynomial of degree $2n + 1$ or less, satisfying $p(x_k) = y_k$, $p'(x_k) = y_k'$ provided:

(a) $U_i(x)$ and $V_i(x)$ are polynomials of degree $2n + 1$
(b) $U_i(x_k) = \delta_{ik}$, $V_i(x_k) = 0$
(c) $U_i'(x_k) = 0$, $V_i'(x_k) = \delta_{ik}$

where

$$\delta_{ik} = \begin{cases} 0 & \text{for } i \neq k \\ 1 & \text{for } i = k \end{cases}$$

▌ The degree issue is obvious, since an additive combination of polynomials of given degree is a polynomial of the same or lower degree. Substituting $x = x_k$ we have

$$p(x_k) = U_k(x_k) y_k + 0 = y_k$$

and similarly substituting $x = x_k$ into $p'(x)$,

$$p'(x_k) = V_k'(x_k) y_k' = y_k'$$

all other terms being zero.

5.3 Recalling that the Lagrangian multiplier $L_i(x)$ satisfies $L_i(x_k) = \delta_{ik}$, show that

$$U_i(x) = [1 - 2L_i'(x_i)(x - x_i)][L_i(x)]^2 \qquad V_i(x) = (x - x_i)[L_i(x)]^2$$

meet the requirements listed in Problem 5.2.

▌ Since $L_i(x)$ is of degree n, its square has degree $2n$ and both $U_i(x)$ and $V_i(x)$ are of degree $2n + 1$. For the second requirement we note that $U_i(x_k) = V_i(x_k) = 0$ for $k \neq i$, since $L_i(x_k) = 0$. Also, substituting $x = x_i$,

$$U_i(x_i) = [L_i(x_i)]^2 = 1 \qquad V_i(x_i) = 0$$

so that $U_i(x_k) = \delta_{ik}$ and $V_i(x_k) = 0$. Next calculate the derivatives

$$U_i'(x) = [1 - 2L_i'(x_i)(x - x_i)]2L_i'(x)L_i(x) - 2L_i'(x_i)[L_i(x)]^2$$
$$V_i'(x) = (x - x_i)2L_i(x)L_i'(x) + [L_i(x)]^2$$

At once $U_i'(x_k) = 0$ and $V_i'(x_k) = 0$ for $k \neq i$ because of the $L_i(x_k)$ factor, and for $x = x_i$, $U_i'(x_i) = 2L_i'(x_i) -$

$2L_i'(x_i) = 0$ since $L_i(x_i) = 1$. Finally, $V_i'(x_i) = [L_i(x_i)]^2 = 1$. The Hermite formula is therefore

$$p(x) = \sum_{i=0}^{n} [1 - 2L_i'(x_i)(x - x_i)][L_i(x)]^2 y_i + (x - x_i)[L_i(x)]^2 y_i'$$

5.4 A switching path between parallel railroad tracks is to be a cubic polynomial joining positions $(0,0)$ and $(4,2)$ and tangent to the lines $y = 0$ and $y = 2$, as shown in Fig. 5-1. Apply Hermite's formula to produce this polynomial.

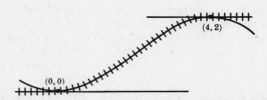

(4, 2)

(0, 0)

Fig. 5-1

▮ The specifications ask for a cubic polynomial matching the data

x_k	y_k	y_k'
0	0	0
4	2	0

With $n = 1$, we have

$$L_0(x) = \frac{x - x_1}{x_0 - x_1} \qquad L_1(x) = \frac{x - x_0}{x_1 - x_0} \qquad L_0'(x) = \frac{1}{x_0 - x_1} \qquad L_1'(x) = \frac{1}{x_1 - x_0}$$

and substituting into Hermite's formula (only the y_1 term need be computed since $y_0 = y_0' = y_1' = 0$),

$$p(x) = \left[1 - 2\frac{x - 4}{4 - 0}\right]\left[\frac{x - 0}{4 - 0}\right]^2 \cdot 2 = \frac{1}{16}(6 - x)x^2$$

The significance of this switching path is, of course, that it provides a smooth journey. Being tangent to both of the parallel tracks, there are no sudden changes of direction, no corners. Since $p''(0)$ and $p''(4)$ are not zero, there are, however, discontinuities in curvature. (But see Problem 5.8.)

5.5 Obtain a formula for the difference between $y(x)$ and its polynomial approximation $p(x)$.

▮ The derivation is very similar to that for the simpler collocation polynomial. Since $y(x) = p(x)$ and $y'(x) = p'(x)$ at the arguments $x_0, \ldots, x_n$ we anticipate a result of the form

$$y(x) - p(x) = C[\pi(x)]^2$$

where $\pi(x) = (x - x_0) \cdots (x - x_n)$ as before. Accordingly we define the function

$$F(x) = y(x) - p(x) - C[\pi(x)]^2$$

which has $F(x_k) = F'(x_k) = 0$ for $k = 0, \ldots, n$. By choosing any new argument x_{n+1} in the interval between x_0 and x_n, and making

$$C = [y(x_{n+1}) - p(x_{n+1})] / [\pi(x_{n+1})]^2$$

we also make $F(x_{n+1}) = 0$. Since $F(x)$ now has $n + 2$ zeros at least, $F'(x)$ will have $n + 1$ zeros at intermediate points. It also has zeros at $x_0, \ldots, x_n$, making $2n + 2$ zeros in all. This implies that $F''(x)$ has $2n + 1$ zeros at least. Successive applications of Rolle's theorem now show that $F'''(x)$ has $2n$ zeros at least, $F'''(x)$ has $2n - 1$ zeros and so on to $F^{(2n+2)}(x)$, which is guaranteed at least one zero in the interval between x_0 and x_n, say at $x = \xi$.

Calculating this derivative, we get

$$F^{(2n+2)}(\xi) = y^{(2n+2)}(\xi) - C(2n+2)! = 0$$

which can be solved for C. Substituting back,

$$y(x_{n+1}) - p(x_{n+1}) = \frac{y^{(2n+2)}(\xi)}{(2n+2)!}\left[\pi(x_{n+1})\right]^2$$

Recalling that x_{n+1} can be any argument other than $x_0, \ldots, x_n$ and noticing that this result is even true for $x_0, \ldots, x_n$ (both sides being zero), we replace x_{n+1} by the simpler x:

$$y(x) - p(x) = \frac{y^{(2n+2)}(\xi)}{(2n+2)!}\left[\pi(x)\right]^2$$

5.6   Prove that only one polynomial can meet the specifications of Problem 5.2.

❚   Suppose there were two. Since they must share common y_k and y_k' values at the arguments x_k, we may choose one of them as the $p(x)$ of Problem 5.5 and the other as the $y(x)$. In other words, we may view one polynomial as an approximation to the other. But since $y(x)$ is now a polynomial of degree $2n+1$, it follows that $y^{(2n+2)}(\xi)$ is zero. Thus $y(x)$ is identical with $p(x)$ and our two polynomials are actually one and the same.

5.7   How can a polynomial be found that matches the following data?

$$\begin{array}{cccc} x_0 & y_0 & y_0' & y_0'' \\ x_1 & y_1 & y_1' & y_1'' \end{array}$$

In other words, at two arguments the values of the polynomial and its first two derivatives are specified.

❚   Assume for simplicity that $x_0 = 0$. If this is not true, then a shift of argument easily achieves it. Let

$$p(x) = y_0 + xy_0' + \tfrac{1}{2}x^2 y_0'' + Ax^3 + Bx^4 + Cx^5$$

with A, B and C to be determined. At $x = x_0 = 0$ the specifications have already been met. At $x = x_1$ they require

$$Ax_1^3 + Bx_1^4 + Cx_1^5 = y_1 - y_0 - x_1 y_0' - \tfrac{1}{2}x_1^2 y_0''$$

$$3Ax_1^2 + 4Bx_1^3 + 5Cx_1^4 = y_1' - y_0' - x_1 y_0''$$

$$6Ax_1 + 12Bx_1^2 + 20Cx_1^3 = y_1'' - y_0''$$

These three equations determine A, B and C uniquely.

5.8   A switching path between parallel railroad tracks is to join positions $(0,0)$ and $(4,2)$. To avoid discontinuities in both direction and curvature, the following specifications are made:

x_k	y_k	y_k'	y_k''
0	0	0	0
4	2	0	0

Find a polynomial that meets these specifications.

❚   Applying the procedure of Problem 5.7,

$$p(x) = Ax^3 + Bx^4 + Cx^5$$

the quadratic portion vanishing entirely. At $x_1 = 4$ we find

$$64A + 256B + 1024C = 2 \qquad 48A + 256B + 1280C = 0 \qquad 24A + 192B + 1280C = 0$$

from which $A = 40/128$, $B = -15/128$ and $C = 24/128$. Substituting, $p(x) = \tfrac{1}{256}(80x^3 - 30x^4 + 3x^5)$.

5.9   Apply Hermite's formula to find a cubic polynomial that meets the specifications

x_k	y_k	y'_k
0	0	0
1	1	1

This can be viewed as a switching path between nonparallel tracks.

▌ As in Problem 5.4, we have $n = 1$. This time there are two terms to generate, the y_1 and y'_1 terms:

$$p(x) = [1 - 2(x - 1)](x - 0)^2 + (x - 1)x^2 = 2x^2 - x^3$$

It is easy to verify that the four given specifications are met.

5.10   Apply Hermite's formula to find a polynomial that meets the specification

x_k	y_k	y'_k
0	0	0
1	1	0
2	0	0

▌ Now $n = 2$, but once again there is only one term to generate, the y_1 term. The Lagrange multiplier needed is

$$L_1(x) = x(x - 2)/(-1)$$

with derivative $L'_1(x) = -2x + 2$. Substituting we find

$$p(x) = [1 - 2 \cdot 0 \cdot (x - 1)]x^2(x - 2)^2 = x^2(x - 2)^2$$

5.11   Apply the method of Problem 5.7 to find a fifth-degree polynomial that meets the specifications

x_k	y_k	y'_k	y''_k
0	0	0	0
1	1	1	0

This is a smoother switching path than that of Problem 5.9.

▌ Direct substitution of these specifications into the equations of the earlier problem leads to the system

$$
\begin{aligned}
a + \quad b + \quad c &= 1 \\
3a + \quad 4b + \quad 5c &= 1 \\
6a + 12b + 20c &= 0
\end{aligned}
$$

The coefficients are then determined to be $a = 6$, $b = -8$, $c = 3$ and the osculating polynomial follows:

$$p(x) = 6x^3 - 8x^4 + 3x^5$$

5.12   Find two second-degree polynomials, one having $p_1(0) = p'_1(0) = 0$, the other having $p_2(4) = 2$, $p'_2(4) = 0$, both passing through $(2, 1)$, as shown in Fig. 5-2. Show that $p'_1(2) = p'_2(2)$ so that a pair of parabolic arcs also serves as a switching path between parallel tracks, as well as the cubic of Problem 5.4.

▌ A method of undetermined coefficients seems indicated. With

$$p_1(x) = ax^2 \qquad p_2(x) = 2 + b(x - 4)^2$$

the endpoint conditions are already satisfied. Forcing both of the quadratics through $(2, 1)$ determines $a = \frac{1}{4}$ and $b = -\frac{1}{4}$. The symmetry of the situation suggests matching slopes at the join, and calculating $p'_1(x) = x/2$ and $p'_2(x) = -(x - 4)/2$ we find both equaling 1 at $x = 2$.

5.13 Find two fourth-degree polynomials, one having $p_1(0) = p_1'(0) = p_1''(0) = 0$, the other having $p_2(4) = 2$, $p_2'(4) = p_2''(4) = 0$, both passing through $(2,1)$ with $p_1''(2) = p_2''(2) = 0$. This is another switching path for which direction and curvature are free of discontinuities, like the fifth-degree polynomial of Problem 5.11. Verify this by showing that first and second derivatives agree on both sides of $(0,0)$, $(2,1)$ and $(4,2)$ where the four pieces of track are butted together.

▮ Again using the method of undetermined coefficients, we set

$$p_1(x) = ax^3 + bx^4 \qquad p_2(x) = 2 + c(x-4)^3 + d(x-4)^4$$

and have satisfied all conditions at the two endpoints. Forcing both polynomials to enter $(2,1)$ leads to the equations

$$1 = 8a + 16b \qquad 1 = 2 - 8c + 16d$$

and from the right hand pair of

$$p_1'(x) = 3ax^2 + 4bx^3 \qquad\qquad p_1''(x) = 6ax + 12bx^2$$
$$p_2'(x) = 3c(x-4)^2 + 4d(x-4)^3 \qquad p_2''(x) = 6c(x-4) + 12d(x-4)^2$$

comes this second pair,

$$0 = 12a + 48b \qquad 0 = -12c + 48d$$

The four coefficients are then determined in pairs as $a = \frac{1}{4}$, $b = -\frac{1}{16}$, $c = \frac{1}{4}$, $d = \frac{1}{16}$ after which it is a relatively short road to the two polynomials

$$p_1(x) = x^3(4-x)/16 \qquad p_2(x) = 2 - x(4-x)/16$$

As for the smoothness of first derivatives at $(2,1)$ both prove to be 1.

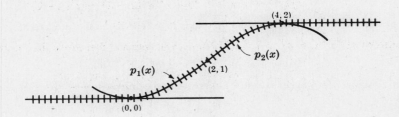

Fig. 5-2

5.14 From Hermite's formula for two point osculation derive the midpoint formula

$$p_{1/2} = \tfrac{1}{2}(y_0 + y_1) + \tfrac{1}{8}L(y_0' - y_1')$$

where $L = x_1 - x_0$.

▮ At the midpoint of the (x_0, x_1) interval we have $x - x_0 = h/2$ and $x - x_1 = -h/2$ where h is the length of the interval. Also the Lagrange multiplier function $L_0(x)$ for the given x becomes

$$L_0(x) = (x - x_1)/(x_0 - x_1) = \tfrac{1}{2}$$

with $L_1(x)$ taking the same value. For the derivatives we find $L_0' = -1/h = -L_1'$. Substituting these various values into the Hermite formula yields the desired result.

5.15 Find a polynomial of degree 4 that meets the conditions

x_k	y_k	y_k'
0	1	0
1	0	—
2	9	24

Note that one of the y_k' values is not available.

▮ The polynomial $p(x) = 1 + cx^2 + dx^3 + ex^4$ meets the two conditions at $x = 0$. Since $p'(x) = 2cx + 3dx^2 + 4ex^3$, the other three conditions become

$$
\begin{aligned}
0 &= 1 + c + d + e \\
9 &= 1 + 4c + 8d + 16e \\
24 &= 4c + 12d + 32e
\end{aligned}
$$

with solution $c = -2$, $d = 0$, $e = 1$. The required polynomial is then

$$p(x) = 1 - 2x^2 + x^4 = (1 - x^2)^2$$

5.16 Find a polynomial of degree 4 that meets the conditions

x_k	y_k	y_k'	y_k''
0	1	−1	0
1	2	7	—

▮ The polynomial $p(x) = 1 - x + ax^3 + bx^4$ meets all the conditions set at $x = 0$. With $p'(x) = -1 + 3ax^2 + 4bx^3$ the other two conditions reduce to

$$2 = a + b \qquad 7 = -1 + 3a + 4b$$

with solution $a = 0$, $b = 2$, so $p(x) = 1 - x + 2x^4$.

5.17 Find a polynomial of degree 3 that meets the conditions

x_k	y_k	y_k''
0	1	−2
1	1	4

▮ Here we have no information about the first derivative. The method of undetermined coefficients suggests the beginning

$$p(x) = 1 + ax - x^2 + bx^3$$

with the conditions at $x = 0$ again satisfied. Differentiation brings $p''(x) = -2 + 6bx$ and leads to the little system

$$a + b = 1 \qquad 6b = 6$$

and then to $p(x) = 1 - x^2 + x^3$.

5.18 In Problem 4.91 we observed that the divided difference of first order

$$y(x_0, x_1) = (y_1 - y_0)/(x_1 - x_0)$$

approaches the derivative $y'(x_0)$ if x_1 approaches x_0. This can be related to the idea of osculation, with Newton's divided-difference formula providing the link. The appropriate segment of the difference table is

$$
\begin{array}{llll}
x_0 & y_0 & & \\
 & & y'(x_0) & \\
x_0 & y_0 & & y(x_0, x_0, x_2) \\
 & & y(x_0, x_2) & \\
x_2 & y_2 & &
\end{array}
$$

in which the entries x_2, y_2 indicate the continuation of the table along normal lines. Now suppose that x_2 also approaches x_0. Show that the second difference in this table has the limit $y''(x_0)/2$.

▮ We have

$$y(x_0, x_0, x_2) = \frac{y(x_0, x_2) - y'(x_0)}{x_2 - x_0}$$

and since

$$y(x_2) = y(x_0) + y'(x_0)(x_2 - x_0) + y''(x_0)(x_2 - x_0)^2/2 + \cdots$$

we find that

$$y(x_0, x_2) = y'(x_0) + y''(x_0)(x_2 - x_0)/2 + \cdots$$

so that $y(x_0, x_0, x_2)$ converges to $y''(x_0)/2$ as expected.

A corresponding result holds for higher-order differences, that is,

$$\lim y(x_0, x_0, x_0, x_3) = y'''(x_0)/3!$$

as x_3 approaches x_0 and, in general,

$$\lim y(x_0, x_0, \ldots, x_0, x_k) = y^{(k)}(x_0)/k!$$

This is proved by referring to Problem 4.97 and letting x and all the x_k approach zero. It may also be convenient to reduce n by 1, obtaining $y(x_0, \ldots, x_0) = y^{(n)}(x_0)/n!$.

5.19 Illustrate the use of Newton's divided-difference formula for solving an osculation problem such as 5.4.

▮ Specifying derivative values, as in this problem, is what is called osculation. It is also known as higher-order contact of the two curves in question, here the switching path and a given track. To develop this higher-order contact we use each point of collocation twice. The difference table begins as shown in the left half of Table 5.1. Note that the given values of the derivatives have been entered for $y(0,0)$ and $y(4,4)$. The right half of the figure then shows the completion of the table as usual.

The table's top diagonal is now used in the familiar way,

$$p(x) = 0 + 0(x - 0) + \tfrac{1}{8}(x - 0)(x - 0) - \tfrac{1}{16}(x - 0)(x - 0)(x - 4)$$

which reduces to the same $\tfrac{1}{16}(6 - x)x^2$ found earlier.

TABLE 5.1

x	y		x	y			
0	0		0	0			
		0			0		
0	0		0	0		1/8	
					1/2		-1/16
4	2		4	2		-1/8	
		0			0		
4	2		4	2			

5.20 Rework Problem 5.15 for which we found the solution $p(x) = (1 - x^2)^2$ by the method of undetermined coefficients. This time use the Newton formula.

▮ The difference table is initialized as shown in the left half of Table 5.2. The right half is the completed table. The osculating polynomial is

$$p(x) = 1 + 0 \cdot x - x^2 + 3x^2(x - 1) + x^2(x - 1)(x - 2)$$

TABLE 5.2

x	y		x	y			
0	1		0	1			
		0			0		
0	1		0	1		-1	
					-1		3
1	0		1	0		5	
					9		5
2	9		2	9		15	1
		24			24		
2	9		2	9			

which reduces to the earlier solution. Note how in this example there is no double contact at $x = 1$, no specified value for the derivative and so this data line is entered only once.

5.21 Rework Problem 5.16 by the current method.

▮ As before, the left half of Table 5.3 shows the steps taken to arrange multiple contacts, this time a triple and a double. The entry $y''(0)/2$ is, of course, zero. Note that it is based on three identical x, y pairs.

The required polynomial is, using the Newton formula,

$$p(x) = 1 - x + 0 \cdot x^2 + 2x^3 + 2x^3(x - 1) = 1 - x + 2x^4$$

as before.

TABLE 5.3

x	y				x	y				
0	1				0	1				
		−1					−1			
0	1		0		0	1		0		
		−1					−1		2	
0	1				0	1		2		2
							1		4	
1	2				1	2		6		
		7					7			
1	2				1	2				

5.22 Obtain a polynomial of degree 5 or less that agrees in value with $\sin(x)$ and its derivative at $x = 0, \pi/2, \pi$.

▮ This is precisely the problem solved by Hermite's formula, but suppose we use Newton's instead. The difference table is initiated at the left in Table 5.4 and completed at the right.

If we denote the six entries on the top diagonal a, b, c, d, e, f, then the Newton formula offers

$$p(x) = bx + cx^2 + dx^2\left(x - \frac{\pi}{2}\right) + ex^2\left(x - \frac{\pi}{2}\right)^2$$

omitting the zero terms. (See also the following problem.)

TABLE 5.4

x	sin x		x	sin x					
0	0		0	0					
		1			1				
0	0		0	0		−.2313350			
					.6366198		−.1107398		
$\pi/2$	1		$\pi/2$	1		−.4052847		.0352495	
		0			0		0		0
$\pi/2$	1		$\pi/2$	1		−.4052847		.0352495	
					−.6366198		.1107398		
π	0		π	0		−.2313350			
		−1			−1				
π	0		π	0					

5.23 Proceed by undetermined coefficients to solve the sine osculation problem.

▮ Assume the polynomial in Newton's form

$$p(x) = a + bx + cx^2 + dx^2\left(x - \frac{\pi}{2}\right) + ex^2\left(x - \frac{\pi}{2}\right)^2 + fx^2\left(x - \frac{\pi}{2}\right)^2(x - \pi)$$

and substitute the six specifications. A diagonal system of equations is found,

$$0 = a$$
$$1 = b$$
$$1 = (\pi/2)b + (\pi^2/4)c$$
$$0 = 1 + \pi c + \tfrac{1}{4}\pi^2 d$$
$$0 = \pi b + \pi^2 c + \tfrac{1}{2}\pi^3 d + \tfrac{1}{4}\pi^4 e$$
$$-1 = 1 + 2\pi c + 2\pi^2 d + \tfrac{3}{2}\pi^3 e + \tfrac{1}{4}\pi^4 f$$

with the solutions

$$a = 0 \qquad d = \frac{4}{\pi^2} - \frac{16}{\pi^3}$$
$$b = 1 \qquad e = \frac{16}{\pi^4} - \frac{4}{\pi^3}$$
$$c = \frac{4}{\pi^2} - \frac{2}{\pi} \qquad f = 0$$

These are the exact values for which seven decimal place approximations were found in the preceding problem.

5.24 Develop a formula for a quadratic polynomial with given y_0, y_1 and y_0'.

$\blacksquare$ A solution by undetermined coefficients could begin with

$$p(x) = y_0 + y_0'(x - x_0) + a(x - x_0)^2$$

and determine the coefficient a from $p(x_1) = y_1$. However, the table

$$
\begin{array}{lll}
x_0 & y_0 & \\
 & & y_0' \\
x_0 & y_0 & & y(x_0, x_0, x_1) \\
 & & y(x_0, x_1) \\
x_1 & y_1 &
\end{array}
$$

and Newton's formula offer us

$$p(x) = y_0 + y_0'(x - x_0) + y(x_0, x_0, x_1)(x - x_0)^2$$

so that in fact $a = y(x_0, x_0, x_1)$.

If $y(0) = 1$, $y'(0) = -1$ and $y(1) = 2$ we have the little table

$$
\begin{array}{lll}
0 & 1 & \\
 & & -1 \\
0 & 1 & & 2 \\
 & & 1 \\
1 & 2 &
\end{array}
$$

and

$$p(x) = 1 - x + 2x^2$$

5.2 THE TAYLOR POLYNOMIAL

5.25 What is the Taylor polynomial?

$\blacksquare$ In a way the Taylor polynomial is the ultimate osculation. For a single argument x_0 the values of the polynomial and its first n derivatives are required to match those of a given function $y(x)$:

$$p^{(i)}(x_0) = y^{(i)}(x_0) \quad \text{for } i = 0, 1, \ldots, n$$

5.26   Derive a formula for the Taylor polynomial.

▮ A polynomial of degree n can be written

$$p(x) = a_0 + a_1(x - x_0) + \cdots + a_n(x - x_0)^n$$

Successive differentiations produce

$$p^{(1)}(x) = a_1 + 2a_2(x - x_0) + \cdots + na_n(x - x_0)^{n-1}$$
$$p^{(2)}(x) = 2a_2 + 3 \cdot 2a_3(x - x_0) + \cdots + n(n-1)a_n(x - x_0)^{n-2}$$
$$\vdots$$
$$p^{(n)}(x) = n!a_n$$

The specifications then require

$$p(x_0) = a_0 = y_0 \qquad p^{(1)}(x_0) = a_1 = y_0^{(1)} \qquad p^{(2)}(x_0) = 2a_2 = y_0^{(2)} \qquad \cdots \qquad p^{(n)}(x_0) = n!a_n = y_0^{(n)}$$

Solving for the a_n coefficients and substituting,

$$p(x) = y_0 + y_0^{(1)}(x - x_0) + \cdots + \frac{1}{n!}y_0^{(n)}(x - x_0)^n = \sum_{i=0}^{n} \frac{1}{i!} y_0^{(i)}(x - x_0)^i$$

5.27   Find a polynomial $p(x)$ of degree n, such that, at $x_0 = 0$, $p(x)$ and e^x agree in value together with their first n derivatives.

▮ Since for e^x derivatives of all orders are also e^x,

$$y_0 = y_0^{(1)} = y_0^{(2)} = \cdots = y_0^{(n)} = 1$$

The Taylor polynomial can then be written

$$p(x) = \sum_{i=0}^{n} \frac{1}{i!} x^n = 1 + x + \frac{1}{2}x^2 + \frac{1}{6}x^3 + \cdots + \frac{1}{n!}x^n$$

5.28   Consider a second function $y(x)$ also having the specifications of Problem 5.26. We shall think of $p(x)$ as a polynomial approximation to $y(x)$. Obtain a formula for the difference $y(x) - p(x)$ in integral form, assuming $y^{(n+1)}(x)$ continuous between x_0 and x.

▮ Here it is convenient to use a different procedure from that which led us to error estimates for the collocation and osculating polynomials. We start by temporarily calling the difference R,

$$R = y(x) - p(x)$$

or in full detail

$$R(x, x_0) = y(x) - y(x_0) - y'(x_0)(x - x_0) - \frac{1}{2}y''(x_0)(x - x_0)^2 - \cdots - \frac{1}{n!}y^{(n)}(x_0)(x - x_0)^n$$

This actually defines R as a function of x and x_0. Calculating the derivative of R relative to x_0, holding x fixed, we find

$$R'(x, x_0) = -y'(x_0) + y'(x_0) - y''(x_0)(x - x_0) + y''(x_0)(x - x_0)$$
$$- \frac{1}{2}y'''(x_0)(x - x_0)^2 + \cdots - \frac{1}{n!}y^{(n+1)}(x_0)(x - x_0)^n$$
$$= -\frac{1}{n!}y^{(n+1)}(x_0)(x - x_0)^n$$

since differentiation of the second factor in each product cancels the result of differentiating the first factor in the previous product. Only the very last term penetrates through. Having differentiated relative to x_0, we reverse direction and integrate relative to x_0 to recover R:

$$R(x, x_0) = -\frac{1}{n!}\int_x^{x_0} y^{(n+1)}(u)(x-u)^n \, du + \text{constant}$$

By the original definition of R, $R(x_0, x_0) = 0$ and the constant of integration is 0. Reversing the limits,

$$R(x, x_0) = \frac{1}{n!}\int_{x_0}^{x} y^{(n+1)}(u)(x-u)^n \, du$$

which is known as an integral form of the error.

5.29 Obtain Lagrange's form of the error from the integral form.

$\blacksquare$ Here we use a mean value theorem of calculus, which says that if $f(x)$ is continuous and $w(x)$ does not change sign in the interval (a, b) then

$$\int_a^b f(x)w(x) \, dx = f(\xi)\int_a^b w(x) \, dx$$

where ξ is between a and b. Choosing $w(x) = (x - x_0)^n$, we easily get

$$R(x, x_0) = \frac{1}{(n+1)!}y^{(n+1)}(\xi)(x - x_0)^{n+1}$$

where ξ is between x_0 and x but otherwise unknown. This form of the error is very popular because of its close resemblance to the terms of the Taylor polynomial. Except for a ξ in place of an x it would be the term which produced the Taylor polynomial of next higher degree.

5.30 Estimate the degree of a Taylor polynomial for the function $y(x) = e^x$, with $x_0 = 0$, which guarantees approximations correct to three decimal places for $-1 < x < 1$. Perform the estimate to six decimal places also.

$\blacksquare$ By the Lagrange formula for the error,

$$|e^x - p(x)| = |R| \le \frac{e}{(n+1)!}$$

For three place accuracy this should not exceed .0005, a condition which is satisfied for $n = 7$ or higher. The polynomial

$$p(x) = \sum_{i=0}^{7} \frac{1}{i!} x^i$$

is therefore adequate. Similarly, for six place accuracy $|R|$ should not exceed .0000005, which will be true for $n = 10$.

5.31 The *operator D* is defined by $D = h(\partial/\partial x)$. What is the result of applying the successive powers of D to $y(x)$?

$\blacksquare$ We have at once $D^i y(x) = h^i y^{(i)}(x)$.

5.32 Express the Taylor polynomial in operator symbolism.

$\blacksquare$ Let $x - x_0 = kh$. This is the symbolism we have used earlier, with x_k now abbreviated to x. Then direct substitution into the Taylor polynomial of Problem 5.26 brings

$$p(x) = \sum_{i=0}^{n} \frac{1}{i!} y_0^{(i)}(x - x_0)^i = \sum_{i=0}^{n} \frac{1}{i!} y_0^{(i)} k^i h^i = \sum_{i=0}^{n} \frac{1}{i!} k^i D^i y(x_0)$$

A common way of rewriting this result is

$$p(x) = \left[\sum_{i=0}^{n} \frac{1}{i!} k^i D^i\right] y(x_0)$$

or in terms of the integer variable k alone as

$$p_k = \left[\sum_{i=0}^{n} \frac{1}{i!} k^i D^i \right] y_0$$

where as usual $p(x_k) = p_k$.

5.33 A function $y(x)$ is called *analytic* on the interval $|x - x_0| \le r$ if as $n \to \infty$,

$$\lim R(x, x_0) = 0$$

for all arguments x in the interval. It is then customary to write $y(x)$ as an infinite series, called a *Taylor series*:

$$y(x) = \lim p(x) = \sum_{i=0}^{\infty} \frac{1}{i!} y_0^{(i)} (x - x_0)^i$$

Express this in operator form.

▮ Proceeding just as in Problem 5.32, we find $y(x_k) = [\sum_{i=0}^{\infty}(1/i!)k^i D^i] y_0$. This is our first infinite series operator. The arithmetic of such operators is not so easy to justify as was the case with the simpler operators used earlier.

5.34 The operator e^{kD} is defined by $e^{kD} = \sum_{i=0}^{\infty}(1/i!)k^i D^i$. Write the Taylor series using this operator.

▮ We have at once $y(x_k) = e^{kD} y_0$.

5.35 Prove $e^D = E$.

▮ By Problem 5.34 with $k = 1$ and the definition of E, $y(x_1) = y_1 = Ey_0 = e^D y_0$ making $E = e^D$.

5.36 Develop the Taylor series for $y(x) = \ln(1 + x)$, using $x_0 = 0$.

▮ The derivatives are $y^{(i)}(x) = (-1)^{i+1}(i-1)!/(1+x)^i$ so that $y^{(i)}(0) = (-1)^{i+1}(i-1)!$. Since $y(0) = \ln 1 = 0$, we have

$$y(x) = \ln(1 + x) = \sum_{i=1}^{\infty} \frac{(-1)^{i+1}}{i} x^i = x - \frac{1}{2}x^2 + \frac{1}{3}x^3 - \frac{1}{4}x^4 + \cdots$$

The familiar ratio test shows this to be convergent for $-1 < x < 1$. It does not, however, prove that the series equals $\ln(1 + x)$. To prove this let $p(x)$ represent the Taylor polynomial, of degree n. Then by the Lagrange formula for the error,

$$|\ln(1 + x) - p(x)| \le \frac{1}{(n+1)!} \cdot \frac{n!}{(1+\xi)^{n+1}} \cdot x^{n+1}$$

For simplicity consider only the interval $0 \le x < 1$. The series is applied mostly to this interval anyway. Then the error can be estimated by replacing ξ by 0 and x by 1 to give $|\ln(1 + x) - p(x)| \le 1/(n + 1)$ and this does have limit 0. Thus $\lim p(x) = \ln(1 + x)$, which was our objective.

5.37 Estimate the degree of a Taylor polynomial for the function $y(x) = \ln(1 + x)$, with $x_0 = 0$, which guarantees three decimal place accuracy for $0 < x < 1$.

▮ By the Lagrange formula for the error,

$$|\ln(1 + x) - p(x)| \le \frac{1}{(n+1)!} \cdot \frac{n!}{(1+\xi)^n} \cdot x^{n+1} \le \frac{1}{n+1}$$

Three place accuracy requires that this not exceed .0005, which is satisfied for $n = 2000$ or higher. A polynomial of degree 2000 would be needed. This is an example of a slowly convergent series.

5.38 Express the operator D in terms of the operator Δ.

▮ From $e^D = E$ we find $D = \ln E = \ln(1 + \Delta) = \Delta - \frac{1}{2}\Delta^2 + \frac{1}{3}\Delta^3 - \frac{1}{4}\Delta^4 + \cdots$.
    The validity of this calculation is surely open to suspicion, and any application of it must be carefully checked. It suggests that the final series operator will produce the same result as the operator D.

5.39 Express $y(x) = (1 + x)^p$ as a Taylor series.

▮ For p a positive integer this is the binomial theorem of algebra. For other values of p it is the *binomial series*. Its applications are extensive. We easily find

$$y^{(i)}(x) = p(p - 1) \cdots (p - i + 1)(1 + x)^{p-i} = p^{(i)}(1 + x)^{p-i}$$

where $p^{(i)}$ is again the factorial polynomial. Choosing $x_0 = 0$

$$y^{(i)}(0) = p^{(i)}$$

and substituting into the Taylor series,

$$y(x) = \sum_{i=0}^{\infty} \frac{p^{(i)}}{i!} x^i = \sum_{i=0}^{\infty} \binom{p}{i} x^i$$

where $\binom{p}{i}$ is the generalized binomial coefficient. The convergence of this series to $y(x)$ for $-1 < x < 1$ can be demonstrated.

5.40 Use the binomial series to drive the *Euler transformation*.

▮ The Euler transformation is an extensive rearrangement of the alternating series $S = a_0 - a_1 + a_2 - a_3 + \cdots$ which we rewrite as

$$S = [1 - E + E^2 - E^3 + \cdots] a_0 = [1 + E]^{-1} a_0$$

by the binomial theorem with $p = -1$. The operator $[1 + E]^{-1}$ may be interpreted as the inverse operator of $1 + E$. A second application of the binomial theorem now follows:

$$S = [1 + E]^{-1} a_0 = [2 + \Delta]^{-1} a_0 = \frac{1}{2} \left[1 + \frac{\Delta}{2} \right]^{-1} a_0$$

$$= \frac{1}{2} \left[1 - \frac{\Delta}{2} + \frac{\Delta^2}{4} - \frac{\Delta^3}{8} + \cdots \right] a_0 = \frac{1}{2} \left[a_0 - \frac{1}{2} \Delta a_0 + \frac{1}{4} \Delta^2 a_0 - \frac{1}{8} \Delta^3 a_0 + \cdots \right]$$

Our derivation of this formula has been a somewhat optimistic application of operator arithmetic. No general, easy-to-apply criterion for insuring its validity exists, but see Problem 5.59.

5.41 The *Bernoulli numbers* are defined to be the numbers B_i in the following series

$$y(x) = \frac{x}{e^x - 1} = \sum_{i=0}^{\infty} \frac{1}{i!} B_i x^i$$

Find $B_0, \ldots, B_{10}$.

▮ The Taylor series requires that $y^{(i)}(0) = B_i$, but it is easier in this case to proceed differently. Multiplying by $e^x - 1$ and using the Taylor series for e^x, we get

$$x = \left(x + \tfrac{1}{2} x^2 + \tfrac{1}{6} x^3 + \cdots \right) \left(B_0 + B_1 x + \tfrac{1}{2} B_2 x^2 + \tfrac{1}{6} B_3 x^3 + \cdots \right)$$

Now comparing the coefficients of the successive powers of x,

$$B_0 = 1 \qquad B_1 = -\tfrac{1}{2} \qquad B_2 = \tfrac{1}{6} \qquad B_3 = 0 \qquad B_4 = -\tfrac{1}{30} \qquad B_5 = 0$$

$$B_6 = \tfrac{1}{42} \qquad B_7 = 0 \qquad B_8 = -\tfrac{1}{30} \qquad B_9 = 0 \qquad B_{10} = \tfrac{5}{66}$$

The process could be continued in an obvious way.

5.42 Suppose $\Delta F_k = y_k$. Then an *inverse operator* Δ^{-1} can be defined by

$$F_k = \Delta^{-1} y_k$$

This inverse operator is indefinite in that for given y_k the numbers F_k are determined except for an arbitrary additive constant. For example, in the following table the numbers y_k are listed as first differences. Show that the number F_0 can be chosen arbitrarily and that the other F_k numbers are then determined.

F_k	F_0	.	.	.	.	.	.	.	.
y_k	y_0	y_1	y_2	y_3	y_4	.	.	.	

■ We have at once

$$F_1 = F_0 + y_0 \qquad F_2 = F_1 + y_1 = F_0 + y_0 + y_1 \qquad F_3 = F_2 + y_2 = F_0 + y_0 + y_1 + y_2$$

and in general $F_k = F_0 + \sum_{i=0}^{k-1} y_i$. The requirements plainly hold for an arbitrary F_0 and the analogy with indefinite integration is apparent.

5.43 Obtain a formula for Δ^{-1} in terms of the operator D.

■ The result $e^D = 1 + \Delta$ suggests

$$\Delta^{-1} = (e^D - 1)^{-1} = D^{-1}\left[D(e^D - 1)^{-1} \right]$$

where D^{-1} is an *indefinite integral operator*, an inverse of D. From the definition of Bernoulli numbers,

$$\dot{\Delta}^{-1} = D^{-1} \sum_{i=0}^{\infty} (1/i!) B_i D^i$$

$$= D^{-1}\left[1 - \tfrac{1}{2}D + \tfrac{1}{12}D^2 - \tfrac{1}{720}D^4 + \cdots \right] = D^{-1} - \tfrac{1}{2} + \tfrac{1}{12}D - \tfrac{1}{720}D^3 + \cdots$$

As always with the indefinite integral (and here we also have an indefinite summation), the presence of an additive constant may be assumed.

5.44 Derive the Euler–Maclaurin formula operationally.

■ Combining the results of the previous two problems, we have

$$F_k = \Delta^{-1} y_k = F_0 + \sum_{i=0}^{k-1} y_i$$

$$F_k = \left[D^{-1} - \tfrac{1}{2} + \tfrac{1}{12}D - \tfrac{1}{720}D^3 + \cdots \right] y_k$$

From the first of these,

$$F_n - F_0 = \sum_{i=0}^{n-1} y_i$$

while from the second,

$$F_n - F_0 = \frac{1}{h} \int_{x_0}^{x_n} y(x)\, dx - \frac{1}{2}(y_n - y_0) + \frac{h}{12}(y_n' - y_0') - \frac{h^3}{720}(y_n^{(3)} - y_0^{(3)}) + \cdots$$

so that finally,

$$\sum_{i=0}^{n-1} y_i = \frac{1}{h} \int_{x_0}^{x_n} y(x)\, dx - \frac{1}{2}(y_n - y_0) + \frac{h}{12}(y_n' - y_0') + \cdots$$

which is the Euler–Maclaurin formula. The operator arithmetic used in this derivation is clearly in need of supporting logic, but the result is useful in spite of its questionable pedigree and in spite of the fact that the series obtained is usually *not convergent*.

5.45 Show that the series for $\sin x$,

$$\sin x = x - x/3! + x/5! - \cdots$$

is convergent to $\sin x$ for any value of x.

❚ The Lagrange form of the error of the Taylor polynomial shows this error to be bounded by $|x|^{n+1}/(n+1)!$ if the degree of the polynomial is n. For $n \to \infty$ this has limit zero for any x. The same argument proves the corresponding result for $\cos x$.

5.46 For what value of n will the Taylor polynomial approximate $\sin x$ correctly to three decimal places for $0 < x < \pi/2$?

❚ Again using the Lagrange error formula, we will need

$$(\pi/2)^{n+1}/(n+1)! < .0005$$

which holds if n is at least 7. Easy computations also show that for $n = 9, 10$ and 11 we would obtain four, five and six correct places. Of course, since all terms in the sine series are of odd degree, the $n = 10$ term would have a zero coefficient.

5.47 Rework the preceding problem for the cosine function.

❚ The same computations apply. Since only terms of even degree are now present, the odd values of n can be reduced by 1 for the given accuracy.

5.48 Express the operator Δ as a series operator in D.

❚
$$\Delta = E - 1 = e^D - 1 = D + D^2/2! + D^3/3! + \cdots$$

5.49 The functions $\sinh x$ and $\cosh x$ are defined by

$$\sinh x = \frac{e^x - e^{-x}}{2} \qquad \cosh x = \frac{e^x + e^{-x}}{2}$$

Show that their Taylor series are

$$\sinh x = \sum_{i=0}^{\infty} \frac{1}{(2i+1)!} x^{2i+1} \qquad \cosh x = \sum_{i=0}^{\infty} \frac{1}{(2i)!} x^{2i}$$

❚ Adding the two exponential series

$$e^x = 1 + x + x^2/2! + x^3/3! + x^4/4! + \cdots$$
$$e^{-x} = 1 - x + x^2/2! - x^3/3! + x^4/4! - \cdots$$

term by term and then dividing by 2 produces

$$(e^x + e^{-x})/2 = 1 + x^2/2! + x^4/4! + \cdots$$

the odd power terms having been eliminated. This is the $\cosh x$ series. Term by term subtraction leads in a similar way to the series for $\sinh x$.

5.50 Show by operator arithmetic that $\delta = 2 \sinh \frac{1}{2} D$ and $\mu = \cosh \frac{1}{2} D$.

❚ We have $E = e^D$, $E^{1/2} = e^{D/2}$ and $E^{-1/2} = e^{-D/2}$ so that

$$\delta = E^{1/2} - E^{-1/2} = e^{D/2} - e^{-D/2} = 2 \sinh\left(\tfrac{1}{2} D\right)$$

with an almost parallel proof for μ.

5.51 Use the binomial series to express $\Delta = \frac{1}{2}\delta^2 + \delta\sqrt{1 + \frac{1}{4}\delta^2}$ as a series in powers of δ, through the term in δ^7.

❚ This expression for Δ was found in Problem 4.29. The binomial theorem converts it to

$$\Delta = \tfrac{1}{2}\delta^2 + \delta\left(1 + \tfrac{1}{8}\delta^2 - \tfrac{1}{8}\cdot\tfrac{1}{16}\delta^4 + \tfrac{1}{16}\cdot\tfrac{1}{64}\delta^6 + \cdots\right)$$
$$= \delta + \tfrac{1}{2}\delta^2 + \tfrac{1}{8}\delta^3 - \tfrac{1}{128}\delta^5 + \tfrac{1}{1024}\delta^7 + \cdots$$

through the seventh power term.

5.52 Develop this series representation of D in terms of δ:

$$D = \delta - \frac{1^2}{2^2 \cdot 3!}\delta^3 + \frac{1^2 \cdot 3^2}{2^4 \cdot 5!}\delta^5 - \frac{1^2 \cdot 3^2 \cdot 5^2}{2^6 \cdot 7!}\delta^7 + \cdots$$

▮ There are several approaches. One might start from

$$D = \Delta - \tfrac{1}{2}\Delta^2 + \tfrac{1}{3}\Delta^3 - \tfrac{1}{4}\Delta^4 + \cdots$$

and

$$\Delta = \delta + \tfrac{1}{2}\delta^2 + \tfrac{1}{8}\delta^3 - \tfrac{1}{128}\delta^5 + \tfrac{1}{1024}\delta^7 + \cdots$$

substituting the latter series into the former. This involves a moderate slog starting with

$$\Delta^2 = \delta^2 + \delta^3 + \tfrac{1}{2}\delta^4 + \tfrac{1}{8}\delta^5 - \tfrac{1}{128}\delta^7 + \cdots$$

and

$$\Delta^3 = \delta^3 + \tfrac{3}{2}\delta^4 + \tfrac{9}{8}\delta^5 + \tfrac{1}{2}\delta^6 + \tfrac{15}{128}\delta^7 + \cdots$$

up through $\Delta^7 = \delta^7 + \cdots$, followed by substitutions into the series for D and collection of corresponding power terms.

A more interesting trail begins with

$$\delta = E^{1/2} - E^{-1/2} = e^{D/2} - e^{-D/2} = 2\sinh\left(\tfrac{1}{2}D\right)$$

which solves for D in the form

$$D = 2\sinh^{-1}(\delta/2)$$

The known series for the inverse hyperbolic sine

$$\sinh^{-1} x = x - \frac{1}{2}\frac{x^3}{3} + \frac{1 \cdot 3}{2 \cdot 4}\frac{x^5}{5} - \frac{1 \cdot 3 \cdot 5}{2 \cdot 4 \cdot 6}\frac{x^7}{7} + \cdots$$

is then available and leads directly to

$$D = \delta - \tfrac{1}{24}\delta^3 + \tfrac{3}{640}\delta^5 - \tfrac{5}{7168}\delta^7 + \cdots$$

which can be used to estimate the derivative of a tabulated function. (But see also Problem 5.54.)

5.53 Verify these terms of a Taylor series for D^2:

$$D^2 = \delta^2 - \tfrac{1}{12}\delta^4 + \tfrac{1}{90}\delta^6 - \tfrac{1}{560}\delta^8 + \cdots$$

by squaring the series of the preceding problem and collecting together the various powers of δ.

▮ We must, of course, square each term and take twice the product of each pair. Certainly the leadoff term in D^2 will be δ^2. There can be no odd power terms in this series and only the product $-2/2 \cdot 3! = -1/12$ contributes to a δ^4 term. For δ^6 we have two contributions, $1/16 \cdot 36$ and $2 \cdot 9/16 \cdot 120$, which combine to produce the indicated $1/90$. As for the δ^8 term, two product terms contribute $2 \cdot 9 \cdot 25/64 \cdot 5040$ and $2 \cdot 9/64 \cdot 6 \cdot 25$ for a total of $1/560$.

5.54 The formula of Problem 5.52, if applied to y_0, would require unlisted data such as $y_{1/2}$, $y_{3/2}$, etc. Modify this formula by multiplying by $\mu/\sqrt{1 + \tfrac{1}{4}\delta^2}$, which is 1, to obtain

$$D = \mu\left(\delta - \frac{1^2}{3!}\delta^3 + \frac{1^2 \cdot 2^2}{5!}\delta^5 - \frac{1^2 \cdot 2^2 \cdot 3^2}{7!}\delta^7 + \cdots\right)$$

which may be applied directly to y_0.

▮ By the binomial theorem

$$\left(1 + \frac{1}{4}\delta^2\right)^{-1/2} = 1 - \frac{1}{8}\delta^2 + \frac{3}{128}\delta^4 - \frac{15}{48 \cdot 64}\delta^6 + \cdots$$

and our series for D is

$$D = \delta - \frac{1}{24}\delta^3 + \frac{9}{16 \cdot 5!}\delta^5 - \frac{9 \cdot 25}{64 \cdot 7!}\delta^7 + \cdots$$

so multiplying the two and introducing the factor μ,

$$D = \mu\left[\delta - \frac{1}{6}\delta^3 + \left(\frac{3}{128} + \frac{1}{8 \cdot 4!} + \frac{9}{16 \cdot 5!}\right)\delta^5 - \left(\frac{15}{48 \cdot 64} + \frac{3}{128 \cdot 4!} + \frac{9}{128 \cdot 5!} + \frac{9 \cdot 25}{64 \cdot 7!}\right)\delta^7 + \cdots\right]$$

which will be found to agree with the stated result.

5.55 Prove $D^3 = \mu(\delta^3 - \frac{1}{4}\delta^5 + \frac{7}{120}\delta^7 - \cdots)$.

▌ Multiplying our series of the preceding two problems

$$D^3 = \mu\left[\delta^3 - \left(\frac{1}{12} + \frac{1}{6}\right)\delta^5 + \left(\frac{1}{90} + \frac{1}{72} + \frac{1}{30}\right)\delta^7 - \cdots\right]$$

which quickly reduces to the required result.

5.56 Prove $D^4 = \delta^4 - \frac{1}{6}\delta^6 + \frac{7}{240}\delta^8 - \cdots$.

▌ Squaring the series for D^2

$$D^4 = \delta^4 - \frac{2}{12}\delta^6 + \left(\frac{1}{144} + \frac{2}{90}\right)\delta^8 - \cdots$$

which is equivalent to the theorem.

5.57 Find the operator series for $\mu\delta$ in terms of D.

▌ Observing that

$$\mu\delta = \tfrac{1}{2}(E^1 - E^{-1}) = \tfrac{1}{2}(e^D - e^{-D}) = \sinh D$$

the series for this hyperbolic function is what is needed.

5.58 Apply Taylor polynomials to find an approximation to a root of $e^{-x} = A \sin x$ near zero.

▌ Using first-degree polynomials for both sides, $1 - x = Ax$ and we have

$$x = 1/(A + 1)$$

Moving to quadratic polynomials,

$$1 - x + \tfrac{1}{2}x^2 = Ax \quad \text{or} \quad x^2 - 2(A + 1)x + 2 = 0$$

with solution

$$x = (A + 1) - \sqrt{(A + 1)^2 - 2}$$

Here are a few numerical results:

A	1	2	3	4
degree 1	.500	.333	.250	.200
degree 2	.586	.354	.258	.204
correct	.5885327	.3573274	.2599482	.2050800

The correct values were found by Newton's method, which is well known and will appear in a later chapter. The point here is that a very simple effort based on Taylor polynomials, even a mental calculation, may be satisfactory for some purposes.

5.59 Rederive Euler's formula, finding an error term.

▌ Consider the finite sum

$$S_n = \sum_{k=0}^{n-1} a_k t^k = \sum_{k=0}^{n-1} a_k \Delta v_k$$

where $v_k = (1 - t^k)/(1 - t)$. Summation by parts leads to

$$S_n = a_n \left(\frac{1 - t^n}{1 - t} \right) - \frac{1}{1 - t} \sum_{k=0}^{n-1} \Delta a_k + \frac{t}{1 - t} \sum_{k=0}^{n-1} t^k \Delta a_k$$

and since the first sum on the right is simply $a_n - a_0$,

$$S_n = \frac{a_0}{1 - t} - \frac{a_n t^n}{1 - t} + \frac{t}{1 - t} \sum_{k=0}^{n-1} t^k \Delta a_k$$

Notice that the last term has the same form as the original sum, with Δa_k in place of a_k. Apply summation by parts to this last term to obtain

$$S_n = \frac{a_0}{1 - t} + \frac{t \Delta a_0}{(1 - t)^2} + \frac{t^2}{(1 - t)^2} \sum_{k=0}^{n-1} t^k \Delta^2 a_k - \frac{a_n t^n}{1 - t} - \frac{1}{1 - t} \frac{\Delta a_n t^{n+1}}{1 - t}$$

Continuing through r such summation by parts,

$$S_n = \frac{1}{1 - t} \sum_{i=0}^{r-1} \left(\frac{t}{1 - t} \right)^i \Delta^i a_0 + \left(\frac{t}{1 - t} \right)^r \sum_{k=0}^{n-1} t^k \Delta^r a_k - \frac{t^n}{1 - t} \sum_{i=0}^{r-1} \left(\frac{t}{1 - t} \right)^i \Delta^i a_n$$

If for n tending to infinity we set $\lim S_n = S$, the last term has limit zero, making

$$S = \frac{1}{1 - t} \sum_{i=0}^{r-1} \left(\frac{t}{1 - t} \right)^i \Delta^i a_0 + \left(\frac{t}{1 - t} \right)^r \sum_{k=0}^{\infty} t^k \Delta^r a_k$$

If now r tends to infinity and *if it is assumed that the final term has limit zero*, then a generalized Euler formula appears. Put $t = -1$ to obtain the special case derived in Problem 5.40. We have here a test for the validity of the Euler formula, namely, the convergence of the original sums to S and the vanishing in the limit of the final term exhibited. Unfortunately the latter is not always easy to decide. Moreover, the Euler formula has also been found helpful when $\lim S_n$ fails to exist.

5.3 PIECEWISE APPROXIMATION

5.60 What is meant by piecewise approximation?

▮ Up to now we have used a single polynomial to represent the data or given function over the entire interval of interest. To improve the accuracy of approximation our recourse has been to try polynomials of higher and higher degree. Often enough this is quite satisfactory, but it has been observed that there are occasions when a numerical instability enters and the true values are overwhelmed by error.

Instead of using a single polynomial we may instead join together several polynomial segments, each of fairly low degree. The classic example of this is a set of line segments, each fitted to the data over a single subinterval. Such an approximation is continuous but has a first derivative with discontinuities at the interval endpoints, or *knots* (Fig. 5-3). It is the basis for elementary interpolations and for the trapezoidal rule method of numerical integration. The implicit assumption that between data points the background function is almost linear may be reasonable if the knots are close enough together.

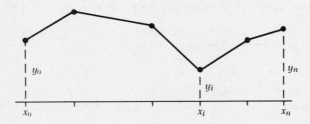

Fig. 5-3

Segments of higher-degree polynomials, usually degree 2 or 3, are also used for piecewise approximation. Parabolic segments are used to develop Simpson's rule for numerical integration. Cubic segments are the standard for constructing spline approximations. In the opposite direction, step functions (see Fig. 5-4), which use zero-degree polynomial segments, have played important roles in mathematical analysis, the definition of the integral being one notable example and the statistician's histogram another.

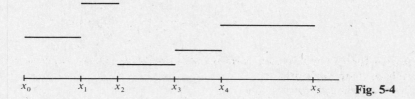

Fig. 5-4

5.61 How can the broken line (or piecewise linear function) of Fig. 5-5 be represented?

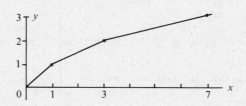

x	0	1	3	7
y	0	1	2	3

Fig. 5-5

▌ The most natural way would seem to be by a linear formula for each subinterval:

$$\text{on } (0,1) \qquad p(x) = x$$
$$\text{on } (1,3) \qquad p(x) = 1 + \tfrac{1}{2}(x-1)$$
$$\text{on } (3,7) \qquad p(x) = 2 + \tfrac{1}{4}(x-3)$$

To compute the function for a given x, a two step process is then involved. First the subinterval must be located and then the appropriate formula computed. For $x = 2$ we find the second subinterval active and correctly figure $p(2)$ to be 1.5, while for $x = 5$ the active subinterval is the third and $p(5) = 2.5$. This representation of $p(x)$ is easy to program and for purposes of computation is quite satisfactory.

5.62 Why have other representations of piecewise polynomial functions been developed?

▌ For theoretical purposes the "separate formula for each subinterval" representation has often been found troublesome and various attempts have been made to obtain single formula representations such as

$$p(x) = p_0(x) + c_1 b_1(x) + \cdots + c_k b_k(x)$$

in which the functions $b_i(x)$ are called basis functions and $p_0(x)$ is a sort of initializer. It is then possible to obtain more compact expressions for the derivative or integral of $p(x)$ by exploiting the features of the basis functions.

5.63 What are the truncated power functions?

▌ These are the familiar powers of x truncated at 0, the left half replaced by 0 and the whole then shifted right or left to an arbitrary position t. For degree 0 to 3 they are pictured in Fig. 5-6.

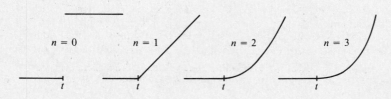

Fig. 5-6

5.64   Use truncated power functions of degree 1 as a basis for the piecewise linear function of Problem 5.61.

▌ Recalling that the data points were $(0,0)$, $(1,1)$, $(3,2)$ and $(7,3)$, we combine four such functions

$$p(x) = c_{-1}\, \text{TP}(x+1) + c_0\, \text{TP}(x) + c_1\, \text{TP}(x-1) + c_2\, \text{TP}(x-2)$$

with each TP function zero for negative arguments. Substituting the given data determines the coefficients

$$
\begin{aligned}
0 &= c_{-1} & c_{-1} &= 0 \\
1 &= 2c_{-1} + c_0 & c_0 &= 1 \\
2 &= 4c_{-1} + 3c_0 + 2c_1 & c_1 &= -\tfrac{1}{2} \\
3 &= 8c_{-1} + 7c_0 + 6c_1 + 4c_2 & c_2 &= -\tfrac{1}{4}
\end{aligned}
$$

and we have the required representation.

$$p(x) = \text{TP}(x) - \tfrac{1}{2}\,\text{TP}(x-1) - \tfrac{1}{4}\,\text{TP}(x-3)$$

This is a relatively favorable example for truncated powers, due to the monotone function. With data that bounce up and down a great deal each truncated power may be obliged to compensate for the effect of its predecessor, the coefficients c_i swinging wildly and a numerical instability developing.

5.65   Find a pair of quadratic segments fitting this slight extension of the data of the preceding problem:

x	0	1	3	7	15
y	0	1	2	3	4

▌ Either the Lagrange or the Newton formula will serve the purpose. Choosing Newton we have the pair

$$p(x) = 0 + x - \tfrac{1}{6}x(x-1) \quad \text{on } (0,3)$$
$$p(x) = 2 + \tfrac{1}{4}(x-3) - \tfrac{1}{96}(x-3)(x-7) \quad \text{on } (3,15)$$

The point is, a quadratic can certainly be fitted to any three data points but if done in this way there will probably be corners at the joins, where two quadratic segments meet; no effort has been made to avoid this. Here the first derivative jumps from $\tfrac{1}{6}$ to $\tfrac{7}{24}$ as argument $x = 3$ is crossed.

5.66   What is a quadratic spline?

▌ A quadratic spline is a set of quadratic segments joined in such a way that the first derivative of the composite function is continuous. Let data $(x_i,\ y_i)$ be given. We seek a quadratic segment over each interval $x_i,\ x_{i+1}$. Suppose we were to generate these segments from left to right beginning at x_0. Since each segment will have three coefficients and the first has only to match the data values y_0 and y_1, we have one initial degree of freedom. Choose y_0' arbitrarily for the moment. The first segment is then determined. Moving along to the interval x_1, x_2 the next segment will have to match y_1 and y_2 and also duplicate the y' value of the first segment at x_1. So the second quadratic is completely determined. Clearly the process now continues in this way, each segment determined by its own two y_i values and the responsibility to keep the first derivative continuous.

We see from this approach that a quadratic spline is not entirely determined by the given data and the requirement of a continuous derivative. One extra condition can be imposed, the initial slope or whatever is appropriate.

5.67   Returning to the data of Problem 5.65, find a quadratic spline using a basis of truncated power functions.

▌ With five data points and one free condition we set up our spline in the form

$$p(x) = c_{-2}\,\text{TP}(x+2)^2 + c_{-1}\,\text{TP}(x+1)^2 + c_0\,\text{TP}\,x^2 + c_1\,\text{TP}(x-1)^2 + c_2\,\text{TP}(x-3)^2 + c_3\,\text{TP}(x-7)^2$$

and have six coefficients available to assist in satisfying the six specifications. Note that $\text{TP}(x-15)^2$ would be useless to us since it is zero on the entire interval $(0,15)$. Also note that the first two truncated powers listed are both active in some part of this interval and so can play a role. There is a shred of evidence that the derivative at $x = 0$ may be approximately 1, so let us proceed on this assumption. First

$$p(0) = 4c_{-2} + c_{-1} = 0 \qquad p'(0) = 4c_{-2} + 2c_{-1} = 1$$

and we are led to $c_{-2} = -\frac{1}{4}$, $c_{-1} = 1$. In passing we observe that derivatives of truncated power functions are available by the familiar power rule, that is,

$$\text{TP}'(x-t)^n = n \cdot \text{TP}(x-t)^{n-1}$$

for $x \geq t$. For $x < t$ all derivatives are, of course, zero. Continuing,

$$p(1) = 9c_{-2} + 4c_{-1} + c_0 = 1$$
$$p(3) = 25c_{-2} + 16c_{-1} + 9c_0 + 4c_1 = 2$$
$$p(7) = 81c_{-2} + 64c_{-1} + 49c_0 + 36c_1 + 16c_2 = 3$$
$$p(15) = 289c_{-2} + 256c_{-1} + 225c_0 + 196c_1 + 144c_2 + 64c_3 = 4$$

from which one eventually learns that

$$c_0 = -\tfrac{3}{4} \qquad c_1 = -\tfrac{1}{4} \qquad c_2 = \tfrac{5}{16} \qquad c_3 = -\tfrac{7}{64}$$

and the spline is in hand.

5.68   What other splines are available for the data of the preceding problem? Which is best?

▮   Over the initial interval $(0,1)$ the function $ax + (1-a)x^2$ will fit the two data points $(0,0)$ and $(1,1)$ with initial slope a. The spline just found corresponds to $a = 1$. It is not hard to verify that $a = \frac{7}{6}$ leads to the spline

$$p(x) = \tfrac{7}{6}x - \tfrac{1}{6}x^2 + \tfrac{3}{16}\text{TP}(x-3)^2 - \tfrac{3}{64}\text{TP}(x-7)^2$$

the first two terms of which are familiar from Problem 5.65. Taking $a = 0$ and making a suitable followup we find

$$p(x) = x^2 - \tfrac{7}{4}\text{TP}(x-1)^2 + \tfrac{17}{16}\text{TP}(x-3)^3 - \tfrac{31}{64}\text{TP}(x-7)^2$$

which is plotted in Fig. 5-7 along with the "true" function $\log(x+1)$. Though a spline, its credentials are unimpressive, and the result of Problem 5.67 was surely better.

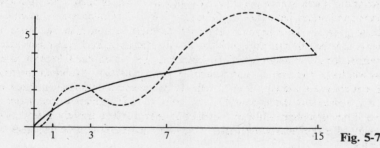

Fig. 5-7

5.69   What is a spline of degree n?

▮   It is a piecewise polynomial, each segment a polynomial of degree n, matching given data (x_i, y_i) and having all derivatives up through the $(n-1)$th continuous at the knots. The argument just employed to discover 1 degree of freedom for splines of degree 2 is easily reworked to show that there will be $n-1$ such degrees of freedom for splines of degree n.

5.70   What are the *hat* functions?

▮   These are an alternative to the truncated linear functions in providing a basis for piecewise representation of data. The hat function corresponding to consecutive knots t_{i-1}, t_i, t_{i+1} is a set of line segments forming an inverted V, equaling 0 at t_{i-1} and t_{i+1}, equaling 1 at t_i and identically zero elsewhere. See Fig. 5-8.

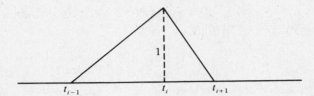

Fig. 5-8

5.71 How can the hat functions be used to represent a broken line function or piecewise linear polynomial? (Note that such a function can now be described as a spline, since as defined in Problem 5.69 no continuity requirements on derivatives are set when $n = 1$.)

▮ Let the given data be (x_i, y_i) for $i = 0, 1, \ldots, n$. The knots are to be at x_1 to x_{n-1}. Let $b_i(x)$ be the hat function with peak at x_i. Then

$$p(x) = \sum_{i=0}^{n} c_i b_i(x)$$

will represent the broken line function if we choose $c_i = y_i$. To see this, note that for each x_i the function $b_i(x_i)$ will have the value 1 while all other basis functions are 0. On each of the intervals two of the $b(x)$ functions will cross and because the sum of two linear functions is linear, we have the required broken line. Since only two basis functions are active on any interval the computation of $p(x)$ is a two term computation.

5.72 Develop basis functions similar to the hat functions of the preceding problem but suitable for quadratic splines.

▮ The idea is to find a set of "basic" quadratic splines from which a general quadratic spline can be formed by superposition. A moment's thought recommends the model of Fig. 5-9, which is active on three intervals, is zero together with its first derivative at both end knots and has a continuous derivative at the interior knots. It will also prove possible to maintain symmetry if the intervals have equal length.

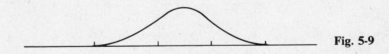

Fig. 5-9

The model just described can be implemented as follows for the case of equal intervals. First define

$$\phi(t) = \begin{cases} at^2 & \text{on } (0,1) \\ M - b\left(t - \frac{3}{2}\right)^2 & \text{on } (1,2) \\ a(3 - t)^2 & \text{on } (2,3) \\ 0 & \text{elsewhere} \end{cases}$$

For continuity at $t = 1$ of both $\phi(t)$ and its first derivative, we need $a = M - \frac{1}{4}b$ and $2a = b$, which together imply $M = \frac{3}{4}a$. This means that the value of a, the initial curvature, is open. The shape of $\phi(t)$ is determined but its scale is not. Choosing $a = \frac{1}{2}$, we have $b = 1$ and $M = \frac{3}{4}$.

For basis functions we now let $x = x_i + th$ and have the shifted versions of the $\phi(t)$ function

$$B_i(x) = \phi\left(\frac{x - x_i}{h}\right)$$

$$= \begin{cases} \dfrac{\frac{1}{2}(x - x_i)^2}{h^2} & \text{on } (x_i, x_{i+1}) \\ M - \left(\dfrac{x - x_i}{h} - \dfrac{3}{2}\right)^2 & \text{on } (x_{i+1}, x_{i+2}) \\ \dfrac{1}{2}\left(3 - \dfrac{x - x_i}{h}\right)^2 & \text{on } (x_{i+2}, x_{i+3}) \\ 0 & \text{elsewhere} \end{cases}$$

5.73 How can the basis functions of the preceding problem be used to represent a spline?

▌ Let the spline be formulated over (x_0, x_n) as

$$p(x) = \sum_{i=-2}^{n-1} c_i B_i(x)$$

observing that $n + 2$ coefficients are available to accommodate the values y_0 to y_n and the 1 degree of freedom that we have come to expect. The conditions $p(x_i) = y_i$ reduce to

$$\tfrac{1}{2} c_{i-2} + \tfrac{1}{2} c_{i-1} = y_i \quad \text{for } i = 0, 1, \ldots, n$$

since only the two basis functions B_{i-1} and B_{i-2} are nonzero at x_i, both equaling $\tfrac{1}{2}$. Suppose we choose $p_0'(x) = y_0'$ as the additional specification. This reduces to

$$B'_{-2}(x_0) c_{-2} + B'_{-1}(x_0) c_{-1} = \frac{1}{h} c_{-1} - \frac{1}{h} c_{-2} = y_0'$$

and the set of $n + 2$ equations for the various c_i can be expressed in the matrix form

$$\begin{bmatrix} -1 & 1 & & & & \\ 1 & 1 & & & & \\ & 1 & 1 & & & \\ & & 1 & 1 & & \\ & & & & \ddots & \\ & & & & 1 & 1 \end{bmatrix} \begin{bmatrix} c_{-2} \\ c_{-1} \\ c_0 \\ c_1 \\ \vdots \\ c_{n-1} \end{bmatrix} = \begin{bmatrix} hy_0' \\ 2y_0 \\ 2y_1 \\ 2y_2 \\ \vdots \\ 2y_n \end{bmatrix}$$

5.74 Apply the procedure of Problem 5.73 to this data, which correspond to the function $y(x) = x^3$:

x	0	1	2	3
y	0	1	8	27

▌ Since $h = 1$ and $y_0' = 0$ the linear system is

$$\begin{bmatrix} -1 & 1 & & & \\ 1 & 1 & & & \\ & 1 & 1 & & \\ & & 1 & 1 & \\ & & & 1 & 1 \end{bmatrix} \begin{bmatrix} c_{-2} \\ c_{-1} \\ c_0 \\ c_1 \\ c_2 \end{bmatrix} = \begin{bmatrix} 0 \\ 0 \\ 2 \\ 16 \\ 54 \end{bmatrix}$$

One sees at once that $c_{-2} = c_{-1} = 0$ after which c_0, c_1 and c_2 are found to be 2, 14 and 40. The spline we seek is

$$p(x) = 2B_0(x) + 14B_1(x) + 40B_2(x)$$

It may be helpful to check the specifications of this spline and to begin by noting the active parts of our basis functions on the three intervals. Since $B_{-2}(x)$ and $B_{-1}(x)$ have zero coefficients we have only the parts

	on $(0,1)$	on $(1,2)$	on $(2,3)$
$B_0(x)$	$\tfrac{1}{2} x^2$	$\tfrac{3}{4} - \left(x - \tfrac{3}{2}\right)^2$	$\tfrac{1}{2}(3 - x)^2$
$B_1(x)$	0	$\tfrac{1}{2}(x - 1)^2$	$\tfrac{3}{4} - \left(x - \tfrac{5}{2}\right)^2$
$B_2(x)$	0	0	$\tfrac{1}{2}(x - 2)^2$

At $x = 0$ only B_0 is active and the specifications $y_0 = y_0' = 0$ are met. At $x = 1$ either of the first two parts of B_0 can be used to produce the required 1 value. At $x = 2$,

$$2B_0(2) + 14B_1(2) = 1 + 7 = 8$$

while at $x = 3$

$$14B_1(3) + 40B_2(3) = 7 + 20 = 27$$

since the basis functions equal $\frac{1}{2}$ at the nodes.

5.75 Why are cubic splines the most commonly used?

❚ A principal reason for this is that with cubics the second derivative can be made continuous. Second derivatives have at least two important areas of application. They affect the curvature of trajectories and they are involved with the physical ideas of force and acceleration. A jump in curvature is likely to be noticed by a traveler and is caused by an acceleration of some sort. Any sudden change in force is also likely to be noticed. It is useful to have a continuous second derivative.

5.76 The cubic spline can be developed using the common form

$$p(x) = c_0 + c_1 x + c_2 x^2 + c_3 x^3$$

for each segment. Let $I = (a, b)$ be the interval in question and suppose it divided as usual into subintervals as

$$a = x_0, x_1, x_2, \ldots, x_n = b$$

The spline is to assume the values y_i at arguments x_i, with first and second derivatives continuous at the knots x_1 to x_{n-1}. By counting available coefficients and required specifications, confirm that 2 degrees of freedom are available. What are some common choices for the extra specifications?

❚ We will have n cubic segments, one for each of the subintervals. Each segment will have four coefficients for a total of $4n$. We now have four specifications at each knot, namely, the segment on each side must reach this point, and the first two derivatives have to agree. This comes to $4n - 4$ conditions. At the two endpoints we ask only for collocation, two more conditions, making a grand total of $4n - 2$. As anticipated, there are two free specifications at our disposal. Sometimes these are used to make the second derivatives zero at the endpoints, leading to what is called the natural spline. Alternatively one may ask the end segments to match the end derivative values of the given function, if these are known or can be approximated. A third option, reducing the specifications at knots x_1 and x_{n-1}, has also been used.

5.77 Let the subintervals of Problem 5.76 be called I_1 to I_n, so that $I_i = (x_{i-1}, x_i)$. Also define $h_i = x_i - x_{i-1}$, noting that the subintervals need not be of equal length. If $S_i(x)$ is the spline segment on I_i, show that

$$S_i''(x) = C_{i-1} \frac{x_i - x}{h_i} + C_i \frac{x - x_{i-1}}{h_i}$$

for constants C_i and $i = 1, \ldots, n$.

❚ On I_i the spline segment is cubic, so its first derivative will be quadratic and the second derivative linear. It remains to verify the continuity at each knot x_k for $k = 1, \ldots, n - 1$. The segment S_k touches this knot at its right end while S_{k+1} touches it at its left end. The required derivatives are thus

$$S_k''(x_k^-) = C_{k-1} \frac{x_k - x_k}{h_k} + C_k \frac{x_k - x_{k-1}}{h_k}$$

and

$$S_{k+1}''(x_k^+) = C_k \frac{x_{k+1} - x_k}{h_{k+1}} + C_{k+1} \frac{x_k - x_k}{h_{k+1}}$$

both of which reduce to C_k. Continuity is thus assured and we discover that the constants C_k are in fact the common values of spline second derivatives.

5.78 Integrate the result of the preceding problem twice to obtain the spline segments and then impose the requirement that segments pass through appropriate knots to determine the constants of integration.

❚ The two integrations manage

$$S_i(x) = C_{i-1}\frac{(x_i - x)^3}{6h_i} + C_i\frac{(x - x_{i-1})^3}{6h_i} + c_i(x_i - x) + d_i(x - x_{i-1})$$

the last two terms being the linear function introduced by the constants of integration. For collocation at the knots, we must have $S_i(x_{i-1}) = y_{i-1}$ and $S_i(x_i) = y_i$. These conditions fix c_i and d_i and lead to

$$S_i(x) = C_{i-1}\frac{(x_i - x)^3}{6h_i} + C_i\frac{(x - x_{i-1})^3}{6h_i} + \left(y_{i-1} - \frac{C_{i-1}h_i^2}{6}\right)\frac{x_i - x}{h_i} + \left(y_i - \frac{C_i h_i^2}{6}\right)\frac{x - x_{i-1}}{h_i}$$

as may be verified by inserting x_{i-1} and x_i.

5.79   It remains to ensure the continuity of the first derivatives. To arrange this, differentiate the result of the preceding problem and compare adjoining values as in Problem 5.77.

❚ Differentiating

$$S_i'(x) = -C_{i-1}\frac{(x_i - x)^2}{2h_i} + C_i\frac{(x - x_{i-1})^2}{2h_i} + \frac{y_i - y_{i-1}}{h_i} - \frac{C_i - C_{i-1}}{6}h_i$$

so the required derivatives at knot x_k are

$$S_k'(x_k^-) = \frac{h_k}{6}C_{k-1} + \frac{h_k}{3}C_k + \frac{y_k - y_{k-1}}{h_k}$$

and

$$S_{k+1}'(x_k^+) = -\frac{h_{k+1}}{3}C_k - \frac{h_{k+1}}{6}C_{k+1} + \frac{y_{k+1} - y_k}{h_{k+1}}$$

Since these are to be equal, we have, for $k = 1, \ldots, n-1$,

$$\frac{h_k}{6}C_{k-1} + \frac{h_k + h_{k+1}}{3}C_k + \frac{h_{k+1}}{6}C_{k+1} = \frac{y_{k+1} - y_k}{h_{k+1}} - \frac{y_k - y_{k-1}}{h_k}$$

which is a linear system of $n - 1$ equations for the constants C_0 to C_n. As observed earlier, the system is underdetermined. We are two equations short.

There is an interesting way to include two additional equations in the linear system, keeping our options open and preserving the general character of the matrix. First let

$$\alpha_i = \frac{h_{i+1}}{h_i + h_{i+1}}$$

$$\beta_i = 1 - \alpha_i = \frac{h_i}{h_i + h_{i+1}}$$

$$d_i = \frac{6}{h_i + h_{i+1}}\left(\frac{y_{i+1} - y_i}{h_{i+1}} - \frac{y_i - y_{i-1}}{h_i}\right)$$

for $i = 1, \ldots, n-1$. The system can then be rewritten, still for $i = 1, \ldots, n-1$, as

$$\beta_i C_{i-1} + 2C_i + \alpha_i C_{i+1} = d_i$$

Now take two additional conditions in the form

$$2C_0 + \alpha_0 C_1 = d_0 \qquad \beta_n C_{n-1} + 2C_n = d_n$$

with α_0, d_0, β_n and d_n at our disposal. The combined system then takes this shape:

$$\begin{bmatrix} 2 & \alpha_0 & 0 & & & & \\ \beta_1 & 2 & \alpha_1 & & & & \\ 0 & \beta_2 & 2 & & & & \\ & & & \ddots & & & \\ & & & & 2 & \alpha_{n-2} & 0 \\ & & & & \beta_{n-1} & 2 & \alpha_{n-1} \\ & & & & 0 & \beta_n & 2 \end{bmatrix} \begin{bmatrix} C_0 \\ C_1 \\ C_2 \\ \vdots \\ C_{n-2} \\ C_{n-1} \\ C_n \end{bmatrix} = \begin{bmatrix} d_0 \\ d_1 \\ d_2 \\ \vdots \\ d_{n-2} \\ d_{n-1} \\ d_n \end{bmatrix}$$

The coefficient matrix is triple diagonal, all other elements being zero.

5.80 How can the linear system of the preceding problem be used to find a natural spline?

❚ Choose α_0, d_0, β_n and d_n as zero. The top and bottom equations then force C_0 and C_n to be zero also and this is what identifies the natural spline. The system is reduced to order $n-1$ for determining the remaining C_1 to C_{n-1}.

5.81 Similarly, how can we arrange that the end conditions

$$S_1'(x_0) = y_0' \qquad S_n'(x_n) = y_n'$$

be met?

❚ Borrowing appropriate formulas from Problem 5.79, we have

$$S_1'(x_0^+) = -\frac{h_1}{3}C_0 - \frac{h_1}{6}C_1 + \frac{y_1 - y_0}{h_1} = y_0'$$

and

$$S_n'(x_n^-) = \frac{h_n}{6}C_{n-1} + \frac{h_n}{3}C_n + \frac{y_n - y_{n-1}}{h_n} = y_n'$$

which are easily converted to

$$2C_0 + C_1 = \frac{6}{h_1}\left(\frac{y_1 - y_0}{h_1} - y_0'\right)$$

and

$$C_{n-1} + 2C_n = \frac{6}{h_n}\left(y_n' - \frac{y_n - y_{n-1}}{h_n}\right)$$

Now comparing with the first and last equations of the linear system, namely $2C_0 + \alpha_0 C_1 = d_0$ and $\beta_n C_{n-1} + 2C_n = d_n$, suggests the choices

$$\alpha_0 = 1 = \beta_n \qquad d_0 = \frac{6}{h_1}\left(\frac{y_1 - y_0}{h_1} - y_0'\right) \qquad d_n = \frac{6}{h_n}\left(y_n' - \frac{y_n - y_{n-1}}{h_n}\right)$$

which will, in fact, provide the required end values.

5.82 Fit cubic spline segments to the function $f(x) = \sin x$ on the interval $(0, \pi)$. Use just the two interior points $\pi/3$ and $2\pi/3$.

❚ The corresponding data set is

x_i	0	$\pi/3$	$2\pi/3$	π
y_i	0	$\sqrt{3}/2$	$\sqrt{3}/2$	0

with $i = 0, \ldots, 3$ and all $h_i = \pi/3$. There are three cubic segments to find. The uniform h_i values at once make α_1, α_2, β_1 and β_2 all equal to $\frac{1}{2}$. Then

$$d_1 = \frac{3}{h}\left(\frac{0}{h} - \frac{\sqrt{3}/2}{h}\right) = -\frac{27\sqrt{3}}{2\pi^2}$$

with the same result for d_2. This leads us to the equations

$$\frac{1}{2}C_0 + 2C_1 + \frac{1}{2}C_2 = \frac{-27\sqrt{3}}{2\pi^2}$$

$$\frac{1}{2}C_1 + 2C_2 + \frac{1}{2}C_3 = \frac{-27\sqrt{3}}{2\pi^2}$$

and to the matter of end conditions. The natural spline is certainly appropriate here because the sine function does have zero second derivatives at the endpoints. So we set C_0 and C_3 to zero. The remaining system then quickly yields $C_1 = C_2 = -27\sqrt{3}/5\pi^2$. Substituting into the formulas of Problem 5.78 finally produces the spline segments, which after simplifications are

$$S_1(x) = \left(\frac{-27\sqrt{3}}{10\pi^3}\right)x^3 + \left(\frac{9\sqrt{3}}{5\pi}\right)x$$

$$S_2(x) = \left(\frac{-27\sqrt{3}}{10\pi^3}\right)\left(\frac{2\pi}{3} - x\right)^3 + \left(x - \frac{\pi}{3}\right)^3 + \frac{3\sqrt{3}}{5}$$

$$S_3(x) = \left(\frac{-27\sqrt{3}}{10\pi^3}\right)(\pi - 3)^3 + \left(\frac{9\sqrt{3}}{5\pi}\right)(\pi - x)$$

These three cubics can be verified by checking all the conditions imposed on them. The simplicity of the example has allowed exact values to be carried throughout. Notice also that the central "cubic" segment is actually quadratic.

5.83 Again fit cubic segments to the sine function, this time asking that endpoint first derivatives equal sine derivatives.

▌ The new endpoint conditions are $S_1'(0) = 1$ and $S_3'(\pi) = -1$. From Problem 5.81 we find

$$\alpha_0 = \beta_n = 1 \qquad d_0 = d_3 = \left(\frac{18}{\pi}\right)\left(\frac{3\sqrt{3}}{2\pi} - 1\right)$$

so the new linear system is

$$2C_0 + C_1 = \left(\frac{18}{\pi}\right)\left(\frac{3\sqrt{3}}{2\pi} - 1\right)$$

$$\frac{1}{2}C_0 + 2C_1 + \frac{1}{2}C_2 = \frac{-27\sqrt{3}}{2\pi^2}$$

$$\frac{1}{2}C_1 + 2C_2 + \frac{1}{2}C_3 = \frac{-27\sqrt{3}}{2\pi^2}$$

$$C_2 + 2C_3 = \left(\frac{18}{\pi}\right)\left(\frac{3\sqrt{3}}{2\pi} - 1\right)$$

and has the solution

$$C_0 = C_3 = \frac{18\sqrt{3}}{\pi^2} - \frac{10}{\pi}$$

$$C_1 = C_2 = \frac{2}{\pi} - \frac{9\sqrt{3}}{\pi^2}$$

Substituting into the $S_i(x)$ formulas of Problem 5.78, we again have the cubic segments. Verification that these segments meet all conditions imposed on them is direct, and it may also be found that the end values of $S''(x)$ are not zero.

5.84 A third way to obtain a well-determined system for spline approximation is to relax our requirements slightly. For example, omitting the segments $S_1(x)$ and $S_n(x)$, we can ask $S_2(x)$ and $S_{n-1}(x)$ to take care of the endpoint collocations. This also eliminates continuity requirements at x_1 and x_{n-1}, which are no longer knots. Show that the resulting problem will have just as many conditions to be met as coefficients available to meet them.

▮ There will now be $n - 2$ instead of n cubic segments, with $4n - 8$ coefficients available. But there will be only $n - 3$ rather than $n - 1$ knots. With four requirements per knot, this makes $4n - 12$ conditions to be satisfied. Since collocation is also required at x_0, x_1, x_{n-1} and x_n, the count of conditions climbs to $4n - 8$.

5.85 Modify the developments in Problem 5.77 to 5.74 to meet the requirements suggested in Problem 5.84.

▮ A careful rereading of the problems mentioned will show that a great deal can be saved. The center $n - 3$ equations of our linear system, as presented in Problem 5.74, are still valid because they refer to knots x_2 to x_{n-2} where no changes are being made. These already provide $n - 3$ equations for the $n - 1$ coefficients C_1 to C_{n-1}. The other two needed equations will make $S_2(x_0) = y_0$ and $S_{n-1}(x_n) = y_n$. Returning to the $S_i(x)$ formula given in Problem 5.73, these conditions can be implemented. After some algebraic manipulation they can be induced to take the form

$$2C_1 + \alpha_1 C_2 = d_1 \qquad \beta_{n-1} C_{n-2} + 2C_{n-1} = d_{n-1}$$

with the definitions

$$\alpha_1 = \frac{2(h_1 h_2^2 - h_1^3)}{(h_1 + h_2)^3 - (h_1 + h_2)h_2^2}$$

$$\beta_{n-1} = \frac{2(h_{n-1}^2 h_n - h_n^3)}{(h_{n-1} + h_n)^3 - (h_{n-1} + h_n)h_{n-1}^2}$$

$$d_1 = \left[12h_2\left(y_0 - \frac{h_1 + h_2}{h_2}y_1 + \frac{h_1}{h_2}y_2\right)\right] \Big/ \left[(h_1 + h_2)^3 - (h_1 + h_2)h_2^2\right]$$

$$d_{n-1} = \left[12h_{n-1}\left(y_n - \frac{h_{n-1} + h_n}{h_{n-1}}y_{n-1} + \frac{h_n}{h_{n-1}}y_{n-2}\right)\right] \Big/ \left[(h_{n-1} + h_n)^3 - (h_{n-1} + h_n)h_{n-1}^2\right]$$

The final form of the system is then

$$\begin{bmatrix} 2 & \alpha_1 & 0 & & & & \\ \beta_2 & 2 & \alpha_2 & & & & \\ 0 & \beta_3 & 2 & & & & \\ & & & \ddots & & & \\ & & & & 2 & \alpha_{n-3} & 0 \\ & & & & \beta_{n-2} & 2 & \alpha_{n-2} \\ & & & & 0 & \beta_{n-1} & 2 \end{bmatrix} \begin{bmatrix} C_1 \\ C_2 \\ C_3 \\ \vdots \\ C_{n-3} \\ C_{n-2} \\ C_{n-1} \end{bmatrix} = \begin{bmatrix} d_1 \\ d_2 \\ d_3 \\ \vdots \\ d_{n-3} \\ d_{n-2} \\ d_{n-1} \end{bmatrix}$$

again triple diagonal, all other elements being zero.

5.86 Apply the method just developed to $f(x) = \sin x$ on the interval $(0, \pi)$ using three equally spaced interior points.

▮ There are four subintervals, with spline segments to be found for the inner two. The one knot will be at $x_2 = \pi/2$. This makes it clear why we are not continuing the earlier example, which had one fewer interval. There would be no knots at all and a single cubic would interpolate the four given points. The present data set is

x_i	0	$\pi/4$	$\pi/2$	$3\pi/4$	π
y_i	0	$\sqrt{2}/2$	1	$\sqrt{2}/2$	0

with all $h_i = \pi/4$. The formulas for α_i and β_i now apply only at the knot x_2 and yield $\alpha_2 = \beta_2 = \frac{1}{2}$. We also find $d_2 = 48(\sqrt{2} - 2)/\pi^2$ and then the one equation

$$\frac{1}{2}C_1 + 2C_2 + \frac{1}{2}C_3 = \frac{48(\sqrt{2} - 2)}{\pi^2}$$

Turning to more recent formulas, $\alpha_1 = 0$, $\beta_3 = 0$ and

$$d_1 = d_3 = \frac{32(1 - \sqrt{2})}{\pi^2}$$

so our linear system is

$$2C_1 \qquad\qquad = d_1$$

$$\frac{1}{2}C_1 + 2C_2 + \frac{1}{2}C_3 = \left(\frac{3\sqrt{2}}{2}\right)d_1$$

$$2C_3 = d_1$$

Solving, and appealing again to Problem 5.78, we come to the two spline segments

$$S_2(x) = \frac{16(1-\sqrt{2})(\pi - 2x)^3 + (4\sqrt{2}-7)(4x-\pi)^3}{12\pi^3} + \frac{(8\sqrt{2}-2)(2\pi - 4x) + (19 - 4\sqrt{2})(4x-\pi)}{12\pi}$$

$$S_3(x) = \frac{16(1-\sqrt{2})(2x-\pi)^3 + (4\sqrt{2}-7)(3\pi-4x)^3}{12\pi^3} + \frac{(8\sqrt{2}-2)(4x-2\pi) + (19 - 4\sqrt{2})(3\pi-4x)}{12\pi}$$

With a little patience it can be verified that S_2 joins the first three points, S_3 the last three and that they make a proper knot at x_2. This is all that was required. Bonuses such as $S_2'(0) = 1$ or $S_2''(\pi/2) = -1$ would have been nice, but there is no point in being greedy. The approximations 1.05 and -1.09 will have to do.

5.87 What is the error of a cubic spline approximation?

▮ It can be shown that

$$\max|f(x) - S(x)| \le \tfrac{5}{384} \max|f^4(x)|H^4$$

where H is the largest of the h_i and the maxima are on the interval I.

5.88 Apply the error bound of Problem 5.87 to the spline of Problem 5.82.

▮ The fourth derivative of $\sin x$ is, of course, bounded by 1 and $H = \pi/3$. Thus

$$\max|\sin x - S(x)| \le \frac{5}{384}\frac{\pi^4}{81} = .016$$

5.89 How well does a cubic spline approximate the derivative $f'(x)$?

▮ It can be shown that

$$\max|f'(x) - S'(x)| \le \frac{\max|f^4(x)|H^3}{24}$$

5.90 Apply the formula of Problem 5.89 to the spline of Problem 5.87.

▮ We find $H^3/24 = .05$ approximately. Generally speaking, splines are quite good approximations to derivatives.

5.91 What is meant by saying that a spline is a global approximation to $f(x)$?

▮ The segments of the spline are not determined independently of each other. Each is linked with all the others. The set of coefficients C_i that identify the segments is determined by one linear system. By way of contrast, one could fit a cubic polynomial to the first four points, x_0 to x_3, then another to set x_3 to x_6 and so on across the interval I. Each segment would then be found independently of the others, but the continuity properties of the spline at knots would almost surely be absent.

5.92 Show that the natural spline on (a, b) uniquely minimizes

$$\int_a^b f''(x)^2\, dx$$

among all functions $f(x)$ that have continuous second derivatives and satisfy $f(x_i) = y_i$ at the knots.

▮ First note that

$$\int_a^b f''(x)^2 \, dx - \int_a^b S''(x)^2 \, dx = \int_a^b [f''(x) - S''(x)]^2 \, dx + 2\int_a^b S''(x)[f''(x) - S''(x)] \, dx$$

with $S(x)$ the cubic spline. Integration by parts over each subinterval converts the last integral as

$$\int_{x_{i-1}}^{x_i} S_i''(x)[f''(x) - S_i''(x)] \, dx = S_i''(x)[f'(x) - S_i'(x)]\big|_{x_{i-1}}^{x_i} - \int_{x_{i-1}}^{x_i} [f'(x) - S_i'(x)]S_i^{(3)}(x) \, dx$$

$$= S_i''(x)[f'(x) - S_i'(x)]\big|_{x_{i-1}}^{x_i} - S_i^{(3)}(x)[f(x) - S_i(x)]\big|_{x_{i-1}}^{x_i}$$

$$+ \int_{x_{i-1}}^{x_i} [f(x) - S_i(x)]S_i^{(4)}(x) \, dx$$

The last two terms vanish since $f(x) = S_i(x)$ at the knots and $S_i^{(4)}(x) = 0$. Summing what is left for $i = 1, \ldots, n$ there is cancellation of all interior values leaving

$$S''(b)[f'(b) - S'(b)] - S''(a)[f'(a) - S'(a)]$$

which also vanishes since S is the natural spline. Notice that this remnant would still vanish if we assumed instead that f' and S' agree at the endpoints. In either case, reordering the original equation just slightly,

$$\int_a^b S''(x)^2 \, dx = \int_a^b f''(x)^2 \, dx - \int_a^b [f''(x) - S''(x)]^2 \, dx$$

which does make the first integral smaller than the second.

5.93 Fit a cubic spline to the data

x_i	0	2	2.5	3	3.5	4	4.5	5	6
y_i	0	2.9	3.5	3.8	3.5	3.5	3.5	2.6	0

▮ Choosing the natural spline, the system of Problem 5.79 provides seven equations for the seven interior C_i. Their solution, rounded to two places, is

i	1	2	3	4	5	6	7
C_i	$-.23$	$-.72$	-4.08	2.65	$.69$	-5.40	$-.70$

A plot of the nine data points and the spline segments appears in Fig. 5-10. Recalling that the C_i are the second-derivative values at the data points, with C_0 and C_8 zero, it is reassuring to observe their behavior across the interval, particularly the large values more or less where expected.

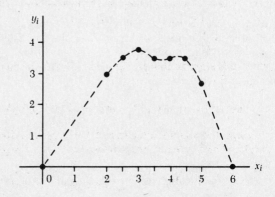

Fig. 5-10

5.94 Find the natural cubic spline for the data

x	0	1	2	3	4
y	0	0	1	0	0

▮ For the natural spline $C_0 = C_4 = 0$. With the equal spacing all the h_i are 1 making $\alpha_i = \beta_i = \frac{1}{2}$ for all i. We then find d_1, d_2, d_3 to be $3, -6, 3$. The linear system for the C_i becomes

$$2C_1 + \tfrac{1}{2}C_2 \qquad = 3$$
$$\tfrac{1}{2}C_1 + 2C_2 + \tfrac{1}{2}C_3 = -6$$
$$\tfrac{1}{2}C_2 + 2C_3 = 3$$

with solution $C_1 = C_3 = 18/7$, $C_2 = -30/7$. Substituting into the formula for the $S_i(x)$, we have our spline

$$S_1(x) = \tfrac{3}{7}(x^3 - x) \qquad\qquad\text{on } (0,1)$$
$$S_2(x) = \tfrac{3}{7}(2-x)^3 - \tfrac{5}{7}(x-1)^3 - \tfrac{3}{7}(2-x) + \tfrac{12}{7}(x-1) \qquad \text{on } (1,2)$$
$$S_3(x) = -\tfrac{5}{7}(3-x)^3 + \tfrac{3}{7}(x-2)^3 + \tfrac{12}{7}(3-x) - \tfrac{3}{7}(x-2) \qquad \text{on } (2,3)$$
$$S_4(x) = \tfrac{3}{7}\left[(4-x)^3 - (4-x)\right] \qquad\qquad \text{on } (3,4)$$

With a little patience all the specifications can be checked.

5.95 The case in which all data points fall on a straight line is hardly one that calls for a spline, but is worth a moment's attention. Recall that the constants C_i are values of the second derivative, so in this case must all be zero. How does our linear system manage this?

▮ The d_i will all be zero.

5.96 What happens to our linear system if all data points fall on a parabola?

▮ The d_i are six times the second divided difference of $y(x)$, which is a constant. All equations except the end conditions reduce to $3C = d_i$.

5.97 To develop a set of basis functions for cubic splines, first find a function $\phi(t)$ corresponding to the one found in Problem 5.72 for quadratic splines.

▮ We will seek a spline of the shape shown in Fig. 5-11, nonzero only between $t = -2$ and $t = 2$. For continuity of $\phi(x)$ and its first two derivatives at $t = -2$ all three of these values must be zero, forcing

$$\phi(t) = a(t+2)^3 \quad \text{on } (-2, -1)$$

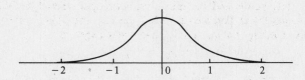

Fig. 5-11

with a arbitrary. At $t = -1$ this segment has $\phi(-1) = a$, $\phi'(-1) = 3a$ and $\phi''(-1) = 6a$, so the next segment must retain these values.

$$\phi(t) = a + 3a(t+1) + 3a(t+1)^2 + b(t+1)^3 \quad \text{on } (-1, 0)$$

Moving along to $t = 0$ we find the present cubic segment taking the values $\phi(0) = 7a + b$, $\phi'(0) = 9a + 3b$ and $\phi''(0) = 6a + 6b$. The two familiar degrees of freedom can now be exploited. Requiring symmetry at $t = 0$ we set the first derivative to zero,

$$9a + 3b = 0$$

To choose the scale of our function we ask that the values of ϕ at the three interior nodes sum to 1,

$$9a + b = 1$$

and discover that we have chosen $a = \frac{1}{6}$, $b = -\frac{1}{2}$. These values, together with the symmetry, now lead to

$$\phi(t) = \begin{cases} \frac{1}{6}(t+2)^3 & \text{on } (-2,-1) \\ \frac{1}{6} + \frac{1}{2}(t+1) + \frac{1}{2}(t+1)^2 - \frac{1}{2}(t+1)^3 & \text{on } (-1,0) \\ \frac{1}{6} + \frac{1}{2}(1-t) + \frac{1}{2}(1-t)^2 - \frac{1}{2}(1-t)^3 & \text{on } (0,1) \\ \frac{1}{6}(2-t)^3 & \text{on } (1,2) \end{cases}$$

with $\phi(t) = 0$ elsewhere.

5.98 Find the first and second derivatives of $\phi(t)$.

▮ Both will, of course, be zero outside the interval $(-2,2)$. Inside this interval the differentiations are elementary:

$$\phi'(t) = \begin{cases} \frac{1}{2}(t+2)^2 & \text{on } (-2,-1) \\ \frac{1}{2} + (t+1) - \frac{3}{2}(t+1)^2 & \text{on } (-1,0) \\ -\frac{1}{2} - (1-t) + \frac{3}{2}(1-t)^2 & \text{on } (0,1) \\ -\frac{1}{2}(2-t)^2 & \text{on } (1,2) \end{cases}$$

$$\phi''(t) = \begin{cases} t+2 & \text{on } (-2,-1) \\ 1 - 3(t+1) & \text{on } (-1,0) \\ 1 - 3(1-t) & \text{on } (0,1) \\ 2-t & \text{on } (1,2) \end{cases}$$

5.99 Define the basis functions.

▮ As earlier, it is simply a matter of changing variables. Let $x = x_i + th$, with h the equal spacing between the arguments x_i. (Basis functions can also be defined for unequal spacing.) Then define

$$B_i(x) = \phi((x - x_i)/h)$$

making each basis function a translation of $\phi(t)$ accompanied by a change of scale for the independent variable. Each $B_i(x)$ has the same shape and features as $\phi(t)$ itself.

An explicit formulation is

$$B_i(x) = \frac{1}{6h^3} \begin{cases} (x - x_{i-2})^3 & \text{on } (x_{i-2}, x_{i-1}) \\ h^3 + 3h^2(x - x_{i-1}) + 3h(x - x_{i-1})^2 - 3(x - x_{i-1})^3 & \text{on } (x_{i-1}, x_i) \\ h^3 + 3h^2(x_{i+1} - x) + 3h(x_{i+1} - x)^2 - 3(x_{i+1} - x)^3 & \text{on } (x_i, x_{i+1}) \\ (x_{i+2} - x)^3 & \text{on } (x_{i+1}, x_{i+2}) \end{cases}$$

with zero value elsewhere. It will be convenient to tabulate its values and those of its derivatives at the key knots. The derivatives of the parent function $\phi(t)$ are already in hand and $B_i'(x) = \phi'(t)/h$, $B_i''(x) = \phi''(t)/h$. This is all that is needed to complete Table 5.5.

TABLE 5.5

	x_{i-2}	x_{i-1}	x_i	x_{i+1}	x_{i+2}
$B_i(x)$	0	$1/6$	$2/3$	$1/6$	0
$B_i'(x)$	0	$1/2h$	0	$-1/2h$	0
$B_i''(x)$	0	$1/h^2$	$-2/h^2$	$1/h^2$	0

5.100   Develop the algorithm for representing a spline in terms of the basis functions just defined.

▮ The representation will have the form

$$S(x) = \sum_{i=-1}^{n+1} c_i B_i(x)$$

providing enough coefficients to accommodate the $n + 1$ data values and the two additional specifications at our disposal. Taking the data values first, at $x = x_i$ only three terms will be nonzero, as Table 5.5 shows. [Recall that outside the interval x_{i-2}, x_{i+2}, $B_i(x)$ is a solid zero.] This leads to

$$c_{i-1}B_{i-1}(x_i) + c_i B_i(x_i) + c_{i+1}B_{i+1}(x_i) = y_i$$

or, taking the B values from the table, for $i = 0, \ldots, n$,

$$c_{i-1} + 4c_i + c_{i+1} = 6y_i$$

Suppose the additional specifications set the derivative values at the endpoints x_0 and x_n. Since

$$S'(x_0) = \sum_{i=-1}^{n+1} c_i B_i'(x_0) = c_{-1}(-1/2h) + c_1(1/2h)$$

(see the table again for the last step), we have

$$-c_{-1} + c_1 = 2hy_0'$$

and by an almost identical effort,

$$-c_{n-1} + c_{n+1} = 2hy_n'$$

The full set of $n + 3$ equations for the $n + 3$ coefficients c_i can be expressed in a convenient matrix form.

$$
\begin{bmatrix}
-1 & 0 & 1 & & & & \\
1 & 4 & 1 & & & & \\
& 1 & 4 & 1 & & & \\
& & & \ddots & & & \\
& & & & 1 & 4 & 1 \\
& & & & -1 & 0 & 1
\end{bmatrix}
\begin{bmatrix}
c_{-1} \\ c_0 \\ c_1 \\ \vdots \\ c_n \\ c_{n+1}
\end{bmatrix}
=
\begin{bmatrix}
2hy_0' \\ 6y_0 \\ 6y_1 \\ \vdots \\ 6y_n \\ 2hy_n'
\end{bmatrix}
$$

The matrix can be put into a three diagonal form (tridiagonal) by row operations at top and bottom, but this is not necessary.

5.101   Apply the method just developed to the data function

x	0	1	2	3	4
y	0	0	1	0	0

used in Problem 5.94. Assume end derivatives of zero.

▮ The end derivatives are, of course, being estimated crudely from the available data, but will serve to illustrate the procedure. The linear system of the preceding problem becomes

$$
\begin{bmatrix}
-1 & 0 & 1 & & & & \\
1 & 4 & 1 & & & & \\
& 1 & 4 & 1 & & & \\
& & 1 & 4 & 1 & & \\
& & & 1 & 4 & 1 & \\
& & & & 1 & 4 & 1 \\
& & & & & -1 & 0 & 1
\end{bmatrix}
\begin{bmatrix}
c_{-1} \\ c_0 \\ c_1 \\ c_2 \\ c_3 \\ c_4 \\ c_5
\end{bmatrix}
=
\begin{bmatrix}
0 \\ 0 \\ 0 \\ 6 \\ 0 \\ 0 \\ 0
\end{bmatrix}
$$

and has the solution vector

$$(c_{-1}, c_0, c_1, c_2, c_3, c_4, c_5) = (-.5, .25, -.5, 1.75, -.5, .25, -.5)$$

The indicated spline is

$$\sum c_i B_i(x)$$

It may be helpful to compute a few values of this spline by hand using the following slightly extended table of $B_i(x)$ values:

x	$i-2$	$i-1.5$	$i-1$	$i-.5$	i	$i+.5$	$i+1$	$i+1.5$	$i+2$
$B_i(x)$	0	1/48	1/6	23/48	2/3	23/48	1/6	1/48	0

For example, at $x = .5$,

$$S(.5) = -\frac{1}{2} \cdot \frac{1}{48} + \frac{1}{4} \cdot \frac{23}{48} - \frac{1}{2} \cdot \frac{23}{48} + \frac{7}{4} \cdot \frac{1}{48} + 0 + 0 + 0 = -\frac{3}{32} = -.09375$$

while at $x = 1.5$,

$$S(1.5) = -\frac{1}{2} \cdot 0 + \frac{1}{4} \cdot \frac{1}{48} - \frac{1}{2} \cdot \frac{23}{48} + \frac{7}{4} \cdot \frac{23}{48} - \frac{1}{2} \cdot \frac{1}{48} + 0 + 0 = \frac{19}{32} = .59375$$

and then, of course,

$$S(2) = -\frac{1}{2} \cdot 0 + \frac{1}{4} \cdot 0 - \frac{1}{2} \cdot \frac{1}{6} + \frac{7}{4} \cdot \frac{2}{3} - \frac{1}{2} \cdot \frac{1}{6} + 0 + 0 = 1$$

at the center point.

5.102 Also apply the method of basis functions to the small scale sine approximation of Problem 5.82, again finding the natural spline.

■ For the natural spline we will need the second derivative

$$S''(x_0) = \sum c_i B_i''(x_0) = \frac{1}{h^2} c_{-1} - \frac{2}{h^2} c_0 + \frac{1}{h^2} c_1$$

with the last inequality following from Table 5.5 and the fact that only three basis functions will be active for any x_0. The natural spline will require this derivative to be zero. At the opposite end of the interval the similar condition

$$S''(x_n) = \frac{1}{h^2} c_{n-1} - \frac{2}{h^2} c_n + \frac{1}{h^2} c_{n+1} = 0$$

must also hold. These conditions replace the first derivative conditions found in Problem 5.100. The linear system is thus

$$\begin{vmatrix} 1 & -2 & 1 & & & \\ 1 & 4 & 1 & & & \\ & 1 & 4 & 1 & & \\ & & 1 & 4 & 1 & \\ & & & 1 & 4 & 1 \\ & & & & 1 & -2 & 1 \end{vmatrix} \begin{vmatrix} c_{-1} \\ c_0 \\ c_1 \\ c_2 \\ c_3 \\ c_4 \end{vmatrix} = \begin{vmatrix} 0 \\ 0 \\ 3\sqrt{3} \\ 3\sqrt{3} \\ 0 \\ 0 \end{vmatrix}$$

and using the figure 5.19615 for $3\sqrt{3}$ leads to the solution

$$(-c, 0, c, c, 0, -c)$$

where $c = 1.03923$. The indicated solution is

$$S(x) = 1.03923[-B_{-1}(x) + B_1(x) + B_2(x) - B_4(x)]$$

which receives a partial vote of confidence from the direct computation of S for the arguments

x	0	$\pi/3$	$\pi/2$	$2\pi/3$	π
$S(x)$	0	.866025	.996	.866025	0

all values correct to six places with one obvious exception.

5.103 Rework Problem 5.101 with the end derivatives no longer prescribed and the natural spline conditions substituted.

❚ The linear system appears as Figure 5-12 and differs from the earlier one only in the top and bottom rows. Solving the system we have the coefficients c_{-1} to c_5:

$$(.42857, 0, -.42857, 1.71429, -.42857, 0, .42857)$$

$$
\begin{vmatrix}
1 & -2 & 1 & 0 & 0 & 0 & 0 \\
1 & 4 & 1 & 0 & 0 & 0 & 0 \\
0 & 1 & 4 & 1 & 0 & 0 & 0 \\
0 & 0 & 1 & 4 & 1 & 0 & 0 \\
0 & 0 & 0 & 1 & 4 & 1 & 0 \\
0 & 0 & 0 & 0 & 1 & 4 & 1 \\
0 & 0 & 0 & 0 & 1 & -2 & 1
\end{vmatrix}
\begin{vmatrix}
c_{-1} \\ c_0 \\ c_1 \\ c_2 \\ c_3 \\ c_4 \\ c_5
\end{vmatrix}
=
\begin{vmatrix}
0 \\ 0 \\ 0 \\ 6 \\ 0 \\ 0 \\ 0
\end{vmatrix}
$$

Fig. 5-12

The required spline is thus

$$S(x) = \sum c_i B_i(x)$$

from which the following values may be computed:

x	0	.5	1	1.5	2
$S(x)$	0	$-.16$	0	.61	1

Values from $x = 2$ to $x = 4$ are symmetric.

5.104 Apply the method of cubic basis splines to the factorial polynomial function $y(x) = x^{(4)} = x(x-1)(x-2)(x-3)$. Use only the data points

x	0	1	2	3
y	0	0	0	0

and the end derivative values $y_0' = -6$, $y_3' = 6$.

❚ The question is, of course, just how much any method can extract from this exotic collection. Our system is

$$
\begin{vmatrix}
-1 & 0 & 1 & 0 & 0 & 0 \\
1 & 4 & 1 & 0 & 0 & 0 \\
0 & 1 & 4 & 1 & 0 & 0 \\
0 & 0 & 1 & 4 & 1 & 0 \\
0 & 0 & 0 & 1 & 4 & 1 \\
0 & 0 & 0 & -1 & 0 & 1
\end{vmatrix}
\begin{vmatrix}
c_{-1} \\ c_0 \\ c_1 \\ c_2 \\ c_3 \\ c_4
\end{vmatrix}
=
\begin{vmatrix}
-12 \\ 0 \\ 0 \\ 0 \\ 0 \\ 12
\end{vmatrix}
$$

with solution $(38, -10, 2, 2, -10, 38)/3$. Computing $\sum c_i B_i(x)$ we find the values

x	0	.5	1	1.5	2	2.5	3
$S(x)$	0	-1	0	.5	0	-1	0

The correct intermediate values are $-.94$ and .56 to two places.

5.105 Repeat the preceding problem but requiring end values of the second derivative to match those of the given function.

▋ We abandon the first derivatives and compute $y_0'' = y_3'' = 22$. The top and bottom rows of our linear system become

$$
\begin{array}{cccccccc}
1 & -2 & 1 & 0 & 0 & 0 & c_{-1} & 22 \\
0 & 0 & 0 & 1 & -2 & 1 & c_4 & 22
\end{array}
$$

with no changes in the middle four rows. The coefficients prove to be $(13.9333, -3.6777, .7333, .7333, -3.6777, 13.9333)$ and the spline $\sum c_i B_i(x)$ yields the values

x	0	.5	1	1.5	2	2.5	3
$S(x)$	0	-1.1	0	.55	0	-1.1	0

5.4 HERMITE CUBICS

5.106 What is meant by the term Hermite cubics?

▋ These are osculating polynomials as developed beginning with Problem 5.1. At each data point we ask that the polynomial take the given value and also that its derivative take a given value. For the cubic case there will be only two such data points.

The following results will be borrowed from those earlier problems. Let the data values be $(x_{i-1}, y_{i-1}, z_{i-1})$ and (x_i, y_i, z_i), with z denoting the derivative. Then with $h = x_i - x_{i-1}$,

$$L_{i-1}(x) = (x_i - x)/h \qquad L_i(x) = (x - x_{i-1})/h \qquad -L_{i-1}' = 1/h = L_i'$$

$$U_{i-1}(x) = \left[1 + \frac{2}{h}(x - x_{i-1})\right] \cdot L_{i-1}^2(x) \qquad U_i(x) = \left[1 - \frac{2}{h}(x - x_i)\right] \cdot L_i^2(x)$$

$$V_{i-1}(x) = (x - x_{i-1}) \cdot L_{i-1}^2(x) \qquad V_i(x) = (x - x_i) \cdot L_i^2(x)$$

$$p(x) = U_{i-1}(x) y_{i-1} + U_i(x) y_i + V_{i-1}(x) z_{i-1} + V_i(x) z_i$$

where $p(x)$ is the Hermite cubic osculating polynomial. The four terms correspond to indices $n = 0$ and $n = 1$ of the earlier sum.

5.107 Apply Hermite cubics to the data of Problem 5.82 for the sine function, now augmented with the corresponding derivative values

x	0	$\pi/3$	$2\pi/3$	π
y	0	$\sqrt{3}/2$	$\sqrt{3}/2$	0
z	1	1/2	1/2	-1

▋ The point is that a cubic will now be fitted to the appropriate interval, depending on the x argument of interest. The skeleton of an algorithm for this purpose follows:

(1) For $i = 0$ to n read the given x, y, z data.
(2) Read the x argument of interest.
(3) Locate the appropriate interval: For $i = 1$ to n, if $x \Leftarrow x_i$ then exit (the interval will be x_{i-1}, x_i).
(4) Compute $L_{i-1}, L_i, U_{i-1}, U_i, V_{i-1}, V_i$ and $p(x)$ from the formulas of Problem 5.106.
(5) Print x and $p(x)$.
(6) Return to step 2 if necessary.

For the present data the output generated was

x	0	$\pi/3$	$\pi/2$	$2\pi/3$	π
$p(x)$	0	.866025	.997	.866025	0

and is almost identical with that of the natural spline found in Problem 5.102.

5.108 Also apply Hermite cubics to the data of Problem 5.104 for the factorial polynomial $y(x) = x^{(4)}$, augmented with the required derivative values

x	0	1	2	3
y	0	0	0	0
z	-6	2	-2	6

The algorithm of the preceding problem was used to produce the values

x	0	.5	1	1.5	2	2.5	3
$p(x)$	0	-1	0	.5	0	-1	0

With the simple numbers involved it is easy to trace through the execution. For $x = 0$, .5 or 1 the search for the appropriate interval will terminate with $i = 1$, indicating that the interval is $(0, 1)$. With $h = 1$ the various Lagrange and Hermite coefficient functions become as listed in Table 5.6, along with their values for the arguments of interest. Since all y values are zero the U functions are not involved.

TABLE 5.6

	$x = 0$	$x = .5$	$x = 1$
$L_0(x) = 1 - x$	1	$1/2$	0
$L_1(x) = x - 0 = x$	0	$1/2$	1
$U_0(x) = [1 + 2x](1 - x)^2$	1	$1/2$	0
$U_1(x) = [1 - 2(x - 1)]x^2$	0	$1/2$	1
$V_0(x) = x(1 - x)^2$	0	$1/8$	0
$V_1(x) = (x - 1)x^2$	0	$-1/8$	0
$p(x) = -6V_0(x) + 2V_1(x)$	0	-1	0

5.109 Apply Hermite cubics to the data that correspond to the function $y(x) = 1/(1 + x^2)$:

x	-2	-1	0	1	2
y	.2	.5	1	.5	.2
z	.16	.5	0	$-.5$	$-.16$

▮ The algorithm produced the following values, with the true values listed for comparison:

x	-2	-1.5	-1	$-.5$	0
Hermite	.2	.3075	.5	.8125	1
True	.2	.3077	.5	.8000	1

5.110 Compare the spline and Hermite cubic representations of the familiar data

x	0	1	2	3	4
y	0	0	1	0	0

▮ The spline found in Problem 5.101 is plotted in Fig. 5-13(a) and the set of four Hermite cubics in Fig. 5-13(b). Both are, of course, obliged to pass through the five data points. Both have also been assigned zero derivatives at the endpoints, the Hermite cubics having zero slope at the interior points as well. The outer Hermite segments are flat, while each of the inner two segments runs between the two zero slope positions of a cubic curve. This means that there will be discontinuities in the curvature at $x = 1$ and $x = 3$. The spline, however, is obliged to preserve continuity of both the first and second derivatives, and to do this has to dip below the horizontal axis for its outer segments. This gives it a steeper start for the climb to the peak. (Note that in both of these graphs the vertical scale

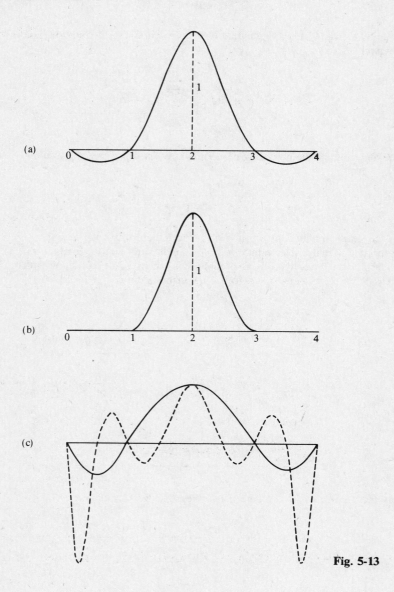

Fig. 5-13

has been doubled for emphasis.) For further comparisons, the polynomial found in Problem 3.144 is plotted in Fig. 5-13(c), the solid curve, along with a polynomial of degree 8 (shown dotted) found by adding data points midway between the original five. Clearly the second of these is quite uncomfortable with the given function, which, in fact, is not an easy function to represent in any of these forms. Each curve is a response to different specifications.

5.111 Compare various polynomial representations of the step function in Fig. 5-14.

▮ The basis splines, assuming end derivatives of zero and using the method of Problem 5.100, have the coefficients

$$(.03846154, -.01923077, .03846154, -.1346154, .5, 1.1346154, .9615384, 1.019231, .9615384)$$

A plot of this six piece spline appears in Fig. 5-14(a), the solid curve. Once again the need to maintain continuity of the first and second derivatives requires the curve to oscillate slightly (the size of these oscillations has been exaggerated for emphasis) at the two ends. The dotted curve is a 12 part spline. Over the flatter stretches it has values of only a few thousandths, but the jump is difficult for any polynomial to match. In Fig. 5-14(b) we have two

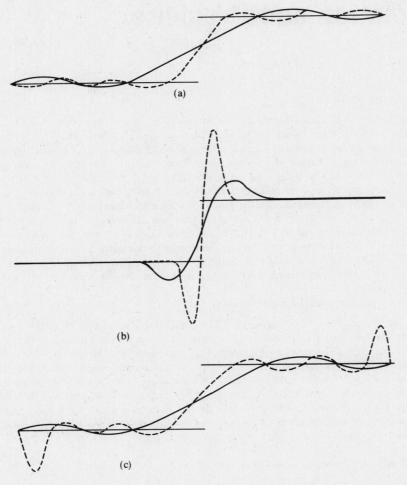

(a)

(b)

(c)

Fig. 5-14

sets of Hermite cubics. The solid curve corresponds to the specifications

x	0	1	2	3	4	5	6
y	0	0	0	.5	1	1	1
y'	0	0	0	5	0	0	0

The step function has an infinite slope at the central point. The artificial value of 5 causes the center pair of the six Hermite segments to overshoot in both directions. A larger value would increase the overshoot. The dotted curve consists of 12 Hermite cubics found by inserting new data points midway between the originals and using a central slope of 10. The increased overshoot is conspicuous. Figure 5-14(c) shows (solid curve) the polynomial of degree 6 fitted to the seven data points listed above, as well as the polynomial of degree 12 passing through the six midway points as well. The latter is showing signs of strain in its outermost parts. Unquestionably the step function is a hard one to handle with these devices.

CHAPTER 6
Interpolation and Prediction

6.1   What is interpolation?

▌ It is the process of estimating values of a function $y(x)$ for arguments between the data points $x_0, x_1, \ldots, x_n$ at which the values of the function are known to be $y_0, y_1, \ldots, y_n$. Inverse interpolation proceeds in the opposite direction. Subtabulation is the systematic interpolation of many values between each pair of data points, which reduces the spacing of a table of values perhaps from h to $h/10$. Prediction requires estimating a value of $y(x)$ outside the interval in which the data arguments fall.

All these operations were substantial computational hurdles before the arrival of high-speed computers, which explains the variety of formulas devised by our mathematical ancestors for dealing with them under a variety of circumstances. Many of the formulas of this chapter bear the names of prominent mathematicians of a century or more ago. Their place in our subject is now partly historical. It is interesting to see how the computational hurdles of an earlier time were surmounted. Taking the trouble to rearrange a formula without changing its substance may have been more important when all the arithmetic had to be done with more primitive equipment. But it is quite likely that more interpolations are made now than then, so the subject is far from extinct.

6.2   What methods are available for interpolation?

▌ The possibilities include all the methods of representing a set of data by polynomials as developed in preceding chapters. The principal idea is to have something that is easy to compute, provided it approximates $y(x)$ with sufficient accuracy. The formulas of Newton, Lagrange, Stirling, Hermite and Gauss, as well as those for splines, are all available. It was realized at an early stage that using data from both sides of the interpolation argument x made sense and led to better values or briefer computations. Many methods were motivated by this reasoning and a study of the errors involved provides logical support. At the ends of a table of data this could not be done and the original formulas of Newton had their turn. It was unnecessary to choose the degree of the approximating polynomial in advance, simply to continue fitting differences into their proper places so long as the results seemed to warrant.

6.3   What are the error sources in interpolation?

▌ It is customary to identify three main error sources.

(1) *Input errors* arise when the data values $y_0, \ldots, y_n$ are inexact, as experimental or computed values usually are.
(2) *Truncation error* is the difference between the given function $y(x)$ and the approximating function $p(x)$, which we accept the moment we decide to use a polynomial as an approximation. This error was found earlier to be

$$y(x) - p(x) = \frac{\pi(x)}{(n+1)!} y^{(n+1)}(\xi)$$

Though ξ is unknown, this formula can still be used at times to obtain error bounds. Truncation error is a type of algorithm error. In prediction problems this error can be substantial, since the factor $\pi(x)$ becomes very large outside the interval in which the data arguments x_0 and x_n fall.
(3) *Roundoff errors* occur since computers operate with a fixed number of digits and any excess digits produced in multiplications or divisions are lost. They are another type of algorithm error.

6.4   Predict the two missing values of y_k.

$k = x_k$	0	1	2	3	4	5	6	7
y_k	1	2	4	8	15	26		

▌ This is a simple example, but it will serve to remind us that the basis on which applications are to be made is polynomial approximation. Calculate some differences:

$$
\begin{array}{ccccccccc}
1 & & 2 & & 4 & & 7 & & 11 \\
& 1 & & 2 & & 3 & & 4 & \\
& & 1 & & 1 & & 1 & &
\end{array}
$$

Presumably the missing y_k values might be any numbers at all, but the evidence of these differences points strongly toward a polynomial of degree 3, suggesting that the six y_k values given and the two to be predicted all belong to such a polynomial. Accepting this as the basis for prediction, it is not even necessary to find this collocation polynomial. Adding two more 1s to the row of third differences, we quickly supply a 5 and 6 to the row of second differences, a 16 and 22 as new first differences and then predict $y_6 = 42$, $y_7 = 64$. This is the same data used in Problem 3.121 where the cubic collocation polynomial was found.

6.5 Values of $y(x) = \sqrt{x}$ are listed in Table 6.1, rounded off to four decimal places, for arguments $x = 1.00(.01)1.06$. (This means that the arguments run from 1.00 to 1.06 and are equally spaced with $h = .01$.) Calculate differences to Δ^6 and explain their significance.

▎ The differences are also listed in Table 6.1.

For simplicity, leading zeros are often omitted in recording differences. In this table all differences are in the fourth decimal place. Though the square root function is certainly not linear, the first differences are almost constant, suggesting that over the interval tabulated and to four-place accuracy this function may be accurately approximated by a linear polynomial. The entry Δ^2 is best considered a unit roundoff error, and its effect on higher differences follows the familiar binomial coefficient pattern observed in Problem 3.33. In this situation one would ordinarily calculate only the first differences. Many familiar functions such as $\sqrt{x}$, $\log x$, $\sin x$, etc., have been tabulated in this way, with arguments so tightly spaced that first differences are almost constant and the function can be accurately approximated by a linear polynomial.

TABLE 6.1

x	$y(x) = \sqrt{x}$	Δ	Δ^2	Δ^3	Δ^4	Δ^5	Δ^6
1.00	1.0000						
		50					
1.01	1.0050		0				
		50		−1			
1.02	1.0100		−1		2		
		49		1		−3	
1.03	1.0149		0		−1		4
		49		0		1	
1.04	1.0198		0		0		
		49		0			
1.05	1.0247		0				
		49					
1.06	1.0296						

6.6 Apply Newton's forward formula with $n = 1$ to interpolate for $\sqrt{1.005}$.

▎ Newton's formula reads

$$p_k = y_0 + \binom{k}{1} \Delta y_0 + \binom{k}{2} \Delta^2 y_0 + \cdots + \binom{k}{n} \Delta^n y_0$$

Choosing $n = 1$ for a linear approximation we find, with $k = (x - x_0)/h = (1.005 - 1.00)/.01 = \frac{1}{2}$,

$$p_k = 1.0000 + \tfrac{1}{2}(.0050) = 1.0025$$

This is hardly a surprise. Since we have used a linear collocation polynomial, matching our $y = \sqrt{x}$ values at arguments 1.00 and 1.01, we could surely have anticipated this midway result.

6.7 What would be the effect of using a higher-degree polynomial for the interpolation of Problem 6.6?

▎ An easy computation shows the next several terms of the Newton formula, beginning with the second difference term, to be approximately .00001. They would not affect our result at all.

6.8 Values of $y(x) = \sqrt{x}$ are listed in Table 6.2, rounded off to five decimal places, for arguments $x = 1.00(.05)1.30$. Calculate differences to Δ^6 and explain their significance.

▮ The differences are listed in the table. Here the error pattern is more confused by the fluctuations of plus and minus signs in the last three columns are reminiscent of the effects produced in Problems 3.33 and 3.34. It may be best to view these three columns as error effects, not as useful information for computing the square root function.

TABLE 6.2

x	$y(x) = \sqrt{x}$	Δ	Δ^2	Δ^3	Δ^4	Δ^5	Δ^6
1.00	1.00000						
		2470					
1.05	1.02470		-59				
		2411		5			
1.10	1.04881		-54		-1		
		2357		4		-1	
1.15	1.07238		-50		-2		4
		2307		2		3	
1.20	1.09544		-48		1		
		2259		3			
1.25	1.11803		-45				
		2214					
1.30	1.14017						

6.9 Use the data of Problem 6.8 to interpolate for $\sqrt{1.01}$.

▮ Newton's forward formula is convenient for interpolations near the top of a table. With $k = 0$ at the top entry $x_0 = 1.00$, this choice usually leads to diminishing terms and makes the decision of how many terms to use almost automatic. Substituting into the formula as displayed in Problem 6.6, with $k = (x - x_0)/h = (1.01 - 1.00)/.05 = \frac{1}{5}$, we find

$$p_k = 1.00000 + \tfrac{1}{5}(.02470) - \tfrac{2}{25}(-.00059) + \tfrac{6}{125}(.00005)$$

stopping with this term since it will not affect the fifth decimal place. Notice that this last term uses the highest-order difference which we felt, in Problem 6.8, to be significant for square root computations. We have not trespassed into columns which were presumably only error effects. The value p_k reduces to

$$p_k = 1.000000 + .004940 + .000048 + .000002 = 1.00499$$

which is correct to five places. (It is a good idea to carry an extra decimal place during computations, if possible, to control algorithm errors described in Chapter 2. In machine computations, of course, the number of digits is fixed anyway, so this remark would not apply.)

6.10 Use the data of Problem 6.8 to interpolate for $\sqrt{1.28}$.

▮ Here Newton's backward formula is convenient and most of the remarks made in Problem 6.9 again apply. With $k = 0$ at the bottom entry $x_0 = 1.30$, we have $k = (x - x_0)/h = (1.28 - 1.30)/.05 = -\frac{2}{5}$. Substituting into the backward formula (Problem 4.15),

$$p_k = y_0 + k \nabla y_0 = \frac{k(k+1)}{2} \nabla^2 y_0 + \frac{k(k+1)(k+2)}{3!} \nabla^3 y_0 + \cdots + \frac{k(k+1)\cdots(k+n-1)}{n!} \nabla^n y_0$$

we obtain

$$p_k = 1.14017 + \left(-\tfrac{2}{5}\right)(.02214) + \left(-\tfrac{3}{25}\right)(-.00045) + \left(-\tfrac{8}{125}\right)(.00003)$$
$$= 1.140170 - .008856 + .000054 - .000002 = 1.13137$$

which is correct to five places.

6.11 The previous two problems have treated special cases of the interpolation problem, working near the top or near the bottom of a table. This problem is more typical in that data will be available on both sides of the point of interpolation. Interpolate for $\sqrt{1.12}$ using the data of Problem 6.8.

❚ The central difference formulas are now convenient since they make it easy to use data more or less equally from both sides. In Problem 6.22 we will see that this also tends to keep the truncation error small. Everett's formula will be used:

$$p_k = \binom{k}{1} y_1 + \binom{k+1}{3} \delta^2 y_1 + \binom{k+2}{5} \delta^4 y_1 + \cdots - \binom{k-1}{1} y_0 - \binom{k}{3} \delta^2 y_0 - \binom{k+1}{5} \delta^4 y_0 + \cdots$$

where higher-order terms have been omitted since we will not need them in this problem. Choosing $k = 0$ at $x_0 = 1.10$, we have $k = (x - x_0)/h = (1.12 - 1.10)/.05 = \frac{2}{5}$. Substituting into Everett's formula,

$$p_k = \left(\frac{2}{5}\right)(1.07238) + \left(-\frac{7}{125}\right)(-.00050) + \left(\frac{168}{5^6}\right)(-.00002)$$

$$-\left(-\frac{3}{5}\right)(1.04881) - \left(\frac{8}{125}\right)(-.00054) - \left(-\frac{182}{5^6}\right)(-.00001)$$

$$= .428952 + .000028 + .629286 + .000035$$

the two highest-order terms contributing nothing (as we hoped, since these are drawn from the error effects columns). Finally $p_k = 1.05830$, which is correct to five places. Notice that the three interpolations made in Table 6.2 have all been based on collocation polynomials of degree 3.

6.12 The laboratory's newest employee has been asked to "look up" the value $y(.3333)$ in table NBS-AMS 52 of the National Bureau of Standards Applied Mathematics Series. On the appropriate page of this extensive volume abundant information is found, a small part of which is reproduced in Table 6.3. Apply Everett's formula for the needed interpolation.

❚ Choosing $x = 0$ and $x_0 = .33$, we have $k = (x - x_0)/h = (.3333 - .33)/.01 = .33$. Writing Everett's formula through second differences in the form

$$p_k = k y_1 + (1 - k) y_0 + E_1 \delta^2 y_1 - E_0 \delta^2 y_0$$

where $E_1 = \binom{k+1}{3}$ and $E_0 = \binom{k}{3}$, the interpolator will find all ingredients available in tables. For $k = .33$, we find $E_1 = -.0490105$ and $E_0 = .0615395$. Then

$$p_k = (.33)(.13545218) + (.67)(3.1105979) + (-.0490145)(.00002349) - (.0615395)(.00002365)$$

$$= .13250667$$

❚ This table was prepared with Everett's formula in mind.

TABLE 6.3

x	$y(x)$	δ^2
.31	.12234609	2392
.32	.12669105	2378
.33	.13105979	2365
.34	.13545218	2349
.35	.13986806	2335

6.13 Use the following extract from the table of the arctangent function, NBS-AMS 26, to obtain arctan 2.682413 to an exotic 12 decimal places:

x	arctan x	δ^2
2.682	1.213906583322	-79909
2.683	1.214028596946	-79833

❚ We quickly find $k = .413$, look up $E_1 = .0641230$, $E_0 = -.0570925$ and compute

$$E_1 \delta^2 y_1 - E_0 \delta^2 y_0 = .0000000096819$$

leading to $p_k = 1.213956984631$.

6.14 The problem of *inverse interpolation* reverses the roles of x_k and y_k. We may view the y_k numbers as arguments and the x_k as values. Clearly the new arguments are not usually equally spaced. Given that $\sqrt{x} = 1.05$, use the data of Table 6.2 to find x.

▌ Since we could easily find $x = (1.05)^2 = 1.1025$ by a simple multiplication, this is plainly another test case of our available algorithms. Since it applies to unequally spaced arguments, suppose we use Lagrange's formula. Interchanging the roles of x and y,

$$p = \frac{(y - y_1)(y - y_2)(y - y_3)}{(y_0 - y_1)(y_0 - y_2)(y_0 - y_3)} x_0 + \frac{(y - y_0)(y - y_2)(y - y_3)}{(y_1 - y_0)(y_1 - y_2)(y_1 - y_3)} x_1$$

$$+ \frac{(y - y_0)(y - y_1)(y - y_3)}{(y_2 - y_0)(y_2 - y_1)(y_2 - y_3)} x_2 + \frac{(y - y_0)(y - y_1)(y - y_2)}{(y_3 - y_0)(y_3 - y_1)(y_3 - y_2)} x_3$$

With the same four x_k, y_k pairs used in Problem 6.13, this becomes

$$p = (-.014882)1.05 + (.97095)1.10 + (.052790)1.15 + (-.008858)1.20 = 1.1025$$

as expected.

6.15 Apply Everett's formula to the inverse interpolation problem just solved.

▌ Since the Everett formula requires equally spaced arguments, we return x and y to their original roles. Writing Everett's formula as

$$1.05 = k(1.07238) + \binom{k+1}{3}(-.00050) + \binom{k+2}{5}(-.00002)$$

$$+ (1 - k)(1.04881) - \binom{k}{3}(-.00054) - \binom{k+1}{5}(-.00001)$$

we have a fifth-degree polynomial equation in k. This is a problem treated extensively in a later chapter. Here a sample, iterative procedure can be used. First neglect all differences and obtain a first approximation by solving

$$1.05 = k(1.07238) + (1 - k)(1.04881)$$

The result of this linear inverse interpolation is $k = .0505$. Insert this value into the δ^2 terms, still neglecting the δ^4 terms, and obtain a new approximation from

$$1.05 = k(1.07238) + \binom{1.0505}{3}(-.00050) + (1 - k)(1.04881) - \binom{.0505}{3}(.00054)$$

This proves to be $k = .0501$. Inserting this value into both the δ^2 and δ^4 terms then produces $k = .0500$. Reintroduced into the δ^2 and δ^4 terms this last value of k reproduces itself, so we stop. The corresponding value of x is 1.1025 to four places.

6.16 Interpolate for $\sqrt{1.125}$ and $\sqrt{1.175}$ in Table 6.2.

▌ For these arguments, which are midway between tabulated arguments, Bessel's formula has a strong appeal. First choose $k = 0$ at $x_0 = 1.10$, making $k = (1.125 - 1.10)/.05 = \frac{1}{2}$. The Bessel formula (Problem 4.49) is

$$p_k = \mu y_{1/2} + \binom{k}{2}\mu \delta^2 y_{1/2} + \binom{k+1}{4}\mu \delta^4 y_{1/2}$$

if we stop at degree 4. The odd difference terms disappear entirely because of the factor $k - \frac{1}{2}$. Substituting

$$p_k = 1.06060 + \left(-\tfrac{1}{8}\right)(-.00052) + \left(\tfrac{3}{128}\right)(-.000015) = 1.06066$$

with the δ^4 term again making no contributions. Similarly in the second case, with $k = 0$ now at $x_0 = 1.15$, we again have $k = \frac{1}{2}$ and find $p_k = 1.08397$. By finding all such midway values, the size of a table may be doubled. This is a special case of the problem of *subtabulation*.

6.17 Apply the Lagrange formula to obtain $\sqrt{1.12}$ from the data of Table 6.2.

❚ The Lagrange formula does not require equally spaced arguments. It can of course be applied to such arguments as a special case, but there are difficulties. The degree of the collocation polynomial must be chosen at the outset. With the Newton, Everett or other difference formulas the degree can be determined by computing terms until they no longer appear significant. Each term is an additive correction to terms already accumulated. But with the Lagrange formula a change of degree involves a completely new computation of all terms. In Table 6.2 the evidence is strong that a third-degree polynomial is suitable. On this basis we may proceed to choose $x_0 = 1.05, \ldots, x_3 = 1.20$ and substitute into

$$p = \frac{(x - x_1)(x - x_2)(x - x_3)}{(x_0 - x_1)(x_0 - x_2)(x_0 - x_3)} y_0 + \frac{(x - x_0)(x - x_2)(x - x_3)}{(x_1 - x_0)(x_1 - x_2)(x_1 - x_3)} y_1$$
$$+ \frac{(x - x_0)(x - x_1)(x - x_3)}{(x_2 - x_0)(x_2 - x_1)(x_2 - x_3)} y_2 + \frac{(x - x_0)(x - x_1)(x - x_2)}{(x_3 - x_0)(x_3 - x_1)(x_3 - x_2)} y_3$$

to produce

$$p = \frac{-8}{125}(1.02470) + \frac{84}{125}(1.04881) + \frac{56}{125}(1.07238) + \frac{-7}{125}(1.09554) = 1.05830$$

This agrees with the result of Problem 6.11.

6.18 Apply the Lagrange formula to interpolate for the $y(1.50)$ using some of the following values of the normal error function, $y(x) = e^{-x^2/2}/\sqrt{2\pi}$:

x_k	1.00	1.20	1.40	1.60	1.80	2.00
y_k	.2420	.1942	.1497	.1109	.0790	.0540

❚ An algorithm for computing the Lagrange formula might run as follows:

Initialize: set the value of n; read the data pairs $x(0)$, $y(0)$ to $x(n)$, $y(n)$; read the interpolation argument x; set the polynomial value to $p = 0$.
For $i = 0$ to n
$\quad l = 1$
$\quad$ for $j = 0$ to $i - 1$ and $j = i + 1$ to n
$\quad\quad l = l * (x - x(j))/(x(i) - x(j))$
$\quad p = p + l * y(i)$.

For the given data the algorithm generates the values

$$n = 1 \quad \text{(center pair of points)} \quad p = .1303$$
$$n = 3 \quad \text{(center four points)} \quad p = .1295$$
$$n = 5 \quad \text{(all six points)} \quad p = .1295$$

The correct value is .1295.

6.19 Apply Bessel's formula to the same data.

❚ This formula was derived in Problem 4.49 and is especially useful at midway points since the terms with factor $(k - \frac{1}{2})$ are then zero. For example

$$p(1.5) = \frac{1}{2}(.1497 + .1109) + 0 - \frac{1}{8}(.0063) + 0 - \frac{3}{128}(-.0012) + 0$$

which is .1295 as before. Notice that the attempt to calculate $p(1.3)$ in the same way,

$$p(1.3) = .1942 + 0 - \frac{1}{8}(.0045) + 0$$

is abbreviated at this point due to missing data. The terms in hand do, however, yield the correct .1714. It seems that cubic representation of the data is adequate for four place work. In the same way we find

$$p(1.7) = .09495 - .00086 = .0941$$

which is also correct to four places.

6.20 How accurately can the argument x corresponding to $y = .1300$ be estimated from the data of the preceding problem?

▮ Reversing the roles of x and y and choosing $y = .1300$, the algorithm produces $x = .1.497465$. Taking a primitive approach we again reverse roles and after a moment's search find the pairs

x	1.497	1.498
y	.1301	.1301

suggesting a continued search. One more such step finds all arguments from 1.4971 to 1.4976 managing .1300 to four places.

6.21 The Aitken procedure has an advantage over Lagrange's formula. Like the difference formulas, it gives an indication of what degree polynomial to choose. Apply this method to find $\sqrt{1.12}$.

▮ Proceeding as described in Problem 4.71 we obtain the results in Table 6.4.

The entries on the upper diagonal serve as successive approximations to the result, so that we may stop when we have the accuracy anticipated. Here the value 1.05830 once again appears.

TABLE 6.4

x	y				
1.05	1.02470				−.07
1.10	1.04881	1.05845			−.02
1.15	1.07238	1.05808	1.05830		.03
1.20	1.09544	1.05771	1.05830	1.05830	.08

6.22 The problem of *subtabulation* can be approached by means of a new difference operator Δ_α associated with an interval αh, where h is the spacing of the given table. (Often $\alpha = 1/10$.) Define $\Delta_\alpha = E^\alpha - 1$ and then show that

$$\Delta_\alpha = \alpha\Delta + \frac{\alpha(\alpha - 1)}{2}\Delta^2 + \frac{\alpha(\alpha - 1)(\alpha - 2)}{6}\Delta^3 + \cdots$$

Also compute Δ_α^i as a series operator in powers of Δ.

▮ Since $\Delta_\alpha = E^\alpha - 1 = (1 + \Delta)^\alpha - 1$, the result for Δ_α follows quickly by the binomial theorem. Factoring out $\alpha\Delta$, the binomial theorem may again be applied. For example,

$$\Delta_\alpha^2 = \alpha^2\Delta^2\left[1 + (\alpha - 1)\Delta + \frac{(\alpha - 1)(5\alpha - 7)}{6}\Delta^2 + \cdots\right]$$

$$\Delta_\alpha^3 = \alpha^3\Delta^3\left[1 + \frac{3(\alpha - 1)}{2}\Delta + \cdots\right]$$

$$\Delta_\alpha^4 = \alpha^4\Delta^4[1 + \cdots]$$

only terms through fourth differences being explicitly shown. As usual, the validity of these series operators remains uncertain and results obtained from them must be inspected with care.

6.23 Apply Problem 6.22 to subtabulate Table 6.2 for the arguments $x = 1.00(.01)1.05$.

▮ Apply the operators Δ_α and Δ to $y_0 = 1.00000$, with $\alpha = \frac{1}{5}$. We find, stopping at cubic terms,

$$\Delta_\alpha y_0 = \tfrac{1}{5}(.02470) + \tfrac{-2}{25}(-.00059) + \tfrac{6}{125}(.00005) = .00499$$

$$\Delta_\alpha^2 y_0 = (.04)[(-.00059) - (.9)(.00005)] = -.000024$$

$$\Delta_\alpha^3 y_0 = (.008)[.00005] = .0000004$$

We now have the layout shown in Table 6.5, higher differences being zero.

The second difference column may be filled with -2.4 entries, after which first differences may be obtained, with the y_k values following. The completed result is Table 6.6

The fact that $y(1.05)$ is incorrect by one unit in the last place shows that in subtabulation work it is preferable to have the entries in the master table computed to one extra decimal place beyond what is ultimately required of the completed table.

TABLE 6.5

x_k	y_k	Δ_α	Δ_α^2
1.00	1.00000		
		499	
1.01			-2.4
1.02			
1.03			
1.04			
1.05			

TABLE 6.6

1.00	1.00000		
		499	
1.01	1.00499		-2.4
		497	
1.02	1.00996		-2.4
		494	
1.03	1.01490		-2.4
		492	
1.04	1.01982		-2.4
		489	
1.05	1.02471		

6.24 In using a collocation polynomial $p(x)$ to compute approximations to a function $y(x)$, we accept what is called a *truncation error*, $y(x) - p(x)$. Estimate this error for our interpolations in Table 6.1.

▮ The formula for truncation error of a collocation polynomial was derived in Problem 3.9 and is

$$y(x) - p(x) = \frac{\pi(x)}{(n+1)!} y^{(n+1)}(\xi)$$

when the polynomial approximation is of degree n. For Table 6.1 we found $n = 1$ suitable. The collocation points may be called x_0 and x_1, leading to this error estimate for linear interpolation:

$$y(x) - p(x) = \frac{(x - x_0)(x - x_1)}{2} y^{(2)}(\xi) = \frac{k(k-1)}{2} h^2 y^{(2)}(\xi)$$

Since $h = .01$ and $y^{(2)}(x) = -\frac{1}{4}x^{-3/2}$, we have

$$|y(x) - p(x)| \le \frac{k(k-1)}{8}(.0001)$$

For k between 0 and 1, which we arrange for any interpolation by our choice of x_0, the quadratic $k(k-1)$ has a maximum size of $\frac{1}{4}$ at the midpoint $k = \frac{1}{2}$ (see Fig. 6-1). This allows us to complete our truncation error estimate,

$$|y(x) - p(x)| \le \frac{1}{32}(.0001)$$

and we discover that it cannot affect the fourth decimal place. Table 6.1 was prepared with linear interpolation in mind. The interval $h = .01$ was chosen to keep truncation error this small.

Fig. 6-1

6.25 Estimate truncation errors for our computations in Table 6.2.

▐ Here for the most part we used Everett's formula for a cubic polynomial. For other cubic formulas the same error estimate follows. Assuming equally spaced collocation arguments x_{-1}, x_0, x_1 and x_2,

$$y(x) - p(x) = \frac{(x - x_{-1})(x - x_0)(x - x_1)(x - x_2)}{4!}y^{(4)}(\xi)$$

$$= \frac{(k+1)k(k-1)(k-2)h^4 y^{(4)}\xi}{24}$$

The polynomial $(k+1)k(k-1)(k-2)$ has the general shape of Fig. 6-2. Outside the interval $-1 < k < 2$ it climbs sensationally. Inside $0 < k < 1$ it does not exceed $\frac{9}{16}$ and this is the appropriate part for interpolation. We now have, for the maximum error in cubic interpolation,

$$|y(x) - p(x)| \le \frac{9}{16} \cdot \frac{1}{24}h^4 |y^{(4)}(\xi)| = \frac{3}{128}h^4 |y^{(4)}(\xi)|$$

For this example $h = .05$ and $y^{(4)}(x) = -\frac{15}{16}x^{-7/2}$, and hence $|y(x) - p(x)| \le \frac{1}{64}(.00005)$ so that truncation error has not affected our five-decimal calculations.

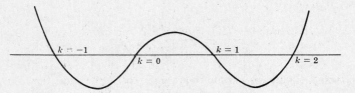

Fig. 6-2

6.26 How large could the interval length h be made in a table of $\sqrt{x}$ with a cubic formula still giving five-place accuracy? (Assume $1 \le x$.)

▐ This sort of question is naturally of interest to table makers. Our truncation error formula can be written as

$$|y(x) - p(x)| \le \left(\frac{9}{16}\right)h^4\left(\frac{15}{16}\right)\left(\frac{1}{24}\right)$$

To keep this less than .000005 requires $h^4 < .000228$, or very closely $h < \frac{1}{8}$. This is somewhat larger than the $h = .05$ used in Table 6.1, but other errors enter our computations and it pays to be on the safe side.

6.27 The previous problem suggests that Table 6.2 may be abbreviated to half length if Everett's cubic polynomial is to be used for interpolations. Find the second differences needed in this Everett formula.

▮ The result is Table 6.7, in which first differences may be ignored.

TABLE 6.7

x_k	y_k	δ	δ^2
1.00	1.00000		
		4881	
1.10	1.04881		-217
		4664	
1.20	1.09544		-191
		4473	
1.30	1.14017		

6.28 Use Table 6.7 to interpolate for $y(1.15)$.

▮ With Everett's formula and $k = \frac{1}{2}$,

$$p_k = \tfrac{1}{2}(1.09544) - \tfrac{1}{16}(-.00191) + \tfrac{1}{2}(1.04881) - \tfrac{1}{16}(-.00217) = 1.07238$$

as listed in Table 6.2. This confirms Problem 6.26 in this instance.

6.29 In a table of the function $y(x) = \sin x$ to four decimal places, what is the largest interval h consistent with linear interpolation?

▮ Borrowing our work of Problem 6.24 and keeping truncation error below .00005, we will need

$$\tfrac{1}{8}h^2 < .00005$$

leading to the estimate $h < .02$.

6.30 Rework the preceding problem but require five decimal place accuracy.

▮ We now need $\tfrac{1}{8}h^2 < .000005$, which will hold if $h < .006$. A standard table of the sine function to five places uses the interval $h = .01$. Our approach here has been conservative.

6.31 If Everett's cubic formula were used for interpolations, rather than a linear polynomial, how large an interval h could be chosen for a four place table of the sine function? For a five place table?

▮ Borrowing our result of Problem 6.25, we will need, for the four place table

$$\tfrac{9}{16}\tfrac{1}{24}h^4 < .00005$$

leading to $h < .22$. For a five place table we, of course, put .000005 instead of .00005 and emerge with $h < .12$. These intervals corresponding to about 12° and 7°.

6.32 In quadratic approximation with Newton's formula, the function $k(k-1)(k-2)$ appears in the truncation error estimate. Plot this function and find its maximum absolute value on the interval $(0,2)$.

▮ The derivative of this cubic is $3k^2 - 6k + 2$ and becomes zero at the points $k = 1 \pm \sqrt{3}/3$. Evaluation of the cubic at these points produces the absolute value $2\sqrt{3}/9$. The plot is given in Fig. 6-3.

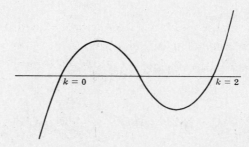

Fig. 6-3

6.33   e function $k(k^2 - 1)(k^2 - 4)$ appears in the truncation error estimate for Stirling's formula. Diagram this for $-2 < k < 2$ and estimate its maximum absolute value for $-\frac{1}{4} < k < \frac{1}{4}$, which is the interval to which use of this formula is usually limited.

▮ The graph appears as Fig. 6-4. The maximum absolute value for $-\frac{1}{4} < k < \frac{1}{4}$ is about .92, which we may conservatively call 1.

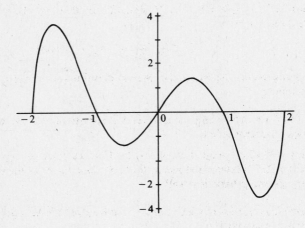

Fig. 6-4

6.34   Show that the relative maxima and minima of the polynomials

$$k(k^2 - 1)(k^2 - 4) \qquad k(k^2 - 1)(k^2 - 9)$$

increase in magnitude as their distance from the interval $-1 < k < 1$ increases. These polynomials appear in the truncation error for Stirling's formula. The implication is that this formula is most accurate in the center of the range of collocation.

▮ It will be enough to compute these functions for a suitable range of values of k. The first has just been plotted and we see that its relative maxima and minima are of sizes 1.4 and 3.6, more or less, as it moves away from $k = 0$. For the second there are three such turning points on either side of $k = 0$, where the function has magnitudes 12, 22 and 92 to the nearest whole number.

6.35   Show that the relative maxima and minima of the polynomials

$$(k + 1)k(k - 1)(k - 2) \qquad (k + 2)(k + 1)k(k - 1)(k - 2)(k - 3)$$

increase in magnitude with distance from the interval $0 < k < 1$. These polynomials appear in the truncation error for Everett's or Bessel's formula. The implication is that these formulas are most accurate over this central interval.

▮ The first of these has extremes of about

$$-.94 \text{ on } (-1, 0) \qquad .56 \text{ on } (0, 1) \qquad -.94 \text{ on } (1, 2)$$

while for the second we find

$$-16 \text{ on } (-2, -1) \qquad 5 \text{ on } (-1, 0) \qquad -3.5 \text{ on } (0, 1)$$

with symmetric values to the right.

6.36   Estimate the truncation error for a fifth-degree formula.

▮ Assume the collocation arguments equally spaced and at $k = -2, -1, \ldots, 3$ as in Everett's formula (the position is actually immaterial):

$$y(x) - p(x) = \frac{\pi(x)}{(n + 1)!} y^{(n+1)}(\xi) = \frac{(k + 2)(k + 1)k(k - 1)(k - 2)(k - 3)}{720} h^6 y^{(6)}(\xi)$$

The numerator factor, for $0 < k < 1$, takes a maximum absolute value of $\frac{225}{64}$ at $k = \frac{1}{2}$, as may easily be verified, making

$$|y(x) - p(x)| \leq \frac{1}{720} \cdot \frac{225}{64} \cdot h^6 y^{(6)}(\xi)$$

6.37 For the function $y(x) = \sqrt{x}$, $1 \leq x$, how large an interval h is consistent with five-place accuracy if Everett's fifth-degree formula is to be used in interpolations?

∎ For this function, $y^{(6)}(x) = \frac{945}{64} x^{-11/2} \leq \frac{945}{64}$. Substituting this into the result of the previous problem and requiring five-place accuracy.

$$\frac{1}{720} \cdot \frac{225}{64} \cdot h^6 \cdot \frac{945}{64} \leq .000005$$

leading to $h \leq \frac{1}{5}$ approximately. Naturally the interval permitted with fifth-degree interpolation exceeds that for third-degree interpolation.

6.38 For the function $y(x) = \sin x$, how large an interval h is consistent with five-place accuracy if Everett's fifth-degree formula is to be used in interpolations?

∎ For this function $y^{(6)}(x)$ is bounded absolutely by 1, so we need $\frac{1}{720} \cdot \frac{225}{64} \cdot h^6 \leq .000005$, leading to $h \leq .317$. This is the equivalent of $18°$ intervals and means that only four values of the sine function, besides $\sin 0$ and $\sin 90°$ are needed to cover this entire basic interval!

6.39 How large an interval h is consistent with interpolation by Everett's fifth-degree formula if the function is $\log x$ and five-place accuracy is required?

∎ As in the preceding two problems it is the sixth derivative that distinguishes between functions. For $\log x$ this proves to be $-120/x^6$ so, assuming $x > 1$, we will need

$$\frac{1}{720} \cdot \frac{225}{64} h^6 (120) < .000005$$

which reduces to $h < .15$. The assumption $x > 1$ is no restriction since for $x < 1$ we may use $\log x = -\log 1/x$. Notice that of the three functions considered in Problems 6.37 to 6.39 the sine permits the largest spacing and the log the smallest. In tables designed for linear interpolation to five places, all three will have $h = .01$.

6.40 A second source of error in the use of our formulas for the collocation polynomial (the first source being truncation error) is the presence of *inaccuracies in the data values*. The numbers y_k, for example, if obtained by physical measurement will contain inaccuracy due to the limitations imposed by equipment, and if obtained by computations probably contain roundoff errors. Show that linear interpolation does not magnify such errors.

∎ The linear polynomial may be written in Lagrangian form,

$$p = ky_1 + (1 - k) y_0$$

where the y_k are as usual the actual data values. Suppose these values are inaccurate. With Y_1 and Y_0 denoting the exact but unknown values, we may write

$$Y_0 = y_0 + e_0 \qquad Y_1 = y_1 + e_1$$

where the numbers e_0 and e_1 are the errors. The exact result desired is therefore

$$P = kY_1 + (1 - k) Y_0$$

making the error of our computed result

$$P - p = ke_1 + (1 - k) e_0$$

If the errors e_k do not exceed E in magnitude, then

$$|P - p| \leq kE + (1 - k) E = E$$

for $0 < k < 1$. This means that the error in the computed value p does not exceed the maximum data error. No magnification of error has occurred.

6.41   Estimate the magnification of data inaccuracies due to second-degree interpolation.

▮ Using the Lagrangian form of the polynomial, but assuming equally spaced arguments for convenience, we have the parallel expressions

$$p = L_{-1}y_{-1} + L_0 y_0 + L_1 y_1 \qquad P = L_{-1}Y_{-1} + L_0 Y_0 + L_1 Y_1$$

with the L_i the Lagrange coefficient functions.

$$L_{-1} = k(k-1)/2 \qquad L_0 = (k+1)(k-1)/(-1) \qquad L_1 = (k+1)k/2$$

For $0 < k < 1$ the first coefficient will be negative and the others positive, so that by changing the sign of the first term,

$$|P - p| \le \left[(k - k^2)/2 + (1 - k^2) + (k^2 + k)/2 \right] E = (1 + k - k^2) E$$

The quadratic has a maximum value of $\frac{5}{4}$ on the interval $0 < k < 1$ making this the maximum possible magnification of E, the common bound of the e_i. A similar argument leads to an identical result for the interval $-1 < k < 0$. The closer k comes to the zero position the smaller the magnificiation will be, but in any event it does not seem to be a crippling factor.

6.42   Estimate the magnification of data inaccuracies due to cubic interpolation.

▮ Again using the Lagrangian form but assuming equally spaced arguments at $k = -1, 0, 1, 2$, the cubic can be written as

$$p = \frac{k(k-1)(k-2)}{-6} y_{-1} + \frac{(k+1)(k-1)(k-2)}{2} y_0 + \frac{(k+1)k(k-2)}{-2} y_1 + \frac{(k+1)k(k-1)}{6} y_2$$

As in Problem 6.40, we let $Y_k = y_k + e_k$, with Y_k denoting the exact data values. If P again stands for the exact result desired, then the error is

$$P - p = \frac{k(k-1)(k-2)}{-6} e_{-1} + \frac{(k+1)(k-1)(k-2)}{2} e_0 + \frac{(k+1)k(k-2)}{-2} e_1 + \frac{(k+1)k(k-1)}{6} e_2$$

Notice that for $0 < k < 1$ the errors e_{-1} and e_2 have negative coefficients, while the other two have positive coefficients. This means that if the errors do not exceed E in magnitude,

$$|P - p| \le E \left[\frac{k(k-1)(k-2)}{6} + \frac{(k+1)(k-1)(k-2)}{2} + \frac{(k+1)k(k-2)}{-2} + \frac{(k+1)k(k-1)}{-6} \right]$$

which simplifies to

$$|P - p| \le (-k^2 + k + 1) E = m_k E$$

Not surprisingly the quadratic magnification factor m_k takes its maximum at $k = \frac{1}{2}$ (Fig. 6-5) and so $|P - p| \le \frac{5}{4} E$. The data error E may be magnified by as much as $\frac{5}{4}$. This is, of course, a pessimistic estimate. In certain cases errors may even annul one another, making the computed value p more accurate than the data y_k.

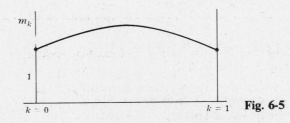

Fig. 6-5

6.43 Estimate the magnification of data inaccuracies due to fourth-degree interpolation. Assume $0 < k < 1$.

I The device used in the three preceding problems is still available and leads to

$$P - p = L_{-2}e_{-2} + L_{-1}e_{-1} + L_0 e_0 + L_1 e_1 + L_2 e_2$$

with coefficients

$$L_{-2} = (k+1)k(k-1)(k-2)/24 \qquad L_{-1} = (k+2)k(k-1)(k-2)/(-6)$$

$$L_0 = (k+2)(k+1)(k-1)(k-2)/4 \qquad L_1 = (k+2)(k+1)k(k-2)/(-6)$$

$$L_2 = (k+2)(k+1)k(k-1)/24.$$

On the interval $0 < k < 1$ the first three of these are positive and the last two negative, which leads after some simplifications to the expression for the absolute error,

$$|P - p| \le (6k^4 - 12k^3 - 30k^2 + 36k + 24) E/24$$

Computation of this quartic finds its maximum size on the given interval to be 33.375 at $k = .5$, making the magnification factor no greater than 1.4.

6.44 What other source of error is there in an interpolation?

I One source that is very important to keep in mind, even though it is often entirely out of one's control, is the continual necessity to make roundoffs during the carrying out of the algorithm. Working to a limited number of digits, this cannot be avoided. Our various formulas, even when they represent exactly the same collocation polynomial, process the data involved in differing ways. In other words, they represent different algorithms. Such formulas accept the same input error (data inaccuracies) and may have the same truncation error, but still differ in the way algorithm roundoffs develop.

6.45 Describe how Taylor's series may be used for interpolation.

I Consider the function $y = e^x$. By Taylor's series,

$$e^{x+t} = e^x \cdot e^t = e^x\left(1 + t + \tfrac{1}{2}t^2 + \cdots\right)$$

Assume the factor e^x is known. Truncating the series after the t^2 term means an error (inside the parentheses) of at most $\frac{1}{6}(h/2)^3$ where h is the interval at which arguments are spaced in the table. This assumes that interpolation will always be based on the nearest tabular entry. If $h = .05$ this error is $(\frac{125}{48})10^{-6}$, or $(2.6)10^{-6}$. This means that, stopping at the t^2 term, accuracy to five digits (not decimal places) will be obtained in the computed value of e^{x+t}. For example, using the data of Table 6.8 the interpolation for $e^{2.718}$ runs as follows. With $t = .018$, $1 + t + \tfrac{1}{2}t^2 = 1.01816$ and

$$e^{2.718} = e^{2.70}(1.01816) = (14.880)(1.01816) = 15.150$$

which is correct to its full five digits. Our collocation polynomials would also produce this result.

TABLE 6.8

x	2.60	2.65	2.70	2.75	2.80
$y = e^x$	13.464	14.154	14.880	15.643	16.445

6.46 Apply the Lagrange formula to the data of the preceding problem. How many points does it need to duplicate the value produced by the Taylor series?

I Using the three center points the Lagrange formula offers us a quadratic polynomial with value $p(2.718) = 15.1504$. Taking all five data points it becomes a fourth-degree polynomial with value 15.1503. Both agree with the Taylor series value to the tabulated three decimal places.

6.47 How can Taylor series interpolation be used for the function $y(x) = \sin x$?

▮ Since $\sin x$ and $\cos x$ are usually tabulated together, we may express

$$\sin(x \pm t) = \sin x \pm t \cos x - \tfrac{1}{2}t^2 \sin x$$

Here, of course, t is measured in radians. If the tabular interval is $h = .0001$, as it is in NBS-AMS 36, of which Table 6.9 is a brief extract, then the above formula will give accuracy to nine digits, since $\tfrac{1}{6}(h/2)^3$ is out beyond the twelfth place.

TABLE 6.9

x	$\sin x$	$\cos x$
1.0000	.841470985	.540302306
1.0001	.841525011	.540218156
1.0002	.841579028	.540134001
1.0003	.841633038	.540049840

6.48 Compute $\sin 1.00005$ by the Taylor series interpolation.

▮ With $x = 1$ and $t = .00005$,

$$\sin 1.00005 = .841470985 + (.00005)(.540302306) - \left(\tfrac{1}{8}\right)(10^{-8})(.841470985) = .841497999$$

6.49 Apply Newton's backward formula to the *prediction* of $\sqrt{1.32}$ in Table 6.2.

▮ With $k = 0$ at $x_0 = 1.30$ we find $k = (1.32 - 1.30)/.05 = .4$. Substituting into the Newton formula,

$$p = 1.14017 + (.4)(.02214) + (.28)(-.00045) + (.224)(.00003) = 1.14891$$

which is correct as far as it goes. Newton's backward formula seems the natural choice for such prediction problems, since the supply of available differences is greatest for this formula and one may introduce difference terms until they do not contribute to the decimal places retained. This allows the degree of the approximating polynomial to be chosen as the computation progresses.

6.50 Use Newton's backward formula to predict the square roots of 1.35, 1.40, 1.45, 1.50 and 1.55 from the data of Table 6.2. Compare with the correct values and observe the beginning of an error explosion.

▮ For $x = 1.35$ the value of k will be $(1.35 - 1.30)/.05$ or 1. For the other listed values k will equal 2, 3, 4 and 5. The Newton formula manages

$$p(1) = 1.14017 + .02214 - .00045 + .00003 = 1.16189$$
$$p(2) = 1.14017 + 2(.02214) - 3(.00045) + 4(.00003) = 1.18322$$
$$p(3) = 1.14017 + 3(.02214) - 6(.00045) + 10(.00003) = 1.20419$$
$$p(4) = 1.14017 + 4(.02214) - 10(.00045) + 20(.00003) = 1.22483$$
$$p(5) = 1.14017 + 5(.02214) - 15(.00045) + 35(.00003) = 1.24517$$

which are to be compared with the correct values

$$1.16189 \qquad 1.18322 \qquad 1.20416 \qquad 1.22474 \qquad 1.24499$$

The errors are 0, 0, 3, 9 and 18 in the fifth decimal place, and it is easy to believe that worse lies ahead as the next problem will also suggest.

6.51 Analyze the truncation error in prediction.

▮ The truncation error of the collocation polynomial can be expressed as

$$\frac{k(k+1)\cdots(k+n)}{(n+1)!}h^{n+1}y^{(n+1)}(\xi)$$

where the collocation points are at $k = 0, -1, \ldots, -n$ as is the case when Newton's backward formula is used. For prediction, k is positive. The numerator factors grow rapidly with increasing k, more rapidly for large n, as Fig. 6-6 suggests. This indicates that truncation error will not be tolerable beyond a certain point and that prediction far beyond the end of a table is dangerous, as might be anticipated. The truncation error of a collocation polynomial is oscillatory between the points of collocation, but once outside the interval of these points it becomes explosive.

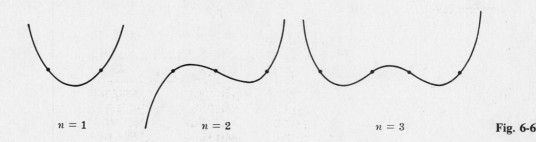

$n = 1$ $n = 2$ $n = 3$ **Fig. 6-6**

6.52   Illustrate the ideas of *modified differences* and *throwback*.

     I   Three typical terms of Everett's fifth-degree formula are

$$\binom{k}{1} y_1 + \binom{k+1}{3} \delta^2 y_1 + \binom{k+2}{5} \delta^4 y_1 = k y_1 + \frac{k(k^2-1)}{6} \left[\delta^2 y_1 - \frac{4-k^2}{20} \delta^4 y_1 \right]$$

For k between 0 and 1 the factor $(4 - k^2)/20$ varies only from .15 to .20. If this factor is approximated by a constant C, then a *modified second* difference may be defined as

$$\delta_m^2 y_1 = \delta^2 y_1 - C \delta^4 y_1$$

The other three terms of the Everett formula lead to a similar modified second difference,

$$\delta_m^2 y_0 = \delta^2 y_0 - C \delta^4 y_0$$

This is also described as *throwback* of the fourth difference upon the second. The same idea may be applied to any difference and to any formula, but we continue with Everett's of degree 5.

6.53   Consider the modified Everett formula

$$p_k = k y_1 + (1-k) y_0 + \binom{k+1}{3} \delta_m^2 y_1 - \binom{k}{3} \delta_m^2 y_0$$

and evaluate the error made in using this in place of Everett's formula of degree 5.

     I   The difference between the two is

$$e_k = \left[\binom{k+2}{5} + C \binom{k+1}{3} \right] \delta^4 y_1 - \left[\binom{k+1}{5} + C \binom{k}{3} \right] \delta^4 y_0$$

6.54   Assuming fourth differences constant, simplify the error formula of the previous problem and discuss error behavior for $0 < k < 1$.

     I   Denoting both fourth differences by $\delta^4 y$, we find after a slight effort,

$$e_k = \frac{k(k-1)}{24} \left[k^2 - k - 2 + 12C \right] \delta^4 y$$

and denoting $12C - 2$ by α,

$$e_k = k(k-1)(k^2 - k + \alpha) \delta^4 y / 24 = F_k \delta^4 y / 24$$

For small values of α the factor F_k has the behavior shown in Fig. 6-7 for $0 < k < 1$. There are two minima and one maximum. The three extreme values of $|F_k|$ can be equalized by a proper choice of α, and it is not hard to show that in this way the maximum of $|F_k|$ is made as small as possible. By the usual method the center maximum is found to be of height $(1/16 - \alpha/4)$ and the two minima of depth $\alpha^2/4$. Equating these leads to

$$\alpha = \tfrac{1}{2}(\sqrt{2} - 1) \qquad C = (3 + \sqrt{2})/24 = .1839$$

making $C = .1839$ approximately. With this choice for C and still assuming fourth differences constant, we find

$$|e_k| \le |\max F_k| \, \delta^4 y/24 = \left((\sqrt{2} - 1)^2/384\right) \delta^4 y$$

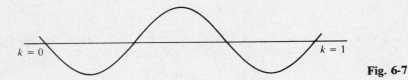

$k = 0 \qquad\qquad\qquad\qquad\qquad\qquad k = 1$

Fig. 6-7

6.55 If fourth differences are not constant, which is what we expect, then the value of C suggested by the previous problem may still be as good a choice as any. Find the error e_k in such a case.

▮ The formula of Problem 6.53 still applies, and a direct evaluation shows that

$$|e_k| \le .00122 \max\left(|\delta^4 y_0|, |\delta^4 y_1|\right)$$

Thus if the fourth differences are absolutely less than 400 units in the last decimal plane used, this error will be smaller than half a unit in that place. The values given by the modified Everett formula will not then differ significantly from those given by the fifth-degree formula.

6.56 Prepare a table of $y(x) = \sin x$ with modified second differences suitable for five place accuracy.

▮ Problem 6.38 suggests the interval $h = 18°$ for fifth-degree interpolation, but to keep fourth differences nearer to the level recommended by Problem 6.55, we use the slightly more conservative interval of $15°$. This is also a little more convenient. Values of $\sin x$ at this interval are given in Table 6.10. A few extra values are included at the ends to fill out the fourth difference column. They are consequences of the symmetry of the sine function.

Modified second differences are now computed from

$$\delta_m^2 y = \delta^2 y - .1839 \, \delta^4 y$$

and suppressing the first and third differences we obtain Table 6.11.

TABLE 6.10

		25882			−1764	
0	.00000		0			0
		25882			−1764	
15	.25882		−1764			121
		24118			−1643	
30	.50000		−3407			231
		20711			−1412	
45	.70711		−4819			329
		15892			−1083	
60	.86603		−5902			402
		9990			−681	
75	.96593		−6583			450
		3407			−231	
90	1.00000		−6814			462
		−3407			231	

TABLE 6.11

x	$\sin x$	δ_m^2
0	.00000	0
15	.25882	-1786
30	.50000	-3449
45	.70711	-4880
60	.86603	-5976
75	.96593	-6656
90	1.00000	-6919

6.57 Use Table 6.11 to interpolate for $\sin 80°$.

▮ Using Everett's cubic formula with the modified second differences and choosing $k = 0$ at $x_0 = 75°$, we find $k = \frac{1}{3}$ at $x = 80°$, and so

$$\sin 80° = \tfrac{1}{3}(1.00000) + \tfrac{2}{8}(.96593) - \tfrac{4}{81}(-.06919) - \tfrac{5}{81}(-.06656) = .98481$$

which is correct to five places.

6.58 Apply Hermite's formula to interpolate for $y(1.05)$ from the following data.

x	y	y'
1.00	1.00000	.50000
1.10	1.04881	.47673

▮ It is not uncommon for experimental work to yield measured values of both y and y'. (See the railroad switching Problem 5.4.) Some computed tables also list both y and y'. Hermite's formula is appropriate in such cases. With $n = 1$ in this formula we need

$$L_0(x) = \frac{x - x_1}{x_0 - x_1} = \frac{1.05 - 1.10}{1.00 - 1.10} \qquad L_1(x) = \frac{x - x_0}{x_1 - x_0} = \frac{1.05 - 1.00}{1.10 - 1.00}$$

$$L_0'(x) = \frac{1}{x_0 - x_1} = \frac{1}{-.10} \qquad L_1'(x) = \frac{1}{x_1 - x_0} = \frac{1}{.10}$$

Substituting into Hermite's formula,

$$p(x) = \sum_{i=0}^{n} \left[1 - 2L_i'(x_i)(x - x_i)\right]\left[L_i(x)\right]^2 y_i + (x - x_i)\left[L_i(x)\right]^2 y_i'$$

$$= \left[1 - 2\left(\tfrac{1}{-.10}\right)(.05)\right]\left(\tfrac{1}{2}\right)^2(1) + (.05)\left(\tfrac{1}{2}\right)^2(.5)$$

$$+ \left[1 - 2\left(\tfrac{1}{.10}\right)(-.05)\right]\left(\tfrac{1}{2}\right)^2(1.04881) + (-.05)\left(\tfrac{1}{2}\right)^2(.47673)$$

$$= 1.02470$$

Since the original data were taken from the square root function which has been so prominent in these numerical test runs, it is reassuring to have once again recovered $\sqrt{1.05}$ correct to five places.

6.59 Apply the Hermite, Lagrange and spline methods to interpolate $y(1.05)$ from the following data.

x	.90	1.00	1.10	1.20
$y = \sin x$	.78333	.84147	.89121	.93204
$y' = \cos x$	.62161	.54030	.45360	.36236

What parts of the data does each method actually use? How do the results compare?

I Suppose we apply the Hermite method to the center two data points, for arguments 1.00 and 1.10. This means a Hermite cubic matching the y and y' values in those columns. An algorithm for this cubic was developed beginning with Problem 5.106, which for the present data yields the value $y(1.05) = .86742$, which is the correct sine value to five places.

The Lagrange method ignores y' and produces a cubic matching the four given y values. The details have been provided in the set of problems culminating in 4.56. The interpolated value in this case is the same .86742 provided by Hermite's cubic.

As for a spline, there is no point in limiting its domain to the center columns of the table since this would reproduce the Hermite cubic. Instead let us ask it to reproduce the four data values of y and, as the two supplementary specifications, to have the two endpoint y' values. A method of representing this spline in terms of basis functions was developed in Chapter 5 and a linear system for determining the coefficients appears in Problem 5.100. The resulting spline once again offers us the value .86742 for sin(1.05). In Problem 6.31 it was estimated that for a cubic polynomial to interpolate the sine function to five-place accuracy the spacing $h = .12$ might be advisable. Here we have managed quite well with $h = .10$, or about 6° between data points.

6.60 Strictly in a sporting spirit, apply the same three methods of interpolation used in the preceding problem to the following values of the sine function, the spacing between data points being $h = .5$ or approximately 30°.

x	.5	1.0	1.5	2.0
y	.47943	.84147	.99749	.90930
y'	.87758	.54030	.07074	$-.41615$

How much accuracy is achieved and is there any noticeable difference in performance?

I Here are the interpolated values at four arguments within the center interval, along with the correct five-place sines:

x	1.1	1.2	1.3	1.4
Hermite	.89114	.93190	.96341	.98538
Lagrange	.89038	.93074	.96225	.98460
spline	.89112	.93187	.96339	.98537
correct	.89121	.93204	.96356	.98545

All the interpolations are on the low side, the Hermite results averaging 11 under, the Lagrange 107 and the spline 13.

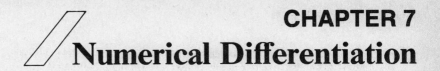
7.1 What are the three primitive numerical differentiation formulas and what is their geometric interpretation?

▌ They are

$$y(x) \simeq \frac{y(x+h) - y(x)}{h} \qquad y'(x) \simeq \frac{y(x+h) - y(x-h)}{2h} \qquad y'(x) \simeq \frac{y(x) - y(x-h)}{h}$$

and their geometric background is apparent in Fig. 7-1 as the slopes of the three chords of the curve. The tangent line to the curve at x_0 is also shown and we are to view the slopes of the chords as approximations to the slope of the tangent line. It seems clear that the middle formula will provide the best of the three approximations, since its chord uses information from both sides of x_0, and this suspicion will receive support when we compute the truncation errors in Problem 7.8. Because of their simplicity these differentiation formulas are frequently applied in complex problems, where anything more sophisticated might lead to overwhelming arithmetical difficulties. (See in particular the chapter on partial differential equations.) In the next several problems these formulas will be derived as first approximations to presumably more accurate ones.

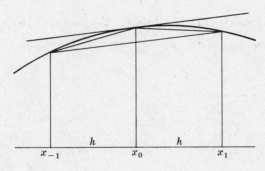

$$x_{-1} \qquad h \qquad x_0 \qquad h \qquad x_1 \qquad \textbf{Fig. 7-1}$$

7.2 Differentiate Newton's forward formula,

$$p_k = y_0 + \binom{k}{1} \Delta y_0 + \binom{k}{2} \Delta^2 y_0 + \binom{k}{3} \Delta^3 y_0 + \binom{k}{4} \Delta^4 y_0 + \cdots$$

▌ The Stirling numbers may be used to express the factorials as powers, after which an easy computation produces derivatives relative to k. With the operator D continuing to represent such derivatives, $Dp_k, D^2 p_k, \ldots$, we use the familiar $x = x_0 + kh$ to obtain derivatives relative to the argument x:

$$p'(x) = (Dp_k)/h \qquad p^{(2)}(x) = (D^2 p_k)/h^2 \qquad \cdots$$

The results are

$$p'(x) = \frac{1}{h}\left(\Delta y_0 + \left(k - \frac{1}{2}\right)\Delta^2 y_0 + \frac{3k^2 - 6k + 2}{6}\Delta^3 y_0 + \frac{2k^3 - 9k^2 + 11k - 3}{12}\Delta^4 y_0 + \cdots\right)$$

$$p^{(2)}(x) = \frac{1}{h^2}\left(\Delta^2 y_0 + (k-1)\Delta^3 y_0 + \frac{6k^2 - 18k + 11}{12}\Delta^4 y_0 + \cdots\right)$$

$$p^{(3)}(x) = \frac{1}{h^3}\left(\Delta^3 y_0 + \frac{2k-3}{2}\Delta^4 y_0 + \cdots\right)$$

$$p^{(4)}(x) = \frac{1}{h^4}\left(\Delta^4 y_0 + \cdots\right) \quad \text{and so on}$$

7.3 Apply the formulas of Problem 7.2 to produce $p'(1)$, $p^{(2)}(1)$ and $p^{(3)}(1)$ from the data of Table 7.1 (This is the same as Table 6.2, with the differences beyond the third suppressed. Recall that those differences were written off as error effects. The table is reproduced here for convenience.)

▮ With $h = .05$ and $k = 0$ at $x_0 = 1.00$, our formulas produce

$$p'(1) = 20(.02470 + .000295 + .000017) = .50024$$
$$p^{(2)}(1) = 400(-.00059 - .00005) = -.256$$
$$p^{(3)}(1) = 8000(.00005) = .4$$

The correct results are, since $y(x) = \sqrt{x}$, $y'(1) = \frac{1}{2}$, $y^{(2)}(1) = -\frac{1}{4}$ and $y^{(3)}(1) = \frac{3}{8}$.

Though the input data are accurate to five decimal places, we find $p'(1)$ correct to only three places, $p^{(2)}(1)$ not quite correct to two places and $p^{(3)}(1)$ correct to only one. Obviously algorithm errors are prominent.

TABLE 7.1

x	$y(x) = \sqrt{x}$			
1.00	1.00000			
		2470		
1.05	1.02470		-59	
		2411		5
1.10	1.04881		-54	
		2357		4
1.15	1.07238		-50	
		2307		2
1.20	1.09544		-48	
		2259		3
1.25	1.11803		-45	
		2214		
1.30	1.14017			

7.4 Differentiate Stirling's formula

$$p_k = y_0 + \binom{k}{1} \delta\mu y_0 + \frac{k}{2}\binom{k}{1} \delta^2 y_0 + \binom{k+1}{3} \delta^3 \mu y_0 + \frac{k}{4}\binom{k+1}{3} \delta^4 y_0 + \cdots$$

▮ Proceeding as in Problem 7.2, we find

$$p'(x) = \frac{1}{h}\left(\delta\mu y_0 + k\,\delta^2 y_0 + \frac{3k^2 - 1}{6} \delta^3 \mu y_0 + \frac{2k^3 - k}{12} \delta^4 y_0 + \cdots \right)$$

$$p^{(2)}(x) = \frac{1}{h^2}\left(\delta^2 y_0 + k\,\delta^3 \mu y_0 + \frac{6k^2 - 1}{12} \delta^4 y_0 + \cdots \right)$$

$$p^{(3)}(x) = \frac{1}{h^3}\left(\delta^3 \mu y_0 + k\,\delta^4 y_0 + \cdots \right)$$

$$p^{(4)}(x) = \frac{1}{h^4}\left(\delta^4 y_0 + \cdots \right) \quad \text{and so on}$$

7.5 Apply the formulas of Problem 7.4 to produce $p'(1.10)$, $p^{(2)}(1.10)$ and $p^{(3)}(1.10)$ from the data of Table 7.1.

▮ With $k = 0$ at $x_0 = 1.10$, our formulas produce

$$p'(1.10) = 20\left[\frac{.02411 + .02357}{2} + 0 - \frac{1}{6}\left(\frac{.00005 + .00004}{2} \right) \right] = .4766$$

$$p^{(2)}(1.10) = 400(-.00054 + 0) = -.216$$

$$p^{(3)}(1.10) = 8000(.000045) = .360$$

The correct results are $y'(1.10) = .47674$, $y^{(2)}(1.10) = -.2167$ and $y^{(3)}(1.10) = .2955$.

The input data were correct to five places, but our approximations to these first three derivatives are correct to roughly four, three and one place, respectively.

7.6   Differentiate Newton's backward formula,

$$p_k = y_0 + k\nabla y_0 + \frac{k(k+1)}{2!}\nabla^2 y_0 + \cdots + \frac{k\cdots(k+n-1)}{n!}\nabla^n y_0$$

▌ Again as in Problem 7.2,

$$p'(x) = \frac{1}{h}\left(\nabla y_0 + \left(k+\frac{1}{2}\right)\nabla^2 y_0 + \frac{3k^2+6k+2}{6}\nabla^3 y_0 + \frac{4k^3+18k^2+22k+6}{24}\nabla^4 y_0 + \cdots\right)$$

$$p''(x) = \frac{1}{h^2}\left(\nabla^2 y_0 + (k+1)\nabla^3 y_0 + \frac{12k^2+36k+22}{24}\nabla^4 y_0 + \cdots\right)$$

$$p'''(x) = \frac{1}{h^3}\left(\nabla^3 y_0 + \left(k+\frac{3}{2}\right)\nabla^4 y_0 + \cdots\right)$$

$$p''''(x) = \frac{1}{h^4}\left(\nabla^4 y_0 + \cdots\right)$$

up through fourth difference terms.

7.7   Apply the formulas of the preceding problem to approximate the first three derivatives of $y(x)$ at $x = 1.30$, still using the data of Table 7.1.

▌ With $k = 0$ at $x = 1.30$, we find

$$p'(1.30) = 20(.02214 - .00022 + .00001) = .4386$$
$$p''(1.30) = 400(-.00045 + .00003) = -.168$$
$$p'''(1.30) = 8000(.00003) = .24$$

which are to be compared with the more accurate values

$$y'(1.30) = .43853 \qquad y''(1.30) = -.1687 \qquad y'''(1.30) = .19$$

The errors are about .0007, .0007 and .05, so the deterioration is once again evident. No fourth derivative was requested since differences beyond the third have been deemed to consist largely of error. However, as a bit of computing nonsense we may refer back to the earlier Table 6.2 and note that the appropriate difference of order 4 had value .00001. This suggests, but not very strongly, the computation

$$p^{(4)}(1.30) = 160000(.00001) = 1.6$$

which may be compared with

$$y^{(4)}(1.30) = -.37$$

to see that skepticism was surely appropriate.

7.8   Problems 7.2, 7.4, and 7.6 suggest three approximations to $y'(x_0)$ using only first differences,

$$\frac{y_1 - y_0}{h} \qquad \frac{y_1 - y_{-1}}{2h} \qquad \frac{y_0 - y_{-1}}{h}$$

Interpreted geometrically, these are the slopes of three lines shown in Fig. 7-1. The tangent line at x_0 is also shown. It would appear that the middle approximation is closest to the slope of the tangent line. Confirm this by computing the truncation errors of the three formulas.

▌ Newton's forward formula, truncated after the first difference term, leaves the truncation error

$$y(x) - p(x) = \frac{h^2}{2}\left[k(k-1)y^{(2)}(\xi)\right]$$

with $x = x_0 + kh$ as usual. It is helpful here to consider k as a continuous argument, no longer restricting it to integer values. Assuming $y^{(2)}(\xi)$ continuous, we then find the error of our derivative formula (by the chain rule) for $k = 0$:

$$y'(x_0) - p'(x_0) = -(h/2)\, y^{(2)}(\xi_0)$$

Note that for $k = 0$ the derivative of the troublesome $y^{(2)}(\xi)$ factor is not involved. Similarly for Newton's backward formula,

$$y'(x_0) - p'(x_0) = (h/2)\, y^{(2)}(\xi_0)$$

With Stirling's formula we receive an unexpected bonus. Retaining even the *second* difference term in our approximation we find that at $k = 0$ it disappears from $p'(x)$. (See Problem 7.4.) Thus we may consider the middle approximation under discussion as arising from a second-degree polynomial approximation. The truncation error is then

$$y(x) - p(x) = \frac{h^3}{6}\left[(k+1)k(k-1)\, y^{(3)}(\xi)\right]$$

leading to

$$y'(x_0) - p'(x_0) = \frac{-h^2}{6}\, y^{(3)}(\xi)$$

It is true that the symbol ξ probably represents three distinct unknown numbers in these three computations. But since h is usually small, the appearance of h^2 in the last result, compared with h in the others, suggests that this truncation error is the smallest, by an order of magnitude. This confirms the geometrical evidence.

7.9 Apply the middle formula of Problem 7.8 to approximate $y'(1.10)$ for the data of Table 7.1. Find the actual error of this result and compare with the truncation error estimate of Problem 7.8.

▮ This approximation is actually the first term computed in Problem 7.5: $y'(1.10) \approx .4768$. The actual error is, to five places,

$$y'(1.10) - .4768 = .47674 - .47680 = -.00006$$

The estimate obtained in Problem 7.8 was $-h^2 y^{(3)}(\xi)/6$. Since $y^{(3)}(x) = \frac{3}{8}x^{-5/2}$, we exaggerate only slightly by replacing the unknown ξ by 1, obtaining $-h^2 y^{(3)}(\xi)/6 \sim -(.05)^2(\frac{1}{16}) = -.00016$. This estimate is generous, though not unrealistic.

7.10 Convert the formula for $p'(x_0)$ obtained in Problem 7.4 to a form that exhibits the y_k values used rather than the differences.

▮ We have $k = 0$ for this case, making

$$p'(x_0) = \frac{1}{h}\left[\frac{1}{2}(y_1 - y_{-1}) - \frac{1}{12}(y_2 - 2y_1 + 2y_{-1} - y_{-2})\right] = \frac{1}{12h}(y_{-2} - 8y_{-1} + 8y_1 - y_2)$$

7.11 Estimate the truncation error in the formula of Problem 7.10.

▮ Since the formula was based on Stirling's fourth-degree polynomial,

$$y(x) - p(x) = h^5(k^2 - 4)(k^2 - 1)ky^{(5)}(\xi)/120$$

Differentiating as in Problem 7.8 and putting $k = 0$, $y'(x_0) - p'(x_0) = h^4 y^{(5)}(\xi)/30$.

7.12 Compare the estimate of Problem 7.11 with the actual error of the computed result in Problem 7.5.

▮ To five places the actual error is

$$y'(1.10) - p'(1.10) = .47674 - .47660 = .00014$$

while the formula of Problem 7.11, with $y^{(5)}(1)$ substituting for the unknown $y^{(5)}(\xi)$ and causing a slight exaggeration, yields

$$h^4 y^{(5)}(\xi)/30 \sim (.05)^4(7/64) = .0000007$$

Surely this is disappointing. Though the truncation error has been essentially eliminated by using differences of higher order, the actual error is greater. Clearly another source of error is dominant in these algorithms. It proves to be the input errors of the y_i values and how the algorithm magnifies them. For brevity we shall include this in the term roundoff error.

7.13 Estimate the roundoff error behavior for the formula $(y_1 - y_{-1})/2h$.

▌ As before, let Y_1 and Y_{-1} be the exact (unknown) data values. Then $Y_1 = y_1 + e_1$ and $Y_{-1} = y_{-1} + e_{-1}$ with e_1 and e_{-1} representing data errors. The difference

$$\frac{Y_1 - Y_{-1}}{2h} - \frac{y_1 - y_{-1}}{2h} = \frac{e_1 - e_{-1}}{2h}$$

is then the error in our output due to input inaccuracies. If e_1 and e_{-1} do not exceed E in magnitude, then this output error is at worst $2E/2h$, making the maximum roundoff error E/h.

7.14 Apply the estimate of Problem 7.13 to the computation of Problem 7.9.

▌ Here $h = .05$ and $E = .000005$, making $E/h = .00010$. Thus roundoff error in the algorithm may influence the fourth place slightly.

7.15 Estimate roundoff error behavior for the formula of Problem 7.10.

▌ Proceeding just as in Problem 7.12, we find $1/12h(e_{-2} - 8e_{-1} + 8e_1 - e_2)$ for the error in the output due to input inaccuracies. If the e_k do not exceed E in magnitude, then this output error is at worst $18E/12h$, i.e., maximum roundoff error $= (3/2h)E$. The factor $(3/2h)$ is the magnification factor, as $(1/h)$ was in Problem 7.13. Note that for small h, which we generally associate with high accuracy, this factor is large and roundoff errors in the input information become strongly magnified.

7.16 Apply the estimate of Problem 7.15 to the computation of Problem 7.5. Then compare with various errors associated with our efforts to compute $y'(1.10)$.

▌ With $h = .05$ and $E = .000005$, $(3/2h)E = .00015$. The various errors are grouped in Table 7.2.
In the first case roundoff error has helped, but in the second case it has hurt. Plainly, the high magnification of such errors makes low truncation error pointless, except for extremely accurate data.

TABLE 7.2

Formula	Actual error	Est. trunc. error	Max. R.O. error
$(y_1 - y_{-1})/2h$	$-.00006$	$-.00016$	$\pm.00010$
$(y_{-2} - 8y_{-1} + 8y_1 - y_2)/12h$	$.00014$	$.0000007$	$\pm.00015$

7.17 Differentiate Bessel's formula, obtaining derivatives up through the fifth in terms of differences through the fifth.

▌ Bessel's formula is

$$p_k = \mu y_{1/2} + \left(k - \tfrac{1}{2}\right)\delta y_{1/2} + \binom{k}{2}\mu\delta^2 y_{1/2} + \left(\tfrac{1}{3}\right)\left(k - \tfrac{1}{2}\right)\binom{k}{2}\delta^3 y_{1/2}$$

$$+ \cdots + \binom{k+n-1}{2n}\mu\delta^{2n} y_{1/2} + (1/[2n+1])\left(k - \tfrac{1}{2}\right)\binom{k+n-1}{2n}\delta^{2n+1} y_{1/2}$$

and there is no difficulty in taking its derivatives relative to k. Then multiplying by $1/h$,

$$hp^1 = \delta y_{1/2} + \left(k - \frac{1}{2}\right)\mu\,\delta^2 y_{1/2} + \frac{6k^2 - 6k + 1}{12}\,\delta^3 y_{1/2}$$

$$+ \frac{4k^3 - 6k^2 - 2k + 2}{24}\,\mu\,\delta^4 y_{1/2} + \frac{5k^4 - 10k^3 + 5k - 1}{120}\,\delta^5 y_{1/2}$$

$$h^2 p^{(2)} = \mu\,\delta^2 y_{1/2} + \left(k - \frac{1}{2}\right)\delta^3 y_{1/2} + \frac{12k^2 - 12k - 2}{24}\,\mu\,\delta^4 y_{1/2} + \frac{4k^3 - 6k^2 + 1}{24}\,\delta^5 y_{1/2}$$

$$h^3 p^{(3)} = \delta^3 y_{1/2} + \left(k - \frac{1}{2}\right)\mu\,\delta^4 y_{1/2} + \frac{1}{2}(k^2 - k)\,\delta^5 y_{1/2}$$

$$h^4 p^{(4)} = \mu\,\delta^4 y_{1/2} + \left(k - \frac{1}{2}\right)\delta^5 y_{1/2}$$

$$h^5 p^{(5)} = \delta^5 y_{1/2}$$

7.18 Apply the formulas of the preceding problem to find derivatives of the square root function at the point $x = 1.125$ from the data in Table 7.1.

∥
$$p' = 20\left[.02357 + 0 - \tfrac{1}{24}(.00004)\right] = .4714$$

$$p'' = 400(-.00052) = -.208$$

$$p''' = 8000(.00004) = .32$$

The fact that we are working at a point midway between arguments listed in the table makes $k = \frac{1}{2}$ and reduces many of the formula terms to zero. The correct values are $.4714045$, $-.2095$ and $.28$, each given to enough places to indicate the actual error sizes.

7.19 Express the formula for p' at $k = \frac{1}{2}$ found in Problem 7.17 in terms of the data values y_i, using only differences through the third. Then estimate the error caused by inaccuracies in the y_i values.

∥ We find for $k = \frac{1}{2}$,

$$hp' = y_1 - y_0 - \tfrac{1}{24}(y_2 - 3y_1 + 3y_0 - y_{-1})$$

$$p' = (-y_2 + 27y_1 - 27y_0 + y_{-1})/24h$$

This is not a particularly attractive formula, but estimating its potential for magnifying data inaccuracies is easy enough. The procedure used in Problem 7.13 serves here just as well and leads quickly to the $56E/24h$ or $7E/3h$. With $E = .000005$ and $h = \frac{1}{20}$ this can be applied to the p' of Problem 7.18 for which it finds a possible error of $.0000666$, which may be compared with the actual error of only $.0000045$. We appear to be safely below the maximum.

7.20 Express the formula for p'' found in Problem 7.17 in terms of the data values y_i, using only differences through the third. Then estimate the possibilities for magnification of inaccuracies in the data values.

∥ In fact even the third difference will not be active since at $k = \frac{1}{2}$ it has a zero coefficient. So for this k,

$$h^2 p'' = \tfrac{1}{2}\left[(y_2 - 2y_1 + y_0) + (y_1 - 2y_0 + y_{-1})\right]$$

$$p'' = (y_2 - y_1 - y_0 + y_{-1})/2h^2$$

The maximum error due to the y_i values is $4E/2h^2$ or $2E/h^2$. For the p'' calculation of Problem 7.18 this comes to $(400)(.00001)$ or $.004$. The actual error was found to be $.0015$.

7.21 If Stirling's formula up to sixth differences is differentiated, what will be added to the p' result of Problem 7.4?

∥ There will be two new terms,

$$\frac{5k^4 - 15k^2 + 4}{120}\,\delta^5\mu\,y_0 + \frac{6k^5 - 20k^3 + 8k}{720}\,\delta^6 y_0$$

At $k = 0$ the second will be zero and the result

$$p' = \frac{1}{h}\left(\delta\mu\, y_0 - \frac{1}{6}\delta^3\mu\, y_0 + \frac{1}{30}\delta^5\mu\, y_0 \right)$$

is indicated, the last term being the addition.

7.22 Express the formula of the preceding problem in terms of the y_i values. What is the maximum possible error due to roundoff, or magnification of input errors?

❚ The details are a small nuisance but no intellectual hurdle:

$$hp' = \tfrac{1}{2}(y_1 - y_{-1}) - \tfrac{1}{12}(y_2 - 3y_1 + 3y_0 - y_{-1} + y_1 - 3y_0 + 3y_{-1} - y_{-2})$$
$$+ \tfrac{1}{60}(y_3 - 5y_2 + 10y_1 - 10y_0 + 5y_{-1} - y_{-2} + y_2 - 5y_1 + 10y_0 - 10y_{-1} + 5y_{-2} - y_{-3})$$

reducing to

$$p' = (y_3 - 9y_2 + 45y_1 - 45y_0 + 9y_{-1} - y_{-2})/60h$$

Totalling the numerator coefficients, we find a maximum roundoff error of $110E/60h$ or $11E/6h$.

7.23 Estimate the truncation error of the formula

$$y^{(2)}(x_0) \approx \frac{1}{h^2}\delta^2 y_0 = \frac{1}{h^2}(y_1 - 2y_0 + y_{-1})$$

obtainable from Problem 7.4 by stopping after the second difference term.

❚ Here it may be convenient to follow a different route to the truncation error, using Taylor series. In particular

$$y_1 = y_0 + hy_0' + \tfrac{1}{2}h^2 y_0^{(2)} + \tfrac{1}{6}h^3 y_0^{(3)} + \tfrac{1}{24}h^4 y^{(4)}(\xi_1)$$
$$y_{-1} = y_0 - hy_0' + \tfrac{1}{2}h^2 y_0^{(2)} - \tfrac{1}{6}h^3 y_0^{(3)} + \tfrac{1}{24}h^4 y^{(4)}(\xi_2)$$

so that adding these up and then subtracting $2y_0$ we find

$$\delta^2 y_0 = h^2 y_0^{(2)} + \tfrac{1}{24}h^4\left[y^{(4)}(\xi_1) + y^{(4)}(\xi_2) \right]$$

Unfortunately ξ_1 is probably not the same as ξ_2, but for an estimate of truncation error suppose we replace both fourth derivatives by a number $y^{(4)}$ which remains open for our choice. For complete safety we could choose $y^{(4)} = \max|y^{(4)}(x)|$ over the interval involved, leading to an upper bound for the magnitude of truncation error, but conceivably other choices might be possible. We now have

$$\text{truncation error} = y_0^{(2)} - \frac{1}{h^2}\delta^2 y_0 = -\frac{h^2}{12}y^{(4)}$$

7.24 Apply the estimate in Problem 7.23 to the computation of Problem 7.5.

❚ The computation of $p^{(2)}(1.10)$ in Problem 7.5 was actually made by the formula

$$p^{(2)}(1.10) = \delta^2 y_0/h^2 = -.21600$$

since higher difference terms contributed nothing. The result has already been compared with the correct $y''(1.10) = -.21670$. The truncation error estimate of Problem 7.23, with

$$y^{(4)}(x) = -(15/16)x^{-7/2} \sim -15/16$$

suggests a slight exaggeration:

$$\text{truncation error} \sim 1/5120 = .00020$$

The actual error is $-.00070$, again indicating that truncation is not the major error source.

7.25 Estimate the roundoff error of the formula $\delta^2 y_0 / h^2$:

▌ Proceeding as before, we find the output error due to input inaccuracies to be $(1/h^2)(e_1 - 2e_0 + e_{-1})$ where the e_k are the input errors. If these do not exceed E in magnitude, then this can be at worst $(4/h^2)E$; thus the maximum roundoff error is $(4/h^2)E$.

7.26 Apply the formula of Problem 7.25 to the computation of Problem 7.5 and compare the actual error of our approximation to $y^{(2)}(1.10)$ with truncation and roundoff estimates.

▌ As before $h = .05$ and $E = .000005$, making $(4/h^2)E = .00800$.

The magnification factor $(4/h^2)$ has a powerful effect. Our results confirm that roundoff has been the principal error source in our approximation of $y^{(2)}(1.10)$ and that it has contributed only about 90 of a potential 800 units.

Actual error	Est. truncation error	Max. R.O. error
− .00070	.00020	±.00800

7.27 Estimate roundoff error for the formula $y^{(4)}(x_0) \sim \delta^4 y_0 / h^4$ obtained in Problem 7.4.

▌ In terms of y_k values this formula becomes $(y_{-2} - 4y_{-1} + 6y_0 - 4y_1 + y_2)/h^4$ and involves an error due to data inaccuracies of amount $(e_{-2} - 4e_{-1} + 6e_0 - 4e_1 + e_2)/h^4$. If the e_k do not exceed E in magnitude, this cannot exceed $(16/h^4)E$.

We made no attempt to use this formula in Table 7.1 because fourth differences were only error effects. With $h = .05$ and E the usual .000005, roundoff error might have come to 12.8 anyway, completely obscuring any meaningful result. To approximate fourth derivatives excessive data accuracy is required.

7.28 Find a minimum value of $y(x)$ given the data in Table 7.3

▌ First we compute the differences that are also shown in the table.

A polynomial of degree 2 seems to be indicated. Stirling's formula with $k = 0$ at $x_0 = .70$ becomes

$$p_k = .6318 + k(.00075) + \tfrac{1}{2}k^2(.0049)$$

The derivative relative to k is $Dp_k = .00075 + k(.0049)$ and becomes zero at $k = -.153$. Inserted into the polynomial, the minimum value is found to be .6137. The corresponding argument is $x = .70 - (.153)(.05) = .692$. These values $y(x)$ actually come from $y(x) = e^x - 2x$, which has a minimum of close to .6137 at $x = \log 2 = .693$.

7.29 By Problems 7.23 and 7.25 we find the combined truncation and roundoff errors of the approximation

$$y^{(2)}(x_0) \approx (1/h^2)(y_1 - 2y_0 + y_{-1})$$

to have the form $Ah^2 + 4E/h^2$ where $A = |y^{(4)}(\xi)/12|$. What choice of h will minimize this combination?

▌ The derivative relative to h is $2Ah - 8E/h^3$. This is zero for $h^4 = 4E/A$, or $h = (4E/A)^{1/4}$. For the square root function and five place accuracy, this recommends $h = .13$ so that a *wider* spacing than that of Table 7.1 would be more suitable for this formula. Of course, the combination we have minimized does not represent the exact error,

TABLE 7.3

x	$y(x)$		
.60	.6221		
		− 66	
.65	6155		49
		− 17	
.70	.6138		49
		32	
.75	.6170		

only an approximation to it, but this theoretical result certainly comes as a surprise. Actual computations yield the results in Table 7.4.

It is at least clear that the accuracy does not improve indefinitely as h diminishes. At $h = .08$ we find a perfect result, after which roundoff errors begin to obscure things.

TABLE 7.4

h	$y^{(2)}(1) \sim (1/h^2)(y_1 - 2y_0 + y_{-1})$
.01	$-.2000$
.05	$-.2480$
.08	$-.2500$
.10	$-.2510$
.13	$-.2509$
.15	$-.2520$

7.30   In the right circumstances (accurate data and an interval h not too small) a more sophisticated formula for numerical differentiation may be justified. Apply Problems 5.52 to 5.56 to approximate the first four derivatives of $y(x) = \sin x$ at $x = \pi/4$ from the data of Table 7.5.

▮ First we use Problem 5.54, with $k = 0$ at $x = \pi/4$:

$$y'(\pi/4) = (1/h) \, Dy_0 = (12/\pi)(.183020 + .002078 + .000028 - .000003) = .70711$$

the .000003 actually being important! Next, by Problem 5.53,

$$y^{(2)}(\pi/4) = (1/h^2) \, D^2 y_0 = (12/\pi)^2 (-.048190 - .000273 - .000002) = -.70719$$

Then using Problem 5.55,

$$y^{(3)}(\pi/4) = (1/h^3) \, D^3 y_0 = (12/\pi)^3 (-.012470 - .000211 - .000002) = -.70683$$

Finally, by Problem 5.56,

$$y^{(4)}(\pi/4) = (1/h^4) \, D^4 y_0 = (12/\pi)^4 (.003280 + .000035) = .70568$$

Since all results should be .70711 apart from sign, diminishing returns are again apparent.

TABLE 7.5

x	$\sin x$	δ	δ^2	δ^3	δ^4	δ^5	δ^6	δ^7	δ^8
0	.00000		0000		0				
		25882		-1764		120			
$\pi/12$	.25882		-1764		120		-7		
		24118		-1644		113			
$2\pi/12$	.50000		-3407		233		-18		
		20711		-1411		95		-3	
$3\pi/12$	.70711		-4819		328		-21		-2
		15892		-1083		74		-5	
$4\pi/12$	.86603		-5902		402		-26		
		9990		-681		48			
$5\pi/12$	.96593		-6583		450		-36		
		3407		-231		12			
$\pi/2$	1.00000		-6814		462				

7.31 Find the argument that makes $y' = 0$ in Table 7.6 using the Lagrange formula and inverse cubic interpolation. Then find the corresponding y value by direct interpolation.

▮ With y' and x playing the roles of x and y in our standard Lagrange formula (see Problem 4.56), the inverse interpolation finds that $y' = 0$ occurs for $x = 1.57080$. (The data are an extract, which may have been suspected, from the sine function, so the correct zero belongs to $x = \pi/2$ or 1.57078 to five places.) The direct interpolation from column one to column two of the table then manages $y(1.57080) = .999994$, so the cubic approximation has not done so badly at determining the maximum point.

TABLE 7.6

x	y	y'
1.4	.98545	.16997
1.5	.99749	.07074
1.6	.99957	−.02920
1.7	.99166	−.12884

7.32 Ignoring the top and bottom lines of Table 7.6, apply Hermite's formula to find a cubic polynomial fitting the remaining data. Where does the derivative of this cubic equal zero? Compare with the result of the preceding problem.

▮ Hermite's formula (see Problem 5.2) is in this simplest case

$$p(x) = U_0(x) y_0 + U_1(x) y_1 + V_0(x) y_0' + V_1(x) y_1'$$

with

$$U_0(x) = [1 + 2/h(x - x_0)](x - x_i)^2/h^2$$
$$U_1(x) = [1 - 2/h(x - x_1)](x - x_0)^2/h^2$$
$$V_0(x) = (x - x_0)(x - x_1)^2/h^2$$
$$V_1(x) = (x - x_1)(x - x_0)^2/h^2$$

and $h = x_1 - x_0$. It follows that the derivative of $h^2 p(x)$ involves the four coefficients

$$h^2 U_0'(x) = [1 + 2/h(x - x_0)]2(x - x_1) + 2/h(x - x_1)^2$$
$$h^2 U_1'(x) = [1 - 2/h(x - x_1)]2(x - x_0) - 2/h(x - x_0)^2$$
$$h^2 V_0'(x) = 2(x - x_0)(x - x_1) + (x - x_1)^2$$
$$h^2 V_1'(x) = 2(x - x_1)(x - x_0) + (x - x_G)^2$$

each multiplying its appropriate y or y' value. While it is true that this is a quadratic and its zero is available by the familiar formula, it is still a minor nuisance to organize the various terms. The alternative of direct computation, though pedestrian and itself not without nuisance value, was chosen. A search between 1.56 and 1.58 showed the zero to be between 1.57 and 1.571, after which a narrowing search pinned it down at 1.57082 to five places. (Compare with the 1.57080 found by the Lagrange interpolation and the correct 1.57078.)

7.33 The normal distribution function $y(x) = (1/\sqrt{2\pi})e^{-x^2/2}$ has an inflection point ($y'' = 0$) exactly at $x = 1$. How closely could this be determined from each of the four place data sets provided in Table 7.7?

TABLE 7.7

x	y	x	y
.50	.3521	.98	.2468
.75	.3011	.99	.2444
1.00	.2420	1.00	.2420
1.25	.1827	1.01	.2396
1.50	.1295	1.02	.2371

■ For the left side of the table the differences are (all entries are in the fifth decimal place)

$$
\begin{array}{cccc}
-510 & & & \\
& -81 & & \\
-591 & & 79 & \\
& -2 & & -16 \\
-593 & & 63 & \\
& 61 & & \\
-532 & & &
\end{array}
$$

Stirling's formula after two differentiations, developed in Problem 7.4, becomes in this case

$$
h^2 p'' = -.00002 + .00071 \cdot k + \frac{6k^2 - 1}{12}(-.00016)
$$

which simplifies to

$$
-10^5 h^2 p'' = 8k^2 - 71k + .67
$$

and is to be zero at the point of inflection. The quadratic formula provides $k = .0094466$ leading to

$$
x = 1 + .25k = 1.002
$$

The right side of the table has the differences

$$
\begin{array}{cccc}
-.0024 & & & \\
& 0 & & \\
-.0024 & & 0 & \\
& 0 & & -.0001 \\
-.0024 & & -.0001 & \\
& -.0001 & & \\
-.0025 & & &
\end{array}
$$

One glance at the column of first differences makes one wonder what lies ahead. The data are almost linear. Perservering we find the fairly flat parabola

$$
h^2 p'' = 0 - .00005 \cdot k - .0001 \frac{6k^2 - 1}{12}
$$

and then the quadratic equation

$$
5k^2 + 5k + .83 = 0
$$

with solution $k = -.21$ or $x = 1 - .002 = .998$. The denser packing of the data has not brought a better estimate of the inflection point.

7.34 From Problems 7.8 and 7.13 we find the combined truncation and roundoff errors of the approximation

$$
y'(x_0) = (y_1 - y_{-1})/2h
$$

to be of the form $Ah^2 + E/h$ with $A = |y^{(3)}(\xi)/6|$. What is the interval h for which this is a minimum?

■ Setting the derivative relative to h to zero

$$
2Ah = E/h^2 \qquad h = (E/2A)^{1/3}
$$

For Table 7.1, in which $E = .000005$ and A is bounded by $\frac{1}{16}$, this works out to $h = .03$. This is slightly smaller than the interval used in the table. Computations using a five-place table of the square root function produced these results for $x = 1.10$. The correct value is .47673:

h	.04	.03	.02	.01
$(y_1 - y_{-1})/2h$	.47688	.47683	.47675	.47650

The closest estimate occurs for $h = .02$.

7.35 From Problems 7.11 and 7.15 we find the combined truncation and roundoff errors of the approximation

$$y'(x_0) = (y_{-2} - 8y_{-1} + 8y_1 - y_2)/12h$$

to have the form $Ah^4 + 3E/2h$ where $A = |y^{(5)}(\xi)|/30|$. For what interval h will this be a minimum?

$$4Ah^3 = 3E/2h^2 \qquad h = (3E/8A)^{1/5}$$

For Table 7.1 this proves to be about .11, somewhat larger than the spacing of the table. Here are a few computed results:

h	.2	.1	.05	.01
approximate $y'(1.10)$	.47663	.47668	.47667	.47642

The correct value is still .47673, so this time the prediction of $h = .11$ as the optimum looks good.

7.36 Show that the truncation error of the formula

$$y^{(4)}(x_0) = \delta^4 y_0/h^4$$

is approximately $h^2 y^{(6)}(x_0)/6$.

▌ Applying the Taylor series method

$$y_2 = y_0 + 2hy_0' + 2h^2 y_0'' + \tfrac{4}{3}h^3 y_0^{(3)} + \tfrac{2}{3}h^4 y_0^{(4)} + \tfrac{4}{15}h^5 y_0^{(5)} + \tfrac{4}{45}h^6 y_0^{(6)} + \cdots$$
$$y_{-2} = y_0 - 2hy_0' + 2h^2 y_0'' - \tfrac{4}{3}h^3 y_0^{(3)} + \tfrac{2}{3}h^4 y_0^{(4)} - \tfrac{4}{15}h^5 y_0^{(5)} + \tfrac{4}{45}h^6 y_0^{(6)} - \cdots$$

which combine to the sum

$$y_2 + y_{-2} = 2y_0 + 4h^2 y_0'' + \tfrac{4}{3}h^4 y_0^{(4)} + \tfrac{8}{45}h^6 y_0^{(6)} + \cdots$$

Next

$$y_1 = y_0 + hy_0' + \tfrac{1}{2}h^2 y_0'' + \tfrac{1}{6}h^3 y_0^{(3)} + \tfrac{1}{24}h^4 y_0^{(4)} + \tfrac{1}{120}h^5 y_0^{(5)} + \tfrac{1}{720}h^6 y_0^{(6)} + \cdots$$
$$y_{-1} = y_0 - hy_0' + \tfrac{1}{2}h^2 y_0'' - \tfrac{1}{6}h^3 y_0^{(3)} + \tfrac{1}{24}h^4 y_0^{(4)} - \tfrac{1}{120}h^5 y_0^{(5)} + \tfrac{1}{720}h^6 y_0^{(6)} - \cdots$$

which combine to make this sum.

$$-4(y_1 + y_{-1}) = -8y_0 - 4h^2 y_0'' - \tfrac{1}{3}h^4 y_0^{(4)} - \tfrac{1}{90}h^6 y_0^{(6)} + \cdots$$

The two partial sums together with $6y_0$ produce a grand total of

$$h^4 y_0^{(4)} + \left(\tfrac{8}{45} - \tfrac{1}{90}\right)h^6 y_0^{(6)} + \cdots$$

making the truncation error $h^6 y_0^{(6)}/6 + \cdots$. The leading term of this series serves as an approximation.

7.37 Illustrate the method of undetermined coefficients by finding a numerical differentiation formula of the sort

$$y'(0) = a_1 y(h) + a_2 y(0) + a_3 y(-h)$$

▌ The coefficients of this three term formula will be found by making it exact for the simple polynomials $y(x) = 1$, x and x^2. For these functions the formula becomes

$$0 = a_1 + a_2 + a_3$$
$$1 = a_1 h - a_3 h$$
$$0 = a_1 h^2 + a_3 h^2$$

in turn, and the system has the familiar solution

$$a_1 = -a_3 = 1/2h \qquad a_2 = 0$$

The formula is $[y(h) - y(-h)]/2h$.

7.38   Apply the method of undetermined coefficients to find a formula of the sort

$$y'(0) = a_2 y(2h) + a_1 y(h) + a_0 y(0) + a_{-1} y(-h) + a_{-2} y(-2h)$$

▮ This time there are five coefficients to be determined, so we require the formula to be exact for $y(x) = 1, x, x^2, x^3, x^4$ and we have the five equations

$$0 = a_2 + a_1 + a_0 + a_{-1} + a_{-2}$$
$$1 = 2ha_2 + ha_1 - ha_{-1} - 2ha_{-2}$$
$$0 = 4h^2 a_2 + h^2 a_1 + h^2 a_{-1} + 4h^2 a_{-2}$$
$$0 = 8h^3 a_2 + h^3 a_1 - h^3 a_{-1} - 8h^3 a_{-2}$$
$$0 = 16h^4 a_2 + h^4 a_1 + h^4 a_{-1} + 16h^4 a_{-2}$$

In one way or another (reference to problem 7.10 perhaps) the solution

$$(a_2, a_1, a_0, a_{-1}, a_{-2}) = (-1, 8, 0, -8, 1)/12h$$

may be found. The formula of the earlier problem has been rediscovered.

7.39   Use undetermined coefficients to produce a differentiation formula of the unsymmetric sort, with $x_1 < 0 < x_2$,

$$y'(0) = a_1 y(x_1) + a_0 y(0) + a_2 y(x_2)$$

▮ There are many possible specifications for determining the coefficients, but suppose we again ask for exactness for the polynomials $y(x) = 1, x$ and x^2:

$$0 = a_1 + a_0 + a_2$$
$$1 = x_1 a_1 + x_2 a_2$$
$$0 = x_1^2 a_1 + x_2^2 a_2$$

The solution of this system is easily found:

$$a_1 = x_2/x_1(x_2 - x_1) \qquad a_0 = -(x_1 + x_2)/x_1 x_2 \qquad a_2 = -x_1/x_2(x_2 - x_1)$$

7.40   Apply the formula of the preceding problem to the exponential function at $x = 0$ to observe the effect of asymmetry.

▮ Choosing several values for x_1 and keeping $x_2 = x_1 + 2$, the following approximate derivatives were found:

x_1	$-.2$	$-.6$	-1.0	-1.4	-1.8	-1.99
p'	1.10	1.18	1.18	1.12	1.04	1.00

The symmetric case of $x_1 = -1$ gives one of the poorest values. It is an exercise of elementary calculus to show that the limit for x_1 approaching -2 will be the correct derivative, which is 1.

7.41   Use the method of undetermined coefficients is to find a formula of the sort, for the approximation of the second derivative,

$$y''(0) = a_1 y(h) + a_0 y(0) + a_{-1} y(-h)$$

▮ As before we require exactness for $y(x) = 1, x$ and x^2, and we have good reason to anticipate a second difference expression on the right. In fact the three conditions,

$$0 = a_1 + a_0 + a_{-1}$$
$$0 = ha_1 - ha_{-1}$$
$$2 = h^2 a_1 + h^2 a_{-1}$$

are easily solved for $a_1 = a_{-1} = 1/h^2$ and $a_0 = -2/h^2$ and we have

$$[y(h) - 2y(0) + y(-h)]/h^2$$

as anticipated.

7.42 How can the method of undetermined coefficients be extended to estimate truncation error?

▌ The idea is to see how badly the formula does for the first power function for which it is not exact. Our basic truncation error formula contains the factors $y^{(n+1)}(\xi)/(n+1)!$, which, for the power function $y = x^{n+1}$, is exactly 1. We must find the other factor of the error. For example, returning to

$$[y(h) - y(-h)]/2h$$

which is exact (see Problem 7.37) for $y(x) = 1, x, x^2$, we evaluate for $y(x) = x^3$ and have

$$h^3 - (-h)^3/2h = h^2$$

The truncation error estimate is then $0 - h^2$, where zero is the correct derivative at $x = 0$. For a different function $y(x)$ we now suspect the truncation error to be

$$-y^{(n+1)}(\xi)h^2/(n+1)!$$

which is precisely the estimate found in Problem 7.8.
  In the same way, the formula

$$[-y(2h) + 8y(h) - 8y(-h) + y(-2h)]/12h$$

was found in Problem 7.38 to yield $y'(0)$ exactly for $y(x) = 1, x, x^2, x^3$ and x^4. For x^5 it produces

$$[-32h^5 + 8h^5 + 8h^5 - 32h^5]/12h = -4h^4$$

leading us to the truncation error estimate

$$4h^4 y^{(5)}(\xi)/120 = y^{(5)}(\xi)h^4/30$$

the same estimate obtained in Problem 7.11.
  As a further example take the second derivative formula

$$[y(h) - 2y(0) + y(-h)]/h^2$$

derived in the preceding problem by requiring exactness for the power functions $1, x$ and x^2. With $y(x) = x^3$ it manages

$$h^3 = (-h)^3 = 0$$

so we have a bonus level of exactness. Now try x^4:

$$\left[h^4 + (-h)^4\right]/h^2 = 2h^2$$

leading to the estimate

$$-2h^2 y^{(4)}(\xi)/4! = -y^{(4)}(\xi)h^2/12$$

which agrees with the result of Problem 7.23.

7.43 The blend of function and first derivative values,

$$a_1 y(h) + b_1 y'(h) + a_2 y(-h) + b_2 y'(-h)$$

seems to offer reasonable prospects as a formula for the second derivative $y''(0)$. Proceed by undetermined coefficients.

▌ Exactness for $y(x) = 1, x, x^2, x^3$ requires

$$0 = a_1 + a_2$$
$$0 = ha_1 + b_1 - ha_2 + b_2$$
$$2 = h^2 a_1 + 2hb_1 + h^2 a_2 - 2hb_2$$
$$0 = h^3 a_1 + 3h^2 b_1 - h^3 a_2 + 3h^2 b_2$$

which responds to the usual algebraic manipulations and yields $a_1 = a_2 = 0$ and $b_1 = -b_2 = 1/2h$. The resulting formula is

$$[y'(h) - y'(-h)]/2h$$

and is a slight disappointment, not being the expected blend at all but simply a restatement of an earlier approximation to the first derivative.

7.44 What should be expected of an attempt to approximate the third derivative of $y(x)$ at $x = 0$ by the formula

$$a_1 y(3h/2) + a_2 y(h/2) + a_3 y(-h/2) + a_4 y(-3h/2)?$$

What does the method of undetermined coefficients produce?

▮ A four term approximation to the third derivative using equally spaced arguments may be expected to be essentially the third difference. Proceeding by undetermined coefficients,

$$0 = a_1 + a_2 + a_3 + a_4$$
$$0 = \tfrac{3}{2}ha_1 + \tfrac{1}{2}ha_2 - \tfrac{1}{2}ha_3 - \tfrac{3}{2}ha_4$$
$$0 = \tfrac{9}{4}h^2 a_1 + \tfrac{1}{4}h^2 a_2 + \tfrac{1}{4}h^2 a_3 + \tfrac{9}{4}h^2 a_4$$
$$6 = \tfrac{27}{8}h^3 a_1 + \tfrac{1}{8}h^3 a_2 - \tfrac{1}{8}h^3 a_3 - \tfrac{27}{8}h^3 a_4$$

which with a little persistence reduces to

$$a_1 = 1/h^3 \qquad a_2 = -3/h^3 \qquad a_3 = 3/h^3 \qquad a_4 = -1/h^3$$

bringing the central difference into view:

$$[y(3h/2) - 3y(h/2) + 3y(-h/2) - y(3h/2)]/h^3$$

7.45 Estimate the truncation error of the third central difference as an approximate third derivative by testing it for the next higher power function.

▮ With $y(x) = x^4$ we quickly find exactness. Turning to x^5,

$$h^5(243/32 - 3/32 - 3/32 + 243/32) = 15h^5$$

and our estimate is

$$-15h^5 y^{(5)}(\xi)/5! = -h^5 y^{(5)}(\xi)/8$$

This is consistent with the y''' result in Problem 7.17 where a different notation uses $k = \tfrac{1}{2}$ for the same midway position.

7.46 How can the Richardson extrapolation method be applied to numerical differentiation?

▮ As usual, information about the error in an approximation formula is used to make a correction. As an illustration, take the central formula

$$y'(x) = \frac{y(x+h) - y(x-h)}{2h} + T$$

where T is the truncation error. An easy calculation using Taylor series finds

$$T = a_1 h^2 + a_2 h^4 + a_3 h^6 + \cdots$$

Making two applications, using h and $h/2$, we have

$$y'(x) = F(h) + a_1 h^2 + a_2 h^4 + \cdots$$
$$y'(x) = F\left(\frac{h}{2}\right) + \frac{a_1 h^2}{4} + \frac{a_2 h^4}{16} + \cdots$$

with $F(h)$ and $F(h/2)$ denoting the approximate derivatives and where we assume that the a_i do not change much for small h. Eliminating the a_1 terms leads to

$$y'(x) = \frac{4F(h/2) - F(h)}{3} + b_1 h^4 + O(h^6)$$

so that in

$$F_1\left(\frac{h}{2}\right) = \frac{4F(h/2) - F(h)}{3}$$

we have an approximate differentiation formula of fourth-order accuracy, obtained by combining two results from a formula of second-order accuracy.

The argument can now be repeated, beginning with

$$y'(x) = F_1\left(\frac{h}{2}\right) + b_1 h^4 + O(h^6)$$

$$y'(x) = F_1\left(\frac{h}{4}\right) + \frac{b_1 h^4}{16} + O(h^6)$$

and eliminating the b_1 term to produce an approximation

$$F_2\left(\frac{h}{2}\right) = \frac{16F_1(h/4) - F_1(h/2)}{15}$$

with sixth-order accuracy. Clearly further repetitions are possible, the overall process being known as extrapolation to the limit.

The set of approximations calculated during an extrapolation to the limit is usually displayed as

	F	F_1	F_2	F_3
h	$F(h)$			
$h/2$	$F(h/2)$	$F_1(h/2)$		
$h/4$	$F(h/4)$	$F_1(h/4)$	$F_2(h/4)$	
$h/8$	$F(h/8)$	$F_1(h/8)$	$F_2(h/8)$	$F_3(h/8)$

more entries being added as needed. The general formula is

$$F_m\left(\frac{h}{2^k}\right) = F_{m-1}\left(\frac{h}{2^k}\right) + \frac{F_{m-1}(h/2^k) - F_{m-1}(h/2^{k-1})}{2^{2m} - 1}$$

It is not hard to modify the process just sketched so that the step size is reduced in some other way, perhaps $h_i = r^{i-1}h_1$ with h_1 the initial h. An arbitrary sequence of h_i could even be handled at little cost. Examples exist to show that sometimes these variations can be profitable.

7.47   Apply the Richardson extrapolation to the function $y(x) = -1/x$ to find $y'(.05)$, the exact value being 400.

▮ The computations are summarized in Table 7.8. The original formula of Problem 7.46 produced the column headed F (with all entries reduced by 400 for simplicity), so its best effort, for $h = .0001$, was off in the third

TABLE 7.8 (Entries reduced by 400)

h	F	F_1	F_2	F_3
.0128	28.05289			
.0064	6.66273	$-.46732$		
.0032	1.64515	$-.02737$	.00196	
.0016	.41031	$-.00130$	.00043	.00041
.0008	.10250	$-.00010$	$-.00002$	$-.00002$
.0004	.02625	.00084	.00090	.00091
.0002	.00750	.00125	.00127	.00127
.0001	.00500	.00417	.00436	.00441
.00005	.01000	.01166	.01215	.01227

decimal place, after which roundoff error took over. Looking elsewhere in the table one sees that values almost correct to five places appear, so the method of extrapolation has made a definite contribution.

7.48 Using the F functions of the preceding problems, show that

$$F_1(h) = \frac{4F(h) - F(2h)}{3} = \frac{-y_2 + 8y_1 - 8y_{-1} + y_{-2}}{12h}$$

once again recovering the formula first found in Problem 7.10 and having its origin in the differentiation of Stirling's formula.

▮ We have

$$\frac{4}{3}F(h) - \frac{1}{3}F(2h) = \frac{1}{6h}(4y_1 - 4y_{-1}) - \frac{1}{12h}(y_2 - y_{-2})$$

which reduces at once to the indicated result.

7.49 Show that

$$y'(x_0) = \frac{F(nh) - n^2 F(h)}{1 - n^2} + O(h^4)$$

▮
$$F(nh) = y_0' - An^2h^2 + O(h^4)$$
$$F(h) = y_0' - Ah^2 + O(h^4)$$

so that

$$F(nh) - n^2 F(h) = (1 - n^2) y_0' + O(h^4)$$

from which it is one step to what was required.

7.50 Apply the splines of Problem 5.82 and 5.83 to find approximate derivatives of the sine function.

▮ In Problem 5.82 we found the natural spline, having zero second derivatives at the endpoints. Since the sine itself has these end derivatives, the natural spline is appropriate in this case. Taking the center point first, we find the derivative of the center spline segment S_2 to be

$$S_2'(x) = -\frac{27\sqrt{3}}{10\pi^3}(2\pi x - \pi^2)$$

which is precisely zero at $x = \pi/2$. Clearly the symmetry has been helpful. A fairer test may be made at $x = \pi/3$, which was one of the knots, where we find S_2' to be .496. The error of .4% may be judged keeping in mind that only three spline segments were used over the interval $(0, \pi)$.

In Problem 5.83 we found the spline that matched the endpoint first derivatives of the sine function. For the center section we found

$$S_2'(x) = \frac{2\pi - 9\sqrt{3}}{2\pi^3}(2\pi x - \pi^2)$$

which is again zero at $x = \pi/2$. At $x = \pi/3$, it manages $(9\sqrt{3} - 2\pi)/6\pi$ or .494.

For the second derivative the anticipated deterioration again appears. The natural spline predicts $S_2'' = -.948$ for the entire center interval, where the true second derivative ranges from $-.866$ to -1.

7.51 The representation of splines in terms of basis functions can be used to compute derivatives. Develop the needed formulas by differentiating the basis spline $B_i(x)$ of Problem 5.99.

▮ The four active parts of $B_i(x)$ have the derivatives [with $B_i'(x)$ equal to zero elsewhere]

$$B_i'(x) = \frac{1}{6h^3}\begin{cases} 3(x - x_{i-2})^2 & \text{on } (x_{i-2}, x_{i-1}) \\ 3h^2 + 6h(x - x_{i-1}) - 9(x - x_{i-1})^2 & \text{on } (x_{i-1}, x_i) \\ -3h^2 - 6h(x_{i+1} - x) + 9(x_{i+1} - x)^2 & \text{on } (x_i, x_{i+1}) \\ -3(x_{i+2} - x)^2 & \text{on } (x_{i+1}, x_{i+2}) \end{cases}$$

Since $B_i(x)$ is itself a cubic spline, each part of its derivative is quadratic. The linear segments of its second derivative will be

$$B_i''(x) = \frac{1}{6h^3} \begin{cases} 6(x - x_{i-2}) & \text{on } (x_{i-2}, x_{i-1}) \\ 6h - 18(x - x_{i-1}) & \text{on } (x_{i-1}, x_i) \\ 6h - 18(x_{i+1} - x) & \text{on } (x_i, x_{i+1}) \\ 6(x_{i+2} - x) & \text{on } (x_{i+1}, x_{i+2}) \end{cases}$$

7.52 Apply the formulas for the derivatives of basis splines, found in the preceding problem, to the data function

x	0	1	2	3
y	0	0	0	0

for which coefficients were found in Problem 5.104 using the additional specification that endpoint derivatives must agree with those of the parent function $y(x) = x(x - 1)(x - 2)(x - 3)$.

❚ The endpoint derivatives were -6 and 6. Since the values of the coefficients are in hand, it is only a matter of applying them to the derivatives of appropriate segments. A program for computing the spline itself is easily modified for this purpose; simply replace the spline segments by their first or second derivatives. A few computed results appear in Table 7.9 along with the correct values. Considering the character of the data set, it is encouraging to see the good results for the first derivative, where there was agreement to five decimal places or more. The second derivative has been imitated qualitatively but not quantitatively.

TABLE 7.9

x	0	.5	1	1.5	2	2.5	3
y'	-6	1	2	0	-2	-1	6
s'			the same to five places				
s''	20	8	-4	-4	-4	8	20
y''	22	7	-2	-5	-2	7	22

7.53 How well does the spline of Problem 5.102 represent the first derivative of the sine function?

❚ Recall that this spline was based on a four point data set with arguments 0, $\pi/3$, $2\pi/3$ and π. The coefficients of the required basis functions were obtained in the earlier problem. We now apply them to the segment derivatives instead of to the segments themselves and have the computed values

x	0	$\pi/6$	$\pi/3$	$\pi/2$	$2\pi/3$	$5\pi/6$	π
s'	1.04	.91	.52	0	$-.52$	$-.91$	-1.04
true	1	.87	.50	0	$-.50$	$-.87$	-1

The errors range from 4% at the ends to essentially zero at the center.

7.54 Apply the method of spline numerical differentiation to the "spike" function of Problem 5.101:

x	0	1	2	3	4
y	0	0	1	0	0

❚ This is another rough function for approximation by cubic or other polynomials. The coefficients of a representation using our basis splines were worked out in the earlier problem, using the further specification of zero end derivatives. Here are a few computed derivatives. The exact values are unknown since no parent function has been indicated. These values should be matched with the plot of this spline given as Fig. 5-12(a):

x	0	.5	1	1.5	2	2.5	3	3.5	4
s'	0	$-.2$	.75	1.3	0	-1.3	$-.75$	.2	0

7.55 Find a six part spline for the following data drawn from the square root function.

x	.85	.90	.95	1.00	1.05	1.10	1.15
$y = \sqrt{x}$	.92195	.94868	.97468	1.00000	1.02470	1.04881	1.07238

▌ Choosing as the extra specifications the end values of the first derivative, which are .5423261 and .4662524, the coefficients of the nine basis functions will satisfy the system shown as Fig. 7-2. To six places they are

$$(.894572, .922087, .948805, .974792, 1.000104, 1.024792, 1.048899, 1.072465, 1.095524)$$

Applying these coefficients to the first and second derivatives of the spline segments we get the values shown in Table 7.10.

The correct values also appear. It will be noted that the first derivatives are accurate to five or six places, while the second derivatives are dependable to only three or four.

$$
\begin{bmatrix}
-1 & 0 & 1 & 0 & 0 & 0 & 0 & 0 & 0 \\
1 & 4 & 1 & 0 & 0 & 0 & 0 & 0 & 0 \\
0 & 1 & 4 & 1 & 0 & 0 & 0 & 0 & 0 \\
0 & 0 & 1 & 4 & 1 & 0 & 0 & 0 & 0 \\
0 & 0 & 0 & 1 & 4 & 1 & 0 & 0 & 0 \\
0 & 0 & 0 & 0 & 1 & 4 & 1 & 0 & 0 \\
0 & 0 & 0 & 0 & 0 & 1 & 4 & 1 & 0 \\
0 & 0 & 0 & 0 & 0 & 0 & 1 & 4 & 1 \\
0 & 0 & 0 & 0 & 0 & 0 & -1 & 0 & 1
\end{bmatrix}
\begin{bmatrix}
c_{-1} \\ c_0 \\ c_1 \\ c_2 \\ c_3 \\ c_4 \\ c_5 \\ c_6 \\ c_7
\end{bmatrix}
=
\begin{bmatrix}
.05423261 \\ 5.53170 \\ 5.69208 \\ 5.84808 \\ 6 \\ 6.14820 \\ 6.29286 \\ 6.43428 \\ .04662524
\end{bmatrix}
$$

Fig. 7-2

TABLE 7.10

x	s'	y'	s''	y''
.85	.542326	.542326	$-.3188$	$-.3190$
.90	.527044	.527046	$-.2924$	$-.2928$
.95	.512988	.512989	$-.2698$	$-.2700$
1.00	.500002	.500000	$-.2497$	$-.2500$
1.05	.487950	.487950	$-.2324$	$-.2324$
1.10	.476728	.476731	$-.2165$	$-.2167$
1.15	.466248	.466252	$-.2028$	$-.2027$

7.56 For the step function of Problem 5.111 all derivatives should certainly be zero, except at the step. How well does the six part spline obtained in that problem oblige?

▌ Borrowing the coefficients from the earlier problem we have the figures

x	0	1	2	3	4	5	6
s'	0	$-.1$	.2	.6	.2	$-.1$	0
s''	.1	$-.2$	.8	0	$-.8$	.2	$-.1$

which are perhaps as good as we should have expected from this rough function.

7.57 The previous problems suggest that approximate differentiation is an inaccurate affair. Illustrate this further by comparing the function $y(x) = e \sin(x/e^2)$ with the polynomial approximation $p(x) = 0$.

▌ The two functions collocate at the equally spaced arguments $x = ie^2\pi$ for integers i. For a very small number e, the approximation is extremely accurate, $y(x) - p(x)$ never exceeding e. However, since $y'(x) = (1/e)\cos(x/e^2)$ and $p'(x) = 0$, the difference in derivatives is enormous. This example shows that accurate approximation of a function should not be expected to mean accurate approximation of its derivative. See Fig. 7-3.

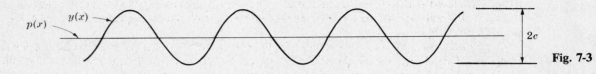

Fig. 7-3

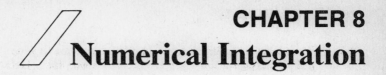

CHAPTER 8
Numerical Integration

8.1 What are some of the types of methods available for approximation of integrals?

❚ *A single polynomial* $p(x)$ can be used to approximate $y(x)$ over the interval of integration. The integral of the polynomial then substitutes for the integral of $y(x)$. The resulting Newton–Cotes formulas are the ingredients from which a variety of useful and popular methods have been developed, but are not often used as they come.

Composite methods such as the trapezoidal and Simpson rules use numerous segments of low degree polynomials to cover the interval of integration, rather than a single polynomial of high degree. This has numerous advantages, including simplicity and avoidance of the oscillatory character of high-degree polynomials which has already been noted in interpolation problems.

Extrapolation methods use information about the truncation error to accelerate convergence toward the true integral. The Romberg method, for example, uses the fact that the trapezoidal rule has a truncation error proportional to h^2. Two applications of the rule, with say h and $h/2$, may therefore be expected to reduce the error by a factor of $\frac{1}{4}$. Comparing the two results leads to an estimate of the error remaining. The estimate then leads to an acceleration device. Romberg's method systematically exploits this idea.

Correction terms for the trapezoidal and Simpson rules can be found by integrating the approximating polynomials over less than the full range of collocation. For example, integration of the Stirling formula of degree 6, which provides collocation at the arguments $x_{-3}, \ldots, x_3$, over just the center two intervals x_{-1} to x_1 then leads to a composite formula consisting of Simpson's rule with correction terms. Similarly, starting with Bessel's formula of degree 5 and integrating over just the one center interval leads to the trapezoidal rule with correction terms.

Series methods, particularly Taylor series, may be applied to express the integrand as a power series, after which term by term integration sometimes leads to a feasible computation. It is also possible to modify the series being used step by step as one advances across the interval of integration, the integral being treated as a simplified differential equation.

Adaptive methods have been devised to deal with the fact that many functions are harder to integrate accurately over some parts of the interval than over others. A particularly difficult section might, for example, force the use of a very small h value in Simpson's rule and cause a great deal of unnecessary computation elsewhere. Adaptive methods use a finer grid only where it is actually needed.

Special devices may be required for special situations. The formula of Filon works well for integrals involving periodic functions. If a function oscillates rapidly it will be hard to handle by the usual devices. The Filon formula is an attempt to build the oscillations into the procedure. Singular integrals also require special devices, especially those with infinite singularities.

Gaussian methods can be derived using osculating polynomials, the Hermite formula being useful for this purpose. The formulas obtained prove to be very accurate, as might be anticipated from the higher order of the polynomial approximations. The relations between Gaussian methods and orthogonal polynomials provide some attractive mathematics, but the importance of these methods is not esthetic but utilitarian.

8.2 What are the important sources of error in numerical integration?

❚ All the usual sources are present. However, input errors in the data values $y_0, \ldots, y_n$ are not magnified by most integration formulas, so this source of error is not nearly so troublesome as it is in numerical differentiation. Truncation error, which is

$$\int_a^b [y(x) - p(x)] \, dx$$

for our simplest formulas, and a composite of similar pieces for most of the others, is now the major contributor. A variety of methods for estimating this error has been developed.

A related question is that of *convergence*. This asks whether as continually higher-degree polynomials are used, or as continually smaller intervals h between data points are used with $\lim h = 0$, a sequence of approximations is generated for which the limit of truncation error is zero. In many cases, including the trapezoidal and Simpson rules, convergence can be proved.

Roundoff errors also have a strong effect. A small interval h means a lot of computing and many roundoffs. Examples will be offered in which the effect of roundoff will be very evident. It usually limits the accuracy attainable for a given integral and obscures the convergence that should theoretically occur. It is found in practice that decreasing h below a certain level leads to larger errors rather than smaller, truncation error becoming negligible while roundoff error accumulates.

8.1 COLLOCATION METHODS

8.3 Integrate Newton's formula for a collocation polynomial of degree n. Use the limits x_0 and x_n, which are the outside limits of collocation. Assume equally spaced arguments.

▮ This involves integrating a linear function from x_0 to x_1, or a quadratic from x_0 to x_2 and so on. See Fig. 8-

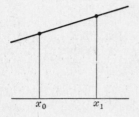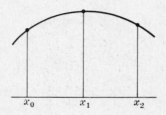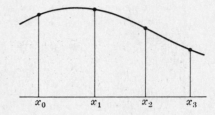

Fig. 8-1

The linear function certainly leads to $\frac{1}{2}h(y_0 + y_1)$. For the quadratic

$$p_k = y_0 + k\,\Delta y_0 + \tfrac{1}{2}k(k-1)\,\Delta^2 y_0$$

and easy computation produces, since $x = x_0 + kh$,

$$\int_{x_0}^{x_2} p(x)\,dx = h\int_0^2 p_k\,dk = h\left(2y_0 + 2\,\Delta y_0 + \frac{1}{3}\,\Delta^2 y_0\right) = \frac{h}{3}(y_0 + 4y_1 + y_2)$$

For the cubic polynomial a similar calculation produces

$$\int_{x_0}^{x_3} p(x)\,dx = h\int_0^3 p_k\,dk = h\int_0^3 \left[y_0 + k\,\Delta y_0 + \binom{k}{2}\Delta^2 y_0 + \binom{k}{3}\Delta^3 y_0\right]dk$$

$$= h\left(3y_0 + \frac{9}{2}\,\Delta y_0 + \frac{9}{4}\,\Delta^2 y_0 + \frac{3}{8}\,\Delta^3 y_0\right) = \frac{3h}{8}(y_0 + 3y_1 + 3y_2 + y_3)$$

Results for higher-degree polynomials can also be obtained in the same form,

$$\int_{x_0}^{x_n} p(x)\,dx = Ch(c_0 y_0 + \cdots + c_n y_n)$$

and values of C and c_i for the first few values of n are given in Table 8.1. Such formulas are called the *Cotes formulas*. Higher-degree formulas are seldom used, partly because simpler and equally accurate formulas are available and partly because of the somewhat surprising fact that higher-degree polynomials do not always mean improved accuracy.

TABLE 8.1

n	C	c_0	c_1	c_2	c_3	c_4	c_5	c_6	c_7	c_8
1	1/2	1	1							
2	1/3	1	4	1						
3	3/8	1	3	3	1					
4	2/45	7	32	12	32	7				
6	1/140	41	216	27	272	27	216	41		
8	4/14175	989	5888	-928	10496	-4540	10496	-928	5888	989

8.4   Estimate the truncation error of the $n = 1$ formula.

▮ For this simple case we can integrate the formula

$$y(x) - p(x) = \tfrac{1}{2}(x - x_0)(x - x_1)y^{(2)}(\xi)$$

directly and apply the mean value theorem as follows, obtaining the exact error

$$\int_{x_0}^{x_1} y(x)\,dx - \tfrac{1}{2}h(y_0 + y_1) = \int_{x_0}^{x_1} \tfrac{1}{2}(x - x_0)(x - x_1)y^{(2)}(\xi)\,dx$$

$$= y^{(2)}(\xi)\int_{x_0}^{x_1} \tfrac{1}{2}(x - x_0)(x - x_1)\,dx = -\tfrac{1}{12}h^3 y^{(2)}(\xi)$$

where $h = x_1 - x_0$. The application of the mean value theorem is possible because $(x - x_0)(x - x_1)$ does not change sign in (x_0, x_1). The continuity of $y^{(2)}(\xi)$ is also involved. For $n > 1$ a sign change prevents a similar application of the mean value theorem and many methods have been devised to estimate truncation error, most having some disadvantages. We now illustrate one of the oldest methods, using the Taylor series, for the present simple case $n = 1$. First we have

$$\tfrac{1}{2}h(y_0 + y_1) = \tfrac{1}{2}h\left[y_0 + \left(y_0 + hy_0' + \tfrac{1}{2}h^2 y_0^{(2)} + \cdots \right) \right]$$

Using an indefinite integral $F(x)$, where $F'(x) = y(x)$, we can also find

$$\int_{x_0}^{x_1} y(x)\,dx = F(x_1) - F(x_0) = hF'(x_0) + \tfrac{1}{2}h^2 F^{(2)}(x_0) + \tfrac{1}{6}h^3 F^{(3)}(x_0) + \cdots = hy_0 + \tfrac{1}{2}h^2 y_0' + \tfrac{1}{6}h^3 y_0^{(2)} + \cdots$$

and subtracting,

$$\int_{x_0}^{x_1} y(x)\,dx - \frac{1}{2}h(y_0 + y_1) = -\frac{h^3}{12}y_0^{(2)} + \cdots$$

presenting the truncation error in series form. The first term may be used as an error estimate. It should be compared with the actual error as given by $-(h^3/12)y^{(2)}(\xi)$ where $x_0 < \xi < x_1$.

8.5   Estimate the truncation error of the $n = 2$ formula.

▮ Proceeding as in the previous problem, we find first

$$\tfrac{1}{3}h(y_0 + 4y_1 + y_2) = \tfrac{1}{3}h\left[y_0 + 4\left(y_0 + hy_0' + \tfrac{1}{2}h^2 y_0^{(2)} + \tfrac{1}{6}h^3 y_0^{(3)} + \tfrac{1}{24}h^4 y_0^{(4)} + \cdots \right) \right.$$

$$\left. + \left(y_0 + 2hy_0' + 2h^2 y_0^{(2)} + \tfrac{4}{3}h^3 y_0^{(3)} + \tfrac{2}{3}h^4 y_0^{(4)} + \cdots \right) \right]$$

$$= \tfrac{1}{3}h\left(6y_0 + 6hy_0' + 4h^2 y_0^{(2)} + 2h^3 y_0^{(3)} + \tfrac{5}{6}h^4 y_0^{(4)} + \cdots \right)$$

The integral itself is

$$\int_{x_0}^{x_2} y(x)\,dx = F(x_2) - F(x_0)$$

$$= 2hF'(x_0) + \tfrac{1}{2}(2h)^2 F^{(2)}(x_0) + \tfrac{1}{6}(2h)^3 F^{(3)}(x_0) + \tfrac{1}{24}(2h)^4 F^{(4)}(x_0) + \tfrac{1}{120}(2h)^5 F^{(5)}(x_0) + \cdots$$

$$= 2hy_0 + 2h^2 y_0 + \tfrac{4}{3}h^3 y_0^{(2)} + \tfrac{2}{3}h^4 y_0^{(3)} + \tfrac{4}{15}h^5 y_0^{(4)} + \cdots$$

and subtracting,

$$\int_{x_0}^{x_2} y(x)\,dx - \tfrac{1}{3}h(y_0 + 4y_1 + y_2) = -\tfrac{1}{90}h^5 y_0^{(4)} + \cdots$$

we again have the truncation error in series form. The first term will be used as an approximation. It can also be shown that the error is given by $-(h^5/90)y^{(4)}(\xi)$ where $x_0 < \xi < x_2$. (See Problem 8.6.)

A similar procedure applies to the other formulas. Results are presented in Table 8.2, the first term only being shown.

TABLE 8.2

n	Truncation error	n	Truncation error
1	$-(h^3/12)\,y^{(2)}$	4	$-(8h^7/945)\,y^{(6)}$
2	$-(h^5/90)\,y^{(4)}$	6	$-(9h^9/1400)\,y^{(8)}$
3	$-(3h^5/80)\,y^{(4)}$	8	$-(2368h^{11}/467{,}775)\,y^{(10)}$

Notice that formulas for odd n are comparable with those for the next smaller integer. (Of course, such formulas do cover one more interval of length h, but this does not prove to be significant. The even formulas are superior.)

8.6 Derive the exact expression for the error of the $n = 2$ formula as given in the preceding problem.

▮ Let

$$F(h) = \int_{-h}^{h} y(x)\,dx - \frac{h}{3}[\,y(-h) + 4y(0) + y(h)\,]$$

Differentiate three times relative to h, using the theorem on "differentiating under the integral sign,"

$$\frac{d}{dh}\int_{a(h)}^{b(h)} y(x,h)\,dx = \int_{a}^{b}\frac{\partial y}{\partial h}\,dx + y(b,h)\,b'(h) - y(a,h)\,a'(h)$$

to obtain

$$F^{(3)}(h) = -\frac{h}{3}\Big[\,y^{(3)}(h) - y^{(3)}(-h)\,\Big]$$

Notice that $F'(0) = F^{(2)}(0) = F^{(3)}(0) = 0$. Assuming $y^{(4)}(x)$ continuous, the mean value theorem now produces

$$F^{(3)}(h) = -\tfrac{2}{3}h^2 y^{(4)}(\theta h)$$

where θ depends on h and falls between -1 and 1. We now reverse direction and recover $F(h)$ by integration. It is convenient to replace h by t (making θ a function of t). We then have

$$F(h) = -\tfrac{1}{3}\int_{0}^{h}(h-t)^2 t^2 y^{(4)}(\theta t)\,dt$$

as may be verified by differentiating three times relative to h to recover the preceding $F^{(3)}(h)$. Because this formula also makes $F(0) = F'(0) = F''(0)$, it is the original $F(h)$. Now apply the mean value theorem

$$\int_{a}^{b} f(t)\,g(t)\,dt = g(\xi)\int_{a}^{b} f(t)\,dt$$

with $a < \xi < b$, which is valid for continuous functions provided $f(t)$ does not change sign between a and b. These conditions do hold here with $f(t) = -t^2(h-t)^2/3$. The result is

$$F(h) = y^{(4)}(\xi)\int_{0}^{h} f(t)\,dt = -\frac{h^5}{90}y^{(4)}(\xi)$$

The early parts of this proof, in which we maneuver from $F(h)$ to its third derivative and back again, have as their goal a representation of $F(h)$ to which the mean value theorem can be applied. This is often the main difficulty in obtaining a truncation error formula of the sort just achieved.

8.7 The Newton–Cotes formulas can also be derived by a method of undetermined coefficients. Illustrate this for $n = 2$.

▮ It is convenient to make the change of variable $k = (x - x_1)/h$ converting the integral to the form

$$I = \int_{x_0}^{x_2} y(x)\,dx = h\int_{-1}^{1} y_k\,dk = c_{-1}y_{-1} + c_0 y_0 + c_1 y_1$$

Asking that this hold exactly for $y = 1$, k and k^2 then lead to the equations

$$2h = \quad c_{-1} + c_0 + c_1$$
$$0 = -c_{-1} + c_1$$
$$2h/3 = \quad c_{-1} + c_1$$

from which come $c_{-1} = c_1 = h/3$ and $c_0 = 4h/3$ as predicted.

8.8 Show that the truncation error estimate of Problem 8.5 can also be found by testing the formula for the next higher powers of k.

▮ We know it to be exact for powers of $(x - x_1)/h$ up through the second. For power 3 we find

$$I = h \int_{-1}^{1} k^3 \, dk = 0 \qquad A = \frac{h}{3}\left[(-1)^3 + 4 \cdot 0 + 1^3\right] = 0$$

so that the integral I and its approximation A agree. For power 4, however,

$$I = h \int_{-1}^{1} k^4 \, dk = \frac{2h}{5} \qquad A = \frac{h}{3}\left[(-1)^4 + 4 \cdot 0 + 1^4\right] = \frac{2h}{3}$$

making $I - A = -4h/15$. For this power function, the fourth derivative relative to k will be 24, making the corresponding derivative relative to $x = 24/h^4$. This leads to

$$\frac{(I - A)}{y^{(4)}} = \frac{-4h}{15} \cdot \frac{h^4}{24} = \frac{-h^5}{90}$$

and assuming this to hold for other functions $y(x)$, we have the required estimate:

$$\text{truncation error} = -h^5 y^{(4)}/90$$

It is clear that this derivation involves a bit of optimism, but the details are so simple that the routine is seductive.

8.9 Use the method of undetermined coefficients to derive the $n = 4$ Newton–Cotes formula.

▮ Let $k = (x - x_2)/h$ and obtain the integral in the form

$$I = \int_{x_0}^{x_4} y(x) \, dx = h \int_{-2}^{2} y_k \, dk = c_{-2}y_{-2} + c_{-1}y_{-1} + c_0 y_0 + c_1 y_1 + c_2 y_2$$

We require equality for powers of k up through 4:

$$4h = \quad c_{-2} + c_{-1} + c_0 + c_1 + c_2$$
$$0 = -2c_{-2} - c_{-1} \quad\quad + c_1 + 2c_2$$
$$16h/3 = \quad 4c_{-2} + c_{-1} \quad\quad + c_1 + 4c_2$$
$$0 = -8c_{-2} - c_{-1} \quad\quad + c_1 + 8c_2$$
$$64h/5 = \quad 16c_{-2} + c_{-1} \quad\quad + c_1 + 16c_2$$

Solving we first find $c_{-2} = c_2 = 14h/45$ and then $c_{-1} = c_1 = 64h/45$ after which $c_0 = 24h/45$ completes the set. The $n = 4$ formula is thus

$$A = (2h/45)(7y_{-2} + 32y_{-1} + 12y_0 + 32y_1 + 7y_2)$$

as Table 8.1 announces.

8.10   Obtain the truncation error estimate of Table 8.2 for the case $n = 4$ by testing the formula for the next higher-power function.

▌ Taking first the fifth power of k,

$$I = h \int_{-2}^{2} k^5 \, dk = 0$$

$$A = (2h/45)\left[7(-2)^5 + 32(-1)^5 + 12 \cdot 0 + 32 + 7(2)^5\right] = 0$$

and we again have agreement. Moving to the sixth power,

$$I = h \int_{-2}^{2} k^6 \, dk = h \cdot 2^8/7$$

$$A = (2h/45)\left[7(-2)^6 + 32 + 0 + 32 + 7(2)^6\right] = -384h/9$$

making $I - A = -128h/21$. Now the sixth derivative of k^6 is 720 so the sixth derivative of $y(x)$ relative to x will be $720/h^6$. The ratio of $I - A$ to this derivative works out to $-8h^7/945$ and if it is assumed that this will hold for other $y(x)$, then we have the estimate

$$\text{truncation error} = -8h^7 y^{(6)}/945$$

as Table 8.2 suggests.

8.11   Use the Taylor series method to estimate the truncation error of the $n = 3$ Newton–Cotes formula.

▌ First we treat the approximation

$$A = \frac{3h}{8}(y_0 + 3y_1 + 3y_2 + y_3)$$

$$= \frac{3h}{8}\left[y_0 + 3\left(y_0 + hy_0' + \frac{h^2}{2}y_0'' + \frac{h^3}{6}y_0''' + \frac{h^4}{24}y_0'''' + \cdots \right) \right.$$

$$+ 3\left(y_0 + 2hy_0' + \frac{4h^2}{2}y_0'' + \frac{8h^3}{6}y_0''' + \frac{16h^4}{24}y_0'''' + \cdots \right)$$

$$\left. + \left(y_0 + 3hy_0' + \frac{9h^2}{2}y_0'' + \frac{27h^3}{6}y_0''' + \frac{81h^4}{24}y_0'''' + \cdots \right) \right]$$

$$= \frac{3h}{8}\left[8y_0 + 12hy_0' + 12h^2 y_0'' + 9h^3 y_0''' + \frac{11}{2}h^4 y_0'''' + \cdots \right]$$

and then the integral itself, with $F(x)$ some indefinite integral:

$$I = F(x_3) - F(x_0) = F'(x_0)(3h) + F''(x_0)(3h)^2/2 + \cdots$$

$$= 3hy_0 + \tfrac{9}{2}h^2 y_0' + \tfrac{9}{2}h^3 y_0'' + \tfrac{27}{8}h^4 y_0''' + \tfrac{81}{40}h^5 y_0'''' + \cdots$$

Comparisons show that terms through the third derivative match, after which a little arithmetic manages

$$I - A = -3h^5 y''''/80 + \cdots$$

the first term of which is listed in Table 8.2.

8.12   Develop the $n = 6$ formula using undetermined coefficients.

▌ Choosing $k = (x - x_3)/h$,

$$\int_{x_0}^{x_6} y(x) \, dx = h \int_{-3}^{3} y_k \, dk = c_{-3} y_{-3} + c_{-2} y_{-2} + \cdots + c_3 y_3$$

which we plan to make exact for powers of k through the sixth. The resulting system of equations is

$$6h = c_{-3} + c_{-2} + c_{-1} + c_0 + c_1 + c_2 + c_3$$
$$0 = -3c_{-3} - 2c_{-2} - c_{-1} + c_1 + 2c_2 + 3c_3$$
$$2 \cdot 3^3 h/3 = 9c_{-3} + 4c_{-2} + c_{-1} + c_1 + 4c_2 + 9c_3$$
$$0 = -27c_{-3} - 8c_{-2} - c_{-1} + c_1 + 8c_2 + 27c_3$$
$$2 \cdot 3^5 h/5 = 81c_{-3} + 16c_{-2} + c_{-1} + c_1 + 16c_2 + 81c_3$$
$$0 = -243c_{-3} - 32c_{-2} - c_{-1} + c_1 + 32c_2 + 243c_3$$
$$2 \cdot 3^7 h/7 = 729c_{-3} + 64c_{-2} + c_{-1} + c_1 + 64c_2 + 729c_3$$

and seems formidable. However, if we risk the assumption that the coefficients are symmetric, $c_{-3} = c_3$, $c_{-2} = c_2$ and $c_{-1} = c_1$, then the system is reduced to

$$6h = c_0 + 2c_1 + 2c_2 + 2c_3$$
$$18h = 2c_1 + 8c_2 + 18c_3$$
$$2 \cdot 3^5 h/5 = 2c_1 + 32c_2 + 162c_3$$
$$2 \cdot 3^7 h/7 = 2c_1 + 128c_2 + 1458c_3$$

which is manageable and eventually gives up $c_3 = 41h/140$ and the other coefficients listed in Table 8.1. Substituting back, it can be verified that the seven c_i do in fact satisfy the original system of seven equations, so the assumption of symmetry proves to have been justified.

8.13 Derive the *trapezoidal rule*.

▌ This ancient formula still finds application and illustrates very simply how the formulas of Problem 8.3 may be stretched to cover many intervals. The trapezoidal rule applies our $n = 1$ formula to successive intervals up to x_n:

$$\tfrac{1}{2}h(y_0 + y_1) + \tfrac{1}{2}h(y_1 + y_2) + \tfrac{1}{2}h(y_2 + y_3) + \cdots + \tfrac{1}{2}h(y_{n-1} + y_n)$$

This leads to

$$\int_{x_0}^{x_n} y(x)\, dx \sim \tfrac{1}{2}h[y_0 + 2y_1 + \cdots + 2y_{n-1} + y_n]$$

which is the trapezoidal rule.

8.14 Apply the trapezoidal rule to the integration of $\sqrt{x}$ between the arguments 1.00 and 1.30. Use the data of Table 8.1. Compare with the correct value of the integral.

▌ We easily find

$$\int_{1.00}^{1.30} \sqrt{x}\, dx \sim \tfrac{.05}{2}[1 + 2(1.02470 + \cdots + 1.11803) + 1.14017] = .32147$$

The correct value is $\tfrac{2}{3}[(1.3)^{3/2} - 1] = .32149$ to five places, making the actual error .00002.

8.15 Derive an estimate of the truncation error of the trapezoidal rule.

▌ The result of Problem 8.4 may be applied to each interval, producing a total truncation error of about

$$(-h^3/12)[y_0^{(2)} + y_1^{(2)} + \cdots + y_{n-1}^{(2)}]$$

Assuming the second derivative bounded, $m < y^{(2)} < M$, the sum in brackets will be between nm and nM. Also assuming this derivative continuous allows the sum to be written as $ny^{(2)}(\xi)$ where $x_0 < \xi < x_n$. This is because $y^{(2)}(\xi)$ then assumes all values intermediate to m and M. It is also convenient to call the ends of the interval of integration $x_0 = a$ and $x_n = b$, making $b - a = nh$. Putting all this together, we have

$$\text{truncation error} \sim -\frac{(b-a)h^2}{12} y^{(2)}(\xi)$$

8.16   Apply the estimate of Problem 8.15 to our square root integral.

▌ With $h = .05$, $b - a = .30$ and $y^{(2)}(x) = -x^{-3/2}/4$, truncation error $\sim .000016$, which is slightly less than the actual error of $.00002$. However, rounding to five places and adding this error estimate to our computed result does produce $.32149$, the correct result.

8.17   Estimate the effect of inaccuracies in the y_k values on results obtained by the trapezoidal rule.

▌ With Y_k denoting the true values, as before, we find $\frac{1}{2}h[e_0 + 2e_1 + \cdots + 2e_{n-1} + e_n]$ as the error due to inaccuracies $e_k = Y_k - y_k$. If the e_k do not exceed E in magnitude, this output error is bounded by $\frac{1}{2}h[E + 2(n-1)E + E] = (b - a)E$.

8.18   Apply the preceding information to the square root integral of Problem 8.14.

▌ We have $(b - a)E = (.30)(.000005) = .0000015$, so that this source of error is negligible.

8.19   Derive *Simpson's rule*.

▌ This may be the most popular of all integration formulas. It involves applying our $n = 2$ formula to successive pairs of intervals up to x_n, obtaining the sum

$$(h/3)(y_0 + 4y_1 + y_2) + (h/3)(y_2 + 4y_3 + y_4) + \cdots + (h/3)(y_{n-2} + 4y_{n-1} + y_n)$$

which simplifies to

$$(h/3)[y_0 + 4y_1 + 2y_2 + 4y_3 + \cdots + 2y_{n-2} + 4y_{n-1} + y_n]$$

This is Simpson's rule. It requires n to be an even integer.

8.20   Apply Simpson's rule to the integral of Problem 8.14.

▌
$$\int_{1.00}^{1.30} \sqrt{x}\, dx = (.05/3)[1.0000 + 4(1.02470 + 1.07238 + 1.11803) + 2(1.04881 + 1.09544) + 1.14017]$$
$$= .32149$$

which is correct to five places.

8.21   Estimate the truncation error of Simpson's rule.

▌ The result of Problem 8.5 may be applied to each pair of intervals, producing a total truncation error of about

$$-(h^5/90)\left[y_0^{(4)} + y_2^{(4)} + \cdots + y_{n-2}^{(4)}\right]$$

Assuming the fourth derivative continuous allows the sum in brackets to be written as $(n/2)y^{(4)}(\xi)$ where $x_0 < \xi < x_n$. (The details are almost the same as in Problem 8.15.) Since $b - a = nh$,

$$\text{truncation error} \sim -\frac{(b-a)h^4}{180} y^{(4)}(\xi)$$

8.22   Apply the estimate of Problem 8.21 to our square root integral.

▌ Since $y^{(4)}(x) = -(15/16)x^{-7/2}$, truncation error $\sim .00000001$ which is minute.

8.23   Estimate the effect of data inaccuracies on results computed by Simpson's rule.

▌ As in Problem 8.17, this error is found to be

$$\frac{1}{3}h[e_0 + 4e_1 + 2e_2 + 4e_3 + \cdots + 2e_{n-2} + 4e_{n-1} + e_n]$$

and if the data inaccuracies e_k do not exceed E in magnitude, this output error is bounded by

$$\tfrac{1}{3}hE\left[1 + 4\left(\tfrac{1}{2}n\right) + 2\left(\tfrac{1}{2}n - 1\right) + 1\right] = (b - a)E$$

exactly as for the trapezoidal rule. Applying this to the square root integral of Problem 8.20 we obtain the same .0000015 as in Problem 8.18, so that once again this source of error is negligible.

8.24 Compare the results of applying Simpson's rule with intervals $2h$ and h and obtain a new estimate of truncation error.

$\blacksquare$ Assuming data errors negligible, we compare the two truncation errors. Let E_1 and E_2 denote these errors for the intervals $2h$ and h, respectively. Then

$$E_1 = -\frac{(b-a)(2h)^4}{180}y^{(4)}(\xi_1) \qquad E_2 = -\frac{(b-a)h^4}{180}y^{(4)}(\xi_2)$$

so that $E_2 \sim E_1/16$. The error is reduced by a factor of 16 by halving the interval h. This may now be used to get another estimate of the truncation error of Simpson's rule. Call the correct value of the integral I and call the two Simpson approximations A_1 and A_2. Then

$$I = A_1 + E_1 = A_2 + E_2 \sim A_1 + 16E_2$$

Solving for E_2, the truncation error associated with interval h is $E_2 \sim (A_2 - A_1)/15$.

8.25 Use the estimate of Problem 8.24 to correct the Simpson's rule approximation.

$\blacksquare$ This is an elementary but very useful idea. We find

$$I = A_2 + E_2 \sim A_2 + (A_2 - A_1)/15 = (16A_2 - A_1)/15$$

8.26 Apply the trapezoidal, Simpson and $n = 6$ formulas to compute the integral of $\sin x$ between 0 and $\pi/2$ from the seven values provided in Table 8.3. Compare with the correct value of 1.

$\blacksquare$ The trapezoidal rule produces .99429. Simpson manages 1.00003. The $n = 6$ formula leads to

$$\frac{\pi}{140 \times 12}\left[41(0) + 216(.25882) + 27(.5) + 272(.70711) + 27(.86603) + 216(.96593) + 41(1)\right] = 1.000003$$

Clearly the $n = 6$ rule performs best for this fixed data supply.

8.27 Show that to obtain the integral of the previous problem correct to five places by using the trapezoidal rule would require an interval h of approximately .006 rad. By contrast, Table 8.3 has $h = \pi/12 \sim .26$.

TABLE 8.3

x	0	$\pi/12$	$2\pi/12$	$3\pi/12$	$4\pi/12$	$5\pi/12$	$\pi/2$
$\sin x$	.00000	.25882	.50000	.70711	.86603	.96593	1.00000

$\blacksquare$ The truncation error of Problem 8.15 suggests that we want

$$\frac{(b-a)h^2}{12}y^{(2)}(\xi) \leq \frac{(\pi/2)h^2}{12} < .000005$$

which will occur provided $h < .006$.

8.28 What interval h would be required to obtain the integral of Problem 8.26 correct to five places using Simpson's rule?

❚ The truncation error of Problem 8.21 suggests

$$\frac{(b-a)h^4}{180}y^{(4)}(\xi) \le \frac{(\pi/2)h^4}{180} < .000005$$

or $h < .15$ approximately.

8.29 Prove that the trapezoidal and Simpson's rules are *convergent*.

❚ If we assume truncation to be the only source of error, then in the case of the trapezoidal rule

$$I - A = -\frac{(b-a)h^2}{12}y^{(2)}(\xi)$$

where I is the exact integral and A the approximation. If $\lim h = 0$, then assuming $y^{(2)}$ bounded, $\lim(I - A) = 0$. (This is the definition of convergence.)

For Simpson's rule we have the similar result

$$I - A = -\frac{(b-a)h^4}{180}y^{(4)}(\xi)$$

If $\lim h = 0$ then assuming $y^{(4)}$ bounded, $\lim(I - A) = 0$. Multiple use of higher-degree formulas also leads to convergence.

8.30 Apply Simpson's rule to the integral $\int_0^{\pi/2} \sin x \, dx$, continually halving the interval h in the search for greater accuracy.

❚ Machine computations, carrying eight digits, produce the results in Table 8.4.

TABLE 8.4

h	Approx. integral	h	Approx. integral
$\pi/8$	1.0001344	$\pi/128$	.99999970
$\pi/16$	1.0000081	$\pi/256$	.99999955
$\pi/32$	1.0000003	$\pi/512$	.99999912
$\pi/64$	.99999983 (best)	$\pi/1024$	.99999870

8.31 The computations of Problem 8.30 indicate a durable error source that does not disappear as h diminishes and actually increases as work continues. What is this error source?

❚ For very small intervals h the truncation error is small and, as seen earlier, data inaccuracies have little impact on Simpson's rule for any interval h. But small h means much computing, with the prospect of numerous computational roundoffs. This error source has not been a major factor in the much briefer algorithms encountered in interpolation and approximate differentiation. Here it has become dominant and limits the accuracy obtainable, even though our algorithm is convergent and the effect of data inaccuracies is small (we are saving eight decimal places). This problem emphasizes the importance of continuing the search for briefer algorithms.

8.32 Apply the device of Problem 8.24 to the trapezoidal rule, that is, develop a correction term based on the trapezoidal results for intervals $2h$ and h.

❚ We begin with the truncation error estimate of Problem 8.15 simplified to

$$E = I - A = Ch^2$$

with I the exact integral and A the trapezoidal approximation. For the two intervals chosen this becomes

$$E_1 = I - A = C(2h)^2 \qquad E_2 = I - A = Ch^2$$

from which it follows that $E_1 = 4E_2$. Since $I = A_1 + E_1 = A_2 + E_2$, we then have $A_1 + 4E_2 = A_2 + E_2$ making

$$E_2 = (A_2 - A_1)/3$$

This now allows the new approximation

$$I = A_2 + (A_2 - A_1)/3 = (4A_2 - A_1)/3$$

For a further extension of this same idea see the development of Romberg's method in Problem 8.47.

8.33 Apply several of our formulas for numerical integration to the data supply

x	1	5/4	3/2	7/4	2
y	1	4/5	2/3	4/7	1/2

▮ The trapezoidal rule manages

$$(1/8)(1 + 8/5 + 4/3 + 8/7 + 1/2) = 1171/1680 = .69702$$

to five places. Simpson's rule produces

$$(1/12)(1 + 16/5 + 4/3 + 16/7 + 1/2) = 1747/2520 = .69325$$

while the $n = 4$ formula yields .69317. The parent function of these data is rather clearly $y(x) = 1/x$, so the exact integral is the natural logarithm of 2, which is .693147 to six places. The $n = 4$ rule has come the closest.

8.34 Suppose the data supply of the preceding problem doubled, that is, we now have the values of $y(x) = 1/x$ for $x = 1$ to $x = 2$ with a spacing of $h = 1/8$. Again apply various integration rules.

▮ The trapezoidal rule now produces .6941, still a point off in the third decimal place. Suppose we apply the correction process of Problem 8.32, with

$$A_1 = .6970 \quad \left(\text{found in Problem 8.33 using } h = \tfrac{1}{4}\right)$$

$$A_2 = .6941 \quad \left(\text{just found using } h = \tfrac{1}{8}\right).$$

We obtain

$$(4A_2 - A_1)/3 = (2.7764 - .6970)/3 = .69313$$

which is off by only two points in the fifth place.

Turning to Simpson's rule we find the value .69315 which is correct to all five places. The Newton–Cotes $n = 4$ formula also produces this value, making two applications, one over the left half of the interval and one over the right half. So a set of four quadratic polynomial segments (Simpson's rule) agrees with a pair of quartic polynomial segments in this example.

8.35 From the following data estimate $\int_0^2 y(x)\,dx$ as best you can:

x	0	.25	.50	.75	1.00	1.25	1.50	1.75	2
$y(x)$	1.000	1.284	1.649	2.117	2.718	3.490	4.482	5.755	7.389

How much confidence do you place in your results? Do you believe them correct to three places?

▮ The trapezoidal rule, with $h = .5$ and skipping over the values in between, yields 6.52175. With $h = .25$ and using the full data set it manages 6.42237. The correction procedure of Problem 8.32 then makes $(4A_2 - A_1)/3 =$ 6.38924 to five places.

Simpson's rule with $h = .5$ and using the five alternate data values produces 6.39150. With $h = .25$ and the full data set its value is 6.38925. The corresponding figures for the Newton–Cotes $n = 4$ formula are 6.38958 (half the data) and 6.38910 (full data).

Looking back over the evidence it seems that the integral has a value near 6.3892.

8.36 The data of the preceding problem come from the parent function $y(x) = e^x$. The correct integral is, therefore,

$$\int_0^2 e^x \, dx = e^2 - 1 = 6.389056$$

to six places. Our previous result is correct to three places but it is optimistic to expect more than these four correct digits (6.389) from four digit input information.

8.37 From the following data, estimate $\int_1^5 y(x) \, dx$ as best as you can:

x	1	1.5	2	2.5	3	3.5	4	4.5	5
$y(x)$	0	.41	.69	.92	1.10	1.25	1.39	1.50	1.61

How much confidence do you place in your results?

 Here we face a similar situation, but the data are offered to only two decimal places and two or three significant digits. If we appeal first to the trapezoidal rule, it offers 3.9850 from alternate data points alone and then 4.0325 from the full set. The now familiar correction process then provides 4.0483, which almost surely is not correct to all four places. Simpson's rule yields 4.0433 from a half data set and 4.0483 from the full set, in total agreement with the trapezoidal correction. Turning to Newton–Cotes $n = 4$ we obtain 4.0458 from the half set and then 4.0487 from the full. If no further information is available on the integrand we may be wise to settle for the estimate 4.05, but, of course, there is a followup problem.

8.38 The data of the preceding problem were drawn from the parent function $y(x) = \log x$. The correct integral is, therefore,

$$\int_1^5 \log x \, dx = 5 \log 5 - 4 = 4.0472$$

to four places. The estimate of 4.05 was thus correct, but even though our various formulas appear to agree more or less to one more decimal place, it is not wise to trust them. The data supply has its limitations.

8.39 Use the truncation error estimate for Simpson's rule obtained in Problem 8.21 to predict how small an interval h would be needed to produce $\log 2$ correct to four places. Also to eight places.

 We will need for four place accuracy

$$(b - a) h^4 y^{(4)} / 180 < h^4 24 / 180 < .00005$$

which requires h^4 to be no larger than $3.75 \cdot 10^{-4}$ or $h < .14$. For eight place accuracy an additional 10^{-4} appears on the right, so after taking the fourth root we end up with .014 instead. The influence of the particular function $1/x$ is limited here to the size of its fourth derivative, in this case 24 maximum on the interval $(1, 2)$. In Problems 8.33 and 8.34, $h = \frac{1}{4}$ proved to be almost adequate for this level of accuracy and $h = \frac{1}{8}$ actually produced a correct five place result. Estimates obtained as in this problem are strictly conservative.

8.40 Rework the preceding problem assuming that the trapezoidal rule is to be used rather than Simpson's.

 We now need

$$(b - a) h^2 y'' / 12 < h^2 2 / 12 < .00005$$

since y'' does not exceed 2 on the interval $(1, 2)$. This means $h^2 < 3 \cdot 10^{-4}$ or $h < .017$ more or less. For eight place accuracy another 10^{-4} must be included on the right side, and after taking the square root we have $h < .00017$, which means quite a lot of computing to get this precision from a brute force application of the trapezoidal rule.

8.41 Compare truncation error estimates for the interval h required to get four place accuracy from Simpson's rule for the functions $y(x) = 1/x$, e^x and $\log x$ used in the recent examples.

 As just observed, the differences are due to the differences in fourth derivatives, although the length of the interval of integration clearly has some influence. For $1/x$ this derivative is $24x^{-5}$ and will not exceed 24 on the interval $(1, 2)$. For e^x the derivative is also e^x and on $(0, 2)$ will not exceed e^2. The interval h is then limited by

$$2h^4 e^2 / 180 < h^4 / 12 < .00005$$

or $h < .15$. For $y(x) = \log x$ the fourth derivative is bounded in size by 6, so on the interval $(1, 5)$ of our example we need

$$4h^4 6/180 < .00005$$

or $h < .14$. It seems that the three integrals that have been used as illustrations all have about the same needs. Rechecking we see that the intervals actually used in the logarithm and exponential examples were considerably greater than .15 and that in fact four place accuracy was not achieved in either case.

8.42   Calculate $\int_0^1 dx/(1 + x^2)$ correct to seven places by any of our approximate methods. The correct value is $\pi/4$, or to seven places .7853982.

■ Here are the results of applying Simpson's rule with 4, 8 and 16 intervals:

$$.7853921 \qquad .7853982 \qquad .7853982$$

After this, with continued doubling of the number of intervals, the last two digits behaved as follows:

$$82 \quad 81 \quad 83 \quad 82 \quad 83 \quad 82 \quad 81 \quad 90 \quad 75$$

The last two of these suggest the beginning of error growth, but the true value has long since been found. The Newton–Cotes $n = 4$ formula applied successively to sets of four intervals gave the figures

$$.7855294 \qquad .7853985 \qquad .7853982$$

the first again corresponding to just four intervals with continued doubling to follow. The last decimal place eventually showed error problems much as in the case of Simpson's rule.

8.43   Calculate the elliptic integral

$$\int_0^{\pi/2} \sqrt{1 - .25 \sin^2 t}\ dt$$

■ Both Simpson's rule and the $n = 4$ rule managed 1.467462 using eight intervals. This figure persisted until the number of intervals had been doubled to 512, after which the final digit drifted between 1 and 4 for a while and the computation was terminated.

8.44   Calculate the elliptic integral

$$\int_0^{\pi/2} \sqrt{1 - .99 \sin^2 t}\ dt$$

■ The change in value of the parameter seemed to slow down the convergence somewhat. Simpson's rule managed $(4, 8, \ldots$ intervals)

$$1.020319 \qquad 1.016805 \qquad 1.016056 \qquad 1.015994$$

after which the final digit drifted between 3 and 4 for a while.

8.45   Use one of our approximate integration formulas to verify

$$\int_0^{\pi/2} \frac{dx}{\sin^2 x + \frac{1}{4} \cos^2 x} = 3.1415927$$

the exact value being π.

■ The $n = 4$ rule with $4, 8, 16, \ldots$ intervals produced

$$3.125501 \qquad 3.142912 \qquad 3.141613 \qquad 3.141592 \qquad 3.141593$$

after which the final two digits drifted as follows:

$$93 \quad 93 \quad 90 \quad 89 \quad 89 \quad 92 \quad 96$$

Simpson's rule vascillated briefly between the 93 and 92 closings before becoming unstable.

8.46   Compute the total arclength of the ellipse $x^2 + y^2/4 = 1$.

▮ The ellipse is conveniently represented by

$$x = \cos A \qquad y = 2 \sin A$$

making the arclength

$$s = 4 \int_0^{\pi/2} \sqrt{\sin^2 A + 4 \cos^2 A} \; dA = 4 \int_0^{\pi/2} \sqrt{1 + 3 \cos^2 A} \; dA$$

Simpson's rule generated the sequence of approximations

$$9.691322 \qquad 9.688460 \qquad 9.688449 \qquad 9.688447 \qquad 9.688447$$

after which the last two digits drifted to 51, 44 and 58. Our best value may be the closing 47.

It is amusing to note that the truncation error estimate for this famous rule has it more or less proportional to h^4. If this were strictly true, then each halving of h would reduce the error by a factor of $1/16$. Two halvings would make a reduction by $1/256$ and three by $1/4096$. In the present example the original value of 9.691322 was off by some 2875 points in the final place, assuming the accepted value of 9.688447 to be at least close to correct. Three steps later this error may then be only $1/4096$ of this, or less than 1. We are hoping that this is true.

8.47   Develop the idea of Problems 8.25 and 8.32 into Romberg's method of approximate integration.

▮ Suppose that the error of an approximate formula is proportional to h^n. Then two applications of the formula, with intervals h and $2h$, involve errors

$$E_1 \simeq C(2h)^n \qquad E_2 \simeq Ch^n$$

making $E_2 \simeq E_1/2^n$. With $I = A_1 + E_1 = A_2 + E_2$ as before, we soon find the new approximation

$$I \simeq A_2 + \frac{A_2 - A_1}{2^n - 1} = \frac{2^n A_2 - A_1}{2^n - 1}$$

For $n = 4$ this duplicates Problem 8.25. For $n = 2$ it applies to the trapezoidal rule in which the truncation error is proportional to h^2. It is not hard to verify that for $n = 2$ our last formula duplicates Simpson's rule and that for $n = 4$ it duplicates the Cotes $n = 4$ formula. It can be shown that the error in this formula is proportional to h^{n+2} and this suggests a recursive computation. Apply the trapezoidal rule several times, continually halving h. Call the results $A_1, A_2, A_3, \dots$. Apply our foregoing formula with $n = 2$ to each pair of consecutive A_i. Call the results $B_1, B_2, B_3, \dots$. Since the error is now proportional to h^4 we may reapply the formula, with $n = 4$, to the B_i. The results may be called $C_1, C_2, C_3, \dots$. Continuing in this fashion an array of results is obtained:

$$
\begin{array}{cccc}
A_1 & A_2 & A_3 & A_4 \quad \cdots \\
 & B_1 & B_2 & B_3 \quad \cdots \\
 & & C_1 & C_2 \quad \cdots \\
 & & & D_1 \quad \cdots
\end{array}
$$

The computation is continued until entries at the lower right of the array agree within the required tolerance.

8.48   Show that the B row of the Romberg method theoretically contains values produced by Simpson's rule.

▮ Let y_0, y_1, y_2 be values corresponding to arguments with equal spacing h. Then the two trapezoidal values of interest are

$$A_1 = \frac{2h}{2}(y_0 + y_2) \qquad A_2 = \frac{h}{2}(y_0 + 2y_1 + y_2)$$

making

$$4A_2 - A_1 = \frac{h}{2}(2y_0 + 8y_1 + 2y_2)$$

so that dividing by 3 we have the expected equivalence. It will often be found that minor differences occur between the computed values of the two sides of this last equation. The explanation is that the values y_i are processed in different orders, so that roundoffs may not be the same.

8.49 Show that the C row of the Romberg method is equivalent to the Newton–Cotes $n = 4$ rule.

▌ Let $y_0, y_1, \ldots, y_4$ be the active values. Then the two Simpson results of interest are

$$B_1 = \frac{2h}{3}(y_0 + 4y_2 + y_4) \qquad B_2 = \frac{h}{3}(y_0 + 4y_1 + 2y_2 + 4y_3 + y_4)$$

making

$$16B_2 - B_1 = \frac{h}{3}(14y_0 + 64y_1 + 24y_2 + 64y_3 + 14y_4)$$

after which division by 15 leads quickly to the equivalence.

8.50 Apply Romberg's method to $\int_0^{\pi/2} \sin x \, dx$.

▌ The various results are

Points used	4	8	16	32
Trapezoidal result	.987116	.996785	.999196	.999799
		1.000008	1.000000	1.000000
			.999999	1.000000
				1.000000
				1.000000

For example,

$$[4(.996785) - .987116]/3 = 1.000008$$
$$[16(1) - 1.000008]/15 = .999999$$

Convergence to the correct value of 1 is apparent. The integral is an easy one for approximate methods because the sine function is a smooth function, all its derivatives existing and being bounded by 1.

8.51 Apply the Romberg method to $\int_0^9 \sqrt{x} \, dx$.

▌ The exact value is 18. Table 8.5 depicts a substantial struggle to achieve three place accuracy. All values in the table are reduced by 17 and so should converge to 1.

TABLE 8.5

− .704	.369	.770	.917	.970	.989	.996	.999
	.726	.903	.966	.988	.996	.998	.999
		.915	.970	.989	.996	.999	.999
			.971	.990	.996	.999	.99954
				.990	.996	.999	.99955
					.996	.999	.99955
						.999	.99955
							.99955

The column at the right is certainly the place to look for the best results of the Romberg method, containing the best of the trapezoidal, Simpson, $n = 4$, $n = 6$ and other rules. It should be noted that Romberg entries are obtained with little computing, one multiplication, subtraction and division for each. In fact, the trapezoidal row may require more time than all the others combined, and the total display provides a fine view of the convergence, with a chance to form opinions of the ultimate error. It should also be remarked that the square root integrand is not an easy one for polynomials to approximate, because of the infinite slope at $x = 0$.

8.52 Apply Romberg's method to $\int_0^1 - x \log x \, dx$.

▌ Here again we have an infinite slope at $x = 0$, but Table 8.6 shows convergence to about five places, the exact value being .25.

TABLE 8.6

.17329	.22723	.24341	.24813	.24948	.24985
	.24521	.24880	.24970	.24992	.24998
		.24904	.24976	.24994	.24998
			.24977	.24994	.24999
				.24994	.24999
					.24999

8.53 Apply Romberg's method to $\int_0^\pi x \sin(20x)\, dx$.

▮ The integrand oscillates fairly rapidly, making 10 complete sine oscillations in the interval. The correct value is $-\pi/20$ or $-.15708$ to five places. Table 8.7 shows the strain of the rapid oscillations, with its mix of positive and negative values at upper left. However, the column at the right is reassuring.

TABLE 8.7

.00002	.00000	−.61685	.12776	−.10304	−.14426	−.15391
	.00000	−.82247	.37596	−.17997	−.15799	−.15713
		−.87730	.45585	−.21703	−.15653	−.15707
			.47701	−.22772	−.15557	−.15708
				−.23048	−.15529	−.15709
					−.15521	−.15709
						−.15709

8.54 Try the Romberg method on $\int_0^1 x \sin(50x)\, dx$. Does it outperform the Simpson rule?

▮ The Romberg table has the seemingly chaotic beginning

$$
\begin{array}{cccc}
0 & -1.2 & -1.5 & .1 \\
  & -1.6 & -1.6 & .6 \\
  &   & -1.6 & .7
\end{array}
$$

the correct value being $-\pi/50$ or $-.0628318$ to seven places. In spite of the rough start, however, the eighth and ninth columns of the table finish with the entries

$$
\begin{array}{cc}
-.06286 & -.0628317 \\
        & -.0628317
\end{array}
$$

the last being one point low in the last place. Simpson's rule produced the same value on the twelfth approximation, though with a very noticeable increase of computer time.

8.55 More accurate integration formulas may be obtained by integrating a polynomial over less than the full range of collocation. Integrate Stirling's formula over the two center intervals.

▮ Up through sixth differences Stirling's formula is

$$
p_k = y_0 + k\mu\, \delta y_0 + \frac{1}{2}k^2\, \delta^2 y_0 + \frac{k(k^2-1)}{6}\mu\, \delta^3 y_0 + \frac{k^2(k^2-1)}{24}\, \delta^4 y_0
$$

$$
+ \frac{k(k^2-1)(k^2-4)}{120}\mu\, \delta^5 y_0 + \frac{k^2(k^2-1)(k^2-4)}{720}\, \delta^6 y_0
$$

Integration brings, since $x - x_0 = kh$ and $dx = h\, dk$,

$$
\int_{x_0-h}^{x_0+h} p(x)\, dx = h \int_{-1}^{1} p_k\, dk = h\left[2y_0 + \tfrac{1}{3}\, \delta^2 y_0 - \tfrac{1}{90}\, \delta^4 y_0 + \tfrac{1}{756}\, \delta^6 y_0 \right]
$$

More terms are clearly available by increasing the degree of the polynomial. Stopping with the second difference term leaves us once again with the starting combination of Simpson's rule, in the form $(h/3)(y_{-1} + 4y_0 + y_1)$. In this case the integration has extended over the full range of collocation, as in Problem 8.3. With the fourth difference term we integrate over only half the range of collocation (Fig. 8-2).

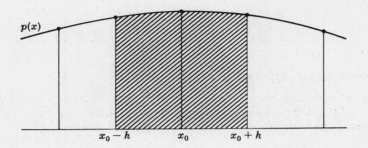

Fig. 8-2

As more differences are used, $y(x)$ and $p(x)$ collocate at additional arguments, but the integration is extended over the center two intervals only. Since these are the intervals where Stirling's formula has the smallest truncation error, it can be anticipated that an integration formula obtained in this way will be more accurate. This extra accuracy is, however, purchased at a price: In application such formulas require y_k values outside the interval of integration.

The truncation error of this formula may be estimated by the Taylor series method used in Problem 8.4, and proves to be approximately $-(23h^9/113400)y_0^{(8)} + \cdots$.

8.56 Use the result of Problem 8.55 to develop the Simpson rule with correction terms.

▌ We make $n/2$ applications centered at $x_1, x_3, \ldots, x_{n-1}$, where n is even. The result is

$$\int_{x_0}^{x_n} p(x)\, dx = \frac{h}{3}\left[y_0 + 4y_1 + 2y_2 + \cdots + 4y_{n-1} + y_n \right] - \frac{h}{90}\left[\delta^4 y_1 + \delta^4 y_3 + \cdots + \delta^4 y_{n-1} \right]$$

$$+ \frac{h}{756}\left[\delta^6 y_1 + \delta^6 y_3 + \cdots + \delta^6 y_{n-1} \right]$$

This can be extended to higher differences if desired.

The truncation error of the result will be approximately $n/2$ times that of the previous problem and can be written as $-[23(x_n - x_0)h^8/226800]y_0^{(8)} + \cdots$.

8.57 Apply Problem 8.56 to the data of Table 7.5.

▌ Since Simpson's rule was applied to the data in Problem 8.26 and gave 1.00003, we need only the correction terms. The fourth differences yield $\pi/12 \cdot -1/90(.00120 + .00328 + .00450)$ which come to $-.000026$, and sixth differences prove to be negligible. Applying this correction makes

$$\int_0^{\pi/2} \sin x\, dx = 1.00000$$

which is correct. Notice that a number of values of $\sin x$ outside the interval of integration contribute to this result.

8.58 Integrate Bessel's formula over the center interval.

▌ Writing Bessel's formula as

$$p_k = \mu y_{1/2} + \left(k - \tfrac{1}{2}\right)\delta y_{1/2} + \binom{k}{2}\mu\,\delta^2 y_{1/2} + \binom{k}{2}\left(k - \tfrac{1}{2}\right)\tfrac{1}{3}\delta^3 y_{1/2} + \binom{k+1}{4}\mu\,\delta^4 y_{1/2} + \cdots$$

integration leads to

$$\int_{x_0}^{x_0+h} p(x)\, dx = h\int_0^1 p_k\, dk$$

$$= \frac{h}{2}\left[(y_0 + y_1) - \frac{1}{12}\left(\delta^2 y_0 + \delta^2 y_1\right) + \frac{11}{720}\left(\delta^4 y_0 + \delta^4 y_1\right) - \frac{191}{60480}\left(\delta^6 y_0 + \delta^6 y_1\right) \right]$$

Again more terms are available if wanted. Stopping before the second difference term leaves us once again with the starting combination of the trapezoidal rule, $(y_0 + y_1)/2h$. In this case integration extends over the full range of collocation, from x_0 to x_1. With the second and fourth difference terms included, we integrate over only one-fifth the range of collocation (Fig. 8-3).

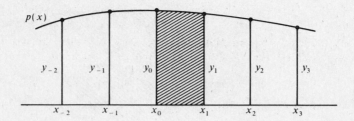

Fig. 8-3

If more differences are used $y(x)$ and $p(x)$ would collocate at additional arguments, but the integration would extend over the center interval only. Since this is where Bessel's formula has the smallest truncation error (Problem 6.35), we may expect an accurate result. Again, however, values of y_k outside the interval of integration are the price we pay.

The truncation error can be found to be approximately $.0007 h^9 y_0^{(8)} + \cdots$.

8.59   Develop the trapezoidal rule with correction terms.

I   There are several formulas meeting this description. To obtain one of them we apply the previous problem n times from x_0 to x_n:

$$\int_{x_0}^{x_n} p(x)\, dx = \frac{h}{2} [y_0 + 2y_1 + 2y_2 + \cdots + 2y_{n-1} + y_n]$$

$$- \frac{h}{24} [\delta^2 y_0 + 2\delta^2 y_1 + 2\delta^2 y_2 + \cdots + 2\delta^2 y_{n-1} + \delta^2 y_n]$$

$$+ \frac{11h}{1440} [\delta^4 y_0 + 2\delta^4 y_1 + 2\delta^4 y_2 + \cdots + 2\delta^4 y_{n-1} + \delta^4 y_n]$$

and higher difference terms are available if desired. The last result may be simplified by summing the differences. One easily finds that

$$\delta^2 y_0 + 2\delta^2 y_1 + \cdots + 2\delta^2 y_{n-1} + \delta^2 y_n = 2\mu\, \delta y_n - 2\mu\, \delta y_0$$

with similar expressions for the other differences. As a result,

$$\int_{x_0}^{x_n} p(x)\, dx = \frac{h}{2} [y_0 + 2y_1 + 2y_2 + \cdots + 2y_{n-1} + y_n]$$

$$- \frac{h}{12} [\mu\, \delta y_n - \mu\, \delta y_0] + \frac{11h}{720} [\mu\, \delta^3 y_n - \mu\, \delta^3 y_0] - \frac{191h}{60480} [\mu\, \delta^5 y_n - \mu\, \delta^5 y_0]$$

The truncation error if fifth difference terms are used is approximately $.0007(x_n - x_0) h^8 y_0^{(8)}$.

8.60   Apply the previous problem to the data of Table 7.5.

I   The trapezoidal rule was used in Problem 8.26 and managed .99429, so we need only the correction terms. They are

$$- \frac{h}{12} \left[\frac{-.03407 + .03407}{2} - \frac{.25882 + .25882}{2} \right] \qquad \frac{11h}{720} \left[\frac{.00231 - .00231}{2} - \frac{-.01764 - .01764}{2} \right]$$

which simplify to .005647 and .000070, respectively. Apply these corrections to get

$$\int_0^{\pi/2} \sin x\, dx = 1.00001$$

8.61 Derive Gregory's formula.

▮ This is another form of the trapezoidal rule with correction terms and can be derived in many ways. One way begins with the Euler–Maclaurin formula (Problem 5.44) in the form

$$\int_{x_0}^{x_n} y(x)\,dx = \frac{h}{2}[\,y_0 + 2y_1 + \cdots + 2y_{n-1} + y_n\,] - \frac{h^2}{12}(y_n' - y_0') + \frac{h^4}{720}(y_n^{(3)} - y_0^{(3)}) - \frac{h^6}{30240}(y_n^{(5)} - y_0^{(5)})$$

more terms being available if needed. Now express the derivatives at x_n in terms of backward differences and the derivatives at x_0 in terms of forward differences (Problems 7.2 and 7.6).

$$hy_0' = \left(\Delta - \tfrac{1}{2}\Delta^2 + \tfrac{1}{3}\Delta^3 - \tfrac{1}{4}\Delta^4 + \tfrac{1}{5}\Delta^5 - \cdots\right)y_0$$
$$hy_n' = \left(\nabla + \tfrac{1}{2}\nabla^2 + \tfrac{1}{3}\nabla^3 + \tfrac{1}{4}\nabla^4 + \tfrac{1}{5}\nabla^5 + \cdots\right)y_n$$
$$h^3 y_0^{(3)} = \left(\Delta^3 - \tfrac{3}{2}\Delta^4 + \tfrac{7}{4}\Delta^5 - \cdots\right)y_0$$
$$h^3 y_n^{(3)} = \left(\nabla^3 + \tfrac{3}{2}\nabla^4 + \tfrac{7}{4}\nabla^5 + \cdots\right)y_n$$
$$h^5 y_0^{(5)} = \left(\Delta^5 - \cdots\right)y_0$$
$$h^5 y_n^{(5)} = \left(\nabla^5 + \cdots\right)y_n$$

The result of substituting these expressions is

$$\int_{x_0}^{x_n} p(x)\,dx = \frac{h}{2}[\,y_0 + 2y_1 + \cdots + 2y_{n-1} + y_n\,] - \frac{h}{12}(\nabla y_n - \Delta y_0) - \frac{h}{24}(\nabla^2 y_n + \Delta^2 y_0)$$
$$- \frac{19h}{720}(\nabla^3 y_n - \Delta^3 y_0) - \frac{3h}{160}(\nabla^4 y_n + \Delta^4 y_0) - \frac{863h}{60480}(\nabla^5 y_n - \Delta^5 y_0)$$

and again more terms can be computed if needed. This is Gregory's formula. It does not require y_k values outside the interval of integration.

8.62 Apply Gregory's formula to the data of Table 7.5.

▮ The trapezoidal rule produced .99429. The correction terms generate

$$-\frac{h}{12}(-.22475) - \frac{h}{24}(-.08347) - \frac{19h}{720}(.00963) - \frac{3h}{160}(.00635)$$

fifth difference terms being negligible. The total correction is .00572 and added to .99429 again gives us $\int_0^{\pi/2} \sin x\,dx = 1.00001$.

8.63 Apply the Euler–Maclaurin formula itself to the same integral.

▮ As given at the start of Problem 8.61, this formula adds to the trapezoidal rule various end corrections in the form of derivatives. These correction terms are, in the present example,

$$-\frac{h^2}{12}(0 - 1) + \frac{h^4}{720}(0 + 1)$$

and bring the same .00572 total as in Problem 8.62.

8.64 Use the Euler–Maclaurin formula to improve the trapezoidal result obtained in Problem 8.34 for $\int_1^2 1/x\,dx = .6931472$ to seven places.

▮ The trapezoidal rule itself, which is the first part of the Euler–Maclaurin formula, produces .6941218, which was rounded to .6941 in the earlier problem. The correction terms are now

$$-(h^2/12)(-\tfrac{1}{4} + 1) = -.0009765$$
$$(h^4/720)(-\tfrac{3}{8} + 6) = .0000019$$

with the fifth derivative term zero to seven places. Putting the pieces together we have the correct .6931472.

8.2 TAYLOR METHODS

8.65 Apply Taylor's theorem to evaluate the error function integral

$$H(x) = \frac{2}{\sqrt{\pi}} \int_0^x e^{-t^2} \, dt$$

for $x = .5$ and $x = 1$, correct to four decimal places.

▮ The series $e^{-t^2} = 1 - t^2 + t^4/2 - t^6/6 + t^8/24 - t^{10}/120 + \cdots$ leads to

$$H(x) = \frac{2}{\sqrt{\pi}} \left[x - \frac{x^3}{3} + \frac{x^5}{10} - \frac{x^7}{42} + \frac{x^9}{216} - \frac{x^{11}}{1320} + \cdots \right]$$

For $x = .5$ this produces .5205 and for $x = 1$ we find .8427. The character of this series assures that the error made in truncating it does not exceed the last term used, so we can be confident in our results. The series method has performed very well here, but it becomes clear that if more decimal places are wanted or if larger upper limits x are to be used, then many more terms of this series will become involved. In such cases it is usually more convenient to proceed as in the next problem.

8.66 Tabulate the error function integral for $x = 0(.1)4$ to six decimal places:

$$H(x) = \frac{2}{\sqrt{\pi}} \int_0^x e^{-t^2} \, dt$$

▮ We adopt the method that was used to prepare the 15 place table of this function, NBS-AMS 41. The derivatives needed are

$$H'(x) = \left(2/\sqrt{\pi}\right) e^{-x^2} \qquad H^{(2)}(x) = -2xH'(x) \qquad H^{(3)}(x) = -2xH^{(2)}(x) - 2H'(x)$$

and in general

$$H^{(n)}(x) = -2xH^{(n-1)}(x) - 2(n-2)H^{(n-2)}(x)$$

The Taylor series may be written as

$$H(x + h) = H(x) + hH'(x) + \cdots + \frac{h^n}{n!} H^{(n)}(x) + R$$

where the remainder is the usual $R = h^{n+1}H^{(n+1)}(\xi)/(n+1)!$. Notice that if M denotes the sum of even power terms and N the sum of odd power terms, then

$$H(x + h) = M + N \qquad H(x - h) = M - N$$

For six place accuracy we use terms of the Taylor series that affect the eighth place, because the length of the task ahead makes substantial roundoff error growth a possibility. With $H(0) = 0$, the computation begins with

$$H(.1) = \frac{2}{\sqrt{\pi}}(.1) - \frac{2}{3\sqrt{\pi}}(.1)^3 + \frac{1}{5\sqrt{\pi}}(.1)^5 = .11246291$$

only the odd powers contributing. Next we put $x = .1$ and find

$$H'(.1) = \left(2/\sqrt{\pi}\right) e^{-.01} = 1.1171516$$

$$H^{(2)}(.1) = -.2H'(.1) = -.22343032$$

$$H^{(3)}(.1) = -.2H^{(2)}(.1) - 2H'(.1) = -2.1896171$$

$$H^{(4)}(.1) = -.2H^{(3)}(.1) - 4H^{(2)}(.1) = 1.3316447$$

$$H^{(5)}(.1) = -.2H^{(4)}(.1) - 6H^{(3)}(.1) = 12.871374$$

$$H^{(6)}(.1) = -.2H^{(5)}(.1) - 8H^{(4)}(.1) = -13.227432$$

leading to

$$M = .11246291 - .00111715 + .00000555 - .00000002 = .11135129$$
$$N = .11171516 - .00036494 + .00000107 = .11135129$$

Since $H(x - h) = M - N$, we rediscover $H(0) = 0$, which serves as a check on the correctness of the computation. We also obtain

$$H(.2) = H(x + h) = M + N = .22270258$$

The process is now repeated to obtain a check on $H(.1)$ and a prediction of $H(.3)$. Continuing in this way one eventually reaches $H(4)$. The last two decimal places can then be rounded off. Correct values to six places are given in Table 8.8 for $x = 0(.5)4$. In NBS-AMS 41 computations were carried to 25 places, then rounded to 15. Extensive subtabulations were then made for small x arguments.

TABLE 8.8

x	.5	1.0	1.5	2.0	2.5	3.0	3.5	4.0
$H(x)$	.520500	.842701	.966105	.995322	.999593	.999978	.999999	1.000000

8.67 Use a Taylor series to evaluate $\int_0^1 \sqrt{x} \sin x \, dx$ to six places:

$$\sqrt{x} \sin x = x^{3/2} - \frac{1}{6} x^{7/2} + \frac{1}{120} x^{11/2} - \frac{1}{5040} x^{15/2} + \frac{1}{362880} x^{17/2} + \cdots$$

▌ Integrating and inserting the limits, we have the terms

$$2/5 - 1/27 + 2/(13)(120) - 2/(17)(5040) + 2/(19)(362880) - \cdots$$
$$= .4 - .037037 + .001282 - .0000233 + .0000002$$
$$= .364222$$

8.68 Use the Taylor series method to evaluate the sine integral

$$\text{Si}(x) = \int_0^x (\sin t)/t \, dt$$

for $x = 0(.2)1$ to six places.

▌ The integrand is $1 - t^2/6 + t^4/120 - t^6/5040 + t^8/362880$ through the first five terms of its series. This integrates to

$$x - x^3/18 + x^5/600 - x^7/35280 + x^9/3265920$$

from which this brief table can be verified:

x	.2	.4	.6	.8	1.0
$\text{Si}(x)$	.199556	.3964615	.5881288	.7720958	.9460830

8.69 How well does the series method work for the following integral?

$$\int_0^{.5} 1/\sqrt{1 + x^4} \, dx$$

▌ The binomial series is called for and offers us

$$\int_0^{.5} \left(1 - x^4/2 + 3x^8/8 - 15x^{12}/48 + 105x^{16}/384 - \cdots\right) dx$$

which integrates to

$$x - x^5/10 + x^9/24 - 15x^{13}/624 + 105x^{17}/6528 - \cdots$$

For the given upper limit this computes to

$$.5 - .0031250 + .0000813 - .0000029 + .0000001 + \cdots$$

or .4969535. As the upper limit nears 1, however, the binomial series converges very slowly and does not provide a viable path to the integral.

8.70 How does Simpson's rule handle the integral of the preceding problem?

❚ It converges quickly to .496954 and then roundoff error blocks further improvement. However, for upper limit 1 it just as quickly finds the value .927037.

For most integrals there are many ways to find good numerical approximations. The series method may be particularly useful if the integral is needed as a function of its upper limit.

8.71 Add the term $(h/140)\delta^6 y_3$ to the Newton–Cotes $n = 6$ formula and so derive the once popular Weddle's rule:

$$\int_{x_0}^{x_6} y(x)\, dx = (3h/10)(y_0 + 5y_1 + y_2 + 6y_3 + y_4 + 5y_5 + y_6)$$

❚ The addition does somewhat increase the truncation error but the simplicity of the resulting coefficients made the rule convenient when hand computing was the only way to go.

The coefficients of the $n = 6$ formula, apart from the factor of $h/140$, are

$$41 \quad 216 \quad 27 \quad 272 \quad 27 \quad 216 \quad 41$$

while those of the sixth difference are

$$1 \quad -6 \quad 15 \quad -20 \quad 15 \quad -6 \quad 1$$

so in combination we have

$$42 \quad 210 \quad 42 \quad 252 \quad 42 \quad 210 \quad 42$$

which, after reintroducing the $h/140$, gives the rule as stated.

Applying Weddle's rule to the familiar data

x	0	$\pi/12$	$\pi/6$	$\pi/4$	$\pi/3$	$\pi/12$	$\pi/2$
$\sin x$	0	.25882	.5	.70711	.86603	.96593	1

we find it producing 1.0000035 for the integral of $\sin x$ over this interval. In Problem 8.26 the $n = 6$ rule itself yielded the figure 1.000003, so little seems to have been lost by making the simplification. In any event, both manage the integral correct to five places from the five place data.

8.3 SOME SPECIAL FORMULAS

8.72 Apply the method of undetermined coefficients to derive a formula of the type

$$\int_0^h y(x)\, dx = h(a_0 y_0 + a_1 y_1) + h^2(b_0 y_0' + b_1 y_1')$$

❚ With four coefficients available, we try to make the formula exact when $y(x) = 1$, x, x^2 and x^3. This leads to the four conditions

$$1 = a_0 + a_1$$
$$\tfrac{1}{2} = \quad a_1 + b_0 + b_1$$
$$\tfrac{1}{3} = \quad a_1 \quad + 2b_1$$
$$\tfrac{1}{4} = \quad a_1 \quad + 3b_1$$

which yield $a_0 = a_1 = \frac{1}{2}$ and $b_0 = -b_1 = \frac{1}{12}$. The resulting formula is

$$\int_0^h y(x)\, dx = \frac{h}{2}(y_0 + y_1) + \frac{h^2}{12}(y_0' - y_1')$$

which reproduces the first terms of the Euler–Maclaurin formula. (Compare with Problem 8.61.)

8.73 Use the method of undetermined coefficients to derive a formula of the form

$$\int_{-h}^h y(x)\, dx = h(a_{-1}y_{-1} + a_0 y_0 + a_1 y_1) + h^2(b_{-1}y_{-1}' + b_0 y_0' + b_1 y_1')$$

which is exact for polynomials of as high a degree as possible.

▌ Substituting the power functions $y(x) = 1, x, \ldots, x^5$ leads to this system of equations for the six coefficients:

$$2 = a_{-1} + a_0 + a_1$$
$$0 = -a_{-1} + a_1 + b_{-1} + b_0 + b_1$$
$$\tfrac{2}{3} = a_{-1} + a_1 - 2b_{-1} + 2b_1$$
$$0 = -a_{-1} + a_1 + 3b_{-1} + 3b_1$$
$$\tfrac{2}{5} = a_{-1} + a_1 - 4b_{-1} + 4b_1$$
$$0 = -a_{-1} + a_1 + 5b_{-1} + 5b_1$$

A little algebra then leads to

$$a_0 = \tfrac{16}{15} \qquad a_{-1} = a_1 = \tfrac{7}{15} \qquad b_0 = 0 \qquad b_{-1} = -b_1 = \tfrac{1}{15}$$

and we have another integration formula,

$$\int_{-h}^h y(x)\, dx = \frac{h}{15}(7y_{-1} + 16y_0 + 7y_1) + \frac{h^2}{15}(y_{-1}' - y_1')$$

8.74 Use the method of undetermined coefficients to derive the formula

$$\int_0^h y(x)\, dx = \frac{h}{2}(y_0 + y_1) - \frac{h^3}{24}(y_0^{(2)} + y_1^{(2)})$$

which should be compared with the result of Problem 8.59.

▌ We ask that

$$\int_0^h y(x)\, dx = h(a_0 y_0 + a_1 y_1) + h^3(b_0 y_0'' + b_1 y_1'')$$

be exact for $y(x) = 1, x, x^2, x^3$:

$$1 = a_0 + a_1 \qquad\qquad \tfrac{1}{2} = a_1$$
$$\tfrac{1}{3} = a_1 + 2b_0 + 2b_1 \qquad \tfrac{1}{4} = a_1 + 6b_1$$

This little system has the solution

$$a_0 = a_1 = \tfrac{1}{2} \qquad b_0 = b_1 = -\tfrac{1}{24}$$

as anticipated.

8.75 Use the method of undetermined coefficients to derive

$$\int_0^h y(x)\,dx = \frac{h}{2}(y_0 + y_1) + \frac{h^2}{10}(y_0' - y_1') + \frac{h^3}{120}(y_0^{(2)} + y_1^{(2)})$$

proving it exact for polynomials of degree up through 5.

▌ The usual process generates the linear system

$$1 = a_0 + a_1 \qquad\qquad \tfrac{1}{2} = a_1 + b_0 + b_1$$
$$\tfrac{1}{3} = a_1 + 2b_1 + 2c_0 + 2c_1 \qquad \tfrac{1}{4} = a_1 + 3b_1 + 6c_1$$
$$\tfrac{1}{5} = a_1 + 4b_1 + 12c_1 \qquad \tfrac{1}{6} = a_1 + 5b_1 + 20c_1$$

with solution $a_0 = a_1 = \tfrac{1}{2}$, $b_0 = -b_1 = \tfrac{1}{10}$ and $c_0 = c_1 = \tfrac{1}{120}$.

8.76 Integrals involving oscillatory functions often require special treatment. Develop the Filon formula for $\int_a^b f(x)\sin x\,dx$.

▌ The method of undetermined coefficients may be applied. As a simple example, just to illustrate the method, we choose

$$\int_0^{2\pi} y(x)\sin x\,dx = A_1 y(0) + A_2 y(\pi) + A_3 y(2\pi)$$

Requiring that this be exact for $y(x) = 1$, x and x^2, we obtain

$$0 = A_1 + A_2 + A_3 \qquad -2\pi = \pi A_2 + 2\pi A_3 \qquad -4\pi^2 = \pi^2 A_2 + 4\pi^2 A_3$$

Solving for the coefficients, $A_1 = 1$, $A_2 = 0$ and $A_3 = -1$. Then

$$\int_0^{2\pi} y(x)\sin x\,dx = y(0) - y(2\pi)$$

Filon has developed the more general result

$$\int_a^b y(x)\sin kx\,dx \sim h[Ay(a)\cos ka - Ay(b)\cos kb + BS + CT]$$

where $2nh = b - a$ and

$$A = \frac{1}{kh} + \frac{\sin 2kh}{2k^2h^2} - \frac{2\sin^2 kh}{k^3h^3} \qquad B = \frac{1 + \cos^2 kh}{k^2h^2} - \frac{\sin 2kh}{k^3h^3} \qquad C = \frac{4\sin kh}{k^3h^3} - \frac{4\cos kh}{k^2h^2}$$

$$S = -y(a)\sin ka - y(b)\sin kb + 2\sum_{i=0}^{n} y(a + 2ih)\sin(ka + 2ikh)$$

$$T = \sum_{i=1}^{n} y[a + (2i-1)h]\sin[ka + (2i-1)kh]$$

The truncation error proves to be

$$R_n = \frac{h^3}{12}(b-a)\left[1 - \frac{1}{16\cos(kh/4)}\right]\sin\left(\frac{kh}{2}\right)\cdot y^{(4)}(\xi)$$

8.77 Apply Filon's formula to $\int_0^{2\pi}\log(1+x)\sin 10x\,dx$.

▌ Results using from 4 to 256 points are listed in Table 8.9. Corresponding results for Simpson's rule are included for comparison. Note that the Filon formula with four points excels the Simpson formula with 128 and that Filon with 16 points wins out over Simpson with 256 points. Clearly there is something to be said for giving oscillatory

TABLE 8.9

Points	Filon	Simpson
4	−.19800636	
8	−.19815543	.00000155
16	−.19764683	−.95075040
32	−.19762639	.44885465
64	−.19762670	−.22532862
128	−.19762681	−.19876029
256	−.19762755	−.19769189
4096		−.19762765

integrals special attention. Note also the fluctuation in sign of the early Simpson approximations. This is due to the fact that so few points are within each period of $\sin 10x$ that their location has a major effect on the result. The correct result to eight places is $-.19762761$.

8.4 ADAPTIVE METHODS

8.78 What is the basic idea of adaptive integration methods?

❚ It may best be explained with an example. Suppose we apply the trapezoidal rule to this trivial integral

$$\int_0^1 x^4 \, dx$$

and require the error to be no greater than .005, the exact value being, of course, .2. Using eight equal intervals we would find the value .2052, which almost meets the specifications. It is interesting to see the contribution that each of the eight trapezoids makes to the error of 52 points. In the fourth decimal place they are

$$0 \quad 1 \quad 2 \quad 4 \quad 6 \quad 9 \quad 13 \quad 17$$

Almost all the error enters at the right, which is understandable since the error of the trapezoidal rule is governed by the second derivative, which grows like $12x^2$. In a brute force application of the rule we might next halve the interval, from $\frac{1}{8}$ to $\frac{1}{16}$, and try another run. The basic idea of adaptive integration is that this would be wasteful at the left. We are already doing very well there and could easily afford to *increase* the length of the interval in the spirit of economy of effort. The acceptable error of 50 points in the fourth place would be achieved if we made errors of 6 points for each of eight trapezoids, spreading the total error equally over the set. After just a bit of trial and error we encountered the set of endpoints

$$x_i = 0 \quad .28 \quad .42 \quad .54 \quad .65 \quad .75 \quad .84 \quad .92 \quad 1$$

The eight trapezoids built on these foundations have a total area of .2035, putting us well under the error limitation, but it is the error contributions of the trapezoids that is of interest:

$$5 \quad 3 \quad 4 \quad 5 \quad 5 \quad 5 \quad 4 \quad 5$$

The technique has been casual but the idea of spreading out the error has been implemented. Figure 8-4 gives the picture.

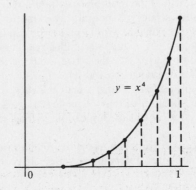

Fig. 8-4

8.79   Repeat the experiment of the preceding problem using

$$\int_0^1 x^8\, dx$$

and seeking the same level of accuracy.

▌ With eight equal intervals the trapezoidal rule manages .1214, which is to be compared with the correct .1111 to four places. A breakdown of the 103 point error by trapezoids looks like

$$0 \quad 0 \quad 0 \quad 1 \quad 3 \quad 10 \quad 27 \quad 63$$

and is hardly a fair split. This power function has even greater curvature growth than the last and the error imbalance shows it. Once again some informal but scientific guesswork leads to a more equitable arrangement. The endpoints

$$x_i = 0 \quad .48 \quad .64 \quad .74 \quad .82 \quad .89 \quad .95 \quad 1$$

provide foundations for only seven, not eight, trapezoids, but the new sum of trapezoidal areas is .1151. The reduced error is 40 points, with this distribution over the seven trapezoids

$$5 \quad 6 \quad 5 \quad 5 \quad 6 \quad 6 \quad 5$$

The arithmetical discrepancy is due to roundoffs, but the point is that seven trapezoids have done better than eight. Figure 8-5 gives the picture.

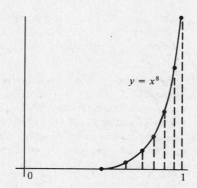

$y = x^8$

Fig. 8-5

8.80   How can the idea of adaptive integration be developed more systematically?

▌ There are numerous ways. An interesting possibility, perhaps more picturesque than practical, is illustrated in Fig. 8-6(a).

The combined areas of the four trapezoids is clearly less than the area under the curve, which is concave downward over this interval. (The idea involved can be adapted to cover convexity as well.) If the shaded areas were included it also seems clear that the resulting area would then be too large, so the total shaded area serves as an error measure. The adaptive strategy is now to order the four intervals according to the sizes of the corresponding error triangles. The interval with largest error measure (shaded triangle) is then subdivided and a pair of trapezoids replaces the original. Figure 8-6(b) illustrates how the error measure is reduced by this effort, and the interval list is modified to properly place these two smaller error triangles. The process continues in this way until the total area of all the error triangles is less than a predetermined tolerance.

The formulas of trigonometry required to implement this idea are somewhat messy, but if implemented usually lead to rapid error reduction. The disadvantage of having to continually sort the interval list so that the interval having the largest error triangle will be at the top can be serious, particularly since this list will grow very long if high accuracy is demanded.

An alternate procedure avoids the problem of maintaining a priority list by taking intervals from left to right, subdividing if necessary and discarding intervals for which the error bound is less than a proportional share of the total tolerance,

$$\text{bound} < \text{tolerance} * (\text{interval length})/(b - a)$$

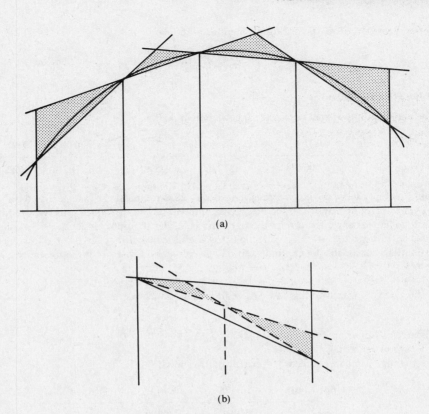

(a)

(b)

Fig. 8-6

where (a, b) is the total interval of integration. The number of active intervals eventually diminishes and the algorithm stops when it reaches zero.

Tests have found that this adaptive method requires something like a third as many trapezoids for smooth integrands and perhaps only a twentieth as many for troublesome integrands, having large curvature in places. For such integrals it often outperforms the Simpson rule.

8.81 Develop an adaptive integration method based on the error estimator of Problem 8.32.

$\blacksquare$ In that problem the errors of two successive applications of the trapezoidal rule, with intervals h and $2h$, were compared. The error of the former was found to be approximately

$$e = \tfrac{1}{3}(A_2 - A_1)$$

with A_2 corresponding to the interval h. The following algorithm is then available.

(1) Define the integral and the tolerance, set the approximation initially to zero and choose $2h$ initially as the entire interval (a, b).
(2) Compute A_1 and A_2 for the active subinterval.
(3) Apply the error estimator $e = \tfrac{1}{3}(A_2 - A_1)$.
(4) If the estimate is unacceptable, but the active interval is the result of doubling an interval for which a satisfactory error has been found, then accumulate the earlier result for the left half and return to step 2 with the right half interval. (Too much acceleration.) Otherwise, return to step 2 with the left half interval.
If the estimate is acceptable, and the endpoint b has been reached, accumulate this result, print the total and stop. If the active interval is the result of halving an earlier interval for which an unsatisfactory error was found, then accumulate this result and return to step 2 with the next interval, keeping the same value of h. Otherwise, save the present result but double the interval and return to step 2. (The attempt to move ahead more rapidly.)

The algorithm is nothing more than an attempt to use the widest trapezoids possible consistent with the error tolerance specified at the outset. Various embellishments can be attached, such as intermediate output to observe the speed of advance or a minimum value of h to avoid endless looping.

8.82 Apply the algorithm of the preceding problem to the integral

$$\int_0^2 x^7/128\, dx = .25$$

with a tolerance of .0001.

▮ The following output was obtained, somewhat rounded:

Upper limit	1.219	1.516	1.672	1.828	1.977	2.000
Integral	.00478	.02723	.05965	.12187	.22755	.25005

Each output step, except the last, records the work of 20 trapezoids; the last step used only 6. Note that the first 20 covered an interval of length 1.2, the next 20 an interval of length about .3, the next about .16 and so on. The integrand has increasing curvature at the right and is harder to approximate there. Also note that the tolerance of .0001 has been respected. In this connection it is useful to recall that the error estimator being used is only an approximation, not an absolute bound, so it may sometimes be wise to set the tolerance a bit less than the accuracy required.

8.83 Apply the same algorithm to this integral, requiring the same tolerance:

$$\int_0^2 x^9/1024\, dx = .1$$

▮ These output values were recorded, somewhat rounded:

Upper limit	1.4375	1.7031	1.8594	2.0000
Integral	.00376	.02008	.04827	.10004

Again each output step records the work of 20 trapezoids, except the last which needed only 18. The rapid progress at the left is conspicuous. Almost half the trapezoids were needed between 1.7 and 2. The tolerance has again been respected.

8.84 Apply the same algorithm to

$$\int_0^2 x^{11}/1024\, dx = \tfrac{1}{3}$$

requiring an error no greater than .0001.

▮ The upper limits reached by successive sets of 20 trapezoids somewhat rounded, were

$$1.31 \quad 1.52 \quad 1.66 \quad 1.74 \quad 1.82 \quad 1.90 \quad 1.98 \quad 1.99 \quad 2.00$$

The pattern is similar to that found in preceding problems and the resulting approximation of .33337 is satisfactory.

8.85 Apply the same algorithm to the integral

$$\int_{-1}^1 \exp(-100x^2)\, dx$$

with an error tolerance of .0001.

▮ The integral has a sharp peak at $x = 0$ and drops quickly to near zero on both sides. The algorithm should move quickly from the left limit of -1, slow down substantially in the vicinity of the peak and then accelerate again for the finish. Here are the upper limits reached by successive sets of 20 trapezoids, rounded as usual:

$$-.203 \quad -.156 \quad -.117 \quad -.062 \quad -.025 \quad -.006 \quad .014 \quad .039 \quad .090 \quad .129 \quad .168 \quad .238$$

and finally 1. The integrand is, of course, symmetric about $x = 0$ but the algorithm does not quite exploit this. Even so, both of the outer intervals are of length about .8 and the heavier going near the center is clear. The approximation .17730 was obtained and may be compared with the correct $\sqrt{\pi}/10$ or .17725. A minimum h value of 1/512 had been set but proved to be too conservative. The change to 1/1024 led to the given results.

8.86 How can the idea of the adaptive algorithm just illustrated be modified for use with the Simpson rule?

▌ The modification is direct, using the error estimator of Problem 8.24,

$$\text{error} = (A_2 - A_1)/15$$

It may be useful to summarize the underlying idea in other words. Each part of the interval of integration is to be subdivided just finely enough for it to contribute only its proportionate share of the overall permissible error. Let that error be E. We apply Simpson's rule to the interval and then use the error estimator. If the error is less than E we are finished. If not, attention shifts to the left half interval. If the new error estimate is less than $E/2$ we are finished with that half interval. Otherwise the halving process continues until eventually an interval of length $(b-a)/2^k$ is reached, (a, b) being the original interval, where the estimator predicts an error less than $E/2^k$. Forward progress then resumes beginning at the right edge of the accepted interval.

The current integrator is to be

$$A_2 = (h/3)(y_0 + 4y_1 + 2y_2 + 4y_3 + y_4)$$

while the simpler

$$A_1 = (2h/3)(y_0 + 4y_2 + y_4)$$

is needed for the error estimator. The approximation A_2 is then accepted when

$$(A_2 - A_1) \le 15E/2^k$$

and is accumulated into the sum of previously accepted results to its left. The process ends when the fragments cover (a, b).

8.87 Apply the adaptive integration method of the preceding problem to the integral

$$\int_0^8 x^5 \, dx$$

▌ A few runs were made with differing tolerances and slight changes in the upper limit. The following abbreviated output is typical. Note especially the values of k, which started at 1 (not printed) and rose to 7. An effort to raise the upper limit further found k skyrocketing:

x	$x^6/6$	Computed	k
2	10.667	10.667	4
4	682.667	682.667	5
6	7776.000	7775.99	6
8	43690.67	43690.58	7

8.88 Apply adaptive integration as in the preceding problem to the arcsine integral

$$\int_0^1 \frac{dx}{\sqrt{1-x^2}}$$

▌ Tension is generated by the infinite discontinuity at the upper limit, which suggests a diminishing step size near this end much as in the preceding problem. Values of k confirmed this suspicion, climbing steadily as the computation progressed and reaching 15 with the result

$$\text{upper limit} = .9999$$
$$\text{integral} = 1.5573$$

At this point the correct arcsine value is 1.5575.

8.5 GAUSSIAN METHODS

8.89 What is the underlying idea of Gaussian integration?

▌ The main idea behind Gaussian integration is that in the selection of a formula

$$\int_a^b y(x) \, dx \approx \sum_{i=1}^{n} A_i y(x_i)$$

it may be wise not to specify that the arguments x_i be equally spaced. All our earlier formulas assume equal spacing, and if the values $y(x_i)$ are obtained experimentally this will probably be true. Many integrals, however, involve familiar analytic functions that may be computed for any argument and to great accuracy. In such cases it is useful to ask what choice of the x_i and A_i together will bring maximum accuracy. It proves to be convenient to discuss the slightly more general formula

$$\int_a^b w(x)\, y(x)\, dx \simeq \sum_{i=1}^n A_i\, y(x_i)$$

in which $w(x)$ is a weighting function to be specified later. When $w(x) = 1$ we have the original, simpler formula.

One approach to such Gaussian formulas is to ask for perfect accuracy when $y(x)$ is one of the power functions $1, x, x^2, \ldots, x^{2n-1}$. This provides $2n$ conditions for determining the $2n$ numbers x_i and A_i. In fact,

$$A_i = \int_a^b w(x)\, L_i(x)\, dx$$

where $L_i(x)$ is the Lagrange multiplier function. The arguments $x_1, \ldots, x_n$ are the zeros of the nth-degree polynomial $p_n(x)$ belonging to a family having the orthogonality property

$$\int_a^b w(x)\, p_n(x)\, p_m(x)\, dx = 0 \quad \text{for } m \neq n$$

These polynomials depend on $w(x)$. The weighting function therefore influences both the A_i and the x_i but does not appear explicitly in the Gaussian formula.

Hermite's formula for an osculating polynomial provides another approach to Gaussian formulas. Integrating the osculating polynomial leads to

$$\int_a^b w(x)\, y(x)\, dx \simeq \sum_{i=1}^n \left[A_i\, y(x_i) + B_i\, y'(x_i) \right]$$

but the choice of the arguments x_i as the zeros of a member of an orthogonal family makes all $B_i = 0$. The formula then reduces to the prescribed type. This suggests, and we proceed to verify, that a simple collocation polynomial at these unequally spaced arguments would lead to the same result.

8.90 What are some of the types of Gaussian formulas?

❚ Particular types of Gaussian formulas may be obtained by choosing $w(x)$ and the limits of integration in various ways. Occasionally one may also wish to impose constraints, such as specifying certain x_i in advance. A number of particular types are presented.

(1) *Gauss–Legendre formulas* occur when $w(x) = 1$. This is the prototype of the Gaussian method and we discuss it in more detail than the other types. It is customary to normalize the interval (a, b) to $(-1, 1)$. The orthogonal polynomials are then the Legendre polynomials

$$P_n(x) = \frac{1}{2^n n!} \frac{d^n}{dx^n} (x^2 - 1)^n$$

with $P_0(x) = 1$. The x_i are the zeros of these polynomials and the coefficients are

$$A_i = \frac{2(1 - x_i^2)}{n^2 \left[P_{n-1}(x_i) \right]^2}$$

Tables of the x_i and A_i are available to be substituted directly into the Gauss–Legendre formula

$$\int_a^b y(x)\, dx \simeq \sum_{i=1}^n A_i\, y(x_i)$$

Various properties of Legendre polynomials are required in the development of these results, including

$$\int_{-1}^{1} x^k P_n(x)\, dx = 0 \quad \text{for } k = 0, \ldots, n-1$$

$$\int_{-1}^{1} x^n P_n(x)\, dx = \frac{2^{n+1}(n!)^2}{(2n+1)!}$$

$$\int_{-1}^{1} [P_n(x)]^2\, dx = \frac{2}{2n+1}$$

$$\int_{-1}^{1} P_m(x) P_n(x)\, dx = 0 \quad \text{for } m \neq n$$

$$P_n(x) \text{ has } n \text{ real zeros in } (-1,1)$$

$$(n+1) P_{n+1}(x) = (2n+1) x P_n(x) - n P_{n-1}(x)$$

$$(t-x) \sum_{i=0}^{n} (2i+1) P_i(x) P_i(t) = (n+1)[P_{n+1}(t) P_n(x) - P_n(t) P_{n+1}(x)]$$

$$\int_{-1}^{1} \frac{P_n(x)}{x - x_k}\, dx = \frac{-2}{(n+1) P_{n+1}(x_k)}$$

$$(1 - x^2) P_n'(x) + n x P_n(x) = n P_{n-1}(x)$$

(2) *Gauss–Laguerre formulas* take the form

$$\int_{0}^{\infty} e^{-x} y(x)\, dx \simeq \sum_{i=1}^{n} A_i y(x_i)$$

the arguments x_i being the zeros of the nth Laguerre polynomial

$$L_n(x) = e^x \frac{d^n}{dx^n}(e^{-x} x^n)$$

and the coefficients A_i being

$$A_i = \frac{(n!)^2}{x_i [L_n'(x_i)]^2}$$

The numbers x_i and A_i are available in tables.

The derivation of Gauss–Laguerre formulas parallels that of Gauss–Legendre very closely, using properties of the Laguerre polynomials.

(3) *Gauss–Hermite formulas* take the form

$$\int_{-\infty}^{\infty} e^{-x^2} y(x)\, dx \simeq \sum_{i=1}^{n} A_i y(x_i)$$

the arguments x_i being the zeros of the nth Hermite polynomial

$$H_n(x) = (-1)^n e^{x^2} \frac{d^n}{dx^n}(e^{-x^2})$$

and the coefficients A_i being

$$A_i = \frac{2^{n+1} n! \sqrt{\pi}}{[H_n'(x_i)]^2}$$

The numbers x_i and A_i are available in tables.

(4) *Gauss–Chebyshev formulas* take the form

$$\int_{-1}^{1} \frac{y(x)}{\sqrt{1-x^2}}\, dx \simeq \frac{\pi}{n} \sum_{i=1}^{n} y(x_i)$$

the arguments x_i being the zeros of the nth Chebyshev polynomial $T_n(x) = \cos(n \arccos x)$.

(5) *Gauss–Lobatto formulas* take the form

$$\int_{-1}^{1} y(x)\, dx \sim \frac{2}{n(n-1)}[y(-1)+y(1)] + \sum_{i=1}^{n-2} A_i y(x_i)$$

the x_i being the zeros of $P'_{n-1}(x)$ and where

$$A_i = \frac{2}{n(n-1)[P_{n-1}(x_i)]^2}$$

Note that the endpoints $x = \pm 1$ have been prescribed as two of the $n+1$ arguments.

(6) *Chebyshev formulas* take the form

$$\int_{-1}^{1} y(x)\, dx \sim (2/n) \sum_{i=1}^{n} y(x_i)$$

and have equal coefficients.

(7) *The two-point formula*

$$\int_{-1}^{1} y(x)\, dx \sim \frac{2^n n!}{(2n)!} \sum_{i=0}^{n-1} \frac{(n+i)!}{(n-i)! 2^i i!} Y_i$$

where $Y_i = y^{(n-i-1)}(-1) + (-1)^{n-i-1} y^{(n-i-1)}(1)$, uses values of $y(x)$ and its derivatives *only at the endpoints* of the interval of integration.

8.91 Integrate Hermite's formula for an osculating polynomial approximation to $y(x)$ at arguments x_1 to x_n.

▮ Here it is convenient to delete the argument x_0 in our osculating polynomial. This requires only minor changes in our formulas of Chapter 5. The Hermite formula itself becomes

$$p(x) = \sum_{i=1}^{n} [1 - 2L'_i(x_i)(x - x_i)][L_i(x)]^2 y_i + (x - x_i)[L_i(x)]^2 y'_i$$

where $L_i(x) = F_i(x)/F_i(x_i)$ is the Lagrange multiplier function and $F_i(x)$ is the product $F_i(x) = \prod_{k \neq i}(x - x_k)$. Integrating, we find

$$\int_{a}^{b} w(x) p(x)\, dx = \sum_{i=1}^{n} (A_i y_i + B_i y'_i)$$

where

$$A_i = \int_{a}^{b} w(x)[1 - 2L'_i(x_i)(x - x_i)][L_i(x)]^2\, dx \qquad B_i = \int_{a}^{b} w(x)(x - x_i)[L_i(x)]^2\, dx$$

8.92 Find the truncation error of the formula in Problem 8.91.

▮ Surprisingly enough, this comes more easily than for formulas obtained from simple collocation polynomials, because the mean value theorem applies directly. The error of Hermite's formula (Problem 5.5), with n in place of $n+1$ because we have deleted one argument, becomes

$$y(x) - p(x) = \frac{y^{(2n)}(\xi)}{(2n)!}[\pi(x)]^2$$

Multiplying by $w(x)$ and integrating,

$$\int_a^b w(x)[y(x)-p(x)]\,dx = \int_a^b w(x)\frac{y^{(2n)}(\xi)}{(2n)!}[\pi(x)]^2\,dx$$

Since $w(x)$ is to be chosen a nonnegative function and $[\pi(x)]^2$ is surely positive, the mean value theorem at once yields

$$E = \int_a^b w(x)[y(x)-p(x)]\,dx = \frac{y^{(2n)}(\theta)}{(2n)!}\int_a^b w(x)[\pi(x)]^2\,dx$$

for the truncation error. Here $a < \theta < b$, but as usual θ is not otherwise known. Notice that if $y(x)$ were a polynomial of degree $2n-1$ or less, this error term would be exactly 0. Our formula will be exact for all such polynomials.

8.93 Show that all the coefficients B_i will be 0 if

$$\int_a^b w(x)\pi(x)x^k\,dx = 0 \quad \text{for } k = 0,1,\ldots,n-1$$

▮ By Problem 4.57, $(x-x_i)L_i(x) = \pi(x)/\pi'(x_i)$. Substituting this into the formula for B_i,

$$B_i = \frac{1}{\pi'(x_i)}\int_a^b w(x)\pi(x)L_i(x)\,dx$$

But $L_i(x)$ is a polynomial in x of degree $n-1$ and so

$$B_i = \frac{1}{\pi'(x_i)}\int_a^b w(x)\pi(x)\sum_{k=0}^{n-1}\alpha_k x^k\,dx = \frac{1}{\pi'(x_i)}\sum_{k=0}^{n-1}\alpha_k\int_a^b w(x)\pi(x)x^k\,dx = 0$$

8.94 Define *orthogonal functions* and restate the result of Problem 8.93 in terms of orthogonality.

▮ Functions $f_1(x)$ and $f_2(x)$ are called orthogonal on the interval (a,b) with weight function $w(x)$ if

$$\int_a^b w(x)f_1(x)f_2(x)\,dx = 0$$

The coefficients B_i of our formula will be zero if $\pi(x)$ is orthogonal to x^p for $p = 0,1,\ldots,n-1$. By addition $\pi(x)$ will then be orthogonal to any polynomial of degree $n-1$ or less, including the Lagrange multiplier functions $L_i(x)$. Such orthogonality depends on and determines our choice of the collocation arguments x_k and is assumed for the remainder of this chapter.

8.95 Prove that with all the $B_i = 0$, the coefficients A_i reduce to $A_i = \int_a^b w(x)[L_i(x)]^2\,dx$ and are therefore positive numbers,

▮ $$A_i = \int_a^b w(x)[L_i(x)]^2\,dx - 2L_i'(x_i)B_i \quad \text{reduces to the required form when} \quad B_i = 0$$

8.96 Derive the simpler formula $A_i = \int_a^b w(x)L_i(x)\,dx$.

▮ The result follows if we can show that $\int_a^b w(x)L_i(x)[L_i(x)-1]\,dx = 0$. But $L_i(x)-1$ must contain $(x-x_i)$ as a factor, because $L_i(x_i)-1 = 1-1 = 0$. Therefore

$$L_i(x)[L_i(x)-1] = \frac{\pi(x)}{\pi'(x_i)(x-x_i)}[L_i(x)-1] = \pi(x)p(x)$$

with $p(x)$ of degree $n-1$ at most. Problem 8.93 then guarantees that the integral is zero.

8.97   The integration formula of this section can now be written as

$$\int_a^b w(x)\,y(x)\,dx \simeq \sum_{i=1}^n A_i\,y(x_i)$$

where $A_i = \int_a^b w(x)L_i(x)\,dx$ and the arguments x_i are to be chosen by the orthogonality requirements of Problem 8.93. This formula was obtained by integration of an osculating polynomial of degree $2n-1$ determined by the y_i and y_i' values at arguments x_i. Show that the same formula is obtained by integration of the simpler collocation polynomial of degree $n-1$ determined by the y_i values alone. (This is one way of looking at Gaussian formulas; they extract high accuracy from polynomials of relatively low degree.)

▮   The collocation polynomial is $p(x) = \sum_{i=1}^n L_i(x)\,y(x_i)$ so that integration produces

$$\int_a^b w(x)\,p(x)\,dx = \sum_{i=1}^n A_i\,y(x_i)$$

as suggested. Here $p(x)$ represents the collocation polynomial. In Problem 8.91 it stood for the more complicated osculating polynomial. Both lead to the same integration formula. (For a specific example of this, see Problem 8.115.)

8.6   GAUSS–LEGENDRE FORMULAS

8.98   The special case $w(x) = 1$ leads to Gauss–Legendre formulas. It is the custom to use the interval of integration $(-1,1)$. As a preliminary exercise, determine the arguments x_k directly from the conditions of Problem 8.93,

$$\int_{-1}^1 \pi(x)\,x^k\,dx = 0 \quad k = 0,1,\dots,n-1$$

for the value $n = 3$.

▮   The polynomial $\pi(x)$ is then cubic, say $\pi(x) = a + bx + cx^2 + x^3$. Integrations produce

$$2a + \tfrac{2}{3}c = 0 \qquad \tfrac{2}{3}b + \tfrac{2}{5} = 0 \qquad \tfrac{2}{3}a + \tfrac{2}{5}c = 0$$

which lead quickly to $a = c = 0$, $b = -\tfrac{3}{5}$. This makes

$$\pi(x) = x^3 - \tfrac{3}{5}x = \left(x + \sqrt{\tfrac{3}{5}}\right)x\left(x - \sqrt{\tfrac{3}{5}}\right)$$

The collocation arguments are therefore $x_k = -\sqrt{\tfrac{3}{5}}, 0, \sqrt{\tfrac{3}{5}}$.

Theoretically this procedure would yield the x_k for any value of n but it is quicker to use a more sophisticated approach.

8.99   The Legendre polynomial of degree n is defined by

$$P_n(x) = \frac{1}{2^n n!}\frac{d^n}{dx^n}(x^2 - 1)^n$$

with $P_0(x) = 1$. Prove that for $k = 0,1,\dots,n-1$,

$$\int_{-1}^1 x^k P_n(x)\,dx = 0$$

making $P_n(x)$ also orthogonal to any polynomial of degree less than n.

▮   Apply integration by parts k times:

$$\int_{-1}^1 x^k \frac{d^n}{dx^n}(x^2 - 1)^n\,dx = \underbrace{\left[x^k \frac{d^{n-1}}{dx^{n-1}}(x^2 - 1)^n\right]_{-1}^1}_{= 0} - \int_{-1}^1 kx^{k-1}\frac{d^{n-1}}{dx^{n-1}}(x^2 - 1)^n\,dx$$

$$= \cdots = (-1)^k k!\int_{-1}^1 \frac{d^{n-k}}{dx^{n-k}}(x^2 - 1)^n\,dx = 0$$

8.100 Prove $\int_{-1}^{1} x^n P_n(x)\, dx = (2^{n+1}(n!)^2)/(2n+1)!$.

$\blacksquare$ Taking $k = n$ in the preceding problem,

$$\int_{-1}^{1} x^n \frac{d^n}{dx^n}(x^2-1)^n\, dx = (-1)^n n! \int_{-1}^{1}(x^2-1)^n\, dx$$

$$= 2n! \int_{0}^{1}(1-x^2)^n\, dx = 2n! \int_{0}^{\pi/2} \cos^{2n+1} t\, dt$$

This last integral responds to the treatment

$$\int_{0}^{\pi/2} \cos^{2n+1} t\, dt = \underbrace{\left[\frac{\cos^{2n} t \sin t}{2n+1}\right]_{0}^{\pi/2}}_{=0} + \frac{2n}{2n+1}\int_{0}^{\pi/2}\cos^{2n-1} t\, dt$$

$$= \cdots = \frac{2n(2n-2)\cdots 2}{(2n+1)(2n-1)\cdots 3}\int_{0}^{\pi/2}\cos t\, dt$$

so that

$$\int_{-1}^{1} x^n \frac{d^n}{dx^n}(x^2-1)^n\, dx = 2n! \frac{2n(2n-2)\cdots 2}{(2n+1)(2n-1)\cdots 3}$$

Now multiply top and bottom by $2n(2n-2)\cdots 2 = 2^n n!$ and recall the definition of $P_n(x)$ to obtain, as required,

$$\int_{-1}^{1} x^n P_n(x)\, dx = \frac{1}{2^n n!} 2n! \frac{2^n n! 2^n n!}{(2n+1)!} = \frac{2^{n+1}(n!)^2}{(2n+1)!}$$

8.101 Prove $\int_{-1}^{1}[P_n(x)]^2\, dx = 2/(2n+1)$.

$\blacksquare$ Splitting off the highest power of x in one $P_n(x)$ factor,

$$\int_{-1}^{1}[P_n(x)]^2\, dx = \int_{-1}^{1}\left[\frac{1}{2^n n!}\frac{(2n)!}{n!}x^n + \cdots\right]P_n(x)\, dx$$

Powers below x^n make no contribution, by Problem 8.99. Using the preceding problem, we have

$$\int_{-1}^{1}[P_n(x)]^2\, dx = \frac{(2n)!}{2^n(n!)^2}\frac{2^{n+1}(n!)^2}{(2n+1)!} = \frac{2}{2n+1}$$

8.102 Prove that for $m \neq n$, $\int_{-1}^{1} P_m(x) P_n(x)\, dx = 0$.

$\blacksquare$ Writing out the lower-degree polynomial, we find each power in it orthogonal to the higher-degree polynomial. In particular with $m = 0$ and $n \neq 0$ we have the special case $\int_{-1}^{1} P_n(x)\, dx = 0$.

8.103 Prove that $P_n(x)$ has n real zeros between -1 and 1.

$\blacksquare$ The polynomial $(x^2-1)^n$ is of degree $2n$ and has multiple zeros at ± 1. Its derivative therefore has one interior zero, by Rolle's theorem. This first derivative is also zero at ± 1, making three zeros in all. The second derivative is then guaranteed two interior zeros by Rolle's theorem. It also vanishes at ± 1, making four zeros in all. Continuing in this way we find that the nth derivative is guaranteed n interior zeros, by Rolle's theorem. Except for a constant factor, this derivative is the Legendre polynomial $P_n(x)$.

8.104 Show that for the weight function $w(x) = 1$, $\pi(x) = [2^n(n!)^2/(2n)!]P_n(x)$.

$\blacksquare$ Let the n zeros of $P_n(x)$ be called $x_1, \ldots, x_n$. Then

$$\left[\frac{2^n n!^2}{(2n)!}\right]P_n(x) = (x - x_1)\cdots(x - x_n)$$

The only other requirement on $\pi(x)$ is that it be orthogonal to x^k for $k = 0, 1, \ldots, n-1$. But this follows from Problem 8.99.

8.105 Calculate the first several Legendre polynomials directly from the definition, noticing that only even or only odd powers can occur in any such polynomial.

▮ $P_0(x)$ is defined to be 1. Then we find

$$P_1(x) = \frac{1}{2}\frac{d}{dx}(x^2 - 1) = x \qquad P_3(x) = \frac{1}{48}\frac{d^3}{dx^3}(x^2 - 1)^3 = \frac{1}{2}(5x^3 - 3x)$$

$$P_2(x) = \frac{1}{8}\frac{d^2}{dx^2}(x^2 - 1)^2 = \frac{1}{2}(3x^2 - 1) \qquad P_4(x) = \frac{1}{16\cdot 24}\frac{d^4}{dx^4}(x^2 - 1)^4 = \frac{1}{8}(35x^4 - 30x^2 + 3)$$

Similarly,

$$P_5(x) = \frac{1}{8}(63x^5 - 70x^3 + 15x) \qquad P_7(x) = \frac{1}{16}(429x^7 - 693x^5 + 315x^3 - 35x)$$

$$P_6(x) = \frac{1}{16}(231x^6 - 315x^4 + 105x^2 - 5) \qquad P_8(x) = \frac{1}{128}(6435x^8 - 12012x^6 + 6930x^4 - 1260x^2 + 35)$$

and so on. Since $(x^2 - 1)^n$ involves only even powers of x, the result of differentiating n times will contain only even or only odd powers.

8.106 Show that x^n can be expressed as a combination of Legendre polynomials up through $P_n(x)$. The same is then true of any polynomial of degree n.

▮ Solving in turn for successive powers, we find

$$1 = P_0(x) \qquad x = P_1(x) \qquad x^2 = \tfrac{1}{3}[2P_2(x) + P_0(x)]$$

$$x^3 = \tfrac{1}{5}[2P_3(x) + 3P_1(x)] \qquad x^4 = \tfrac{1}{35}[8P_4(x) + 20P_2(x) + 7P_0(x)]$$

and so on. The fact that each $P_k(x)$ begins with a nonzero term in x^k allows this procedure to continue indefinitely.

8.107 Prove the recursion for Legendre polynomials,

$$(n + 1)P_{n+1}(x) = (2n + 1)xP_n(x) - nP_{n-1}(x)$$

▮ The polynomial $xP_n(x)$ is of degree $n + 1$, and so can be expressed as the combination (see Problem 8.106)

$$xP_n(x) = \sum_{i=0}^{n+1} c_i P_i(x)$$

Multiply by $P_k(x)$ and integrate to find

$$\int_{-1}^{1} xP_k(x)P_n(x)\,dx = c_k\int_{-1}^{1} P_k^2(x)\,dx$$

all other terms on the right vanishing since Legendre polynomials of different degrees are orthogonal. But for $k < n - 1$ we know $P_n(x)$ is also orthogonal to $xP_k(x)$, since this product then has degree at most $n - 1$. (See Problem 8.99.) This makes $c_k = 0$ for $k < n - 1$ and

$$xP_n(x) = c_{n+1}P_{n+1}(x) + c_n P_n(x) + c_{n-1}P_{n-1}(x)$$

Noticing that, from the definition, the coefficient of x^n in $P_n(x)$ will be $(2n)!/2^n(n!)^2$, we compare coefficients of x^{n+1} in the preceding equations to find

$$\frac{(2n)!}{2^n(n!)^2} = c_{n+1}\frac{(2n + 2)!}{2^{n+1}[(n + 1)!]^2}$$

from which $c_{n+1} = (n + 1)/(2n + 1)$ follows. Comparing the coefficients of x^n, and remembering that only alternate powers appear in any Legendre polynomial, brings $c_n = 0$. To determine c_{n-1} we return to our integrals. With

$k = n - 1$ we imagine $P_k(x)$ written out as a sum of powers. Only the term in x^{n-1} needs to be considered, since lower terms, even when multiplied by x, will be orthogonal to $P_n(x)$. This leads to

$$\frac{(2n-2)!}{2^{n-1}[(n-1)!]^2} \int_{-1}^{1} x^n P_n(x)\, dx = c_{n-1} \int_{-1}^{1} P_{n-1}^2(x)\, dx$$

and using the results of Problems 8.100 and 8.101, one easily finds $c_{n-1} = n/(2n+1)$. Substituting these coefficients into our expression for $xP_n(x)$ now brings the required recursion. As a bonus we also have the integral

$$\int_{-1}^{1} x P_{n-1}(x) P_n(x)\, dx = \frac{n}{2n+1} \frac{2}{2n-1} = \frac{2n}{4n^2-1}$$

8.108   Illustrate the use of the recursion formula.

■   Taking $n = 5$, we find

$$P_6(x) = \tfrac{11}{6} x P_5(x) - \tfrac{5}{6} P_4(x) = \tfrac{1}{16}(231x^6 - 315x^4 + 105x^2 - 5)$$

and with $n = 6$,

$$P_7(x) = \tfrac{13}{7} x P_6(x) - \tfrac{6}{7} P_5(x) = \tfrac{1}{16}(429x^7 - 693x^5 + 315x^3 - 35x)$$

confirming the results obtained in Problem 8.105. The recursion process is well suited to automatic computation of these polynomials, while the differentiation process of Problem 8.105 is not.

8.109   Derive Christoffel's identity

$$(t - x) \sum_{i=0}^{n} (2i + 1) P_i(x) P_i(t) = (n + 1)[P_{n+1}(t) P_n(x) - P_n(t) P_{n+1}(x)]$$

■   The recursion formula of Problem 8.107 can be multiplied by $P_i(t)$ to obtain

$$(2i + 1) x P_i(x) P_i(t) = (i + 1) P_{i+1}(x) P_i(t) + i P_{i-1}(x) P_i(t)$$

Writing this also with arguments x and t reversed (since it is true for any x and t) and then subtracting, we have

$$(2i + 1)(t - x) P_i(x) P_i(t) = (i + 1)[P_{i+1}(t) P_i(x) - P_i(t) P_{i+1}(x)] - i[P_i(t) P_{i-1}(x) - P_{i-1}(t) P_i(x)]$$

Summing from $i = 1$ to $i = n$, and noticing the "telescoping effect" on the right, we have

$$(t - x) \sum_{i=1}^{n} (2i + 1) P_i(x) P_i(t) = (n + 1)[P_{n+1}(t) P_n(x) - P_n(t) P_{n+1}(x)] - (t - x)$$

The last term may be transferred to the left side where it may be absorbed into the sum as an $i = 0$ term. This is the Christoffel identity.

8.110   Use the Christoffel identity to evaluate the integration coefficients for the Gauss–Legendre case, proving $A_k = 2/(n P_n'(x_k) P_{n-1}(x_k))$.

■   Let x_k be a zero of $P_n(x)$. Then the preceding problem, with t replaced by x_k, makes

$$\frac{(n + 1) P_{n+1}(x_k) P_n(x)}{x - x_k} = - \sum_{i=0}^{n} (2i + 1) P_i(x) P_i(x_k)$$

Now integrate from -1 to 1. By a special case of Problem 8.102 only the $i = 0$ term survives on the right, and we have

$$\int_{-1}^{1} \frac{P_n(x)}{x - x_k}\, dx = \frac{-2}{(n + 1) P_{n+1}(x_k)}$$

The recursion formula with $x = x_k$ makes $(n+1)P_{n+1}(x_k) = -nP_{n-1}(x_k)$ which allows us the alternative

$$\int_{-1}^{1} \frac{P_n(x)}{x - x_k} \, dx = \frac{2}{nP_{n-1}(x_k)}$$

By Problems 8.96 and 8.104 we now find

$$A_k = \int_{-1}^{1} L_k(x) \, dx = \int_{-1}^{1} \frac{\pi(x)}{\pi'(x_k)(x - x_k)} \, dx = \int_{-1}^{1} \frac{P_n(x)}{P_n'(x_k)(x - x_k)} \, dx$$

leading at once to the result stated.

8.111   Prove that $(1 - x^2)P_n'(x) + nxP_n(x) = nP_{n-1}(x)$, which is useful for simplifying the result of Problem 8.110.

❚   We first notice that the combination $(1 - x^2)P_n' + nxP_n$ is at most of degree $n + 1$. However, with A representing the leading coefficient of $P_n(x)$, it is easy to see that x^{n+1} comes multiplied by $-nA + nA$ and so is not involved. Since P_n contains no term in x^{n-1}, our combination also has no term in x^n. Its degree is at most $n - 1$ and by Problem 8.106 it can be expressed as

$$(1 - x^2)P_n'(x) + nxP_n(x) = \sum_{i=0}^{n-1} c_i P_i(x)$$

Proceeding as in our development of the recursion formula, we now multiply by $P_k(x)$ and integrate. On the right only the kth term survives, because of the orthogonality, and we obtain

$$\frac{2}{2k+1} c_k = \int_{-1}^{1} (1 - x^2)P_n'(x) P_k(x) \, dx + n\int_{-1}^{1} xP_n(x) P_k(x) \, dx$$

Integrating the first integral by parts, the integrated piece is zero because of the factor $(1 - x^2)$. This leaves

$$\frac{2}{2k+1} c_k = -\int_{-1}^{1} P_n(x) \frac{d}{dx}\left[(1 - x^2) P_k(x)\right] dx + n\int_{-1}^{1} xP_n(x) P_k(x) \, dx$$

For $k < n - 1$ both integrands have $P_n(x)$ multiplied by a polynomial of degree $n - 1$ or less. By Problem 8.99 all such c_k will be zero. For $k = n - 1$ the last integral is covered by the Problem 8.107 bonus. In the first integral only the leading term of $P_{n-1}(x)$ contributes (again because of Problem 8.99) making this term

$$\int_{-1}^{1} P_n(x) \frac{d}{dx}\left\{ x^2 \frac{(2n-2)!}{2^{n-1}[(n-1)!]^2} x^{n-1} \right\} dx$$

Using Problem 8.100, this now reduces to

$$\frac{(2n-2)!}{2^{n-1}[(n-1)!]^2}(n+1)\frac{2^{n+1}(n!)^2}{(2n+1)!} = \frac{2n(n+1)}{(2n+1)(2n-1)}$$

Substituting these various results, we find

$$c_{n-1} = \frac{2n-1}{2}\left[\frac{2n(n+1)}{(2n+1)(2n-1)} + \frac{2n^2}{(2n+1)(2n-1)} \right] = n$$

which completes the proof.

8.112   Apply Problem 8.111 to obtain $A_k = (2(1 - x_k^2))/(n^2[P_{n-1}(x_k)]^2)$.

❚   Putting $x = x_k$, a zero of $P_n(x)$, we find $(1 - x_k^2)P_n'(x_k) = nP_{n-1}(x_k)$. The derivative factor can now be replaced in our result of Problem 8.110, producing the required result.

8.113 The Gauss–Legendre integration formula can now be expressed as

$$\int_{-1}^{1} y(x)\,dx \simeq \sum_{i=1}^{n} A_i y(x_i)$$

where the arguments x_k are the zeros of $P_n(x)$ and the coefficients A_k are given in Problem 8.112. Tabulate these numbers for $n = 2, 4, 6, \ldots, 16$.

▮ For $n = 2$ we solve $P_2(x) = \frac{1}{2}(3x^2 - 1) = 0$ to obtain $x_k = \pm\sqrt{\frac{1}{3}} = \pm.57735027$. The two coefficients prove to be the same. Problem 8.112 makes $A_k = 2(1 - \frac{1}{3})/[4(\frac{1}{3})] = 1$.

For $n = 4$ we solve $P_4(x) = \frac{1}{8}(35x^4 - 30x^2 + 3) = 0$ to find $x_k^2 = (15 \pm 2\sqrt{30})/35$, leading to the four arguments $x_k = \pm[(15 \pm 2\sqrt{30})/35]^{1/2}$.

Computing these and inserting them into the formula of Problem 8.112 produces the x_k, A_k pairs given in Table 8.10. The results for larger integers n are found in the same way, the zeros of the high-degree polynomials being found by methods of successive approximations.

TABLE 8.10

n	x_k	A_k	n	x_k	A_k
2	±.57735027	1.00000000	14	±.98628381	.03511946
4	±.86113631	.34785485		±.92843488	.08015809
	±.33998104	.65214515		±.82720132	.12151857
6	±.93246951	.17132449		±.68729290	.15720317
	±.66120939	.36076157		±.51524864	.18553840
	±.23861919	.46791393		±.31911237	.20519846
				±.10805495	.21526385
8	±.96028986	.10122854	16	±.98940093	.02715246
	±.79666648	.22238103		±.94457502	.06225352
	±.52553241	.31370665		±.86563120	.09515851
	±.18343464	.36268378		±.75540441	.12462897
10	±.97390653	.06667134		±.61787624	.14959599
	±.86506337	.14945135		±.45801678	.16915652
	±.67940957	.21908636		±.28160355	.18260342
	±.43339539	.26926672		±.09501251	.18945061
	±.14887434	.29552422			
12	±.98156063	.04717534			
	±.90411725	.10693933			
	±.76990267	.16007833			
	±.58731795	.20316743			
	±.36783150	.23349254			
	±.12533341	.24914705			

8.114 Apply the two-point formula to $\int_0^{\pi/2} \sin t\,dt$.

▮ The change of argument $t = \pi(x+1)/4$ converts this to our standard interval as

$$\int_{-1}^{1} \frac{\pi}{4} \sin\frac{\pi(x+1)}{4}\,dx$$

and the Gaussian arguments $x_k = \pm.57735027$ lead to $y(x_1) = .32589$, $y(x_2) = .94541$. The two-point formula now generates $(\pi/4)(.32589 + .94541) = .99848$ which is correct to almost three places. The *two*-point Gaussian formula has produced a better result than the trapezoidal rule with *seven* points (Problem 8.26). The error is two-tenths of 1%.

It is amusing to see what a *one*-point formula could have done. For $n = 1$ the Gauss–Legendre result is, as one may easily verify, $\int_{-1}^{1} y(x)\,dx \simeq 2y(0)$. For the sine function this becomes

$$\int_{-1}^{1} \frac{\pi}{4} \sin\frac{\pi(x+1)}{4}\,dx \simeq \frac{\pi}{4}\sqrt{2} \simeq 1.11$$

which is correct to within about 10%.

8.115 Explain the accuracy of the extremely simple formulas used in Problem 8.114 by exhibiting the polynomials on which the formulas are based.

▮ The $n = 1$ formula can be obtained by integrating the collocation polynomial of *degree zero*, $p(x) = y(x_1) = y(0)$. However, it can also be obtained, and this is the idea of the Gaussian method, from the osculating polynomial of degree $2n - 1 = 1$, which by Hermite's formula is $y(0) + xy'(0)$. Integrating this linear function between -1 and 1 produces the same $2y(0)$, the derivative term contributing zero. The zero-degree collocation polynomial produces the same integral as a first-degree polynomial, because the point of collocation was the Gaussian point (Fig. 8-7).

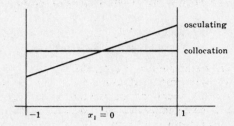

Fig. 8-7

Similarly, the $n = 2$ formula can be obtained by integrating the collocation polynomial of *degree* 1, the points of collocation being the Gaussian points

$$\int_{-1}^{1}\left(\frac{x-r}{-2r}y_1 + \frac{x+r}{2r}y_2\right) dx = y_1 + y_2$$

where $r = \sqrt{\frac{1}{3}}$. This same formula is obtained by integrating the osculating polynomial of degree 3, since

$$\int_{-1}^{1}\left[\left(1 + \frac{x+r}{r}\right)\frac{3}{4}(x-r)^2 y_1 + \left(1 - \frac{x-r}{r}\right)\frac{3}{4}(x+r)^2 y_2 + \frac{3}{4}(x^2 - r^2)(x-r) y_1' + \frac{3}{4}(x^2 - r^2)(x+r) y_2'\right] dx$$
$$= y_1 + y_2$$

The polynomial of degree 1 performs so well because the points of collocation were the Gaussian points (Fig. 8-8).

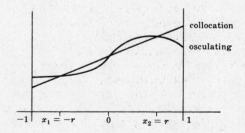

Fig. 8-8

8.116 Apply the Gaussian four-point formula to the integral of Problem 8.114.

▮ Using the same change of argument, the four-point formula produces $\sum_{i=1}^{4} A_i y_i = 1.000000$, correct to six places. Comparing with the Simpson 32-point result of 1.0000003 and the Simpson 64-point result of .99999983, we find it superior to either.

8.117 Adapt the truncation error estimate of Problem 8.92 to the special case of Gauss–Legendre approximation.

▮ Combining Problems 8.92, 8.101 and 8.104, we find the error to be

$$E = \frac{y^{(2n)}(\theta)}{(2n)!}\left[\frac{2^n(n!)^2}{(2n)!}\right]^2 \frac{2}{2n+1} = \frac{2^{2n+1}(n!)^4}{(2n+1)[(2n)!]^3} y^{(2n)}(\theta)$$

This is not an easy formula to apply if the derivatives of $y(x)$ are hard to compute. Some further idea of the accuracy of Gaussian formulas is, however, available by computing the coefficient of $y^{(2n)}$ for small n:

$$n = 2 \qquad E = .0074\,y^{(4)}$$
$$n = 4 \qquad E = .0000003\,y^{(8)}$$
$$n = 6 \qquad E = 1.5(10^{-12})\,y^{(12)}$$

8.118 Apply the error estimates of Problem 8.117 to the integral of Problem 8.114 and compare with the actual errors.

▌ After the change of argument that brings this integral to our standard form, we find

$$\left|y^{(4)}(x)\right| < \left(\frac{\pi}{4}\right)^5 \qquad \left|y^{(8)}(x)\right| < \left(\frac{\pi}{4}\right)^9$$

For $n = 2$ this makes our error estimate $E = (.0074)(.298) = .00220$, while for $n = 4$ we find $E = (.0000003)(.113) = .00000003$. The actual errors were .00152 and, to six places, zero. So our estimates are consistent with our results.

This example offers a favorable situation. The sine function is easy to integrate, even by approximate methods, because its derivatives are all bounded by the same constant, namely, 1. The powers of $\pi/4$ do enter with the change of argument, but they actually help in this case. The next example deals with a familiar function whose derivatives do not behave so favorably.

8.119 Apply the Gauss–Legendre formula to $\int_0^{\pi/2} \log(1 + t)\,dt$.

▌ The correct value of this integral is

$$\left(1 + \frac{\pi}{2}\right)\left[\log\left(1 + \frac{\pi}{2}\right) - 1\right] + 1 = .856590$$

to six places. The change of argument $t = \pi(x + 1)/4$ converts the integral to

$$\int_{-1}^{1} \frac{\pi}{4}\log\left[1 + \frac{\pi(x + 1)}{4}\right]dx$$

The fourth derivative of the new integrand is $(\pi/4)^5[-6/(1 + t)^4]$. In the interval of integration this cannot exceed $6(\pi/4)^5$, so the truncation error cannot exceed $6(\pi/4)^5(.0074)$ if we use the two-point Gaussian formula. This is six times the corresponding estimate for the integral of the sine function. Similarly, the eighth derivative is $(\pi/4)^9[-7!/(1 + t)^8]$. This means a truncation error of at most $(\pi/4)^9 \cdot 7!(.0000003)$ which is 7! times the corresponding estimate for the integral of the sine function. While the successive derivatives of the sine function remain bounded by 1, those of the logarithm function increase as factorials. The difference has an obvious impact on the truncation errors of any of our formulas, perhaps especially on Gaussian formulas where especially high derivatives are involved. Even so, these formulas perform well. Using just two points we obtain .858, while four points manage .856592 which is off by just two units in the last place. The six-point Gaussian formula scores a bull's-eye to six places, even though its truncation error term involves $y^{(12)}(x)$, which is approximately of size 12!. For contrast, Simpson's rule requires 64 points to produce this same six place result.

The function $\log(1 + t)$ has a singularity at $t = -1$. This is not on the interval of integration, but it is close, and even a complex singularity nearby could produce the slow kind of convergence in evidence here.

8.120 How does the length of the interval of integration affect the Gaussian formulas?

▌ For an integral over the interval $a \le t \le b$, the change of argument $t = a + (b - a)/2(x + 1)$ produces the standard interval $-1 \le x \le 1$. It also makes

$$\int_a^b y(t)\,dt = \int_{-1}^{1} \frac{b - a}{2}\,y\left[a + \frac{b - a}{2}(x + 1)\right]dx$$

The effect on truncation error is in the derivative factor, which is

$$\left(\frac{b - a}{2}\right)^{2n+1} y^{(2n)}(t)$$

In the examples just given $b - a$ was $\pi/2$ and this interval length actually helped to reduce error, but with a longer interval the potential of the powers of $b - a$ to magnify error is clear.

8.121 Apply the Gaussian method to $(2/\sqrt{\pi})\int_0^4 e^{-t^2}\,dt$.

■ The higher derivatives of this error function are not easy to estimate realistically. Proceeding with computations, one finds the $n = 4, 6, 8, 10$ formulas giving these results:

n	4	6	8	10
Approximation	.986	1.000258	1.000004	1.000000

For larger n the results agree with that for $n = 10$. This suggests accuracy to six places. We have already computed this integral by a patient application of Taylor's series (Problem 8.66) and found it to equal 1, correct to six places. For comparison, the Simpson formula requires 32 points to achieve six-place accuracy.

8.122 Apply the Gaussian method to $\int_0^4 \sqrt{1 + \sqrt{t}}\,dt$.

■ The $n = 4, 8, 12, 16$ formulas give the results

n	4	8	12	16
Approximation	6.08045	6.07657	6.07610	6.07600

This suggests accuracy to four places. The exact integral can be found, by a change of argument, to be $\frac{8}{5}(2\sqrt{3} + \frac{1}{3})$, which is 6.07590 correct to five places. Observe that the accuracy obtained here is inferior to that of the previous problem. The explanation is that our square root integrand is not as smooth as the exponential function. Its higher derivatives grow very large, like factorials. Our other formulas also feel the influence of these large derivatives. Simpson's rule for instance produces the values

No. of points	16	64	256	1024
Simpson values	6.062	6.07411	6.07567	6.07586

Even with a thousand points it has not managed the accuracy achieved in the previous problem with just 32 points.

8.123 Derive the Lanczos estimate for the truncation error of Gaussian formulas.

■ The relation $\int_{-1}^1 [xy(x)]'\,dx = y(1) + y(-1)$ holds exactly. Let I be the approximate integral of $y(x)$ obtained by the Gaussian n-point formula and I^* be the corresponding result for $[xy(x)]'$. Since $[xy(x)]' = y(x) + xy'(x)$,

$$I^* = I + \sum_{i=1}^n A_i x_i y'(x_i)$$

so that the error in I^* is

$$E^* = y(1) + y(-1) - I - \sum_{i=1}^n A_i x_i y'(x_i)$$

Calling the error in I itself E, we know that

$$E = C_n y^{(2n)}(\theta_1) \qquad E^* = C_n (xy)^{(2n+1)}(\theta_2)$$

for suitable θ_1 and θ_2 between -1 and 1. Suppose $\theta_1 = \theta_2 = 0$. On the one hand, $(xy)^{(2n+1)}(0)/(2n)!$ is the coefficient of x^{2n} in the Taylor series expansion of $(xy)'$, while on the other hand,

$$y(x) = \cdots + \frac{y^{(2n)}(0)\,x^{2n}}{(2n)!} + \cdots$$

leading directly to

$$[xy(x)]' = \cdots + \frac{(2n+1)\,y^{(2n)}(0)\,x^{2n}}{(2n)!} + \cdots$$

from which we deduce

$$(xy)^{(2n+1)}(0) = (2n+1)\,y^{(2n)}(0)$$

Thus $E^* = (2n+1)E$ approximately, making

$$E \simeq \frac{1}{2n+1}\left[y(1) + y(-1) - I - \sum_{i=1}^n A_i x_i y'(x_i) \right]$$

This involves applying the Gaussian formula to $xy'(x)$ as well as to $y(x)$ itself, but it avoids the often troublesome calculation of $y^{(2n)}(x)$. Putting $\theta_1 = \theta_2 = 0$ is the key move in deducing this formula. This has been found to be more reasonable for smooth integrands such as that of Problem 8.121, than for integrands with large derivatives, which seems reasonable since $y^{(2n)}(\theta_1)/y^{(2n)}(\theta_2)$ should be nearly 1 when $y^{(2n+1)}$ is small.

8.124 Apply the error estimate of the previous problem to the integral of Problem 8.121.

▮ For $n = 8$ the Lanczos estimate is .000004 and is identical with the actual error. For $n = 10$ and above, the Lanczos estimate correctly predicts a six-place error of zero. If applied to the integral of Problem 8.122, however, in which the integrand is very unsmooth, the Lanczos estimate proves to be too conservative to be useful. The limits to the usefulness of this error formula are still to be determined.

8.125 Prove $P_n'(x) = xP_{n-1}'(x) + nP_{n-1}(x)$.

▮ This and the next six problems explore some additional properties of the Legendre polynomials. From their definition

$$P_n'(x) = \frac{1}{2^n n!} \frac{d^n}{dx^n}\left[n(x^2-1)^{n-1}(2x) \right]$$

Applying the theorem on the nth derivative of a product,

$$P_n'(x) = \frac{n}{2^n n!} \frac{d}{dx}\left[2x\frac{d^{n-1}}{dx^{n-1}}(x^2-1)^{n-1} + 2(n-1)\frac{d^{n-2}}{dx^{n-2}}(x^2-1)^{n-1} \right]$$

$$= \frac{d}{dx}\left[xP_{n-1}(x) \right] + (n-1)P_{n-1}(x)$$

from which the given theorem follows at once.

8.126 Prove the classic differential equation for Legendre polynomials

$$(1-x^2)P_n^{(2)}(x) - 2xP_n'(x) + n(n+1)P_n(x) = 0$$

▮ Let $z = (x^2-1)^n$. Then $z' = 2nx(x^2-1)^{n-1}$ and $(x^2-1)z' = 2nxz$. Repeatedly differentiate this equation, obtaining

$$(x^2-1)z^{(2)} - (2n-2)xz' - 2nz = 0$$

$$(x^2-1)z^{(3)} - (2n-4)xz^{(2)} - [2n+(2n-2)]z' = 0$$

$$(x^2-1)z^{(4)} - (2n-6)xz^{(3)} - [2n+(2n-2)+(2n-4)]z^{(2)} = 0$$

and ultimately

$$(x^2-1)z^{(n+2)} - (2n-2n-2)xz^{(n+1)} - [2n+(2n-2)+(2n-4)+\cdots+(2n-2n)]z^{(n)} = 0$$

which simplifies to

$$(x^2-1)z^{(n+2)} + 2xz^{(n+1)} - n(n+1)z^{(n)} = 0$$

Since $P_n(x) = z^{(n)}/2^n n!$, the required result soon follows.

8.127 Prove $xP_n'(x) - P_{n-1}'(x) = nP_n(x)$.

▮ In Problem 8.111 we found

$$(1-x^2)P_n'(x) + nxP_n(x) = nP_{n-1}(x)$$

Differentiating,

$$(1-x^2)P_n''(x) + (n-2)xP_n'(x) + nP_n(x) = nP_{n-1}'(x)$$

Subtract the result of the preceding problem to obtain

$$nxP_n'(x) - nP_{n-1}'(x) = nP_n(x)$$

and cancel n. We now have several results that can be viewed as recursions for the derivatives of $P_n(x)$.

8.128 Prove that for all n, $P_n(1) = 1$ and $P_n(-1) = (-1)^n$.

   ❚ From Problem 8.111 with $x = 1$ we see that $P_n(1) = P_{n-1}(1)$ for positive integers n. Since $P_0(1) = 1$ it follows that $P_n(1) = 1$. Again from Problem 8.111 with $x = -1$ we see that $P_n(-1) = -P_{n-1}(-1)$. Since $P_0(-1) = 1$ it follows that the other polynomials alternate in sign at this argument.

8.129 Determine the first derivative values of the Legendre polynomials at 1 and -1.

   ❚ From Problem 8.125 we have

$$P_n'(1) = P_{n-1}'(1) + nP_{n-1}(1) = P_{n-1}'(1) + n$$

and since $P_0'(1) = 0$ the next several values are

n	0	1	2	3	4	5	6	$\cdots$
$P'(1)$	0	1	3	6	10	15	21	$\cdots$

Clearly we are once again summing the positive integers and have $P_n'(1) = n(n+1)/2$.
   Also from Problem 8.125 we have

$$P_n'(-1) = -P_{n-1}'(-1) + n(-1)^{n-1}$$

and since $P_0'(-1) = 0$ the next several values are

n	0	1	2	3	4	5	6	$\cdots$
$P'(-1)$	0	1	-3	6	-10	15	-21	$\cdots$

The result $P_n'(-1) = (-1)^{n+1} P_n'(1)$ is indicated.

8.130 Following up on the preceding problem, also find the values of the higher derivatives of the Legendre polynomials at 1 and -1.

   ❚ Once again we start from the result of Problem 8.125,

$$P_n'(x) = xP_{n-1}'(x) + nP_{n-1}(x)$$

Successive differentiations manage

$$P_n''(x) = xP_{n-1}''(x) + (n+1)P_{n-1}'(x)$$
$$P_n^{(3)}(x) = xP_{n-1}^{(3)}(x) + (n+2)P_{n-1}''(x)$$
$$P_n^{(k)}(x) = xP_{n-1}^{(k)}(x) + (n+k-1)P_{n-1}^{(k-1)}(x)$$

Setting $x = 1$ in the first of these,

$$P_n''(1) = P_{n-1}''(1) + (n+1)P_{n-1}'(1) = P_{n-1}''(1) + (n+1)n(n-1)/2$$

since $P_{n-1}'(1) = n(n-1)/2$ by the preceding problem. If we think of $P_n''(1)$ as y_n, then we are faced with the finite integration problem

$$y_n - y_{n-1} = (n+1)^{(3)}/2$$

with both y_0 and $y_1 = 0$. Integration of the factorial polynomial quickly leads to

$$y_n = P_n''(1) = (n+2)^{(4)}/2 \cdot 4 = (n+2)(n+1)n(n-1)/2 \cdot 4$$

   We are then in a position to deal with the third derivatives. From our preceding differentiations we have, setting $x = 1$,

$$P_n^{(3)}(1) = P_{n-1}^{(3)}(1) + (n+2)(n+1)^{(4)}/2 \cdot 4$$

and using z_n in place of $P_n^{(3)}(1)$,

$$z_n - z_{n-1} = (n+2)^{(5)}/2 \cdot 4$$

and finitely integrating the factorial polynomial,

$$z_n = P_n^{(3)}(1) = (n+3)^{(6)}/2 \cdot 4 \cdot 6$$

The general result is

$$P_n^{(k)}(1) = (n+k)^{(2k)}/2^k k!$$

and an inductive proof is indicated. Since Legendre polynomials are either even or odd functions, it also follows that

$$P_n^{(k)}(-1) = (-1)^{n+k} P_n^{(k)}(1)$$

8.131 Prove

$$P_{n+1}'(x) - P_{n-1}'(x) = (2n+1) P_n(x)$$

| From Problems 8.125 and 8.127 we have

$$P_{n+1}'(x) = x P_n'(x) + (n+1) P_n(x)$$
$$= P_{n-1}'(x) + n P_n(x) + (n+1) P_n(x) = P_{n-1}'(x) + (2n+1) P_n(x)$$

and by comparing the extremes we find the stated result.

8.132 The leading coefficient in $P_n(x)$ is, as we know,

$$A_n = (2n)!/2^n(n!)^2$$

but show that it can also be written as

$$1 \cdot 3 \cdot 5 \cdots (2n-1)/n!$$

| It is only necessary to separate the even and odd factors of the $(2n)!$. The odd ones penetrate through, while for the even ones we have $2 \cdot 4 \cdot 6 \cdots (2n) = 2^n n!$ cancelling a part of the given denominator.

8.133 Compute the exact Gauss–Legendre arguments and coefficients for the case $n = 3$.

| The argument are the roots of $P_3(x) = \frac{1}{2}(5x^3 - 3x) = 0$, which are 0 and $\pm \sqrt{\frac{3}{5}}$. For the coefficients the derivative $P_3'(x)$ will also be needed, and is $\frac{1}{2}(15x^2 - 3)$. By Problem 8.110,

$$A_k = 2/n P_n'(x_k) P_{n-1}(x_k)$$

so that corresponding to $x = 0$

$$A_0 = \frac{2}{3}\left(-\frac{3}{2}\right)\left(-\frac{1}{2}\right) = \frac{8}{9}$$

while for the other two arguments

$$A_k = \frac{2}{3}(3)\left(\frac{2}{5}\right) = \frac{5}{9}$$

8.134 Apply the three point formula of the preceding problem to the integral of the sine function.

$$\int_0^{\pi/2} \sin t \, dt$$

| The change of variable $t = \pi/4(x+1)$ converts the limits to the required form

$$\int_0^{\pi/2} \sin t \, dt = \frac{\pi}{4} \int_{-1}^1 \sin \frac{\pi}{4}(x+1) \, dx$$

The three point approximation is then

$$I = \frac{\pi}{4}\left[\frac{8}{9}\sin\frac{\pi}{4} + \frac{5}{9}\sin\frac{\pi}{4}\left(1 + \sqrt{\frac{3}{5}}\right) + \frac{5}{9}\sin\frac{\pi}{4}\left(1 - \sqrt{\frac{3}{5}}\right) \right] = 1.000008$$

to six places. In Problem 8.26, using seven points, we found the trapezoidal rule producing .99429, Simpson's rule 1.00003, and the Newton–Cotes $n = 6$ formula 1.000003.

8.135 Show that the Gauss–Legendre $n = 5$ arguments and coefficients can be obtained exactly, though in a somewhat messy form.

▮ The arguments are, of course, the roots of $P_5(x) = 0$, or

$$63x^5 - 70x^3 + 15x = 0$$

Removing the $x = 0$ root and letting $u = x^2$ leads to

$$63u^2 - 70u + 15 = 0$$

from which the familiar formula extracts

$$u = x^2 = (70 \pm 4\sqrt{70})/126$$

With the exact arguments in hand, however messy, our formula of Problem 8.110 would yield exact values for the corresponding coefficients. The easy one is

$$A_0 = 2/5(15/8)(3/8) = 128/225$$

As for the others, approximations are more convenient and prove to be

Arguments	Coefficient
± .53846931	.47862867
± .90617985	.23692689

8.136 Apply the Gauss–Legendre two point formula to $\int_{-1}^{1}[1/(1 + x^2)]\,dx$ and compare with the exact value $\pi/2$.

· ▮ In Problem 8.113 the arguments were found to be the square roots of $\frac{1}{3}$, with coefficients of 1. Accordingly, the two point approximation is $1(\frac{3}{4}) + 1(\frac{3}{4}) = 1.5$. The correct value to four places is 1.5708, so we have a relative error of less than 5%.

8.137 Apply other Gauss–Legendre formulas to the same integral.

▮ The programs are easily written and we find these results,

n	4	6	8	16
Integral	1.5686	1.57073	1.570794	1.570796

the last of which is correct to six places. An adaptive method based on the trapezoidal rule, and seeking six place accuracy, produced 1.570797 using well over 400 trapezoids, but Romberg's method from a trapezoidal start managed this result from a table of size 4×4.

8.138 Apply Gauss–Legendre methods to the integral

$$\int_0^{\pi/2} \frac{\cos x}{1 + x}\,dx$$

and compare with the work of the trapezoidal and Simpson rules.

▮ An adaptive application of the trapezoidal rule, seeking six place accuracy, managed the figure .673622 using well over 600 trapezoids. Simpson's rule took 32 points to achieve .6736212. A 5×5 Romberg table yielded

.6736211. The Gauss–Legendre values were

n	2	4	6	8
Integral	.6718	.673615	.6736211	.6736211

The power of the Gaussian method is conspicuous.

8.139 How well do our Gaussian formulas deal with $\int_0^1 x^x \, dx$?

❚ Here are the figures up through $n = 16$:

n	2	4	6	8	16
Integral	.7746	.7827	.78327	.78338	.78343

The integrand here seems to be more stubborn than in the earlier problems, but we appear to be closing in on five place accuracy. The adaptive trapezoidal rule, for comparison, produced .78346 from several hundreds of trapezoids. Simpson's rule managed the figure .78345 using 64 points and .78343 from 256 points. A Romberg table of size 7×7 duplicated this value, while a 9×9 table managed .7834306.

8.140 Apply various methods to $\int_0^{\pi/2} e^{\sin x} \, dx$.

❚ Gauss–Legendre, using 2, 4, 6, and 8 points in turn, produced 3.1095, 3.104398, 3.104380, and 3.104379. Simpson agrees with this final value using the same number of points.

8.141 Evaluate the integral $\int_0^{\pi/2} \sqrt{1 - \frac{1}{4} \sin^2 t} \, dt$.

❚ Gauss–Legendre yields 1.477404 from six points and 1.467462 from both eight points and sixteen. Simpson's rule also needs only eight points to produce this figure, as does the trapezoidal rule. This appears to be an easy integral.

8.142 Similarly evaluate $\int_0^{\pi/2} \sqrt{1 - \frac{1}{2} \sin^2 t} \, dt$.

❚ Simpson's rule, the trapezoidal rule and the Gauss–Legendre formula, all using eight points, agree on the value 1.350644.

8.143 Evaluate $\int_0^{\pi/2} dx/(\sin^2 x + \frac{1}{4} \cos^2 x)$.

❚ The correct value is π. Gauss–Legendre used 16 points and Simpson's rule used 32 to produce 3.141593, which is correct to its full six places.

8.144 The integral $\int_0^1 (1/\sqrt{-\log x}) \, dx$ has a singularity at the upper limit. (At the lower limit the integrand can be defined as zero.) How can it be approximated?

❚ One way is to shift the limits to $-1, 1$ by the variable change $t = \frac{1}{2}(x + 1)$ and use Gauss–Legendre methods, the points at which the function is to be evaluated being well within the interval so the singularity does not enter directly. Unfortunately, the results obtained from just a few points are not very satisfactory:

n	6	8	16
Integral	1.639	1.670	1.720

The correct value is 1.772 to three places. For another attack on this integral see Problem 8.248.

8.145 The integral

$$\int_0^{\pi/2} \frac{\cos x}{1 - x/1000} \left(1 - e^{-100x}\right) dx$$

is not singular but has near singular behavior at $x = 0$. Try our approximation methods on this integral.

▎ Gauss–Legendre methods managed the values

n	6	8	16	16 + 16
Integral	1.04	.997	.99066	.99057

the last by applying the 16 point formula to each half of the interval of integration. Simpson's rule duplicated this final value by using 256 points. A 9×9 Romberg table extended this to .9905719.

8.7 A MORE GENERAL APPROACH

8.146 What is the Rodrigues formula?

▎ In this and the next several problems we retrace some of the track followed in Problems 8.91 to 8.97, obtaining a slightly different view of certain details of Gaussian integration and orthogonal functions. Consider a sequence of polynomials $\phi_n(x)$ with ϕ_n of degree n and orthogonal on the interval (a, b) to all polynomials of lesser degree, in the sense

$$\int_a^b w(x)\phi_n(x)q(x)\,dx = 0$$

with $q(x)$ any polynomial of degree $n - 1$ or less. The function $w(x)$ is assumed to be nonnegative throughout the interval. The sequence of Legendre polynomials $P_n(x)$ is such a sequence with $w(x) = 1$.

Now let

$$w(x)\phi_n(x) = \frac{d^n}{dk^n} U_n(x)$$

and consider the integral

$$I = \int_a^b U_n^{(n)}(x)q(x)\,dx$$

in which the superscript denotes the nth derivative. Integrating by parts

$$I = U_n^{(n-1)}(x)q(x)\Big|_a^b - \int_a^b U_n^{(n-1)}(x)q'(x)\,dx$$

and the integrated piece will vanish if the boundary values of $U_n^{(n-1)}$ are both zero. Making this assumption and again integrating by parts we have

$$I = -U_n^{(n-2)}(x)q'(x)\Big|_a^b + \int_a^b U_n^{(n-2)}(x)q''(x)\,dx$$

where again the integrated piece will vanish if the boundary values of $U_n^{(n-2)}$ are both zero. Assuming such, and continuing in this way, we come eventually to

$$I = (-1)^{n+1}U_n(x)q^{(n-1)}(x)\Big|_a^b \pm \int_a^b U_n(x)q^{(n)}(x)\,dx$$

with the remaining integral vanishing since $q(x)$ is of degree at most $n - 1$. We deduce that the integral I will be zero if $U_n(x)$ is zero at the boundaries together with its first $n - 1$ derivatives. But this establishes the required orthogonality property of the polynomial $\phi_n(x)$.

In summary, the sequence of function $\phi_n(x)$ is available in the form

$$\phi_n(x) = \left(\frac{1}{w(x)}\right)\frac{d^n}{dx^n} U_n(x)$$

with $U_n(x)$ satisfying the differential equation

$$\frac{d^{n+1}}{dx^{n+1}}\left(\frac{1}{w(x)}\right)\frac{d^n}{dx^n} U_n(x) = 0$$

[since $\phi_n(x)$ is of degree n] subject to the boundary conditions imposed. This is the Rodrigues formula.

8.147 Develop a representation of the important integral

$$c_n = \int_a^b w(x) \phi_n^2(x) \, dx$$

in terms of the function $U_n(x)$.

▎ First express $\phi_n(x)$ as

$$\phi_n(x) = a_n x^n + q(x)$$

with $q(x)$ of degree $n-1$ at most. Then

$$c_n = \int_a^b w(x) \phi_n^2(x) \, dx = \int_a^b w(x) \phi_n(x) [a_n x^n + q(x)] \, dx$$

$$= a_n \int_a^b x^n w(x) \phi_n(x) \, dx$$

since the other integral is zero by the orthogonality of $\phi_n(x)$ with q. Now introducing $U_n(x)$,

$$c_n = a_n \int_a^b x^n U_n^{(n)}(x) \, dx$$

and integrating by parts

$$c_n = a_n x^n U_n^{(n-1)}(x) \Big|_a^b - a_n \int_a^b n x^{n-1} U_n^{(n-1)}(x) \, dx$$

and the integrated piece vanishes because of the boundary values assumed by $U_n^{(n-1)}$. After a total of n integrations by parts, with the integrated piece vanishing each time, we come eventually to

$$c_n = (-1)^n n! a_n \int_a^b U_n(x) \, dx$$

As a byproduct we also have

$$\int_a^b x^n w(x) \phi_n(x) \, dx = c_n / a_n$$

8.148 Show that if $w(x)$ does not change sign in (a, b), then the zeros of $\phi_n(x)$ are real, distinct and fall inside (a, b).

▎ Let $x_1, x_2, \ldots, x_k$ be the points within (a, b) where $\phi_n(x)$ has changes of sign. Then

$$(x - x_1) \cdots (x - x_k) \phi_n(x)$$

will not change sign in this interval. Now if k were less than n, then the integral

$$\int_a^b w(x)(x - x_1) \cdots (x - x_k) \phi_n(x) \, dx$$

would be zero by the orthogonality of $\phi_n(x)$ to all polynomials of degree no greater than $n-1$. But the integrand does not change sign in (a, b) so the integral cannot possibly be zero. It follows that $k = n$, which proves the n zeros of $\phi_n(x)$ to be real, distinct and inside (a, b).

8.149 Develop a three term recursion for the polynomials $\phi_n(x)$.

▎ Let

$$\alpha_n = a_{n+1} / a_n$$

the ratio of leading coefficients (highest degree terms) in ϕ_{n+1} and ϕ_n. Then $\phi_{n+1}(x) - \alpha_n x \phi_n(x)$ will have a zero coefficient of x^{n+1} and we can write

$$\phi_{n+1}(x) - \alpha_n x \phi_n(x) = d_n \phi_n(x) + \cdots + d_0 \phi_0(x)$$

[Here it is assumed that none of the leading coefficients a_n is zero, permitting the polynomial on the left side to be written as a combination of the lower degree $\phi_j(x)$.] We now use the orthogonality property to show that $d_0 = d_1 = \cdots = d_{n-2} = 0$ so that the recursion is reduced to

$$\phi_{n+1}(x) = (\alpha_n x + d_n)\phi_n(x) + d_{n-1}\phi_{n-1}(x)$$

For example, multiply throughout by $w(x)\phi_0(x)$ and integrate. All terms will be zero by the orthogonality except for

$$d_0 \int_a^b w(x)\phi_0^2(x)\, dx$$

which forces $d_0 = 0$. The same argument applies with $\phi_1, \ldots, \phi_{n-2}$ replacing ϕ_0. With ϕ_{n-1}, however, the term

$$\alpha_n x \phi_{n-1}(x)\phi_n(x)$$

no longer yields to the orthogonality argument since $x\phi_{n-1}$ is no longer of degree $n-1$ or less. This stops the term by term annihilation process and leaves us with a three term recursion.

8.150 Using the notation

$$\phi_n(x) = a_n x^n + b_n x^{n-1} + q$$

to define the second leading coefficient in $\phi_n(x)$ as b_n, with q now of degree at most $n-2$, show that the coefficient d_n of the recursion is given by

$$d_n = \alpha_n \left(\frac{b_{n+1}}{a_{n+1}} - \frac{b_n}{a_n} \right)$$

❚ It is sufficient to compare the x^n terms in the recursion. On the left it is simply b_{n+1} by definition. On the right we find two such terms, with multipliers $\alpha_n b_n$ and $d_n a_n$. Thus

$$b_{n+1} = \alpha_n b_n + d_n a_n$$

which may be solved for d_n to obtain the stated result.

8.151 Also show that the remaining coefficient in the recursion is

$$d_{n-1} = -a_{n+1}a_{n-1}c_n / a_n^2 c_{n-1}$$

❚ Recall our earlier notations

$$c_n = \int_a^b w(x)\phi_n^2(x)\, dx \qquad c_n/a_n = \int_a^b x^n w(x)\phi_n(x)\, dx$$

and observe that

$$\int_a^b x w(x)\phi_n(x)\phi_{n+1}(x)\, dx = \int_a^b x w(x)(a_n x^n + q)\phi_{n+1}(x)\, dx$$

$$= a_n \int_a^b x w(x) x^n \phi_{n+1}(x)\, dx = a_n c_{n+1}/a_{n+1}$$

the term involving q vanishing since ϕ_{n+1} is orthogonal to xq. Now multiply the three term recursion by $w(x)\phi_{n+1}(x)$ and integrate to obtain

$$c_{n+1} = \alpha_n a_n c_{n+1}/a_{n+1}$$

and similarly multiply by $w(x)\phi_{n-1}(x)$ and integrate to obtain

$$0 = \alpha_n a_{n-1} c_n / a_n + d_{n-1} c_{n-1}$$

The latter two equations may be solved for d_{n-1}.

8.152 Derive the Christoffel identity from the three term recursion.

    ❚ Begin by multiplying through the recursion by $\phi_k(y)$:

$$\phi_k(y)\phi_{k+1}(x) = (\alpha_k x + d_k)\phi_k(y)\phi_k(x) + d_{k-1}\phi_k(y)\phi_{k-1}(x)$$

Rewrite with x and y reversed, possible since the result is true for any such pair of values

$$\phi_k(x)\phi_{k+1}(y) = (\alpha_k y + d_k)\phi_k(x)\phi_k(y) + d_{k-1}\phi_k(x)\phi_{k-1}(y)$$

Now divide both equations by $\alpha_k c_k$, subtract one from the other, and use the result of the preceding problem. The details are elementary and lead to the result

$$(x-y)\frac{\phi_x(x)\phi_k(y)}{c_k} = \frac{\phi_{k+1}(x)\phi_k(y) - \phi_k(x)\phi_{k+1}(y)}{\alpha_k c_k} - \frac{\phi_k(x)\phi_{k-1}(y) - \phi_{k-1}(x)\phi_k(y)}{\alpha_{k-1}c_{k-1}}$$

Notice that on the right side the terms are the same except for the shift down in the value of k. In other words, this side is in the form of a finite difference. Summing from $k=0$ to $k=n$ we will have a telescoping sum, only the end values surviving. If we artificially insist that $\phi_{-1}(x)$ be zero, then the final form of the summation will be

$$\sum_{k=0}^{n} \frac{\phi_k(x)\phi_k(y)}{c_k} = \frac{\phi_{n+1}(x)\phi_n(y) - \phi_n(x)\phi_{n+1}(y)}{\alpha_n c_n(x-y)}$$

This is the Christoffel identity.

8.153 Use the Christoffel identity to show that for $j = 1, \ldots, n$,

$$A_j = \frac{-a_{n+1}c_n}{a_n \phi_{n+1}(x_j)\phi_n'(x_j)}$$

provides an alternative formula for the coefficients A_j.

    ❚ Set $y = x_j$ in the Christoffel identity, where x_j is a zero of $\phi_n(x)$, reducing it to

$$\sum_{k=0}^{n-1} \frac{\phi_k(x)\phi_k(x_j)}{c_k} = -\frac{\phi_n(x)\phi_{n+1}(x_j)}{\alpha_n c_n(x - x_j)}$$

Multiply through by $w(x)\phi_0(x)$ and integrate:

(∗) $$\frac{\phi_0(x_j)c_0}{c_0} = \frac{-\phi_{n+1}(x_j)}{\alpha_n c_n}\int_a^b \frac{w(x)\phi_0(x)\phi_n(x)}{x - x_j}\,dx$$

Since the Lagrange multiplier function $L_j(x)$ is given by

$$L_j(x) = \pi_n(x)/(x - x_j)\pi_n'(x_j) = \phi_n(x)/(x - x_j)\phi_n'(x_j)$$

and since ϕ_0 is a constant, we can rewrite (∗) as

$$1 = \frac{-\phi_{n+1}(x_j)}{\alpha_n c_n} \int_a^b \frac{w(x)\phi_n(x)}{x - x_j}\, dx$$

$$= \frac{-\phi_{n+1}(x_j)\phi_n'(x_j)}{\alpha_n c_n} \int_a^b w(x) L_j(x)\, dx = \frac{-\phi_{n+1}(x_j)\phi_n'(x_j)}{\alpha_n c_n} A_j$$

which is equivalent to the stated result. [Note that in this problem the function denoted earlier as $\pi(x)$ has been called $\pi_n(x)$ instead.]

8.154 Redevelop the recursion for the Legendre polynomials by following the track laid out in the preceding problems.

❚ We have $w(x) = 1$, $(a, b) = (-1, 1)$, and $U_n(x) = (x - 1)^n/2^n n!$. The leading coefficient is still

$$a_n = (2n)!/2^n n!^2$$

making $\alpha_n = a_{n+1}/a_n = (2n + 1)/(n + 1)$. Since $U_n(x)$ is a function of x^2 it follows that alternate terms in the polynomials are zero, so that $b_n = 0$. We have also computed the integral $c_n = 2/(2n + 1)$ so that all the needed ingredients are in hand. Problem 8.150 at once shows that $d_n = 0$. The formula of Problem 8.151 then leads to a brief exercise with factorials which culminates with the anticipated $d_{n-1} = -n/(n + 1)$. Slipping α_n and d_{n-1} into the three term formula of Problem 8.149 we are back once again to the familiar

$$(n + 1)P_{n+1}(x) = (2n + 1)xP_n(x) - nP_{n-1}(x)$$

8.155 What formula for the coefficient A_j is obtained by the current approach and how does it compare with the one found in Problem 8.110?

❚ From Problem 8.153 we now have

$$\frac{-a_{n+1}c_n}{a_n P_{n+1}(x_j) P_n'(x_j)} = \frac{-(2n + 1)}{n + 1} \frac{2}{2n + 1} \frac{1}{P_{n+1}(x_j) P_n'(x_j)}$$

$$= \frac{-2}{(n + 1)P_{n+1}(x_j) P_n'(x_j)}$$

whereas our earlier result was

$$A_j = \frac{2}{nP_n'(x_j) P_{n-1}(x_j)}$$

To reconcile the two we have only to let $x = x_j$ in the recursion, making $P_n(x_j) = 0$ and leaving us with

$$(n + 1)P_{n+1}(x_j) = -nP_{n-1}(x_j)$$

8.8 GAUSS–LAGUERRE FORMULAS

8.156 The Laguerre polynomials can be developed by taking $w(x) = e^{-x}$, $(a, b) = (0, \infty)$ and $U_n(x) = e^{-x}x^n$ in Problem 8.146. Find the first few of these polynomials directly from the definition.

❚ The traditional notation for these orthogonal polynomials is $L_n(x)$, which will be used in place of the $\phi_n(x)$ of the earlier problem. Thus

$$L_n(x) = e^x \frac{d^n}{dx^n} e^{-x} x^n$$

so that

$$L_0(x) = 1$$
$$L_1(x) = -x + 1$$
$$L_2(x) = x^2 - 4x + 2$$

after which it may be easier to proceed by the recursion.

8.157 Evaluate

$$c_n = \int_a^b w(x)\phi_n^2(x)\,dx = \int_0^\infty e^{-x}L_n^2(x)\,dx$$

▌ By Problem 8.147 we have the alternative computation

$$c_n = (-1)^n n! a_n \int_a^b U_n(x)\,dx$$

$$= (-1)^n n!(-1)^n \int_0^\infty e^{-x}x^n\,dx$$

since the leading coefficient in $L_n(x)$ is easily found to be $a_n = (-1)^n$. But this final integral is well known, or may be evaluated by the technique of repeated integrations by parts, its value being $n!$. Thus $c_n = n!^2$.

8.158 Develop the recursion for the Laguerre polynomials.

▌ The leading coefficient (of the x^n term) in $L_n(x)$ has been found to be $a_n = (-1)^n$, so that $\alpha_n = a_{n+1}/a_n = -1$. It is not much harder to follow the coefficient of the x^{n-1} term:

$$U = e^{-x}x^n$$
$$U' = e^{-x}(-x^n + nx^{n-1})$$
$$U'' = e^{-x}(x^n - 2nx^{n-1} + \cdots)$$
$$U^{(3)} = e^{-x}(-x^n + 3nx^{n-1} + \cdots)$$
$$U^{(4)} = e^{-x}(x^n - 4nx^{n-1} + \cdots)$$

and discover that $b_n = (-1)^{n+1}n^2$. The necessary ingredients now being in hand, we find from Problem 8.151

$$d_{n-1} = -(-1)^{n+1}(-1)^{n+1}c_n/c_{n-1} = -n^2$$

while from Problem 8.150 we find

$$d_n = -\frac{(-1)^n(n+1)^2}{(-1)^{n+1}} - \frac{(-1)^{n+1}n^2}{(-1)^n}$$

$$= (n+1)^2 - n^2 = 2n+1$$

The recursion is then, by Problem 8.149,

$$L_{n+1}(x) = (-x + 2n + 1)L_n(x) - n^2 L_{n-1}(x)$$

8.159 Generate a few of the Laguerre polynomials using $L_0(x) = 1$ and $L_1(x) = -x + 1$ as starters for the recursion:

$$L_2(x) = (-x+3)(-x+1) - 1 = x^2 - 4x + 2$$
$$L_3(x) = (-x+5)(x^2 - 4x + 2) - 4(-x+1) = -x^3 + 9x^2 - 18x + 6$$
$$L_4(x) = (-x+7)(-x^3 + 9x^2 - 18x + 6) - 9(x^2 - 4x + 2)$$
$$= -x^4 - 16x^3 + 72x^2 - 96x + 24$$

and so on.

8.160 Find the zeros of the first few Laguerre polynomials.

▌ The zero of $L_1(x)$ is clearly at $x = 1$. The zeros of $L_2(x)$ are quickly found to be $2 \pm \sqrt{2}$. For $n > 2$ the zeros are obtained by approximation methods and are listed in Table 8.11 for even n up to 14.

TABLE 8.11

n	x_k	A_k	n	x_k	A_k
2	.58578644	.85355339	12	.11572212	.26473137
	3.41421356	.14644661		.61175748	.37775928
4	.32254769	.60315410		1.51261027	.24408201
	1.74576110	.35741869		2.83375134	.09044922
	4.53662030	.03888791		4.59922764	.02010238
	9.39507091	.00053929		6.84452545	.00266397
6	.22284660	.45896467		9.62131684	.00020323
	1.18893210	.41700083		13.00605499	.00000837
	2.99273633	.11337338		17.11685519	.00000017
	5.77514357	.01039920		22.15109038	.00000000
	9.83746742	.00026102		28.48796725	.00000000
	15.98287398	.00000090		37.09912104	.00000000
8	.17027963	.36918859	14	.09974751	.23181558
	.90370178	.41878678		.52685765	.35378469
	2.25108663	.17579499		1.30062912	.25873461
	4.26670017	.03334349		2.43080108	.11548289
	7.04590540	.00279454		3.93210282	.03319209
	10.75851601	.00009077		5.82553622	.00619287
	15.74067864	.00000085		8.14024014	.00073989
	22.86313174	.00000000		10.91649951	.00005491
10	.13779347	.30844112		14.21080501	.00000241
	.72945455	.40111993		18.10489222	.00000006
	1.80834290	.21806829		22.72338163	.00000000
	3.40143370	.06208746		28.27298172	.00000000
	5.55249614	.00950152		35.14944366	.00000000
	8.33015275	.00075301		44.36608171	.00000000
	11.84378584	.00002826			
	16.27925783	.00000042			
	21.99658581	.00000000			
	29.92069701	.00000000			

8.161 Evaluate the coefficients A_j for the Gauss–Laguerre case.

▌ Referring once again to Problem 8.153 and introducing the values now available for a_{n+1}/a_n and c_n,

$$A_j = \frac{-a_{n+1} c_n}{a_n \phi_{n+1}(x_j) \phi_n'(x_j)} = \frac{n!^2}{L_{n+1}(x_j) L_n'(x_j)}$$

For $n = 1$ and $x_1 = 1$ this becomes

$$A_1 = 1/L_2(1) L_1'(1) = 1$$

For $n = 2$ it is

$$A_j = 4/L_3(x_j) L_2'(x_j)$$

and choosing the zero $x_1 = 2 + \sqrt{2}$,

$$A_1 = 1/(2\sqrt{2} + 4)$$

with a similar result for $x_2 = 2 - \sqrt{2}$, namely, $A_2 = 1/(-2\sqrt{2} + 4)$.

8.162 Prove

$$L_{n+1}(x) = xL_n'(x) + (n+1-x)L_n(x)$$

▌ By definition

$$L_n(x) = e^x \frac{d^n}{dx^n} e^{-x} x^n$$

and differentiating

(∗)
$$L_n'(x) = L_n(x) + e^x \frac{d^{n+1}}{dx^{n+1}} e^{-k} x^n$$

To compute this remaining derivative consider

$$\frac{d^{n+1}}{dx^{n+1}} e^{-x} x^{n+1} = \frac{d^{n+1}}{dx^{n+1}} (e^{-x} x^n) x$$

and use the theorem on higher derivatives of a product to find

$$\frac{d^{n+1}}{dx^{n+1}} e^{-x} x^{n+1} = \left(\frac{d^{n+1}}{dx^{n+1}} e^{-x} x^n \right) x + (n+1) \frac{d^n}{dx^n} (e^{-x} x^n) 1$$

all other terms vanishing since derivatives of x beyond the first are zero. Translating this last equation

$$x \frac{d^{n+1}}{dx^{n+1}} (e^{-x} x^n) = e^{-x} L_{n+1}(x) - (n+1) e^{-x} L_n(x)$$

and substituting this into (∗)

$$xL_n'(x) = xL_n(x) + L_{n+1}(x) - (n+1)L_n(x)$$

which rearranges into the stated result.

8.163 Apply the result in the preceding problem to show that

$$A_j = n!^2 / x_j \left[L'(x_j) \right]^2$$

providing yet another form for these coefficients.

▌ Setting $x = x_j$, a zero of $L_n(x)$,

$$L_{n+1}(x_j) = x_j L_n'(x_j)$$

and making this replacement in the formula of Problem 8.161 we obtain the alternative formula.

8.164 What are the Gauss–Laguerre integration formulas?

▌ Summarizing the work in the preceding problems, these are of the form

$$\int_0^\infty e^{-x} y(x) \, dx \approx \sum_{j=1}^n A_j y(x_j)$$

the arguments x_j being the zeros of the nth Laguerre polynomial

$$L_n(x) = e^x \frac{d^n}{dx^n} (e^{-x} x^n)$$

and the coefficients A_j being

$$A_j = n!^2 / x_j \left[L_n'(x_j) \right]^2$$

The weight function $w(x) = e^{-x}$ does not appear explicitly in the approximating summation. Table 8.11 provides values of the x_j and the A_j.

8.165 What is the truncation error of the Gauss–Laguerre formula?

▮ Problem 8.92 provides the general result

$$E = \frac{y^{(2n)}(\theta)}{(2n)!} \int_a^b w(x)[\pi(x)]^2\, dx$$

But $\phi_n(x) = a_n\pi(x)$ or in the present case

$$L_n(x) = (-1)^n\pi(x)$$

so the integral is simply c_n or $n!^2$. Thus

$$E = n!^2 y^{(2n)}(\theta)/(2n)!$$

8.166 Apply the Gauss–Laguerre one point formula to the integration of e^{-x} itself.

▮ From Problem 8.161 we have the one point formula

$$\int_0^\infty e^{-x}y(x)\, dx \approx y(1)$$

since $x_1 = 1$ and $A_1 = 1$. For the case in question $y(x) = 1$ so the formula manages the value 1 for the integral. This happens to be the exact value, which comes as no surprise since with $n = 1$ we are guaranteed exact results for $y(x)$ any polynomial of degree 1 or less. In fact with $y(x) = ax + b$ the formula produces

$$\int_0^\infty e^{-x}(ax + b)\, dx = y(1) = a + b$$

which is the correct value.

8.167 By Problem 8.165 the Gauss–Laguerre two point formula should give exact values for $y(x)$ any polynomial of degree 3 or less. By Problem 8.161 this formula is

$$\int_0^\infty e^{-x}y(x)\, dx = \frac{y(2 + \sqrt{2})}{4 + \sqrt{2}\,2} + \frac{y(2 - \sqrt{2})}{4 - \sqrt{2}\,2}$$

Apply it to the case $y(x) = x^2$,

$$\int_0^\infty e^{-x}x^2\, dx$$

obtaining the exact value of 2!.

▮ A little arithmetic manages $y(2 + \sqrt{2}) = 6 + 4\sqrt{2}$ while $y(2 - \sqrt{2}) = 6 - 4\sqrt{2}$. The formula then offers

$$\frac{6 + 4\sqrt{2}}{4 + 2\sqrt{2}} + \frac{6 - 4\sqrt{2}}{4 - 2\sqrt{2}}$$

which reduces to 2.

8.168 Similarly apply the two point formula to the case $y(x) = x^3$,

$$\int_0^\infty e^{-x}x^3\, dx$$

obtaining the exact 3!.

▮ A little more arithmetic achieves $y(2 + \sqrt{2}) = 20 + 14\sqrt{2}$ while $y(2 - \sqrt{2}) = 20 - 14\sqrt{2}$. The two point formula then offers

$$\frac{20 + 14\sqrt{2}}{4 + 2\sqrt{2}} + \frac{20 - 14\sqrt{2}}{4 - 2\sqrt{2}}$$

which does work out to 6.

8.169 The correct value of the integral

$$\int_0^\infty e^{-x} x^7 \, dx$$

is 7! or 5040. Since our error formula contains as a factor the derivative $y^{(2n)}(\theta)$, the exact value of this integral should be available by choosing the four point Gauss–Laguerre formula, the eighth derivative of $y(x) = x^7$ being zero. How good a result is obtained by using $n = 2$?

$\blacksquare$ The computation

$$.85355339(.58578644)^7 + .14644661(3.41421356)^7$$

yields 792 to the nearest integer, not an especially good result. The three point formula (coefficients not included in Table 8.11)

$$.711093(.415775)^7 + .278518(2.294280)^7 + .010389(6.289945)^7$$

manages the somewhat better 4140, but the error is still around 20%. Apart from roundoff error, the four point formula does score a bull's-eye.

Certainly we ought not to expect too much accuracy from just two or three points, particularly when the integration is over an interval of infinite length. It seems almost miraculous that the exact value is available from four points. Note, however, that the factor $y^{(2n)}(\theta)$ in the error estimator would be $840 \cdot \theta^3$ for $n = 2$, making the estimate $140 \cdot \theta^3$, with θ anywhere between 0 and ∞. For $n = 3$ the corresponding derivative is $7! \cdot \theta$ and the error estimate $252 \cdot \theta$. The infinite range makes it impossible in this case to bound the error. In such situations it may be unwise to go ahead blindly

8.170 Apply the Gauss–Laguerre method to $\int_0^x e^{-x} \sin x \, dx$.

$\blacksquare$ The exact value of this integral is easily found to be $\frac{1}{2}$. The smoothness of $\sin x$, by which is meant the boundedness of its derivatives, suggests that our formulas will perform well. The error estimate of $(n!)^2/(2n)!$, which replaces $y^{(2n)}$ by its maximum of 1, reduces to $1/924$ for $n = 6$ and suggests about three place accuracy. Actually substituting into $\sum_{i=1}^n A_i \sin x_i$ brings the results

n	2	6	10	14
Σ	.43	.50005	.5000002	.50000000

so that our error formula is somewhat pessimistic.

8.171 Apply the Gauss–Laguerre method to $\int_1^x (e^{-1}/t) \, dt$.

$\blacksquare$ The unsmoothness of $y(t) = 1/t$, meaning that its nth derivative

$$y^{(n)}(t) = (-1)^n n! t^{-(n+1)}$$

increases rapidly with n, does not suggest overconfidence in approximation formulas. Making the change of argument $t = x + 1$, this integral is converted into our standard interval as

$$\int_0^x e^{-x} \frac{1}{e(x+1)} \, dx$$

and the error formula becomes

$$E = \left[(n!)^2/(2n)! \right] \left[(2n)!/e(\theta+1)^{2n+1} \right]$$

which reduces to $(n!)^2/e(\theta+1)^{2n+1}$. If we replaced θ by 0 to obtain the maximum derivative this would surely be discouraging, and yet no other choice nominates itself. Actual computations with the formula

$$\frac{1}{e} \sum_{i=1}^n A_i/(x_i + 1)$$

bring the results

n	2	6	10	14
Approximation	.21	.21918	.21937	.21938

Since the correct value to five places is .21938 we see that complete pessimism was unnecessary. The elusive argument θ appears to increase with n. A comparison of the actual and theoretical errors allows θ to be determined.

n	2	6	10
θ	1.75	3.91	5.95

In this example the function $y(x)$ has a singularity at $x = -1$. Even a complex singularity near the interval of integration can produce the slow convergence in evidence here. (Compare with Problem 8.119.) The convergence is more rapid if we move away from the singularity. For example, integration of the same function by the same method over the interval from 5 to ∞ brings the results

n	2	6	10
Approximation	.001147	.0011482949	.0011482954

The last value is almost correct to 10 places.

8.172 Apply various Gauss–Laguerre formulas to the smooth integral

$$\int_0^\infty e^{-x} \cos x \, dx$$

comparing the convergence with that for the case $y(x) = \sin x$.

∎ Here are the numbers:

n	6	10	14
Approximation	.4997	.5000004	.5000001

The convergence is very similar, the present results having been obtained on a smaller computer.

8.173 Apply several Gauss–Laguerre formulas to the "unsmooth" integral

$$\int_0^\infty e^{-x} \log(1 + x) \, dx$$

∎ The correct value to five decimal places is .59634 and the added difficulty due to the unsmoothness of the integrand can be seen in the numbers

n	6	10	14
Approximation	.59650	.596355	.596348

8.174 How well does the integral

$$\int_0^\infty e^{-(x+1/x)} \, dx$$

respond to the Gauss–Laguerre method?

▮ The convergence is even slower than in the preceding problem:

n	6	10	14
Approximation	.275	.282	.281

The correct value to four places is .2797.

8.175 Which of the following two integrals responds more quickly to the method of Gauss–Laguerre?

$$I_1 = \int_0^\infty e^{-x-x^2}\, dx \qquad I_2 = \int_0^\infty e^{-x}\sqrt{1+x}\, dx$$

▮ One might speculate or test some higher derivatives of the two functions $y(x)$, but the numbers leave little doubt.

n	6	8	10
I_1	.5382	.5465	.5455
I_2	1.37896	1.378937	1.378937

The second integral has converged to seven digits while the first is still struggling in the third decimal place.

8.176 Use the recursion of Problem 8.162 to generate some of the early Laguerre polynomials.

▮ With $L_0(x) = 1$ as the starter and

$$L_{n+1}(x) = xL_n'(x) + (n+1-x)L_n(x)$$

as recursion we find

$$L_1(x) = 1 - x$$
$$L_2(x) = x(-1) + (2-x)(1-x) = x^2 - 4x + 2$$
$$L_3(x) = x(2x-4) + (3-x)(x^2 - 4x + 2) = -x^3 + 9x^2 - 18x + 6$$
$$L_4(x) = x(-3x^2 + 18x - 18) + (4-x)(-x^3 + 9x^2 - 18x + 6)$$
$$= x^4 - 16x^3 + 72x^2 - 96x + 24$$

with the next step producing

$$L_5(x) = -x^5 + 25x^4 - 200x^3 + 600x^2 - 600x + 120$$

and an endless sequence beyond.

8.177 The classical differential equation of the Laguerre polynomials is

$$xL_n''(x) + (1-x)L_n'(x) + nL_n(x) = 0$$

Verify that the first few polynomials satisfy this equation.

▮ For $n = 0$ the constant $L_0(x) = 1$ clearly does the trick. For $n = 1$ the linear $L_1(x) = 1 - x$ reduces the equation to

$$0 + (x-1) + (1-x) = 0$$

while the $n = 2$ the substitution of $L_2(x) = x^2 - 4x + 2$ manages

$$2x + (1-x)(2x-4) + 2(x^2 - 4x + 2) = 0$$

8.9 GAUSS–HERMITE FORMULAS

8.178 The Hermite polynomials can be developed by taking $w(x) = e^{-x^2}$ and $(a, b) = (-\infty, \infty)$ with $U_n(x) = (-1)^n e^{-x^2}$ in Problem 8.146. Find the first few of these polynomials directly from the definition.

■ These polynomials will be denoted $H(x)$. Thus

$$H_n(x) = e^{x^2} \frac{d^n}{dx^n} (-1)^n e^{-x^2}$$

which for $n = 0$ is taken to mean $H_0(x) = 1$. Then

$$H_1(x) = 2x$$
$$H_2(x) = 4x^2 - 2$$
$$H_3(x) = 8x^3 - 12x$$

and so on.

8.179 Develop the three term recursion for the Hermite polynomials.

■ The leading coefficient (of the x^n term) in $H_n(x)$ is easily found to be

$$a_n = 2^n$$

so that $\alpha_n = a_{n+1}/a_n = 2$. Because $U_n(x)$ is a function of x^2 it follows that its various derivatives will have alternate terms of zero. This implies that the next leading coefficient will be

$$b_n = 0$$

As for the integral

$$c_n = \int_{-\infty}^{\infty} e^{-x^2} \phi_n^2(x)\, dx = \int_{-\infty}^{\infty} e^{-x^2} H_n^2(x)\, dx$$

we again have from Problem 8.147 the alternative computation

$$c_n = (-1)^n n! 2^n \int_{-\infty}^{\infty} (-1)^n e^{-x^2}\, dx = n! 2^n \sqrt{\pi}$$

These ingredients now determine the coefficients of the recursion as, from Problem 8.151,

$$d_{n-1} = -\frac{2^{n+1} 2^{n-1} n! 2^n \sqrt{\pi}}{2^{2n}(n-1)! 2^{n-1} \sqrt{\pi}} = -2n$$

while by Problem 8.150 we have $d_n = 0$ since the bs are zero. The recursion is therefore

$$H_{n+1}(x) = 2x H_n(x) - 2n H_{n-1}(x)$$

8.180 Generate a few of the Hermite polynomials from the recursion.

■ Taking $H_0(x) = 1$ and $H_1(x) = 2x$ as starters,

$$H_2(x) = 2x(2x) - 2(1) = 4x^2 - 2$$
$$H_3(x) = 2x(4x^2 - 2) - 4(2x) = 8x^3 - 12x$$
$$H_4(x) = 2x(8x^3 - 12x) - 6(4x^2 - 2) = 16x^4 - 48x^2 + 12$$

and so on.

8.181   Find the zeros of the first few Hermite polynomials.

▮ For $n = 1$ there is the single zero at $x = 0$. For $n = 2$ we quickly find $x = \pm\sqrt{2}/2$. At $n = 3$ there is $x = 0$ and also $x = \pm\sqrt{6}/2$. For $n = 4$, $x^2 = (3 \pm \sqrt{6})/2$ from which the zeros follow. Even $n = 5$ allows an exact determination, but this will be omitted.

8.182   Evaluate the coefficient A_j for Gauss–Hermite quadrature.

▮ From Problem 8.153 and using the available values of α_n and c_n we have

$$A_j = \frac{-a_{n+1}c_n}{a_n H_{n+1}(x_j) H_n'(x_j)} = -\frac{2^{n+1}n!2^n\sqrt{\pi}}{2^n H_{n+1}(x_j) H_n'(x_j)}$$

$$= -\frac{2^{n+1}n!\sqrt{\pi}}{H_{n+1}(x_j) H_n'(x_j)}$$

For example, with $n = 1$ the coefficient corresponding to the zero at $x_1 = 0$ will be

$$A_1 = -4\sqrt{\pi}/(-2)(2) = \sqrt{\pi}$$

For $n = 2$ the arguments are at $x^2 = \frac{1}{2}$ so

$$A_j = -16\sqrt{\pi}/H_3(x) H_2'(x) = -16\sqrt{\pi}/4x(2x^2 - 3)8x = \sqrt{\pi}/2$$

for each.

8.183   Prove that $H_n'(x) = -H_{n+1}(x) + 2xH_n(x)$.

▮ Much as in Problem 8.162 we have

$$H_n(x) = e^{x^2} \frac{d^n}{dx^n}(-1)^n e^{-x^2}$$

so that

$$H_n' = e^{x^2} \frac{d^{n+1}}{dx^{n+1}}\left[(-1)^n e^{-x^2}\right] + 2xH_n(x)$$

$$= -e^{x^2} \frac{d^{n+1}}{dx^{n+1}}\left[(-1)^{n+1} e^{-x^2}\right] + 2xH_n(x)$$

$$= -H_{n+1}(x) + 2xH_n(x)$$

8.184   Use the result of the preceding problem to modify the formula for A_j as given in Problem 8.182.

▮ For a zero x_j of $H_n(x)$ the preceding result simplifies to

$$H_n'(x_j) = -H_{n+1}(x_j)$$

and the coefficient A_j becomes

$$A_j = 2^{n+1}n!\sqrt{\pi}/H_n'(x_j)^2$$

8.185   Summarize the Gauss–Hermite quadrature formulas.

▮ Using the several preceding problems we have

$$\int_{-\infty}^{\infty} e^{-x^2}y(x)\, dx \approx \sum_{j=1}^{n} A_j y(x_j)$$

the arguments x_j being the zeros of the nth Hermite polynomial

$$H_n(x) = e^{x^2} \frac{d^n}{dx^n}(-1)^n e^{-x^2}$$

and the coefficients being

$$A_j = 2^{n+1} n! \sqrt{\pi} / H_n'(x_j)^2$$

The weight function $w(x)$ does not appear explicitly in the approximating summation. Table 8.12 provides values of the x_j and A_j.

TABLE 8.12

n	x_k	A_k	n	x_k	A_k
2	$\pm .70710678$	.88622693	12	$\pm .31424038$	.57013524
				$\pm .94778839$	.26049231
4	$\pm .52464762$	.80491409		± 1.59768264	.05160799
	± 1.65068012	.08131284		± 2.27950708	.00390539
6	$\pm .43607741$	.72462960		± 3.02063703	.00008574
	± 1.33584907	.15706732		± 3.88972490	.00000027
	± 2.35060497	.00453001	14	$\pm .29174551$	.53640591
8	$\pm .38118699$	.66114701		$\pm .87871379$	.27310561
	± 1.15719371	.20780233		± 1.47668273	.06850553
	± 1.98165676	.01707798		± 2.09518326	.00785005
	± 2.93063742	.00019960		± 2.74847072	.00035509
				± 3.46265693	.00000472
10	$\pm .34290133$	.61086263		± 4.30444857	.00000001
	± 1.03661083	.24013861			
	± 1.75668365	.03387439			
	± 2.53273167	.00134365			
	± 3.43615912	.00000764			

8.186 What is the truncation error of the Gauss–Hermite formulas?

▮ Problem 8.92 provides the general result

$$E = \frac{y^{(2n)}(\theta)}{(2n)!} \int_a^b w(x) [\pi(x)]^2 \, dx$$

With $\phi_n(x) = a_n \pi(x)$ or, in the present case,

$$H_n(x) = 2^n \pi(x)$$

$$E = \frac{y^{(2n)}(\theta)}{(2n)! 2^{2n}} \int_{-\infty}^{\infty} e^{-x^2} H_n^2(x) \, dx = \frac{y^{(2n)}(\theta)}{(2n)! 2^{2n}} c_n$$

$$= \frac{y^{(2n)}(\theta) \, n! 2^n \sqrt{\pi}}{(2n)! 2^{2n}} = \frac{n! \sqrt{\pi} \, y^{(2n)}(\theta)}{2^n (2n)!}$$

8.187 What is the one point Gauss–Hermite formula?

▮ The zero of $H_1(x)$ is at $x = 0$ and the corresponding coefficient $A_1 = \sqrt{\pi}$ was found in Problem 8.182. The one point formula is

$$\int_{-\infty}^{\infty} e^{-x^2} y(x) \, dx = \sqrt{\pi} \, y(0)$$

Applied to the integral

$$\int_{-\infty}^{\infty} e^{-x^2} \, dx$$

in which $y(x)$ is identically 1, the formula produces the correct value of $\sqrt{\pi}$. For $y(x) = x$ and the integral

$$\int_{-\infty}^{\infty} x e^{-x^2} \, dx = 0$$

it also produces the exact value. Since the error is a multiple of $y^{(2n)}(\theta)$ and $n = 1$, an exact result for any linear $y(x)$ is to be expected.

8.188   What is the Gauss–Hermite two point formula? Apply it to

$$\int_{-\infty}^{\infty} x^2 e^{-x^2} \, dx \quad \text{and} \quad \int_{-\infty}^{\infty} x^3 e^{-x^2} \, dx$$

for which exact results can be anticipated.

▌ Arguments and coefficients were given in Problem 8.182. The two point formula is

$$\int_{-\infty}^{\infty} e^{-x^2} y(x) \, dx \sim \frac{\sqrt{\pi}}{2}\left[y\left(\frac{\sqrt{2}}{2}\right) + y\left(-\frac{\sqrt{2}}{2}\right) \right]$$

With $y(x) = x^2$ this becomes $\int_{-\infty}^{\infty} e^{-x^2} x^2 \, dx = \sqrt{\pi}/2$, which is the exact value of the integral. With $y(x) = x^3$ it yields the obviously correct value of zero.

8.189   Apply the two point formula to the integral

$$\int_{-\infty}^{\infty} x^4 e^{-x^2} \, dx$$

for which the correct value is $3\sqrt{\pi}/4$.

▌ The result $\sqrt{\pi}/4$ is instant. The error formula of Problem 8.186 suggests an error of $\sqrt{\pi}/2$ and is absolutely correct.

8.190   Develop the three point formula and apply it to the integral of the preceding problem.

▌ The arguments have been found to be 0 and $\pm\sqrt{6}/2$. The coefficients are given by

$$A_j = 2^{n+1} n! \sqrt{\pi} / H_n'(x_j)^2$$

and with $H_3(x) = 8x^3 - 12x$, making $H_3'(x) = 24x^2 - 12$,

$$A_0 = 96\sqrt{\pi}/144 = 2\sqrt{\pi}/3$$

while for the other two arguments

$$A_j = 96\sqrt{\pi}/144(2x_j^2 - 1)^2 = \sqrt{\pi}/6$$

The three point formula is thus

$$\int_{-\infty}^{\infty} e^{-x^2} y(x) \, dx \approx \frac{2\sqrt{\pi}}{3} y(0) + \frac{\sqrt{\pi}}{6} y\left(\frac{\sqrt{6}}{3}\right) + y\left(\frac{-\sqrt{6}}{3}\right)$$

For $y(x) = x^4$ it produces the exact value of $3\sqrt{\pi}/4$, as it should. It also manages the correct value of 0 when $y(x) = x^5$.

8.191   Evaluate correct to six places $\int_{-\infty}^{\infty} e^{-x^2} \sin^2 x \, dx$.

▌ The Gauss–Hermite formula produces the results

n	2	4	6	8	10
Approximation	.748	.5655	.560255	.560202	.560202

This appears to suggest six place accuracy and the result is actually correct to six places, the exact integral being $\sqrt{\pi}(1 - e^{-1})/2$ which is .56020226 to eight places.

8.192 Evaluate correct to three places, $\int_{-\infty}^{\infty} [e^{-x^2}/\sqrt{1+x^2}]\, dx$.

▌ The square root factor is not so smooth as the sine function of the preceding problem, so we should not expect quite so rapid convergence, and do not get it:

n	2	4	6	8	10	12
Approximation	.145	.151	.15202	.15228	.15236	.15239

The value .152 seems to be indicated.

8.193 How closely do the four, eight and twelve point formulas come to the result

$$\int_{-\infty}^{\infty} e^{-x^2} \cos x\, dx = \sqrt{\pi}\, e^{-1/4}$$

which is 1.3804 to four places?

▌ The values

$$1.38033 \qquad 1.380388 \qquad 1.380388$$

were obtained using a seven place computer. Even the four point formula has come close to the figure given.

8.194 Which of the following two integrals responds better to the four, eight and twelve point Gauss–Hermite formulas?

$$\int_0^{\infty} e^{-x^2 - 1/x^2}\, dx = \frac{\sqrt{\pi}}{2e^2} \approx .11994 \qquad \int_{-\infty}^{\infty} \left[e^{-x^2}/(1+x^2) \right] dx \approx 1.343$$

▌ For the first the values

$$.0776 \qquad .1126 \qquad .1238$$

were found, and for the second

$$1.306 \qquad 1.3392 \qquad 1.3426$$

so while we may have the second to a pair of decimal places and three figures, it seems unlikely that the first has reached this point.

8.195 Of these two integrals, which seems to respond better to the four, eight and twelve point formulas?

$$\int_{-\infty}^{\infty} e^{-x^2}\sqrt{1+x}\, dx \qquad \int_{-\infty}^{\infty} e^{-x^2} \log(1+x^2)\, dx$$

▌ Here are the numbers. For the first integral

$$2.1318 \qquad 2.1278 \qquad 2.1276$$

and for the second

$$.6053 \qquad .5879 \qquad .5868$$

so perhaps the first has proved somewhat more cooperative, at least up to this point.

8.196 Show that the Hermite polynomials appear as coefficients in the power series,

$$F(x, t) = e^{-t^2 + 2tx} = e^{x^2} e^{-(t-x)^2} = \sum_{n=0}^{\infty} H_n(x) t^n/n!$$

known as the generator of these polynomials.

▌ The position occupied by $H_n(x)$ is that belonging to the nth derivative of F relative to t, computed at $t = 0$. But this derivative will be computed in the factor $e^{-(t-x)^2}$ where for $t = 0$ it will be equivalent to computing $(d^n/dx^n)e^{-x^2}$, apart from a factor of $(-1)^n$. This means that the required derivative is actually $e^{x^2}(-1)^n(d^n/dx^n)e^{-x^2}$, which is $H_n(x)$.

8.197 Prove that $H_n'(x) = 2nH_{n-1}(x)$.

   ∎ From the preceding problem it will be seen that

$$\partial F(x,t)/\partial x = 2tF(x,t)$$

or, in series form,

$$\sum H_n'(x)t^n/n! = \sum 2H_n(x)t^{n+1}/n!$$

and comparing the coefficients of the t^n terms, we have at once the required proof.

8.198 Rederive the recursion of Problem 8.179 by using the $F(x,t)$ generator function.

   ∎ Note that

$$\partial F(x,t)/\partial t = -2(t-x)F(x,t)$$

or, in series form,

$$\sum H_n(x)nt^{n-1}/n! + 2\sum H_n(x)t^{n+1}/n! - 2x\sum H_n(x)t^n/n! = 0$$

Extracting the t^n term on the left, we find its coefficient to be

$$[H_{n+1}(x) + 2nH_{n-1}(x) - 2xH_n(x)]/n!$$

which must be zero.

8.199 Prove the classical differential equation of the family of Hermite polynomials

$$H_n''(x) - 2xH_n'(x) + 2nH_n(x) = 0$$

   ∎ By the preceding problem

$$H_n''(x) = 2nH_{n-1}'(x) = 2n \cdot 2(n-1)H_{n-2}(x)$$

while

$$-2xH_n'(x) = -4nxH_{n-1}(x)$$

Combining these two, including $2nH_n(x)$, and cancelling a factor of $2n$ we arrive at the differential equation.

8.10 OTHER GAUSSIAN FORMULAS

8.200 In a later chapter, on least-squares approximation, the Chebyshev polynomials will be defined as

$$T_n(x) = \cos(n \cdot \arccos x)$$

or, with $A = \arccos x$, $T_n(x) = \cos nA$. It follows quickly that $T_0(x) = 1$, $T_1(x) = x$, $T_2(x) = 2x^2 - 1$ and so on. The orthogonal feature of these function is

$$\int_{-1}^{1} T_m(x)T_n(x)/\sqrt{1-x^2}\,dx = 0$$

with $1/\sqrt{1-x^2}$ playing the role of the weight function $w(x)$. The integral

$$c_n = \int_{-1}^{1} T_n^2(x)/\sqrt{1-x^2}\,dx$$

will be shown to have the value $\pi/2$ for $n \geq 1$. For $n = 0$ it has the value π. The three term recursion

$$T_{n+1}(x) - 2xT_n(x) - T_{n-1}(x) = 0$$

will also be proved. Using these results, find the values of the various parameters introduced in Problems 8.146 to 8.151.

❚ From the recursion we have

$$a_0 = 1 \qquad a_n = 2^{n-1} \quad \text{for } n \geq 1$$

because of the $2x$ factor multiplying $T_n(x)$. The nature of the recursion also shows that only alternate powers of x will appear in any one $T_n(x)$. This means that the second coefficient b_n will be zero. Moreover, $\alpha_n = a_{n+1}/a_n = 2$ for $n \geq 1$. From Problem 8.150 we then have $d_n = 0$, while from 8.151

$$d_{n-1} = -\alpha_n c_n / \alpha_{n-1} c_{n-1} = -1$$

Introducing these coefficients into the recursion as it appears in Problem 8.149 we then have

$$T_{n+1}(x) - 2xT_n(x) + T_{n-1}(x) = 0$$

agreeing with the results stated at the outset. The recursion holds for $n \geq 1$.

8.201 Show that the quadrature coefficients A_j for the n point formula are given by

$$A_j = \pi/n$$

❚ From Problem 8.153,

$$A_j = -\alpha_n c_n / T_{n+1}(x_j) T_n'(x_j) = -\pi / T_{n+1}(x_j) T_n'(x_j)$$

where x_j is a zero of $T_n(x)$. These zeros are easily seen to be

$$x_j = \cos(2j-1)\pi/2n$$

for $j = 1, \ldots, n$. Using the notation $A = \arccos x$, so that $\cos nA = 0$, it also follows that $\sin nA = \pm 1$. Thus

$$T_{n+1}(x_j) = \cos(n+1)A = \cos(nA)x_j - \sin(nA)\sin A$$
$$= \mp\sqrt{1-x_j^2}$$

Also

$$T_n'(x_j) = \sin(nA)n/\sqrt{1-x_j^2} = \pm n/\sqrt{1-x_j^2}$$

and putting these pieces together, $A_j = \pi/n$.

8.202 Summarize the Gauss–Chebyshev formulas for quadrature.

❚ These are of the form

$$\int_{-1}^{1} y(x)/\sqrt{1-x^2} \, dx = \sum_{j=1}^{n} A_j y(x_j)$$

the arguments x_j being the zeros of the nth Chebyshev polynomial

$$x_j = \cos(2j-1)\pi/2n$$

and the coefficients $A_j = \pi/n$. The weight function $1/\sqrt{1-x^2}$ does not appear explicitly in the approximating summation.

8.203 What is the error of the Gauss–Chebyshev formulas?

❚ From Problem 8.92 we still have the general result

$$E = \frac{y^{(2n)}(\theta)}{(2n)!} \int_a^b w(x)\pi^2(x) \, dx$$

which, with $\pi(x) = T_n(x)/a_n^2$, becomes

$$E = \frac{y^{(2n)}(\theta)}{(2n)!} 2^{2-2n} \int_{-1}^{1} T_n^2(x)/\sqrt{1-x^2} \, dx$$

The remaining integral is c_n for which the value $\pi/2$ is in hand. The error formula then takes the form

$$E = \frac{2\pi y^{(2n)}(\theta)}{(2n)! 2^{2n}}$$

8.204 What is the one point formula?

   ▌ The zero of $T_1(x)$ is at $x_1 = 0$, and the corresponding coefficient works out to $A_1 = \pi$. So the formula is

$$\int_{-1}^{1} y(x)/\sqrt{1-x^2} \, dx = \pi y(0)$$

The error formula of the preceding problem indicates that this will be exact for $y(x)$ any linear function. In fact it makes

$$\int_{-1}^{1} 1/\sqrt{1-x^2} \, dx = \pi \qquad \int_{-1}^{1} x/\sqrt{1-x^2} \, dx = 0$$

as it should.

8.205 Develop the $n = 2$ formula.

   ▌ The zeros of $T_2(x)$ are at $x^2 = \frac{1}{2}$ making the two point formula

$$\int_{-1}^{1} y(x)/\sqrt{1-x^2} \, dx = \pi/2 \left[y(\sqrt{2}/2) + y(-\sqrt{2}/2) \right]$$

This should be exact for $y(x)$ any polynomial of degree 3 or less. In particular, it finds the correct value

$$\int_{-1}^{1} x^2/\sqrt{1-x^2} \, dx = \pi/2 \left(\frac{1}{2} + \frac{1}{2} \right) = \pi/2$$

For $y(x) = x^4$, however, it fails as it should:

$$\int_{-1}^{1} x^4/\sqrt{1-x^2} \, dx \approx \pi/2 \left(\frac{1}{4} + \frac{1}{4} \right) = \pi/4$$

For this latter integral the error formula offers the perfect estimate $E = \pi/\theta$, which may be added to our $\pi/4$ to produce the correct value.

8.206 Develop the three point formula.

   ▌ The zeros of $T_3(x)$ are at $x = 0$ and $x^2 = \frac{3}{4}$. The formula is

$$\int_{-1}^{1} y(x)/\sqrt{1-x^2} \, dx = \pi/3 \left[y(0) + y(\sqrt{3}/2) + y(-\sqrt{3}/2) \right]$$

It is exact for $y(x)$ a polynomial of degree up to 6. Applying it to the case $y(x) = x^4$, it lives up to its reputation:

$$\pi/3[0 + 9/16 + 9/16] = 3\pi/8$$

(Elementary methods reduce this integral to that of $\sin^4 t$.)

8.207 Use Gauss–Chebyshev methods to approximate the integral

$$\int_{-1}^{1} e^x/\sqrt{1-x^2} \, dx$$

■ The derivatives of e^x on $(-1,1)$ are all bounded by 1, making our error estimate at most

$$E = 2\pi e / 2^{2n} (2n)!$$

For $n = 2$, 4 and 6 this amounts to about .044, .0000017 and 5×10^{-9}, respectively. From these three formulas we may anticipate one, five and eight correct decimal places in turn. The numbers are

n	2	4	6
Approximation	3.960	3.977463	3.977463

the last pair being correct as far as they go.

8.208 The integral

$$\int_{-1}^{1} \cos(x) / \sqrt{1 - x^2} \, dx = \pi J_0(1)$$

where $J_0(x)$ is a Bessel function, is 2.40394 to five places. How large a value of n would be needed to obtain this value?

■ Since the derivatives of $\cos(x)$ are also bounded by 1 we can expect accuracy similar to that obtained in the previous problem and we are not disappointed:

n	3	4	5	6
Approximation	2.40407	2.403939	2.403940	2.403940

8.209 Evaluate the integral

$$\int_{-1}^{1} \frac{dx}{\sqrt{1 - x^4}} = \int_{-1}^{1} \frac{dx}{\sqrt{1 - x^2} \sqrt{1 + x^2}}$$

■ Here are four numerical results:

4	6	8	12
2.620809	2.622027	2.622057	2.622058

8.210 A generating function for the Chebyshev polynomials is available in the form

$$F(x, t) = (1 - t^2) / 2(1 - 2xt + t^2) = \tfrac{1}{2} T_0(x) + \sum_{j=1}^{\infty} T_j(x) t^j + \cdots$$

Verify the first few terms.

■ The geometric series

$$1 / (1 - 2xt + t^2) = 1 + (2xt - t^2) + (2xt - t^2)^2 + (2xt - t^2)^3 + \cdots$$

is to be multiplied by $(1 - t^2)/2$. The result is

$$\tfrac{1}{2} + xt + (2x^2 - 1) t^2 + (4x^3 - 3x) t^3 + \cdots$$

from which the Chebyshev polynomials up to $T_3(x)$ can be plucked.

8.211 The classical differential equation of the Chebyshev polynomials is

$$(1 - x^2) T_n''(x) - x T_n'(x) + n^2 T_n(x) = 0$$

Verify that the first few do satisfy the equation.

■ For $n = 0$ all terms are zero. For $n = 1$ the left side becomes $-x + x$. For $n = 2$ it is $(1 - x^2)(4) - x(4x) + 4(2x^2 - 1)$ which does equal zero.

8.11 PRESCRIBED ARGUMENTS OR COEFFICIENTS

8.212 Derive an integration formula of the form

$$\int_{-1}^{1} y(x)\,dx \sim A_1 y(x_1) + A_2 y(x_2) + A_3 y(-1) + A_4 y(1)$$

that will be exact for as high-degree polynomials as possible.

▮ A procedure similar to that used for our other Gaussian formulas would also suffice here, but the method of undetermined coefficients will be used instead. The method is capable of generating any of the Gaussian formulas and serves as an alternate approach. The present example illustrates the general procedure. Here we have six unknown arguments and coefficients, so we ask exactness for $y(x) = 1, x, \ldots, x^5$:

$$2 = A_1 + A_2 + A_3 + A_4$$
$$0 = A_1 x_1 + A_2 x_2 - A_3 + A_4$$
$$\tfrac{2}{3} = A_1 x_1^2 + A_2 x_2^2 + A_3 + A_4$$
$$0 = A_1 x_1^3 + A_2 x_2^3 - A_3 + A_4$$
$$\tfrac{2}{5} = A_1 x_1^4 + A_2 x_2^4 + A_3 + A_4$$
$$0 = A_1 x_1^5 + A_2 x_2^5 - A_3 + A_4$$

These equations are linear in the A_i but nonlinear in x_1 and x_2.

Introduce the familiar polynomial

$$\pi(x) = (x - x_1)(x - x_2)(x + 1)(x - 1) = x^4 + C_1 x^3 + C_2 x^2 + C_3 x + C_4$$

First notice that $\pi(1) = \pi(-1) = 0$ because of the prescribed arguments. This implies

$$1 + C_1 + C_2 + C_3 + C_4 = 0$$
$$1 - C_1 + C_2 - C_3 + C_4 = 0$$

Next multiply the first five equations above by $C_4, C_3, C_2, C_1, 1$ and add. Since $\pi(x_1) = \pi(x_2) = 0$, we obtain $\tfrac{2}{5} + \tfrac{2}{3}C_2 + 2C_4 = 0$. Using the same multipliers on the last five equations brings $\tfrac{2}{5}C_1 + \tfrac{2}{3}C_3 = 0$. We now have four linear equations for the C_i and solve them to get $C_1 = C_3 = 0$, $C_2 = -\tfrac{6}{5}$ and $C_4 = \tfrac{1}{5}$ making

$$\pi(x) = \tfrac{1}{5}(5x^4 - 6x^2 + 1) = \tfrac{1}{5}(x - 1)(x + 1)(5x^2 - 1)$$

The missing arguments are therefore $x_1 = \sqrt{1/5}$ and $x_2 = -\sqrt{1/5}$. With these in our hands, the original set of six equations is linear. Solving the first four for the coefficients A_i, we find $A_1 = A_2 = \tfrac{5}{6}$ and $A_3 = A_4 = \tfrac{1}{6}$. Finally we may write our formulas as

$$\int_{-1}^{1} y(x)\,dx \sim \tfrac{5}{6}\big[y(x_1) + y(x_2) \big] + \tfrac{1}{6}\big[y(-1) + y(1) \big]$$

8.213 What are the Gauss–Lobatto formulas?

▮ These have the form

$$\int_{-1}^{1} y(x)\,dx \sim \frac{2}{n(n-1)}\big[y(-1) + y(1) \big] + \sum_{i=1}^{n-2} A_i y(x_i)$$

where the x_i are the zeros are $P'_{n-1}(x) = 0$ and the coefficients are

$$A_i = 2/n(n-1)\big[P_{n-1}(x_i) \big]^2$$

The previous example produces the $n = 4$ formula. The truncation error can be shown to be

$$E = -\frac{n(n-1)^3 2^{2n-1}\big[(n-2)!\big]^4}{(2n-1)\big[(2n-2)!\big]^3} y^{(2n-2)}(\theta)$$

8.214   Derive an integration formula of the form

$$\int_{-1}^{1} y(x)\, dx = A[\, y(x_1) + y(x_2) + y(x_3)\,]$$

which will be exact for polynomials of degree up to 3.

▌   Again we illustrate the method of undetermined coefficients. Requiring exactness for $y(x) = 1,\, x,\, x^2,\, x^3$ leads to

$$2 = 3A \qquad 0 = A(x_1 + x_2 + x_3) \qquad \tfrac{2}{3} = A\big(x_1^2 + x_2^2 + x_3^2\big) \qquad 0 = A\big(x_1^3 + x_2^3 + x_3^3\big)$$

With

$$\pi(x) = (x - x_1)(x - x_2)(x - x_3) = x^3 + C_1 x^2 + C_2 x + C_3$$

we multiply the preceding equations by $C_3, C_2, C_1, 1$ and add to get $2C_3 + \tfrac{2}{3}C_1 = 0$ since $\pi(x_i) = 0$. Also, by comparing the two forms of $\pi(x)$,

$$C_1 = -(x_1 + x_2 + x_3) \qquad C_2 = x_1 x_2 + x_1 x_3 + x_2 x_3 \qquad C_3 = -x_1 x_2 x_3$$

It is now clear that $A = \tfrac{2}{3}$ and $C_1 = 0$. For the rest it is convenient to use the classical relationships between the various symmetric functions. For example,

$$2(x_1 x_2 + x_1 x_3 + x_2 x_3) = (x_1 + x_2 + x_3)^2 - \big(x_1^2 + x_2^2 + x_3^2\big)$$

$$3x_1 x_2 x_3 = (x_1 + x_2 + x_3)(x_1 x_2 + x_1 x_3 + x_2 x_3) - \big(x_1^2 + x_2^2 + x_3^2\big)(x_1 + x_2 + x_3) + \big(x_1^3 + x_2^3 + x_3^3\big)$$

$$= 0$$

become in this example

$$2C_2 = C_1^2 - 1 \qquad -3C_3 = -C_1 C_2 + C_1$$

so that $C_2 = -\tfrac{1}{2}$ and $C_3 = 0$. This makes $\pi(x) = x^3 - \tfrac{1}{2}x$ and the required arguments are $0, \pm \sqrt{1/2}$. The integration formula is then

$$\int_{-1}^{1} y(x)\, dx \sim 2/3\big[\, y\big(-\sqrt{1/2}\,\big) + y(0) + y\big(\sqrt{1/2}\,\big)\big]$$

8.215   What are the Chebyshev formulas?

▌   These are of the form

$$\int_{-1}^{1} y(x)\, dx \sim \frac{2}{n} \sum_{i=1}^{n} y(x_i)$$

and are to be exact for polynomials of degree n or less. Notice that all values of the function $y(x)$ receive the *same weight*, $2/n$. The arguments x_i are not so easily described. A few are listed in Table 8.13. For $n = 8$ and $n > 10$ no formula of this type exists, the analysis leading to fewer than n real zeros for $\pi(x)$.

TABLE 8.13

n	x_i	n	x_i	n	x_i
3	0	7	0	9	0
	$\pm .70710678$		$\pm .32391181$		$\pm ..16790618$
5	0		$\pm .52965678$		$\pm .52876178$
	$\pm .37454141$		$\pm .88386170$		$\pm .60101866$
	$\pm .83249749$				$\pm .91158931$

8.216 If all temperatures are given equal weight and five readings are to be used in a 24 h period stretching from midnight to midnight, what times should be chosen for the most accurate determination of the mean daily temperature?

$\blacksquare$ Choosing the formula of the preceding problem with $n = 5$, it is enough to shift the interval $(-12, 12)$ to $(-1, 1)$ which makes the five arguments $0, \pm .375, \pm .832$ equivalent to the hours $0, \pm 4.5, \pm 10$, more or less. Translating to more conventional timekeeping, the temperatures should be taken at roughly 0200, 0730, 1200, 1630 and 2200, which may be further translated if necessary.

If the measurements at these hours were, for example,

$$40 \quad 60 \quad 80 \quad 90 \quad 70 \quad \text{degrees}$$

then the mean temperature for the day would be the ordinary average taken at these five times, since using $x = t/12$,

$$\text{mean temp} = \frac{1}{24} \int_{-12}^{12} T(t) \, dt = \frac{1}{24} \int_{-1}^{1} T(x) 12 \, dx \approx \frac{12}{24} \frac{2}{n} \sum T_j = \frac{1}{n} \sum T_j$$

The average is, of course, $68°$.

8.217 Redo the preceding problem if only three readings of temperature can be afforded.

$\blacksquare$ The appropriate formulas uses arguments 0 and $\pm \sqrt{1/2}$. On a -12 to 12 time scale this means $\pm 12(.707)$ or ± 8.5 h, more or less. The optimal times would appear to be 0350, 1200 and 2050.

8.218 Derive an integration formula that uses *only the endpoints* of the interval of integration, values of the integrand and its derivatives being combined at these points.

$\blacksquare$ Such a formula may be obtained by using a "two point Taylor theorem" that provides an approximating polynomial of degree $2n - 1$, which agrees with $y(x)$ and its first $n - 1$ derivatives at ± 1. The following alternate derivation using Legendre polynomials is also of interest. It begins with a familiar succession of integrations by parts, the result of which is

$$\int_{-1}^{1} y(x) v^{(n)}(x) \, dx = \sum_{i=0}^{n-1} y^{(n-i-1)}(x) v^{(i)}(x) (-1)^{n-i-1} \Big|_{-1}^{1} + (-1)^n \int_{-1}^{1} v(x) y^{(n)}(x) \, dx$$

The Legendre polynomials $P_n(x)$ now play another useful role. Taking $v(x) = 2^n n! P_n(x)/(2n)!$ we find $v^{(n)}(x) = 1$, so that

$$\int_{-1}^{1} y(x) \, dx \sim \sum_{i=0}^{n-1} y^{(n-i-1)}(x) \frac{2^n n!}{(2n)!} P_n^{(i)}(x) (-1)^{n-i-1} \Big|_{-1}^{1}$$

with truncation error

$$E = \frac{(-1)^n 2^n n!}{(2n)!} \int_{-1}^{1} P_n(x) y^{(n)}(x) \, dx$$

In Problem 8.130 a proof that

$$P_n^{(i)}(1) = (-1)^{n+i} P_n^{(i)}(-1) = (n+i)!/(n-i)! 2^i i!$$

is given. Using these values

$$\int_{-1}^{1} y(x) \, dx \sim \frac{2^n n!}{(2n)!} \sum_{i=1}^{n-1} \frac{(n+1)!}{(n-i)! 2^i i!} Y_i$$

where $Y_i = y^{(n-i-1)}(-1) + (-1)^{n-i-1} y^{(n-i-1)}(1)$.

8.219 Simplify the truncation error formula of the preceding problem.

▮ After n integrations by parts, we have

$$\int_{-1}^{1} \frac{d^n}{dx^n} (x^2 - 1)^n y^{(n)}(x)\, dx = (-1)^n \int_{-1}^{1} (x^2 - 1)^n y^{(2n)}(x)\, dx$$

since all integrated terms vanish at both limits. Since $(x^2 - 1)^n$ does not change sign in the interval of integration, the mean value theorem may be applied. We find

$$E = [2^n n! / (2n)!]\, y^{(2n)}(\theta) \int_{-1}^{1} (x^2 - 1)^n\, dx$$

The remaining integral was evaluated in Problem 8.100, and we have

$$E = \frac{(-1)^n 2^{3n+1} (n!)^3}{(2n)!(2n+1)!}\, y^{(2n)}(\theta)$$

Our formula is exact for polynomials of degree $2n - 1$ or less, which is reasonable since $2n$ values of $y(x)$ and its derivatives are being used.

8.220 Show that for $n = 1$ the endpoint formula becomes

$$\int_{-1}^{1} y(x)\, dx \approx y(-1) + y(1)$$

and that this corresponds to integration of the linear function $p(x)$ for which $p(-1) = y(-1)$ and $p(1) = y(1)$.

▮ For $n = 1$ the endpoint formula reduces to a single term Y_0, which from its definition is $y(-1) + y(1)$. The line through the endpoints is, of course,

$$p(x) = y(-1) + \tfrac{1}{2}[y(1) - y(-1)](x + 1)$$

and integrating this does yield the stated formula.

8.221 Show that for $n = 2$ the endpoint formula becomes

$$\int_{-1}^{1} y(x)\, dx \approx y(-1) + y(1) + \tfrac{1}{3}y'(-1) - \tfrac{1}{3}y'(1)$$

and that this corresponds to integration of a polynomial $p(x)$ which matches the endpoint values of $y(x)$ and $y'(x)$.

▮ With $n = 2$ the formula reduces to two terms, namely,

$$\tfrac{1}{3} Y_0 + Y_1$$

and working out the values of these two expressions does lead to the stated result.

The polynomial $p(x)$ is the Hermite cubic corresponding to the endpoints. Borrowing from our earlier work

$$U_0(x) = \left[1 - 2\left(-\tfrac{1}{2}\right)(x+1)\right](x-1)^2/4$$

$$U_1(x) = \left[1 - 2\left(\tfrac{1}{2}\right)(x-1)\right](x+1)^2/4$$

$$V_0(x) = (x+1)(x-1)^2/4$$

$$V_1(x) = (x-1)(x+1)^2/4$$

$$p(x) = U_0 y_0 + U_1 y_1 + V_0 y_0' + V_1 y_1'$$

and elementary calculations show the integrals of the two Us and the two Vs to be 1, 1, $\tfrac{1}{3}$ and $-\tfrac{1}{3}$, respectively.

8.222 Apply the $n = 2$ endpoint formula to $\int_0^{\pi/2} \sin t\,dt$.

▮ The familiar transformation $t = \pi/4(x + 1)$ converts the integral to

$$\frac{\pi}{4} \int_1^1 \sin \frac{\pi}{4}(x + 1)\,dx$$

which is then approximated by

$$\frac{\pi}{4}\left[0 + 1 + \frac{1}{3}\frac{\pi}{4}(1 - 0)\right] = .99$$

approximately. The correct value is, of course, 1.

8.223 Apply the $n = 3$ formula to $\int_0^{\pi/2} \sin t\,dt$.

▮ Computing the coefficients, we have the formula

$$\int_{-1}^1 y(x)\,dx \sim y(-1) + y(1) + \tfrac{2}{5}[y'(-1) - y'(1)] + \tfrac{1}{15}[y^{(2)}(-1) + y^{(2)}(1)]$$

The usual change of argument $t = \pi(x + 1)/4$ presents us once again with

$$\int_{-1}^1 \frac{\pi}{4}\sin\frac{\pi}{4}(x + 1)\,dx$$

making $y(x) = (\pi/4)\sin(\pi/4)(x + 1)$. We easily find $Y_0 = -(\pi/4)^3$, $Y_1 = (\pi/4)^2$ and $Y_2 = \pi/4$ making

$$\int_0^{\pi/2} \sin t\,dt \sim \frac{1}{15}\left[-\left(\frac{\pi}{4}\right)^3 + 6\left(\frac{\pi}{4}\right)^2 + 15\left(\frac{\pi}{4}\right)\right]$$

which reduces to .99984. The error formula of Problem 8.219 produces $E = 32\sin\theta/525$, and if the maximum of $\sin\theta$ is used we obtain a very conservative estimate compared with the actual error of .00016. The value of θ in this case is actually very near zero.

8.224 Show that the $n = 3$ formula just used corresponds to integration of a polynomial $p(x)$ which matches the $y(x)$, $y'(x)$ and $y^{(2)}(x)$ values at 1 and -1.

▮ It is not hard to verify that the required polynomial is

$$\begin{aligned}
p(x) = \tfrac{1}{16}\big[&(x + 1)^3(3x^2 - 9x + 8)\,y_1 - (x - 1)^3(3x^2 + 9x + 8)\,y_{-1} \\
&- (x + 1)^3(x - 1)(3x - 5)\,y_1' - (x + 1)(x - 1)^3(3x + 5)\,y_{-1}' \\
&+ (x + 1)^3(x - 1)^2\,y_1^{(2)} + (x + 1)^2(x - 1)^3\,y_{-1}^{(2)}\big]
\end{aligned}$$

Integration of $p(x)$ brings the formula of the previous problem.

8.12 SINGULAR INTEGRALS

8.225 Why should one be cautious in applying the earlier methods of this chapter to the approximation of singular integrals?

▮ It is unwise to apply the earlier formulas of this chapter blindly in the case of singular integrals. They are all based on the assumption that the function being integrated can be conveniently approximated by a polynomial $p(x)$. When this is not true those formulas may produce poor, if not completely deceptive, results. It would be comforting to be sure that the application

$$\int_1^2 \frac{dx}{x^2 - 2} \sim \left(\frac{1}{6}\right)\left[-1 + 4(4) + \frac{1}{2}\right] = \frac{31}{12}$$

of Simpson's rule will never be made, but less obvious singular points have probably been temporarily missed. Not quite so serious are the efforts to apply polynomial-based formulas to functions having singularities in their derivatives. Since polynomials breed endless generations of smooth derivatives they are not ideally suited to such functions and poor results are usually obtained.

8.226 What procedures are available for singular integrals?

▌ There is a variety of methods, both for singular integrands and for infinite range of integration.

(1) *Ignoring the singularity* may even be successful. Under certain circumstances it is enough to use more and more arguments x_i until a satisfactory result is obtained.

(2) *Series expansions* of all or part of the integrand, followed by term by term integration, is a popular procedure provided convergence is adequately fast.

(3) *Subtracting the singularity* amounts to splitting the integral into a singular piece that responds to the classical methods of analysis and a nonsingular piece to which our approximate integration formulas may be applied without anxiety.

(4) *Change of argument* is one of the most powerful weapons of analysis. Here it may exchange a difficult singularity for a more cooperative one or it may remove the singularity completely.

(5) *Differentiation relative to a parameter* involves imbedding the given integral in a family of integrals and then exposing some basic property of the family by differentiation.

(6) *Gaussian methods* also deal with certain types of singularity.

(7) *Asymptotic series* are also relevant, but this procedure is treated in the following chapter.

8.227 Compare the results of applying Simpson's rule to the integration of $\sqrt{x}$ near 0 and away from 0.

▌ Take first the interval between 1 and 1.30 with $h = .05$, since we made this computation earlier. Simpson's rule gave a correct result to five places. Even the trapezoidal rule gave an error of only .00002. Applying Simpson's rule now to the interval between 0 and .30, which has the same length but includes a singular point of the derivative of $\sqrt{x}$, we obtain $\int_0^{0.3} \sqrt{x}\, dx \sim .10864$. Since the correct figure is .10954, our result is not quite correct to three places. The error is more than a hundred times greater.

8.228 What is the effect of ignoring the singularity in the derivative of $\sqrt{x}$ and applying Simpson's rule with successively smaller intervals h?

▌ Polya has proved that for functions of this type (continuous with singularities in derivatives), Simpson's rule and others of similar type should converge to the correct integral. Computations show these results:

$1/h$	8	32	128	512
$\int_0^1 \sqrt{x}\, dx$	.663	.6654	.66651	.66646

The convergence to $\frac{2}{3}$ is slow but does appear to be occurring.

8.229 Determine the effect of ignoring the singularity and applying Simpson's rule to the integral $\int_0^1 (1/\sqrt{x})\, dx = 2$.

▌ Here the integrand itself has a discontinuity, and an infinite one, but Davis and Rabinowitz have proved (*SIAM Journal*, 1965) that convergence should occur. They also found Simpson's rule producing these results, which show that ignoring the singularity is sometimes successful:

$1/h$	64	128	256	512	1024	2048
Approx. integral	1.84	1.89	1.92	1.94	1.96	1.97

The convergence is again slow but does appear to be occurring. At current computing speeds slow convergence may not be enough to rule out a computing algorithm. There is, however, the usual question of how much roundoff error will affect a lengthy computation. For this same integral the trapezoidal rule with $h = 1/4096$ managed 1.98, while application of the Gaussian 48 point formula to quarters of the interval (192 points in all) produced 1.99.

8.230 Determine the result of ignoring the singularity and applying the Simpson and Gauss rules to the integral $\int_0^1 (1/x)\sin(1/x)\, dx \sim .6347$.

▌ Here the integrand has an infinite discontinuity and is also highly oscillatory. The combination can be expected to produce difficulty in numerical computation. Davis and Rabinowitz (see preceding problem) found Simpson's rule failing:

$1/h$	64	128	256	512	1024	2048
Approx. integral	2.31	1.69	$-.60$	1.21	.72	.32

and the Gaussian 48 point formula doing no better. So the singularity cannot always be ignored.

8.231 Evaluate to three places the singular integral $\int_0^1 (e^x/\sqrt{x})\,dx$.

    ▌ Direct use of the Taylor series leads to

$$\int_0^1 \left(\frac{e^x}{\sqrt{x}}\right) dx = \int_0^1 \left(\frac{1}{\sqrt{x}} + x^{1/2} + \frac{1}{2}x^{3/2} + \frac{1}{6}x^{5/2} + \cdots\right) dx$$

$$= 2 + \frac{2}{3} + \frac{1}{5} + \frac{1}{21} + \frac{1}{108} + \frac{1}{660} + \frac{1}{4680} + \frac{1}{37,800} + \cdots = 2.925$$

After the first few terms the series converges rapidly and higher accuracy is easily achieved if needed. Note that the singularity $1/\sqrt{x}$ has been handled as the first term of the series. (See also the next problem.)

8.232 Apply the method of "subtracting the singularity" to the integral of Problem 8.231.

    ▌ Calling the integral I, we have

$$I = \int_0^1 \frac{1}{\sqrt{x}}\,dx + \int_0^1 \frac{e^x - 1}{\sqrt{x}}\,dx$$

The first integral is elementary and the second has no singularity. However, since $(e^x - 1)/\sqrt{x}$ behaves like $\sqrt{x}$ near zero, it does have a singularity in its first derivative. This is enough, as we saw in Problem 8.227, to make approximation integration inaccurate.

    The subtraction idea can be extended to push the singularity into a higher derivative. For example, our integral can also be written as

$$I = \int_0^1 \frac{1 + x}{\sqrt{x}}\,dx + \int_0^1 \frac{e^x - 1 - x}{\sqrt{x}}\,dx$$

Further terms of the series for the exponential function may be subtracted if needed. The first integral here is $\frac{8}{3}$, and the second could be handled by our formulas, though the series method still seems preferable in this case.

8.233 Evaluate the integral of Problem 8.231 by a change of argument.

    ▌ The change of argument, or substitution, may be the most powerful device in integration. Here we let $t = \sqrt{x}$ and find $I = 2\int_0^1 e^{t^2}\,dt$, which has no singularity of any kind, even in its derivatives. This integral may be evaluated by any of our formulas or by a series development.

8.234 Evaluate correct to six decimal places $\int_0^1 (\cos x)(\log x)\,dx$.

    ▌ Here a procedure like that of Problem 8.231 is adopted. Using the series for $\cos x$, the integral becomes

$$\int_0^1 \left(1 - \frac{x^2}{2!} + \frac{x^4}{4!} - \frac{x^6}{6!} + \cdots\right) \log x\,dx$$

Using

$$\int_0^1 x^i \log x\,dx = \frac{x^{i+1}}{i+1}\left(\log x - \frac{1}{i+1}\right)\bigg|_0^1 = -\frac{1}{(i+1)^2}$$

the integral is replaced by the series

$$-1 + \frac{1}{3^2 2!} - \frac{1}{5^2 4!} + \frac{1}{7^2 6!} - \frac{1}{9^2 8!} + \cdots$$

which reduces to $-.946083$.

8.235 Evaluate $\int_1^\infty (1/t^2)\sin(1/t^2)\,dt$ by a change of variable which converts the infinite interval of integration into a finite interval.

    ▌ Let $x = 1/t$. Then the integral becomes $\int_0^1 \sin(x^2)\,dx$, which can be computed by various approximate methods. Choosing a Taylor series expansion leads to

$$\int_0^1 \sin(x^2)\,dx = \frac{1}{3} - \frac{1}{42} + \frac{1}{1320} - \frac{1}{75600} + \cdots$$

which is $.310268$ to six places, only four terms contributing.

8.236 Show that the change of variable used in Problem 8.235 converts $\int_1^\infty (\sin t/t)\, dt$ into a badly singular integral, so that reducing the interval of integration to finite length may not always be a useful step.

▮ With $x = 1/t$ we obtain the integral $\int_0^1 (1/x)\sin(1/x)\, dx$ encountered in Problem 8.230, which oscillates badly near zero, making numerical integration nearly impossible. The integral of this problem may best be handled by asymptotic methods to be discussed in the next chapter.

8.237 Compute $\int_1^\infty (1/x^5)\sin \pi x\, dx$ by direct evaluation between the zeros of $\sin x$, thus developing part of an alternating series.

▮ Applying the Gaussian eight point formula to each of the successive intervals $(1, 2)$, $(2, 3)$, and so on, these results are found:

Interval	Integral	Interval	Integral
(1, 2)	−.117242	(2, 3)	.007321
(3, 4)	−.001285	(4, 5)	.000357
(5, 6)	−.000130	(6, 7)	.000056
(7, 8)	−.000027	(8, 9)	.000014
(9, 10)	−.000008		

The total is −.11094, which is correct to five places.

   This method of direct evaluation for an interval of finite length resembles in spirit the method of ignoring a singularity. The upper limit is actually replaced by a finite substitute, in this case 10, beyond which the contribution to the integral may be considered zero to the accuracy required.

8.238 Compute $\int_0^\infty e^{-x^2 - 1/x^2}\, dx$ by differentiation relative to a parameter.

▮ This problem illustrates still another approach of the problem of integration. We begin by imbedding the problem in a family of similar problems. For t positive, let

$$F(t) = \int_0^\infty e^{-x^2 - t^2/x^2}\, dx$$

Since the rapid convergence of this singular integral permits differentiation under the integral sign, we next find

$$F'(t) = -2t \int_0^\infty \frac{1}{x^2} e^{-x^2 - t^2/x^2}\, dx$$

Now introduce the change of argument $y = t/x$, which allows the attractive simplification

$$F'(t) = -2 \int_0^\infty e^{-y^2 - t^2/y^2}\, dy = -2F(t)$$

Thus $F(t) = Ce^{-2t}$ and the constant C may be evaluated from the known result

$$F(0) = \int_0^\infty e^{-x^2}\, dx = \frac{\sqrt{\pi}}{2}$$

The result is

$$\int_0^\infty e^{-x^2 - t^2/x^2}\, dx = \tfrac{1}{2}\sqrt{\pi}\, e^{-2t}$$

For the special case $t = 1$, this produces .119938 correct to six digits.

8.239 What is the result of applying Simpson's rule to the following integral?

$$\int_0^1 x \log(x)\, dx$$

▮ There is no actual singularity, only an apparent one, the value of the integrand at $x = 0$ being zero. Simpson's rule, using four subintervals to start and then doubling a few times yields the values

$$-.245 \qquad -.2488 \qquad -.2497 \qquad -.2499$$

The convergence toward the correct $-\frac{1}{4}$ is not especially rapid but definitely seems to be occurring.

8.240 Evaluate to three places, by using the series for $\sin x$,

$$I = \int_0^1 (\sin x)/x^{3/2}\, dx$$

■ This would surely seem to be the best way to proceed. The series puts the integral into the form

$$\int_0^1 \left(x^{-1/2} - \tfrac{1}{6} x^{3/2} + \tfrac{1}{12} x^{7/2} - \tfrac{1}{5040} x^{11/2} + \cdots \right) dx$$

$$= 2\sqrt{x} - \tfrac{1}{15} x^{5/2} + \tfrac{1}{540} x^{9/2} - \tfrac{1}{32760} x^{1/2} + \cdots \Big|_0^1$$

$$= 2 + .0018 - .0667 + .00003$$

$$= 1.935$$

with more places easily available.

8.241 Apply the method of subtracting the singularity to the integral of the preceding problem, obtaining an elementary integral and an integral that involves no singularity until the second derivative of its integrand.

■ Since

$$\frac{\sin x}{x^{3/2}} = \frac{x}{x^{3/2}} + \frac{\sin x - x}{x^{3/2}}$$

we have

$$I = \int_0^1 x^{-1/2}\, dx + \int_0^1 (\sin x - x)/x^{3/2}\, dx$$

with the first integral equal to 2. The second is nonsingular and so can be treated by various methods. Simpson's rule, for example, produced these values using from 4 to 32 subintervals:

$$-.06492 \qquad -.06486 \qquad -.064847 \qquad -.064845$$

Together the two pieces reproduce the 1.935 obtained by the series method.

8.242 Apply the methods of Simpson and Romberg to the integral of the preceding two problems.

■ Since both methods use the endpoint values of the integrand, and the integrand is infinite at $x = 0$, the interval of integration is reduced to $(L, 1)$ with L a small positive number. Here are some results using single precision arithmetic:

L	.001	.0001	.00001
Simpson	1.87	1.92	2.00
Romberg	1.87	1.94	2.07

Clearly things are not going especially well. The very large function values near the singularity have brought instability to the computation.

8.243 Evaluate $\int_0^1 e^{-x} \log x\, dx$ correct to four places by using the series for the exponential function.

■ The integral becomes

$$\int_0^1 \left(1 - x + \tfrac{1}{2} x^2 - \tfrac{1}{6} x^3 + \tfrac{1}{24} x^4 - \tfrac{1}{120} x^5 + \cdots \right) \log x\, dx$$

and since $\int_0^1 x^i \log(x)\, dx = -1/(i+1)^2$ we have the series

$$-\left(1 - \tfrac{1}{4} + \tfrac{1}{18} - \tfrac{1}{96} + \tfrac{1}{600} - \tfrac{1}{4320} + \cdots \right)$$

which works out to $-.7966$ to four places.

8.244 How well do the Simpson, Romberg and Gauss formulas handle the integral of the preceding problem?

▮ The Gauss–Legendre formula, after a change of variables to shift the lower limit, produced $-.788$ from eight points and $-.794$ from sixteen. Simpson and Romberg must avoid the lower limit, and replacing it by L led to the results

L	.0001	.00001	.000001	.0000001
Simpson	$-.796$	$-.797$	$-.798$	$-.799$
Romberg	$-.796$	$-.798$	$-.800$	$-.801$

Once again it seems that pressing the singularity too hard has led to an instability in the computation.

8.245 Use a series to evaluate $I = \int_0^1 \log(x)/(1-x)\,dx$.

▮ The geometric series for $1/(1-x)$ leads at once to

$$I = \int_0^1 \log(x)(1 + x + x^2 + x^3 + \cdots)\,dx$$
$$= -\sum_{i=0}^{\infty} 1/(1+i)^2$$

which is a familiar series for $-\pi^2/6$.
   In exactly the same way we find

$$\int_0^1 \log(x)/(1+x)\,dx = -\left(1 - \tfrac{1}{4} + \tfrac{1}{9} - \tfrac{1}{16} + \cdots\right)$$

which will be shown in a later chapter on trigonometric approximations to be $-\pi^2/12$. As one further example

$$\int_0^1 \log(x)/(1-x^2)\,dx = -\left(1 + \tfrac{1}{9} + \tfrac{1}{25} + \tfrac{1}{49} + \cdots\right)$$

which will be shown to be $-\pi^2/8$. The series can, of course, be approximated without this supporting theory. (See the following chapter.)

8.246 Verify that to four decimal places $\int_0^\infty e^{-x^2}/(1+x^2)\,dx = .6716$.

▮ The form is right for the Gauss–Hermite formulas, and using 4, 8, 12 and finally 14 points, these produced

$$.653 \qquad .6696 \qquad .6713 \qquad .6715$$

so we are certainly on the doorstep. Applying the Romberg and Simpson methods over various intervals $(0, L)$, we found

L	2	3	5	10
Romberg value	.6709	.67164	.6716465	.6716465

with the Simpson value corresponding to $L = 10$, .6716468. It seems safe to say that the four place figure has been verified.

8.247 Verify that to four places $\int_0^\infty e^{-x} \log x\,dx = -.5772$.

▮ Here the Gauss–Leguerre formulas appear to be indicated. In Problem 8.243, however, the same integrand was treated over the interval $(0, 1)$ using a series method, and the value $-.7966$ found. This disposes of the infinite singularity at $x = 0$ and leaves us with the problem of the infinite range of integration. So we do have alternatives.
   Direct application of the Gauss–Laguerre 10 and 14 point formulas managed the somewhat disappointing figures of $-.515$ and $-.532$ so the presence of the singularity appears to be noticed. Retrieving the fruit of our earlier labors, the $-.7966$, we turn the Romberg and Simpson formulas loose on the interval $(1, L)$ and get the output

L	10	12	15
Romberg	.21928	.21937	.2193832

with Simpson managing .2193831 for $L = 15$. Combining the two parts of the integral now achieves $-.5772$ as predicted.

8.248   Verify that to four places $\int_0^1 1/\sqrt{-\log x}\ dx = 1.7725$.

▮ The Gauss–Legendre 16 point formula comes up with the estimate 1.720, which is not impressive. The infinite discontinuity at the right end of the interval is not helpful. Turning to the Romberg and Simpson rules brings questionable improvement, values for upper limit U being

U	.999	.9999	.99999
Romberg	1.710	1.775	1.906
Simpson	1.709	1.761	1.835

Perhaps a change of variable will help. Let $u = -\log x$. The integral is exchanged for

$$\int_0^\infty e^{-u}/\sqrt{u}\ du$$

which is well known to have the value $\sqrt{\pi}$, approximately 1.7725 as forecast at the outset. But to continue with our efforts at approximation, the infinite discontinuity now being at the lower limit, suppose we dispose of it first. Over $(0, 1)$ replace the exponential function by its series to obtain, after a step or two,

$$\int_0^1 \frac{e^{-u}}{\sqrt{u}}\ du = 2\sqrt{u} - \frac{2}{3}u^{3/2} + \frac{1}{5}u^{5/2} - \frac{1}{21}u^{7/2} + \frac{1}{108}u^{9/2} - \frac{1}{660}u^{11/2}$$
$$+ \frac{1}{13 \cdot 360}u^{13/2} - \frac{1}{15 \cdot 2520}u^{15/2} + \cdots$$

which upon insertion of the limits gives up the value 1.4937. Simpson's rule with 1024 points then easily handles the remaining interval $(1, U)$, offering .2788 for $U = 10$ to 14. Putting the two pieces together we have $1.4937 + .2788 = 1.7725$, a bull's-eye, but only after a bit of scattershot.

8.249   The evidence of a number of the preceding problems is that using Simpson's rule, for example, over successive intervals such as $(L, 1)$ or $(0, U)$ to gradually approach an infinite discontinuity is somewhat treacherous. Try this approach on the trivial integrals $\int_0^1 (1/\sqrt{x})\ dx = 2$ and $\int_0^1 \log x\ dx = -1$.

▮ Here are the numbers. For the first integral

L	.001	.0001	.00001
Simpson	1.937	1.989	2.063
Romberg	1.937	2.003	2.134

while for the second

L	.0001	.00001
Simpson	$-.9992$	-1.0006
Romberg	$-.9995$	-1.0015

the correct value being overshot in both cases when the singularity is pressed too hard. The general lesson to be learned is that singular integrals are not to be treated casually.

8.250   Try a few of our methods on $\int_0^\infty e^{-x}\sqrt{x}\ dx$ which equals .8862 to four places.

▮ Gauss–Laguerre generates .8895 from 10 points and .8882 from 14. There being no singularity apart from the upper limit of integration, the ordinary quadrature formulas can be applied. Simpson's rule suggests .8861 for the interval $(0, 15)$ but (in single precision) smaller values for longer intervals, while Romberg's method yields a best value of .8859. Comparing with the elementary integral $\int_0^\infty e^{-x}x\ dx$, it will be found that it is unnecessary to go beyond the upper limit 15.

8.251   Apply the method of differentiation relative to a parameter to the integral

$$F(t) = \int_0^\infty e^{-x^2} \cos tx\ dx$$

and eventually obtain $F(\pi)$.

▌ Differentiating

$$F'(t) = -\int_0^\infty e^{-x^2} x \sin tx \, dx$$

which an integration by parts converts to

$$F'(t) = -(t/2)\int_0^\infty e^{-x^2} \cos tx \, dx = -(t/2)F(t)$$

since the integrated term is zero at both limits. It follows that $F(t) = Ce^{-t^2/4}$ with C a constant. But for $t = 0$ the value of the integral is known to be $\sqrt{\pi}/2$, so

$$F(t) = \left(\sqrt{\pi}/2\right)e^{-t^2/4}$$

into which $t = \pi$ can be substituted to obtain

$$\int_0^\infty e^{-x^2} \cos \pi x \, dx = .075156$$

to six places.

8.252 Evaluate $I = \int_0^\infty (\sin \pi x)/(1 + x^2) \, dx$.

▌ Suppose we adopt the idea of Problem 8.237 and integrate between the zero of the sine function to develop an alternating series. Some of the terms are

$$I = .50440 - .20096 + .08923 - .04851 + .03015 - .02046 + .01477 - .01115 + .00871$$
$$- .00699 + .00573 - .00478 + .00405 - .00348 + .00302 - .00264 + \cdots$$

The convergence of these terms toward zero is depressingly slow, but at least we know that the error made in stopping at a given term is no greater than the next. Summing them all except the last one we, therefore, discover the approximation .362 with error on more than .003 apart from possible roundoff errors.

A more interesting computation uses the Euler transformation of Problem 5.40, which will be further illustrated in the coming chapter. Its role will be to accelerate the convergence. It may be useful to sum a few of the early members of our sequence directly, say the first six, obtaining .35385. From the others we form a difference table. In fact, the segment in Table 8.14 will prove to be sufficient. The differences needed appear along the top diagonal. The Euler transformation now blends these together as

$$\tfrac{1}{2}a_0 - \tfrac{1}{4}\Delta a_0 + \tfrac{1}{8}\Delta^2 a_0 - \tfrac{1}{16}\Delta^3 a_0 + \tfrac{1}{32}\Delta^4 a_0 - \tfrac{1}{64}\Delta^5 a_0$$
$$= \tfrac{1}{2}(.01477) - \tfrac{1}{4}(-.00362) + \tfrac{1}{8}(.00118) - \tfrac{1}{16}(-.00046)$$
$$+ \tfrac{1}{32}(.00020) - \tfrac{1}{64}(-.00009)$$
$$= .00847$$

to five places, the last two terms having little impact. Putting our two pieces together we have .36232 which it may be wise to trim to .3623.

TABLE 8.14 (All entries are in the fifth decimal place)

1477					
	−362				
1115		118			
	−244		−46		
871		72		20	
	−172		−26		−9
699		46		11	
	−126		−15		
573		31			
	−95				
478					

8.253   Evaluate the similar integral $\int_0^\infty \sin(\pi x)/(1 + x^4)\,dx$.

❚   The fourth power of x means more rapid convergence of the integrand to zero for large x, so appeal to the Euler transformation may not be necessary. Some of the terms of the alternating series obtained by integrating between the zeros of the sine function (Romberg's method was used) are

$$.57041 - .11880 + .01708 - .00438 + .00159 - .00071 + .00036 - .00020 + .00012$$
$$- .00008 + .00005 - .00004 + .00003 - .00002 + \cdots$$

This supply seems adequate for a four place result, and together they come to .46541.

8.254   Evaluate $\int_1^\infty dx/(1 + x^3)$ by comparing it with the simpler and elementary integral $\int_1^\infty dx/x^3$.

❚
$$\int_1^\infty \frac{dx}{1 + x^3} = \int_1^\infty \frac{dx}{x^3} - \int_1^\infty \left[\frac{1}{(1 + x^3)} - \frac{1}{x^3} \right] dx = \frac{1}{2} - \int_1^\infty \frac{dx}{x^3(1 + x^3)}$$

The point is, we are now faced with an integrand that converges to zero like $1/x^6$, rather than like $1/x^3$, reducing the amount of computation to be done and the roundoff errors. For example, the original integrand is dominated by $1/x^3$ and

$$\int_U^\infty dx/x^3 = 1/2U^2$$

which equals .005 at $U = 10$ and .00125 at $U = 20$. This means that such an integration must proceed well beyond $x = 20$ to achieve even three place accuracy. Over $(1, 25)$ Simpson's rule manages .373. But

$$\int_U^\infty dx/x^6 = 1/5U^5$$

which at $U = 10$ is already down to .000002. Applying the Romberg method to $\int_1^U dx/x^3(1 + x^3)$ led to the values

U	5	10	12
Romberg	.12639	.1264473	.1264486

Doubt remains in the sixth decimal place, as the preceding estimate of .000002 suggested. The given integral appears to equal

$$.5 - .12645 = .37355$$

to five places.

If there is interest in reducing the lower limit to zero, then a direct computation over $(0, 1)$ can be made. Almost any of our methods will produce .83505 more or less, which adds to the preceding result to make 1.20860. The idea of computing a more difficult integral by comparing it with an easier one will be borrowed in the next chapter for the evaluation of series.

8.255   Treat the integral

$$\int_1^\infty dx/(1 + x^4)$$

by the comparison method of the preceding problem.

❚   The details are only slightly different, the basic idea being identical:

$$\int_1^\infty \frac{dx}{1 + x^4} = \int_1^\infty \frac{dx}{x^4} + \int_1^\infty \left[\frac{1}{1 + x^4} - \frac{1}{x^4} \right] dx = \frac{1}{3} - \int_1^\infty \frac{dx}{x^4(1 + x^4)}$$

The last integral is found to six places by integrating to the upper limit 8, its value being .089585, and subtracting from $\frac{1}{3}$ brings .24373 if we round. A direct attack on the original integral using Simpson's rule managed .2436 from $(1, 15)$ and .2437 from $(1, 20)$.

CHAPTER 9
Sums and Series

9.1 The representation of numbers and functions as finite or infinite sums is, of course, a large part of mathematics. What part does numerical analysis play?

■ The role of numerical analysis is in the efficient evaluation of the sums and series. Among the devices available for carrying out this role are:

(1) *The finite integral.* Some preliminary examples of this have already been given in Chapter 3 and others will be added.
(2) *Acceleration methods.* The Euler transformation and the Euler–Maclaurin formula are two prominent examples.
(3) *Asymptotic series.* These will be illustrated in the computation of integrals from x to ∞.
(4) *Special functions.* The Bernoulli and Euler polynomials are examples that have made contributions.

9.1 SOME FINITE INTEGRALS

9.2 Evaluate $\sum_{2}^{\infty} \log[(i-1)/i]$.

■ This is another telescoping sum. We easily find

$$\sum_{i=2}^{n} \log \frac{i-1}{i} = \sum_{i=2}^{n} \left[\log(i-1) - \log i \right] = -\log n$$

The telescoping method is of course the summation of differences. The sum $\sum y_i$ can be easily evaluated if y_i can be expressed as a difference, for then $\sum_{i=a}^{b} y_i = \sum_{i=a}^{b} \Delta Y_i = Y_{b+1} - Y_a$.

9.3 Evaluate the power sum $\sum_{i=1}^{n} i^4$.

■ Since powers can be expressed in terms of factorial polynomials, which in turn can be expressed as differences, any such power sum can be telescoped. In the present example

$$\sum_{i=1}^{n} i^4 = \sum_{i=1}^{n} \left[i^{(1)} + 7i^{(2)} + 6i^{(3)} + i^{(4)} \right] = \sum_{i=1}^{n} \Delta \left[\tfrac{1}{2}i^{(2)} + \tfrac{7}{3}i^{(3)} + \tfrac{6}{4}i^{(4)} + \tfrac{1}{5}i^{(5)} \right]$$

$$= \tfrac{1}{2}(n+1)^{(2)} + \tfrac{7}{3}(n+1)^{(3)} + \tfrac{6}{4}(n+1)^{(4)} + \tfrac{1}{5}(n+1)^{(5)} = \tfrac{1}{30}n(n+1)(2n+1)(3n^2+3n-1)$$

Other power sums are treated in similar fashion.

9.4 Evaluate $\sum_{i=1}^{n}(i^2 + 3i + 2)$.

■ Since power sums may be evaluated by summing differences, sums of polynomial values are easy bonuses. For example,

$$\sum_{i=1}^{n} i^2 + 3\sum_{i=1}^{n} i + \sum_{i=1}^{n} 2 = \frac{n(n+1)(2n+1)}{6} + \frac{3n(n+1)}{2} + 2n$$

9.5 Evaluate $\sum_{i=1}^{n} 1/[i(i+1)(i+2)]$.

■ This can also be written as a sum of differences. Recalling the factorial polynomials with negative exponent in Chapter 3, we find

$$\frac{1}{2i(i+1)} - \frac{1}{2(i+1)(i+2)} = \frac{1}{i(i+1)(i+2)}$$

and it follows that the given sum telescopes to

$$\frac{1}{4} - \frac{1}{2(n+1)(n+2)}$$

In this example the infinite series is convergent and

$$\sum_{i=1}^{\infty} \frac{1}{i(i+1)(i+2)} = \frac{1}{4}$$

9.6 Evaluate $\sum_{i=1}^{n} 3/[i(i+3)]$.

▌ Simple rational functions such as this (and in Problem 9.5) are easily summed. Here

$$\sum_{i=1}^{n} \frac{3}{i(i+3)} = \sum_{i=1}^{n} \left(\frac{1}{i} - \frac{1}{i+3} \right) = 1 + \frac{1}{2} + \frac{1}{3} - \frac{1}{n+1} - \frac{1}{n+2} - \frac{1}{n+3}$$

The infinite series converges to $\sum_{i=1}^{\infty} 3/[i(i+3)] = \frac{11}{6}$.

9.7 Evaluate the sums $\sum_{k}^{n} i^{(k)}$.

▌ The solution is instant by finite integration.

$$\sum_{k}^{n} i^{(k)} = i^{(k+1)}/(k+1) \Big|_{k}^{n+1} = (n+1)^{(k+1)}/(k+1)$$

We have, as examples, the sums

$$2 \cdot 1 + 3 \cdot 2 + 4 \cdot 3 + \cdots + n(n-1) = (n+1)n(n-1)/3$$
$$3 \cdot 2 \cdot 1 + 4 \cdot 3 \cdot 2 + \cdots + n(n-1)(n-2) = (n+1)n(n-1)(n-2)/4$$
$$4 \cdot 3 \cdot 2 \cdot 1 + 5 \cdot 4 \cdot 3 \cdot 2 + \cdots + n(n-1)(n-2)(n-3) = (n+1)n(n-1)(n-2)(n-3)/5$$

and so on.

9.8 Evaluate $\sum_{1}^{n} i^5$.

▌ Conversion to factorial polynomials is again in order, as in the case of a few lower powers of i. Using Stirling's numbers

$$i^5 = i^{(5)} + 10i^{(4)} + 25i^{(3)} + 15i^{(2)} + i^{(1)}$$

and finite integration, the sum becomes

$$\left[\tfrac{1}{6}i^{(6)} + 2i^{(5)} + \tfrac{25}{4}i^{(4)} + 5i^{(3)} + \tfrac{1}{2}i^{(2)} \right] \Big|_{1}^{n+1}$$

which after inserting the limits can be expressed in the form

$$(n+1)n\left[\tfrac{1}{2} + (n-1)\left[5 + (n-2)\left[\tfrac{25}{4} + (n-3)\left[2 + (n-4)\left[\tfrac{1}{6} \right] \right] \right] \right] \right]$$

At first glance this seems awkward, but its nested form is often preferred for the computation of a polynomial. A little algebra will convert it to

$$(n+1)^2 n^2 (2n^2 + 2n - 1)/12$$

which also has its devotees. For large n the sum's dominant term is $n^6/6$.

9.9 Generalizing the idea of factorial polynomial in the sense

$$x^{(n)} = x(x-h)(x-2h) \cdots (x-(n-1)h)$$

which makes our earlier version the case $h = 1$, show that

$$\Delta x^{(n)} = nx^{(n-1)}h \qquad \Delta^{-1}x^{(n)} = x^{(n+1)}/(n+1)h$$

▮ The proofs imitate those given earlier:

$$\Delta x^{(n)} = (x+h)^{(n)} - x^{(n)}$$

$$= (x+h)x \cdots (x-(n-2)h) - x(x-h) \cdots (x-(n-1)h)$$

$$= x(x-h) \cdots (x-(n-2)h)[(x+h)-(x-(n-1)h)]$$

$$= x^{(n-1)}(nh)$$

as anticipated. The second result follows from the first by the definition of the finite integral Δ^{-1} and a shift from n to $n+1$.

9.10 Continuing the generalization begun in the preceding problem, let

$$x^{(-n)} = 1/(x+h)(x+2h) \cdots (x+nh)$$

and show that

$$\Delta x^{(-n)} = -nhx^{(-(n+1))}$$

▮ Again the proof is direct:

$$\Delta x^{(-n)} = (x+h)^{(-n)} - x^{(-n)}$$

$$= \frac{1}{(x+2h) \cdots (x+(n+1)h)} - \frac{1}{(x+h) \cdots (x+nh)}$$

$$= \frac{1}{(x+2h) \cdots (x+nh)} \left[\frac{1}{x+(n+1)h} - \frac{1}{x+n} \right]$$

$$= x^{(-(n+1))}(-nh)$$

9.11 Find a closed formula for the sum

$$\frac{1}{1 \cdot 3 \cdot 5} + \frac{1}{3 \cdot 5 \cdot 7} + \frac{1}{5 \cdot 7 \cdot 9} + \cdots + \frac{1}{(2n-1)(2n+1)(2n+3)}$$

▮ By the definition in the preceding problem,

$$(x-h)^{(-3)} = 1/x(x+h)(x+2h)$$

and with $h=2$ the given sum is

$$\sum_{x=1}^{x=2n-1} (x-2)^{(-3)} = \Delta^{-1}(x-2)^{(-3)} \Big|_1^{2n+1} = \frac{(x-2)^{(-2)}}{(-2)(2)} \Big|_1^{2n+1}$$

$$= \frac{(2n-1)^{(-2)}}{-4} - \frac{(-1)^{(-2)}}{-4}$$

$$= \left[-\frac{1}{4} \cdot \frac{1}{(2n+1)(2n+3)} \right] - \left[-\frac{1}{4} \frac{1}{3} \right]$$

$$= \frac{1}{12} - \frac{1}{4(2n+1)(2n+3)}$$

9.12 Find the value of the infinite series

$$\frac{1}{1 \cdot 3 \cdot 5} + \frac{1}{3 \cdot 5 \cdot 7} + \frac{1}{5 \cdot 7 \cdot 9} + \cdots$$

▮ Letting n go to infinity in the preceding result we find at once the value $\frac{1}{12}$.

9.13 Evaluate the series

$$\frac{1}{1 \cdot 4} + \frac{1}{4 \cdot 7} + \frac{1}{7 \cdot 10} + \frac{1}{10 \cdot 13} + \cdots$$

❚ The series can be represented as $\Sigma(x-3)^{(-2)}$ with the understanding that in the factorial polynomial the value of h is 3. That is,

$$(x-h)^{(-2)} = (x-3)^{(-2)} = 1/x(x+h) = 1/x(x+3)$$

Suppose we first find the nth partial sum of the series, noting that its nth term is $1/(3n-2)(3n+1)$:

$$\sum_{x=1}^{x=3n-2} (x-3)^{(-2)} = \frac{(x-3)^{(-1)}}{-h} \Bigg|_{x=1}^{x=3n+1}$$

$$= \frac{(3n-2)^{(-1)}}{-3} - \frac{(-2)^{(-1)}}{-3}$$

$$= \frac{1}{3} - \frac{1}{3(3n+1)}$$

For $n=1$ and $n=2$ this does give the correct sums of $\frac{1}{4}$ and $\frac{2}{7}$. Putting it to its ultimate purpose we let n go to infinity and have the series values of $\frac{1}{3}$.

9.14 Find the sum $1^2 + 4^2 + 7^2 + \cdots + (3n-2)^2$ by finite integration.

❚ First note that for generalized factorial polynomials,

$$x^{(2)} = x(x-h) = x^2 - hx$$

$$x^{(2)} + hx^{(1)} = x^2$$

a result which reduces to our earlier one if $h=1$. With this in hand our sum becomes (recall that $h=3$)

$$\sum_{x=1}^{3n-2} x^2 = \sum (x^{(2)} + 3x^{(1)}) = \frac{x^{(3)}}{3 \cdot 3} + \frac{3x^{(2)}}{2 \cdot 3} \Bigg|_{1}^{3n+1}$$

$$= \frac{(3n+1)(3n-2)(3n-5)}{9} + \frac{(3n+1)(3n-2)}{2} - \frac{(-2)(-5)}{9} - \frac{-2}{2}$$

which eventually reduces to $n(6n^2 - 3n - 1)/2$.

In much the same way one can show that

$$2^2 + 5^2 + 8^2 + \cdots + (3n-1)^2 = n(6n^2 + 3n - 1)/2$$

The third sum in this trilogy is

$$3^2 + 6^2 + 9^2 + \cdots + (3n)^2 = 9(1^2 + 2^2 + 3^2 + \cdots + n^2)$$

and using Problem 3.114 it equals $9n(n+1)(2n+1)/6$.

9.15 Compute the sum $\sin A + \sin 2A + \cdots + \sin nA$ by using finite integration.

❚ As a preliminary we observe that

$$\Delta \cos x = \cos(x+h) - \cos x = -2\sin(x + \tfrac{1}{2}h)\sin\tfrac{1}{2}h$$

or shifting the argument x to $x - \tfrac{1}{2}h$,

$$\Delta \cos(x - \tfrac{1}{2}h) = -2\sin\tfrac{1}{2}h \sin x$$

Rewritten in finite integral form this is

$$\Delta^{-1}\sin x = \cos\left(x - \tfrac{1}{2}h\right) / \left(-2\sin\tfrac{1}{2}h\right)$$

Returning to the summation, and using the step size $h = A$,

$$\sum_{x=A}^{x=nA}\sin x = \frac{\cos\left(x - \tfrac{1}{2}A\right)}{-2\sin\tfrac{1}{2}A}\Bigg|_{A}^{(n+1)A}$$

$$= \frac{\cos\left(n + \tfrac{1}{2}\right)A}{-2\sin\tfrac{1}{2}A} + \frac{\cos\tfrac{1}{2}A}{2\sin\tfrac{1}{2}A} = \frac{\cos\left(n + \tfrac{1}{2}\right)A - \cos\tfrac{1}{2}A}{-2\sin\tfrac{1}{2}A}$$

In much the same way one finds

$$\frac{1}{2} + \cos A + \cos 2A + \cdots + \cos nA = \frac{\sin\left(n + \tfrac{1}{2}\right)A}{2\sin\tfrac{1}{2}A}$$

9.16   Find the sum

$$\frac{1}{1\cdot 4} + \frac{1}{2\cdot 5} + \frac{1}{3\cdot 6} + \cdots + \frac{1}{n(n+3)}$$

▌ The simple idea used in Problem 3.126 is available, particularly for the sum to infinity, which was shown in Problem 3.132 to be $1/n$ times the sum of terms of the harmonic series. For the finite sum the following procedure is useful. The sum can be written as

$$\sum_{x=1}^{x=n} 1/x(x+3)$$

and a finite integral can be found in terms of factorial polynomials. First note that

$$\frac{1}{x(x+3)} = \frac{(x+1)(x+2)}{x(x+1)(x+2)(x+3)} = \frac{(x+1)(x+2)}{(x+3)^{(4)}}$$

and then seek coefficients c_0 to c_2 such that

$$(x+1)(x+2) = c_0 + c_1(x+3)^{(1)} + c_2(x+3)^{(2)}$$

$$= c_0 + c_1(x+3) + c_2(x+3)(x+2)$$

It is easy enough to discover that to make this an identity we must choose $c_0 = 2$, $c_1 = -2$ and $c_2 = 1$. Thus

$$\frac{1}{x(x+3)} = \frac{2 - 2(x+3) + (x+3)(x+2)}{x(x+1)(x+2)(x+3)}$$

$$= \frac{2}{x(x+1)(x+2)(x+3)} - \frac{2}{x(x+1)(x+2)} + \frac{1}{x(x+1)}$$

$$= 2(x-1)^{(-4)} - 2(x-1)^{(-3)} + (x-1)^{(-2)}$$

and a finite integral is in hand:

$$\Delta^{-1}\left[\frac{1}{x(x+3)}\right] = \frac{2(x-1)^{(-3)}}{-3} - \frac{2(x-1)^{(-2)}}{-2} + \frac{(x-1)^{(-1)}}{-1}$$

Finally it is time to compute the sum.

$$\sum_{1}^{n} \frac{1}{x(x+3)} = \left\{ \frac{2(x-1)^{(-3)}}{-3} - \frac{2(x-1)^{(-2)}}{-2} + \frac{(x-1)^{(-1)}}{-1} \right\}\Bigg|_{1}^{n+1}$$

$$= -\frac{2}{3(n+1)(n+2)(n+3)} + \frac{1}{(n+1)(n+2)} - \frac{1}{n+1} + \frac{2}{3(1)(2)(3)} - \frac{1}{(1)(2)} + \frac{1}{1}$$

$$= \frac{11}{18} - \frac{1}{n+1} + \frac{1}{(n+1)(n+2)} - \frac{2}{3(n+1)(n+2)(n+3)}$$

If the sum to infinity is wanted it again works out to $\frac{11}{18}$, as did

$$\tfrac{1}{3}\left(\tfrac{1}{1} + \tfrac{1}{2} + \tfrac{1}{3}\right)$$

9.17 Rework the sum of Problem 9.14 using our earlier factorial polynomials instead of the generalized ones.

$\blacksquare$
$$1^2 + 4^2 + 7^2 + \cdots + (3n-2)^2 = \sum_{1}^{n} (3k-2)^2 = \sum_{1}^{n} (9k^2 - 12k + 4)$$

$$= \sum_{1}^{n} (9k^{(2)} - 3k^{(1)} + 4) = \left(3k^{(3)} - \tfrac{3}{2}k^{(2)} + 4k\right)\Bigg|_{1}^{n+1}$$

leading at last to the familiar $n(6n^2 - 3n - 1)/2$.

9.18 Find a closed form for the sum

$$1 \cdot 3 \cdot 5 + 3 \cdot 5 \cdot 7 + 5 \cdot 7 \cdot 9 + \cdots + (2n-1)(2n+1)(2n+3)$$

$\blacksquare$ A preliminary computation may be useful:

$$\Delta(2k+3)^{(4)} = (2(k+1)+3)^{(4)} - (2k+3)^{(4)}$$
$$= (2k+5)(2k+3)(2k+1)(2k-1) - (2k+3)(2k+1)(2k-1)(2k-3)$$
$$= (2k+3)^{(3)}[(2k+5) - (2k-3)] = 4(2k+3)^{(3)}(2)$$

The parallel with a simple case of the *chain rule* of calculus will be noticed. In terms of finite integrals, this result is

$$\Delta^{-1}(2k+3)^{(3)} = (2k+3)^{(4)}/(4)(2)$$

It is important to observe that while the step size for k is $h = 1$ the factors in the polynomials will still differ by 2, since k is multiplied by this factor. We are now ready for the sum:

$$1 \cdot 3 \cdot 5 + 3 \cdot 5 \cdot 7 + \cdots + (2n-1)(2n+1)(2n+3) = \sum_{1}^{n} (2k-1)(2k+1)(2k+3) = \sum (2k+3)^{(3)}$$

$$= \frac{(2k+3)^{(4)}}{8}\Bigg|_{1}^{n+1} = \frac{(2n+5)^{(4)}}{8} - (5)(3)(1)\frac{-1}{8}$$

$$= \frac{(2n+5)(2n+3)(2n+1)(2n-1) + 15}{8}$$

Here is a variation of the preceding procedure, differing in notation only. With step size 2 for the summation,

$$\sum_{x=5}^{x=2n+3} x^{(3)} = x^{(4)}/(4)(2)\Big|_{5}^{2n+5}$$

and now substituting the limits leads to exactly the same finish as before.

9.19   Find a closed form for

$$1 \cdot 2 \cdot 3 + 2 \cdot 3 \cdot 4 + 3 \cdot 4 \cdot 5 + \cdots + n(n+1)(n+2)$$

▮ Here the $h = 1$ factorial polynomials will do nicely:

$$\sum_1^n k(k+1)(k+2) = \sum (k+2)^{(3)} = (k+2)^{(4)}/4 \Big|_1^{n+1}$$

$$= (n+3)(n+2)(n+1)n/4$$

It now follows that

$$2 \cdot 4 \cdot 6 + 4 \cdot 6 \cdot 8 + 6 \cdot 8 \cdot 10 + \cdots + 2n(2n+2)(2n+4)$$

$$= 8[1 \cdot 2 \cdot 3 + 2 \cdot 3 \cdot 4 + \cdots + n(n+1)(n+2)]$$

$$= 2(n+3)(n+2)(n+1)n$$

and combining the results for odd and even factors,

$$1 \cdot 3 \cdot 5 + 2 \cdot 4 \cdot 6 + 3 \cdot 5 \cdot 7 + \cdots + 2n(2n+2)(2n+4)$$

can be summed.

9.20   Evaluate $1^2 + 3^2 + 5^2 + \cdots + (2n-1)^2$.

▮ Following the same track used in Problem 9.14, with step size $h = 2$ in the summation,

$$\sum_{x=1}^{x=2n-1} x^2 = \sum_{x=1}^{x=2n-1} [x^{(2)} + 2x^{(1)}] = \frac{x^{(3)}}{3 \cdot 2} + \frac{2x^{(2)}}{2 \cdot 2} \Bigg|_{x=1}^{x=2n+1}$$

$$= \frac{(2n+1)(2n-1)(2n-3)}{6} + \frac{(2n+1)(2n-1)}{2} - \frac{1(-1)(-3)}{6} - \frac{1(-1)}{2}$$

$$= \frac{2n(4n^2 - 1)}{6}$$

Of course, the same result is available by appealing to Problem 3.114 in which we found the well known

$$1^2 + 2^2 + \cdots + n^2 = n(n+1)(2n+1)/6$$

Extending this to

$$1^2 + 2^2 + \cdots + (2n)^2 = 2n(2n+1)(4n+1)/6$$

and extracting the even squares in the form

$$2^2 + 4^2 + \cdots + (2n)^2 = 4(1^2 + 2^2 + \cdots + n^2)$$

$$= 4n(n+1)(2n+1)/6$$

the sum of odd powers can be found by subtraction,

$$1^2 + 3^2 + \cdots + (2n-1)^2 = \tfrac{1}{6}2n(2n+1)(4n+1) - \tfrac{1}{6}4n(n+1)(2n+1)$$

$$= \tfrac{1}{6}2n(2n+1)[(4n+1) - 2(n+1)]$$

$$= \tfrac{1}{6}2n(2n+1)(2n-1)$$

as expected.

9.21 The method of Problems 3.126 and 3.132 can be used to show that

$$\frac{1}{1 \cdot 3} + \frac{1}{2 \cdot 4} + \frac{1}{3 \cdot 6} + \cdots + \frac{1}{n(n+2)} = \frac{1}{2}\left[\frac{1}{1} + \frac{1}{2} - \frac{1}{n+1} - \frac{1}{n+2}\right]$$

$$= \frac{3}{4} - \frac{1}{2}\frac{2n+3}{(n+1)(n+2)}$$

Derive this same result by the more general method of Problem 9.16.

$\blacksquare$ First note that

$$\frac{1}{x(x+2)} = \frac{x+1}{x(x+1)(x+2)} = \frac{-1+(x+2)}{x(x+1)(x+2)}$$

$$= \frac{-1}{x(x+1)(x+2)} + \frac{1}{x(x+1)}$$

$$= -(x-1)^{(-3)} + (x-1)^{(-2)}$$

so that

$$\Delta^{-1}\frac{1}{x(x+2)} = \frac{-(x-1)^{(-2)}}{-2} + \frac{(x-1)^{(-1)}}{-1}$$

With this in hand, the sum will be equal to

$$\frac{(x-1)^{(-2)}}{2} + \frac{(x-1)^{(-1)}}{-1}\bigg|_1^{n+1} = \frac{1}{2(n+1)(n+2)} - \frac{1}{n+1} - \frac{1}{4} + \frac{1}{1}$$

$$= \frac{3}{4} - \frac{2n+3}{2(n+1)(n+2)}$$

9.2 RAPIDLY CONVERGENT SERIES

9.22 How many terms of the Taylor series for $\sin x$ in powers of x are needed to provide eight place accuracy for all arguments between 0 and $\pi/2$?

$\blacksquare$ Since the series $\sin x = \sum_{i=0}^{\infty}(-1)^i x^{2i+1}/(2i+1)!$ is alternating with steadily decreasing terms, the truncation error made by using only n terms will not exceed the $(n+1)$th term. This important property of such series makes truncation error estimation relatively easy. Here we find $(\pi/2)^{15}/15! \simeq 8 \cdot 10^{-10}$ so that seven terms of the sine series are adequate for eight place accuracy over the entire interval.

This is an example of a rapidly convergent series. Since other arguments may be handled by the periodicity feature of this function, all arguments are covered. Notice, however, that a serious loss of significant digits can occur in argument reduction. For instance, with $x \simeq 31.4$ we find

$$\sin x \simeq \sin 31.4 = \sin(31.4 - 10\pi) \simeq \sin(31.4 - 31.416) = \sin(-.016) \simeq -.016$$

In the same way $\sin 31.3 \simeq -.116$ while $\sin 31.5 \simeq .084$. This means that although the input data 31.4 is known to three significant figures, the output is not certain even to one significant figure. Essentially it is the number of digits to the right of the decimal point in the argument x which determines the accuracy obtainable in $\sin x$.

9.23 How many terms of the Taylor series for e^x in powers of x are needed to provide eight place accuracy for all arguments between 0 and 1?

$\blacksquare$ The series is the familiar $e^x = \sum_{i=0}^{\infty} x^i/i!$. Since this is not an alternating series, the truncation error may not be less than the first omitted term. Here we resort to a simple comparison test. Suppose we truncate the series after the x^n terms. Then the error is

$$\sum_{i=n+1}^{\infty} \frac{x^i}{i!} = \frac{x^{n+1}}{(n+1)!}\left[1 + \frac{x}{n+2} + \frac{x^2}{(n+2)(n+3)} + \cdots\right]$$

and since $x < 1$, this error will not exceed

$$\frac{x^{n+1}}{(n+1)!}\left[1 + \frac{1}{n+2} + \frac{1}{(n+2)^2} + \cdots\right] = \frac{x^{n+1}}{(n+1)!}\frac{1}{1-1/(n+2)} = \frac{x^{n+1}}{(n+1)!}\frac{n+2}{n+1}$$

so that it barely exceeds the first omitted term. For $n = 11$ this error bound becomes about $2 \cdot 10^{-9}$ so that a polynomial of degree 11 is indicated. For example, at $x = 1$ the successive terms are

1.00000000	.50000000	.04166667	.00138889	.00002480	.00000028
1.00000000	.16666667	.00833333	.00019841	.00000276	.00000003

and their total is 2.71828184. This is wrong by one unit in the last place because of roundoff errors.
The error could also have been estimated using Lagrange's form (Problem 5.29), which gives

$$E = \frac{1}{(n+1)!}e^{\xi}x^{n+1} \quad \text{with } 0 < \xi < x$$

9.24 Compute e^{-10} to six significant digits.

❚ This problem illustrates an important difference. For six places we could proceed as in Problem 9.23, with $x = -10$. The series would however converge very slowly and there is trouble of another sort. In obtaining this small number as a difference of larger numbers we lose digits. Working to eight places we would obtain $e^{-10} \simeq .00004540$ which has only four significant digits. Such loss is frequent with alternating series. Occasionally double-precision arithmetic (working to twice as many places) overcomes the trouble. Here, however, we simply compute e^{10} and then take the reciprocal. The result is $e^{-10} \simeq .0000453999$ which is correct to the last digit.

9.25 In Problem 8.65 the integral $(2/\sqrt{\pi})\int_0^x e^{-t^2}\, dt$ was calculated by the Taylor series method for $x = 1$. Suppose the series is used for larger x, but to avoid roundoff error growth no more than 20 terms are to be summed. How large can x be made, consistent with four place accuracy?

❚ The nth term of the integrated series is $2x^{2n-1}/\sqrt{\pi}(2n-1)(n-1)!$ apart from the sign. Since this series alternates, with steadily decreasing terms, the truncation error will not exceed the first omitted term.
Using 20 terms we require that $(2/\sqrt{\pi})x^{41}/41 \cdot 20! < 5 \cdot 10^{-5}$. This leads to $x < 2.5$ approximately. For such arguments the series converges rapidly enough to meet our stipulations. For larger arguments it does not.

9.26 How many terms of the cosine series

$$\cos x = 1 - x^2/2! + x^4/4! - \cdots$$

would be needed to guarantee eight place accuracy for x between 0 and $\pi/2$?

❚ As with the sine series the terms are of alternating sign and steadily decreasing (after the first two). The error cannot be greater than the first omitted term. Since the nth term involves x to power $2n - 2$, we have a truncation error bound in

$$x^{2n}/(2n)!$$

which a little arithmetic, or a few logarithms, will show to be less than the required $5 \cdot 10^{-9}$ if x does not exceed $\pi/2$ and n is at least 8. Eight terms should be enough.

9.27 For how large an argument x will 20 terms of the series

$$\arctan x = x - \tfrac{1}{3}x^3 + \tfrac{1}{5}x^5 - \tfrac{1}{7}x^7 + \cdots$$

produce six place accuracy?

❚ This is another alternating series, so we will need

$$x^{41}/41 \le 5 \cdot 10^{-7}$$

since the 21st term will have power 41. The inequality holds for x up to .7.

9.28 The logarithm series

$$\log(1 + x) = x - \tfrac{1}{2}x^2 + \tfrac{1}{3}x^3 - \tfrac{1}{4}x^4 + \cdots$$

converges rather slowly. For how large an x will 20 terms produce four place accuracy?

▮ Assuming x to be positive we once more have an alternating series. The 21st term will be $x^{21}/21$ and

$$x^{21}/21 \le 5 \cdot 10^{-5}$$

will be assured if x does not exceed .7. The arctan series of the preceding problem is not a fast converger, but manages two more correct decimal places than the logarithm series for the same x and from the same number of terms.

9.29 The series for the hyperbolic sine is

$$\sinh x = x + \frac{x^3}{3!} + \frac{x^5}{5!} + \frac{x^7}{7!} + \cdots$$

and converges rapidly because of the factorials. Estimate the truncation error in terms of the first omitted term, using the method of Problem 9.23. (The series is not alternating.) For how large an argument x will 20 terms be enough for eight place accuracy?

The remainder after n terms will be

$$\frac{x^{2n+1}}{(2n+1)!} + \frac{x^{2n+3}}{(2n+3)!} + \cdots = \frac{x^{2n+1}}{(2n+1)!}\left[1 + \frac{x^2}{(2n+2)(2n+3)} + \frac{x^4}{(2n+4)(2n+5)} + \cdots\right]$$

$$\le \frac{x^{2n+1}}{(2n+1)!}\left[1 + \frac{x^2}{(2n+2)^2} + \frac{x^4}{(2n+2)^4} + \cdots\right]$$

$$= \frac{x^{2n+1}}{(2n+1)!}\left[1\Big/\left(1 - \frac{x^2}{(2n+2)^2}\right)\right]$$

For $n = 20$ and eight place accuracy what is needed is

$$x^{41}/41!\left(1 - x^2/42^2\right) \le 5 \cdot 10^{-9}$$

which proves to be true for x up to 10.

9.3 ACCELERATION METHODS

9.30 Not all series converge as rapidly as those of the previous problems. From the binomial series

$$\frac{1}{1 + x^2} = 1 - x^2 + x^4 - x^6 + \cdots$$

one finds by integrating between 0 and x that

$$\arctan x = x - \tfrac{1}{3}x^3 + \tfrac{1}{5}x^5 - \tfrac{1}{7}x^7 + \cdots$$

At $x = 1$ this gives the Leibnitz series

$$\frac{\pi}{4} = 1 - \frac{1}{3} + \frac{1}{5} - \frac{1}{7} + \cdots$$

How many terms of this series would be needed to yield four place accuracy?

▮ Since the series is alternating with steadily decreasing terms, the truncation error cannot exceed the first term omitted. If this term is to be .00005 or less, we must use terms out to about 1/20000. This comes to 10000 terms. In summing so large a number of terms we can expect roundoff errors to accumulate to 100 times the maximum individual roundoff. But the accumulation *could* grow to 10000 times that maximum if we were unbelievably unlucky. At any rate this series does not lead to a pleasant algorithm for computing $\pi/4$.

9.31   Apply the Euler transformation of Problem 5.40 to the series of the preceding problem to obtain four place accuracy.

▮   The best procedure is to sum the early terms and apply the transformation to the rest. For example, to five places,

$$1 - \tfrac{1}{3} + \tfrac{1}{5} - \cdots - \tfrac{1}{19} = .76046$$

The next few reciprocals and their differences are

$$
\begin{array}{ccccc}
.04762 & & & & \\
& -414 & & & \\
.04348 & & 66 & & \\
& -348 & & -14 & \\
.04000 & & 52 & & 3 \\
& -296 & & -11 & \\
.03704 & & 41 & & \\
& -255 & & & \\
.03448 & & & &
\end{array}
$$

The Euler transformation is

$$y_0 - y_1 + y_2 - y_3 + \cdots = \sum_{i=0}^{n} \frac{(-1)^i \Delta^i y_0}{2^{i+1}} = \tfrac{1}{2} y_0 - \tfrac{1}{4} \Delta y_0 + \tfrac{1}{8} \Delta^2 y_0 - \cdots$$

which applied to our table produces

$$.02381 + .00104 + .00008 + .00001 = .02494$$

Finally we have

$$\frac{\pi}{4} = 1 - \frac{1}{3} + \frac{1}{5} - \frac{1}{7} + \cdots = .76046 + .02494 = .78540$$

which is correct to five places. In all, 15 terms of the original series have seen action rather than 10000. The Euler transformation often produces superb acceleration like this, but it can also fail.

9.32   Compute $\pi/4$ from the formula

$$\frac{\pi}{4} = 2 \arctan \frac{1}{5} + \arctan \frac{1}{7} + 2 \arctan \frac{1}{8}$$

working to eight digits.

▮   This illustrates how special properties of the function involved may be used to bring accelerated convergence. The series

$$\arctan x = x - \tfrac{1}{3} x^3 + \tfrac{1}{5} x^5 - \tfrac{1}{7} x^7 + \cdots$$

converges quickly for the arguments now involved. We find, using no more than five terms of the series,

$$2 \arctan \tfrac{1}{5} = .39479112 \qquad \arctan \tfrac{1}{7} = .14189705 \qquad 2 \arctan \tfrac{1}{8} = .24870998$$

with a total of .78539815. The last digit should be a 6.

9.33   How many terms of $\sum_{i=1}^{\infty} 1/(i^2 + 1)$ would be needed to evaluate the series correct to three places?

▮   Terms beginning with $i = 45$ are all smaller than .0005, so that none of these individually affects the third decimal place. Since all terms are positive, however, it is clear that collectively the terms from $i = 45$ onward will affect the third place, perhaps even the second. Stegun and Abramowitz (*Journal of SIAM*, 1956) showed that 5745 terms are actually required for three place accuracy. This is a good example of a slowly convergent series of positive terms.

9.34 Evaluate the series of Problem 9.33 by the *comparison method*, correct to three places. (This method is analogous to the evaluation of singular integrals by subtracting out the singularity.)

▐ The comparison method involves introducing a known series of the same rate of convergence. For example,

$$\sum_{i=1}^{\infty} \frac{1}{i^2+1} = \sum_{i=1}^{\infty} \frac{1}{i^2} - \sum_{i=1}^{\infty} \frac{1}{i^2(i^2+1)}$$

We will prove later that the first series on the right is $\pi^2/6$. The second converges more rapidly than the others, and we find

$$\sum_{i=1}^{\infty} \frac{1}{i^2(i^2+1)} = \frac{1}{2} + \frac{1}{20} + \frac{1}{90} + \frac{1}{272} + \frac{1}{650} + \frac{1}{1332} + \frac{1}{2450} + \cdots \simeq .56798$$

with just 10 terms being used. Subtracting from $\pi^2/6 \simeq 1.64493$ makes a final result of 1.07695, which can be rounded to 1.077.

9.35 Verify that the result obtained in Problem 9.34 is correct to at least three places.

▐ The truncation error of our series computation is

$$E = \sum_{i=11}^{\infty} \frac{1}{i^2(i^2+1)} < \sum_{i=11}^{\infty} \frac{1}{i^4} = \sum_{i=1}^{\infty} \frac{1}{i^4} - \sum_{i=1}^{10} \frac{1}{i^4}$$

The first series on the right will later be proved to be $\pi^4/90$, and the second comes to at least 1.08200. This makes $E < 1.08234 - 1.08200 = .00034$. Roundoff errors cannot exceed $11 \cdot 5 \cdot 10^{-6}$ since 11 numbers of five place accuracy have been summed. The combined error therefore does not exceed .0004, making our result correct to three places.

9.36 Apply the comparison method to $\sum_{i=1}^{\infty} 1/[i^2(i^2+1)]$.

▐ This series was summed directly in the preceding problem. To illustrate how the comparison method may be reapplied, however, notice that

$$\sum_{i=1}^{\infty} \frac{1}{i^2(i^2+1)} = \sum_{i=1}^{\infty} \frac{1}{i^4} - \sum_{i=1}^{\infty} \frac{1}{i^4(i^2+1)}$$

Direct evaluation of the last series brings $\frac{1}{2} + \frac{1}{80} + \frac{1}{810} + \frac{1}{4352} + \frac{1}{16250} + \cdots$ which comes to .51403. Subtracting from $\pi^4/90$ we find

$$\sum_{i=1}^{\infty} \frac{1}{i^2(i^2+1)} \simeq 1.08234 - .51403 = .56831$$

which agrees nicely with the results of the previous two problems, in which this same sum was computed to be .56798 with an estimated error of .00034. The error estimate was almost perfect.

9.37 Evaluate $\sum_{i=1}^{\infty} 1/i^3$ to four places.

▐ The series converges a little too slowly for comfort. Applying the comparison method,

$$\sum_{i=1}^{\infty} \frac{1}{i^3} = 1 + \sum_{i=2}^{\infty} \frac{1}{(i-1)i(i+1)} - \sum_{i=2}^{\infty} \frac{1}{i^2(i^3-i)}$$

The first series on the right is telescoping and was found in Problem 9.5 to be exactly $\frac{1}{4}$. The last may be summed directly,

$$\frac{1}{24} + \frac{1}{216} + \frac{1}{960} + \frac{1}{3000} + \frac{1}{7560} + \frac{1}{16464} + \cdots$$

and comes to .04787. Subtracting from 1.25, we have finally $\sum_{i=1}^{\infty} 1/i^3 = 1.20213$ which is correct to four places.

9.38   Apply the comparison method to $\sum_1^\infty i^2/(i^4 + 1)$.

▌ The terms are similar to $1/i^2$, which we have already seen do not approach zero fast enough to make for a comfortable and safe computation. But

$$\sum \frac{i^2}{i^4 + 1} = \sum \frac{1}{i^2} + \sum \left(\frac{i^2}{i^4 + 1} - \frac{1}{i^2} \right) = \frac{\pi^2}{6} + \sum \frac{-1}{i^2(i^4 + 1)}$$

$$= \frac{\pi^2}{6} - \left(\frac{1}{2} + \frac{1}{68} + \frac{1}{728} + \frac{1}{4112} + \cdots \right)$$

and the acceleration is apparent. For the remaining series we find the partial sums

Number of terms	10	15	20
Sum	.5164046	.5164059	.5164060

with another 10 terms not changing this last result. Adding this to $\pi^2/6$, which is 1.644934 to six places, the figure 1.128528 is obtained.

9.39   Compute the sum $\sum_1^\infty 1/i^2$ by a comparison method.

▌ We know this sum to be $\pi^2/6$ or 1.644934 to six places, but it would take thousands of terms to achieve this sum directly, if roundoff errors did not ruin the computation along the way. The following effort is to be viewed as an exercise with comparisons rather than as the definitive way to handle the sum. Begin with

$$\sum \frac{1}{i^2} = \sum \frac{1}{i(i+1)} + \sum \left(\frac{1}{i^2} - \frac{1}{i(i+1)} \right) = 1 + \sum \frac{1}{i^2(i+1)}$$

to obtain a substitute series with terms like $1/i^3$ instead of the original $1/i^2$. The convergence has been accelerated but the rate is still uncomfortably slow. A second comparison leads to

$$1 + \sum \frac{1}{i^3} - \sum \frac{1}{i^3(i+1)}$$

with $\sum 1/i^3$ having been found in Problem 9.37 to be 1.2021. (The improved value 1.202057 will be found in Problem 9.88.) One more comparison gets us to

$$2.2020569 - \sum \frac{1}{i^4} + \sum \frac{1}{i^4(i+1)}$$

involving the sum $\sum 1/i^4 = \pi^2/90 = 1.082324$ to six places. (See Problem 9.66.) The remaining series responds to direct calculation in this way:

Number of terms	25	30	35
Sum	.5251997	.5252000	.5252001

Another 10 terms do not change this final value. Putting the various pieces together the sum of 1.644933 is obtained, differing from the true value by one unit in the last place. Twenty-five terms of a substitute series have managed almost six place accuracy, where thousands of terms of the original series would have been needed to achieve three places.

9.40   Compute $\sum 1/(i^3 + 1)$ to three places.

▌ The rate of convergence is not impressive but the accuracy demanded is not excessive. Direct summation brings the numbers

Number of terms	10	20	30	40	50
Sum	.682	.685	.6860	.6862	.6863

and the correct .6866 may be only a hundred or so terms away. But the comparison,

$$\sum \frac{1}{i^3 + 1} = \sum \frac{1}{i^3} - \sum \left(\frac{1}{i^3} - \frac{1}{i^3 + 1} \right) = 1.2021 - \sum \frac{1}{i^3(i^3 + 1)}$$

leads to a more comfortable computation. For the series that remains, six place accuracy seems to be easy:

Number of terms	10	15	20
Sum	.5155520	.5155533	.5155534

In fact five terms are enough to reach .5155, which subtracts from 1.2021 to produce .687, correct to the required three places.

9.41 Evaluate the series $\sum 1/T_2(i) = \sum 1/(2i^2 - 1)$ to four places.

▮ The polynomial $T_2(x)$ may be recognized as the Chebyshev polynomial of degree 2. The series has terms like $1/i^2$, which means relatively slow convergence,

$$1 + \tfrac{1}{7} + \tfrac{1}{17} + \tfrac{1}{31} + \cdots$$

Direct summation brings the numbers

Terms	20	40	60	80	100
Sum	1.321	1.333	1.337	1.339	1.3405

The comparison

$$\sum \frac{1}{2i^2 - 1} = \sum \frac{1}{2i^2} + \sum \frac{1}{(2i^2 - 1)2i^2}$$

offers a first term of $\pi^2/12$ and a second that sums much more comfortably, 10 terms offering a correct .5230. With $\pi^2/12$ equaling .8225 to four places we emerge with the sum 1.3454.

9.42 The three series

$$\sum \frac{1}{i^2} = \frac{\pi^2}{6} = 1.635 \qquad \sum \frac{1}{i^4} = \frac{\pi^4}{90} = 1.082322 \qquad \sum \frac{1}{i^6} = \frac{\pi^6}{945} = 1.017344$$

clearly converge at different rates. Using direct summation, how many correct decimal places are found from 20, 50 and 100 terms?

▮ The following table gives the answers. Six place accuracy was the maximum attempted.

Terms	20	50	100
$\sum 1/i^2$	1	2	3
$\sum 1/i^4$	4	6	6
$\sum 1/i^6$	6	6	6

The improvement in convergence rates is obvious.

9.43 Apply the Euler transformation to the series

$$1 - 1/\sqrt{2} + 1/\sqrt{3} - 1/\sqrt{4} + \cdots$$

▮ Summing the first 20 terms directly finds a total of .4944918. For the next five terms the difference table appears as Table 9.1.

From the top diagonal the Euler transformation produces

$$.1091089 + .0012543 + .0000414 + .0000022 + .0000002$$

or .1104069. Adding this to the sum of the first 20 terms brings the estimate .6648987 for the series.

TABLE 9.1

.2182179				
	− 50172			
.2132007		3309		
	− 46863		− 349	
.2085144		2960		49
	− 43903		− 299	
.2041241		2662		
	− 41241			
.2000000				

9.44   Apply the Euler transformation to

$$\log 2 = 1 - \tfrac{1}{2} + \tfrac{1}{3} - \tfrac{1}{4} + \cdots$$

How many terms are need to achieve six place accuracy?

❚   The six place value is .693147. Summing the first 20 terms directly we have .668771, but by the theorem on alternating series we know that it would take almost a million terms to reach the required accuracy by this brute force technique. The table of differences for the next four terms appears as Table 9.2.

The Euler transformation generates four terms from the top row of differences:

$$.023810 + .000541 + .000024 + .000001 = .024376$$

Combined with the .668771 this duplicates the correct value given previously. It has taken only 4 terms beyond the original 24 to achieve this.

TABLE 9.2

.047619			
	− 2165		
.045455		188	
	− 1976		− 24
.043478		165	
	− 1812		
.041667			

9.45   The series $1 - \tfrac{1}{9} + \tfrac{1}{25} - \tfrac{1}{49} + \cdots$ converges more rapidly than those of the two preceding problems, but acceleration may still be useful. Seven place accuracy would require several thousands of terms of direct summation. Suppose the first 10 terms are summed, with total .9147248. From the difference table of the next seven terms (not reproduced here) the Euler transformation squeezes

$$.00113379 + .00009430 + .00001086 + .00000155 + .00000026 + .00000005 = .0012408$$

Combined with the earlier .9147248 this produces .9159656, which is correct to all seven places. Seventeen terms in all have been used.

9.46   Treat the very slowly convergent series

$$1 - 1/\log 2 + 1/\log 3 - 1/\log 4 + \cdots$$

by the Euler transformation.

▮ Following the now familiar track we sum the first 20 terms and have $-.08981$ to five places. From the next three terms the difference table

$$
\begin{array}{ccc}
.32846 & & \\
& -494 & \\
.32352 & & 36 \\
& -459 & \\
.31893 & & \\
\end{array}
$$

is constructed. The Euler transformation then yields

$$.16423 + .00124 + .00004 = .16551$$

and subtracting the earlier $.08981$ we have $.0757$, which is the correct value to four places.

9.47 Show that power series for logarithms converge slowly for large arguments.

▮ The familiar identity

$$\frac{1}{1-t^2} = 1 + t^2 + t^4 + \cdots + t^{2n-2} + \frac{t^{2n}}{1-t^2}$$

can be integrated from 0 to x, with the result

$$\frac{1}{2}\log\frac{1+x}{1-x} = x + \frac{1}{3}x^3 + \frac{1}{5}x^5 + \cdots + \frac{1}{2n-1}x^{2n-1} + R_n \quad \text{where} \quad R_n = \int_0^x \frac{t^{2n}}{1-t^2}\, dt$$

For $x^2 < 1$ we find $1/(1-t^2) \le 1/(1-x^2)$, making $|R_n| \le |x|^{2n+1}/(2n+1)(1-x^2)$. As n increases, $\lim R_n = 0$ and the series obtained does represent the logarithm function. Moreover, for $-1 < x < 1$ the quotient $(1+x)/(1-x)$ assumes all positive values, so that theoretically any real logarithm is computable from the series. Using $x = \frac{1}{3}$, the series produces

$$\log 2 = 2\left(\tfrac{1}{3} + \tfrac{1}{81} + \tfrac{1}{1215} + \tfrac{1}{15309} + \cdots\right) \simeq .693147$$

six terms being adequate for six place accuracy since $|R_6| \le \tfrac{9}{8}|\tfrac{1}{3}|^{13}/13 \sim 5 \cdot 10^{-8}$.

Computations were carried to eight digits, finally rounded to six, so that roundoff errors could not possibly influence the result. For $|x| < \frac{1}{3}$ the series is rapidly convergent. However, for $x = \frac{2}{3}$, which leads to $\log 5$, almost 20 terms are needed, and as x nears 1 the series begins to resemble the divergent $1 + \frac{1}{3} + \frac{1}{5} + \frac{1}{7} + \cdots$. This corresponds to the fact that logarithms grow without bound as their arguments increase. The series converges very slowly for such arguments and roundoff error accumulation becomes a serious factor. Other series, such as the one for $\log(1 + x)$, are slower still.

9.48 Devise an accelerated method for computing logarithms of large arguments.

▮ Let $(1+x)/(1-x) = p^2/(p^2 - 1)$. Then

$$\log p = \frac{1}{2}\log(p^2 - 1) + \frac{1}{2}\log\frac{1+x}{1-x}$$

and since $x = 1/(2p^2 - 1)$, we can use the series of Problem 9.47 to get

$$\log p = \frac{1}{2}\log(p-1) + \frac{1}{2}\log(p+1) + \frac{1}{2p^2 - 1} + \frac{1}{3(2p^2 - 1)^3} + \cdots$$

If we restrict x to the interval $0 < x < 1$, which costs us nothing since negative x lead to logarithms of reciprocals, then $p^2 > 1$. If p is a prime greater than 2, this series expresses $\log p$ as a combination of logarithms of smaller integers (since $p + 1$ will be even and can be factored) plus a rapidly convergent series. The truncation error of this series can be estimated by comparison with a geometric series. The remainder beyond the term $1/n(2p^2 - 1)^n$ is

$$R_n < \frac{1}{(n+2)(2p^2 - 1)^{n+2}}\left[1 + \frac{1}{(2p^2 - 1)^2} + \frac{1}{(2p^2 - 1)^4} + \cdots\right] = \frac{1}{(n+2)(2p^2 - 1)^n} \cdot \frac{1}{(2p^2 - 1)^2 - 1}$$

As an example take $p = 3$. Then $2p^2 - 1$ is 17, and using $\log 2 = .693147$,

$$\log 3 = \frac{3}{2}\log 2 + \frac{1}{17} + \frac{1}{3 \cdot 17^3} + \frac{1}{5 \cdot 17^5} + \cdots \simeq 1.098612$$

only these terms contributing since $R_5 < 1/(7 \cdot 288 \cdot 1400000) \sim 3 \cdot 10^{-10}$.

Similar efforts produce $\log 5$, $\log 7$ and so on, the series converging faster as p gets larger. Logarithms of composite integers may be found by additions, and numbers which are not integers may be handled by splitting off the integral part. For example, if $N = I + D$, where I is the integer part of N, then $\log N = \log I + \log(1 + D/I)$. The first logarithm may be found by the method of this problem and the second responds to the series of Problem 9.47.

9.49   Use the series of Problem 9.47 to compute $\log 3$.

▮   To find $\log N$ we set $N = (1 + x)/(1 - x)$ making $x = (N - 1)/(N + 1)$. For $N = 3$ this means $x = 1/2$ so

$$\log 3 = 2\left(\tfrac{1}{2} + \tfrac{1}{24} + \tfrac{1}{160} + \cdots \right)$$

with eight terms accumulating to 1.098611, which is just one unit short in the last place.

9.50   Also find $\log 5$ and $\log 7$ by the same method. How many terms are needed for five place accuracy?

▮   For $N = 5$ and 7 we have $x = \frac{2}{3}$ and $\frac{3}{4}$. A few partial sums appear in Table 9.3. It seems that convergence is still fairly good.

TABLE 9.3

Terms used	$N = 5$	$N = 7$
12	1.60943	1.94578
14	1.60944	1.94588
16	1.60944	1.94590
18	1.60944	1.94591
20	1.60944	1.94591

9.51   Also compute $\log 5$ and $\log 7$ by the acceleration method used in Problem 9.48. How many terms are now needed for the same level of accuracy?

▮   For $p = 5$ the value of x is $1/(2p^2 - 1)$ or $1/49$. Since it is the same series that is to be used, this small value for x means much faster convergence,

$$\log 5 = \frac{1}{2}\log 4 + \frac{1}{2}\log 6 + \frac{1}{49} + \frac{1}{3(49)^2} + \cdots$$

$$= -\log 2 + -\log 3 + \frac{1}{49} + \frac{1}{3(49)^2} + \cdots$$

with the two series terms totaling .020619 and no other terms contributing to these decimal places. Using

$$\log 2 = .693147 \qquad \log 3 = 1.098612$$

we have $\log 5 = 1.609437$. Similarly, for $p = 7$ we have $x = 1/97$ and

$$\log 7 = \frac{1}{2}\log 6 + \frac{1}{2}\log 8 + \frac{1}{97} + \frac{1}{3(97)^2} + \cdots$$

leading easily to $\log 7 = 1.945910$, with one term of the series enough.

9.4   THE BERNOULLI POLYNOMIALS

9.52   The Bernoulli polynomials $B_i(x)$ are defined by

$$e^{xt}\frac{t}{e^t - 1} = \sum_{i=0}^{\infty} \frac{t^i}{i!}B_i(x)$$

Let $B_i(0) = B_i$ and develop a recursion for these B_i numbers.

▮ Replacing x by 0, we have

$$t = (e^t - 1) \sum_{i=0}^{\infty} \frac{t^i B_i}{i!} = \left(\sum_{j=1}^{\infty} \frac{t^i}{j!} \right) \left(\sum_{i=0}^{\infty} \frac{t^i B_i}{i!} \right) = \sum_{k=1}^{\infty} c_k t^k$$

with $c_k = \sum_{i=0}^{k-1} B_i / [i!(k-i)!]$. This makes $k! c_k = \sum_{i=0}^{k-1} \binom{k}{i} B_i$. Comparing the coefficients of t in the preceding series equation, we find that

$$B_0 = 1 \qquad \sum_{i=0}^{k-1} \binom{k}{i} B_i = 0 \quad \text{for } k = 2, 3, \ldots$$

Written out, this set of equations shows how the B_i may be determined one by one without difficulty:

$$B_0 \qquad\qquad = 1$$
$$B_0 + 2B_1 \qquad\quad = 0$$
$$B_0 + 3B_1 + 3B_2 \quad\;\, = 0$$
$$B_0 + 4B_1 + 6B_2 + 4B_3 = 0$$

et cetera. The first several B_i are therefore

$$B_0 = 1 \qquad B_1 = -\tfrac{1}{2} \qquad B_2 = \tfrac{1}{6} \qquad B_3 = 0 \qquad B_4 = -\tfrac{1}{30} \qquad B_5 = 0 \qquad B_6 = \tfrac{1}{42}$$

and so on. The set of equations used can also be described in the form

$$(B + 1)^k - B^k = 0 \quad \text{for } k = 2, 3, \ldots$$

where it is understood that after applying the binomial theorem each *power* B^i is replaced by B_i.

9.53   Find an explicit formula for the Bernoulli polynomials.

▮ From the defining equation and the special case $x = 0$ just treated,

$$\left(\sum_{i=0}^{\infty} \frac{x^i t^i}{i!} \right) \left(\sum_{j=0}^{\infty} \frac{B_j t^j}{j!} \right) = \sum_{k=0}^{\infty} \frac{t^k}{k!} B_k(x)$$

Comparing the coefficients of t^k on both sides makes

$$\frac{1}{k!} B_k(x) = \sum_{i=0}^{k} B_{k-i} \frac{1}{i!(k-i)!} x^i \quad \text{or} \quad B_k(x) = \sum_{i=0}^{k} \binom{k}{i} B_{k-i} x^i$$

The first several Bernoulli polynomials are

$$B_0(x) = 1 \qquad\qquad B_3(x) = x^3 - \tfrac{3}{2} x^2 + \tfrac{1}{2} x$$
$$B_1(x) = x - \tfrac{1}{2} \qquad\quad B_4(x) = x^4 - 2x^3 + x^2 - \tfrac{1}{30}$$
$$B_2(x) = x^2 - x + \tfrac{1}{6} \qquad B_5(x) = x^5 - \tfrac{5}{2} x^4 + \tfrac{5}{3} x^3 - \tfrac{1}{6} x$$

et cetera. The formula can be summarized as $B_k(x) = (x + B)^k$ where once again it is to be understood that the binomial theorem is applied and then each *power* B^i is replaced by B_i.

9.54   Verify that with the conventions just mentioned, the symbolic

$$B_k(x) = (x + B)^k$$

does produce the polynomials listed as $B_2(x)$ and $B_4(x)$ in the preceding problem. Find $B_6(x)$ in the same way.

▮
$$(x + B)^2 = x^2 + 2xB^1 + B^2 = x^2 + 2xB_1 + B_2$$
$$= x^2 - x + \tfrac{1}{6}$$
$$(x + B)^4 = x^4 + 4x^3B^1 + 6x^2B^2 + 4xB^3 + B^4$$
$$= x^4 + 4x^3B_1 + 6x^2B_2 + 4xB_3 + B_4$$
$$= x^4 - 2x^3 + x^2 - \tfrac{1}{30}$$

With the two verifications completed we turn to $B_6(x)$:

$$(x + B)^6 = x^6 + 6x^5B^1 + 15x^4B^2 + 20x^3B^3 + 15x^2B^4 + 6xB^5 + B^6$$

and interpreting the powers of B as subscripts,

$$B_6(x) = x^6 + 3x^5 + \tfrac{5}{2}x^4 - \tfrac{1}{2}x^2 + \tfrac{1}{42}$$

9.55 Prove that $B_i'(x) = iB_{i-1}(x)$.

▮ The defining equation can be written as

$$\frac{te^{xt}}{e^t - 1} = 1 + \sum_{i=1}^{\infty} \frac{t^i B_i(x)}{i!}$$

Differentiating relative to x and dividing through by t,

$$\frac{te^{xt}}{e^t - 1} = \sum_{i=1}^{\infty} \left[\frac{B_i'(x)}{i} \right] t^{i-1}(i-1)!$$

But the defining equations can also be written as

$$\frac{te^{xt}}{e^t - 1} = \sum_{i=1}^{\infty} [B_{i-1}(x)] \left[\frac{t^{i-1}}{(i-1)!} \right]$$

and comparing coefficients on the right, $B_i'(x) = iB_{i-1}(x)$ for $i = 1, 2, \ldots$. Notice also that the same result can be obtained instantly by formal differentiation of $B_i(x) = (x + B)^i$.

9.56 Prove $B_i(x + 1) - B_i(x) = ix^{i-1}$.

▮ Proceeding formally (even though a rigorous proof would not be too difficult) from $(B + 1)^k = B^k$, we find

$$\sum_{k=2}^{i} \binom{i}{k}(B + 1)^k x^{i-k} = \sum_{k=2}^{i} \binom{i}{k} B^k x^{i-k} \quad \text{or} \quad (B + 1 + x)^i - i(B + 1)x^{i-1} = (B + x)^i - iBx^{i-1}$$

From the abbreviated formula for Bernoulli polynomials (Problem 9.53), this converts immediately to $B_i(x + 1) - B_i(x) = ix^{i-1}$.

9.57 Prove $B_i(1) = B_i(0)$ for $i > 1$.

▮ This follows at once from the preceding problem with x replaced by zero.

9.58 Prove that $\int_0^1 B_i(x)\, dx = 0$ for $i = 1, 2, \ldots$.

▮ By the previous problems,

$$\int_0^1 B_i(x)\, dx = \frac{B_{i+1}(1) - B_{i+1}(0)}{i + 1} = 0$$

9.59   The conditions of Problems 9.55 and 9.58 also determine the Bernoulli polynomials, given $B_0(x) = 1$. Determine $B_1(x)$ and $B_2(x)$ in this way.

▮ From $B_1'(x) = B_0(x)$ it follows that $B_1(x) = x + C_1$ where C_1 is a constant. For the integral of $B_1(x)$ to be zero, C_1 must be $-\frac{1}{2}$. Then from $B_2'(x) = 2B_1(x) = 2x - 1$ it follows that $B_2(x) = x^2 - x + C_2$. For the integral of $B_2(x)$ to be zero, the constant C_2 must be $\frac{1}{6}$. In this way each $B_i(x)$ may be determined in its turn.

9.60   Prove $B_{2i-1} = 0$ for $i = 2, 3, \ldots$.

▮ Notice that

$$f(t) = \frac{t}{e^t - 1} + \frac{t}{2} = \frac{t}{2} \cdot \frac{e^t + 1}{e^t - 1} = B_0 + \sum_{i=2}^{\infty} \frac{B_i t^i}{i!}$$

is an even function, that is, $f(t) = (-t)$. All odd powers of t must have zero coefficients, making B_i zero for odd i except $i = 1$.

9.61   Define the Bernoulli numbers b_i.

▮ These are defined as $b_i = (-1)^{i+1} B_{2i}$ for $i = 1, 2, \ldots$. Thus

$$b_1 = \frac{1}{6} \qquad b_4 = \frac{1}{30} \qquad b_7 = \frac{7}{6}$$

$$b_2 = \frac{1}{30} \qquad b_5 = \frac{5}{66} \qquad b_8 = \frac{3617}{510}$$

$$b_3 = \frac{1}{42} \qquad b_6 = \frac{691}{2730} \qquad b_9 = \frac{43867}{798}$$

as is easily verified after computing the corresponding numbers B_i by the recursion formula of Problem 9.52.

9.62   Extend the list of B_i given in Problem 9.52 and so verify the value of b_4 just given.

▮ As a warmup consider the equation

$$B_0 + 5B_1 + 10B_2 + 10B_3 + 5B_4 = 0$$

which determines B_4. Inserting the known values of B_0 to B_3, this becomes

$$1 - \tfrac{5}{2} + \tfrac{5}{3} + 5B_4 = 0$$

and determines the listed $B_4 = -\frac{1}{30}$. To progress to the target B_8 some less familiar binomial coefficients are needed. The Pascal triangle (here truncated to a trapezoid) obliges

			1		5		10		10		5		1		
		1		6		15		20		15		6		1	
	1		7		21		35		35		21		7		1	
1		8		28		56		70		56		28		8		1
1	9		36		84		126		70		126		84		36	9	1

with the resulting equation

$$B_0 + 9B_1 + 36B_2 + 84B_3 + 126B_4 + 126B_5 + 84B_6 + 36B_7 + 9B_8 = 0$$

Assuming B_0 to B_7 previously determined, this becomes

$$1 - \tfrac{9}{2} + 6 - \tfrac{21}{5} + 2 + 9B_8 = 0$$

and leads to $B_8 = -\frac{1}{30}$. Finally $b_4 = -B_8 = \frac{1}{30}$ as given.

9.63 Prove $\int_x^{x+1} B_i(x)\, dx = x^i$.

▮ Using Problems 9.55 and 9.56,

$$\int_x^{x+1} B_i(x)\, dx = \frac{B_{i+1}(x)}{i+1}\bigg|_x^{x+1} = \frac{B_{i+1}(x+1) - B_{i+1}(x)}{i+1}$$

$$= \frac{(i+1)x^i}{i+1} = x^i$$

9.64 Find $B_3(x)$ and $B_4(x)$ by the method of Problem 9.59.

▮
$$B_3'(x) = 3B_2(x) = 3\left(x^2 - x + \tfrac{1}{6}\right) = 3x^2 - 3x + \tfrac{1}{2}$$

$$B_3(x) = x^3 - \tfrac{3}{2}x^2 + \tfrac{1}{2}x + C$$

Since the integral of $B_3(x)$ over $(0,1)$ must be zero, the constant C is determined. Integrating

$$\tfrac{1}{4} - \tfrac{1}{2} + \tfrac{1}{4} + C = 0$$

and $C = 0$. Similarly

$$B_4'(x) = 4B_3(x) = 4x^3 - 6x^2 + 2x$$

$$B_4(x) = x^4 - 2x^3 + x^2 + C$$

Another integration manages $\tfrac{1}{5} - \tfrac{1}{2} + \tfrac{1}{3} + C = 0$ making C the anticipated $-\tfrac{1}{30}$.

9.65 Evaluate the sum of pth powers in terms of Bernoulli polynomials.

▮ Since, by Problem 9.56, $\Delta B_i(x) = B_i(x+1) - B_i(x) = ix^{i-1}$, the Bernoulli polynomials provide "finite integrals" of the power functions. This makes it possible to telescope the power sum,

$$\sum_{x=0}^{n} x^p = \sum_{x=0}^{n} \frac{1}{p+1} \Delta B_{p+1}(x) = \frac{B_{p+1}(n+1) - B_{p+1}(0)}{p+1}$$

9.66 Evaluate the sums of the form $\sum_{k=1}^{\infty} 1/k^{2i}$ in terms of Bernoulli numbers.

▮ It will be proved later (see chapter on trigonometric approximation) that the function

$$F_n(x) = B_n(x) \quad 0 \le x < 1$$

$$F_n(x \pm m) = F_n(x) \quad \text{for } m \text{ an integer}$$

known as a Bernoulli function, having period 1, can be represented as

$$F_n(x) = (-1)^{n/2+1} \cdot n! \cdot \frac{2}{(2\pi)^n} \cdot \sum_{k=1}^{\infty} \frac{\cos 2\pi k x}{k^n}$$

for even n and as

$$F_n(x) = (-1)^{(n+1)/2} \cdot n! \cdot \frac{2}{(2\pi)^n} \cdot \sum_{k=1}^{\infty} \frac{\sin 2\pi k x}{k^n}$$

when n is odd. For even n, say $n = 2i$, we put $x = 0$ and have

$$\sum_{k=1}^{\infty} \frac{1}{k^{2i}} = (-1)^{i+1} \frac{F_{2i}(0)(2\pi)^{2i}}{2(2i)!}$$

But $F_{2i}(0) = B_{2i}(0) = B_{2i} = (-1)^{i+1} b_i$ and so $\sum_{k=1}^{\infty} 1/k^{2i} = b_i(2\pi)^{2i}/2(2i)!$. In particular, $\sum_{k=1}^{\infty} 1/k^2 = \pi^2/6$, $\sum_{k=1}^{\infty} 1/k^4 = \pi^4/90$, et cetera.

9.67 Show that all the Bernoulli numbers are positive and that they become arbitrarily large as i increases.

▮ Noting that $1 < \sum_{k=1}^{\infty} 1/k^{2i} \le \sum_{k=1}^{\infty} 1/k^2 = \pi^2/6 < 2$, we see that

$$\frac{2(2i)!}{(2\pi)^{2i}} < b_i < \frac{4(2i)!}{(2\pi)^{2i}}$$

In particular all the b_i are positive and they grow limitlessly with increasing i.

9.68 Show that as i increases,

$$\lim \frac{(2\pi)^{2i}}{2(2i)!} b_i = 1$$

▮ This also follows quickly from the series of Problem 9.66. All terms except the $k = 1$ term approach zero for increasing i, and because $1/x^p$ is a decreasing function of x,

$$\frac{1}{k^p} < \int_{k-1}^{k} \frac{1}{x^p} \, dx \quad \text{so that, if } p > 1 \quad \sum_{k=2}^{\infty} \frac{1}{k^p} < \int_{1}^{\infty} \frac{1}{x^p} \, dx = \frac{1}{p-1}$$

As p increases (in our case $p = 2i$) this entire series has limit zero, which establishes the required result. Since all terms of this series are positive, it also follows that $b_i > 2(2i)!/(2\pi)^{2i}$.

9.69 Find the much computed sums $\sum_1^n x^2$ and $\sum_1^n x^3$ once again, this time using Problem 9.65.

▮
$$\sum_1^n x^2 = \frac{1}{3}\left[B_3(n+1) - B_3(0) \right] = \frac{1}{3}\left[(n+1)^3 - \frac{3}{2}(n+1)^2 + \frac{1}{2}(n+1) \right]$$

which reduces to $n(n+1)(2n+1)/6$, and

$$\sum_1^n x^3 = \frac{1}{4}\left[B_4(n+1) - B_4(0) \right] = \frac{1}{4}\left[(n+1)^4 - 2(n+1)^3 + (n+1)^2 \right]$$

with the constant terms canceling. The simplification to $n^2(n+1)^2/4$ is routine.

9.70 Apply Problem 9.66 to check the values $\pi^6/945$, $\pi^8/9450$ and $\pi^{10}/93555$ for $\sum_{k=1}^{\infty} 1/k^{2i}$ and $p = 6, 8$ and 10.

▮ Substituting

$$\sum \frac{1}{k^6} = \frac{1}{42} \frac{(2\pi)^6}{2 \cdot 6!} = \frac{\pi^6}{945}$$

$$\sum \frac{1}{k^8} = \frac{1}{30} \frac{(2\pi)^8}{2 \cdot 8!} = \frac{\pi^8}{9450}$$

$$\sum \frac{1}{k^{10}} = \frac{5}{66} \frac{(2\pi)^{10}}{2 \cdot 10!} = \frac{\pi^{10}}{93555}$$

and so on.

9.71 Consider a sequence of polynomials

$$f_0(x), f_1(x), \ldots, f_n(x), \ldots$$

for which $f_n'(x) = f_{n-1}(x)$ and $f_0(x)$ is a constant. Show that

$$f_n(x) = \frac{c_0 x^n}{n!} + \frac{c_1 x^{n-1}}{(n-1)!} + \cdots + c_{n-1} x + c_n$$

with constants c_i independent of n. How does this type of sequence relate to Bernoulli polynomials?

▌ Since each function is the derivative of its successor in the sequence, it follows that each is also an integral of its predecessor. Beginning with $f_0(x) = c_0$,

$$f_1(x) = c_0 x + c_1 \qquad f_2(x) = c_0 x^2/2 + c_1 x + c_2 \qquad \cdots$$

with an obvious continuation to the required result. The c_i are the constants of integration.

To note the connection with the Bernoulli polynomials $B_n(x)$ recall the property

$$B_n'(x) = n B_{n-1}(x)$$

and let $f_n(x)$ be defined by $B_n(x) = n! f_n(x)$. Then

$$n! f_n'(x) = n(n-1)! f_{n-1}(x)$$

and $f'(x) = f_{n-1}(x)$.

9.72 Prove the relationship

$$B_n(x) = \frac{d}{dx} \sum_1^n s_k^{(n)} x^{(k+1)}/(k+1)$$

between Bernoulli polynomials and the Stirling numbers of second kind $s_k^{(n)}$.

▌ From Problem 9.56,

$$B_n(x) = n \, \Delta^{-1} x^{n-1}$$

and shifting from n to $n+1$,

$$B_{n+1}(x) = (n+1) \, \Delta^{-1} x^n = (n+1) \, \Delta^{-1} \sum_1^n s_k^{(n)} x^{(k)}$$

$$= (n+1) \sum_1^n s_k^{(n)} x^{(k+1)}/(k+1)$$

with an arbitrary additive constant ignored. Taking a derivative (which disposes of the additive constant) and again using

$$B_{n+1}'(x) = (n+1) B_n(x)$$

the theorem follows easily. For example, with $n = 2$ it claims that

$$x^2 - x + \frac{1}{6} = \frac{d}{dx} \left(\frac{1}{2} x^{(2)} + \frac{1}{3} x^{(3)} \right)$$

which is verified by doing the differentiations.

9.73 What sequence of polynomials is determined by the conditions

$$Q_n'(x) = n Q_{n-1}(x) \qquad Q_n(0) = 0$$

starting with $Q_0(x) = 1$?

▌ Doing just a few integrations

$$Q_1'(x) = 1 \qquad Q_1(x) = x$$
$$Q_2'(x) = 2x \qquad Q_2(x) = x^2$$
$$Q'(x) = 3x^2 \qquad Q(x) = x^3$$

is enough to solve the mystery. $Q_n(x) = x^n$. The differential equation is the same as for the Bernoulli polynomials but the initial condition $Q_n(0) = 0$ points in a different direction.

9.74 Set $x = 0$ in the generating function of the Bernoulli polynomials (Problem 9.52) and so obtain the B_i numbers by developing the infinite series

$$t/(e^t - 1) = \sum_0^\infty \left[B_i(0)/i! \right] t^i = \sum_0^\infty (B_i/i!) t^i$$

▮ Rewritten this is

$$\left(t + \tfrac{1}{2} t^2 + \tfrac{1}{6} t^3 + \tfrac{1}{24} t^4 + \cdots \right)\left(B_0 + B_1 t + \tfrac{1}{2} B_2 t^2 + \tfrac{1}{6} B_3 t^3 + \cdots \right) = t$$

and carrying out the left side multiplications,

$$B_0 t + \left(B_1 + \tfrac{1}{2} B_0 \right) t^2 + \left(\tfrac{1}{2} B_2 + \tfrac{1}{2} B_1 + \tfrac{1}{6} B_0 \right) t^3 + \left(\tfrac{1}{6} B_3 + \tfrac{1}{4} B_2 + \tfrac{1}{6} B_1 + \tfrac{1}{24} B_0 \right) t^4 + \cdots = t$$

Comparing coefficients of powers of t now forces

$$B_0 = 1 \qquad B_1 = -\tfrac{1}{2} \qquad B_2 = -2\left(\tfrac{1}{6} - \tfrac{1}{4} \right) = \tfrac{1}{6}$$
$$B_3 = -6\left(\tfrac{1}{24} - \tfrac{1}{12} + \tfrac{1}{24} \right) = 0$$

and so on.

9.5 EULER POLYNOMIALS

9.75 Euler polynomials can be defined by their generating function

$$2e^{xt}/(e^t + 1) = \sum_0^\infty E_k(x) t^k$$

with corresponding Euler numbers $E_k = E_k(0)$. Develop a recursion for the Euler numbers.

▮ Rewrite the defining equation with $x = 0$ as

$$2 = \left(1 + \sum_0^\infty t^i/i! \right)\left(\sum_0^\infty E_j t^j \right) = \sum_0^\infty c_k t^k$$

and compare coefficients of t^k.

▮

$$c_k = \sum_0^k E_i/(k - i)! + E_k = \sum_0^k E_{k-i}/i! + E_k$$

The constant terms offer $2 = 2E_0$, making $E_0 = 1$. All other c_k must be zero, which means

$$2E_k + E_{k-1} + E_{k-2}/2! + E_{k-3}/3! + \cdots + E_0/k! = 0$$

In succession we find from this recursion

$$2E_1 + E_0 = 0 \qquad 2E_2 + E_1 + E_0/2! = 0$$
$$2E_3 + E_2 + E_1/2! + E_0/3! = 0$$
$$2E_4 + E_3 + E_2/2! + E_1/3! + E_0/4! = 0$$
$$2E_5 + E_4 + E_3/2! + E_2/3! + E_1/4! + E_0/5! = 0$$

determining $E_1 = -\tfrac{1}{2}$, $E_2 = 0$, $E_3 = \tfrac{1}{24}$, $E_4 = 0$ and $E_5 = -\tfrac{1}{240}$.

9.76 Find an explicit representation of the $E_k(x)$.

▮ Using the result just found,

$$2/(e^t + 1) = \sum_0^\infty E_k t^k$$

the defining equation can be written as

$$\left(\sum_0^\infty x^i t^i/i!\right)\left(\sum_0^\infty E_j t^j\right) = \sum_0^\infty E_k(x) t^k$$

and again comparing coefficients of t^k,

$$E_k(x) = \sum_0^k E_{k-i} x^i/i!$$

$$= x^k/k! + E_1 x^{k-1}/(k-1)! + E_2 x^{k-2}/(k-2)! + \cdots + E_k$$

Successively this gives us

$$E_0(x) = 1$$
$$E_1(x) = x + E_1 = x - \tfrac{1}{2}$$
$$E_2(x) = \tfrac{1}{2}x^2 - \tfrac{1}{2}x$$
$$E_3(x) = \tfrac{1}{6}x^3 - \tfrac{1}{4}x^2 + \tfrac{1}{24}$$
$$E_4(x) = \tfrac{1}{24}x^4 - \tfrac{1}{4}x^3 + \tfrac{1}{24}x$$

with obvious continuation.

9.77 Prove $E_k'(x) = E_{k-1}(x)$.

∎ The generating function can be written

$$2e^{xt}/(e^t + 1) = 1 + \sum_1^\infty E_k(x) t^k$$

and differentiated with respect to x, bringing

$$2te^{xt}/(e^t + 1) = \sum_1^\infty E_k'(x) t^k$$

from which cancellation of t achieves

$$2e^{xt}/(e^t + 1) = \sum_1^\infty E_k'(x) t^{k-1}$$

But the generating function can also be written as

$$2e^{xt}/(e^t + 1) = \sum_1^\infty E_{k-1}(x) t^{k-1}$$

which means that $E_k'(x) = E_{k-1}(x)$. Thus the Euler polynomials form a sequence of the sort described in Problem 9.71.

9.78 Prove $E_k(0) + E_k(1) = 0$ for $k > 0$.

∎ $$E_k(0) + E_k(1) = E_k + \sum_0^k E_{k-i}/i! = E_k - E_k = 0$$

9.79 Show that the differential property $E_n'(x) = E_{n-1}(x)$ together with the supplementary condition

$$\tfrac{1}{2}[E_n(x+1) + E_n(x)] = x^n/n!$$

also determines the Euler polynomials.

$\blacksquare$ For $n = 0$, the new condition alone, assuming $E_0(x)$ a constant, makes $E_0(x) = 1$. Then

$$E_1'(x) = 1 \qquad E_1(x) = x + c$$

and the requirement $\frac{1}{2}(x + 1 + c + x + c) = x$ forces $c = -\frac{1}{2}$. Similarly

$$E_2'(x) = x - \frac{1}{2} \qquad E_2(x) = \frac{1}{2}x^2 - \frac{1}{2}x + c$$

and an easy calculation finds this constant c to be 0. So the first three polynomials are the same as before. An induction is in order and begins by assuming that the sequence determined as suggested in this problem will coincide with the earlier sequence up through $E_k(x)$. So we are still entitled to

$$E_k(x) = x^k/k! + E_1 x^{k-1}/(k-1)! + \cdots + E_k$$

and integrating, we have some function

$$E_{k+1}(x) = x^{k+1}/(k+1)! + E_1 x^k/k! + \cdots + E_k x + c$$

Thus $E_{k+1}(0) = c$ while

$$E_{k+1}(1) = E_0/(k+1)! + E_1/k! + \cdots + E_k + c$$

The supplementary condition at $x = 0$ requires that these add to zero, or

$$E_0/(k+1)! + E_1/k! + \cdots + 2c = 0$$

But this is precisely the recursion of Problem 9.75 where E_{k+1} occupied the place now held by c. So $c = E_{k+1}$ and the induction is complete.

9.6 THE EULER–MACLAURIN FORMULA

9.80 Use the Bernoulli polynomials to derive the Euler–Maclaurin formula with an error estimate. (This formula was obtained in Problem 5.44 by an operator computation, but without an error estimate.)

$\blacksquare$ We begin with an integration by parts, using the facts that $B_1'(t) = B_0(t) = 1$ and $B_1(1) = -B_1(0) = \frac{1}{2}$:

$$\int_0^1 y(t)\, dt = \int_0^1 y(t) B_1'(t)\, dt = \frac{1}{2}(y_0 + y_1) - \int_0^1 y'(t) B_1(t)\, dt$$

Again integrate by parts using $B_2'(t) = 2B_1(t)$ from Problem 9.55 and $B_2(1) = B_2(0) = b_1$ to find

$$\int_0^1 y(t)\, dt = \frac{1}{2}(y_0 + y_1) - \frac{1}{2}b_1(y_1' - y_0') + \frac{1}{2}\int_0^1 y^{(2)}(t) B_2(t)\, dt$$

The next integration by parts brings

$$\frac{1}{2}\int_0^1 y^{(2)}(t) B_2(t)\, dt = \frac{1}{6}y^{(2)}(t) B_3(t)\Big|_0^1 - \frac{1}{6}\int_0^1 y^{(3)}(t) B_3(t)\, dt$$

But since $B_3(1) = B_3(0) = 0$, the integrated term vanishes and we proceed to

$$\frac{1}{2}\int_0^1 y^{(2)}(t) B_2(t)\, dt = -\frac{1}{24}y^{(3)}(t) B_4(t)\Big|_0^1 + \frac{1}{24}\int_0^1 y^{(4)}(t) B_4(t)\, dt$$

$$= \frac{1}{24}b_2(y_1^{(3)} - y_0^{(3)}) + \frac{1}{24}\int_0^1 y^{(4)}(t) B_4(t)\, dt$$

since $B_4(1) = B_4(0) = B_4 = -b_2$. Continuing in this way, we develop the result

$$\int_0^1 y(t)\, dt = \frac{1}{2}(y_0 + y_1) + \sum_{i=1}^k \frac{(-1)^i b_i}{(2i)!}(y_1^{(2i-1)} - y_0^{(2i-1)}) + R_k$$

where

$$R_k = \frac{1}{(2k)!}\int_0^1 y^{(2k)}(t)\,B_{2k}(t)\,dt$$

Integrating R_k by parts the integrated part again vanishes, leaving

$$R_k = \frac{-1}{(2k+1)!}\int_0^1 y^{(2k+1)}(t)\,B_{2k+1}(t)\,dt$$

Corresponding results hold for the intervals between other consecutive integers. Summing, we find substantial telescoping and obtain

$$\sum_{i=0}^n y_i = \int_0^n y(t)\,dt + \frac{1}{2}(y_0 + y_n) - \sum_{i=1}^k \frac{(-1)^i b_i}{(2i)!}\left(y_n^{(2i-1)} - y_0^{(2i-1)}\right)$$

with an error of

$$E_k = \frac{-1}{(2k+1)!}\int_0^n y^{(2k+1)}(t)\,F_{2k+1}(t)\,dt$$

where $F_{2k}(t)$ is the Bernoulli function of Problem 9.66, the periodic extension of the Bernoulli polynomial $B_{2k}(t)$. The same argument may be used between integer arguments a and b rather than 0 and n. We may also allow b to become infinite, provided that the series and the integral we encounter are convergent. In this case we assume that $y(t)$ and its derivatives all become zero at infinity, so that the formula becomes

$$\sum_{i=a}^\infty y_i = \int_a^\infty y(t)\,dt + \frac{1}{2}y_a + \sum_{i=1}^k \frac{(-1)^i b_i}{(2i)!}y_a^{(2i-1)}$$

9.81 Evaluate the power sum $\sum_{i=0}^n i^4$ by use of the Euler–Maclaurin formula.

▎ In this case the function $y(t) = t^4$, so that with $k = 2$ the series of the preceding problem terminates. Moreover, the error E_k becomes zero since $y^{(5)}(t) = 0$. The result is

$$\sum_{i=0}^n i^4 = \tfrac{1}{5}n^5 + \tfrac{1}{2}n^4 + \tfrac{1}{12}(4n^3) - \tfrac{1}{720}(24n) = \tfrac{1}{30}n(n+1)(2n+1)(3n^2 + 3n - 1)$$

as in Problem 9.3. This is an example in which increasing k in the Euler–Maclaurin formula leads to a finite sum. (The method of Problem 9.65 could also have been applied to this sum.)

9.82 Compute Euler's constant

$$C = \lim\left(1 + \frac{1}{2} + \frac{1}{3} + \cdots + \frac{1}{n} - \log n\right)$$

assuming convergence.

▎ Using Problem 9.2, this can be rewritten as

$$C = 1 + \sum_{i=2}^\infty \left(\frac{1}{i} + \log\frac{i-1}{i}\right)$$

The Euler–Maclaurin formula may now be applied with $y(t) = 1/t - \log t + \log(t-1)$. Actually it is more convenient to sum the first few terms directly and then apply the Euler–Maclaurin formula to the rest of the series. To eight places,

$$1 + \sum_{i=2}^9 \left(\frac{1}{i} + \log\frac{i-1}{i}\right) = .63174368$$

Using 10 and ∞ as limits, we first compute

$$\int_{10}^{\infty}\left[\frac{1}{t}-\log t+\log(t-1)\right]dt=(1-t)\log\frac{t}{t-1}\Big|_{10}^{\infty}$$

$$=-1+9\log 10-9\log 9\simeq-.05175536$$

the first term coming from the upper limit by evaluation of the "indeterminate form." Next

$$\tfrac{1}{2}y_{10}=-.00268026\qquad-\tfrac{1}{12}y_{10}'=-.00009259\qquad\tfrac{1}{720}y_{10}^{(3)}=.00000020$$

all values at infinity being zero. Summing the five terms just computed, we have $C\simeq.57721567$. Carrying 10 places and computing only one more term would lead to the better approximation $C\simeq.5772156650$, which is itself one unit too large in the tenth place.

In this example the accuracy obtainable by the Euler–Maclaurin formula is limited. After a point, using more terms (increasing k) leads to poorer approximations to Euler's constant rather than better. In other words, we have used a few terms of a divergent series to obtain our results. To see this we need only note that the ith term of the series is

$$\frac{(-1)^{i+1}b_i}{(2i)(2i-1)}\left(\frac{2i+9}{10^{2i}}-\frac{1}{9^{2i-1}}\right)$$

and that by Problem 9.67 the b_i exceed $2(2i)!/(2\pi)^{2i}$, which guarantees the unlimited growth of this term. Divergence is more typical than convergence for the Euler–Maclaurin series.

9.83 A truck can travel a distance of one "leg" on the maximum load of fuel it is capable of carrying. Show that if an unlimited supply of fuel is available at the edge of a desert, then the truck can cross the desert no matter what its width. Estimate how much fuel would be needed to cross a desert 10 "legs" wide.

▮ On just one load of fuel the truck could cross a desert one leg wide. With two loads available this strategy could be followed: Loading up, the truck is driven out into the desert to a distance of one-third leg. One-third load of fuel is left in a cache and the truck returns to the fuel depot just as its fuel vanishes. On the second load it drives out to the cache, which is then used to fill up. With a full load the truck can then be driven one more leg, thereby cross a desert of width $(1+\tfrac{1}{3})$ legs, as shown in Fig. 9-1. With three loads of fuel available at the depot, two trips can be made to establish a cache of $\tfrac{6}{5}$ loads at a distance of $\tfrac{1}{5}$ leg out into the desert. The third load then brings the truck to the cache with $(\tfrac{4}{5}+\tfrac{6}{5})$ loads available. Repeating the previous strategy then allows a journey of $1+\tfrac{1}{3}+\tfrac{1}{5}$ legs, as shown in Fig. 9-2.

A similar strategy allows a desert of width $(1+\tfrac{1}{3}+\tfrac{1}{5}+\cdots+1/(2n-1))$ to be crossed using n loads of fuel. Since this sum grows arbitrarily large with increasing n, a desert of any width can be crossed if sufficient fuel is available at the depot.

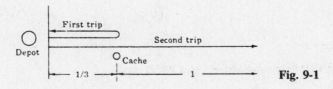

Fig. 9-1

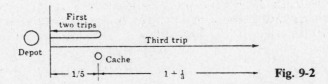

Fig. 9-2

To estimate how much fuel is needed to cross a desert 10 legs wide, we write

$$1 + \frac{1}{3} + \cdots + \frac{1}{2n-1} = \left(1 + \frac{1}{2} + \frac{1}{3} + \cdots + \frac{1}{2n}\right) - \frac{1}{2}\left(1 + \frac{1}{2} + \frac{1}{3} + \cdots + \frac{1}{n}\right)$$

and apply the approximation of Problem 9.82:

$$1 + \frac{1}{3} + \cdots + \frac{1}{2n-1} \simeq \log(2n) + C - \frac{1}{2}(\log n + C)$$

$$= \frac{1}{2}\log n + \log 2 + \frac{1}{2}C \simeq \frac{1}{2}\log n + .98$$

This reaches 10 for n equal to almost 100 million loads of fuel.

9.84 Use the Euler–Maclaurin formula to reproduce the simple and well known results

$$1 + 2 + 3 + \cdots + n = n(n+1)/2$$
$$1^2 + 2^2 + 3^2 + \cdots + n^2 = n(n+1)(2n+1)/6$$

▮ This is more or less like driving tacks with a sledge hammer but the job does get done:

$$\sum_0^n i = \int_0^n t\,dt + \frac{1}{2}n = \frac{1}{2}n^2 + \frac{1}{2}n = \frac{1}{2}n(n+1)$$

$$\sum_0^n i^2 = \int_0^n t^2\,dt + \frac{1}{2}n + \frac{1}{12}(2n) = \frac{1}{3}n^3 + \frac{1}{2}n^2 + \frac{1}{6}n$$

$$= \frac{1}{6}n(n+1)(2n+1)$$

9.85 Compute $\sum_0^n i^3$ by the Euler–Maclaurin formula.

▮ Following the route of Problem 9.81,

$$\sum_0^n i^3 = \int_0^n t^3\,dt + \frac{1}{2}(n^3) + \frac{1}{12}(3n^2) = \frac{1}{4}n^4 + \frac{1}{2}n^3 + \frac{1}{4}n^2$$

$$= \frac{1}{4}n^2(n+1)^2$$

Only the first derivative term is involved since the third derivative is constant and its end values cancel each other.

9.86 Also find $\sum_0^n i^5$ by the Euler–Maclaurin formula.

▮ All such power sums will, of course, terminate. Here the fifth derivative is constant so its term will vanish and, of course, all higher derivatives are zero. This leaves

$$\sum_0^n i^5 = \int_0^n t^5\,dt + \frac{1}{2}(n^5) + \frac{1}{12}(5n^4) + \frac{1}{720}(60n^2)$$

$$= \frac{1}{6}n^5 + \frac{1}{2}n^5 + \frac{5}{12}n^4 - \frac{1}{12}n^2$$

$$= \frac{1}{12}n^2(n+1)^2(2n^2 + 2n - 1)$$

9.87 Apply the Euler–Maclaurin formula to $\sum_0^\infty 1/(i+1)^2$.

▮ Since the function $y(t) = 1/(t+1)^2$ tends to zero at infinity, together with all its derivatives, the formula is

$$\sum_a^\infty y_i = \int_a^\infty y(t)\,dt + \frac{1}{2}y_a - \frac{1}{12}y_a' + \frac{1}{30\cdot 4!}y_a^{(3)} - \frac{1}{42\cdot 6!}y_a^{(5)} + \cdots$$

But $y'(t) = -2(t+1)^{-3}$, $y''(t) = 6(t+1)^{-4}$, $y'''(t) = -24(t+1)^{-5}$ and so on. At $t = 0$ it is evident that the successive odd derivatives will be $-2!, -4!, -6!, \ldots$. Entering these into the formula, with $a = 0$,

$$\sum_0^\infty 1/(i+1)^2 = \int_0^\infty (t+1)^{-2}\,dt + \frac{1}{2} + \frac{1}{6} - \frac{1}{30} + \frac{1}{42} - \frac{1}{30} + \frac{5}{66} - \frac{691}{2730} + \frac{7}{6} - \frac{3617}{510} + \frac{43867}{798} - \cdots$$

and the bad news becomes clear. The Euler–Maclaurin formula does not converge. Computing a few partial sums we find, after 1 (for the integral) and 1.5,

$$1.667 \quad 1.633 \quad 1.657 \quad 1.624 \quad 1.700 \quad 1.446 \quad 2.613 \quad 9.705 \quad 64.676$$

with no need to continue. The correct value of 1.645 to three places has been skirted and then abandoned.

Trying again we sum the first nine terms separately,

$$\sum_0^8 1/(i+1)^2 = 1.5397677$$

to seven places. Applying the Euler–Maclaurin formula to the rest of the series, with $a = 9$,

$$\sum_9^\infty 1/(i+1)^2 = \int_9^\infty (t+1)^{-2}\, dt + \frac{1}{2}\left(\frac{1}{100}\right) - \frac{1}{6}\cdot\frac{1}{2!}\left(\frac{-2}{10^3}\right) + \frac{1}{30}\cdot\frac{1}{4!}\left(\frac{-4!}{10^5}\right) - \frac{1}{42}\cdot\frac{1}{6!}\left(\frac{-6!}{10}\right) + \cdots$$

$$= .1 + .005 + .00016666 - .00000033 + \cdots$$

Together with the sum of the first nine terms this gives a value of 1.6449340 to the series, agreeing with the correct $\pi^2/6$ to at least six places.

9.88 In Problem 9.37 we found $\sum_1^\infty 1/i^3 = 1.2021$ by using a comparison method. Apply the Euler–Maclaurin formula t this series.

❚ Suppose we write it in the form

$$\sum_0^\infty \frac{1}{(i+1)^3} = \sum_0^8 \frac{1}{(i+1)^3} + \int_9^\infty (t+1)^{-3}\, dt + \frac{1}{2}\cdot 10^{-3} - \frac{1}{12}(-3\cdot 10^{-4})$$

$$= 1.1965320 + .005 + .0005 + .000025$$

$$= 1.2020570$$

When the derivatives of $y(t)$ do not change sign in the interval being used it can be proved that the error made by stopping with a particular term does not exceed the first omitted term. Here that term would be $\frac{1}{720}(-60\cdot 10^{-6}) = -.00000008$ to eight places. Our result seems to be secure to six places away.

9.89 The series $1 + 1/3^2 + 1/5^2 + 1/7^2 + \cdots = \pi^2/8 = 1.233701$ to six places. Direct summation leads to the values

Terms used	20	50	100
Sum	1.2218	1.2288	1.2312

How well does the Euler–Maclaurin formula do?

❚ Breaking the series into the parts

$$\sum_0^\infty 1/(2i+1)^2 = \left(\sum_0^4 + \sum_5^\infty\right)1/(2i+1)^2$$

with the first part equaling 1.1838650 to seven places, we note that $y(t) = (2i+1)^{-2}$, $y'(t) = -2\cdot 2(2i+1)^{-3}$, $y'''(t) = -4!2^3(2i+1)^{-5}$ and so on. At $t = 0$ one discovers easily that the odd derivatives are $-2!\cdot 2$, $-4!\cdot 2^3$, $-6!\cdot 2^5$ and so on. So once again there is substantial cancellation of factorials and the second part of the series is replaced by

$$\int_5^\infty y(t)\, dt + \frac{1}{2}\left(\frac{1}{11^2}\right) + \frac{1}{6}\left(\frac{2}{11^3}\right) - \frac{1}{30}\left(\frac{2^3}{11^5}\right) + \frac{1}{42}\left(\frac{2^5}{11^7}\right)$$

with the first neglected term of size $\frac{1}{30}(2^7/11^9)$ and not influencing the seventh decimal place. Putting the various pieces together gives the series a value of 1.2337006.

9.7 WALLIS' INFINITE PRODUCT

9.90 Obtain Wallis' product for π.

▌ Repeated applications of the recursion formula

$$\int_0^{\pi/2} \sin^n x \, dx = \frac{n-1}{n} \int_0^{\pi/2} \sin^{n-2} x \, dx \quad \text{for } n > 1$$

available in integral tables, easily brings the results

$$\int_0^{\pi/2} \sin^{2k} x \, dx = \frac{2k-1}{2k} \cdot \frac{2k-3}{2k-2} \cdots \frac{1}{2} \cdot \int_0^{\pi/2} dx$$

$$\int_0^{\pi/2} \sin^{2k+1} x \, dx = \frac{2k}{2k+1} \cdot \frac{2k-2}{2k-1} \cdots \frac{2}{3} \cdot \int_0^{\pi/2} \sin x \, dx$$

Evaluating the remaining integrals and dividing one result by the other,

$$\frac{\pi}{2} = \frac{2\cdot2\cdot4\cdot4\cdot6\cdot6\cdots 2k\cdot2k}{1\cdot3\cdot3\cdot5\cdot5\cdot7\cdots(2k-1)(2k+1)} \cdot \frac{\int_0^{\pi/2}\sin^{2k}x\,dx}{\int_0^{\pi/2}\sin^{2k+1}x\,dx}$$

The quotient of the two integrals converges to 1 as k increases. This can be proved as follows. Since $0 < \sin x < 1$,

$$0 < \int_0^{\pi/2}\sin^{2k+1}x\,dx \le \int_0^{\pi/2}\sin^{2k}x\,dx \le \int_0^{\pi/2}\sin^{2k-1}x\,dx$$

Dividing by the first integral and using the original recursion formula,

$$1 \le \frac{\int_0^{\pi/2}\sin^{2k}x\,dx}{\int_0^{\pi/2}\sin^{2k+1}x\,dx} \le \frac{2k+1}{2k}$$

so that the quotient does have limit 1. Thus

$$\frac{\pi}{2} = \lim \frac{2\cdot2\cdot4\cdot4\cdot6\cdot6\cdots 2k\cdot2k}{1\cdot3\cdot3\cdot5\cdot5\cdot7\cdots(2k-1)(2k+1)}$$

which is Wallis' infinite product.

9.91 Obtain Wallis' infinite product for $\sqrt{\pi}$.

▌ Since $\lim 2k/(2k+1) = 1$, the result of the previous problem can be written as

$$\frac{\pi}{2} = \lim \frac{2^2\cdot4^2\cdots(2k-2)^2}{3^2\cdot5^2\cdots(2k-1)^2}2k$$

Taking the square root and then filling in missing integers, we find

$$\sqrt{\frac{\pi}{2}} = \lim \frac{2\cdot4\cdots(2k-2)}{3\cdot5\cdots(2k-1)}\sqrt{2k} = \lim \frac{2^{2k}(k!)^2}{(2k)!\sqrt{2k}}$$

from which Wallis' product follows at once in the form

$$\sqrt{\pi} = \lim \frac{2^{2k}(k!)^2}{(2k)!\sqrt{k}}$$

This will be needed in the next problem.

9.8 STIRLING'S SERIES FOR LARGE FACTORIALS

9.92 Derive Stirling's series for large factorials.

▌ In the Euler–Maclaurin formula let $y(t) = \log t$ and use the limits 1 and n. Then

$$\log 1 + \log 2 + \cdots + \log n = n \log n - n + \frac{1}{2} \log n + \sum_{i=1}^{k} \frac{(-1)^i b_i}{(2i)(2i-1)} \left(1 - \frac{1}{n^{2i-1}}\right) - \int_1^n \frac{F_{2k+1}(t)}{(2k+1)t^{2k+1}} \, dt$$

This can be rearranged into

$$\log n! = \left(n + \frac{1}{2}\right) \log n - n + c - \sum_{i=1}^{k} \frac{(-1)^i b_i}{(2i)(2i-1)n^{2i-1}} + \int_n^{\infty} \frac{F_{2k+1}(t)}{(2k+1)t^{2k+1}} \, dt$$

where

$$c = \sum_{i=1}^{k} \frac{(-1)^i b_i}{(2i)(2i-1)} - \int_1^{\infty} \frac{F_{2k+1}(t)}{(2k+1)t^{2k+1}} \, dt$$

To evaluate c let $n \to \infty$ in the previous equation. The finite sum has limit zero. The integral, since F_{2k+1} is periodic and hence bounded, behaves as $1/n^{2k}$ and so also has limit zero. Thus

$$c = \lim \log \frac{n! e^n}{n^{n+1/2}} = \lim \log \alpha_n$$

A simple artifice now evaluates this limit. Since

$$\alpha_n^2 = \frac{(n!)^2 e^{2n}}{n^{2n+1}} \qquad \alpha_{2n} = \frac{(2n)! e^{2n}}{(2n)^{2n+1/2}}$$

we find

$$\lim \alpha_n = \lim \frac{\alpha_n^2}{\alpha_{2n}} = \lim \left[\sqrt{2} \, \frac{(n!)^2 2^{2n}}{\sqrt{n}\,(2n)!}\right] = \sqrt{2\pi}$$

by Wallis' product for $\sqrt{\pi}$. Thus $c = \log\sqrt{2\pi}$. Our result can now be written as the Stirling series

$$\log \frac{n! e^n}{\sqrt{2\pi}\, n^{n+1/2}} \simeq \frac{b_1}{2n} - \frac{b_2}{3 \cdot 4n^3} + \frac{b_3}{5 \cdot 6n^5} - \cdots + \frac{(-1)^{k+1} b_k}{(2k)(2k-1)n^{2k-1}}$$

the error being $E_n = \int_n^{\infty} [F_{2k+1}(t)]/[(2k+1)t^{2k+1}] \, dt$. For large n this means that the logarithm is near zero, making $n! \simeq \sqrt{2\pi}\, n^{n+1/2} e^{-n}$.

9.93 Approximate 20! by Stirling's series.

▌ For $n = 20$ the series itself becomes $1/240 - 1/2880000 + \cdots \simeq .00417$ to five places, only *one* term being used. We now have

$$\log 20! \simeq .00417 - 20 + \log\sqrt{2\pi} + 20.5 \log 20 \simeq 42.33562$$

$$20! \simeq 2.432903 \cdot 10^{18}$$

This is correct to almost seven digits. More terms of the Stirling series could be used for even greater accuracy, but it is important to realize that this series is not convergent. As k is increased beyond a certain point, for fixed n, the terms increase and the error E grows larger. This follows from the fact (see Problem 9.67) that $b_k > 2(2k)!/(2\pi)^{2k}$. As will be proved shortly, the Stirling series is an example of an *asymptotic series*.

The exact value of 20! is 2,432,902,008,176,640,000.

9.94   Also find 5! and 10! by using the Stirling series.

▌ For $n = 5$ the series becomes

$$1/60 - 1/360 \cdot 125 + 1/42 \cdot 30 \cdot 3125 - 1/30 \cdot 56 \cdot 78125 + \cdots$$

which reduces to .0166447. So

$$\log(5!) = .0166447 - 5 + \log(\sqrt{2\pi}) + 5.5\log(5) = 4.78749$$

and $5! = e$ to this power. The computer in use printed a perfect 120.
Turning to 10! the series becomes

$$1/120 - 1/360000 + 1/126000000 - \cdots = .00833612$$

making

$$\log(10!) = .00833612 - 10 + \log(\sqrt{2\pi}) + 10.5\log(10) = 15.10442$$

and $10! = 3628819$. The correct value is 3628800.

9.9   ABEL'S TRANSFORMATION

9.95   Prove

$$\sum_1^n u_k v_k = u_{n+1} \sum_1^n v_k - \sum_1^n \left(\Delta u_k \sum_1^k v_i \right)$$

which is called Abel's transformation.

▌ The theorem resembles the one on summation by parts and its proof begins by applying that earlier result (Problem 3.116):

(∗) $$\sum_1^n f_k \, \Delta g_k = f_{n+1} g_{n+1} - f_1 g_1 - \sum_1^n g_{k+1} \Delta f_k$$

Take $f_k = u_k$ and $\Delta g_k = v_k$ so that

$$\sum_1^{n-1} v_k = g_n - g_1$$

It follows that

$$g_n = g_1 + \sum_1^{n-1} v_k \qquad g_{k+1} = g_1 + \sum_1^k v_i$$

and substituting into (∗),

$$\sum_1^n u_k v_k = u_{n+1} \left[g_1 + \sum_1^n v_k \right] - u_1 g_1 - \sum_1^n \left[\Delta u_k \left(g_1 + \sum_1^k v_i \right) \right]$$

$$= u_{n+1} \sum_1^n v_k - \sum_1^n \left[\Delta u_k \sum_1^k v_i \right] + u_{n+1} g_1 - u_1 g_1 - \sum_1^n g_1 \Delta u_k$$

$$= u_{n+1} \sum_1^n v_k - \sum_1^n \left[\Delta u_k \sum_1^k v_i \right]$$

since the remaining terms equal

$$u_{n+1} g_1 - u_1 g_1 - g_1 \sum_1^n \Delta u_k = u_{n+1} g_1 - u_1 g_1 - g_1 (u_{n+1} - u_1) = 0$$

9.96 Apply the Abel transformation to $\Sigma_1^n i/2^i$.

$\blacksquare$ Take $u_i = i$ and $v_i = 1/2^i$. Then $\Delta u_i = 1$ and $\Sigma_1^n v_i = 1 - (\frac{1}{2})^n$. Substituting

$$\sum_1^n i/2^i = (n+1)\left[1 - (\tfrac{1}{2})^n\right] - \sum_1^n \left[1 - (\tfrac{1}{2})^i\right]$$

$$= n + 1 - (n+1)(\tfrac{1}{2})^n - n + \left[1 - (\tfrac{1}{2})^n\right]$$

$$= 2 - (n+2)(\tfrac{1}{2})^n$$

It also follows that as n tends to infinity the sum approaches 2. (See also Problems 3.117 and 3.118.)

9.97 Apply the Abel transformation to $\Sigma_1^n i^2/2^i$.

$\blacksquare$ Take $u_i = i^2$ and $v_i = 1/2^i$. Then $\Delta u_i = 2i + 1$ while $\Sigma_1^n v_i$ is as in the preceding problem $1 - (\frac{1}{2})^n$. Substituting

$$\sum_1^n i^2/2^i = (n+1)^2\left[1 - (\tfrac{1}{2})^n\right] - \sum_1^n (2i+1)\left[1 - (\tfrac{1}{2})^i\right]$$

$$= (n+1)^2\left[1 - (\tfrac{1}{2})^n\right] - 2\sum_1^n i - n + 2\sum_1^n i/2^i + \sum_1^n (\tfrac{1}{2})^i$$

and using the result of the preceding problem

$$= (n+1)^2\left[1 - (\tfrac{1}{2})^n\right] - n(n+1) - n + 2\left[2 - (n+2)(\tfrac{1}{2})^n\right]$$

$$+ \left[1 - (\tfrac{1}{2})^n\right]$$

which pencil and paper will soon simplify to

$$\sum_1^n i^2/2^i = 6 - (\tfrac{1}{2})^n (n^2 + 4n + 6)$$

As n tends to infinity the sum approaches 6. (See also Problems 3.119 and 3.120.)

9.98 Use Abel's transformation to compute $\Sigma_1^n i^2$.

$\blacksquare$ For this much computed sum we now let $u_i = i$ and $v_i = i$ to find $\Delta u_i = 1$ and $\Sigma_1^n v_i = n(n+1)/2$. Substituting

$$\sum_1^n i^2 = (n+1) \cdot n(n+1)/2 - \sum_1^n i(i+1)/2$$

$$= n(n+1)^2/2 - \tfrac{1}{2}\sum_1^n i^2 - \tfrac{1}{2}\sum_1^n i$$

Combining the two appearances of our "unknown" sum

$$\tfrac{3}{2}\sum_1^n i^2 = n(n+1)^2/2 - n(n+1)/4$$

and eventually

$$\sum_1^n i^2 = n(n+1)(2n+1)/6$$

as anticipated.

9.99   Also compute the sum $\sum_1^n i^3$ by Abel's transformation.

▌ Let $u_i = i^2$ and $v_i = i$. Then $\Delta u_i = 2i + 1$ and $\sum_1^n v_i = n(n+1)/2$. Substituting

$$\sum_1^n i^3 = (n+1)^2 n(n+1)/2 - \sum_1^n (2i+1)i(i+1)/2$$

$$= n(n+1)^3/2 - \sum_1^n i^3 - \tfrac{3}{2}\sum_1^n i^2 - \tfrac{1}{2}\sum_1^n i$$

and combining the two appearances of the target sum

$$2\sum_1^n i^3 = n(n+1)^3/2 - \tfrac{3}{2}n(n+1)(2n+1)/6 - n(n+1)/4$$

which reduces to another familiar result,

$$\sum_1^n i^3 = n^2(n+1)^2/4$$

9.10   ASYMPTOTIC SERIES

9.100   Define an asymptotic series.

▌ Let $S_n(x) = \sum_{i=0}^n a_i x^i$. If for $x \to 0$, $\lim[f(x) - S_n(x)]/x^n = 0$ for any fixed positive integer n, then $f(x)$ is said to be asymptotic to $\sum_{i=0}^\infty a_i x^i$ at zero. This is represented by the symbol

$$f(x) \simeq \sum_{i=0}^\infty a_i x^i$$

With x replaced by $x - x_0$ the same definition applies, the series being asymptotic to $f(x)$ at x_0.
Perhaps the most useful case of all is the asymptotic expansion at infinity. If for $x \to \infty$,

$$\lim x^n[f(x) - S_n(x)] = 0$$

where now $S_n(x) = \sum_{i=0}^n a_i/x^i$, then $f(x)$ has an asymptotic series at infinity and we write

$$f(x) = \sum_{i=1}^\infty \frac{a_i}{x^i}$$

The idea can be further generalized. If, for example,

$$\frac{f(x) - g(x)}{h(x)} \simeq \sum_{i=0}^\infty \frac{a_i}{x^i}$$

then we also say that $f(x)$ has the following asymptotic representation:

$$f(x) \simeq g(x) + h(x) \sum_{i=1}^\infty \frac{a_i}{x^i}$$

Note that none of these series is assumed to converge.

9.101   Obtain an asymptotic series for $\int_x^\infty (e^{-t}/t)\,dt$.

▌ Successive integrations by parts bring

$$f(x) = \int_z^\infty \frac{e^{-t}}{t}\,dt = \frac{e^{-x}}{x} - \int_x^\infty \frac{e^{-t}}{t^2}\,dt = \frac{e^{-x}}{x} - \frac{e^{-x}}{x^2} + 2!\int_x^\infty \frac{e^{-t}}{t^3}\,dt$$

and so on. Ultimately one finds

$$f(x) = \int_x^\infty \frac{e^{-t}}{t}\, dt = e^{-x}\left[\frac{1}{x} - \frac{1}{x^2} + \frac{2!}{x^3} - \frac{3!}{x^4} + \cdots + (-1)^{n+1}\frac{(n-1)!}{x^n}\right] + R_n$$

where $R_n = (-1)^n n! \int_x^\infty (e^{-t}/t^{n+1})\, dt$. Since $|R_n| < n!\, e^{-x}/x^{n+1}$, we have

$$\left| x^n\left[e^x f(x) - \sum_{i=1}^n \frac{(-1)^{i+1}(i-1)!}{x^i}\right]\right| < \frac{n!}{x}$$

so that as $x \to \infty$ this does have limit 10. This makes $e^x f(x)$ asymptotic to the series and by our generalized definition

$$f(x) \simeq e^{-x}\left(\frac{1}{x} - \frac{1}{x^2} + \frac{2!}{x^3} - \frac{3!}{x^4} + \cdots\right)$$

Notice that the series diverges for every value of x.

9.102   Show that the truncation error involved in using the series of the preceding problem does not exceed the first omitted term.

❚   The truncation error is precisely R_n. The first omitted term is $(-1)^{n+2}e^{-x}n!/x^{n+1}$ which is identical with the estimate of R_n occurring in Problem 9.101.

9.103   Use the asymptotic series of Problem 9.101 to compute $f(5)$.

❚   We find

$$e^5 f(5) \simeq .2 - .04 + .016 - .0096 + .00746 - .00746 + \cdots$$

after which terms increase. Since the error does not exceed the first term we omit, only four terms need be used, with the result

$$f(5) \simeq e^{-5}(.166) \simeq .00112$$

with the last digit doubtful. The point is, the series cannot produce $f(5)$ more accurately than this. For larger x amounts the accuracy attainable improves substantially, but is still limited.

9.104   Use the series of Problem 9.101 to compute $f(10)$.

❚   We find, carrying six places,

$$e^{10} f(10) \simeq .1 - .01 + .002 - .0006 + .00024 - .000120 + .000072$$
$$- .000050 + .000040 - .000036 + .000036 - \cdots$$

after which the terms increase. Summing the first nine terms, we have

$$f(10) \simeq e^{-10}(.091582) \simeq .0000041579$$

with the last digit doubtful. In the previous problem two digit accuracy was attainable. Here we have managed four. The essential idea of asymptotic series is that for increasing x arguments the error tends to zero.

9.105   Prove that the Stirling series is asymptotic.

❚   With n playing the role of x and the logarithm the role of $f(x)$ (see Problem 9.92), we must show that

$$\lim n^{2k-1}E_n = \lim n^{2k-1}\int_n^\infty \frac{F_{2k+1}(t)}{(2k+1)t^{2k+1}}\, dt = 0$$

Since $F_{2k+1}(t)$ repeats, with period 1, the behavior of $B_{2k+1}(t)$ in the interval $(0,1)$, it is bounded, say $|F| < M$. Then

$$|n^{2k-1}E_n| < \frac{n^{2k-1}M}{2k(2k+1)n^{2k}}$$

and with increasing n this becomes arbitrarily small.

9.106 Find an asymptotic series for $\int_x^\infty e^{-t^2/2}\,dt$.

∎ The method of successive integrations by parts is again successful. First

$$\int_x^\infty e^{-t^2/2}\,dt = \int_x^\infty -\frac{1}{t}\left(-te^{-t^2/2}\right)\,dt = \frac{1}{x}e^{-x^2/2} - \int_x^\infty \frac{1}{t^2}e^{-t^2/2}\,dt$$

and continuing in this way we find

$$\int_x^\infty e^{-t^2/2}\,dt = e^{-x^2/2}\left[\frac{1}{x} - \frac{1}{x^3} + \frac{1\cdot 3}{x^5} - \cdots + (-1)^{n-1}\frac{1\cdot 3\cdots(2n-3)}{x^{2n-1}}\right] + R_n$$

where $R_n = 1\cdot 3\cdot 5\cdots(2n-1)\int_x^\infty e^{-t^2/2}(1/t^{2n})\,dt$. The remainder can be rewritten as

$$R_n = \frac{1\cdot 3\cdot 5\cdots(2n-1)}{x^{2n+1}}e^{-x^2/2} - R_{n+1}$$

Since both remainders are positive, it follows that

$$R_n < \frac{1\cdot 3\cdot 5\cdots(2n-1)}{x^{2n+1}}e^{-x^2/2}$$

This achieves a double purpose. It shows that the truncation error does not exceed the first omitted term, and since it also makes $\lim e^{x^2/2}x^{2n-1}R_n = 0$, it proves the series asymptotic:

$$\int_x^\infty e^{-t^2/2}\,dt \simeq e^{-x^2/2}\left(\frac{1}{x} - \frac{1}{x^3} + \frac{1\cdot 3}{x^5} - \frac{1\cdot 3\cdot 5}{x^7} + \cdots\right)$$

9.107 Compute $\sqrt{2/\pi}\int_4^\infty e^{-t^2/2}\,dt$ by the series of Problem 9.106.

∎ With $x = 4$ we find

$$\sqrt{\frac{2}{\pi}}\,e^{-8}[.25 - .015625 + .002930 - .000916 + .000401 - .000226$$

$$+ .000155 - .000126 + .000118 - .000125 + \cdots]$$

to the point where terms begin to increase. The result of stopping before the smallest term is

$$\sqrt{\frac{2}{\pi}}\int_4^\infty e^{-t^2/2}\,dt \simeq .0000633266$$

with the 2 digit in doubt.

9.108 Find an asymptotic series for the sine integral.

∎ Once again integration by parts proves useful. First

$$\mathrm{Si}(x) = \int_x^\infty \frac{\sin t}{t}\,dt = \frac{\cos x}{x} - \int_x^\infty \frac{\cos t}{t^2}\,dt$$

after which similar steps generate the series

$$\int_x^\infty \frac{\sin t}{t}\,dt \simeq \frac{\cos x}{x} + \frac{\sin x}{x^2} - \frac{2!\cos x}{x^3} - \frac{3!\sin x}{x^4} + \cdots$$

which can be proved asymptotic as in previous problems.

9.109 Compute Si(10).

▮ Putting $x = 10$ in the previous problem,

$$\text{Si}(10) \simeq -.083908 - .005440 + .001678 + .000326 - .000201$$
$$- .000065 + .000060 + .000027 - .000034 - .000019$$

after which both the cosine and sine terms start to grow larger. The total of these 10 terms rounds to $-.0876$, which is correct to four places.

9.110 Derive the asymptotic series

$$\int_x^\infty \frac{\cos t}{t}\, dt \simeq \sin x \left[-\frac{1}{x} + \frac{2!}{x^3} - \frac{4!}{x^5} + \cdots \right] + \cos x \left[\frac{1}{x^2} - \frac{3!}{x^4} + \frac{5!}{x^6} - \cdots \right]$$

and use it when $x = 10$, obtaining as much accuracy as you can.

▮ Integrating by parts with $u = 1/t$ and $dv = \cos t\, dt$,

$$\int_x^\infty \frac{\cos t}{t}\, dt = \left. \frac{\sin t}{t} \right|_x^\infty + \int_x^\infty \frac{\sin t}{t^2}\, dt$$

with the integrated term reducing to $-\sin x/x$. For the remaining integral let $u = 1/t^2$ and $dv = \sin t\, dt$, leading to

$$\left. \frac{-\cos t}{t^2} \right|_x^\infty - 2\int_x^\infty \frac{\cos t}{t^2}\, dt$$

with the integrated term $\cos x/x^2$. Continuing in this way the indicated series is generated.

9.111 Apply the series of the preceding problem with $x = 10$.

▮ Here are the first six terms of the sine series and the first five of the cosine series:

$$\sin 10(-.1 + .002 - .00016 + .000072 - .000040 + .000036)$$
$$\cos 10(.01 - .0006 + .000120 - .000050 + .000036)$$

The next term of the sine series would include an extra factor of $11 \cdot 12/100$ and so would be greater, with continued increase beyond this point. The last listed term is, therefore, the smallest. In the cosine series the next term would include an extra factor of $10 \cdot 11/100$, so here too the last listed term is the smallest. For greatest accuracy we stop just short of these terms and have

$$\sin 10(-.09813) + \cos 10(.00947) = .04544$$

with the fifth place in doubt. The correct value to five places is $.04546$.

9.112 Derive the asymptotic series

$$\int_x^\infty \sin t^2\, dt \simeq \cos x^2 \left(\frac{1}{2x} - \frac{3}{2^3 x^5} + \frac{3 \cdot 5 \cdot 7}{2^5 x^9} - \cdots \right) + \sin x^2 \left(\frac{1}{2^2 x^3} - \frac{3 \cdot 5}{2^4 x^7} + \frac{3 \cdot 5 \cdot 7 \cdot 9}{2^6 x^{11}} - \cdots \right)$$

▮ With $u = 1/2t$ and $dv = \sin(t^2)2t\, dt$, an integration by parts produces

$$\int_x^\infty \frac{\sin(t^2)2t}{2t}\, dt = \left. \frac{-\cos(t^2)}{2t} \right|_x^\infty - \int_x^\infty \frac{\cos(t^2)}{2t^2}\, dt$$

with the integrated term reducing to $(\cos(x^2))/2x$. For the remaining integral let $u = 1/4t^3$ and $dv = \cos(t^2)2t\, dt$ leading to

$$\left. \frac{-\sin(t^2)}{4t^3} \right|_x^\infty + \int_x^\infty \sin(t^2) \frac{-3}{4t^4}\, dt$$

with the integrated term $(\sin(x^2))/4x^3$. Continuing in this way the indicated series will be generated.

9.113   Apply the asymptotic series of the preceding problem with $x = 10$.

▌ Three terms of the cosine series and two of the sine series are

$$\cos 100\left(\tfrac{1}{2}\cdot 10^{-1} - \tfrac{3}{8}\cdot 10^{-5} + \tfrac{105}{32}\cdot 10^{-9} - \cdots\right)$$

$$\sin 100\left(\tfrac{1}{4}\cdot 10^{-3} - \tfrac{15}{16}\cdot 10^{-7} + \cdots\right)$$

The terms continue to diminish beyond this point, but the last ones listed are already of the order 10^{-7} or less. Stopping just short of these, taking only three terms, we have

$$.8623136(.04999625) + (-.5063747)(.00025) = .042986$$

to six places.

9.114   Also apply the series of Problem 9.112 with $x = 2$.

▌ This will be a rather long reach for an asymptotic series. Here are some of the numbers:

$$\cos 4\left(1/2^2 - 3/2^8 + 3\cdot 35/2^{14} - 3\cdot 35\cdot 99/2^{20} + \cdots\right)$$

$$\sin 4\left(1/2^5 - 15/2^{11} + 15\cdot 63/2^{17} - 15\cdot 63\cdot 143/2^{23} + \cdots\right)$$

In each series it will be noted that the last term listed is larger than the one before it. The third term is the smallest, so we should sum only the first two. From these four terms then we find the integral to be $-.174$, with doubt even of the second decimal place since this is the order of the first omitted term in the sine series.

The integral $\int_0^\infty \sin(t^2)\, dt$ is known to have the value $\sqrt{\pi/8}$, or $.626$ to three places. Suppose we apply a direct power series method to the integral over $(0, 2)$, with the intention of combining the result with our $-.174$ for the interval $(2, \infty)$ and the hope that the sum will agree with $.626$ to one or two digits:

$$\int_0^2 \sin(t^2)\, dt = \int_0^2 \left(t^2 - t^6/3! + t^{10}/5! - t^{14}/7! + \cdots\right) dt$$

$$= \frac{2}{3\cdot 1!} - \frac{2}{7\cdot 3!} + \frac{2}{11\cdot 5!} - \frac{2}{15\cdot 7!}$$

$$= .80478$$

to five places. The integral over $(0, \infty)$ is then, according to our two efforts, $.805 - .174 = .631$. The error of $.005$ must be assigned to the asymptotic series, which was not intended for use with small values of x.

9.115   How accurately does the Stirling series produce $2!$ and at what point do the terms of the series being to increase?

▌ As in the preceding problem, we do not expect an asymptotic series to perform well for such a small argument. The first nine terms are

$$\frac{1}{24} - \frac{1}{360\cdot 8} + \frac{1}{42\cdot 30\cdot 32} - \frac{1}{30\cdot 56\cdot 128} + \frac{5}{66\cdot 90\cdot 512} - \frac{691}{2730\cdot 132\cdot 2048}$$

$$+ \frac{7}{6\cdot 182\cdot 8192} - \frac{3617}{510\cdot 240\cdot 32768} + \frac{43867}{798\cdot 306\cdot 131072}$$

with the last four amounting to 94, 78, 90 and 137 in the eighth decimal place. The smallest is, therefore, the seventh, and its value raises a slight doubt about the sixth decimal place. From the first six terms we find a sum of $.041352$. So

$$\log 2! = .041352 - 2 + \log\sqrt{2\pi} + 2.5\log 2 = .693159$$

the first three digits of which are quite reassuring, and the true final three being 147. We have almost five place accuracy. From this logarithm the factorial is estimated as

$$2! = 2.000023$$

and things have not gone so badly after all.

9.116   The Euler–Maclaurin formula can be used to show that

$$C = \frac{1}{2}\log 2 + \frac{1}{4} + \sum_{i=1}^{k} \frac{(-1)^{i+1}b_i}{(2i)(2i-1)}\left(\frac{2i+1}{2^{2i}} - 1\right) + \text{error term}$$

where C is Euler's constant. At what point do the terms of this asymptotic series begin to increase and how well does the series permit C to be computed?

▌ The first two terms manage .59657 after which, inserting the Bernoulli numbers, the terms

$$-.02083 + .00191 - .00071 + .00057 - .00083 + .00191 - .00640$$

follow and the turning point has clearly been passed. Using the first three we have $C = .57694$ while the correct value to five places is still .57722.

9.11   ADDITIONAL PROBLEMS

9.117   Euler's constant has been defined by

$$C = \lim\left(1 + \frac{1}{2} + \frac{1}{3} + \cdots + \frac{1}{n} - \log n\right)$$

for n tending to infinity. Assuming that

$$1 + \frac{1}{3} + \frac{1}{5} + \cdots + \frac{1}{2n-1} - A\log n$$

approaches a limit for a properly chosen value of A, determine A and the limit.

▌ The sum in question can be rewritten as

$$\left(1 + \frac{1}{2} + \frac{1}{3} + \frac{1}{4} + \cdots + \frac{1}{2n} - \log 2n\right) - \frac{1}{2}\left(1 + \frac{1}{2} + \cdots + \frac{1}{n} - \log n\right)$$

$$+ \log 2n - \frac{1}{2}\log n - A\log n$$

by adding in and subtracting out the even reciprocals. If A is chosen to be $\frac{1}{2}$ the three logarithm terms at the end combine to $\log 2n - \log n = \log 2$. Letting n go to infinity and appealing to the definition of Euler's constant then brings

$$C - \tfrac{1}{2}C + \log 2 = \tfrac{1}{2}C + \log 2$$

as the limit.

9.118   Determine the asymptotic behavior of this sum.

$$1 + \frac{1}{2} + \frac{1}{4} + \frac{1}{5} + \frac{1}{7} + \frac{1}{8} + \cdots + \frac{1}{3n-2} + \frac{1}{3n-1} - A\log n$$

▌ Since it is the reciprocals of the form $1/3i$ that have been omitted, the sum can be rewritten as

$$\left(1 + \frac{1}{2} + \frac{1}{3} + \cdots + \frac{1}{3n} - \log 3n\right) - \frac{1}{3}\left(1 + \frac{1}{2} + \cdots + \frac{1}{n} - \log n\right)$$

$$+ \log 3n - \frac{1}{3}\log n - A\log n$$

in which the choice $A = \frac{2}{3}$ reduces the combination of logarithms at the end to $\log 3$. Letting n go to infinity we find

$$C - \tfrac{1}{3}C + \log 3 = \tfrac{2}{3}C + \log 3$$

as the limit.

9.119 Show that Euler's transformation does accelerate the convergence of the series $\sum_0^\infty (-\frac{1}{2})^i$.

■ Recalling that $\Delta A^i = (A-1)A^i$ so that

$$\Delta(\tfrac{1}{2})^i = (-\tfrac{1}{2})(\tfrac{1}{2})^i \qquad \Delta^2(\tfrac{1}{2})^i = (\tfrac{1}{4})(\tfrac{1}{2})^i \qquad \Delta^3(\tfrac{1}{2})^i = (-\tfrac{1}{8})(\tfrac{1}{2})^i$$

and so on, it follows that at $i = 0$, $\Delta^k(\tfrac{1}{2})^i = (-\tfrac{1}{2})^k$. The Euler transformation

$$y_0 - y_1 + y_2 - y_3 + \cdots = \tfrac{1}{2}y_0 - \tfrac{1}{4}\Delta y_0 + \tfrac{1}{8}\Delta^2 y_0 - \cdots$$

then manages, with $y_i = (\tfrac{1}{2})^i$,

$$\sum_0^\infty (\tfrac{1}{2})^{i+1}(\tfrac{1}{2})^i = \sum_0^\infty (\tfrac{1}{2})^{2i+1}$$

since minus signs match up and leave all terms positive. We now have a choice between the original

$$1 - \tfrac{1}{2} + \tfrac{1}{4} - \tfrac{1}{8} + \cdots$$

and the transformation

$$\tfrac{1}{2} + \tfrac{1}{8} + \tfrac{1}{32} + \tfrac{1}{128} + \cdots$$

with the latter clearly getting the decision for speed.

9.120 Also apply the Euler transformation to $\sum_0^\infty (-\frac{1}{3})^i$. Which series converges faster?

■ With $y = (\tfrac{1}{3})^i$ it follows that

$$\Delta y_i = (-\tfrac{2}{3})(\tfrac{1}{3})^i \qquad \Delta^2 y_i = (-\tfrac{2}{3})^2(\tfrac{1}{3})^i \qquad \Delta^3 y_i = (-\tfrac{2}{3})^3(\tfrac{1}{3})^i$$

and so on. The Euler transformation then becomes

$$\sum_0^\infty (\tfrac{1}{2})^{i+1}(\tfrac{2}{3})^i = \tfrac{1}{2}\sum_0^\infty (\tfrac{1}{3})^i$$
$$= \tfrac{1}{2}(1 + \tfrac{1}{3} + \tfrac{1}{9} + \tfrac{1}{27} + \cdots)$$

which is to be compared with the original

$$1 - \tfrac{1}{3} + \tfrac{1}{9} - \tfrac{1}{27} + \tfrac{1}{81} - \cdots$$

One series alternates and the other does not, but in fact their rates of convergence to the limit of $\frac{3}{4}$ are about equal.

9.121 Finally apply the Euler transformation to $\sum_0^\infty (-\frac{1}{4})^i$. What may be suspected from the evidence of this and the two preceding problems?

■ This time we have $y_i = (\tfrac{1}{4})^i$ making $\Delta^k y_i = (-\tfrac{3}{4})^k(\tfrac{1}{4})^i$ and $\Delta^k y_0 = (-\tfrac{3}{4})^k$. The Euler transformation becomes

$$\sum_0^\infty (\tfrac{1}{2})^{i+1}(\tfrac{3}{4})^i = \tfrac{1}{2}\sum_0^\infty (\tfrac{3}{8})^i$$
$$= \tfrac{1}{2}(1 + \tfrac{3}{8} + \tfrac{9}{64} + \cdots)$$

and is definitely a slower converger than the original series. It requires 15 terms to achieve six place accuracy, which the original manages in 10 terms, and so on.

The evidence suggests that an alternating series that converges fairly rapidly will have an Euler transformation that converges more slowly and vice versa. The breakpoint may be near our middle example.

9.122 Prove

$$y_0 + y_1 + y_2 + \cdots + y_{n-1} = ny_0 + \frac{n(n-1)}{2!}\Delta y_0 + \frac{n(n-1)(n-2)}{3!}\Delta^2 y_0 + \cdots$$

▌ A proof using the operators E and $\Delta = E - 1$ is convenient:

$$\sum_0^{n-1} y_i = y_0 + Ey_0 + E^2 y_0 + \cdots + E^{n-1} y_0$$

$$= (1 + E + E^2 + \cdots + E^{n-1}) y_0$$

$$= \frac{E^n - 1}{E - 1} y_0 = \frac{(1 + \Delta)^n - 1}{\Delta} y_0$$

$$= \frac{1}{\Delta} \left[\frac{n}{1!} \Delta + \frac{n(n-1)}{2!} \Delta^2 + \frac{n(n-1)(n-2)}{3!} \Delta^3 + \cdots \right] y_0$$

$$= \left[n + \frac{n(n-1)}{2!} \Delta + \frac{n(n-1)(n-2)}{3!} \Delta^2 + \cdots \right] y_0$$

which is what was required.

9.123 Apply the theorem of the preceding problem to $\sum_0^{n-1} (i+1)^2$.

▌ This is once again an old favorite only slightly disguised. With $y_i = (i+1)^2$ we find $\Delta y_i = (i+2)^2 - (i+1)^2 = 2i + 3$ and $\Delta^2 y_i = 2$, with all higher differences zero. The theorem then offers this substitute for the original sum

$$n(1) + \frac{n(n-1)}{2} (3) + \frac{n(n-1)(n-2)}{6} (2)$$

the figures in parentheses being the values at $i = 0$ of y_i and its two active differences. The usual $n(n+1)(2n+1)/6$ now follows.

The theorem provides what, in principle, seems a poor exchange of a finite sum for an infinite series, but when as in this case the function y_i is a polynomial in i, the series terminates. It is in such cases that it may be useful.

9.124 Also apply the theorem of Problem 9.122 to $\sum_0^{n-1} (i+1)^3$.

▌ Here, of course, we expect to recover the familiar $n^2(n+1)^2/4$ by still another route. With $y_i = (i+1)^3$ we have

$$\Delta y_i = (i+2)^3 - (i+1)^3 = 3i^2 + 9i + 7$$
$$\Delta^2 y_i = 6i + 12$$
$$\Delta^3 y_i = 6$$

with all higher derivatives zero. The given sum is then equal to

$$n + \frac{n(n-1)}{2} (7) + \frac{n(n-1)(n-2)}{6} (12) + \frac{n(n-1)(n-2)(n-3)}{24} (6)$$

which is acceptable in its present form but can be reduced to the old standard if so desired.

9.125 If $V(x) = \sum_0^\infty v_i x^i$ is a known series, show that a "companion" series $\sum_0^\infty u_i v_i x^i$ can be represented in the form

$$\sum_0^\infty u_i v_i x^i = V(x) u_0 + x V'(x) \Delta u_0 + \frac{1}{2!} x^2 V''(x) \Delta^2 u_0 + \cdots$$

▌ An operator proof is again convenient:

$$\sum_0^\infty u_i v_i x^i = v_0 u_0 + v_1 x E u_0 + v_2 x^2 E^2 u_0 + v_3 x^3 E^3 u_0 + \cdots$$

$$= \left[v_0 + v_1 x E + v_2 x^2 E^2 + v_3 x^3 E^3 + \cdots \right] u_0$$

$$= V(xE) u_0 = V(x + x\Delta) u_0$$

$$= \left[V(x) + V'(x)(x\Delta) + \frac{1}{2!} V''(x)(x\Delta)^2 + \cdots \right] u_0$$

which is equivalent to the stated result. As with the theorem of Problem 9.122, this one is most useful when u_i is a polynomial.

9.126 Prove

$$\sum_{i=0}^{\infty} u_i x^i = \frac{u_0}{1-x} + \frac{x\,\Delta u_0}{(1-x)^2} + \frac{x^2\,\Delta^2 u_0}{(1-x)^3} + \cdots$$

▮ If $v_i = 1$ in the theorem of the preceding problem, then the $V(x)$ series is geometric and equals $1/(1-x)$. This makes

$$V'(x) = 1/(1-x)^2 \qquad V''(x) = 2/(1-x)^3,\ldots$$

and so on, leading to

$$\sum_0^{\infty} u_i x^i = V(x)u_0 + xV'(x)\,\Delta u_0 + \frac{1}{2!}x^2 V''(x)\,\Delta^2 u_0 + \cdots$$

$$= \frac{1}{1-x}u_0 + \frac{x}{(1-x)^2}\Delta u_0 + \frac{x^2}{(1-x)3}\Delta^2 u_0 + \cdots$$

9.127 Sum the series $\sum_0^{\infty} i^2 x^i$.

▮ With $u_i = i^2$ the theorem of the preceding problem can be put to use. The differences $\Delta u_i = 2i + 1$ and $\Delta^2 u_i = 2$ are needed, in particular their values for $i = 0$. Substituting,

$$\sum_0^{\infty} i^2 x^i = 0/(1-x) + x/(1-x)^2 + 2x^2/(1-x)^3 = x(1+x)/(1-x)^3$$

9.128 Also sum the series $\sum_0^{\infty} i^3 x^i$.

▮ With $u_i = i^3$ we have $\Delta u_i = 3i^2 + 3i + 1$, $\Delta^2 u_i = 6i + 6$ and $\Delta^3 u_i = 6$. Substituting,

$$\sum_0^{\infty} i^3 x^i = x(1)/(1-x)^2 + x^2(6)/(1-x)^3 + x^3(6)/(1-x)^4$$

the items in parentheses being the difference values. A little algebra reduces this to $x(x^2 + 4x + 1)/(1-x)^4$.

9.129 Sum the series $\sum_0^{\infty} (i+1)(i+2)x^i$.

▮ With $u_i = (i+1)(i+2)$, which is the factorial polynomial $(i+2)^{(2)}$, it follows that $\Delta u_i = 2(i+2)$ and that $\Delta^2 u_i = 2$. This makes the given series equal to

$$\sum_{i=0}^{\infty} (i+1)(i+2)x^i = \frac{2}{1-x} + \frac{4x}{(1-x)^2} + \frac{2x^2}{(1-x)^3}$$

$$= \frac{2(1-x)^2 + 4x(1-x) + 2x^2}{(1-x)^3}$$

$$= \frac{2}{(1-x)^3}$$

9.130 Compute the two series

$$1\cdot 2 - 2\cdot 3/10 + 3\cdot 4/10^2 - 4\cdot 5/10^3 + \cdots$$

$$1\cdot 2 - 2\cdot 3/2 + 3\cdot 4/2^2 - 4\cdot 5/2^3 + \cdots$$

▮ The first converges at good speed, only eight terms being needed to obtain the six place value of 1.50263, the exact value being, according to the preceding problem with $x = -1/10$, $2(10/11)^3$. With the second, however, things move along more slowly and the direct computation is more tedious. The exact value is, putting $x = -1/2$ in the preceding problem, $2(2/3)^3$ or $16/27$.

9.131   Adapt the theorem of Problem 9.125 to the special case

$$\sum_0^\infty u_i x^i / i!$$

▮ The function v_i is to be identified with $1/i!$ which makes

$$V(x) = \sum_0^\infty x^i / i! = e^x$$

and all the derivatives of $V(x)$ also e^x. Substituting in the general result of the earlier problem then yields

$$\sum_{i=0}^\infty \frac{u_i x^i}{i!} = e^x \left[u_0 + \frac{x\,\Delta u_0}{1!} + \frac{x^2\,\Delta^2 u_0}{2!} + \cdots \right]$$

9.132   Apply the preceding problem to the series $\sum_0^\infty i x^i / i!$.

▮ This is hardly a challenge, the result being available by a simple shifting of the index of summation. However, with $u_i = i$ and $\Delta u_i = 1$ substitutions yield a single term,

$$\sum_0^\infty i x^i / i! = x e^x$$

9.133   Evaluate $\sum_0^\infty i^2 x^i / i!$.

▮ The course is clear. Let $u_i = i^2$, making $\Delta u_i = 2i + 1$ and $\Delta^2 u_i = 2$. Substituting,

$$\sum_0^\infty i^2 x^i / i! = e^x (x + x^2)$$

which can, of course, be verified in several ways.

9.134   Evaluate $\sum_0^\infty P_2(i) x^i / i!$ where $P_2(x) = \frac{3}{2}x^2 - \frac{1}{2}$ is the Legendre polynomial of second degree.

▮ Problem 9.131 can be applied to each term of the polynomial and the outputs combined:

$$\sum_0^\infty P_2(i) x^i / i! = e^x \left(\tfrac{3}{2}x^2 + \tfrac{3}{2}x - \tfrac{1}{2} \right) = e^x \left(P_2(x) + \tfrac{3}{2}x \right)$$

9.135   Evaluate $\sum_0^\infty i^3 x^i / i!$.

▮ Let $u_i = i^3$. Then $\Delta u_i = 3i^2 + 3i + 1$ and $\Delta^2 u_i = 6i + 6$ with $\Delta^3 u_i$ the constant 6. By Problem 9.131,

$$\sum_0^\infty i^3 x^i / i! = e^x (x + 3x^2 + x^3)$$

9.136   Evaluate $\sum_0^\infty P_3(i) x^i / i!$ where $P_3(x) = \frac{5}{2}x^3 - \frac{3}{2}x$ is the Legendre polynomial of third degree.

▮ Combining materials from a few earlier problems,

$$\sum_0^\infty P_3(i) x^i / i! = e^x \left(\tfrac{5}{2}x^3 + \tfrac{15}{2}x^2 + \tfrac{5}{2}x - \tfrac{3}{2}x \right)$$

$$= e^x \left[P_3(x) + \tfrac{5}{2}x(3x + 1) \right]$$

9.137   Show that

$$u_0 + \sum_1^\infty u_i x^i / i = u_0 - u_0 \log(1 - x) + \sum_1^\infty x^i \, \Delta^i u_0 / i(1 - x)^i$$

∎ This is another special case of the result in Problem 9.125. Let $v_0 = 0$ and $v_i = 1/i$ for $i > 0$. Then

$$V(x) = \sum_1^\infty x^i/i = -\log(1-x) \qquad V'(x) = 1/(1-x) \qquad V''(x) = 1/(1-x)^2$$

and in general $V^{(i)}(x) = (i-1)!/(1-x)^i$. Now consider the series

$$u_0 v_0 + u_1 x + u_2 x^2/2 + u_3 x^3/3 + \cdots = \sum_1^\infty u_i v_i x^i$$

which, because of the choice $v_0 = 0$, is less than the original by the amount u_0. By Problem 9.125 this new series equals

$$V(x) u_0 + x V'(x) \Delta u_0 + \frac{1}{2!} x^2 V''(x) \Delta^2 u + \cdots$$

$$= -u_0 \log(1-x) + x \Delta u_0/(1-x) + x^2 \Delta^2 u_0/2(1-x)^2 + \cdots$$

$$= -u_0 \log(1-x) + \sum_1^\infty x^i \Delta^i u_0/i(1-x)^i$$

The original series is then u_0 greater.

9.138 Apply the theorem of the preceding problem with $u_i = Ai + B$.

∎ The series could easily be split into two familiar parts, but following the course set we find $\Delta u_i = A$ and the series equal to

$$B - B\log(1-x) + Ax/(1-x)$$

9.139 Apply the same theorem with $u_i = i^2$.

∎ It will be noted that if u_i is any polynomial with i as a factor, then the denominator i is canceled and the series can be handled as in Problem 9.126. However, to proceed, the present method offers

$$\sum_1^\infty i^2 x^i/i = \sum_1^\infty i x^i = \frac{x}{1-x} + \frac{x^2}{(1-x)^2} = \frac{x}{(1-x)^2}$$

which can be verified in many ways.

9.140 Apply the theorem of Problem 9.137 with $u_i = (i+1)(i+2)$.

∎ The results of the two preceding problems could be combined, using $A = 3$ and $B = 2$. Proceeding directly, however, we have

$$u_i = 2i + 4 \quad \text{and} \quad \Delta u_i = 2$$

so that

$$2 + \frac{2 \cdot 3x}{1} + \frac{3 \cdot 4x^2}{2} + \frac{4 \cdot 5x^3}{3} + \cdots = 2 - 2\log(1-x) + \frac{4x}{1-x} + \frac{x^2}{(1-x)^2}$$

When $x = \frac{1}{2}$ this becomes

$$2 + 3 + \tfrac{3}{2} + \tfrac{5}{6} + \tfrac{15}{32} + \tfrac{21}{80} + \cdots = 7 + 2\log(2)$$

and while the early terms are in no hurry to diminish, the powers of 2 will eventually have their way, producing the correct figure of 8.386294.

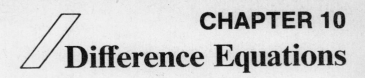

10.1 What is a difference equation and what is meant by the order of such an equation?

▌ The term difference equation might be expected to refer to an equation involving differences. However, an example such as

$$\Delta^2 y_k + 2\,\Delta\,y_k + y_k = 0$$

which quickly collapses to $y_{k+2} = 0$, shows that combinations of differences are not always informative. Instead the term difference equation is used for a relation between the values y_k of a function defined on a discrete set of arguments x_k. Assuming these arguments equally spaced, the argument change $x_k = x_0 + kh$ introduces the integer argument k. An example of a difference equation is

$$y_{k+1} = a_k y_k + b_k$$

in which a_k and b_k are given functions of the integer argument k. This particular example could be rewritten as

$$\Delta y_k = (a_k - 1) y_k + b_k$$

but this is not normally found to be useful.

The order of a difference equation is the difference between the largest and smallest arguments k appearing in it. The first example is of order 1.

10.2 What is a solution of a difference equation?

▌ It is a function y_k for which the equation is true over some range of values of k. The nature of a difference equation allows solution sequences to be found recursively. In the preceding example y_{k+1} can be computed directly if y_k is known. One known value triggers a sequence of values. It is also useful to have a closed formula for the function y_k, and the search for such representations is a large part of the subject.

10.3 In what ways are difference equations analogous to differential equations?

▌ There is a strong analogy. For example, a first-order equation normally has a unique solution satisfying an initial condition such as $y_0 = A$, and a second-order equation normally has exactly one solution satisfying the two initial conditions $y_0 = A$ and $y_1 = B$. Boundary conditions such as $y_0 = A$ and $y_N = B$ lead to second-order boundary value and eigenvalue problems.

Procedures for finding solutions in closed form are similar in the two subjects. First order difference equations can be solved in terms of sums, as the corresponding differential equations are solved in terms of integrals. The theories and solution methods for linear equations are analogous, the case of constant coefficients leading to very similar characteristic equations. A more detailed view of the analogy will appear as we proceed.

10.4 What applications do difference equations have?

▌ There are many social, economic, physical and other problems that have been successfully modeled using difference equations. Also, numerous methods for solving differential equations involve replacing them by difference equation substitutes.

10.1 FIRST-ORDER EQUATIONS

10.5 Compute the first several values of the solution of the equation $y_{k+1} = ky_k + k$, given the initial condition $y_0 = 1$.

▌ It is only necessary to do the additions and subtractions:

$$y_1 = 0 \qquad y_2 = 0 + 1 = 1 \qquad y_3 = 2 + 4 = 6 \qquad y_4 = 18 + 9 = 27$$

with relatively easy continuation. The computational simplicity of difference equations is evident.

10.6 What sort of function satisfies $y_{k+1} = ky_k$ with initial value $y_1 = 1$?

▌ Direct computation finds the next few values to be $1, 2, 6, 24, \cdots$ and the solution function $y_k = (k-1)!$ is suspected, a suspicion that can be confirmed by substitution into the difference equation.

10.7 Show that the gamma function $\Gamma(k) = \int_0^\infty e^{-t} t^{k-1}\, dt$ also satisfies the difference equation of the preceding problem and the initial condition.

▌ One integration by parts achieves the first goal,

$$\Gamma(k+1) = \int_0^\infty e^{-t} t^k\, dt = -t^k e^{-t}\Big|_0^\infty + \int_0^\infty e^{-t} k t^{k-1}\, dt = k\Gamma(k)$$

the integrated part vanishing at both limits. The initial value proves to be

$$\Gamma(1) = \int_0^\infty e^{-t}\, dt = 1$$

as required. Since it is clear that the solution of the difference equation is uniquely determined by the initial value, it may be deduced that

$$\Gamma(k) = (k-1)!$$

for k an integer greater than or equal to 1.

The proof just given is actually valid for any $k > 0$, and as a result the difference equation can be used to extend the gamma function to negative noninteger values, in the form

$$\Gamma(k) = \Gamma(k+1)/k$$

A graph of this function appears as Fig. 10-1.

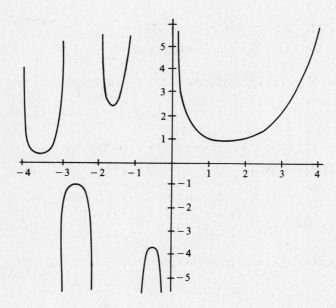

Fig. 10-1

10.8 Solve the equation $y_{k+1} = Ry_k$.

▌ Preliminary computations are probably not necessary to guess the family of solution functions $y_k = R^k y_0$, but the early values

$$y_0, Ry_0, R^2 y_0, \ldots$$

are convincing.

10.9 The general linear homogeneous first-order equation (terminology to be compared with that used with differential equations) is

$$y_{k+1} = a_k y_k$$

What is the solution corresponding to the initial value y_0.

▮ Direct computation finds

$$y_1 = a_0 y_0 \qquad y_2 = a_1 a_0 y_0 \qquad y_3 = a_2 a_1 a_0 y_0, \qquad \cdots$$

and makes the solution obvious:

$$y_n = (a_{n-1} a_{n-2} \cdots a_0) y_0$$

If the initial condition is set at $k = 1$ this can be modified to

$$y_n = (a_{n-1} a_{n-2} \cdots a_1) y_1$$

10.10 Show how logarithms can be introduced to solve the equation of the preceding problem.

▮ For this equation the procedure is far from necessary, but it will be found useful in other cases. We have

$$\log y_{k+1} - \log y_k = \log a_k$$

or, defining L_k to be $\log y_k$,

$$\Delta L_k = L_{k+1} - L_k = \log a_k$$

Finite integration is in order, so we sum from (say) $k = 0$ to $k = n - 1$ to obtain

$$L_n - L_0 = \sum_0^{n-1} \log a_k = \log(a_{n-1} \cdots a_0)$$

Exponentiating both sides then reproduces the earlier result

$$y_n = y_0 (a_{n-1} \cdots a_0)$$

10.11 Given the functions a_k and b_k, what is the solution of the general linear nonhomogeneous first-order difference equation

$$y_{k+1} = a_k y_k + b_k$$

with initial value y_0? (Again note the analogy with terminology used for differential equations.)

▮ Direct computation finds

$$y_1 = a_0 y_0 + b_0$$
$$y_2 = a_1 y_1 + b_1 = a_0 a_1 y_0 + a_1 b_0 + b_1$$
$$y_3 = a_2 y_2 + b_2 = a_0 a_1 a_2 y_0 + a_1 a_2 b_0 + a_2 b_1 + b_2$$

and so on. The general result is clear if not pretty:

$$y_n = a_0 \cdots a_{n-1} y_0 + a_1 \cdots a_{n-1} b_0 + a_2 \cdots a_{n-1} b_1 + \cdots + b_{n-1}$$

By introducing the product function $p_n = a_0 a_1 \cdots a_{n-1}$ things can be made a bit more compact:

$$y_n = p_n \left(A + \frac{b_0}{p_1} + \frac{b_1}{p_2} + \cdots + \frac{b_{n-1}}{p_n} \right)$$

Having the solution in the form of a sum is comparable to having the solution of a differential equation in the form of an integral. It is not entirely satisfactory. In special cases, as has already been seen, still more compact representations can be found.

10.12   Solve $y_{k+1} = Ry_k + 1$ with $y_0 = 1$.

▮ With the coefficient a_k the constant R, we can appeal to the general formula of Problem 10.11, which shows

$$y_n = R^n + R^{n-1} + \cdots + 1 = (R^{n+1} - 1)/(R - 1)$$

10.13   Also apply Problem 10.11 to $y_{k+1} = xy_k + c_{k+1}$ with $y_0 = c_0$.

▮ Here the coefficient a_k is the constant x. Substituting,

$$y_n = c_0 x^n + c_1 x^{n-1} + \cdots + c_n$$

The solution is a polynomial. Horner's method for evaluating this polynomial at argument x involves computing $y_1, y_2, \ldots, y_n$ successively and is equivalent to rearranging the polynomial into

$$y_n = c_n + x(c_{n-1} + \cdots + x(c_3 + x(c_2 + x(c_1 + xc_0))))$$

It is more efficient than building up the powers of x one by one and then computing the standard polynomial form.

10.14   What is the character of the solution of $y_{k+1} = ((k+1)/x)y_k + 1$ with initial value $y_0 = 1$?

▮ Here the p_n of Problem 10.11 becomes $p_n = n!/x^n$, while all $b_k = 1$. The solution is therefore expressible as

$$\frac{y_n}{p_n} = \frac{x^n y_n}{n!} = 1 + x + \frac{1}{2}x^2 + \cdots + \frac{1}{n!}x^n$$

so that for increasing n, $\lim x^n y_n/n! = e^x$.

10.15   What is the character of the solution of $y_{k+1} = [1 - x^2/(k+1)^2]y_k$ with $y_0 = 1$?

▮ Here all the b_k of Problem 10.11 are zero and $A = 1$, making

$$y_n = p_n = (1 - x^2)\left(1 - \frac{x^2}{2^2}\right)\left(1 - \frac{x^2}{3^2}\right)\cdots\left(1 - \frac{x^2}{n^2}\right)$$

This product vanishes for $x = \pm 1, \pm 2, \ldots, \pm n$. For increasing n we encounter the infinite product

$$\lim y_n = \prod_{k=0}^{\infty}\left[1 - \frac{x^2}{(k+1)^2}\right]$$

which can be shown to represent $(\sin \pi x)/\pi x$.

10.16   If a debt is amortized by regular payments of size R and is subject to interest rate i, the unpaid balance is P_k where $P_{k+1} = (1 + i)P_k - R$. The initial debt being $P_0 = A$, show that $P_k = A(1 + i)^k - R[(1 + i)^k - 1]/i$. Also show that to reduce P_k to zero in exactly n payments ($P_n = 0$), we must take $R = Ai/[1 - (1 + i)^{-n}]$.

▮ Substituting,

$$P_{k+1} = A(1 + i)^{k+1} - (R/i)\left[(1 + i)^{k+1} - 1\right]$$

$$(1 + i)P_k = A(1 + i)^{k+1} - (R/i)\left[(1 + i)^{k+1} - (1 + i)\right]$$

with the difference reducing to $-R$ as it should. The initial condition is clearly satisfied. As for the terminal condition, setting $P_n = 0$ brings $A(1 + i)^n - (R/i)[(1 + i)^n - 1] = 0$ which solves for R as described.

10.17   Use a "summing factor" to solve $y_{k+1} = k/(k+1)y_k + k$.

▮ The use of integrating factors in the solution of first-order differential equations will be recalled, often leaving the result in the form of an integral. Here an analogous process leads to a solution in the form of a summation. Multiplying through by the summing factor $k + 1$,

$$(k + 1)y_{k+1} - ky_k = (k + 1)k$$

with the left side in the form of a difference and so ready for telescoping. Summing from 1 to $n-1$,

$$ny_n - y_1 = \sum_1^{n-1} (k+1)^{(2)} = (n+1)n(n-1)/3$$

so that for $n \geq 1$,

$$y_n = \frac{y_1}{n} + \frac{n^2 - 1}{3}$$

with y_1 arbitrary.

The difference equation can also be required to hold for $k = 0$ with the initial value y_0 arbitrary but finite. Then $y_1 = 0$ is forced and $y_n = (n^2 - 1)/3$ otherwise.

10.18 Develop the idea of summing factor for the more general equation $y_{k+1} = a_k y_k + b_k$.

▮ Let p_k be the product $a_k \cdots a_1$. Then the summing factor is $1/p_k$. Multiplying through by this factor,

$$\frac{y_{k+1}}{a_k \cdots a_1} - \frac{a_k y_k}{a_k \cdots a_1} = \frac{b_k}{a_k \cdots a_1}$$

with the left side in the form of a finite difference. (Note the cancellation of a_k in the coefficient of y_k and also that when $k = 1$, this coefficient is 1.) Summing from 1 to $n-1$,

$$\frac{y_n}{a_{n-1} \cdots a_1} = y_1 + \sum_1^{n-1} \frac{b_k}{a_k \cdots a_1}$$

and the solution y_n is available in summation form. At times it may be useful to include a_0 in the summation factor, in which case the summation is from 0 to $n-1$.

10.19 Show that the general summing factor of the preceding problem does become the $k+1$ used in Problem 10.17.

▮ There we had $a_k = k/(k+1)$, making

$$p_k = \frac{k}{k+1} \cdot \frac{k-1}{k} \cdot \frac{k-2}{k-1} \cdots \frac{1}{2}$$

which clearly reduces to $1/(k+1)$.

10.20 Use a summing factor to solve $y_{k+1} = -y_k + 1$ for $k \geq 0$.

▮ Here the coefficient a_k is -1 and it will be convenient to include a_0 in the summing factor, making it $p = (-1)^{k+1}$. It leads to

$$(-1)^{k+1} y_{k+1} - (-1)^k y_k = (-1)^{k+1}$$

which is summed from $k = 0$ to $k = n - 1$ to produce

$$(-1)^n y_n - y_0 = \sum_0^{n-1} (-1)^{k+1}$$

for $n \geq 1$. The sum will be -1 for n odd and 0 for n even, so that

$$y_n = y_0 (-1)^n + A$$

with $A = 1$ for n odd and $A = 0$ for n even. The same result could have been deduced by direct computation.

10.21 Solve $y_{k+1} = ky_k + b_k$ with the aid of a summing factor.

▮ With $a_k = k$ and summing factor $1/(k(k-1) \cdots 1)$ the equation becomes

$$\frac{y_{k+1}}{k!} - \frac{y_k}{(k-1)!} = \frac{b_k}{k!}$$

which is summed from $k = 1$ to $k = n - 1$ to yield

$$\frac{y_n}{(n-1)!} - y_1 = \sum_{1}^{n-1} \frac{b_k}{k!}$$

or

$$y_n = (n-1)! y_1 + (n-1)! \sum_{1}^{n-1} \frac{b_k}{k!}$$

providing the solution in the form of a summation. For our opening example (Problem 10.5) we had $b_k = k^2$, which forces $y_1 = 0$ and makes the solution

$$y_n = (n-1)! \sum_{1}^{n-1} k^2/k!$$

for $n \geq 2$.

10.22   Solve $y_{k+1} = ky_k + 2^k k!$ by using a summing factor.

    ▮   This is a special case of the preceding problem with $b_k = 2^k k!$ and y_1 forced to be 1, so the solution is

$$y_n = (n-1)! \left[1 + \sum_{1}^{n-1} 2^k \right] = (n-1)!(2^n - 1)$$

since the summation is elementary.

10.23   If the "nonhomogeneous term" $2^k k!$ in the equation of the preceding problem is deleted we have the reduced equation $y_{k+1} = ky_k$, with solution $y_n = A(n-1)!$ for any constant A. Show that the given equation, with $b_k = 2^k k!$, can be solved in the form

$$y_k = A(k)(k-1)!$$

The idea involved is known as variation of the parameter A, allowing it to vary with the independent variable k.

    ▮   We find $y_{k+1} = k! A(k+1)$ and $ky_k = k! A(k)$, and substituting into the difference equation,

$$k! [A(k+1) - A(k)] = 2^k k!$$

so that $A(k)$ is determined by $A(k+1) - A(k) = 2^k$ and the forced $A(1) = 1$. Summing from $k = 1$ to $k = n - 1$ now brings

$$A(n) = 1 + (2^n - 2) = 2^n - 1$$

followed by the solution $y_n = (2^n - 1)(n-1)!$ as before.

10.24   Solve the nonlinear equation $y_{k+1} = y_k^2$.

    ▮   The change of variable $z_k = \log y_k$ replaces the equation by $z_{k+1} = 2z_k$, which has the obvious solution $z_k = z_0 2^k$. The original variable y_k can now be recovered by exponentiation. The trivial initial value $y_0 = 1$, which implies that $y_k = 1$ for all k, is accommodated by the logarithms since z_0 will be zero, forcing $z_k = 0$ for all k. More useful initial values are easily handled.

10.25   Solve the nonlinear equation $y_{k+1} = 2y_k^2 - 1$.

    ▮   The change of variable $y_k = \cos A_k$ transforms the equation to $\cos A_{k+1} = 2\cos^2 A_k - 1 = \cos 2A_k$. This can be satisfied by making $A_{k+1} = 2A_k$, which is the same equation just encountered in the preceding problem. Its solution is $A_{k+1} = A_0 2^k$ so that $y_k = \cos(A_0 2^k)$. This difference equation states one of the key properties of the cosine function, assuming that all y_k values are to be in the interval $(-1, 1)$.

    Since the hyperbolic cosine shares this property, the change of variable $y_k = \cosh A_k$ would also have been successful, with the assumption that y_k values will be greater than 1.

10.26   Solve $y_{k+1} = y_k/(1 + y_k)$.

∎   This nonlinear equation does not seem likely to respond to logarithms. A few direct computations may be helpful. For the initial value $y_0 = 1$ we find the successors

$$\tfrac{1}{2} \quad \tfrac{1}{3} \quad \tfrac{1}{4} \quad \tfrac{1}{5} \quad \cdots$$

while $y_0 = 2$ triggers

$$\tfrac{2}{3} \quad \tfrac{2}{5} \quad \tfrac{2}{7} \quad \tfrac{2}{9} \quad \cdots$$

so perhaps a linear denominator is indicated. If $y_k = 1/(a + bk)$, then $y_{k+1} = 1/(a + bk + b)$ and $1 + y_k = (a + bk + 1)/(a + bk)$ so that substitutions into the difference equation lead to

$$\frac{1}{a + bk + b} = \frac{1}{a + bk} \cdot \frac{a + bk}{a + bk + 1}$$

and we find that b must be 1. The solution is $y_k = 1/(a + k)$ with the remaining constant determined by the initial condition.

10.2   THE DIGAMMA FUNCTION

10.27   The method of summing by "telescoping" depends on being able to express a sum as a sum of differences,

$$\sum_{k=0}^{n} b_k = \sum_{k=0}^{n} \Delta y_k = y_{n+1} - y_0$$

That is, it requires solving the first-order difference equation

$$\Delta y_k = y_{k+1} - y_k = b_k$$

Apply this method when $b_k = 1(k + 1)$, solving the difference equation and evaluating the sum.

∎   Start by defining the *digamma function* as $\psi(x) = \sum_{i=1}^{\infty} x/(i(i + x)) - C$ where C is Euler's constant. Directly we find for any $x \neq -i$,

$$\Delta\psi(x) = \psi(x + 1) - \psi(x) = \sum_{i=1}^{\infty}\left[\frac{x+1}{i(i + x + 1)} - \frac{x}{i(i + x)}\right]$$

$$= \sum_{i=1}^{\infty}\left(\frac{1}{i + x} - \frac{1}{i + x + 1}\right) = \frac{1}{x + 1}$$

When x takes integer values, say $x = k$, this provides a new form for the sum of integer reciprocals, since

$$\sum_{k=0}^{n-1}\frac{1}{k + 1} = \sum_{k=0}^{n-1}\Delta\psi(k) = \psi(n) - \psi(0) = \psi(n) + C$$

We may also rewrite this as

$$\psi(n) = \sum_{k=1}^{n}\frac{1}{k} - C$$

so that the digamma function for integer arguments is a familiar quantity. Its behavior is shown in Fig. 10-2, and

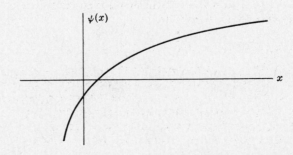

Fig. 10-2

the logarithmic character for large positive x is no surprise when one recalls the definition of Euler's constant. In a sense $\psi(x)$ generalizes from $\psi(n)$ much as the gamma function generalizes factorials.

10.28 Evaluate the sum $\sum_{k=1}^{n} 1/(k+t)$ for arbitrary t.

▮ From Problem 10.27, for any x, $\psi(x+1) - \psi(x) = 1/(x+1)$. Replace x by $k+t-1$ to obtain

$$\psi(k+t) - \psi(k+t-1) = \frac{1}{k+t}$$

Now we have the ingredients of a telescoping sum and find

$$\sum_{k=1}^{n} \frac{1}{k+t} = \sum_{k=1}^{n} [\psi(k+t) - \psi(k+t-1)] = \psi(n+t) - \psi(t)$$

10.29 Evaluate the series $\sum_{k=1}^{\infty} 1/(k+a)(k+b)$ in terms of the digamma function.

▮ Using partial fractions, we find

$$s_n = \sum_{k=1}^{n} \frac{1}{(k+a)(k+b)} = \frac{1}{b-a} \sum_{k=1}^{n} \left(\frac{1}{k+a} - \frac{1}{k+b} \right)$$

Now applying the previous problem, this becomes

$$s_n = \frac{1}{b-a} [\psi(n+a) - \psi(a) - \psi(n+b) + \psi(b)]$$

From the series definition in Problem 10.27 it follows after a brief calculation that

$$\psi(n+a) - \psi(n+b) = (a-b) \sum_{i=1}^{\infty} \frac{1}{(i+n+a)(i+n+b)}$$

so that for $n \to \infty$ this difference has limit zero. Finally,

$$\sum_{k=1}^{\infty} \frac{1}{(k+a)(k+b)} = \lim s_n = \frac{\psi(b) - \psi(a)}{b-a}$$

10.30 Find formulas for $\psi'(x)$, $\psi^{(2)}(x)$, etc., in series form.

▮ Differentiating the series of Problem 10.27 produces $\psi'(x) = \sum_{k=1}^{\infty} 1/(k+x)^2$. Since this converges uniformly in x on any interval not including a negative integer, the computation is valid. Repeating,

$$\psi^{(2)}(x) = \sum_{k=1}^{\infty} \frac{-2!}{(k+x)^3} \qquad \psi^{(3)}(x) = \sum_{k=1}^{\infty} \frac{3!}{(k+x)^4} \quad \text{etc.}$$

In particular, for integer arguments, Problem 9.66 makes $\psi'(0) = \sum_{k=1}^{\infty} 1/k^2 = \pi^2/6$ after which we lose one term at a time to obtain

$$\psi'(1) = \frac{\pi^2}{6} - 1 \qquad \psi'(2) = \frac{\pi^2}{6} - 1 - \frac{1}{4} \quad \text{and in general} \quad \psi'(n) = \frac{\pi^2}{6} - 1 - \frac{1}{4} - \cdots - \frac{1}{n^2}$$

10.31 Evaluate the series

$$\sum_{k=1}^{\infty} \frac{2k+1}{k(k+1)^2}$$

▮ This further illustrates how sums and series involving rational terms in k may be evaluated in terms of the digamma function. Again introducing partial fractions,

$$\sum_{k=1}^{\infty} \frac{2k+1}{k(k+1)^2} = \sum_{k=1}^{\infty} \left[\frac{1}{k} - \frac{1}{k+1} + \frac{1}{(k+1)^2} \right]$$

The first two terms cannot be handled separately since the series would diverge. They can, however, be handled together as in Problem 10.29. The result is

$$\sum_{k=1}^{\infty} \left[\frac{1}{k(k+1)} + \frac{1}{(k+1)^2} \right] = \psi(1) - \psi(0) + \psi'(1) = \frac{\pi^2}{6}$$

Other sums of rational terms may be treated in similar fashion.

10.32 Evaluate the series

$$\sum_{k=1}^{\infty} \frac{1}{1^2 + 2^2 + \cdots + k^2}$$

$\blacksquare$ Summing the squares, we may replace this by

$$\sum_{k=1}^{\infty} \frac{6}{k(k+1)(2k+1)} = \sum_{k=1}^{\infty} \left(\frac{6}{k} + \frac{6}{k+1} - \frac{24}{2k+1} \right)$$

Since no one of these three series is individually convergent, we do not treat each separately. Extending the device used in the problem just solved we may, however, rewrite the combination as

$$\sum_{k=1}^{\infty} \left[\left(\frac{6}{k} - \frac{6}{k} \right) + \left(\frac{6}{k+1} - \frac{6}{k} \right) - \left(\frac{24}{2k+1} - \frac{24}{2k} \right) \right] = \sum_{k=1}^{\infty} \left[\frac{-6}{k(k+1)} + \frac{6}{k(k+\frac{1}{2})} \right]$$

$$= -6[\psi(1) - \psi(0)] + 12 \left[\psi \left(\frac{1}{2} \right) - \psi(0) \right]$$

where Problem 10.29 has been used twice in the last step. Finally,

$$\sum_{k=1}^{\infty} \frac{1}{1^2 + 2^2 + \cdots + k^2} = 12\psi \left(\frac{1}{2} \right) - 6 + 12C$$

10.33 Show that $F(x) = \Gamma'(x+1)/\Gamma(x+1)$ also has the property $\Delta F(x) = 1/(x+1)$, where $\Gamma(x)$ is the gamma function.

$\blacksquare$ The gamma function is defined for positive x by

$$\Gamma(x) = \int_0^{\infty} e^{-t} t^{x-1} \, dt$$

Integration by parts exposes the familiar feature

$$\Gamma(x+1) = x\Gamma(x)$$

and then differentiation brings $\Gamma'(x+1) = x\Gamma'(x) + \Gamma(x)$, or

$$\frac{\Gamma'(x+1)}{\Gamma(x+1)} - \frac{\Gamma'(x)}{\Gamma(x)} = \frac{1}{x}$$

from which the required result follows upon replacing x by $x+1$.

Since $\psi(x+1) - \psi(x) = 1/(x+1)$, we find that

$$\frac{\Gamma'(x+1)}{\Gamma(x+1)} - \psi(x) = A$$

where A is a constant and where x is restricted to a discrete set with unit spacing. The same result can be proved for all x except negative integers, the constant A being zero.

10.34 Compute $\psi(\frac{1}{2})$ from the series definition.

▮ According to the definition,

$$\psi(\tfrac{1}{2}) = \tfrac{1}{2} \sum_1^\infty 1/i(i+\tfrac{1}{2}) - C$$

where $C = .57721567$ is Euler's constant. Earlier problems show that this series will converge very slowly, thousands of terms being needed for even a few correct decimal places. Applying the comparison method (twice) converts the series to

$$\frac{\pi^2}{6} - \frac{1}{2} \sum \frac{1}{i^3} + \frac{1}{4} \sum \frac{1}{i^3(i+\frac{1}{2})}$$

with greatly accelerated convergence and value

$$1.64493 - .60103 + - (.73402) = 1.22740$$

Taking half and subtracting Euler's constant, we find $\psi(\frac{1}{2})$ to be $.0365$ to four places.

10.35 Also compute $\psi(\frac{3}{2})$ and $\psi(-\frac{1}{2})$.

▮ These are easily found from the recursion:

$$\psi(\tfrac{3}{2}) = \psi(\tfrac{1}{2}) + 1/(\tfrac{1}{2}+1) = .7032$$
$$\psi(-\tfrac{1}{2}) = \psi(\tfrac{1}{2}) - 1/(-\tfrac{1}{2}+1) = -1.9635$$

with obvious continuations.

10.3 APPROXIMATING DIFFERENTIAL EQUATIONS

10.36 Develop a difference equation to approximate the differential equation $y'(x) = y(x)$ with initial condition $y(0) = 1$.

▮ Numerous methods will be presented in the next chapter for the approximate solution of differential equations, but this preliminary example will show how a difference equation can be viewed as a discretization of a differential equation.

We consider only those values x_k of the continuous variable x of the form $x_k = x_0 + kh$ with h fixed and $k = 0, 1, 2, \ldots$:

$$\begin{array}{cccc} 0 & 1 & & k \\ \vdash & \vdash & & \vdash \\ x_0 & x_1 & & x_k = x_0 + kh \end{array}$$

In the present example $x_0 = 0$ and $x_k = kh$. Replacing the derivative by a divided difference leads to

$$(y_{k+1} - y_k)/h = y_k \quad \text{or} \quad y_{k+1} = (1+h)y_k$$

The solution of this difference equation satisfying the initial condition is clearly $y_k = (1+h)^k$. To see its relationship with the solution of the differential problem this can be rewritten as

$$y_k = (1+h)^{kh/h} = (1+h)^{x_k/h}$$

If we now let h approach zero, keeping x_k fixed, this approaches $y(x_k) = e^{x_k}$, which we have known from the start to be the solution of the differential problem.

The point is, at least in this example, as h becomes smaller, making the discretization finer, and the computation of the difference solution more laborious, the result does draw closer to the intended target.

10.37 Modify the discretization in the preceding problem to approximate $y' = y + x$ with $y(0) = 0$.

▮ The difference equation

$$(y_{k+1} - y_k)/h = y_k + kh \quad \text{or} \quad y_{k+1} = (1+h)y_k + h^2 k$$

with $y_0 = 0$ is a simple approximation. Its solution can be found using the summing factor $1/(1+h)^{k+1}$, but is also available by combining the solution $C(1+h)^k$ of $y_{k+1} = (1+h)y_k$ with a solution f_k of the given equation, a standard procedure with linear differential equations. Here the trial function $f_k = A + Bk$ can be substituted into the difference equation to find that $A = -1$ and $B = -h$ will produce a solution. So we have

$$y_k = (1 + h)^k - 1 - hk = (1 + h)^{x_k/h} - 1 - x_k$$

which, as h approaches 0, becomes $e^{x_k} - 1 - x_k$. This is the solution of the differential problem.

10.38 For the equation $xy'(x) = 2y(x)$, if $y'(0)$ is required to be finite, it follows that $y(0)$ will have to be zero. Suppose a terminal condition $y(1) = 1$ is also required. The solution $y(x) = x^2$ can quickly be verified, but use this little problem to further illustrate approximation by difference equations.

▌ Following the now familiar track we have

$$(kh)(y_{k+1} - y_k)/h = 2y_k \quad \text{or} \quad ky_{k+1} = (2 + k)y_k$$

as the difference equation. Attempting a solution in the form $y_k = Ak + Bk^2$ (a constant term being useless since $y_0 = 0$ is forced), one finds that $A = B$ is required, making the solution

$$y_k = A(k + k^2)$$

and bringing us to the terminal condition $y(1) = 1$. Suppose the discretization is such that $Nh = 1$, that is, N subintervals are involved in the "grid." Then it is y_N which must be 1, so that

$$1 = y_N = A(N + N^2) \qquad A = 1/(N + N^2) = h^2/(h + 1)$$

Finally

$$y_k = \frac{h^2(k + k^2)}{h + 1} = \frac{hx_k + x_k^2}{h + 1}$$

which does approach x_k^2 as h tends to zero.

10.4 LINEAR SECOND-ORDER EQUATION: HOMOGENEOUS CASE

10.39 The difference equation $y_{k+2} + a_1 y_{k+1} + a_2 y_k = 0$ in which a_1 and a_2 may depend on k is called linear and homogeneous. Prove that if u_k and v_k are solutions, then so are $c_1 u_k + c_2 v_k$ for arbitrary constants c_1 and c_2. (It is this feature that identifies a linear homogeneous equation. The equation is homogeneous because $y_k \equiv 0$ is a solution.)

▌ Since $u_{k+2} + a_1 u_{k+1} + a_2 u_k = 0$ and $v_{k+2} + a_1 v_{k+1} + a_2 v_k = 0$, it follows at once by multiplying the first equation by c_1, the second equation by c_2 and adding that

$$c_1 u_{k+2} + c_2 v_{k+2} + a_1(c_1 u_{k+1} + c_2 v_{k+1}) + a_2(c_1 u_k + c_2 v_k) = 0$$

which was to be proved.

10.40 Show that for a_1 and a_2 constant, two real solutions can be found in terms of elementary functions.

▌ First suppose $a_1^2 > 4a_2$. Then we may take

$$u_k = r_1^k \qquad v_k = r_2^k$$

where r_1 and r_2 are the distinct real roots of the quadratic equation $r^2 + a_1 r + a_2 = 0$. To prove this we verify directly that

$$u_{k+2} + a_1 u_{k+1} + a_2 u_k = r^k(r^2 + a_1 r + a_2) = 0$$

where r is either root. The quadratic equation involved here is known as the characteristic equation.

Next suppose $a_1^2 = 4a_2$. Then the characteristic equation has only one root, say r, and can be rewritten as

$$r^2 + a_1 r + a_2 = (r + \tfrac{1}{2}a_1)^2 = 0$$

Two real solutions are now available in

$$u_k = r^k \qquad v_k = kr^k$$

The solution u_k may be verified exactly as before. As for v_k,

$$(k + 2)r^{k+2} + a_1(k + 1)r^{k+1} + a_2 kr^k = r^k[k(r^2 + a_1 r + a_2) + (2r + a_1)r] = 0$$

since both parentheses are zero.

Finally suppose $a_1^2 < 4a_2$. Then the characteristic equation has complex conjugate roots $\mathrm{Re}^{\pm i\theta}$. Substituting, we find

$$R^2 e^{\pm i2\theta} + a_1 R e^{\pm i\theta} + a_2 = R^2(\cos 2\theta \pm i \sin 2\theta) + a_1 R(\cos\theta \pm i \sin\theta) + a_2$$

$$= (R^2 \cos 2\theta + a_1 R \cos\theta + a_2) \pm i(R^2 \sin 2\theta + a_1 R \sin\theta) = 0$$

This requires that both parentheses vanish:

$$R^2 \cos 2\theta + a_1 R \cos\theta + a_2 = 0 \qquad R^2 \sin 2\theta + a_1 R \sin\theta = 0$$

We now verify that two real solutions of the difference equation are

$$u_k = R^k \sin k\theta \qquad v_k = R^k \cos k\theta$$

For example,

$$u_{k+2} + a_1 u_{k+1} + a_2 u_k = R^{k+2} \sin(k+2)\theta + a_1 R^{k+1} \sin(k+1)\theta + a_2 R^k \sin k\theta$$

$$= R^k(\sin k\theta)(R^2 \cos 2\theta + a_1 R \cos\theta + a_2) + R^k(\cos k\theta)(R^2 \sin 2\theta + a_1 R \sin\theta) = 0$$

since both parentheses vanish. The proof for v_k is almost identical.

It now follows that for a_1 and a_2 constant, the equation $y_{k+2} + a_1 y_{k+1} + a_2 y_k = 0$ always has a family of elementary solutions $y_k = c_1 u_k + c_2 v_k$.

10.41 Solve the difference equation $y_{k+2} - 2A y_{k+1} + y_k = 0$ in terms of power functions, assuming $A > 1$.

Let $y_k = r^k$ and substitute to find that $r^2 - 2Ar + 1 = 0$ is necessary.

This leads to $r = A \pm \sqrt{A^2 - 1} = r_1, r_2$ and $y_k = c_1 r_1^k + c_2 r_2^k = c_1 u_k + c_2 v_k$.

One of these power functions grows arbitrarily large with k and the other tends to zero, since $r_1 > 1$ but $0 < r_2 < 1$. [The fact that $r_2 = A - \sqrt{A^2 - 1} < 1$ follows from $(A-1)^2 = A^2 + 1 - 2A < A^2 - 1$ after taking square roots and transposing terms.]

10.42 Solve the equation $y_{k+2} - 2y_{k+1} + y_k = 0$.

▌ Here we have $a_1^2 = 4a_2 = 4$. The only root of $r^2 - 2r + 1 = 0$ is $r = 1$. This means that $u_k = 1$ and $v_k = k$ are solutions and that $y_k = c_1 + c_2 k$ is a family of solutions. This is hardly surprising in view of the fact that this difference equation may be written as $\Delta^2 y_k = 0$.

10.43 Solve $y_{k+2} - 2A y_{k+1} + y_k = 0$ where $A < 1$.

▌ Now $a_1^2 < 4a_2$. The roots of the characteristic equation become

$$\mathrm{Re}^{\pm i\theta} = A \pm i\sqrt{1 - A^2} = \cos\theta \pm i \sin\theta$$

where $A = \cos\theta$ and $R = 1$. Thus $u_k = \sin k\theta$, $v_k = \cos k\theta$ and the family of solutions

$$y_k = c_1 \sin k\theta + c_2 \cos k\theta$$

is available.

The v_k functions, when expressed as polynomials in A, are known as Chebyshev polynomials. For example,

$$v_0 = 1 \qquad v_1 = A \qquad v_2 = 2A^2 - 1 \qquad \cdots$$

The difference equation of this problem is the recursion for the Chebyshev polynomials.

10.44 Show that if two solutions of $y_{k+2} + a_1 y_{k+1} + a_2 y_k = 0$ agree in value at two consecutive integers k, then they must agree for all integers k. (Assume $a_2 \neq 0$.)

▌ Let u_k and v_k be solutions that agree in value at $k = m$ and $m + 1$. Then their difference $d_k = u_k - v_k$ is a solution (by Problem 10.39) for which $d_m = d_{m+1} = 0$. But then

$$d_{m+2} + a_1 d_{m+1} + a_2 d_m = 0 \qquad d_{m+1} + a_1 d_m + a_2 d_{m-1} = 0$$

from which it follows that $d_{m+2} = 0$ and $d_{m-1} = 0$. In the same way we may prove d_k to be zero for $k > m + 2$ and for $k < m - 1$, taking each integer in its turn. Thus $d_k \equiv 0$ and $u_k \equiv v_k$. (The assumption $a_2 \neq 0$ merely guarantees that we do have a second-order difference equation.)

10.45 Show that any solution of $y_{k+2} + a_1 y_{k+1} + a_2 y_k = 0$ may be expressed as a combination of two particular solutions u_k and v_k,

$$y_k = c_1 u_k + c_2 v_k$$

provided that the Wronskian determinant

$$w_k = \begin{vmatrix} u_k & v_k \\ u_{k-1} & v_{k-1} \end{vmatrix} \neq 0$$

We know that $c_1 u_k + c_2 v_k$ is a solution. By the previous problem it will be identical with the solution y_k if it agrees with y_k for two consecutive integer values of k. In order to obtain such agreement we choose $k = 0$ and $k = 1$ (any other consecutive integers would do) and determine the coefficients c_1 and c_2 by the equations

$$c_1 u_0 + c_2 v_0 = y_0 \qquad c_1 u_1 + c_2 v_1 = y_1$$

The unique solution is $c_1 = (y_1 v_0 - y_0 v_1)/w_1$, $c_2 (y_0 u_1 - y_1 u_0)/w_1$ since $w_1 \neq 0$.

10.46 Show that if the Wronskian determinant is zero for one value of k, it must be identically zero, assuming u_k, v_k to be solutions of the equation of Problem 10.44. Apply this to the particular case of Problem 10.40 to prove $w_k \neq 0$.

▌ We compute the difference

$$\Delta w_k = (u_{k+1} v_k - v_{k+1} u_k) - (u_k v_{k-1} - v_k u_{k-1})$$
$$= v_k(-a_1 u_k - a_2 u_{k-1}) - u_k(-a_1 v_k - a_2 v_{k-1}) - u_k v_{k-1} + v_k u_{k-1}$$
$$= (a_2 - 1) w_k = w_{k+1} - w_k$$

from which it soon follows that $w_k = a_2^k w_0$. Since $a_2 \neq 0$, the only way for w_k to be zero is to have $w_0 = 0$. But then w_k is identically zero.

When w_k is identically zero, it follows that u_k/v_k is the same as u_{k-1}/v_{k-1} for all k, that is, $u_k/v_k = $ constant. Since this is definitely not true for the u_k, v_k of Problem 10.40, w_k cannot be zero there.

10.47 Solve by directly computation the second-order initial-value problem

$$y_{k+2} = y_{k+1} + y_k \qquad y_0 = 0 \qquad y_1 = 1$$

▌ Taking $k = 0, 1, 2, \ldots$ we easily find the successive y_k values $1, 2, 3, 5, 8, 13, 21, 34, 55, 89, 144, \ldots$, which are known as Fibonacci numbers. The computation clearly shows a growing solution but does not bring out its exact character.

10.48 Determine the character of the solution of the previous problem.

▌ Following the historical path mapped in Problems 10.39, 10.40, etc., we consider the characteristic equation $r^2 - r - 1 = 0$.

Since $a_1^2 > 4a_2$, there are two real roots, namely $r_1, r_2 = (1 \pm \sqrt{5})/2$. All solutions can therefore be expressed in the form

$$y_k = c_1 u_k + c_2 v_k = c_1 \left(\frac{1 + \sqrt{5}}{2}\right)^k + c_2 \left(\frac{1 - \sqrt{5}}{2}\right)^k$$

To satisfy the initial conditions, we need $c_1 + c_2 = 0$ and $c_1((1 + \sqrt{5})/2) + c_2((1 - \sqrt{5})/2) = 1$. This makes

$$c_1 = -c_2 = \frac{1}{\sqrt{5}} \quad \text{and} \quad y_k = \frac{1}{\sqrt{5}}\left[\left(\frac{1 + \sqrt{5}}{2}\right)^k - \left(\frac{1 - \sqrt{5}}{2}\right)^k\right]$$

10.49 Show that for the Fibonacci numbers, $\lim(y_{k+1}/y_k) = (1 + \sqrt{5})/2$.

    ▮ For such results it is convenient to know the character of the solution function. Using the previous problem we find, after a brief calculation,

$$\frac{y_{k+1}}{y_k} = \frac{1 + \sqrt{5}}{2} \cdot \frac{1 - \left[(1 - \sqrt{5})/(1 + \sqrt{5})\right]^{k+1}}{1 - \left[(1 - \sqrt{5})(1 + \sqrt{5})\right]^{k}}$$

and $(1 - \sqrt{5})/(1 + \sqrt{5})$ has absolute value less than 1, so that the required result follows.

10.50 The Fibonacci numbers occur in certain problems involving the transfer of information along a communications channel. The capacity C of a channel is defined as $C = \lim(\log y_k)/k$, the logarithm being to base 2. Evaluate this limit.

    ▮ Again the analytic character of the solution y_k is needed. But it is available and we find

$$\log y_k = \log\frac{1}{\sqrt{5}} + \log\left[\left(\frac{1 + \sqrt{5}}{2}\right)^k - \left(\frac{1 - \sqrt{5}}{2}\right)^k\right]$$

$$= \log\frac{1}{\sqrt{5}} + \log\left(\frac{1 + \sqrt{5}}{2}\right)^k + \log\left[1 - \left(\frac{1 - \sqrt{5}}{1 + \sqrt{5}}\right)^k\right]$$

making

$$C = \lim\left\{\frac{\log(1/\sqrt{5})}{k} + \log\frac{1 + \sqrt{5}}{2} + \frac{1}{k}\log\left[1 - \left(\frac{1 - \sqrt{5}}{1 + \sqrt{5}}\right)^k\right]\right\} = \log\frac{1 + \sqrt{5}}{2}$$

10.51 Solve $y_{k+2} + 3y_{k+1} + 2y_k = 0$ with initial conditions $y_0 = 2$ and $y_1 = 1$.

    ▮ The characteristic equation $r^2 + 3r + 2 = 0$ has the roots -1 and -2. The family of solutions is, therefore,

$$y_k = c_1(-1)^k + c_2(-2)^k$$

and the initial conditions determine c_1 and c_2 from

$$c_1 + c_2 = 2 \qquad -c_1 - 2c_2 = 1$$

as $c_1 = 5$, $c_2 = -3$. Either the equation

$$y_k = 5(-1)^k - 3(-2)^k$$

or the difference equation itself can be used to generate a few of the early members of the solution sequence

$$2, 1, -7, 19, -43, \ldots$$

10.52 For what values of the parameter A will the equation

$$y_{k+2} - 2y_{k+1} + (1 - A)y_k = 0$$

have oscillatory solutions?

    ▮ The equation $r^2 - 2r + 1 - A = 0$ will have complex roots if $4 - 4 + 4A$ is negative, which occurs whenever A is negative. This is when the solutions will have oscillatory character.

10.53 What sort of solutions does $y_{k+2} - y_k = 0$ have?

    ▮ It is clear that the sequence of solution values will repeat y_0 and y_1 endlessly. To see how our methods manage to represent this behavior we solve the characteristic equation $r^2 - 1 = 0$ and have the roots 1 and -1. The family of solutions is thus

$$y_k = c_1 + c_2(-1)^k$$

To accommodate the initial values

$$c_1 + c_2 = y_0 \qquad c_1 - c_2 = y_1$$

making $c_1 = \frac{1}{2}(y_0 + y_1)$ and $c_2 = \frac{1}{2}(y_0 - y_1)$.

10.54   For what values of the constant A will all solutions of

$$y_{k+2} - 2Ay_{k+1} + Ay_k = 0$$

tend to zero as k becomes positively infinite?

▮ Since A is constant, solution behavior is governed by the roots of the characteristic equation $r^2 - 2Ar + A = 0$, which are

$$r = A \pm \sqrt{A^2 - A}$$

If $0 < A < 1$ these roots are complex and, apart from oscillations, their behavior depends on the modulus of the roots. Since $R^2 = A^2 + A - A^2 = A$ the modulus R will be less than 1 for all values of A in the interval. Since all solutions include R^k as a factor they will all be driven to zero as k increases.

If $A > 1$ then the root $A + \sqrt{A^2 - A}$ will also exceed 1 and solutions involving powers of this root will "explode" at infinity. Similarly if A is less than -1 then $A - \sqrt{A^2 - A}$ will also be less than -1, and solutions involving powers of this root will experience an infinite oscillation.

There remains the interval $-1 < A < 0$, and a little algebra will reveal that if A is within the $(-\frac{1}{3}, 0)$ part of this range, then both roots are real and of magnitude less than 1, sending all solutions to 0 at infinity.

10.55   Solve $y_{k+2} - 2y_{k+1} + 2y_k = 0$ with initial values $y_0 = 0$ and $y_1 = 1$.

▮ Direct computations produce the first few elements of the solution sequence: $0, 1, 2, 2, 0, -4, \ldots$. To find the closed representation of this function we begin with the characteristic equation $r^2 - 2r + 2 = 0$ and find the roots $1 \pm i = \sqrt{2}\, e^{i\pi/4}$. The solution family is

$$y_k = \sqrt{2}^{\,k} \left[c_1 \sin(\pi k/4) + c_2 \cos(\pi k/4) \right]$$

which puts the initial conditions into the form

$$y_0 = 0 = c_2 \qquad y_1 = 1 = c_1 \sqrt{2}\left(\sqrt{2}/2\right) \quad \text{or} \quad c_1 = 1$$

Finally $y_k = \sqrt{2}^{\,k} \sin(\pi k/4)$.

10.56   Find the Wronskian of the two independent solutions found in Problem 10.55 and verify that it is nonzero.

▮ By the definition in Problem 10.45,

$$
w_k = \begin{vmatrix} \sqrt{2}^{\,k} \sin \pi k/4 & \sqrt{2}^{\,k} \cos \pi k/4 \\ \sqrt{2}^{\,k-1} \sin \pi(k-1)/4 & \sqrt{2}^{\,k-1} \cos \pi(k-1)/4 \end{vmatrix}
$$

$$= \sqrt{2}^{\,2k-1}\left[\sin \pi k/4 \cos \pi(k-1)/4 - \cos \pi k/4 \sin \pi(k-1)/4\right]$$

$$= \sqrt{2}^{\,2k-1} \sin \pi/4 = 2^{k-1}$$

which is definitely nonzero. It can also be noted that with the coefficient $a_2 = 2$, $w_k = a_2^k w_0$ as the earlier problem predicts.

10.57   Find the Wronskian for the two independent solutions found in Problem 10.53 and verify that it is nonzero.

▮ The two solutions were $u_k = 1$ and $v_k = (-1)^k$, making the Wronskian

$$w_k = \begin{vmatrix} 1 & (-1)^k \\ 1 & (-1)^{k-1} \end{vmatrix} = (-1)^{k-1} - (-1)^k = 2(-1)^{k-1}$$

Since $a_2 = -1$ it also follows that $w_k = w_0 a_2^k$.

10.58 Discretize the differential equation $y'' + \omega^2 y = 0$ and show that the difference solution approaches the differential solution as h tends to zero.

▌ The procedure developed in Problem 10.36 will be modified to accommodate the second derivative which is now present. Perhaps the simplest option is

$$\frac{1}{h^2} \Delta^2 y_k + \omega^2 y_k = 0 \quad \text{or} \quad y_{k+2} - 2y_{k+1} + (1 + h^2 w^2) y_k = 0$$

The corresponding characteristic equation is

$$r^2 - 2r + (1 + h^2\omega^2) = 0$$

with roots $r = 1 \pm ih\omega = \sqrt{1 + h^2\omega^2}\, e^{i\theta}$ and $\tan\theta = h\omega$. The solution family is, therefore

$$y_k = \sqrt{1 + h^2\omega^2}^{\,k} \left(c_1 \sin k\theta + c_2 \cos k\theta \right)$$

As h tends to zero, the value of θ will be close to $h\omega$. Taking a shortcut or two we find

$$\lim y_k = \lim\left(1 + \tfrac{1}{2} kh^2\omega^2\right)\left(c_1 \sin kh\omega + c_2 \cos kh\omega \right)$$
$$= c_1 \sin \omega x_k + c_2 \cos \omega x_k$$

since hk is fixed at argument x_k. This is, of course, the family of solutions of the differential equation.

10.59 Do a similar discretization of the equation $y'' - \omega^2 y = 0$ and show that the resulting difference solution converges to the differential solution as h tends to zero.

▌ The difference equation will be

$$y_{k+2} - 2y_{k+1} + (1 - h^2\omega^2) y_k = 0$$

with characteristic equation $r^2 - 2r + (1 - h^2\omega^2) = 0$. The roots are now $r = 1 \pm h\omega$ and lead to the exponential solutions

$$y_k = c_1(1 + h\omega)^k + c_2(1 - h\omega)^k$$
$$= c_1(1 + h\omega)^{kh/h} + c_2(1 - h\omega)^{kh/h}$$

which as h tends to zero, keeping kh fixed at x_k, have the family of differential solutions as limit:

$$c_1 e^{\omega x_k} + c_2 e^{-\omega x_k}$$

10.5 EULER EQUATIONS

10.60 Since $\Gamma(k) = (k-1)\Gamma(k-1)$, it follows that if r is a positive integer

$$\frac{\Gamma(k+r)}{\Gamma(k)} = k(k+1)(k+2) \cdots (k+r-1)$$

which is a factorial polynomial and so has related factorial polynomials as its differences. If r is not an integer, find the first and second differences of this function.

▌
$$\Delta \frac{\Gamma(k+r)}{\Gamma(k)} = \frac{\Gamma(k+r+1)}{\Gamma(k+1)} - \frac{\Gamma(k+r)}{\Gamma(k)} = \frac{(k+r)\Gamma(k+r)}{k\Gamma(k)} - \frac{\Gamma(k+r)}{\Gamma(k)}$$
$$= \frac{r}{k} \frac{\Gamma(k+r)}{\Gamma(k)}$$

$$\Delta^2 \frac{\Gamma(k+r)}{\Gamma(k)} = \frac{r\Gamma(k+r+1)}{(k+1)\Gamma(k+1)} - \frac{r\Gamma(n+r)}{k\Gamma(k)} = \frac{r(k+r)\Gamma(k+r)}{(k+1)k\Gamma(k)} - \frac{r\Gamma(n+r)}{k\Gamma(k)}$$
$$= \frac{\Gamma(k+r)}{\Gamma(k)}\left[\frac{r(k+r)}{(k+1)k} - \frac{r}{k}\right] = \frac{r(r-1)}{k(k+1)} \frac{\Gamma(k+r)}{\Gamma(k)}$$

A more general result can be anticipated and is provable by an induction:

$$\Delta^n \frac{\Gamma(k+r)}{\Gamma(k)} = \frac{r(r-1)(r-2)\cdots(r-n+1)}{k(k+1)(k+2)\cdots(k+n-1)} \frac{\Gamma(k+r)}{\Gamma(k)}$$

10.61   An Euler difference equation of second order is of the form

$$k(k+1)\,\Delta^2 y_k + a_1 k\,\Delta y_k + a_2 y_k = 0$$

and is analogous to an Euler differential equation in which the coefficients are descending powers of x, apart from constants. Show that solutions can be found by trying $y_k = \Gamma(k+r)/\Gamma(k)$ and obtaining a quadratic equation for r.

▮ Substitution leads to

$$\left[k(k+1)\frac{r(r-1)}{k(k+1)} + a_1 k\frac{r}{k} + a_2 \right]\frac{\Gamma(k+r)}{\Gamma(k)} = 0$$

which simplifies to $r(r-1) + a_1 r + a_2 = 0$. This will be called the auxiliary equation.

10.62   Solve $k(k+1)\,\Delta^2 y_k + k\,\Delta y_k - y_k = 0$ by the method of the preceding problem.

▮ With $a_1 = 1$ and $a_2 = -1$ the auxiliary equation becomes $r^2 - 1 = 0$ and has roots 1 and -1. The solution family of the equation is then

$$y_k = c_1 \frac{\Gamma(k+1)}{\Gamma(k)} + c_2 \frac{\Gamma(k-1)}{\Gamma(k)} = c_1 k + c_2 \frac{1}{k-1}$$

10.63   Also solve $k(k+1)\,\Delta^2 y_k + k\,\Delta y_k - \frac{1}{4}y_k = 0$ by the method of Problem 10.61.

▮ Here $a_1 = 1$ and $a_2 = -\frac{1}{4}$ so the roots of the auxiliary equation are $\pm\frac{1}{2}$. The family of solutions is then

$$y_k = c_1 \frac{\Gamma\left(k+\frac{1}{2}\right)}{\Gamma(k)} + c_2 \frac{\Gamma\left(k-\frac{1}{2}\right)}{\Gamma(k)}$$

If the initial conditions $y_1 = 0$ and $y_2 = 1$ are set, then

$$c_1\Gamma\left(\tfrac{3}{2}\right) + c_2\Gamma\left(\tfrac{1}{2}\right) = 0$$
$$c_1\Gamma\left(\tfrac{5}{2}\right) + c_2\Gamma\left(\tfrac{3}{2}\right) = 1$$

and using $\Gamma\left(\tfrac{3}{2}\right) = \left(\tfrac{1}{2}\right)\Gamma\left(\tfrac{1}{2}\right)$ and $\Gamma\left(\tfrac{5}{2}\right) = \left(\tfrac{3}{2}\right)\Gamma\left(\tfrac{3}{2}\right)$ these identify $c_1 = 1$ and $c_2 = -\frac{1}{2}$. The solution satisfying the initial conditions is thus

$$y_k = \frac{\Gamma\left(k+\frac{1}{2}\right)}{\Gamma(k)} - \frac{\Gamma\left(k-\frac{1}{2}\right)}{2\Gamma(k)}$$

10.6   REDUCTION OF THE ORDER

10.64   Suppose a solution u_k is known for the equation

$$y_{k+2} + a_1 y_{k+1} + a_2 y_k = 0$$

in which the coefficients may be functions of k. (Note that the leading coefficient has been set at 1.) Show that a new variable z_k, such that $y_k = u_k z_k$, can be determined by solving a first-order equation. The process has been called reduction of the order of the difference equation.

▮ Since u_k is a solution we have $u_{k+2} + a_1 u_{k+1} + a_2 u_k = 0$, which it is convenient to write as $a_1 u_{k+1} = -u_{k+2} - a_2 u_k$. Now substituting $u_k z_k$ for y_k in the original equation,

$$u_{k+2}z_{k+2} + a_1 u_{k+1}z_{k+1} + a_2 u_k z_k = 0$$

and using the equation for u_k,

$$u_{k+2}(z_{k+2} - z_{k+1}) = a_2(z_{k+1} - z_k)u_k$$

Let $d_k = z_{k+1} - z_k$. The last equation then becomes

$$d_{k+1} = (a_2 u_k / u_{k+2}) d_k$$

which allows d_k to be expressed in a product form. Finite integration then yields z_k, after which a second solution of the given equation is available as $v_k = u_k z_k$. The combination $c_1 u_k + c_2 v_k$ is then the complete solution family.

10.65   As a trivial illustration of order reduction consider the equation $y_{k+2} - y_k = 0$. Let $u_k = 1$ be a known solution.

▮   In Problem 10.53 this equation was solved by the usual method for constant coefficients, the second solution being $(-1)^k$. Here we let $y_k = u_k z_k = z_k$, $d_k = z_{k+1} - z_k$ and obtain

$$d_{k+1} = (a_2 u_k / u_{k+1}) d_k = -d_k$$

with the solution $d_k = (-1)^k$. Then $z_{k+1} - z_k = (-1)^k$ and summing from 0 to $k-1$, $z_k - z_0 = \Sigma_0^{k-1} (-1)^i = \frac{1}{2} - \frac{1}{2}(-1)^k$. The constant term is nothing new, but the $(-1)^k$ is what was expected.

10.66   Use reduction of the order to solve

$$y_{k+2} - \frac{2k+1}{k+1} y_{k+1} + \frac{k-1}{k} y_k = 0$$

given that $u_k = k$ is a solution.

▮   This equation was encountered in Problem 10.62 where the second solution was found to be $1/(k-1)$. Here we let $y_k = u_k z_k = k z_k$ and begin the solution process with

$$d_{k+1} = (a_2 u_k / u_{k+2}) d_k = \frac{k-1}{k+2} d_k$$

after cancellation. But then

$$d_k = \frac{(k-2)(k-3) \cdots 1}{(k+1)k(k-1) \cdots 4} d_2 = \frac{6}{(k+1)k(k-1)} d_2$$

Since an arbitrary multiplicative constant will be attached later, the $6 d_2$ can be ignored and we consider the finite integration problem

$$z_{k+1} - z_k = \frac{1}{(k+1)k(k-1)} = (k-2)^{(-3)}$$

with solution $z_k = (k-2)^{(-2)}/(-2) = -1/2k(k-1)$. Finally

$$y_k = u_k z_k = -3/(k-1)$$

and we have our second solution. The complete family is

$$y = c_1 k + c_2 \frac{1}{k-1}$$

as before.

10.67   Given that $u_k = 1/(k+1)(k+2)$ is a solution of

$$(k+4) y_{k+2} + y_{k+1} - (k+1) y_k = 0$$

find a second solution by reduction of the order.

▮   With $y_k = z_k/(k+1)(k+2)$ and $d_k = z_{k+1} - z_k$ it follows, since $a_2 = -(k+1)/(k+4)$ that

$$d_{k+1} = -\frac{k+3}{k+2} d_k$$

after cancellations. Repeated applications then make

$$d_k = \left(-\frac{k+2}{k+1}\right)\left(-\frac{k+1}{k}\right)\cdots\left(-\frac{3}{2}\right)d_0 = (-1)^k\frac{k+2}{2}d_0$$

after more cancellations. Abandoning the multiplicative constant we face the finite integration

$$z_{k+1} - z_k = (-1)^k(k+2)$$

which can be solved in summation form as

$$z_k = \sum_0^{k-1} (-1)^i(i+2) = 2 - 3 + 4 - 5 + \cdots$$

but for which the representation

$$z_k = \frac{(-1)^{k+1}(2k+3) + 3}{4}$$

is also suitable. With $v_k = u_k z_k$ we again have a complete family of solutions in the form $y_k = c_1 u_k + c_2 v_k$.

10.68 With a known solution $u_k = k^{(2)} = k^2 - k$ of

$$ky_{k+2} - 3ky_{k+1} + (2k+2)y_k = 0$$

solve by reduction of the order.

▮ Since $a_2 = (2k+2)/k$ we begin with $d_{k+1} = 2(k-1)/(k+2)d_k$ after cancellations. The usual repeated applications then manage

$$d_k = \frac{2(k-2)}{k+1}\frac{2(k-3)}{k}\cdots\frac{2\cdot 1}{4}d_2 = \frac{6\cdot 2^{k-2}}{(k+1)k(k-1)}d_2$$

and abandoning multiplicative constants

$$z_{k+1} - z_k = 2^k(k-2)^{(-3)}$$

The summation

$$z_k - z_2 = \sum_2^{k-1} 2^i(i-2)^{(-3)}$$

avoids zero denominators, and the constant z_2 can be skipped since with $y_k = u_k z_k$ the next step, it merely offers a multiple of the solution u_k that was known from the start.

10.69 Let $y_k = \int_0^1 (x^k/(x^2+1))\, dx$. Show that $y_{k+2} + y_k = 1/(k+1)$ and use this recursion to compute various values of y_k:

$$y_{k+2} + y_k = \int_0^1 \frac{x^{k+2} + x^k}{x^2+1}\, dx = \int_0^1 x^k\, dx = \frac{1}{k+1}$$

▮ For $k = 0$ we have an arctan integral with value $\pi/4$, while for $k = 1$ the integral is logarithmic with value $\frac{1}{2}\log 2$. It is true that for any k the integral can be found by elementary means, but with the difference equation in hand it is more convenient to use it instead:

$$y_2 = 1 - y_0 = 1 - \pi/4$$
$$y_3 = \tfrac{1}{2} - y_1 = \tfrac{1}{2} - \tfrac{1}{2}\log 2$$
$$y_4 = \tfrac{1}{3} - y_2 = \tfrac{1}{3} - 1 + \pi/4$$

and so on.

10.70   Solve $y_{k+2} + y_{k+1} + y_k = 1/(k+1)$ in the form of an integral.

▌ With an eye on the preceding problem we let

$$y_k = \int_0^1 \frac{x^k}{x^2 + x + 1}\, dx$$

and readily find it to be a solution of the difference equation. Integral tables lead to $y_0 = \pi/3\sqrt{3}$ and $y_1 = \frac{1}{2}\log 3 - \frac{1}{2}y_0$. It is then a simple matter to run off values of the integral for successive k. In particular, y_2 proves to be $1 - \frac{1}{2}\log 3 - \frac{1}{2}y_0$ and y_3 is $-\frac{1}{2} + y_0$.

10.7 THE NONHOMOGENEOUS CASE

10.71   The equation $y_{k+2} + a_1 y_{k+1} + a_2 y_k = b_k$ is linear and nonhomogeneous. Show that if u_k and v_k are solutions of the associated homogeneous equation (with b_k replaced by 0) with nonvanishing Wronskian and if Y_k is one *particular* solution of the equation as it stands, then every solution can be expressed as $y_k = c_1 u_k + c_2 v_k + Y_k$ where c_1 and c_2 are suitable constants.

▌ With y_k denoting any solution of the nonhomogeneous equation and Y_k the particular solution,

$$y_{k+2} + a_1 y_{k+1} + a_2 y_k = b_k$$
$$Y_{k+2} + a_1 Y_{k+1} + a_2 Y_k = b_k$$

and subtracting,

$$d_{k+2} + a_1 d_{k+1} + a_2 d_k = 0$$

where $d_k = y_k - Y_k$. But this makes d_k a solution of the homogeneous equation, so that $d_k = c_1 u_k + c_2 v_k$. Finally, $y_k = c_1 u_k + c_2 v_k + Y_k$, which is the required result.

10.72   By the previous problem, to find all solutions of a nonhomogeneous equation we may find just one such particular solution and attach it to the solution of the associated homogeneous problem. Follow this procedure for $y_{k+2} - y_{k+1} - y_k = Ax^k$.

▌ When the term b_k is a power function, a solution can usually be found which is itself a power function. Here we try to determine the constant C so that $Y_k = Cx^k$.

Substitution leads to $Cx^k(x^2 - x - 1) = Ax^k$, making $C = A/(x^2 - x - 1)$. All solutions are therefore expressible as

$$y_k = c_1\left(\frac{1 + \sqrt{5}}{2}\right)^k + c_2\left(\frac{1 - \sqrt{5}}{2}\right)^k + \frac{Ax^k}{x^2 - x - 1}$$

Should $x^2 - x - 1 = 0$, this effort fails.

10.73   For the preceding problem, how can a particular solution Y_k be found in the case where $x^2 - x - 1 = 0$?

▌ Try to determine C so that $Y_k = Ckx^k$. Substitution leads to $Cx^k[(k+2)x^2 - (k+1)x - k] = Ax^k$ from which $C = A/(2x^2 - x)$. This makes $Y_k = Akx^k/(2x^2 - x)$.

10.74   For what sort of b_k term may an elementary solution Y_k be found?

▌ Whenever b_k is a power function or a sine or cosine function, the solution Y_k has similar character. Table 10.1 makes this somewhat more precise. If the Y_k suggested in Table 10.1 includes a solution of the associated homogeneous equation, then this Y_k should be multiplied by k until no such solutions are included. Further examples of the effectiveness of this procedure will be given.

TABLE 10.1

b_k	Y_k
Ax^k	Cx^k
k^n	$C_0 + C_1 k + C_2 k^2 + \cdots + C_n k^n$
$\sin A_k$ or $\cos Ak$	$C_1 \sin Ak + C_2 \cos Ak$
$k^n x^k$	$x^k(C_0 + C_1 k + C_2 k^2 + \cdots + C_n k^n)$
$x^k \sin Ak$ or $x^k \cos Ak$	$x^k(C_1 \sin Ak + C_2 \cos Ak)$

10.75 The *national income equation* is $y_{k+2} - 2ay_{k+1} + ay_k = I$ where $0 < a < 1$. Assuming a and I constant, solve this equation and find the limiting national income for increasing k.

$\blacksquare$ The characteristic equation for the associated homogeneous problem is $r^2 - 2ar + a = 0$ and has complex roots

$$r = a \pm \sqrt{a^2 - a} = a \pm i\sqrt{a - a^2} = \sqrt{a} \, e^{\pm i\theta}$$

where $\cos\theta = \sqrt{a}$ and $\sin\theta = \sqrt{1-a}$. The solution of the national income equation is therefore

$$y_k = c_1(\sqrt{a})^k \sin k\theta + c_2(\sqrt{a})^k \cos k\theta + Y_k$$

To determine Y_k we note that $b_k = I$ is constant, so that $Y_k = Y$, also a constant, is suggested by Table 10.1. Substituting we find $Y(1 - a) = I$, and the completed solution is

$$y_k = c_1(\sqrt{a})^k \sin k\theta + c_2(\sqrt{a})^k \cos k\theta + I/(1-a)$$

Since $0 < a < 1$, it follows easily that $\lim y_k = I/(1-a)$ with y_k itself oscillating above and below this limit during the approach.

10.76 Solve the equation $y_{k+2} - 2y_{k+1} + y_k = 1$ with $y_0 = 1$ and $y_1 = 0$.

$\blacksquare$ The associated homogeneous equation was solved earlier, in Problem 10.42. Recalling that result, we may now write $y_k = c_1 + c_2 k + Y_k$.

Since $b_k = 1$ is again a constant, we might anticipate a constant Y_k. However, a constant is included in the homogeneous solution, and so is a constant multiplied by k. Accordingly we try $Y_k = Ck^2$ and substitute to find $C[(k+2)^2 - 2(k+1)^2 + k^2] = 1$ which is true provided $C = \frac{1}{2}$. Thus

$$y_k = c_1 + c_2 k + \tfrac{1}{2}k^2$$

Since our difference equation can be written as $\Delta^2 y_k = 1$, this quadratic could have been guessed at once. The initial conditions lead to $y_k = 1 - \frac{3}{2}k + \frac{1}{2}k^2$.

10.77 Solve $y_{k+2} - 7y_{k+1} + 10y_k = 4^k$ with initial conditions $y_0 = 0$ and $y_1 = 0$.

$\blacksquare$ The homogeneous solution is found first and the roots of the characteristic equation $r^2 - 7r + 10 = 0$ being 2 and 5, proves to be

$$y_k = c_1 2^k + c_2 5^k.$$

For the needed solution of the nonhomogeneous equation we try $Y_k = a \cdot 4^k$. Substituting,

$$a(4^{k+2} - 7 \cdot 4^{k+1} + 10 \cdot 4^k) = 4^k$$

leading to $a = -\frac{1}{2}$. The complete solution of the given equation is then

$$y_k = c_1 \cdot 2^k + c_2 \cdot 5^k - \tfrac{1}{2} \cdot 4^k$$

To satisfy the initial conditions we need

$$c_1 + c_2 = \tfrac{1}{2} \quad \text{and} \quad 2c_1 + 5c_2 = 2$$

so that $c_1 = \frac{1}{6}$ and $c_2 = \frac{1}{3}$. Finally

$$y_k = \tfrac{1}{6} \cdot 2^k + \tfrac{1}{3} \cdot 5^k - \tfrac{1}{2} \cdot 4^k$$

10.78 Solve $y_{k+2} - 7y_{k+1} + 10y_k = 15 \cdot 5^k$ with $y_0 = y_1 = 0$.

$\blacksquare$ The companion homogeneous equation is the same as in the preceding problem, with solution $y_k = c_1 \cdot 2^k + c_2 \cdot 5^k$. But here the term 5^k is itself one of these solutions, so it would be pointless to try $Y_k = a \cdot 5^k$ as a nonhomogeneous solution. Instead we put $Y_k = ak \cdot 5^k$ and substitute:

$$a(k+2)5^{k+2} - 7a(k+1)5^{k+1} + 10ak5^k = 15 \cdot 5^k$$

Cancelling 5^k we find that the k term on the left has a zero coefficient, leaving $15a = 15$ and $a = 1$. The solution family for the given equation is

$$y_k = c_1 \cdot 2^k + c_2 \cdot 5^k + k \cdot 5^k$$

and we are ready for the initial conditions. These require

$$c_1 + c_2 = 0 \quad \text{and} \quad 2c_1 + 5c_2 = -5$$

making $c_1 = -c_2 = \frac{5}{3}$. The final solution is thus

$$y_k = \tfrac{5}{3} \cdot 2^k - \tfrac{5}{3} \cdot 5^k + k \cdot 5^k.$$

10.8 VARIATION OF PARAMETERS

10.79 Develop the procedure known as variation of parameters for finding a particular solution of a nonhomogeneous difference equation.

▌ Let the given equation be

$$y_{k+2} + a_k y_{k+1} + b_k y_k = f_k$$

and suppose the solutions u_k and v_k of the companion homogeneous equation known, making the family of solutions of that equation

$$y_k = c_1 u_k + c_2 v_k$$

The idea of parameter variation is to assume a nonhomogeneous solution in the form

$$Y_k = A_k u_k + B_k v_k$$

with the constants (parameters) c_1 and c_2 replaced by functions of k. Substituting this into the given equation,

$$(*) \qquad A_{k+2} u_{k+2} + B_{k+2} v_{k+2} + a_k A_{k+1} u_{k+1} + a_k B_{k+1} v_{k+1} + b_k A_k u_k + b_k B_k v_k = f_k$$

But $u_{k+2} + a_k u_{k+1} + b_k u_k = 0$ and $v_{k+2} + a_k v_{k+1} + b_k v_k = 0$ since u_k and v_k are homogeneous solutions. Solving these equations for u_{k+2} and v_{k+2} and introducing the results into $(*)$ we have

$$-a_k \big[u_{k+1}(A_{k+2} - A_{k+1}) + v_{k+1}(B_{k+2} - B_{k+1}) \big] - b_k \big[u_k(A_{k+2} - A_k) + v_k(B_{k+2} - B_k) \big] = f_k$$

With two coefficient functions available (A_k and B_k) and only the original difference equation to be satisfied, we have a degree of freedom. It will be used to simplify this last equation by imposing the constraint

$$(**) \qquad u_{k+1}(A_{k+2} - A_{k+1}) + v_{k+1}(B_{k+2} - B_{k+1}) = 0$$

making the coefficient of a_k zero and leaving

$$u_k(A_{k+2} - A_k) + v_k(B_{k+2} - B_k) = -f_k/b_k$$

But the constraint is equivalent to

$$u_k(A_{k+1} - A_k) + v_k(B_{k+1} - B_k) = 0$$

and subtracting the last two equations

$$(**) \qquad u_k(A_{k+2} - A_{k+1}) + v_k(B_{k+2} - B_{k+1}) = -f_k/b_k$$

The two equations marked $(**)$ can now be solved, yielding

$$A_{k+2} - A_{k+1} = v_{k+1} f_k/b_k W \qquad B_{k+2} - B_{k+1} = -u_{k+1} f_k/b_k W$$

where W is the Wronskian $u_{k+1} v_k - v_{k+1} u_k$. The functions A_k and B_k are now found by finite integration and substituted into the formula for Y_k. In summary, the method of variation of parameters reduces the problem of finding a solution of the nonhomogeneous equation to a few finite integrations (two in the second order case), once the companion homogeneous equation has been solved.

10.80 Apply variation of parameters to $y_{k+2} - 5y_{k+1} + 6y_k = k^2$.

▮ Because of the constant coefficients the homogeneous equation is easily solved, two independent solutions being $u_k = 2^k$ and $v_k = 3^k$. This makes $W = u_{k+1}v_k - v_{k+1}u_k = -2^k3^k$. With $f_k = k^2$ and $b_k = 6$, the preceding problem makes

$$A_{k+2} - A_{k+1} = v_{k+1}f_k/b_kW = 3^{k+1}k^2/6(-2^k3^k) = -k^2/2^{k+1}$$
$$B_{k+2} - B_{k+1} = -u_{k+1}f_k/b_kW = -2^{k+1}k^2/6 \cdot 2^k3^k = k^2/3^{k+1}$$

For the finite integrations we recall the result of Problem 3.119:

$$\sum_{i=0}^{\infty} i^2R^i = \frac{n^2R^n}{R-1} - \frac{2R}{R-1}\left[\frac{nR^n}{R-1} + \frac{R(1-R^n)}{(1-R)^2}\right] - \frac{R}{R-1} \cdot \frac{1-R^n}{1-R}$$

With $R = \frac{1}{2}$ this simplifies to $\sum_0^{n-1}i^2/2^i = -(\frac{1}{2})^n(2n^2 + 4n + 6) + 6$ which, applied to our equation for A, where we have k in place of n, yields

$$A_{k+1} - A_1 = -\frac{1}{2}\sum_0^{k-1} i^2/2^i = (\tfrac{1}{2})^k(k^2 + 2k + 3) - 3$$

Both constant terms A_1 and -3 can be abandoned since they would eventually multiply u_k and reproduce part of the homogeneous solution. Similarly, using $R = \frac{1}{3}$, $\sum_0^{n-1}i^2/3^i = -\frac{3}{2}(\frac{1}{3})^n(n^2 + n + 1) + \frac{3}{2}$ and we find

$$B_{k+1} - B_1 = \frac{1}{3}\sum_0^{k-1} i^2/3^i = \frac{1}{3}(-\tfrac{3}{2})(\tfrac{1}{3})^k(k^2 + k + 1) + \frac{1}{2}$$

Replacing k by $k - 1$ and ignoring additive constants, we take

$$A_k = (\tfrac{1}{2})^{k-1}(k^2 + 2) \quad \text{and} \quad B_k = -\frac{1}{2}(\tfrac{1}{3})^{k-1}(k^2 - k + 1)$$

The target solution is thus

$$y_k = (\tfrac{1}{2})^{k-1}(k^2 + 2)2^k - \frac{1}{2}(\tfrac{1}{3})^{k-1}(k^2 - k + 1) \cdot 3^k + c_12^k + c_23^k$$

10.81 Use variation of parameters to solve $y_{k+2} - y_k = f_k$.

▮ Before proceeding it should probably be pointed out that the values of y_k for odd and even arguments k are quite independent of each other. The solution for even k is

$$y_k = y_0 + f_0 + f_2 + f_4 + \cdots + f_{k-2}$$

while that for odd k is

$$y_k = y_1 + f_1 + f_3 + f_5 + \cdots + f_{k-2}$$

So the equation hardly has a challenging character. Even so, it can serve as an easy illustration of various ideas. With $u_k = 1$, $v_k = (-1)^k$, $b_k = -1$ and $W = 2(-1)^k$ we have

$$A_{k+2} - A_{k+1} = (-1)^{k+1}f_k/2(-1)^{k+1} = \frac{1}{2}f_k$$
$$B_{k+2} - B_{k+1} = -f_k/2(-1)^{k+1} = \frac{1}{2}f_k(-1)^k$$

so that, summing from 0 to $k - 2$,

$$A_k - A_1 = \frac{1}{2}\sum_0^{k-2} f_i \quad \text{and} \quad B_k - B_1 = \frac{1}{2}\sum_0^{k-2} (-1)^if_i$$

Abandoning the additive constants A_1 and B_1, the solution family can be written as

$$y_k = c_1 + c_2(-1)^k + A_k(1) + B_k(-1)^k$$
$$= c_1 + c_2(-1)^k + \frac{1}{2}\sum_0^{k-2} f_i + \frac{1}{2}\sum_0^{k-2} (-1)^if_i(-1)^k$$

which is the equivalent of the simpler appearing sums just given. For $k = 0$ and $k = 1$ the summations may be taken to be zero, and the initial values y_0 and y_1 can be accommodated by setting

$$c_1 = \tfrac{1}{2}(y_0 + y_1) \qquad c_2 = \tfrac{1}{2}(y_0 - y_1)$$

10.82 In the preceding problem let $f_k = k \cdot 2^k$ and evaluate the summations in closed form.

▮ From Problem 3.117 we have, after minor changes

$$\sum_0^{k-2} iR^i = \frac{(k-1)R^{k-1}}{R-1} + \frac{R(1-R^{k-1})}{(R-1)^2}$$

With $R = 2$ this determines

$$A_k - A_1 = \tfrac{1}{2}\sum_0^{k-2} i \cdot 2^i = \tfrac{1}{2}\big[(k-1)2^{k-1} + 2(1-2^{k-1})\big] = 2^{k-2}(k-3) + 1$$

while with $R = -2$ it manages

$$B_k - B_1 = \tfrac{1}{2}\sum_0^{k-2} i(-2)^i = \tfrac{1}{2}(-2)^{k-1}\big(-\tfrac{1}{3}k + \tfrac{5}{9}\big) - \tfrac{1}{9}$$

Abandoning the additive terms the solution

$$Y_k = A_k + B_k(-1)^k = 2^{k-2}(k-3) + (-1)^{2k-1}2^{k-2}(-k/3 + \tfrac{5}{9})$$
$$= 2^{k-2}\big(\tfrac{4}{3}k - \tfrac{32}{9}\big)$$

of the nonhomogeneous equation is in hand.

10.83 Solve $(k+4)y_{k+2} + y_{k+1} - (k+1)y_k = 1$ by variation of parameters.

▮ In Problem 10.67 reduction of the order was used to complete the family of solutions of the companion homogeneous equation,

$$u_k = \frac{1}{(k+1)(k+2)} \qquad v_k = \frac{(-1)^{k+1}(2k+3)}{4(k+1)(k+2)}$$

being the independent solutions. Their Wronskian can be found to be $W = (-1)^{k+1}/(k+1)(k+2)(k+3)$. In developing the method of variation of parameters it was assumed that the leading coefficient was 1, so that here we have $a_k = 1/(k+4)$, $b_k = -(k+1)/(k+4)$ and $f_k = 1/(k+4)$. Then, after substantial cancellation,

$$A_{k+2} - A_{k+1} = v_{k+1}f_k/b_kW = (2k+5)/4$$
$$B_{k+2} - B_{k+1} = -u_{k+1}f_k/b_kW = (-1)^{k+1}$$

and summing from 0 to $k - 2$,

$$A_k - A_1 = \tfrac{1}{4}\sum_0^{k-2}(2i+5) = \tfrac{1}{4}k^2 + \tfrac{1}{2}k - \tfrac{3}{4}$$

$$B_k - B_1 = \sum_0^{k-2}(-1)^{i+1} = \tfrac{1}{2}(-1)^{k-1} - \tfrac{1}{2}$$

Abandoning all the additive constants we find the solution for the nonhomogeneous equation:

$$Y_k = \big(\tfrac{1}{4}k^2 + \tfrac{1}{2}k\big)u_k + \tfrac{1}{2}(-1)^{k-1}v_k$$
$$= \frac{k(k+3)}{4(k+1)(k+2)}$$

10.9 BOUNDARY VALUE PROBLEMS

10.84   Show that the equation $y_{k+2} + y_k = 0$ has one solution satisfying the boundary conditions $y_0 = y_N = 0$ if N is odd, and infinitely many if N is even.

  ❚   The solution family is $y_k = c_1 \sin(\pi k/2) + c_2 \cos(\pi k/2)$. The condition $y_0 = 0$ requires $c_2 = 0$. Thus $y_N = c_1 \sin(\pi N/2) = 0$. If N is odd this requires $c_1 = 0$, and $y_k \equiv 0$ becomes the only solution of the boundary value problem. If N is even any constant c_1 serves, and the family of solutions $y_k = c_1 \sin(\pi k/2)$ exists. This alternative is characteristic of homogeneous boundary value problem (which always have the solution $y_k \equiv 0$).

10.85   For the difference equation of the previous problem show that one solution satisfies the boundary conditions $y_0 = A$, $y_N = B$ if N is odd, and that there is no solution at all if N is even, unless $A \cos(\pi N/2) = B$, in which case infinitely many exist.

  ❚   The boundary conditions require $c_2 = A$ and $c_1 \sin(\pi N/2) + A \cos(\pi N/2) = B$. If N is odd, we find $c_1 = B[\sin(\pi N/2)]$ and the solution is uniquely determined. If N is even, then the boundary values A and B must satisfy $A \cos(\pi N/2) = B$ or the condition at $k = N$ cannot be met by any solution. If A and B do meet this requirement, however, any constant c_1 will serve. This alternative is characteristic of nonhomogeneous boundary value problems. The solution will be uniquely determined precisely when the associated homogeneous problem has only the solution $y_k \equiv 0$. This occurs here for N odd (see Problem 10.84) and there will be no solution at all in the case where the associated homogeneous problem has infinitely many solutions (N even), unless the boundary values meet a special requirement, and then both problems have an infinity of solutions.

10.86   Find all solutions of the homogeneous boundary value problem

$$y_{k+2} + (L - 2) y_{k+1} + y_k = 0$$

with $y_0 = y_N = 0$. Assume $0 < L < 4$. (Such problems occur in the study of psychometrics.)

  ❚   The characteristic equation is $r^2 + (L - 2)r + 1 = 0$, and since $0 < L < 4$ the roots are complex,

$$r = \left(1 - \tfrac{1}{2}L\right) \pm \tfrac{1}{2}i\sqrt{L(4 - L)} = e^{\pm i\theta}$$

with $\cos \theta = 1 - \tfrac{1}{2}L$. The solutions are therefore

$$y_k = c_1 \sin k\theta + c_2 \cos k\theta$$

The first boundary condition requires $y_0 = c_2 = 0$. The second makes $y_N = c_1 \sin N\theta = 0$ which can usually be satisfied only by making $c_1 = 0$, leading to the ever-present solution of homogeneous problems, $y_k \equiv 0$.

    The interest lies, however, in the circumstances under which still other solutions will exist. What is required in this example is that $\sin N\theta = 0$, since then c_1 is arbitrary and we have the family of solutions $y_k = c_1 \sin k\theta$. But $\sin N\theta$ is zero only for $N\theta = n\pi$, where n is an integer. This may be converted into a requirement concerning the parameter L which occurs in the difference equation itself. We find

$$L = 2 - 2\cos\theta = 2 - 2\cos(n\pi/N) = 4\sin^2(n\pi/2N)$$

Though n may be any integer, the integers $1, \ldots, N$ exhaust the possibilities and we have shown that the values (known as eigenvalues)

$$L_n = 4\sin^2(n\pi/2N) \qquad n = 1, 2, 3, \ldots, N$$

lead to families of solutions (known as eigenfunctions)

$$y_k^{(n)} = c_1 \sin(n\pi k/N)$$

For other L values, only the solution $y_k \equiv 0$ exists. (See Problem 10.87.)

10.87   Show that the difference equation $y_{k+2} + (L - 2)y_{k+1} + y_k = 0$ with $y_0 = y_N = 0$ has only the solution $y_k = 0$ if L is outside the interval $0 \le L \le 4$.

  ❚   This equation, with L within the interval in question, was treated in Problem 10.86. Here the characteristic equation has the roots $r_1, r_2 = (1 - \tfrac{1}{2}L) \pm \tfrac{1}{2}\sqrt{L(L - 4)}$. The solution family is $y_k = c_1 r_1^k + c_2 r_2^k$, and the boundary conditions become

$$y_0 = c_1 + c_2 \qquad y_N = c_1 r_1^N + c_2 r_2^N$$

which have a unique solution for any y_0, y_N since r_1 and r_2 are distinct. In particular, if both boundary values are set to zero, the constants c_1 and c_2 will also be zero and the solution y_k is identically zero.

10.88 Find the only solution of

$$y_{k+2} + (L-2)\,y_{k+1} + y_k = 1 \qquad k = 0 \text{ to } 8$$

with $L = 1$ and boundary conditions $y_0 = y_{10} = 0$.

▌ Note first that in Problem 10.86 the eigenvalues for this equation were found to be $L_n = 4\sin^2(n\pi/2N)$ for $n = 1$ to N. With $N = 10$ this list does not include the value 1, so there ought to be a unique solution as requested. The needed solution of this nonhomogeneous equation is easily seen to be $Y_k = 1$, making the complete family

$$y_k = c_1 \sin(k\pi/3) + c_2 \cos(k\pi/3) + 1$$

since $\cos\theta = 1 - \frac{1}{2}L = \frac{1}{2}$. The boundary conditions become $y_0 = c_2 + 1 = 0$ and $y_{10} = c_1 \sin(10\theta) + c_2 \cos(10\theta) + 1 = 0$ so that $c_2 = -1$ and $c_1 = (-1 + \cos(10\theta))/\sin(10\theta)$. These are the unique coefficient values and determine the unique solution.

10.89 Show that changing N from 10 to 12 in the preceding problem leads to multiple solutions.

▌ The eigenvalues now become $L_n = 4\sin^2(n\pi/24)$ for $n = 1$ to 12. The $n = 4$ value makes $L = 1$ an eigenvalue and we anticipate multiple solutions. The boundary condition $y_0 = 0$ makes $c_2 = -1$ as before, while $y_{12} = 0$ becomes

$$c_1 \sin(4\pi) - \cos(4\pi) + 1 = 0$$

and holds for any c_1. Any member of the family

$$y_k = c_1 \sin(k\pi/3) - \cos(k\pi/3) + 1$$

will be a solution of this boundary value problem.

10.90 Show that for N even, there is no solution of $y_{k+2} - y_k = 0$ with boundary values $y_0 = 0$ and $y_N = 1$, but that there are infinitely many solutions with both boundary values zero.

▌ This is, of course, a trivial problem, but even so it offers a clear picture of the alternatives. The solution family is most simply written as

$$y_0, y_1, y_0, y_1, y_0, y_1, \ldots$$

and with alternate values the same as y_0 it is abundantly clear that for N even we must also have $y_N = 0$. The first set of boundary value is, therefore, impossible, while the second allows any value for y_1 and so permits an infinity of solution sequences.

10.91 Referring to the preceding problem, what is the situation for N odd?

▌ Specifying the value of y_N is then equivalent to specifying y_1. With y_0 and y_1 in hand we have an initial value problem and a unique solution. If both y_0 and y_N are zero, that solution is identically zero.

10.92 Find the eigenvalues, if any, of the boundary value problem

$$y_{k+2} - y_{k+1} + Ly_k = 0 \qquad y_0 = y_N = 0$$

▌ The characteristic equation has roots $r_1, r_2 = (1 \pm \sqrt{1 - 4L})/2$. If $L < \frac{1}{4}$ these are real and distinct and

$$y_k = c_1 r_1^k + c_2 r_2^k$$

The boundary conditions become

$$c_1 + c_2 = y_0 \qquad c_1 r_1^N + c_2 r_2^N = y_N$$

and have a unique solution for any y_0, y_N. If both of these boundary values are zero, then $y_k = 0$ identically. If $L = \frac{1}{4}$ standard procedure finds the solution

$$y_k = (c_1 + c_2 k)\left(\tfrac{1}{2}\right)^k$$

with the boundary conditions becoming

$$c_1 = y_0 \qquad (c_1 + c_2 N)\left(\tfrac{1}{2}\right)^N = y_N$$

and again having a unique solution.

There remains the case $L > \frac{1}{4}$, with $r_1, r_2 = \frac{1}{2} \pm \frac{1}{2}i\sqrt{4L-1}$, or $Re^{i\theta}$ with $R^2 = L$ and $\tan\theta = \sqrt{4L-1}$. The solution is

$$y_k = R^k(c_1 \sin k\theta + c_2 \cos k\theta)$$

so the boundary conditions become

$$c_2 = y_0 \qquad R^N(c_1 \sin N\theta + c_2 \cos N\theta) = y_N$$

If $\sin N\theta \neq 0$, the coefficients c_1, c_2 are uniquely determined, both being zero in the special case $y_0 = y_N = 0$. However, if $\sin N\theta = 0$, then c_1 is arbitrary and multiple solutions exist. This occurs when $\theta = n\pi/N$ for integer n, and from $\tan^2\theta = 4L - 1$ we obtain the eigenvalues

$$L_n = \frac{1}{4}\sec^2(n\pi/N)$$

The corresponding eigenfunctions are, since $c_2 = 0$,

$$y_k = c_1 R^k \sin(kn\pi/N)$$

with c_1 arbitrary.

10.10 HIGHER-ORDER EQUATIONS

10.93 Solve the difference equation $y_{k+4} - A^4 y_k = f_k$.

∎ As with second-order equations, we first solve the associated homogeneous equation. The search for power functions $y_k = r^k$ quickly leads to the characteristic equation

$$r^4 - A^4 = (r^2 - A^2)(r^2 + A^2) = 0$$

with the possibilities $r = \pm A, \pm Ai$. This suggests the functions

$$y_k = c_1 A^k + c_2(-A)^k + c_3 A^k \sin(\pi k/2) + c_4 A^k \cos(\pi k/2)$$

which can be verified to be solutions of the homogeneous equation. To satisfy the given equation, this may now be augmented by adding one particular solution Y_k. Again following Table 10.1, such a solution can be found by a method of undetermined coefficients. For instance, suppose

$$f_k = P_2(k) = \frac{1}{2}(3k^2 - 1)$$

Then $Y_k = C_1 k^2 + C_2 k + C_3$ will be a solution provided

$$C_1(1 - A^4)k^2 + [8C_1 + C_2(1 - A^4)]k + [16C_1 + 4C_2 + C_3(1 - A^4)] = \frac{3}{2}k^2 - \frac{1}{2}$$

as we find upon substitution. Comparing coefficients leads to

$$Y_k = \frac{3}{2B}k^2 - \frac{12}{B^2}k - \left(\frac{1}{2B} + \frac{24}{B^2} + \frac{48}{B^3}\right)$$

where $B = 1 - A^4$. (For $A = 1$, a higher-degree polynomial is needed.) The functions

$$y_k = c_1 A^k + c_2(-A)^k + c_3 A^k \sin(\pi k/2) + c_4 A^k \cos(\pi k/2) + Y_k$$

can be verified as solutions of the given equation. Four initial conditions would be sufficient to determine the at present arbitrary constants c_i.

10.94 Solve the equation of Problem 10.93 when $f_k = 2^k$.

∎ Let $Y_k = C \cdot 2^k$ and substitute to obtain $C(16 - A^4) = 1$ so that one solution is $Y_k = 2^k/(16 - A^4)$ provided $A \neq 2$. Adding this to the solution of the homogeneous equation already found, we have a four parameter family of solutions, as may easily be verified. The case $A = 2$ responds to the supposition $Y_k = Ck \cdot 2^k$.

10.95 Solve the equation of Problem 10.93 when $f_k = F \cos \omega k$.

▌ Let $Y_k = C_1 \cos \omega k + C_2 \sin \omega k$ and substitute to obtain, after application of a familiar trigonometric identity,

$$(\cos \omega k)\left[C_1(\cos 4\omega - A^4) + C_2 \sin 4\omega\right] + (\sin \omega k)\left[-C_1 \sin 4\omega + C_2(\cos 4\omega - A^4)\right] = F \cos \omega k$$

Matching coefficients of $\cos \omega k$ and $\sin \omega k$ now brings

$$C_1 = F(\cos 4\omega - A^4)/D \qquad C_2 = (F \sin 4\omega)/D$$

where $D = \sin^2 4\omega + (\cos 4\omega - A^4)^2$. Using this C_1 and C_2, the Y_k function may be added to the solution of the homogeneous equation. The case where $D = 0$ must be handled in a slightly different way.

10.96 Solve $y_{k+3} + y_k = 0$.

▌ Of course, the solution sequence is

$$y_0, y_1, y_2, -y_0, -y_1, -y_2, y_0, \ldots$$

but the equation can also be solved in closed form by the usual process. The characteristic equation $r^3 + 1 = 0$ has the roots -1 and $e^{\pm \pi/3}$, making the solution

$$y_k = c_1(-1)^k + c_2 \sin(\pi k/3) + c_3 \cos(\pi k/3)$$

10.97 Solve $y_{k+4} + y_k = 0$ with initial conditions $y_0 = y_1 = y_2 = 0$ and $y_3 = 1$.

▌ The solution sequence is clearly

$$0, 0, 0, 1, 0, 0, 0, -1, 0, \ldots$$

but going for the closed form we solve $r^4 + 1 = 0$ for its roots $e^{\pm i\pi/4}$ and $e^{\pm 3i\pi/4}$. The solution

$$y_k = c_1 \sin(\pi k/4) + c_2 \cos(\pi k/4) + c_3 \sin(3\pi k/4) + c_4 \cos(3\pi k/4)$$

then puts the initial conditions into the form

$$\begin{aligned} c_2 \qquad\quad + c_4 &= 0 \\ c_1 + c_2 + c_3 - c_4 &= 0 \\ c_1 \qquad - c_3 \qquad &= 0 \\ c_1 - c_2 + c_3 + c_4 &= 2/\sqrt{2} \end{aligned}$$

with solution

$$c_1 = -c_2 = c_3 = c_4 = 1/2\sqrt{2}$$

10.98 Solve $y_{k+4} + 5y_{k+2} + 4y_k = 1 + 2^k$.

▌ The characteristic equation has roots $\pm i, \pm 2i$ making the solution of the companion homogeneous equation

$$y_k = c_1 \sin(\pi k/2) + c_2 \cos(\pi k/2) + 2^k\left[c_3 \sin(\pi k/2) + c_4 \cos(\pi k/2)\right]$$

To find a solution Y_k of the nonhomogeneous equation it may be useful to proceed in two steps. The constant 1 on the right suggests a constant C, and mental substitution of C quickly settles its value at $\frac{1}{10}$. The term 2^k suggests $A \cdot 2^k$, and another substitution yields

$$A(16 + 20 + 4) = 1$$

so that $A = \frac{1}{40}$. The complete solution is then found by adding $\frac{1}{10}$ and $2^k/40$ to the preceding y_k.

10.99 Solve $y_{k+4} - 6y_{k+3} + 14y_{k+2} - 14y_{k+1} + 5y_k = 1$.

▌ The procedure is now routine, the only major hurdle being excavation of the roots of the fourth degree equation. Here, as in the preceding problems, there is no great difficulty. The roots of the characteristic equation are 1, 1 and $2 \pm i$. The solution of the companion homogeneous equation is

$$y_k = c_1 + c_2 k + (\sqrt{5})^k\left[c_3 \sin(k\theta) + c_4 \cos(k\theta)\right]$$

where $\theta = \arctan(\frac{1}{2})$.

To find Y_k we observe that the constant term 1 in the difference equation is matched by a constant term in y_k and that there is also the $c_2 k$ term present. Accordingly we try $Y_k = A k^2$, and after substitution and a bit of algebra find $4A = 1$. Adding the term $\frac{1}{4} k^2$ to the preceding y_k produces the complete solution family.

10.100 Solve $y_{k+4} - 2 y_{k+3} + 2 y_{k+2} - 2 y_{k+1} + y_k = k^2$.

▎ Once again we have a cooperative characteristic equation, its roots being 1, 1 and $\pm i$ so that

$$y_k = c_1 + c_2 k + c_3 \sin(\pi k/2) + c_4 \cos(\pi k/2)$$

Here the right-side term of k^2 is not present in the homogeneous solution so we try $Y_k = A k^2$. Substitution leads to the condition $4A = 1$ so we add $\frac{1}{4} k^2$ to the preceding y_k.

10.101 Solve the Euler difference equation

$$k(k+1)(k+2)\, \Delta^3 y_k - k(k+1)\, \Delta^2 y_k - 4k\, \Delta y_k + 4 y_k = 0$$

▎ Following the procedure developed in Problem 10.60 and illustrated in subsequent second-order problems, we let

$$y_k = \Gamma(k+r)/\Gamma(k)$$

and are led to the auxiliary equation

$$r(r-1)(r-2) - r(r-1) - 4r + 4 = 0$$

which could be written directly from the Euler equation itself. The roots prove to be 1, -1 and 4 so

$$y_k = c_1 \frac{\Gamma(k+1)}{\Gamma(k)} + c_2 \frac{\Gamma(k-1)}{\Gamma(k)} + c_3 \frac{\Gamma(k+4)}{\Gamma(k)}$$

and using a familiar property of the gamma function,

$$y_k = c_1 k + c_2 \frac{1}{k-1} + c_3 (k+3)(k+2)(k+1)k$$

10.102 Solve the Euler equation

$$k(k+1)(k+2)(k+3)\, \Delta^4 y_k + \tfrac{15}{16} y_k = 0$$

▎ Because of the missing differences (or zero coefficients) the auxiliary equation is

$$r(r-1)(r-2)(r-3) + \tfrac{15}{16} = 0$$

and a bit of experimentation would probably be enough to discover the root $\frac{1}{2}$ and then the root $\frac{5}{2}$, after which reduction to the quadratic $r^2 - 3r + \frac{3}{4} = 0$ is elementary (also to be reviewed in a later chapter). The last two roots are thus $r_1, r_2 = \frac{1}{2}(3 \pm \sqrt{6})$. The solution

$$y_k = (1/\Gamma(k))[c_1 \Gamma(k+\tfrac{1}{2}) + c_2 \Gamma(k+\tfrac{5}{2}) + c_3 \Gamma(k+r_1) + c_4 \Gamma(k+r_2)]$$

is then available.

10.103 Extend the idea of order reduction to a third degree difference equation.

▎ Let the equation be

$$y_{k+3} + a_1 y_{k+2} + a_2 y_{k+1} + a_3 y_k = 0$$

with the a_i functions of k, and suppose a solution u_k is known. Let $y_k = u_k z_k$ as earlier. Then

(∗) $$u_{k+3} z_{k+3} + a_1 u_{k+2} z_{k+2} + a_2 u_{k+1} z_{k+1} + a_3 u_k z_k = 0$$

But $u_{k+3} + a_1 u_{k+2} + a_2 u_{k+1} + a_3 u_k = 0$ since u_k is a solution, so

$$a_1 u_{k+2} z_{k+2} = -u_{k+3} z_{k+2} - a_2 u_{k+1} z_{k+2} - a_3 u_k z_{k+2}$$

Substituting this into (∗) and rearranging slightly,

$$u_{k+3}(z_{k+3} - z_{k+2}) - a_2 u_{k+1}(z_{k+2} - z_{k+1}) - a_3 u_k(z_{k+2} - z_k) = 0$$

Let $d_k = z_{k+1} - z_k$. The last equation becomes

$$u_{k+3} d_{k+2} - (a_2 u_{k+1} + a_3 u_k) d_{k+1} - a_3 u_k d_k = 0$$

and determines d_k.

10.104 Apply reduction of order to $y_{k+3} - 3y_{k+2} + 3y_{k+1} - y_k = 0$.

▌ The third difference of y_k, which forms the left side, will be recognized. The solution is clearly an arbitrary polynomial of degree 2, so reduction of order is hardly necessary. But to go through the motions, take $u_k = 1$ as the known solutions. Then $y_k = u_k z_k = z_k$. The equation for d_k is not much of a surprise,

$$d_{k+2} - 2d_{k+1} + d_k = 0$$

and has the independent solutions 1 and k. From $d_k = 1$ we have $z_{k+1} - z_k = 1$, making $z_k = y_k = k$. From $d_k = k$ we have, by finite integration, $z_k = \frac{1}{2}k(k-1) = \frac{1}{2}k^2 - \frac{1}{2}k$. Only the k^2 term is new information, and putting the pieces together

$$y_k = c_1 + c_2 k + c_3 k^2$$

as was known from the start.

10.105 Also solve $y_{k+3} - 4y_{k+2} + y_{k+1} + 6y_k = 0$ by order reduction.

▌ Because of the constant coefficients the solution is readily available. However, let $u_k = (-1)^k$ be the known solution and $y_k = (-1)^k z_k$. The equation for d_k is

$$(-1)^{k+3} d_{k+2} - [(-1)^{k+1} + 6(-1)^k] d_{k+1} - 6(-1)^k d_k = 0$$

or

$$-(-1)^k d_{k+2} - 5(-1)^k d_{k+1} - 6(-1)^k d_k = 0$$

simplifying to

$$d_{k+2} + 5d_{k+1} + 6d_k = 0$$

The independent solutions are $(-2)^k$ and $(-3)^k$. From $d_k = (-2)^k$ it follows that $z_k = (-2)^k/(-3)$ and then $y_k = (-1)^k (-2)^k/(-3)$. We choose $y_k = 2^k$. From $d_k = (-3)^k$ there follow $z_k = (-3)^k$ and $y_k = (-1)^k(-3)^k = 3^k$ if we ignore the constant (-4). Putting the pieces together we have the expected solution

$$y_k = c_1(-1)^k + c_2 2^k + c_3 3^k$$

10.106 Solve

$$\frac{y_{k+3}}{k+3} - \frac{y_{k+2}}{k+2} - \frac{y_{k+1}}{k+1} + \frac{y_k}{k} = 0$$

by reduction of the order.

▌ It takes only a moment to see that $u_k = k$ is a solution of this equation, so we can write at once

$$(k+3) d_{k+2} - (k+3) d_k = 0$$

the d_{k+1} term having a zero coefficient. This clearly simplifies to $d_{k+2} = d_k$, which has the solutions 1 and $(-1)^k$. Backtracking from $d_k = 1$ we have

$$z_{k+1} - z_k = 1 \qquad z_k = k \qquad y_k = u_k z_k = k^2$$

while from $d_k = (-1)^k$ come $z_k = (-1)^k/(-2)$ and $y_k = k(-1)^k/(-2)$. The complete solution family of the original difference equation is then

$$y_k = c_1 k + c_2 k^2 + c_3 k(-1)^k$$

10.107 Solve

$$y_{k+3} - \frac{3k+7}{k+2}y_{k+2} + \frac{3k+5}{k+1}y_{k+1} - \frac{k+1}{k}y_k = 0$$

by reduction of the order.

▌ A quick check shows that $u_k = k$ is again a solution. The reduced order equation is thus

$$(k+3)d_{k+2} - 2(k+2)d_{k+1} + (k+1)d_k = 0$$

The variable change $(k+1)d_k = w_k$ may not be entirely necessary but it does replace this by

$$w_{k+2} - 2w_{k+1} + w_k = 0$$

which is clearly satisfied by 1 and k. Backtracking from 1 we encounter

$$d_k = 1/(k+1) = \Delta z_k \qquad z_k = \psi(k) \qquad y_k = k\psi(k)$$

involving the digamma function. Similarly, from k we first find

$$d_k = k/(k+1) = 1 - 1/(k+1)$$

of which the second part has just been treated. Proceeding from the first,

$$\Delta z_k = 1 \qquad z_k = k \qquad y_k = k^2$$

which makes the complete solution family of the given equation

$$y_k = c_1 k + c_2 k^2 + c_3 k\psi(k)$$

10.11 NONLINEAR EQUATIONS

10.108 Solve the nonlinear second-order equation

$$y_{k+2} = y_k y_{k+1}$$

▌ Only some relatively simple nonlinear equations can be solved in a closed form. Here the introduction of logarithms reduces the equation to something familiar. Let $v_k = \log y_k$. Then

$$v_{k+2} = v_{k+1} + v_k$$

which is the equation for Fibonacci numbers. The solution was found in Problem 10.48.

10.109 Solve $y_{k+1}^2 = 1/y_k y_{k+2}$.

▌ Again logarithms lead to a simple linear equation. With v_k denoting $\log y_k$, it follows that $2v_{k+1} = -v_{k+2} - v_k$ or

$$v_{k+2} + 2v_{k+1} + v_k = 0$$

which has the solution $v_k = (c_1 + c_2 k)(-1)^k$. The function y_k is then found by exponentiation.

If initial conditions have been imposed on y_k, it may be simpler to refer them to v_k instead. Thus, $y_0 = 1$ and $y_1 = e$ are converted to $v_0 = 0$ and $v_1 = 1$. This makes $c_1 = 0$ and $c_2 = -1$ so that

$$v_k = k(-1)^{k+1}$$

10.110 Find the solution of $y_{k+2}y_k^2 = y_{k+1}^3$ with initial values $y_1 = 2$ and $y_2 = 1$.

▌ Again letting $v_k = \log y_k$, we soon have

$$v_{k+2} - 3v_{k+1} + 2v_k = 0$$

with initial values $v_1 = \log 2$ and $v_2 = 0$. The solution family

$$v_k = c_1 + c_2 \cdot 2^k$$

accommodates the initial conditions with $c_1 = 2\log 2$ and $c_2 = -\frac{1}{2}\log 2$.

10.111 The equation $y_k y_{k+1} = 2y_k + 1$ can be solved by the change of variable $y_k = c + 1/v_k$ and an appropriate choice of c. Find this c.

▌ Rewriting the equation as $y_{k+1} = 2 + 1/y_k$ and substituting,

$$c + \frac{1}{v_{k+1}} = 2 + \frac{v_k}{v_k c + 1}$$

which, after a bit of algebra, can be altered to

$$(c-2)v_{k+1} + (c^2 - 2c - 1)v_k v_{k+1} + cv_k = -1$$

This will be a linear first-order equation if $c^2 - 2c - 1$ is zero. Choosing $c = 1 + \sqrt{2}$ arranges this and leaves

$$(\sqrt{2} - 1)v_{k+1} + (\sqrt{2} + 1)v_k = -1$$

First-order methods then manage

$$v_k = A\left[\frac{1 + \sqrt{2}}{1 - \sqrt{2}}\right]^k - \frac{1}{2\sqrt{2}}$$

10.112 If $y_{k+1} = \sqrt{2 + y_k}$ with $y_0 = 0$, what is the limiting value of the sequence for k tending to infinity?

▌ Assuming the limit exists and calling it L, we have

$$L = \sqrt{2 + L} \qquad L^2 = 2 + L \qquad L^2 - L - 2 = 0$$

and eventually $L = 2$, the negative root being rejected.

The equation $L = \sqrt{2 + L}$ has an interesting development. If the L on the right is replaced by $\sqrt{2 + L}$, we have $L = \sqrt{2 + \sqrt{2 + L}}$ with an obvious continuation. The ultimate is a representation of 2 in the form

$$2 = \sqrt{2 + \sqrt{2 + \sqrt{2 + \sqrt{2 + \cdots}}}}$$

10.12 MISCELLANEOUS PROBLEMS

10.113 The difference equation

$$L(E)y_k = (E^2 - 5E + 6)y_k = 4^k$$

appears in terms of the operator E. Show how a solution can be found by inverting $L(E)$.

▌ The equation responds to various other procedures, but the inversion of $L(E)$ begins by observing that

$$L(E)R^k = R^{k+2} - 5R^{k+1} + 6R^k = L(R)R^k$$

Rewriting this as

$$\frac{1}{L(E)}R^k = \frac{1}{L(R)}R^k$$

where it is assumed that $L(R)$ is not zero, we see that R^k is a function that responds well to the inverse operator. Applied to the given equation, this result yields

$$y_k = \frac{1}{E^2 - 5E + 6}4^k = \frac{1}{4^2 - 5\cdot 4 + 6}4^k = \frac{1}{2}4^k$$

10.114 Use the equation

$$L(E)y_k = (E^2 - 5E + 6)y_k = f_k$$

to illustrate the method of operator factorization.

∎ Write the equation as $(E-2)(E-3)y_k = f_k$ and then let $(E-3)y_k$ be denoted z_k. Then

$$(E-2)z_k = f_k$$

The summing factor $1/2^k$ converts this to

$$\frac{z_{k+1}}{2^k} - \frac{z_k}{2^{k-1}} = \frac{f_k}{2^k}$$

after which summing from 0 to $k-1$ manages

$$\frac{z_k}{2^{k-1}} - 2z_0 = \sum_0^{k-1} \frac{f_k}{2^k}$$

or, letting $F_k = \sum_0^{k-1} f_i/2^i$,

$$z_k = 2^k z_0 + 2^{k-1}F_k$$

We are now ready for

$$(E-3)y_k = z_k = 2^k z_0 + 2^{k-1}F_k$$

which has the summing factor $1/3^k$. Applying it and then summing from 0 to $k-1$,

$$\frac{y_k}{3^{k-1}} - 3y_0 = z_0 \frac{1 - \left(\frac{2}{3}\right)^k}{1 - \frac{2}{3}} + \sum_0^{k-1} \frac{2^{i-1}F_i}{3^i}$$

from which we find, with a slight change of notation,

$$y_k = c_1 2^k + c_2 3^k + 3^{k-1} \sum_0^{k-1} \frac{2^{i-1}F_i}{3^i}$$

So what has been accomplished? The family of difference equations, with arbitrary f_k, has been "solved" in the form of repeated summations. The first summation produces the function F_k, which in its turn is imbedded in the second summation. It is similar to using repeated integrations to express the solution of a differential equation.

10.115 Test the result of the preceding problem with the case $f_k = 2^k$.

∎ This function seems well chosen. It makes

$$F_k = \sum_0^{k-1} f_i/2^i = \sum_0^{k-1} (1) = k$$

after which the final term in y_k becomes

$$3^{k-1} \sum_0^{k-1} \frac{2^{i-1}F_i}{3^i} = 3^{k-1} \frac{1}{2} \sum_0^{k-1} \left(\frac{2}{3}\right)^i i = 3^{k-1} \frac{1}{2}\left[-3k\left(\frac{2}{3}\right)^k + 6\left(1 - \left(\frac{2}{3}\right)^k\right)\right]$$

where we have used $\sum_0^{k-1} R^i i = kR^k/(R-1) + (R(1-R^k))/((1-R)^2)$. The second of these terms can be referred back to the homogeneous solution, while the first reduces to $-\frac{1}{2}k \cdot 2^k$. This could have been found much more easily by the method of undetermined coefficients.

10.116 Use the factorization method to solve $(E-2)(E-k)y_k = 0$.

∎ The variable coefficient makes things a bit more interesting. Letting $z_k = (E-k)y_k$ we face the almost trivial $(E-2)z_k = 0$, with solution $z_k = z_0 2^k$. So one factor has been handled and we come to

$$(E-k)y_k = z_k = z_0 2^k$$

which has the summing factor $1/(k-1)!$. Suppose, however, we go a different route. The companion homogeneous equation

$$(E-k)y_k = 0 \quad \text{or} \quad y_{k+1} = ky_k$$

has the almost obvious solution $y_k = c(k-1)!$, suggesting a variation of parameters approach. Substituting the trial function $y_k = A_k(k-1)!$ into the original equation leads to

$$A_{k+1}k! - A_k k! = z_0 2^k \quad \text{or} \quad A_{k+1} - A_k = z_0 2^k / k!$$

and then summing from 0 to $k-1$, $A_k - A_0 = z_0 \Sigma_0^{k-1} 2^i / i!$ or

$$y_k = c_1(k-1)! + c_2(k-1)! \sum_0^{k-1} 2^i / i!$$

10.117 Also solve the equation of the preceding problem by reduction of the order.

▌ Let $u_k = (k-1)!$ be the known solution and $y_k = (k-1)! z_k$. The procedure developed in Problem 10.64 then makes

$$d_{k+1} = \frac{2k(k-1)!}{(k+1)!} d_k = \frac{2}{k+1} d_k$$

leading to

$$d_k = \frac{2}{k} \frac{2}{k-1} \frac{2}{k-2} \cdots \frac{2}{1} d_0 = \frac{2^k}{k!} d_0$$

Then $z_{k+1} - z_k = (2^k / k!) d_0$, which sums to

$$z_k = z_0 + d_0 \sum_0^{k-1} 2^i / i!$$

and in one more step we rediscover the solution.

10.118 Now solve the corresponding nonhomogeneous equation

$$(E-2)(E-k) y_k = f_k$$

▌ There are a number of options. Suppose we start with

$$(E-2) z_k = f_k$$

where as before $z_k = (E-k) y_k$. The function $1/2^k$ is a summing factor. Using it leads to

$$(1/2^k) z_{k+1} - (1/2^{k-1}) z_k = f_k / 2^k$$

and summing from 0 to $k-1$,

$$(1/2^{k-1}) z_k - 2z_0 = \sum_0^{k-1} f_i / 2^i$$

or

$$z_k = 2^k z_0 + 2^{k-1} \sum_0^{k-1} f_i / 2^i$$

For the particular case $f_k = 2^k$ this simplifies to

$$z_k = 2^k z_0 + 2^{k-1} k$$

which will be used for the second phase of the procedure. We now have

$$(E-k) y_k = y_{k+1} - ky_k = z_k = z_0 2^k + k \cdot 2^{k-1}$$

for which $1/k!$ is a summing factor. Applying it, summing and rearranging slightly we have our solution:

$$y_k = c_1(k-1)! + c_2(k-1)! \sum_1^{k-1} 2^i / i! + (k-1)! \sum_1^{k-1} 2^{i-1} / (i-1)!$$

10.119 Develop a method of parameter variation for a third-order equation in the form

$$\Delta^3 y_k + a_1 \Delta^2 y_k + a_2 \Delta y_k + a_3 y_k = f_k$$

▌ The idea of parameter variation was introduced in Problem 10.79, using a second-order equation. The procedure worked out could be extended to higher-order equations, but the form of the present equation will lead to something equivalent, though having a different look. Let

$$y_k = c_1 u_k + c_2 v_k + c_3 w_k$$

be the solution family of the companion homogeneous equation and attempt a solution of the given equation in the form

$$y_k = A_k u_k + B_k v_k + C_k w_k$$

Then

$$\Delta y_k = A_k \Delta u_k + B_k \Delta v_k + C_k \Delta w_k + (u_{k+1} \Delta A_k + v_{k+1} \Delta B_k + w_{k+1} \Delta C_k)$$

of which the expression in parentheses will be set to zero. As earlier, the point is that we have three functions A_k, B_k, C_k at our disposal and only one condition to satisfy—the difference equation. The present move proves to be feasible and it does simplify succeeding steps. Again taking differences,

$$\Delta^2 y_k = A_k \Delta^2 u_k + B_k \Delta^2 v_k + C_k \Delta^2 w_k + (\Delta u_{k+1} \Delta A_k + \Delta v_{k+1} \Delta B_k + \Delta w_{k+1} \Delta C_k)$$

and again the expression in parentheses is set to zero. Taking differences one more time,

$$\Delta^3 y_k = A_k \Delta^3 u_k + B_k \Delta^3 v_k + C_k \Delta^3 w_k + (\Delta^2 u_{k+1} \Delta A_k + \Delta^2 v_{k+1} \Delta B_k + \Delta^2 w_{k+1} \Delta C_k)$$

and we must now satisfy the difference equation. Substituting the remaining expressions for y_k and its three differences into the difference equation it is found that the coefficients of A_k, B_k, C_k are all zero (because u_k, v_k, w_k are solutions of the homogeneous equation), leaving the system

$$u_{k+1} \Delta A_k + v_{k+1} \Delta B_k + w_{k+1} \Delta C_k = 0$$
$$\Delta u_{k+1} \Delta A_k + \Delta v_{k+1} \Delta B_k + \Delta w_{k+1} \Delta C_k = 0$$
$$\Delta^2 u_{k+1} \Delta A_k + \Delta^2 v_{k+1} \Delta B_k + \Delta^2 w_{k+1} \Delta C_k = f_k$$

to be solved for $\Delta A_k, \Delta B_k, \Delta C_k$. Finite integrations are then required to obtain these coefficient functions.

10.120 Apply the method of the preceding problem to the equation

$$y_{k+3} - 9 y_{k+2} + 26 y_{k+1} - 24 y_k = 0$$

▌ The characteristic equation has the easy roots 2, 3 and 4 so that we may take $u_k = 2^k$, $v_k = 3^k$ and $w_k = 4^k$. The nonhomogeneous solution is then to be in the form

$$y_k = A_k 2^k + B_k 3^k + C_k 4^k$$

with the coefficients satisfying the system

$$2^{k+1} \Delta A_k + 3^{k+1} \Delta B_k + 4^{k+1} \Delta C_k = 0$$
$$2^{k+1} \Delta A_k + 2 \cdot 3^{k+1} \Delta B_k + 3 \cdot 4^{k+1} \Delta C_k = 0$$
$$2^{k+1} \Delta A_k + 4 \cdot 3^{k+1} \Delta B_k + 9 \cdot 4^{k+1} \Delta C_k = f_k$$

The solution is routine and

$$\Delta A_k = f_k / 2^{k+2} \qquad \Delta B_k = -f_k / 3^{k+1} \qquad \Delta C_k = f_k / 2 \cdot 4^{k+1}$$

Finite integrating and ignoring the additive constant,

$$A_k = \sum_0^{k-1} f_i / 2^{i+2}$$

with similar formulas for the other two.

10.121 The recursion for Legendre polynomials is

$$(n+2)P_{n+2}(x) - (2n+3)xP_{n+1}(x) + (n+1)P_n(x) = 0$$

and for large values of n can be approximated by

$$P_{n+2}(x) - 2xP_{n+1}(x) + P_n(x) = 0.$$

(The 2, 3 and -1 are small in comparison with the accompanying n terms and so are eliminated.) Solve this difference equation.

▮ Consider x as constant and take the case $x > 1$. The characteristic equation will be $r^2 - 2xr + 1 = 0$ with roots $x \pm \sqrt{x^2 - 1}$. Thus

$$P_n(x) = c_1\left(x + \sqrt{x^2-1}\right)^n + c_2\left(x - \sqrt{x^2-1}\right)^n$$

For large n, only the first of these terms need be considered, because it will dominate the other. This makes

$$P_n(x) = c_1\left(x + \sqrt{x^2-1}\right)^n$$

To determine c_1 we can return to the recursion and make a direct computation of both $P_n(x)$ and $(x + \sqrt{x^2-1})^n$ for n increasing. At $x = 2$, for example, the ratio c_1 stabilizes at 7.46 as early as $n = 4$ and becomes a solid 7.4641 a few steps later. At $x = 3$ the value $c_1 = 11.6569$ develops rapidly and, of course, at $x = 1$ all entries have the value 1.

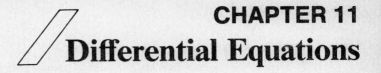

CHAPTER 11
Differential Equations

11.1 THE METHOD OF ISOCLINES

11.1 Use the *method of isoclines* to determine the qualitative behavior of the solutions of $y'(x) = xy^{1/3}$.

▌ This equation can of course be solved by elementary methods, but we shall use it as a test case for various approximation methods. The method of isoclines is based on the family of curves $y'(x) =$ constant which are not themselves solutions but are helpful in determining the character of solutions. In this example the isoclines are the family $xy^{1/3} = M$ where M is the constant value of $y'(x)$. Some of these curves are sketched (dotted) in Fig. 11-1, with M values indicated. Where a solution of the differential equation crosses one of these isoclines, it must have for its slope the M number of that isocline. A few solution curves are also included (solid) in Fig. 11-1. Others can be sketched in, at least roughly.

Accuracy is not the goal of the isocline method, but rather the general character of the solution family. For example, there is symmetry about both axes. One solution through $(0,0)$ and those above it have a U shape. Solutions below this are more unusual. Along $y = 0$ different solutions can come together. A solution can even include a piece of the x axis. One such solution might enter $(0,0)$ on a descending arc, follow the x axis to $(2,0)$ and then start upward again as shown in Fig. 11-2. The possible combinations of line and arc are countless. Information of this sort is often a useful guide when efforts to compute accurate solutions are made.

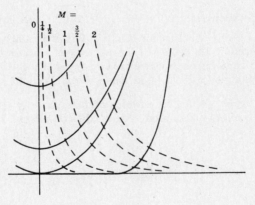

Fig. 11-1

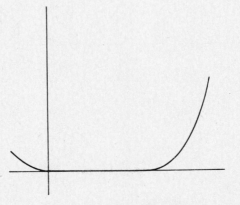

Fig. 11-2

11.2 Apply the method of isoclines to $y'(x) = -xy^2$.

❚ Figure 11-3 shows several isoclines and the solution which passes through $(0, 2)$. Since there is symmetry relative to both axes, only one quadrant is presented. Here it is convenient to also indicate the curve along which $y''(x)$ is zero. It is $y = 1/2x^2$ and, of course, changes in the sign of curvature take place along this curve. Solutions with maxima along the y axis and tending to zero with increasing x, appear to be indicated, although this is not at once obvious from the differential equation itself. (Here again we have a case where elementary methods easily produce the solution and confirm these results.)

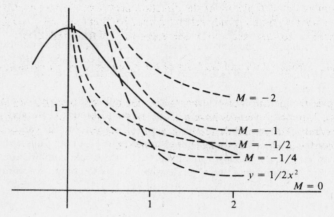

Fig. 11-3

11.3 What is meant by a direction field, and how is it related to the isoclines?

❚ At each point (x, y) where the function $f(x, y)$ of a differential equation $y' = f(x, y)$ is defined there is a prescribed slope or direction. The totality of these directions is called the direction field for that equation. It is the job of a solution function to thread its way through the field in such a way that it has the prescribed slope at each point it transits. The isoclines of the field are curves along which the prescribed slope is constant.

11.4 By considering the direction field, or the isoclines, of the equation $y' = x^2 - y^2$, deduce the qualitative behavior of its solution family. Where will the solutions have maxima or minima? How do they behave as x tends to infinity?

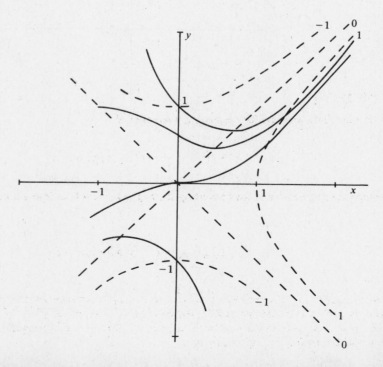

Fig. 11-4

▮ Figure 11-4 shows (dashed) the isoclines for slopes -1, 0 and 1. They are enough to suggest the general picture. Four of the solutions are roughed in (solid). The top one descends rapidly from a region of high negative slopes, rounds to a zero slope on the $y = x$ isocline and then climbs. The next lower solution has a maximum on $y = -x$ (also a part of the zero isocline) and a minimum on $y = x$. The third levels off at the origin and then resumes its climb. When it does cross the (slope 1) isocline, as depicted, it will not be able to get above it again, since this isocline has slope steeper than 1 in this region. It follows that for large positive values of x, solutions will be below the line $y = x$.

11.5 For the equation of the preceding problem, where will the solution through $(-1, 1)$ cross the y axis?

▮ With a much more detailed picture of the direction field it would be possible to give a better answer to this question, but with what we have the only safe reply may be "between the $y = 0$ and $y = 1$ levels." The curve shown in Fig. 11-4 suggests a crossing near $y = .6$. See Problem 11.9 for a followup.

11.6 By considering the direction field and isoclines of the equation $y' = -2xy$, deduce the qualitative behavior of its solutions.

▮ Figure 11-5 gives the picture. The coordinate axes are the zero isocline, and hyperbolic isoclines for slopes -1 and 1 also appear. The solution curves have maxima on the upper y axis or minima on the lower. They settle down to the x axis at infinity. The second derivative is easily found to be $y'' = 2y(2x^2 - 1)$ so the curvature will be zero at $x^2 = \frac{1}{2}$. The lower half plane is symmetric to the upper.

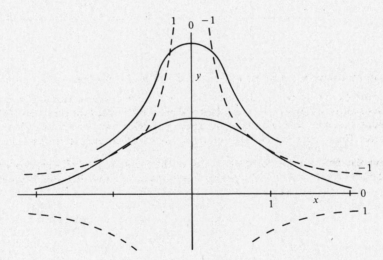

Fig. 11-5

11.2 THE EULER METHOD

11.7 Illustrate the simplest Euler method for computing a solution of

$$y' = f(x, y) = xy^{1/3} \qquad y(1) = 1$$

▮ This is perhaps the original device for converting the method of isoclines into a computational scheme. It uses the formula

$$y_{k+1} - y_k = \int_{x_k}^{x_{k+1}} y' \, dx \sim hy_k'$$

which amounts to considering y' constant between x_k and x_{k+1}. It also amounts to the linear part of a Taylor series, so that if y_k and y_k' were known exactly, the error in y_{k+1} would be $\frac{1}{2}h^2 y^{(2)}(\xi)$. This is called the *local truncation error*, since it is made *in this step* from x_k to x_{k+1}. Since it is fairly large, it follows that rather small increments h would be needed for high accuracy.

The formula is seldom used in practice, but serves to indicate the nature of the task ahead and some of the difficulties to be faced. With x_0, $y_0 = 1$ three applications of this Euler formula, using $h = .01$, bring

$$y_1 \simeq 1 + (.01)(1) = 1.0100$$
$$y_2 \simeq 1.0100 + (.01)(1.01)(1.0033) \simeq 1.0201$$
$$y_3 \simeq 1.0201 + (.01)(1.02)(1.0067) \simeq 1.0304$$

Near $x = 1$ we have $y^{(2)} = y^{1/3} + \frac{1}{3}xy^{-2/3}(xy^{1/3}) \simeq \frac{4}{3}$, which makes the truncation error in each step about .00007. After three such errors, the fourth decimal place is already open to suspicion. A smaller increment h is necessary if we hope for greater accuracy. The accumulation of truncation error is further illustrated in Fig. 11-6 where the computed points have been joined to suggest a solution curve. Our approximation amounts to following successively the tangent lines to various solutions of the equation. As a result the approximation tends to follow the convex side of the solution curve. Notice also that Euler's formula is a nonlinear difference equation of order 1:
$y_{k+1} = y_k + hx_k y_k^{1/3}$.

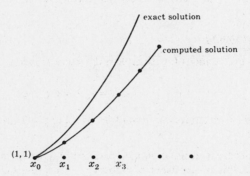

Fig. 11-6

11.8 Illustrate the concept of convergence by comparing the results of applying Euler's method with $h = .10$, .05 and .01 with the correct solution $y = [(x^2 + 2)/3]^{3/2}$.

▌ Convergence refers to the improvement of approximations as the interval h tends to zero. A method that does not converge is of doubtful value as an approximation scheme. Convergence for the various schemes to be introduced will be proved later, but as circumstantial evidence the data of Table 11.1, obtained by Euler's method, are suggestive. Only values for integer x arguments are included, all others being suppressed for brevity.

Notice that across each row there is a reassuring trend toward the exact value. Using smaller intervals means more computing. The value 25.96 in the bottom row, for instance, was obtained in 50 steps, whereas the value 26.89 required 500 steps. The extra labor has brought an improvement, which seems only fair. As h tends to zero the computation grows even longer and we hope that the results approach the exact values as limits. This is the convergence concept. Needless to say, roundoff errors will limit the accuracy attainable, but they are not a part of the convergence issue.

TABLE 11.1

x	$h = .10$	$h = .05$	$h = .01$	Exact
1	1.00	1.00	1.00	1.00
2	2.72	2.78	2.82	2.83
3	6.71	6.87	6.99	7.02
4	14.08	14.39	14.63	14.70
5	25.96	26.48	26.89	27.00

11.9 Referring back to Fig. 11-4 and the solution through $(-1,1)$, how does the Euler method answer the question of its crossing of the y axis?

▌ Using 10 steps at first, and then building to 100, estimates of the crossing point climbed from .58 to .62. This makes the freehand effort shown in the figure look rather good. From this position of strength the method was also asked to follow the solution curve through $(0,0)$, to determine its relationship with the slope 1 isocline. It predicted a crossing between $x = 1.1$ and $x = 1.2$, another boost for freehand solution of differential equations. However, the Euler method then continued on and forecast a second crossing of this same isocline just short of $x = 2$, climbing above the isocline once again, which is clearly impossible. The method is not accurate enough to be trusted on too fine a point.

11.10   Apply the Euler method to $y' = -xy^2$ with $y(0) = 2$. Compute up to $x = 1$ using steps of several sizes. Do the terminal values at $x = 1$ seem to converge toward the exact $y(1) = 1$ as h diminishes?

> ❚ Here are the numbers:

h	.2	.1	.04	.02	.01
$y(1)$	.8	.9	.96	.98	.99

The answer appears to be yes, convergence is occurring. But the pace is slow, with the second decimal place still in doubt.

11.3   AN EXISTENCE THEOREM

11.11   Show that if $f(x, y)$ is a continuous function, then the Euler formula

$$y_{k+1} \simeq y_k + hf(x_k, y_k)$$

permits the construction of an approximate solution $p(x)$, in the form of a polygon chain, for which $|p'(x) - f(x, p(x))| < \epsilon$.

> ❚ The Euler formula produces, from an initial condition $y(x_0) = y_0$, a sequence of values y_k at arguments x_k. Interpreted geometrically this is a sequence of points (x_k, y_k). Connecting these points by line segments, we have a polygon chain $p(x)$ (Fig. 11-7).

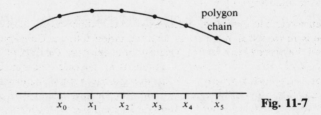

Fig. 11-7

We now show that for at least some local neighborhood of the initial point (x_0, y_0) the condition of the problem can be met. Imagine this initial point the center of a rectangle of width $2a$ and height $2b$. Since $f(x, y)$ is continuous it is also bounded in the rectangle, say $|f(x, y)| < B$. Then if W is the smaller of the two numbers a and b/B, we further restrict our attention to the rectangle R of width $2W$ and height $2BW$ (Fig. 11-8). Choose any positive number ϵ. Since $f(x, y)$ is continuous in R it is uniformly continuous there, which means that there is a

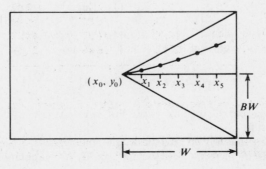

Fig. 11-8

$\delta > 0$ such that $|f(x_1, y_1) - f(x_2, y_2)| < \epsilon$ whenever (x_1, y_1) and (x_2, y_2) are points in R with $|x_1 - x_2| < \delta$ and $|y_1 - y_2| < \delta$. The number δ depends on ϵ but not on the points involved. Now choose h to be the smaller of the two numbers δ and δ/B and apply the Euler method. Noticing that the diagonals of R have slopes of B and $-B$, and recalling $|f(x, y)| < B$, it is clear that no segment of the polygon achieved in this way can be steeper than these diagonals. Accordingly the chain cannot touch these diagonals except at (x_0, y_0). As we follow its progress to the right it must therefore eventually reach the right side of R as shown in the diagram, since the only thing which could stop the Euler method would be to reach a point where $f(x, y)$ is undefined, and this does not happen in R. [The chain can similarly be extended to the left of (x_0, y_0) by using negative h.] Call this chain $p(x)$. Then

$$p'(x) = f(x_k, y_k) \quad \text{for } x_k < x < x_{k+1}$$

and $p'(x)$ fails to exist at the corner points. But now

$$|p'(x) - f(x, p(x))| = |f(x_k, y_k) - f(x, p(x))| < \epsilon$$

for each segment of the chain since

$$|x - x_k| < h < \delta \qquad |p(x) - y_k| < Bh < \delta$$

So the continuity of $f(x, y)$ is enough to guarantee the existence of a function, here a polygon chain, which satisfies the differential equation with accuracy ϵ, at least in the local neighborhood of the initial point. We now proceed to extend this result.

11.12 Show that if $f(x, y)$ also satisfies the Lipschitz condition

$$|f(x, y_1) - f(x, y_2)| < L|y_1 - y_2|$$

L being a positive number, then two functions $Y_1(x)$ and $Y_2(x)$, which satisfy the differential equation with errors ϵ_1 and ϵ_2, respectively,

$$|Y_1'(x) - f(x, Y_1(x))| < \epsilon_1 \qquad |Y_2'(x) - f(x_1 Y_2(x))| < \epsilon_2$$

in the rectangle R of the previous problem, will not differ by more than the following amount for $|x - x_0| \leq W$:

$$|Y_1(x) - Y_2(x)| \leq e^{L|x - x_0|} |Y_1(x_0) - Y_2(x_0)| + \frac{\epsilon_1 + \epsilon_2}{L} [e^{L|x - x_0|} - 1]$$

▮ We focus on the interval $x_0 \leq x \leq x_0 + W$, the argument for the left side of R being similar. Except at a finite set of points where Y_1 and Y_2 may be permitted to have corners, like the polygon chains just produced, we have

$$|Y_1'(x) - Y_2'(x)| \leq |f(x, Y_1(x)) - f(x, Y_2(x))| + \epsilon_1 + \epsilon_2$$
$$\leq L|Y_1(x) - Y_2(x)| + \epsilon_1 + \epsilon_2$$

Let $d(x) = Y_1(x) - Y_2(x)$. Then except at possible corners of Y_1 or Y_2,

$$|d'(x)| \leq L|d(x)| + \epsilon_1 + \epsilon_2$$

First suppose that $d(x)$ is never zero in the interval, say it remains positive. Then

$$d'(x) - Ld(x) \leq \epsilon_1 + \epsilon_2$$

Multiplying by e^{-Lx} and using $d/dx[e^{-Lx} d(x)] = [d'(x) - Ld(x)]e^{-Lx}$, we can integrate between x_0 and x in spite of the finite jumps possible in $d'(x)$, to find

$$e^{-Lx} d(x) - e^{-Lx_0} d(x_0) \leq \frac{\epsilon_1 + \epsilon_2}{L} [e^{-Lx_0} - e^{-Lx}]$$

which easily rearranges into the required result. For $d(x)$ always negative, we may reverse the roles of Y_1 and Y_2 and find the same result.

But it is also possible that $d(x)$ is zero for certain arguments. If it were identically zero then the required result would be true trivially. Suppose that $d(\bar{x})$ is not zero. By its continuity $d(x)$ remains nonzero for some interval about $\bar{x}$ but, since we are concerned with the case that $d(x)$ vanish somewhere, let x^* be its first zero on one side or the other of $\bar{x}$. Since $d(x)$ does not vanish between x^* and $\bar{x}$, we may apply the first case considered with $\bar{x}$ and x^* in place of x and x_0:

$$d(\bar{x}) \le e^{L|\bar{x}-x^*|}|d(x^*)| + \frac{\epsilon_1 + \epsilon_2}{L}[e^{L|\bar{x}-x^*|} - 1] = \frac{\epsilon_1 + \epsilon_2}{L}[e^{L|\bar{x}-x^*|} - 1]$$

This is a stronger inequality than was required, so that the required result holds in all cases.

11.13   Prove that the equation $y' = f(x, y)$ with $y(x_0) = y_0$ has an exact solution for the interval $|x - x_0| \le W$, provided $f(x, y)$ is continuous and satisfies the Lipschitz condition. (Continuity alone guarantees existence but a more strenuous proof is involved.)

❚   This is, of course, the existence theorem. Choose a monotone sequence of positive numbers ϵ_n with $\lim \epsilon_n = 0$. Then by Problem 11.11 we know that a corresponding sequence of polygon chains $p_n(x)$ may be constructed over the indicated interval such that, except at the finite set of corners,

$$|p_n'(x) - f(x, p_n(x))| < \epsilon_n$$

This sequence of functions $p_n(x)$ is uniformly convergent, for since all polygon chains may be started from (x_0, y_0), the inequality of Problem 11.12 makes

$$|p_n(x) - p_m(x)| \le \frac{\epsilon_n + \epsilon_m}{L}[e^{L|x - x_0|} - 1]$$

which is uniformly small for sufficiently large n and m. Since a uniformly convergent sequence of continuous functions has a continuous limit function, we now have

$$\lim p_n(x) = y(x)$$

with $y(x)$ continuous.

Next we show that this $y(x)$ is an exact solution of the differential equation. Notice that

$$|f(x, p_n(x)) - f(x, y(x))| < L|p_n(x) - y(x)|$$

so that the uniform convergence of $p_n(x)$ to $y(x)$ also guarantees

$$\lim f(x, p_n(x)) = f(x, y(x))$$

uniformly. Because of this,

$$\lim \int_{x_0}^{x} f(t, p_n(t))\, dt = \int_{x_0}^{x} f(t, y(t))\, dt$$

Finally we return to

$$|p_n'(x) - f(x, p_n(x))| < \epsilon_n$$

and integrate each side from x_0 to x. Though $p_n(x)$ has corners, its continuity is enough to produce

$$\left| p_n(x) - p_n(x_0) - \int_{x_0}^{x} f(t, p_n(t))\, dt \right| < \epsilon_n W$$

In the limit this becomes

$$y(x) = y_0 + \int_{x_0}^{x} f(t, y(t))\, dt$$

from which

$$y'(x) = f(x, y(x)) \qquad y(x_0) = y_0$$

follow at once. We have now proved that a solution of this initial value problem does exist. We have also proved that the Euler method produces a sequence of functions $p_n(x)$ which converge to this exact solution $y(x)$ as the spacing h between x_k arguments approaches zero (forcing δ and ϵ to zero with it).

11.14 Prove that the exact solution found in Problem 11.13 is unique.

▮ Suppose two solutions existed, say $y_1(x)$ and $y_2(x)$. They could then be considered suitable functions $Y_1(x)$ and $Y_2(x)$ for the inequality of Problem 11.12, with $\epsilon_1 = \epsilon_2 = 0$ and $Y_1(x_0) = Y_2(x_0) = y_0$. Thus $|y_1(x) - y_2(x)| \le 0$ and the two solutions are identical.

11.15 Estimate the difference between $p_n(x)$ and $y(x)$.

▮ By the same inequality, $|p_n(x) - y(x)| \le \epsilon_n/L[e^{L|x-x_0|} - 1]$. Comparing this with $|p'_n(x) - f(x, p_n(x))| \le \epsilon_n$, we are reminded that, quite naturally, there is a difference between how accurately $p_n(x)$ approximates the solution $y(x)$ and how accurately its derivative approximates the function $f(x, y)$.

11.4 THE TAYLOR METHOD

11.16 Apply the local Taylor series method to obtain a solution of $y' = xy^{1/3}$ and $y(1) = 1$ correct to three places for arguments up to $x = 5$.

▮ Generally speaking the method involves using $p(x + h)$ in place of $y(x + h)$, where $p(x)$ is the Taylor polynomial for argument x. We may write directly

$$y(x + h) \simeq y(x) + hy'(x) + \tfrac{1}{2}h^2 y^{(2)}(x) + \tfrac{1}{6}h^3 y^{(3)}(x) + \tfrac{1}{24}h^4 y^{(4)}(x)$$

accepting a local truncation error of amount $E = h^5 y^{(5)}(\xi)/120$.

The higher derivatives of $y(x)$ are computed from the differential equation:

$$y^{(2)}(x) = \tfrac{1}{3}x^2 y^{-1/3} + y^{1/3} \qquad y^{(3)}(x) = -\tfrac{1}{9}x^3 y^{-1} + xy^{-1/3} \qquad y^{(4)}(x) = \tfrac{1}{9}x^4 y^{-5/3} - \tfrac{2}{3}x^2 y^{-1} + y^{-1/3}$$

The initial condition $y(1) = 1$ has been prescribed, so with $x = 1$ and $h = .1$ we find

$$y(1 + .1) \simeq 1 + .1 + \tfrac{2}{3}(.1)^2 + \tfrac{4}{27}(.1)^3 + \tfrac{1}{54}(.1)^4 \simeq 1.10682$$

Next apply the Taylor formula at $x = 1.1$ and find

$$y(1.1 + .1) \simeq 1.22788 \qquad y(1.1 - .1) \simeq 1.00000$$

The second of these serves as an accuracy check since it reproduces our first result to five place accuracy. (This is the same procedure used in Chapter 8 for the error function integral.) Continuing in this way, the results presented in Table 11.2 are obtained. The exact solution is again included for comparison. Though $h = .1$ was used, only values for $x = 1(.5)5$ are listed. Notice that the errors are much smaller than those made in the Euler method with $h = .01$. The Taylor method is a more rapidly convergent algorithm.

TABLE 11.2

x	Taylor result	Exact result	Error
1.0	1.00000	1.00000	—
1.5	1.68618	1.68617	-1
2.0	2.82846	2.82843	-3
2.5	4.56042	4.56036	-6
3.0	7.02123	7.02113	-10
3.5	10.35252	10.35238	-14
4.0	14.69710	14.69694	-16
4.5	20.19842	20.19822	-20
5.0	27.00022	27.00000	-22

11.17 Apply the Taylor method to $y' = -xy^2$ to obtain the solution satisfying $y(0) = 2$.

▮ The procedure of the preceding problem could be applied. Instead, however, an alternative will be illustrated, essentially a method of undetermined coefficients. Assuming convergence at the outset, we write the Taylor series $y(x) = \sum_{i=0}^{\infty} a_i x^i$. Then

$$y^2(x) = \left(\sum_{i=0}^{\infty} a_i x^i \right) \left(\sum_{j=0}^{\infty} a_j x^j \right) = \sum_{k=0}^{\infty} \left(\sum_{i=0}^{k} a_i a_{k-i} \right) x^k \qquad y'(x) = \sum_{i=1}^{\infty} i a_i x^{i-1}$$

Substituting into the differential equation and making minor changes in the indices of summation,

$$\sum_{j=0}^{\infty}(j+1)a_{j+1}x^j = -\sum_{j=1}^{\infty}\left(\sum_{i=0}^{j-1}a_ia_{j-1-i}\right)x^j$$

Comparing coefficients of x^j makes $a_1 = 0$ and

$$(j+1)a_{j+1} = -\sum_{i=0}^{j-1}a_ia_{j-1-i} \quad \text{for } j=1,2,\ldots$$

The initial condition forces $a_0 = 2$ and then we find recursively

$$a_2 = -\tfrac{1}{2}a_0^2 = -2 \qquad\qquad a_6 = -\tfrac{1}{6}\left(2a_0a_4 + 2a_1a_3 + a_2^2\right) = -2$$
$$a_3 = -\tfrac{1}{3}(2a_0a_1) = 0 \qquad\qquad a_7 = -\tfrac{1}{7}(2a_0a_5 + 2a_1a_4 + 2a_2a_3) = 0$$
$$a_4 = -\tfrac{1}{4}\left(2a_0a_2 + a_1^2\right) = 2 \qquad a_8 = -\tfrac{1}{8}\left(2a_0a_6 + 2a_1a_5 + 2a_2a_4 + a_3^2\right) = 2$$
$$a_5 = -\tfrac{1}{5}(2a_0a_3 + 2a_1a_2) = 0$$

and so on. The recursion can be programmed so that coefficients could be computed automatically as far as desired. The indicated series is

$$y(x) = 2(1 - x^2 + x^4 - x^6 + x^8 - \cdots)$$

Since the exact solution is easily found to be $y(x) = 2/(1 + x^2)$, the series obtained is no surprise.

This method sees frequent application. The principle assumption involved is that the solution does actually have a series representation. In this case the series converges only for $-1 < x < 1$. For $-\tfrac{1}{2} < x < \tfrac{1}{2}$ only six terms are needed to give three place accuracy. In the previous problem a new Taylor polynomial was used for each value computed. Here just one such polynomial is enough. The issue is one of range and accuracy required. To proceed up to $x = 5$, for example, the earlier method can be used. In further contrast we may also note that in Problem 11.16 polynomials of fixed degree are used and the convergence issue does not arise explicitly. Here in Problem 11.17 we introduce the entire series into the differential equation, assuming $y(x)$ analytic in the interval of interest.

11.18 Apply the method of the preceding problem to $y' = x^2 - y^2$ with initial condition $y(-1) = 1$.

▌ In Fig. 11-4 a plot of this solution was ventured from a rough picture of the direction field and an estimate of $y = .6$ for its crossing of the y axis was made. In Problem 11.9 the Euler method more or less confirmed this estimate.

Here we assume the solution represented by the series

$$y(x) = \sum_{0}^{\infty}c_i(x+1)^i$$

and compute $y'(x) = \sum_{1}^{\infty}ic_i(x+1)^{i-1}$ and $y^2(x) = \sum_{0}^{\infty}a_k(x+1)^k$ where $a_k = \sum_{0}^{k}c_ic_{k-i}$. Substituting these into the differential equation,

$$\sum_{1}^{\infty}ic_i(x+1)^{i-1} = 1 - 2(x+1) + (x+1)^2 - \sum_{0}^{\infty}a_k(x+1)^k$$

where x^2 has been expressed in terms of $(x+1)$ and its powers. A slight change of index converts this to the more helpful

$$\sum_{0}^{\infty}(k+1)c_{k+1}(x+1)^k = 1 - 2(x+1) + (x+1)^2 - \sum_{0}^{\infty}a_k(x+1)^k$$

Equating coefficients of powers of $(x+1)$ we find at first

$$C_1 = 1 - a_0 \qquad 2c_2 = -2 - a_1 \qquad 3c_3 = 1 - a_2$$

after which the x^2 term no longer has an overt influence and

$$(k+1)c_{k+1} = -a_k \quad \text{for } k \geq 3$$

The initial condition requires $c_0 = 1$, and the next few of these coefficients prove to be $0, -1, 1, -\frac{1}{2}$ making the start of the series

$$y(x) = 1 - (x+1)^2 + (x+1)^3 - \tfrac{1}{2}(x+1)^4 + \cdots$$

Computations with this series produced the numbers

x	$-.8$	$-.6$	$-.4$	$-.2$	0	.2	.4
y	.967	.892	.799	.705	.615	.345	minus

The early values of y are believable and the crossing at $x = 0$ is reassuring, but beyond this point things start to deteriorate.

11.19 Apply the local Taylor series method of Problem 11.16 to the equation of the preceding problem.

▎ The needed derivatives are

$$y' = x^2 - y^2 \quad y'' = 2x - 2yy' \quad y^{(3)} = 2 - 2yy'' - 2(y')^2 \quad y^{(4)} = -2yy^{(3)} - 6y'y''$$

if we truncate the series at this point. These are computed first at the initial point $x = -1$, inserted into the Taylor formula and the value of $y(x+h)$ estimated. The same is then done at $x = x + h$ and so on, a different (truncated) series being used at each step. In runs from $x = -1$ to $x = 2$ results for $h = .1$ and $h = .05$ agreed to three decimal places. Here is a sample of the output:

x	$-.8$	$-.4$	0	.4	.8	1.2	1.6	2.0
y	.967	.799	.620	.516	.556	.788	1.189	1.669

The crossing of the y axis at .620 agrees fairly well with earlier estimates. Comparisons with values of $\sqrt{x^2 - 1}$ (the slope 1 isocline) show the curve passing below it near $x = 1.3$ and then staying below it, as it should.

11.20 Again referring to Fig. 11-4, use the local Taylor series method to track the solutions through $(0, 1)$ and $(0, 0)$. How well do the numbers agree with the rough curves of the figure?

▎ For the initial point $(0, 1)$ the computation found

x	.6	.7	1	1.4	1.6	2.0
y	.684	.680	.750	1.025	1.223	1.679

indicating that the minimum (and the crossing of the 0 isocline) occur between $x = .6$ and $x = .7$, agreeing with the curve of the figure. As for the 1 isocline, its y values at $x = 1.4$ and 1.6 are .980 and 1.249, suggesting a crossing in this interval. No effort was made in the figure to project the curve this far.

For the solution through $(0, 0)$ computations found

x	1	1.2	2
y	.318	.525	1.626

which puts the curve somewhat lower than the sketch, but with the crossing of the isocline between $x = 1$ and $x = 1.2$ as shown.

It should be noted that all solution curves descending across the upper y axis, leveling off as they cross $y = x$, will approach that line again asymptotically. This means a change of curvature from upward to downward. The second derivative being

$$y'' = 2x - 2x^2 y + 2y^3$$

a plot of its zero positions is within reach.

11.21 In applying the local Taylor method to $y' = x^2 - y^2$, as in the preceding problems, what is the computational evidence for convergence?

▎ The runs for the solution through $(0, 0)$ up to $x = 2$ are typical. The terminal values for several step sizes appear in Table 11.3.

TABLE 11.3

h	$y(2)$
1	1.587563
.5	1.613013
.1	1.625859
.05	1.625909
.01	1.625915
.005	1.625916

11.22 What sort of behavior should be expected of the solution through $(0,1)$ of $y' = y^2/(1 - xy)$?

▮ Since an infinite slope is indicated along $xy = 1$, this solution is clearly headed for numerical difficulty. Figure 11-9 shows the appropriate part of the direction field, each dash oriented in the direction of the proper slope. Apart from the infinite slope, it seems that the solution continues upward and backward, as suggested by the solid curve. To the left of the y axis a descent to the $y = 0$ level seems indicated.

The exact solution is available in the form $y = e^{xy}$ since it follows that $y' = e^{xy}(y + xy') = y(y + xy')$ which solves for y' to recover the differential equation. A plot of this curve would confirm the predictions of the direction field.

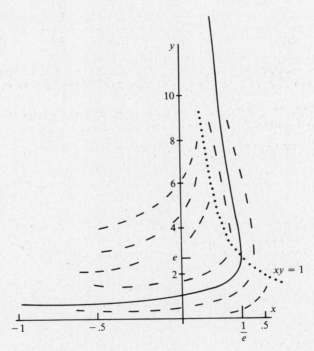

Fig. 11-9

11.23 Find the Taylor series representation of the solution curve just discussed.

▮ Let $y(x) = \sum_0^\infty c_n x^n$, leading to

$$y' = \sum_1^\infty n c_n x^{n-1}$$

$$yy' = \sum_0^\infty a_k x^k \quad \text{with} \quad a_k = \sum_0^k (k - n + 1) c_{k-n+1} c_n$$

$$y^2 = \sum_0^\infty b_k x^k \quad \text{with} \quad b_k = \sum_0^k c_n c_{k-n}$$

Substituting these ingredients into $(1 - xy)y' = y^2$,

$$\sum_{1}^{\infty} nc_n x^{n-1} - \sum_{0}^{\infty} a_k x^{k+1} = \sum_{0}^{\infty} b_k x^k$$

Changes of index are then convenient, bringing

$$\sum_{0}^{\infty} (k+1) c_{k+1} x^k - \sum_{1}^{\infty} a_{k-1} x^k = \sum_{0}^{\infty} b_k x^k$$

in which we must compare coefficients of x^k. For $k = 0$ we must have $c_1 = b_0$, while for all other k,

$$(k+1) c_{k+1} = a_{k-1} + b_k$$

The materials for computing the c_n are now in hand. The initial condition forces $c_0 = 1$. Then follow

$$b_0 = c_1 = 1 \qquad c_1 = b_0 = 1 \qquad b_1 = 2c_0 c_1 = 2 \qquad a_0 = c_0 c_1 = 1$$
$$2c_2 = a_0 + b_1 = 3 \qquad b_2 = c_1^2 + 2c_0 c_2 = 4 \qquad a_1 = 2c_0 c_2 + c_1^2 = 4$$

and so on. The series begins

$$y(x) = 1 + x + \tfrac{3}{2}x^2 + \tfrac{8}{3}x^3 + \tfrac{125}{24}x^4 + \cdots$$

Computations made with this series, using up to 80 terms, put $y(.1) = 1.118$, $y(.2) = 1.296$, $y(.3) = 1.631$ and $y(.4) = 2.519$ at which point the run was stopped, since the last value puts the curve above the infinite isocline. Coefficients of size 10^{30} were noted.

11.24   Apply the local Taylor series method to the preceding problem.

▮ The principal hurdle with this method is to find the several derivatives that are to be used. Starting from $(1 - xy)y' = y^2$ differentiation manages $(1 - xy)y'' + y'(-y - xy') = 2yy'$ or

$$(1 - xy) y'' = 3yy' + x(y')^2$$

Differentiating again,

$$(1 - xy) y^{(3)} + y''(-y - xy') = 3(y')^2 + 3yy'' + (y')^2 + 2xy'y''$$

or

$$(1 - xy) y^{(3)} = 4yy'' + 3xy'y'' + 4(y')^2$$

In the same way we also find

$$(1 - xy) y^{(4)} = 5yy^{(3)} + 15y'y'' + 3x(y'')^2 + 4xy'y^{(3)}$$

These are enough for a fourth degree Taylor polynomial. The initial point being $(0, 1)$, these are computed first at $x = 0$ and prove to be [beginning with $y(0)$ itself] $1, 1, 3, 16, 125$ so that

$$y(x) = 1 + x + \tfrac{3}{2}x^2 + \tfrac{16}{6}x^3 + \tfrac{125}{24}x^4$$

is our first polynomial. The coefficients are, of course, the c_0 to c_4 found in the preceding problem. Evaluating this polynomial at $x = h$ we have the first predicted point and are ready to develop a new polynomial at that point. Each such polynomial is "local" in the sense that it is constructed for the purpose of the next forward step and then abandoned. Here are a few values computed in this way:

x	.1	.2	.3	.35	.36	.37	.38
y for $h = .01$	1.118	1.296	1.631	2.047	2.236	2.720	minus
y for $h = .005$	1.118	1.296	1.631	2.048	2.239	2.980	1.407

Things seem to go along well for a while, but each run shows the stress of approaching the infinite isocline in its own way.

Since the curve $y = e^{xy}$ and the isocline $xy = 1$ are easily seen to intersect at $y = e = 1/x$, the blowup should occur in the vicinity of $x = .368$, $y = 2.718$. At $x = .37$ both runs have overshot this y level.

11.5 RUNGE–KUTTA METHODS

11.25 Find coefficients a, b, c, d, m, n and p in order that the Runge–Kutta formulas

$$k_1 = hf(x, y)$$
$$k_2 = hf(x + mh, y + mk_1)$$
$$k_3 = hf(x + nh, y + nk_2)$$
$$k_4 = hf(x + ph, y + pk_3)$$
$$y(x + h) - y(x) \simeq ak_1 + bk_2 + ck_3 + dk_4$$

duplicate the Taylor series through the term in h^4. Note that the last formula, though not a polynomial approximation, is then near the Taylor polynomial of degree 4.

∎ We begin by expressing the Taylor series in a form that facilitates comparisons. Let

$$F_1 = f_x + ff_y, \qquad F_2 = f_{xx} + 2ff_{xy} + f^2 f_{yy}, \qquad F_3 = f_{xxx} + 3ff_{xxy} + 3f^2 f_{xyy} + f^3 f_{yyy}$$

Then differentiating the equation $y' = f(x, y)$ we find

$$y^{(2)} = f_x + f_y y' = f_x + f_y f = F_1$$
$$y^{(3)} = f_{xx} + 2ff_{xy} + f^2 f_{yy} + f_y(f_x + ff_y) = F_2 + f_y F_1$$
$$y^{(4)} = f_{xxx} + 3ff_{xxy} + 3f^2 f_{xyy} + f^3 f_{yyy} + f_y(f_{xx} + 2ff_{xy} + f^2 f_{yy}) + 3(f_x + ff_y)(f_{xy} + ff_{yy}) + f_y^2(f_x + ff_y)$$
$$= F_3 + f_y F_2 + 3F_1(f_{xy} + ff_{yy}) + f_y^2 F_1$$

which allows the Taylor series to be written as

$$y(x + h) - y(x) = hf + \tfrac{1}{2}h^2 F_1 + \tfrac{1}{6}h^3(F_2 + f_y F_1) + \tfrac{1}{24}h^4 \left[F_3 + f_y F_2 + 3(f_{xy} + ff_{yy}) F_1 + f_y^2 F_1 \right] + \cdots$$

Turning now to the various k values, similar computations produce

$$k_1 = hf$$
$$k_2 = h\left[f + mhF_1 + \tfrac{1}{2}m^2 h^2 F_2 + \tfrac{1}{6}m^3 h^3 F_3 + \cdots \right]$$
$$k_3 = h\left[f + nhF_1 + \tfrac{1}{2}h^2(n^2 F_2 + 2mnf_y F_1) + \tfrac{1}{6}h^3(n^3 F_3 + 3m^2 nf_y F_2 + 6mn^2(f_{xy} + ff_{yy}) F_1) + \cdots \right]$$
$$k_4 = h\left[f + phF_1 + \tfrac{1}{2}h^2(p^2 F_2 + 2npf_y F_1) + \tfrac{1}{6}h^3(p^3 F_3 + 3n^2 pf_y F_2 + 6np^2(f_{xy} + ff_{yy}) F_1 + 6mnpf_y^2 F_1) + \cdots \right]$$

Combining these as suggested by the final Runge–Kutta formula,

$$y(x + h) - y(x) = (a + b + c + d)hf + (bm + cn + dp)h^2 F_1 + \tfrac{1}{2}(bm^2 + cn^2 + dp^2)h^3 F_2$$
$$+ \tfrac{1}{6}(bm^3 + cn^3 + dp^3)h^4 F_3 + (cmn + dnp)h^3 f_y F_1 + \tfrac{1}{2}(cm^2 n + dn^2 p)h^4 f_y F_2$$
$$+ (cmn^2 + dnp^2)h^4(f_{xy} + ff_{yy}) F_1 + dmnph^4 f_y^2 F_1 + \cdots$$

Comparison with the Taylor series now suggests the eight conditions

$$a + b + c + d = 1 \qquad cmn + dnp = \tfrac{1}{6}$$
$$bm + cn + dp = \tfrac{1}{2} \qquad cmn^2 + dnp^2 = \tfrac{1}{8}$$
$$bm^2 + cn^2 + dp^2 = \tfrac{1}{3} \qquad cm^2 n + dn^2 p = \tfrac{1}{12}$$
$$bm^3 + cn^3 + dp^3 = \tfrac{1}{4} \qquad dmnp = \tfrac{1}{24}$$

These eight equations in seven unknowns are actually somewhat redundant. The classical solution set is

$$m = n = \tfrac{1}{2} \qquad p = 1 \qquad a = d = \tfrac{1}{6} \qquad b = c = \tfrac{1}{3}$$

leading to the Runge–Kutta formulas

$$k_1 = hf(x, y) \qquad k_2 = hf\left(x + \tfrac{1}{2}h, y + \tfrac{1}{2}k_1\right) \qquad k_3 = hf\left(x + \tfrac{1}{2}h, y + \tfrac{1}{2}k_2\right)$$

$$k_4 = hf(x + h, y + k_3) \qquad y(x + h) \simeq y(x) + \tfrac{1}{6}(k_1 + 2k_2 + 2k_3 + k_4)$$

It is of some interest to notice that for $f(x, y)$ independent of y this reduces to Simpson's rule applied to $y'(x) = f(x)$.

11.26 What is the advantage of Runge–Kutta formulas over the Taylor method?

▮ Though approximately the same as the Taylor polynomial of degree 4, these formulas do not require prior calculation of the higher derivatives of $y(x)$, as the Taylor method does. Since the differential equations arising in applications are often complicated, the calculation of derivatives can be onerous. The Runge–Kutta formulas involve computation of $f(x, y)$ at various positions instead and this function occurs in the given equation. The method is very extensively used.

11.27 Apply the Runge–Kutta formula to $y' = f(x, y) = xy^{1/3}$, $y(1) = 1$.

▮ With $x_0 = 1$ and $h = .1$ we find

$$k_1 = (.1)f(1, 1) = .1 \qquad\qquad k_3 = (.1)f(1.05, 1.05336) \simeq .10684$$

$$k_2 = (.1)f(1.05, 1.05) \simeq .10672 \qquad k_4 = (.1)f(1.1, 1.10684) \simeq .11378$$

from which we compute

$$y_1 = 1 + \tfrac{1}{6}(.1 + .21344 + .21368 + .11378) \simeq 1.10682$$

This completes one step and we begin another with x_1 and y_1 in place of x_0 and y_0, and continue in this way. Since the method duplicates the Taylor series through h^4, it is natural to expect results similar to those found by the Taylor method. Table 11.4 makes a few comparisons and we do find differences in the last two places. These are partly explained by the fact that the local truncation errors of the two methods are not identical. Both are of the form Ch^5, but the factor C is not the same. Also, roundoff errors usually differ even between algorithms that are algebraically identical, which these are not. Here the advantage is clearly with the Runge–Kutta formulas.

TABLE 11.4

x	Taylor	Runge–Kutta	Exact
1	1.00000	1.00000	1.00000
2	2.82846	2.82843	2.82843
3	7.02123	7.02113	7.02113
4	14.69710	14.69693	14.69694
5	27.00022	26.99998	27.00000

11.28 Illustrate variations of the Runge–Kutta formulas.

▮ It is not hard to verify that

$$y(x + h) = y(x) + hf\left(x + \tfrac{1}{2}h, y + \tfrac{1}{2}hf(x, y)\right)$$

in which y denotes $y(x)$, duplicates the Taylor series through terms of second degree. It is, therefore, known as a Runge–Kutta method of order 2. Similarly,

$$k_1 = hf(x, y)$$

$$k_2 = hf\left(x + \tfrac{1}{2}h, y + \tfrac{1}{2}k_1\right)$$

$$k_3 = hf(x + h, y - k_1 + 2k_2)$$

$$y(x + h) = y(x) + \tfrac{1}{6}(k_1 + 4k_2 + k_3)$$

has order 3. Other methods of order 2 and 3 also exist. The set

$$k_1 = hf(x, y)$$
$$k_2 = hf\left(x + \tfrac{1}{2}h, y + \tfrac{1}{2}k_1\right)$$
$$k_3 = hf\left(x + \tfrac{1}{2}h, y + \tfrac{1}{4}k_1 + \tfrac{1}{4}k_2\right)$$
$$k_4 = hf(x + h, y - k_2 + 2k_3)$$
$$y(x + h) = y(x) + \tfrac{1}{6}(k_1 + 4k_3 + k_4)$$

is an alternate method of order 4, while the more exotic

$$k_1 = hf(x, y)$$
$$k_2 = hf\left(x + \tfrac{1}{2}h, y + \tfrac{1}{2}k_1\right)$$
$$k_3 = hf\left(x + \tfrac{1}{2}h, y + \tfrac{1}{4}k_1 + \tfrac{1}{4}k_2\right)$$
$$k_4 = hf(x + h, y - k_2 + 2k_3)$$
$$k_5 = hf\left(x + \tfrac{2}{3}h, y + \tfrac{7}{27}k_1 + \tfrac{10}{27}k_2 + \tfrac{1}{27}k_4\right)$$
$$k_6 = hf\left(x + \tfrac{1}{5}h, y + \tfrac{28}{625}k_1 - \tfrac{1}{5}k_2 + \tfrac{546}{625}k_3 + \tfrac{54}{625}k_4 - \tfrac{378}{625}k_5\right)$$
$$y(x + h) = y(x) + \tfrac{1}{24}k_1 + \tfrac{5}{48}k_4 + \tfrac{27}{56}k_5 + \tfrac{125}{336}k_6$$

has order 5. The higher the order, the greater is the diversity of possible methods and the lower the truncation error. A method of order n duplicates the Taylor series through terms of nth degree and so has truncation error

$$T = cy^{(n+1)}h^{n+1}$$

which means that for a smooth function $y(x)$ the computation can proceed with a relatively large h and progress more rapidly. The development of high-order methods involves some strenuous algebra and it has been feasible only with the aid of computer programs for doing the manipulations.

11.29 Verify the formula given in the previous problem for a Runge–Kutta method of order 2.

▌ Abbreviating $f(x, y)$ to f, $f_x(x, y)$ to f_x and $f_y(x, y)$ to f_y,

$$f\left(x + \tfrac{1}{2}h, y + \tfrac{1}{2}hf\right) = f + f_x\left(\tfrac{1}{2}h\right) + f_y\left(\tfrac{1}{2}hf\right) + \cdots$$
$$= y'(x) + \tfrac{1}{2}hy''(x) + \cdots$$

where the dots indicate terms with higher powers of h. The method then offers

$$y(x + h) = y(x) + hy'(x) + \tfrac{1}{2}h^2y''(x) + \cdots$$

which is precisely the Taylor series through second degree terms.

11.30 Verify the formula given in Problem 11.28 for a Runge–Kutta method of order 3.

▌ Here things are sufficiently complicated to warrant use of the devices employed in that problem. Of course, $k_1 = hf$ and k_2 is the same as in that problem with $m = \tfrac{1}{2}$, so that

$$k_2 = h\left(f + \tfrac{1}{2}hF_1 + \tfrac{1}{8}h^2F_2 + \cdots\right)$$

where the dots indicate terms involving h to at least power 3. We do have to give k_3 a bit of attention:

$$k_3 = h\left[f + f_x h + f_y\left(-hf + 2hf + h^2F_1 + \cdots\right)\right.$$
$$+ \tfrac{1}{2}f_{xx}h^2 + f_{xy}\left(-h^2f + 2h^2f + \cdots\right)$$
$$+ \tfrac{1}{2}f_{yy}\left(h^2f - 4h^2f^2 + 4h^2f^2 + \cdots\right)$$
$$\left. = hf + h^2F_1 + h^3\left(f_yF_1 + \tfrac{1}{2}F_2\right) + \cdots\right]$$

Finally

$$\tfrac{1}{6}\left(k_1 + 4k_2 + k_3\right) = hf + \tfrac{1}{2}h^2F_1 + \tfrac{1}{6}h^3\left(f_yF_1 + F_2\right) + \cdots$$

and adding in the $y(x)$ term we have the Taylor series up through degree 3.

11.31 Apply the Runge–Kutta method to the equation $y' = x^2 - y^2$.

▐ We have already used the Euler and Taylor methods on this equation, and Fig. 11-4 provides a view of the direction field. For the solution through $(-1,1)$ the crossings of the y axis and the 1 isocline have been our particular targets and a few more relevant numbers appear in Table 11.5. The simpler second-order formula (RK2) is outperformed, but this is expected. It has the crossing at $x = 0$ correct to almost four decimal places and would no doubt do better with a smaller value for h. All the runs put the crossing of the isocline (ordinates attached at the right) just before $x = 1.4$. For a moment the curve and the isocline then draw apart (as they must) but at $x = 3$ they are already closing again. As already observed, the curve will be unable to fully close the gap.

TABLE 11.5

x	RK2 $h = .05$	RK4 $h = .1$	RK4 $h = .05$		
0	.6198	.6199397	.6199403		
1	.6476	.6475110	.6475109		
1.3	.8755	.8752213	.8752202	.83 ⎫	
1.4	.9725	.9722844	.9722829	.98 ⎬	1 isocline
2	1.6689	1.668727	1.668725	1.73 ⎪	
3		2.814095	2.814095	2.83 ⎭	

11.32 Apply the Runge–Kutta method to $y' = y^2/(1 - xy)$.

▐ Table 11.6 provides some relevant information. The second-order method is included but the burden falls primarily on the classic fourth-order procedure. The top three rows of the table follow the uneventful rise of the curve as it prepares for its rendezvous with infinite slope, already viewed in Fig. 11-9. The table shows that even $h = .01$ is too big to pin down this intersection (at $x = 1/e = .3679$, $y = e$) with much accuracy. An additional run with RK4 and $h = .001$ found

$$y(.367) = 2.54 \qquad y(.368) = 2.95$$

so the target is now properly straddled.

TABLE 11.6

x	RK2 $h = .02$	RK2 $h = .01$	RK4 $h = .05$	RK4 $h = .02$	RK4 $h = .01$
.1	1.1182	1.11830	1.118326	1.118325	1.118325
.2	1.2954	1.29574	1.295857	1.295855	1.295855
.3	1.6285	1.63059	1.631371	1.631342	1.631340
.36	2.1864	2.22000		2.239572	2.239214
.38	3.26	2.20		2.413557	2.717391

11.33 How can the Runge–Kutta method be used to find that part of the solution of $y' = y^2/(1 - xy)$, $y(0) = 1$, that lies above $y = e$?

▐ One way is to think of x as a function of y, with initial point $x(1) = 0$ and $x' = (1 - xy)/y^2$. The fourth-order method had no difficulty in locating the maximum x value and carrying on past it, as Table 11.7 documents.

TABLE 11.7

y	x	y	x	y	x	y	x
1	0	2.71	.3678778	2.719	same	3	.3662041
2	.3465737	2.717	.3678795	2.72	same	4	.3465737
2.7	.3678711	2.718	same	2.8	.3677213	5	.3218876

11.34 Try the Runge–Kutta method on the equation $y' = 3y - 4e^{-x}$, with $y(0) = 1$. The equation is solved by elementary means to discover $y = e^{-x}$.

❚ The relative tragedy is displayed in Table 11.8. With $h = .1$ the solution by the fourth-order procedure is already in trouble at $x = 3$, having managed $y(2)$ almost to three decimal places. As h diminishes, the trouble is postponed, but the blowup is relentless.

This is an example of a computational instability. The family of solutions of the differential equation is

$$y(x) = c_1 e^{-x} + c_2 e^{3x}$$

In theory, the initial condition forces c_2 to be zero, eliminating the second term. In practice, each step forward uses the current position as its initial point, shifts to a slightly different trajectory and activates c_2. The growing exponential sooner or later overwhelms the theoretical solution.

TABLE 11.8

x	h = .1	h = .05	h = .02	h = .01	Correct
2	.132	.1352	.1353	.135342	.135335
3	−.007	.0462	.0489	.04993	.049787
4	−1.12	−.054	.0018	.021	.018316
5			−.325	.005	.006738
6				1.14	.002479

11.35 What is the intuitive idea behind Runge–Kutta methods?

❚ The first step, $k_1 = hf(x, y)$, is an Euler method step. But it is clear that the correct slope at (x, y) is not to be counted on for very long, so this step is viewed as a probe, to see where in the plane the curve is headed. The point of arrival for the Euler method would be $(x + h, y + k_1)$ and, as a compromise, the halfway mark at $(x + \frac{1}{2}h, y + \frac{1}{2}k_1)$ is chosen for a new evaluation of the slope. The new estimate is

$$k_2 = hf\left(x + \tfrac{1}{2}h, y + \tfrac{1}{2}k_1\right)$$

and, for the second-order method, it is accepted. For higher-order methods k_2 is viewed as a second probe and other adjustments follow. This is the background intuitive idea. Matching a number of terms of the Taylor series provides a criterion for judging what makes a good adjustment.

11.36 Try the fourth-order Runge–Kutta method on the equation

$$y' = y^2 - 100e^{-100(x-1)^2}$$

with initial point $(.5, 2)$. The final term in this equation is of impulse type, meaning that its effect is strong but only over a short interval, in this case around $x = 1$.

❚ Without the impulse term the solution would be $y = 1/(1 - x)$ and the initial point has been chosen from this curve which, needless to say, has a pole at $x = 1$. Figure 11-10 shows computed solutions corresponding to $h = .1$, .01 and .001. All three start innocently upward along the no-impulse track, but the future is forecast as early as $x = .7$, as the accompanying Table 11.9 indicates, the pure value of $\frac{10}{3}$ not quite prevailing. The climb to the pole is aborted somewhere near $x = .9$ and the solution plunges. Recovering control near $x = 1.1$ or 1.2, the y^2 term of the differential equation then forces the curve upward once again, to gradually become some other member of the family $y = 1/(c - x)$.

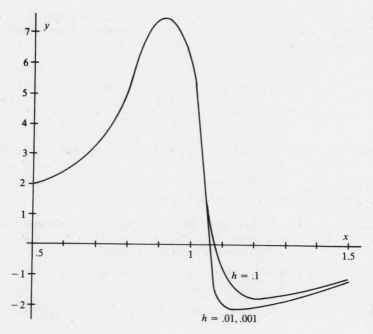

Fig. 11-10

TABLE 11.9

x	$h = .1$	$h = .01$	$h = .001$
.5	2	2	2
.6	2.50	2.5000	2.50000
.7	3.33	3.3331	3.33312
.8	4.95	4.9491	4.94907
.9	7.56	7.5772	7.57747
1.0	5.75	5.6052	5.60726
1.1	− .78	−1.0798	−1.07702
1.2	−1.81	−2.0706	−2.06866
1.3	−1.57	−1.7461	−1.74481
1.4	−1.35	−1.4867	−1.48575
1.5	−1.19	−1.2943	−1.29356

11.37 Discuss the possible error in the computed solutions of the preceding problem.

▮ The local error, or error of each step, in the fourth-order Runge–Kutta method involves the factor h^5. If N steps are made there will be N such local errors, giving us the factor Nh^5 or h^4 times the length of the interval. So globally, to use a popular term, an error involving h^4 is to be anticipated. However, the other factors affecting this truncation error are not constants, particularly the higher derivatives of f, so a direct proportion is unlikely. With this preamble, when h is reduced to $h/10$ as in the example, one might have some hope for four additional correct decimal places. Here that hope appears to be dashed. If the final values are assumed to be close to true, then the reduction from $h = .1$ to $h = .01$ has brought an improvement of perhaps two places, not four. So, rough logic is not always to be trusted.

On the positive side, the impulse term does offer a challenge to accuracy but the results still look good. Why is there not more trouble? Part of the answer is that this impulse depends on x but not on y. Errors in y do not affect it directly. An impulse that was y dependent would cause much more trouble.

11.38 Test the rough error estimator proposed in the preceding problem with the simple equation $y' = y$.

▮ Table 11.10 offers a few relevant numbers, the last column containing the correct values of e, e^2 and e^3 to the places given. Note that the $h = 1$ column is correct to 3, 1 and 2 digits, while the $h = .01$ column is correct to 6, 5 and 6. Here we have gained the four extra digits that rough logic intimated. In the $h = .5$ column there are three correct digits, while in the $h = .05$ column there are seven, so again we have gained four places. It seems that rules of thumb do sometimes work.

TABLE 11.10

x	$h = 1$	$h = .1$	$h = .5$	$h = .05$
1	2.708	2.718280	2.7173	2.718282
2	7.335	7.389046	7.3840	7.389057
3	19.87	20.08549	20.0648	20.08554

11.6 CONVERGENCE OF THE TAYLOR METHOD

11.39 The equation $y' = y$ with $y(0) = 1$ has the exact solution $y(x) = e^x$. Show that the approximate values y_k obtained by the Taylor method converge to this exact solution for h tending to zero and p fixed. (The more familiar convergence concept keeps h fixed and lets p tend to infinity.)

▮ The Taylor method involves approximating each correct value y_{k+1} by

$$Y_{k+1} = Y_k + hY_k' + \frac{1}{2}h^2 Y_k^{(2)} + \cdots + \frac{1}{p!}h^p Y_k^{(p)}$$

For the present problem all the derivatives are the same, making

$$Y_{k+1} = \left(1 + h + \frac{1}{2}h^2 + \cdots + \frac{1}{p!}h^p\right)Y_k = rY_k$$

When $p = 1$ this reduces to the Euler method. In any case it is a difference equation of order 1. Its solution with $Y_0 = 1$ is

$$Y_k = r^k = \left(1 + h + \frac{1}{2}h^2 + \cdots + \frac{1}{p!}h^p\right)^k$$

But by Taylor's polynomial formula,

$$e^h = 1 + h + \frac{1}{2}h^2 + \cdots + \frac{1}{p!}h^p + \frac{h^{p+1}}{(p+1)!}e^{\xi h}$$

with ξ between 0 and 1. Now recalling the identity

$$a^k - r^k = (a - r)(a^{k-1} + a^{k-2}r + \cdots + ar^{k-2} + r^{k-1})$$

we find for the case $a > r > 0$,

$$a^k - r^k < (a - r)ka^{k-1}$$

Choosing $a = e^h$ and r as above, this last inequality becomes

$$0 < e^{kh} - Y_k < \frac{h^{p+1}}{(p+1)!}e^{\xi h}ke^{(k-1)h} < \frac{kh^{p+1}}{(p+1)!}e^{kh}$$

the last step being a consequence of $0 < \xi < 1$. The question of convergence concerns the behavior of values computed for a fixed argument x as h tends to zero. Accordingly, we put $x_k = kh$ and rewrite our last result as

$$0 < e^{x_k} - Y_k < \frac{h^p}{(p+1)!}x_k e^{x_k}$$

Now choose a sequence of step sizes h, in such a way that x_k reoccurs endlessly in the finite argument set of each computation. (The simplest way is to continually halve h.) By the above inequality the sequence of Y_k values obtained at the fixed x_k argument converges to the exact e^{x_k} as h^p. The practical implication is, of course, that the smaller h is chosen the closer the computed result draws to the exact solution. Naturally roundoff errors, which have not been considered in this problem, will limit the accuracy attainable.

11.40 How does the error of the Taylor approximation, as developed in the previous problem, behave for a fixed step size as k increases, in other words as the computation is continued to larger and larger amounts?

I Note that this is not a convergence question, since h is fixed. It is a question of how the error, due to truncation of the Taylor series at the term h^p, accumulates as the computation continues. By the last inequality we see that the error contains the true solution as a factor. Actually it is the relative error that may be more significant, since it is related to the number of significant digits in our computed values. We find

$$\text{relative error} = \left| \frac{e^{x_k} - Y_k}{e^{x_k}} \right| < \frac{h^p}{(p+1)!} x_k$$

which, for fixed h, grows linearly with x_k.

11.41 Prove the convergence of the Taylor method for the general first-order equation $y' = f(x, y)$ with initial condition $y(x_0) = y_0$ under appropriate assumptions on $f(x, y)$.

I This generalizes the result of Problem 11.39. Continuing to use capital Y for the approximate solution, the Taylor method makes

$$Y_{k+1} = Y_k + hY_k' + \frac{1}{2}h^2 Y_k^{(2)} + \cdots + \frac{1}{p!}h^p Y_k^{(p)}$$

where all entries $Y_k^{(i)}$ are computed from the differential equation. For example,

$$Y_k' = f(x_k, Y_k) \qquad Y_k^{(2)} = f_x(x_k, Y_k) + f_y(x_k, Y_k)f(x_k, Y_k) = f'(x_k, Y_k)$$

and suppressing arguments for brevity,

$$Y_k^{(3)} = f_{xx} + 2f_{xy}f + f_{yy}f^2 + (f_x + f_y f)f_y = f''(x_k, Y_k)$$

it being understood that f and its derivatives are evaluated at x_k, Y_k and that Y_k denotes the computed value at arguments x_k. The other $Y_k^{(i)}$ are obtained from similar, but more involved, formulas. If we use $y(x)$ to represent the exact solution of the differential problem, then Taylor's formula offers a similar expression for $y(x_{k+1})$,

$$y(x_{k+1}) = y(x_k) + hy'(x_k) + \frac{1}{2}h^2 y^{(2)}(x_k) + \cdots + \frac{1}{p!}h^p y^{(p)}(x_k) + \frac{h^{p+1}}{(p+1)!}y^{(p+1)}(\xi)$$

provided the exact solution actually has such derivatives. As usual ξ is between x_k and x_{k+1}. In view of $y'(x) = f(x, y(x))$, we have

$$y'(x_k) = f(x_k, y(x_k))$$

and differentiating,

$$y^{(2)}(x_k) = f_x(x_k, y(x_k)) + f_y(x_k, y(x_k))f(x_k, y(x_k)) = f'(x_k, y(x_k))$$

In the same way

$$y^{(3)}(x_k) = f''(x_k, y(x_k))$$

and so on. Subtraction now brings

$$y(x_{k+1}) - Y_{k+1} = y(x_k) - Y_k + h[y'(x_k) - Y_k'] + \frac{1}{2}h^2[y^{(2)}(x_k) - Y_k^{(2)}]$$
$$+ \cdots + \frac{1}{p!}h^p[y^{(p)}(x_k) - Y_k^{(p)}] + \frac{h^{p+1}}{(p+1)!}y^{(p+1)}(\xi)$$

Now notice that if $f(x, y)$ satisfies a Lipschitz condition,

$$|y'(x_k) - Y_k'| = |f(x_k, y(x_k)) - f(x_k, Y_k)| \le L|y(x_k) - Y_k|$$

We will further assume that $f(x, y)$ is such that

$$\left| y^{(i)}(x_k) - Y_k^{(i)} \right| = \left| f^{(i-1)}(x_k, y(x_k)) - f^{(i-1)}(x_k, Y_k) \right| \le L \left| y(x_k) - Y_k \right|$$

This can be proved to be true, for instance, for $i = 1, \ldots, p$ if $f(x, y)$ has continuous derivatives through order $p + 1$. This same condition also guarantees that the exact solution $y(x)$ has continuous derivatives through order $p + 1$, a fact assumed above. Under these assumptions on $f(x, y)$ we now let $d_k = y(x_k) - Y_k$ and have

$$\left| d_{k+1} \right| \le \left| d_k \right| \left(1 + hL + \frac{1}{2}h^2L + \cdots + \frac{1}{p!}h^pL \right) + \frac{h^{p+1}}{(p+1)!}B$$

where B is a bound on $\left| y^{p+1}(x) \right|$. For brevity, this can be rewritten as

$$\left| d_{k+1} \right| \le (1 + \alpha) \left| d_k \right| + \beta$$

where

$$\alpha = L \left(h + \frac{1}{2}h^2 + \cdots + \frac{1}{p!}h^p \right) \qquad \beta = \frac{h^{p+1}}{(p+1)!}B$$

We now prove that

$$\left| d_k \right| \le \beta \frac{e^{k\alpha} - 1}{\alpha}$$

The numbers α and β are positive. Since the exact and approximate solutions both satisfy the initial condition, $d_0 = 0$ and the last inequality holds for $k = 0$. To prove it by induction we assume it for some nonnegative integer k and find

$$\left| d_{k+1} \right| \le (1 + \alpha)\beta \frac{e^{k\alpha} - 1}{\alpha} + \beta = \frac{(1 - \alpha)e^{k\alpha} - 1}{\alpha}\beta < \frac{e^{(k+1)\alpha} - 1}{\alpha}\beta$$

the last step following since $1 + \alpha < e^\alpha$. The induction is therefore valid and the inequality holds for nonnegative integers k. Since $\alpha = Lh + \epsilon h < Mh$ where ϵ tends to zero with h, we can replace L by the slightly larger M and obtain

$$\left| y(x_k) - Y_k \right| \le \frac{h^p B}{(p+1)!} \cdot \frac{e^{M(x_k - x_0)} - 1}{M}$$

with the usual change of argument $x_k = x_0 + kh$, so that convergence is again like h^p.

11.42   What does the result of Problem 11.41 tell about the error for fixed h as the computation continues to larger arguments x_k?

▍ The result is adequate for proving convergence, but since the exact solution is unknown it does not lead at once to an estimate of the relative error. Further error analysis and an extrapolation to the limit process have been explored.

11.43   Are Runge–Kutta methods also convergent?

▍ Since these methods duplicate the Taylor series up to a point (in our example up to the term in h^4), the proof of convergence is similar to that just offered for the Taylor method itself. The details are more complicated.

11.7   THE PREDICTOR–CORRECTOR METHOD

11.44   Derive the modified Euler formula $y_{k+1} \simeq y_k + \frac{1}{2}h(y_k' + y_{k+1}')$ and its local truncation error.

▍ The formula can be produced by applying the trapezoidal rule to the integration of y' as

$$y_{k+1} - y_k = \int_{x_k}^{x_{k+1}} y' \, dx \simeq \frac{1}{2}h(y_k' + y_{k+1}')$$

By Problem 8.4, the error in this application of the trapezoidal rule to y' will be $-h^3 y^{(3)}(\xi)/12$, and this is the local truncation error. (Recall that local truncation error refers to error introduced by the approximation made in the step from x_k to x_{k+1}, that is, in the integration process. Effectively we pretend that y_k and earlier values are known correctly.) Comparing our present result with that for the simpler Euler method, we of course find the present error substantially smaller. This may be viewed as the natural reward for using the trapezoidal rule rather than a still more primitive integration rule. It is also interesting to note that instead of treating y' as constant between x_k and x_{k+1}, so that $y(x)$ is supposed linear, we now treat y' as linear in this interval, so that $y(x)$ is supposed quadratic.

11.45 Apply the modified Euler formula to the problem $y' = xy^{1/3}$, $y(1) = 1$.

▌ Though this method is seldom used for serious computing, it serves to illustrate the nature of the predictor–corrector method. Assuming y_k and y_k' already in hand, the two equations

$$y_{k+1} \simeq y_k + \tfrac{1}{2}h(y_k' + y_{k+1}') \qquad y_{k+1}' = f(x_{k+1}, y_{k+1})$$

are used to determine y_{k+1} and y_{k+1}'. An iterative algorithm much like those to be presented in Chapter 16 for determining roots of equations will be used. Applied successively, beginning with $k = 0$, this algorithm generates sequences of values y_k and y_k'. It is also interesting to recall a remark made in the solution of the previous problem, that we are treating $y(x)$ as though it were quadratic between the x_k values. Our overall approximation to $y(x)$ may thus be viewed as a chain of parabolic segments. Both $y(x)$ and $y'(x)$ will be continuous, while $y''(x)$ will have jumps at the "corner points" (x_k, y_k).

To trigger each forward step of our computation, the simpler Euler formula will be used as a *predictor*. It provides a first estimate of y_{k+1}. Here, with $x_0 = 1$ and $h = .05$ it offers

$$y(1.05) \simeq 1 + (.05)(1) = 1.05$$

The differential equation then presents us with

$$y'(1.05) \simeq (1.05)(1.016) \simeq 1.0661$$

Now the modified Euler formula serves as a corrector, yielding

$$y(1.05) \simeq 1 + (.025)(1 + 1.0661) \simeq 1.05165$$

With this new value the differential equation corrects $y'(1.05)$ to 1.0678, after which the corrector is reapplied and produces

$$y(1.05) \simeq 1 + (.025)(1 + 1.0678) \simeq 1.0517$$

Another cycle reproduces these four place values, so we stop. This iterative use of the corrector formula, together with the differential equation, is the core of the predictor–corrector method. One iterates until convergence occurs, assuming it will. (See Problem 11.58 for a proof.) It is then time for the next step forward, again beginning with a single application of the predictor formula. Since more powerful predictor–corrector formulas are now to be obtained, we shall not continue the present computation further. Notice, however, that the one result we have is only two units too small in the last place, verifying that our corrector formula is more accurate than the simpler Euler predictor, which was barely yielding four place accuracy with $h = .01$. More powerful predictor–corrector combinations will be developed.

11.46 Derive the "predictor" formula $y_{k+1} \simeq y_{k-3} + \tfrac{4}{3}h(2y_{k-2}' - y_{k-1}' + 2y_k')$.

▌ Earlier (Chapter 8) we integrated a collocation polynomial over the entire interval of collocation (Cotes formulas) and also over just a part of that interval (formulas with end corrections). The second procedure leads to more accurate, if more troublesome, results. Now we integrate a collocation polynomial over more than its interval of collocation. Not too surprisingly, the resulting formula will have somewhat diminished accuracy but it has an important role to play nevertheless. The polynomial

$$p_k = y_0' + k\frac{y_1' - y_{-1}'}{2} + k^2\frac{y_1' - 2y_0' + y_{-1}'}{2}$$

satisfies $p_k = y_k'$ for $k = -1, 0, 1$. It is a collocation polynomial for $y'(x)$ in the form of Stirling's formula of degree 2, a parabola. Integrating from $k = -2$ to $k = 2$, we obtain

$$\int_{-2}^{2} p_k \, dk = 4y_0' + \tfrac{8}{3}(y_1' - 2y_0' + y_{-1}') = \tfrac{4}{3}(2y_1' - y_0' + 2y_{-1}')$$

With the usual change of argument $x = x_0 + kh$ this becomes

$$\int_{x_{-2}}^{x_2} p(x)\, dx = \tfrac{4}{3}h(2y_1' - y_0' + 2y_{-1}')$$

Since we are thinking of $p(x)$ as an approximation to $y'(x)$,

$$\int_{x_{-2}}^{x_2} y'(x)\, dx = y_2 - y_{-2} \simeq \tfrac{4}{3}h(2y_1' - y_0' + 2y_{-1}')$$

Since the same argument applies on other intervals, the indices may all be increased by $k - 1$ to obtain the required predictor formula. It is so called because it allows the y_2 to be predicted from data for smaller arguments.

11.47 What is the local truncation error of this predictor?

▮ It may be estimated by the Taylor series method. Using zero as a temporary reference point,

$$y_k = y_0 + (kh)\, y_0' + \tfrac{1}{2}(kh)^2 y_0^{(2)} + \tfrac{1}{6}(kh)^3 y_0^{(3)} + \tfrac{1}{24}(kh)^4 y_0^{(4)} + \tfrac{1}{120}(kh)^5 y_0^{(5)} + \cdots$$

it follows that

$$y_2 - y_{-2} = 4hy_0' + \tfrac{8}{3}h^3 y_0^{(3)} + \tfrac{8}{15}h^5 y_0^{(5)} + \cdots$$

Differentiation also brings

$$y_k' = y_0' + (kh)\, y_0^{(2)} + \tfrac{1}{2}(kh)^2 y_0^{(3)} + \tfrac{1}{6}(kh)^3 y_0^{(4)} + \tfrac{1}{24}(kh)^4 y_0^{(5)} + \cdots$$

from which we find

$$2y_1' - y_0' + 2y_{-1}' = 3y_0' + 2h^2 y_0^{(3)} + \tfrac{1}{6}h^4 y_0^{(5)} + \cdots$$

The local truncation error is therefore

$$(y_2 - y_{-2}) - \tfrac{4}{3}h(2y_1' - y_0' + 2y_{-1}') = \tfrac{14}{45}h^5 y_0^{(5)} + \cdots$$

of which the first term will be used as an estimate. For our shifted interval this becomes

$$E_p \simeq \tfrac{14}{45}h^5 y_{k-1}^{(5)}$$

11.48 Compare the predictor error with that of the "corrector" formula

$$y_{k+1} \simeq y_{k-1} + \tfrac{1}{3}h(y_{k-1}' + 4y_k' + y_{k+1}')$$

▮ This corrector is actually Simpson's rule applied to $y'(x)$. The local truncation error is therefore

$$E_c = \int_{x_{k-1}}^{x_{k+1}} y'(x)\, dx - \tfrac{1}{3}h(y_{k-1}' + 4y_k' + y_{k+1}') \simeq -\tfrac{1}{90}h^5 y_k^{(5)}(\xi)$$

by Problem 8.6. Thus $E_p \simeq -28E_c$ where the difference in the arguments of $y^{(5)}$ has been ignored.

11.49 Show that the error of the corrector formula of Problem 11.48 can be estimated in terms of the difference between predictor and corrector values.

▮ Considering just the local truncation errors made in the step from x_k to x_{k+1}, we have

$$y_{k+1} = P + E_p = C + E_c$$

with P and C denoting the predictor and corrector values. Then

$$P - C = E_c - E_p = 29E_c$$

and

$$E_c = \frac{P - C}{29}$$

more or less. It is not uncommon to apply this estimate as a further correction, yielding

$$y_{k+1} = C + \frac{P - C}{29}$$

and this formula does have truncation error of order h^6. Under some conditions, however, the use of such "mop-up" terms can make a computation unstable.

11.50 The Milne method uses the formula

$$y_{k+1} \simeq y_{k-3} + \tfrac{4}{3}h(2y'_{k-2} - y'_{k-1} - 2y'_k)$$

as a predictor, together with

$$y_{k+1} \simeq y_{k-1} + \tfrac{1}{3}h(y'_{k+1} + 4y'_k + y'_{k-1})$$

as a corrector. Apply this method using $h = .2$ to the problem $y' = -xy^2$, $y(0) = 2$.

$\blacksquare$ The predictor requires four previous values, which it blends into y_{k+1}. The initial value $y(0) = 2$ is one of these. The others must be obtained. Since the entire computation will be based on these starting values, it is worth an extra effort to get them reasonably accurate. The Taylor method or Runge–Kutta method may be used to obtain

$$y(.2) = y_1 \simeq 1.92308 \qquad y(.4) = y_2 \simeq 1.72414 \qquad y(.6) = y_3 \simeq 1.47059$$

correct to five places. The differential equation then yields

$$y'(0) = y'_0 = 0 \qquad y'(.2) = y'_1 - .73964 \qquad y'(.4) = y'_2 \simeq -1.18906 \qquad y'(.6) = y'_3 \simeq -1.29758$$

correct to five places. The Milne predictor then manages

$$y_4 \simeq y_0 + \tfrac{4}{3}(.2)(2y'_3 - y'_2 + 2y'_1) \simeq 1.23056$$

In the differential equation we now find our first estimate of y'_4,

$$y'_4 \simeq -(.8)(1.23056)^2 \simeq -1.21142$$

The Milne corrector then provides the new approximation,

$$y_4 \simeq y_2 + \tfrac{1}{3}(.2)(-1.21142 + 4y'_3 + y'_2) \simeq 1.21808$$

Recomputing y' from the differential equation brings the new estimate $y'_4 \simeq -1.18698$. Reapplying the corrector, we next have

$$y_4 \simeq y_2 + \tfrac{1}{3}(.2)(-1.18698 + 4y'_3 + y'_2) \simeq 1.21971$$

Once again applying the differential equation, we find

$$y'_4 \simeq -1.19015$$

and returning to the corrector,

$$y_4 \simeq y_2 + \tfrac{1}{3}(.2)(-1.19015 + 4y'_3 + y'_2) \simeq 1.21950$$

The next two rounds produce

$$y'_4 \simeq -1.18974 \qquad y_4 \simeq 1.21953 \qquad y'_4 \simeq -1.18980 \qquad y_4 \simeq 1.21953$$

and since our last two estimates of y_4 agree, we can stop. The iterative use of the corrector formula and differential equation has proved to be a convergent process, and the resulting y_4 value is actually correct to four places. In this case four applications of the corrector have brought convergence. If h is chosen too large in a process of this sort, an excessive number of iterative cycles may be needed for convergence or the algorithm may not converge at all. Large differences between predictor and corrector outputs suggest reduction of the interval. On the other hand, insignificant differences between predictor and corrector outputs suggest increasing h and perhaps speeding up the computation. The computation of y_5 and y_5' may now be made in the same way. Results up to $x = 10$ are provided in Table 11.11. Though $h = .2$ was used, only values for integer arguments are printed in the interest of brevity. The exact values are included for comparison.

TABLE 11.11

x	y (Correct)	y (Predictor)	Error	y (Corrector)	Error
0	2.00000				
1	1.00000	1.00037	-37	1.00012	-12
2	.40000	.39970	30	.39996	4
3	.20000	.20027	-27	.20011	-11
4	.11765	.11737	28	.11750	15
5	.07692	.07727	-35	.07712	-20
6	.05405	.05364	41	.05381	14
7	.04000	.04048	-48	.04030	-30
8	.03077	.03022	55	.03041	36
9	.02439	.02500	-61	.02481	-42
10	.01980	.01911	69	.01931	49

11.51 Discuss the error of the previous computation.

▮ Since the exact solution is known for this test case, it is easy to see some things that would usually be quite obscure. The fifth derivative of $y(x) = 2/(1 + x^2)$ has the general behavior shown in Fig. 11-11.

The large fluctuations between 0 and 1 would usually make it difficult to use our truncation error formulas. For example, the local error of the predictor is $14h^5 y^{(5)}/45$ and in our first step (to $x = .8$) we actually find the predictor in error by $-.011$. This corresponds to $y^{(5)} \simeq -100$. The local corrector error is $-h^5 y^{(5)}/90$ and in the same first step the error was actually $-.00002$. This corresponds to $y^{(5)} \simeq 6$. This change of sign in $y^{(5)}$ annuls the anticipated change in sign of error between the predictor and corrector results. It also means that an attempt to use the extrapolation to the limit idea would lead to worse results rather than better, in this case. The oscillating sign of the error as the computation continues will be discussed later.

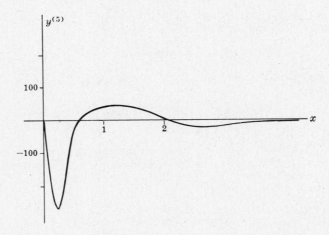

Fig. 11-11

11.52 Derive the Adams predictor formula

$$y_{k+1} = y_k + h\left(y_k' + \tfrac{1}{2}\nabla y_k' + \tfrac{5}{12}\nabla^2 y_k' + \tfrac{3}{8}\nabla^3 y_k' \right)$$
$$= y_k + \tfrac{1}{24}h\left(55y_k' - 59y_{k-1}' + 37y_{k-2}' - 9y_{k-3}' \right)$$

▮ As in Problem 11.46, we obtain this predictor by integrating a collocation polynomial beyond the interval of collocation. The Newton backward formula of degree 3, applied to $y'(x)$, is

$$p_k = y_0' + k\nabla y_0' + \tfrac{1}{2}k(k+1)\nabla^2 y_0' + \tfrac{1}{6}k(k+1)(k+2)\nabla^3 y_0'$$

where as usual $x_k = x_0 + kh$. Integrating from $k = 0$ to $k = 1$ (though the points of collocation are $k =$

$0, -1, -2, -3$), we obtain

$$\int_0^1 p_k \, dk = y_0' + \tfrac{1}{2}\nabla y_0' + \tfrac{5}{12}\nabla^2 y_0' + \tfrac{3}{8}\nabla^3 y_0'$$

In terms of the argument x and using $p(x) \simeq y'(x)$, this becomes

$$\int_{x_0}^{x_1} y'(x) \, dx = y_1 - y_0 \simeq h\left(y_0' + \tfrac{1}{2}\nabla y_0' + \tfrac{5}{12}\nabla^2 y_0' + \tfrac{3}{8}\nabla^3 y_0' \right)$$

Since the same reasoning may be applied between x_k and x_{k+1}, we may raise all indices by k to obtain the first result required. The second then follows by writing out the differences in terms of the y values.

11.53 What is the local truncation error of the Adams predictor?

▮ The usual Taylor series approach leads to $E = 251h^5 y^{(5)}/720$.

11.54 Derive other predictors of the form

$$y_{k+1} = a_0 y_k + a_1 y_{k-1} + a_2 y_{k-2} + h(b_0 y_k' + b_1 y_{k-1}' + b_2 y_{k-2}' + b_3 y_{k-3}')$$

▮ Varying the approach, we shall make this formula exact for polynomials through degree 4. The convenient choices are $y(x) = 1$, $(x - x_k)$, $(x - x_k)^2$, $(x - x_k)^3$ and $(x - x_k)^4$. This leads to the five conditions

$$1 = a_0 + a_1 + a_2 \qquad\qquad 1 = -a_1 - 8a_2 + 3b_1 + 12b_2 + 27b_3$$
$$1 = -a_1 - 2a_2 + b_0 + b_1 + b_2 + b_3 \qquad 1 = a_1 + 16a_2 - 4b_1 - 32b_2 - 108b_3$$
$$1 = a_1 + 4a_2 - 2b_1 - 4b_2 - 6b_3$$

which may be solved in the form

$$a_0 = 1 - a_1 - a_2 \qquad\qquad b_2 = \tfrac{1}{24}(37 - 5a_1 + 8a_2)$$
$$b_0 = \tfrac{1}{24}(55 + 9a_1 + 8a_2) \qquad b_3 = \tfrac{1}{24}(-9 + a_1)$$
$$b_1 = \tfrac{1}{24}(-59 + 19a_1 + 32a_2)$$

with a_1 and a_2 arbitrary. The choice $a_1 = a_2 = 0$ leads us back to the previous problem. Two other simple and popular choices are $a_1 = \tfrac{1}{2}$ and $a_2 = 0$, which leads to

$$y_{k+1} = \tfrac{1}{2}(y_k + y_{k-1}) + \tfrac{1}{48}h(119y_k' - 99y_{k-1}' + 69y_{k-2}' - 17y_{k-3}')$$

with local truncation error $161h^5 y^{(5)}/480$ and $a_1 = \tfrac{2}{3}$ and $a_2 = \tfrac{1}{3}$, which leads to

$$y_{k+1} = \tfrac{1}{3}(2y_{k-1} + y_{k-2}) + \tfrac{1}{72}h(191y_k' - 107y_{k-1}' + 109y_{k-2}' - 25y_{k-3}')$$

with local truncation error $707h^5 y^{(5)}/2160$.

Clearly, one could use these two free parameters to further reduce truncation error, even to order h^7, but another factor to be considered shortly suggests that truncation error is not our only problem. It is also clear that other types of predictor, perhaps using a y_{k-3} term, are possible, but we shall limit ourselves to the abundance we already have.

11.55 Illustrate the possibilities for other corrector formulas.

▮ The possibilities are endless, but suppose we seek a corrector of the form

$$y_{k+1} \simeq a_0 y_k + a_1 y_{k-1} + a_2 y_{k-2} + h(cy_{k+1}' + b_0 y_k' + b_1 y_{k-1}' + b_2 y_{k-2}')$$

for which the local truncation error is of the order h^5. Asking that the corrector be exact for $y(x) = 1$, $(x - x_k), \ldots, (x - x_k)^4$ leads to the five conditions

$$a_0 + a_1 + a_2 = 1 \qquad 13a_1 + 32a_2 - 24b_1 = 5$$
$$a_1 + 24c = 9 \qquad a_1 - 8a_2 + 24b_2 = 1$$
$$13a_1 + 8a_2 - 24b_0 = -19$$

involving seven unknown constants. It would be possible to make this corrector exact for even more powers of x, thus lowering the local truncation error still further. However, the 2 degrees of freedom will be used to bring other desirable features instead to the resulting algorithm. With $a_0 = 0$ and $a_1 = 1$ the remaining constants prove to be those of the Milne corrector:

$$a_2 = 0 \qquad c = \tfrac{1}{3} \qquad b_0 = \tfrac{4}{3} \qquad b_1 = \tfrac{1}{3} \qquad b_2 = 0$$

Another choice, which matches to some extent the Adams predictor, involves making $a_1 = a_2 = 0$, which produces the formula

$$y_{k+1} \simeq y_k + \tfrac{1}{24}h\left(9y'_{k+1} + 19y'_k - 5y'_{k-1} + y'_{k-2}\right)$$

If $a_1 = \tfrac{2}{3}$ and $a_2 = \tfrac{1}{3}$, then we have a formula that resembles another predictor just illustrated:

$$y_{k+1} \simeq \tfrac{1}{3}\left(2y_{k-1} + y_{k-2}\right) + \tfrac{1}{72}h\left(25y'_{k+1} + 91y'_k + 43y'_{k-1} + 9y'_{k-2}\right)$$

Still another formula has $a_0 = a_1 = \tfrac{1}{2}$, making

$$y_{k+1} \simeq \tfrac{1}{2}\left(y_k + y_{k-1}\right) + \tfrac{1}{48}h\left(17y'_{k+1} + 51y'_k + 3y'_{k-1} + y'_{k-2}\right)$$

The various choices differ somewhat in their truncation errors.

11.56 Compare the local truncation errors of the predictor and corrector formulas just illustrated.

❚ The Taylor series method can be applied as usual to produce the following error estimates:

Predictor: $\quad y_{k+1} = y_k + \dfrac{1}{24}h\left(55y'_k - 59y'_{k-1} + 37y'_{k-2} - 9y'_{k-3}\right) + \dfrac{251h^5 y^{(5)}}{720}$

Corrector: $\quad y_{k+1} = y_k + \dfrac{1}{24}h\left(9y'_{k+1} + 19y'_k - 5y'_{k-1} + y'_{k-2}\right) - \dfrac{19h^5 y^{(5)}}{720}$

Predictor: $\quad y_{k+1} = \dfrac{1}{2}\left(y_k + y_{k-1}\right) + \dfrac{1}{48}h\left(119y'_k - 99y'_{k-1} + 69y'_{k-2} - 17y'_{k-3}\right) + \dfrac{161h^5 y^{(5)}}{480}$

Corrector: $\quad y_{k+1} = \dfrac{1}{2}\left(y_k + y_{k-1}\right) + \dfrac{1}{48}h\left(17y'_{k+1} + 51y'_k + 3y'_{k-1} + y'_{k-2}\right) - \dfrac{9h^5 y^{(5)}}{480}$

Predictor: $\quad y_{k+1} = \dfrac{1}{3}\left(2y_{k-1} + y_{k-2}\right) + \dfrac{1}{72}h\left(191y'_k - 107y'_{k-1} + 109y'_{k-2} - 25y'_{k-3}\right) + \dfrac{707h^5 y^{(5)}}{2160}$

Corrector: $\quad y_{k+1} = \dfrac{1}{3}\left(2y_{k-1} + y_{k-2}\right) + \dfrac{1}{72}h\left(25y'_{k+1} + 91y'_k + 43y'_{k-1} + 9y'_{k-2}\right) - \dfrac{43h^5 y^{(5)}}{2160}$

In each case the corrector error is considerably less than that of its predictor mate. It is also of opposite sign, which can be helpful information in a computation. The lower corrector error can be explained by its pedigree. It uses information concerning y'_{k+1} while the predictor must take the leap forward from y_k. This also explains why the burden of the computation falls on the corrector, the predictor being used only as a primer.

For each pair of formulas a mop-up term may be deduced. Take the Adams predictor and the corrector below it, the first pair above. Proceeding in the usual way, considering local truncation errors only and remaining aware that results so obtained must be viewed with some skepticism, we find

$$I = P + E_1 = C + E_2$$

where I is the exact value. Since $19E_1 \simeq -251E_2$, we have $E_2 \simeq \tfrac{19}{270}(P - C)$. This is the mop-up term and $I \simeq C + \tfrac{19}{270}(P - C)$ is the corresponding extrapolation to the limit. Once again it must be remembered that $y^{(5)}$ does not really mean the same thing in both formulas, so that there is still a possibility of sizable error in this extrapolation.

11.57 Apply the Adams method to $y' = -xy^2$ with $y(0) = 2$, using $h = .2$.

❚ The method is now familiar, each step involving a prediction and then an iterative use of the corrector formula. The Adams method uses the first pair of formulas of Problem 11.56 and leads to the results in Table 11.12.

The error behavior suggests that $h = .2$ is adequate for six place accuracy for large x, but that a smaller h (say .1) might be wise at the start. The diminishing error is related to the fact (see Problem 11.89) that for this method the "relative error" remains bounded.

TABLE 11.12

x	y (Correct)	y (Predicted)	Error	y (Corrected)	Error
0	2.000000				
1	1.000000	1.000798	−789	1.000133	−133
2	.400000	.400203	−203	.400158	−158
3	.200000	.200140	−140	.200028	−28
4	.117647	.117679	−32	.117653	−6
5	.076923	.076933	−10	.076925	−2
6	.054054	.054058	−4	.054055	−1
7	.040000	.040002	−2	.040000	
8	.030769	.030770	−1	.030769	
9	.024390	.024391	−1	.024390	
10	.019802	.019802		.019802	

11.58 Prove that, for h sufficiently small, iterative use of a corrector formula does produce a convergent sequence and that the limit of this sequence is the unique value Y_{k+1} satisfying the corrector formula.

▌ We are seeking a number Y_{k+1} with the property

$$Y_{k+1} = hcf(x_{k+1}, Y_{k+1}) + \cdots$$

the dots indicating terms containing only previously computed results, and so independent of Y_{k+1}. Assume as usual that $f(x, y)$ satisfies a Lipschitz condition on y in some region R. Now define a sequence

$$Y^{(0)}, Y^{(1)}, Y^{(2)}, \ldots$$

subscripts $k + 1$ being suppressed for simplicity, by the iteration

$$Y^{(i)} = hcf(x_{k+1}, Y^{(i-1)}) + \cdots$$

and assume all points $(x_{k+1}, Y^{(i)})$ are in R. Subtracting, we find

$$Y^{(i+1)} - Y^{(i)} = hc\left[f(x_{k+1}, Y^{(i)}) - f(x_{k+1}, Y^{(i-1)})\right]$$

Repeated use of the Lipschitz condition then brings

$$|Y^{(i+1)} - Y^{(i)}| \le hcK|Y^{(i)} - Y^{(i-1)}| \le \cdots \le (hcK)^i|Y^{(1)} - Y^{(0)}|$$

Now choose h small enough to make $|hcK| = r < 1$ and consider the sum

$$Y^{(n)} - Y^{(0)} = (Y^{(1)} - Y^{(0)}) + \cdots + (Y^{(n)} - Y^{(n-1)})$$

For n tending to infinity the series produced on the right is dominated (apart from a factor) by the geometric series $1 + r + r^2 + \cdots$ and so converges. This proves that $Y^{(n)}$ has a limit. Call this limit Y_{k+1}.

Now, because of the Lipschitz condition,

$$\left|f(x_{k+1}, Y^{(n)}) - f(x_{k+1}, Y_{k+1})\right| \le K|Y^{(n)} - Y_{k+1}|$$

and it follows that $\lim f(x_{k+1}, Y^{(n)}) = f(x_{k+1}, Y_{k+1})$. We may thus let n tend to infinity in the iteration

$$Y^{(n)} = hcf(x_{k+1}, Y^{(n-1)}) + \cdots$$

and obtain at once, as required,

$$Y_{k+1} = hcf(x_{k+1}, Y_{k+1}) + \cdots$$

To prove uniqueness, suppose Z_{k+1} were another value satisfying the corrector formula at x_{k+1}. Then much as before,

$$|Y_{k+1} - Z_{k+1}| \le hcK|Y_{k+1} - Z_{k+1}| \le \cdots \le (hcK)^{(i)}|Y_{k+1} - Z_{k+1}|$$

for arbitrary i. Since $|hcK| = r < 1$, this forces $Y_{k+1} = Z_{k+1}$. Notice that this uniqueness result proves the correct Y_{k+1} to be independent of $Y^{(0)}$, that is, independent of the choice of predictor formula, at least for small h. The choice of predictor is therefore quite free. It seems reasonable to use a predictor of comparable accuracy, from the local truncation error point of view, with a given corrector. This leads to an attractive mop-up argument as well. The pairings in Problem 11.56 keep these factors and some simple esthetic factors, in mind.

11.8 SOME NUMERICAL EXAMPLES

11.59 In Problem 11.44 the local truncation error of the modified Euler formula was shown to be of the order of h^3. The second-order Runge–Kutta method is also of this order, since it duplicates the first three terms of the Taylor series. Which of these methods performs best on the following problem?

$$y' = -2xy^2 \qquad y(0) = 1$$

▮ The exact solution being $y = 1/(1 + x^2)$ it will be easy enough to identify the winner. The computed solutions, using $h = .1$, appear in Table 11.13, with the exact solution in the center column. Seven digits were carried, not all of which are reproduced. It appears that the modified Euler method has prevailed. (See also the following problem.)

TABLE 11.13

x	Euler	True	Runge–Kutta
1	.500770	.5	.49964
2	.199996	.2	.20036
3	.099949	.1	.10021
4	.058790	.0588236	.05893
5	.038442	.0384616	.03852
10	.0098984	.00990099	.0099076
20	.0024936	.00249376	.0024943

11.60 Compare the relative errors of the two computed solutions in Table 11.13.

▮ The relative error being (computed − true)/true, the figures are easily found (all seven digits being used) and appear below. The modified Euler method dominates, with one early exception.

x	1	2	3	4	5	10	20
Euler	.00154	−.00002	−.00051	−.00056	−.00051	−.00027	−.00006
RK2	−.00072	.00182	.00212	.00185	.00153	.00067	.00024

11.61 Repeat the experiment of Problem 11.59 using the impulse equation

$$y' = y^2 - 100e^{-100(x-1)^2}$$

and the initial condition $y(.5) = 2$.

▮ Using $h = .01$ and running through the impulse from $x = .5$ to 1.5 (as in Problem 11.36) the values listed in Table 11.14 were found. The third column repeats the (presumably) best results found in the earlier problem using a Runge–Kutta fourth-order method and $h = .001$. If we assume these values to be close, the modified Euler method once again gets the nod.

TABLE 11.14

x	Euler	RK4	RK2	Euler	RK2
.6	2.5002	2.5	2.4998	.00006	−.00008
.7	3.3338	3.33312	3.3321	.00022	−.00031
.8	4.9518	4.94907	4.9445	.00055	−.00092
.9	7.5791	7.57747	7.5641	.00021	−.00176
1.0	5.5995	5.60726	5.5850	−.00138	−.00397
1.1	−1.0730	−1.07702	−1.1148	−.00372	.03508
1.2	−2.0709	−2.06866	−2.0906	.00108	.01061
1.3	−1.7468	−1.74481	−1.7595	.00114	.00842
1.4	−1.4872	−1.48575	−1.4964	.00096	.00717
1.5	−1.2946	−1.29356	−1.3017	.00084	.00629

11.62 Compare relative errors of the computed solutions found in the preceding problem, assuming the RK4 column to be correct.

■ These numbers also appear in the table, in the last two columns, and there seems to be no question as to which method wins.

11.63 What is the midpoint formula?

■ This is another of the simpler formulas for approximate solution of differential equations. Replacing $y'(x)$ in the equation $y'(x) = f(x, y)$ by the familiar

$$\frac{y(x+h) - y(x-h)}{2h}$$

leads at once to

$$y(x+h) = y(x-h) + 2hf(x, y)$$

which is the midpoint formula, its name indicating that the point x is midway between the two points used in approximating $y'(x)$. It can be contrasted with the even simpler Euler method in which $y'(x)$ is approximated by $[y(x+h) - y(x)]/h$.

11.64 What is the local truncation error of the midpoint formula?

■ In Problem 7.8 it was found that the error made in replacing $y'(x)$ by the difference quotient just used is of the order h^2. In solving for $y(x+h)$ we multiply through by another h factor, making the error in $y(x+h)$ of the order h^3. This puts the midpoint formula in the same group with the modified Euler formula and RK2.

11.65 Test the midpoint formula using $y' = y^2/(1 - xy)$ and comparing with the results given in Table 11.6 for RK2. Also compare with output from the modified Euler method.

■ The midpoint method, like the predictor–corrector methods, needs help to get started. With the initial condition $y(0) = 1$, and the choice $h = .01$, we need $y(.01)$. This can be found by RK4 with $h = .001$ to be 1.010153. Table 11.15 includes the output of a variety of runs, the last column being the best, its top three entries correct to five places. Among the other three the Euler method had a slight edge, as long as it lasted. But at $x = .36$, it became unable to handle any reasonable tolerance. The modified Euler and Runge–Kutta methods produced nonsense at $x = .38$, as they should.

TABLE 11.15

x	Midpt.	RK2	Euler	RK4
.1	1.11828	1.11830	1.11835	1.11832
.2	1.29564	1.29574	1.29596	1.29586
.3	1.62987	1.63059	1.63209	1.63134
.36	2.20348	2.22000		2.23921
.38	over	2.20		2.71739

11.66 Apply the same three methods to the nose profile of Fig. 11-9, but reversing the roles of the variables as in Problem 11.33. Is any one of these simpler methods superior or inferior?

■ Suppose we choose $h = .01$ again. Since the local truncation error of each formula is like h^3, we may hope for something like four to six place accuracy. The results of runs from the initial point at $(0, 1)$ up to the $y = 10$ level appear in Table 11.16. The true values can be computed by expressing the exact solution in the form $x = (\log y)/y$ and the corresponding figures comprise the last column of the table, the center columns indicating the errors of our three competing methods. Clearly it is the midpoint formula that has faired the worst, though solved to a tolerance compatible with six place accuracy. Between the others things are a standoff, with RK2 running steadily below the true values and modified Euler a bit above.

TABLE 11.16 Errors in Sixth Place

y	RK2	Euler	Midpt.	True Value
2	14	−14	29	.346574
3	10	−11	24	.366204
4	8	−6	20	.346574
5	6	−6	18	.321888
6	5	−5	17	.298627
7	4	−5	16	.277987
8	3	−4	15	.244136
9	3	−4	15	.230258
10	2	−4	15	.149787

11.67 Are the errors just found consistent with local truncation error estimates for these formulas?

❚ The error of the modified Euler formula was found in Problem 11.44 to be $(-h^3/12)$ times the third derivative of y. It follows from Problem 7.8 or from the fact that

$$y(x+h) - y(x-h) = 2hy'(x) + (2h^3/6)y^{(3)}(x) + \cdots$$

that the truncation error of the midpoint formula is $(h^3/3)$ times the same derivative. So we may anticipate that for this pair the errors will be of opposite sign, with the second about 4 times the first. Checking the table we find the opposite sign feature solidly supported, and after the first few entries the factor of 4 gets at least a slight vote of confidence. The truncation errors of Runge–Kutta formulas are more complex, as the next problem will suggest.

11.68 Find the truncation error of the Runge–Kutta formula of order 2.

❚ A slight extension of the effort in Problem 11.29 is in order,

$$f\left(x + \tfrac{1}{2}h,\, y + \tfrac{1}{2}hf\right) = f + f_x\left(\tfrac{1}{2}h\right) + f_y\left(\tfrac{1}{2}hf\right)$$

$$+ \tfrac{1}{2}f_{xx}\left(\tfrac{1}{2}h\right)^2 + f_{xy}\left(\tfrac{1}{2}h\right)\left(\tfrac{1}{2}hf\right) + f_{yy}\left(\tfrac{1}{2}hf\right)^2 + \cdots$$

$$= y' + \tfrac{1}{2}hy'' + \tfrac{1}{8}h^2\left(f_{xx} + 2f_{xy}f + f_{yy}f^2\right)$$

$$= y' + \tfrac{1}{2}hy'' + \tfrac{1}{8}h^2\left(y^{(3)} - f_x f_y - f f_y^2\right)$$

$$= y' + \tfrac{1}{2}hy'' + \tfrac{1}{8}h^2\left(y^{(3)} - f_y y'\right)$$

From this it follows quickly that

$$y(x+h) = y(x) + hf\left(x + \tfrac{1}{2}h,\, y + \tfrac{1}{2}hf\right)$$

$$= y(x) + hy' + \tfrac{1}{2}h^2 y'' + \tfrac{1}{8}h^2\left(y^{(3)} - f_y y'\right)$$

and we have our local truncation error in the final term. Unfortunately it is not the easiest tool to work with and the specific errors for this method as exhibited in Table 11.16 are not so easy to explain.

11.69 With the local truncation error of order h^3, the error over N such steps may be estimated at $Nh^3 = (b-a)h^2$, where $Nh = b - a$ is the length of the interval involved. This suggests that doubling the size of h will quadruple the error. Does this happen in the case under study?

❚ Reruns using $h = .02$ and $h = .04$ were made for the RK2 method, and a sample of the errors follows:

x	2	4	6	8	10
.01 error	14	8	5	3	2
.02 error	54	30	20	15	11
.04 error	217	122	82	61	48

Perfect quadrupling was hardly to be expected, but in general the hypothesis seems to be confirmed. The same treatment of the modified Euler method found the sampling

x	2	4	6	8	10
.01 error	−14	−6	−5	−4	−4
.02 error	−58	−33	−22	−17	−14
.04 error	−234	−133	−89	−68	−54

The same comment seems to be appropriate.

11.70 On the evidence, how small should the step size h be to assure correct values to six places?

▮ It seems that three more halvings of h might do it, even over the early stretch where the curve does more bending. Along the upper part of the "nose" things are already in fair shape and one more halving is probably enough, or one could simply average the results for RK2 and modified Euler. But then, these suggestions are founded upon knowledge of the exact solution. A more useful approach to error control appears in Problem 11.107.

11.71 Apply the Adams predictor–corrector method to the impulse problem. Its local truncation error involves h^5, as does the Runge–Kutta method of order 4. Compare the Adams results with those found in Problem 11.36.

▮ Suppose we choose $h = .01$. For the needed four starting values it is convenient, since they are already in hand, to use output of the Runge–Kutta effort, particularly the $h = .001$ run since it is presumed to be the most accurate. The four values are $y(.5) = 2$, $y(.51) = 2.040816$, $y(.52) = 2.083333$ and $y(.53) = 2.127660$. From them the Adams formulas produce the values in Table 11.17. Corresponding RK4 values ($h = .01$) are repeated for convenience. There is not much difference but, for what it is worth, the RK numbers are a whisker closer to the $h = .001$ results of the earlier problem.

TABLE 11.17

x	Adams	RK4	x	Adams	RK4
.6	2.5000	2.5000	1.1	−1.0802	−1.0798
.7	3.3331	3.3331	1.2	−2.0708	−2.0706
.8	4.9490	4.9491	1.3	−1.7463	−1.7461
.9	7.5771	7.5772	1.4	−1.4868	−1.4867
1.0	5.6052	5.6052	1.5	−1.2944	−1.2943

11.72 Apply the Adams method to the unstable equation of Problem 11.34.

▮ The trouble here lies not with the method of approximation but with the instability of the problem itself, so it is useless to hope that a different method will bring a dramatic rescue. Using starter values drawn from the exact solution, the Adams method actually performed quite a bit worse than Runge–Kutta, if comparisons between blowups are an issue.

11.73 Apply the Adams method to the equation $y' = y^2/(1 - xy)$ which has the nose-shaped solution of Fig. 11-9.

▮ This equation has already been attacked using a Taylor series in Problem 11.23 and by a Runge–Kutta method in Problem 11.32. To trigger the Adams effort, three starter values were computed from the Taylor series and with $h = .01$ the run began, asking for six place convergence in the corrector iteration. The values

$$y(.1) = 1.118325 \qquad y(.2) = 1.295855$$
$$y(.3) = 1.631353 \qquad y(.36) = 2.245515$$

were soon outputted and to a few places agree with earlier results. However, beyond $x = .36$ the iteration "complained" about the required six place tolerance, and reducing it, or increasing the allowed number of iterations, failed to solve the impasse. As applied, the Adams method was not able to climb as close to the position of infinite slope as those used earlier.

11.74 Recalling Problem 11.33, in which an infinite slope was avoided by reversing the roles of the variables, test the predictor–corrector methods of Problem 11.56 by tracking this curve once again.

▮ The first of the three pairs of equations in Problem 11.56 is the Adams set. Call the other two pairs A and B. The local truncation errors listed suggest little difference between these three methods. Using starter values borrowed from the Runge–Kutta run with $h = .001$ and setting a tolerance to achieve six place accuracy, runs were made with $h = .01$. From the initial point at $y = 1$ to an arbitrary finish at $y = 10$ there were no differences in the six places sought. The usual Runge–Kutta fourth-order set was also rerun up to this new finish point and also agreed. The values are given in Table 11.18. The identical pair for $y = 2$ and $y = 4$ is not a misprint.

TABLE 11.18

y	x	y	x
1	0	6	.298627
2	.346574	7	.277987
3	.366204	8	.259930
4	.346574	9	.244136
5	.321888	10	.230259

11.75 What is the behavior of solutions of $y' = 1 - 2xy^2$? Track the one through $(0,1)$.

▮ Figure 11-12 shows the zero slope isocline $y^2 = 1/2x$. To its left slopes must be positive, to its right negative. The computed solution through $(0,1)$, obtained by the Runge–Kutta fourth-order method, is also sketched, threading its way through the invisible direction field. To the right it cannot sink below the zero slope isocline again, since the isocline itself has negative slope. The remaining point of interest is at the left, where the curve drops with increasing steepness. Does it have a pole at some point and approach a vertical asymptote? The computation is neutral on this point. It found a value of the order 10^{19} just short of $x = -2$ and just before overflow was announced, but these events could be explained either with a pole or without one.

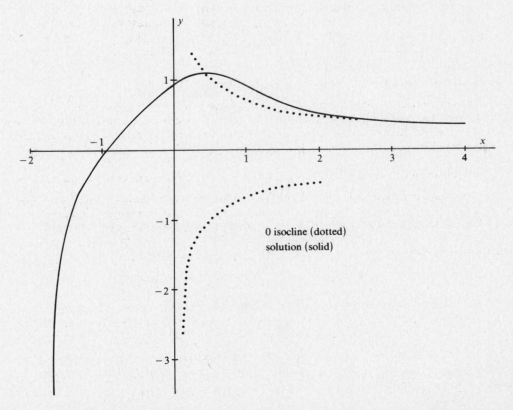

0 isocline (dotted)
solution (solid)

Fig. 11-12

11.76 Show that in the vicinity of the pole, in the preceding problem, the differential equation can be approximated by $y' = -2xy^2$ and solve this new equation.

▌ When x is near -2 and y is large, the term 1 is negligible in comparison with $-2xy^2$. Dropping it leads to the new equation and its elementary solution family $y = 1/(x^2 + C)$ where C is constant. A solution that takes negative y values must have a negative C, and where $x^2 + C$ becomes zero there is a pole. This asymptotic behavior argues strongly that the solution curve of Fig. 11-12 also has a pole.

11.77 Apply the change of variable $y = 1/u$ to the equation of Problem 11.75 and so establish the existence of a pole near $x = -2$. Also determine its position more accurately.

▌ We have $y' = -u'/u^2 = 1 - 2x/u^2$ leading to the equation

$$u' = -u^2 + 2x$$

Rather than integrate back from the initial point $(0,1)$ and face the zero of y near $x = -1$ (and the accompanying singularity of u), we can start at a point such as $(-1.5, -1/.9816)$, the .9816 found during the computation that produced Fig. 11-12. Applying RK4 backward from this point we soon have

$$u(-1.78) = -.0064 \qquad u(-1.79) = .0293$$

straddling a zero of u and a pole of y. An extra run placed it close to -1.782, which would suggest a value near -3.176 for the constant C of the preceding problem.

11.78 How does the solution of $y' = x^2 + y^2$ through $(0,0)$ behave?

▌ The direction field and the isoclines are easily imagined, the latter being concentric circles centered at the origin. The solution through $(0,0)$ has the only zero slope of any solution, but quickly gathers upward momentum and according to RK4 with $h = .01$ takes the values

x	.5	1	1.5	2
y	.0418	.3502	1.5174	279.71

followed by overflow. Does it have a pole near $x = 2$?

Suppose we try the same variable change used a moment ago, $y = 1/u$. From $y' = -u'/u^2 = x^2 + 1/u^2$ we get the equation

$$u' = -1 - x^2 u^2$$

Starting from $(1.5, 1/1.5174)$ we find by RK4

$$u(2.2) = .0057 \qquad u(2.3) = -.0958$$

placing a zero of u and a pole of y close to $x = 2.2$.

11.79 How can the equation of the preceding problem be approximated asymptotically near the pole?

▌ Let $y = x^p u$, with p to be determined. Then

$$y' = x^p u' + p x^{p-1} u = x^{2p} u^2 + x^2$$

or solving for u', $u' = -up/x + x^p u^2 + x^{2-p}$. Choosing $p = 1$ this simplifies to $u' = -u/x + x(u^2 + 1)$. But the second term on the right dominates the first for large x, since $\lim u/x^2(u^2 + 1)$ is zero for x tending to infinity. Abandoning the first we have

$$u' = x(u^2 + 1)$$

which has the obvious solution $u = \tan(\tfrac{1}{2}x^2 + C)$ leading back to

$$y = x \tan(\tfrac{1}{2}x^2 + C)$$

But if this is how the solutions of the original differential equation behave for large x, then they have not just one singular point but an infinite sequence of them. This may be surprising behavior considering the apparent simplicity of the equation.

11.9 CONVERGENCE OF PREDICTOR–CORRECTOR METHODS

11.80 Show that the modified Euler method is convergent.

▌ In this method the simple Euler formula is used to make a first prediction of each y_{k+1} value, but then the actual approximation is found by the modified formula

$$Y_{k+1} = Y_k + \tfrac{1}{2}h\left(Y'_{k+1} + Y'_k\right)$$

The exact solution satisfies a similar relation with a truncation error term. Calling the exact solution $y(x)$ as before, we have

$$y(x_{k+1}) = y(x_k) + \tfrac{1}{2}h\left[y'(x_{k+1}) + y'(x_k)\right] - \tfrac{1}{12}h^3 y^{(3)}(\xi)$$

the truncation error term having been evaluated in Problem 11.44. Subtracting and using d_k for $y(x_k) - Y_k$, we have

$$|d_{k+1}| \leq |d_k| + \tfrac{1}{2}hL\left(|d_{k+1}| + |d_k|\right) + \tfrac{1}{12}h^3 B$$

provided we assume the Lipschitz condition, which makes

$$|y'(x_k) - Y'_k| = |f(x_k, y(x_k)) - f(x_k, Y_k)| \leq L|d_k|$$

with a similar result at argument $k + 1$. The number B is a bound for $|y^{(3)}(x)|$, which we also assume to exist. Our inequality can also be written as

$$\left(1 - \tfrac{1}{2}hL\right)|d_{k+1}| \leq \left(1 + \tfrac{1}{2}hL\right)|d_k| + \tfrac{1}{12}h^3 B$$

Suppose no initial error ($d_0 = 0$) and consider also the solution of

$$\left(1 - \tfrac{1}{2}hL\right)D_{k+1} = \left(1 + \tfrac{1}{2}hL\right)D_k + \tfrac{1}{12}h^3 B$$

with initial value $D_0 = 0$. For purposes of induction we assume $|d_k| \leq D_k$ and find as a consequence

$$\left(1 - \tfrac{1}{2}hL\right)|d_{k+1}| \leq \left(1 - \tfrac{1}{2}hL\right)D_{k+1}$$

so that $|d_{k+1}| \leq D_{k+1}$. Since $d_0 = D_0$ the induction is complete and guarantees $|d_k| \leq D_k$ for positive integers k. To find D_k we solve the difference equation and find the solution family

$$D_k = C\left(\frac{1 + \tfrac{1}{2}hL}{1 - \tfrac{1}{2}hL}\right)^k - \frac{h^2 B}{12L}$$

with C an arbitrary constant. To satisfy the initial condition $D_0 = 0$, we must have $C = (h^2 B/12L)$ so that

$$|y(x_k) - Y_k| \leq \frac{h^2 B}{12L}\left[\left(\frac{1 + \tfrac{1}{2}hL}{1 - \tfrac{1}{2}hL}\right)^k - 1\right]$$

To prove convergence at a fixed argument $x_k = x_0 + kh$ we must investigate the second factor, since as h tends to zero, k will increase indefinitely. But since

$$\left(\frac{1 + \tfrac{1}{2}hL}{1 - \tfrac{1}{2}hL}\right)^k = \left[\frac{1 + L(x_k - x_0)/2k}{1 - L(x_k - x_0)/2k}\right]^k \to \frac{e^{L(x_k - x_0)/2}}{e^{-L(x_k - x_0)/2}} = e^{L(x_k - x_0)}$$

we have

$$y(x_k) - Y_k = O(h^2)$$

Thus as h tends to zero, $\lim Y_k = y(x_k)$, which is the meaning of convergence. Our result also provides a measure of the way truncation errors propagate through the computation.

11.81 Prove the convergence of Milne's method.

▮ The Milne corrector formula is essentially Simpson's rule and provides the approximate values

$$Y_{k+1} = Y_{k-1} + \tfrac{1}{3}h\left(Y'_{k+1} + 4Y'_k + Y'_{k-1}\right)$$

The exact solution $y(x)$ satisfies a similar relation, but with a truncation error term

$$y_{k+1} = y_{k-1} + \tfrac{1}{3}h\left(y'_{k+1} + 4y'_k + y'_{k-1}\right) - \tfrac{1}{90}h^5 y^{(5)}(\xi)$$

with ξ between x_{k-1} and x_{k+1}. Subtracting and using $d_k = y(x_k) - Y_k$,

$$|d_{k+1}| \le |d_{k-1}| + \tfrac{1}{3}hL\left(|d_{k+1}| + 4|d_k| + |d_{k-1}|\right) + \tfrac{1}{90}h^5 B$$

with the Lipschitz condition again involved and B a bound on $y^{(5)}(x)$. Rewriting the inequality as

$$\left(1 - \tfrac{1}{3}hL\right)|d_{k+1}| \le \tfrac{4}{3}hL|d_k| + \left(1 + \tfrac{1}{3}hL\right)|d_{k-1}| + \tfrac{1}{90}h^5 B$$

we compare it with the difference equation

$$\left(1 - \tfrac{1}{3}hL\right)D_{k+1} = \tfrac{4}{3}hLD_k + \left(1 + \tfrac{1}{3}hL\right)D_{k-1} + \tfrac{1}{90}h^5 B$$

Suppose initial errors of d_0 and d_1. We will seek a solution D_k such that $d_0 \le D_0$ and $d_1 \le D_1$. Such a solution will dominate $|d_k|$, that is, it will have the property $|d_k| \le D_k$ for nonnegative integers k. This can be proved by induction much as in the previous problem, for if we assume $|d_{k-1}| \le D_{k-1}$ and $|d_k| \le D_k$ we at once find that $|d_{k+1}| \le D_{k+1}$ also, and the induction is already complete. To find the required solution the characteristic equation

$$\left(1 - \tfrac{1}{3}hL\right)r^2 - \tfrac{4}{3}hLr - \left(1 + \tfrac{1}{3}hL\right) = 0$$

may be solved. It is easy to discover that one root is slightly greater than 1, say r_1, and another in the vicinity of -1, say r_2. More specifically,

$$r_1 = 1 + hL + O(h^2) \qquad r_2 = -1 + \tfrac{1}{3}hL + O(h^2)$$

The associated homogeneous equation is solved by a combination of the kth powers of these roots. The nonhomogeneous equation itself has the constant solution $-h^4 B/180L$ and so we have

$$D_k = c_1 r_1^k + c_2 r_2^k - \frac{h^4 B}{180L}$$

Let E be the greater of the two numbers d_0 and d_1. Then

$$D_k = \left(E + \frac{h^4 B}{180L}\right)r_1^k - \frac{h^4 B}{180L}$$

will be a solution with the required initial features. It has $D_0 = E$ and since $1 < r_1$ it grows steadily larger. Thus

$$|d_k| \le \left(E + \frac{h^4 B}{180L}\right)r_1^k - \frac{h^4 B}{180L}$$

If we make no initial error, then $d_0 = 0$. If also as h is made smaller we improve our value Y_1 (which must be obtained by some other method such as the Taylor series) so that $d_1 = O(h)$, then we have $E = O(h)$ and as h tends to zero so does d_k. This proves the convergence of the Milne method.

11.82 Generalizing the previous problems, prove the convergence of methods based on the corrector formula

$$Y_{k+1} = a_0 Y_k + a_1 Y_{k-1} + a_2 Y_{k-2} + h\left(cY'_{k+1} + b_0 Y'_k + b_1 Y'_{k-1} + b_2 Y'_{k-2}\right)$$

▮ We have chosen the available coefficients to make the truncation error of order h^5. Assuming this to be the case, the difference $d_k = y(x_k) - Y_k$ is found by the same procedure just employed for the Milne corrector to satisfy

$$\left(1 - |c|hL\right)|d_{k+1}| \le \sum_{k=0}^{2}\left(|a_i| + hL|b_i|\right)|d_{k-i}| + T$$

where T is the truncation error term. This corrector requires three starting values, perhaps found by the Taylor series. Call the maximum error of these values E, so that $|d_k| \le E$ for $k = 0,1,2$. Consider also the difference equation

$$(1 - |c|hL) D_{k+1} = \sum_{i=0}^{2} (|a_i| + hL|b_i|) D_{k-i} + T$$

We will seek a solution satisfying $E \le D_k$ for $k = 0,1,2$. Such a solution will dominate $|d_k|$. For, assuming $|d_{k-i}| \le D_{k-i}$ for $i = 0,1,2$ we at once have $|d_{k+1}| \le D_{k+1}$. This completes an induction and proves $|d_k| \le D_k$ for nonnegative integers k. To find the required solution we note that the characteristic equation

$$(1 - |c|hL) r^3 - \sum_{i=0}^{2} (|a_i| + hL|b_i|) r^{2-i} = 0$$

has a real root greater than 1. This follows since at $r = 1$ the left side becomes

$$A = 1 - |c|hL - \sum_{i=0}^{2} (|a_i| + hL|b_i|)$$

which is surely negative since $a_0 + a_1 + a_2 = 1$, while for large r the left side is surely positive if we choose h small enough to keep $1 - |c|hL$ positive. Call the root in question r_1. Then a solution with the required features is

$$D_k = \left(E - \frac{T}{A} \right) r_1^k + \frac{T}{A}$$

since at $k = 0$ this becomes E and as k increases it grows still larger. Thus

$$|y(x_k) - Y_k| \le \left(E - \frac{T}{A} \right) r_1^k + \frac{T}{A}$$

As h tends to zero the truncation error T tends to zero. If we also arrange that the initial errors tend to zero, then $\lim y(x_k) = Y_k$ and convergence is proved.

11.83 The equation $y' = 3xy^{1/3}$ has the solution $y = x^3$ for the initial condition $y(0) = 0$. Why will predictor–corrector methods such as modified Euler and Adams not converge to this solution, given the initial point?

I They will converge to the solution $y = 0$ through this same initial point. Though the function $f(x, y) = 3xy^{1/3}$ is continuous everywhere, its derivative relative to y is infinite along the x axis and the Lipschitz condition fails at these points. So while a solution will exist through any initial point, uniqueness fails when $y = 0$. Our proof of convergence depends on the Lipschitz condition and so fails also for this case. Convergence still does occur, but not to the solution of our choice.

11.10 ERROR AND STABILITY

11.84 What is meant by a stable method for solving differential equations?

I The idea of stability has been described in many ways. Very loosely, a computation is stable if it doesn't "blow up," but this would hardly be appropriate as a formal definition. Stability is sometimes defined as boundedness of the relative error and without question this would be a desirable feature for an algorithm. Gradual deterioration of the relative error means gradual loss of significant digits, which is hardly something to look forward to. The trouble is, over the long run relative error often does deteriorate. An easy example may be useful to gain insight. Consider the modified Euler method.

$$y_{k+1} = y_k + \tfrac{1}{2}h(y'_{k+1} + y'_k)$$

Apply it to the trivial problem

$$y' = Ay \qquad y(0) = 1$$

for which the exact solution is $y = e^{Ax}$. The Euler formula becomes

$$(1 - \tfrac{1}{2}Ah) y_{k+1} = (1 + \tfrac{1}{2}Ah) y_k$$

which is a difference equation of order 1 with solution

$$y_k = r^k = \left(\frac{1 + \frac{1}{2}Ah}{1 - \frac{1}{2}Ah}\right)^k$$

For small h this is close to

$$\left(\frac{e^{(1/2)Ah}}{e^{-(1/2)Ah}}\right)^k = e^{Akh} = e^{Ax}$$

giving us an intuitive proof of convergence. But our goal here lies in another direction. The exact solution satisfies

$$\left(1 - \tfrac{1}{2}Ah\right)y(x_{k+1}) = \left(1 + \tfrac{1}{2}Ah\right)y(x_k) + T$$

where T is the truncation error $-h^3A^3y(\xi)/12$. Subtracting, and using $d_k = y(x_k) - y_k$, we find the similar equation

$$\left(1 - \tfrac{1}{2}Ah\right)d_{k+1} = \left(1 + \tfrac{1}{2}Ah\right)d_k - \tfrac{1}{12}h^3A^3y(\xi)$$

for the error d_k. Now divide by $(1 - \frac{1}{2}Ah)y_{k+1}$ and assume Ah small to obtain

$$R_{k+1} = R_k - \tfrac{1}{12}h^3A^3$$

for the relative error $R_k = y_k/y(x_k)$. Solving

$$R_k = R_0 - \tfrac{1}{12}kh^3A^3 = R_0 - \tfrac{1}{12}x_kh^2A^3$$

suggesting that the relative error grows like x_k, or linearly, as the computation proceeds. This may be far from a blowup, but neither is it a case of relative error remaining bounded.

Taking another view, we will watch the progress of a single error as it penetrates through the solution process, say an initial error d_0. Assuming no other errors committed, we omit T and have

$$d_k = d_0\left(\frac{1 + \frac{1}{2}Ah}{1 - \frac{1}{2}Ah}\right)^k \simeq d_0e^{Akh}$$

which makes the relative error $R_k = d_k/e^{Akh} \simeq d_0$. So the long-range effect of any single error is an imitation of the behavior of the solution itself. If A is positive, the error and the solution grow in the same proportion, while if A is negative, they decay in the same proportion. In both cases the relative error holds firm. That this view is slightly optimistic is suggested by the linear growth predicted above, but at least no blowup is forecast. By some definitions this is enough to consider the Euler algorithm stable. This informal, relaxed usage of the term can be convenient.

There remains the question of how small Ah should be to justify the approximations made in these arguments. Since the true solution is monotone, it seems advisable to keep the value of $(1 + \frac{1}{2}Ah)/(1 - \frac{1}{2}Ah)$ positive. This is true only for Ah between -2 and 2. Prudence suggests keeping one's distance from both of these extremes.

11.85   How well do actual runs using the modified Euler formula and the equations $y' = y$, $y' = -y$ support the theoretical predictions of the preceding problem?

▮   With the first equation, $h = .1$, and a run up to $x = 10$, the relative error climbed from an initial zero to a midway .0019 to a final .0040, about as linear a growth as could be expected. For the second equation a similar run found the relative error to be .0041 midway and .0069 at the end.

11.86   Analyze error behavior in the Milne corrector formula

$$y_{k+1} = y_{k-1} + \frac{h}{3}\left(y'_{k+1} + 4y'_k + y'_{k-1}\right)$$

▮   Again choosing the special equation $y' = Ay$, the error d_k is easily found to satisfy the difference equation of order 2,

$$\left(1 - \tfrac{1}{3}Ah\right)d_{k+1} = \tfrac{4}{3}Ahd_k + \left(1 + \tfrac{1}{3}Ah\right)d_{k-1} + T$$

for which the characteristic equation is

$$\left(1 - \tfrac{1}{3}Ah\right)r^2 - \tfrac{4}{3}Ahr - \left(1 + \tfrac{1}{3}Ah\right) = 0$$

The roots are

$$r_1 = 1 + Ah + O\left(h^2\right) \qquad r_2 = -1 + \tfrac{1}{3}Ah + O\left(h^2\right)$$

which makes

$$d_k \simeq c_1\left(1 + Ah\right)^k + c_2\left(-1 + \tfrac{1}{3}Ah\right)^k$$
$$\simeq c_1 e^{Ahk} + \left(d_0 - c_1\right)\left(-1\right)^k e^{-Ahk/3}$$

Now it is possible to see the long-range effect of the initial error d_0. If A is positive, then d_k behaves very much like the correct solution e^{Ahk}, since the second term tends to zero. In fact, the relative error can be estimated as

$$\frac{d_k}{e^{Ahk}} = c_1 + \left(d_0 - c_1\right)\left(-1\right)^k e^{-4Ahk/3}$$

which approaches a constant. If A is negative, however, the second term does not disappear. Indeed it soon becomes the dominant term. The relative error becomes an unbounded oscillation and the computation degenerates into nonsense beyond a certain point.

The Milne method is said to be stable for A positive and unstable for A negative. In this second case the computed "solution" truly blows up.

11.87 How well do actual runs using the Milne method support the predictions of the preceding problem?

❚ With $y' = y$, $h = .1$ and running up to $x = 10$, the relative error climbed from zero to .0006 and then settled back to zero again. With $y' = -y$, however, it climbed to the third decimal place midway and to .11 at $x = 10$. Anticipating a true blowup, the run was continued to $x = 20$. The relative error responded by skyrocketing to a negative 66000.

11.88 Do the computations reported in Table 11.11 confirm the theoretical predictions?

❚ By referring once again to the table, the relative errors may be computed:

x_k	1	2	3	4	5	6	7	8	9	10
d_k/y_k	$-.0001$	$.0001$	$-.0005$	$.0013$	$-.0026$	$.0026$	$-.0075$	$.0117$	$-.0172$	$.0247$

Although the equation $y' = -xy^2$ is not linear, its solution is decreasing, as that of the linear equation does for negative A. The oscillation in the earlier data was apparent, as is the substantial growth of the relative error.

11.89 Analyze error behavior for the Adams corrector

$$Y_{k+1} = Y_k + \tfrac{1}{24}h\left(9Y'_{k+1} + 19Y'_k - 5Y'_{k-1} + Y'_{k-2}\right)$$

❚ The usual process in this case leads to

$$\left(1 - \tfrac{9}{24}Ah\right)d_{k+1} = \left(1 + \tfrac{19}{24}Ah\right)d_k - \tfrac{5}{24}Ahd_{k-1} + \tfrac{1}{24}Ahd_{k-2} + T$$

Ignoring T we attempt to discover how a solitary error would propagate, in particular what its influence on relative error would be over the long run. The first step is once again to consider the roots of the characteristic equation,

$$\left(1 - \tfrac{9}{24}Ah\right)r^3 - \left(1 + \tfrac{19}{24}Ah\right)r^2 + \tfrac{5}{24}Ahr - \tfrac{1}{24}Ah = 0$$

This has one root near 1, which may be verified to be $r_1 \simeq 1 + Ah$. If this root is removed, the quadratic factor

$$(24 - 9Ah)r^2 - 4Ahr + Ah = 0$$

remains. If Ah were zero this quadratic would have a double root at zero. For Ah nonzero but small the roots, call them r_2 and r_3, will still be near zero. Actually for small positive Ah they are complex with moduli $|r| \simeq \sqrt{Ah/24}$, while for small negative Ah they are real and approximately $\pm \sqrt{-6Ah}/12$. Either way we have

$$|r_2|, |r_3| < 1 + Ah \simeq e^{Ah}$$

for small Ah. The solution of the difference equation can now be written as

$$d_k \simeq c_1(1 + Ah)^k + O(|Ah|^{k/2}) \simeq c_1 e^{Akh} + o(e^{Akh})$$

The constant c_1 depends on the solitary error that has been assumed. Dividing by the exact solution, we find that relative error remains bounded. The Adams corrector is therefore stable for both positive and negative A. A single error will not ruin the computation.

11.90 Do the computations made earlier confirm these theoretical predictions?

▮ Referring once again to Table 11.12, the following relative errors may be computed:

x_k	1	2	3	4	5	6	7 to 10
d_k/y_k	$-.00013$	$-.00040$	$-.00014$	$-.00005$	$-.00003$	$-.00002$	zero

As predicted the errors are diminishing, even the relative error. Once again results obtained for a linear problem prove to be informative about the behavior of computations for a nonlinear problem.

11.91 How does relative error behave for the middle pair of formulas in Problem 11.56?

▮ The analysis is similar, this predictor–corrector pair proving to be stable both for growing and decaying solutions. Direct computations found the relative error diminishing to zero for $y' = y$, growing to .000025 midway and .000040 at the end for $y' = -y$.

11.92 What are parasitic solutions and what is their connection with the idea of computational stability which underlies the preceding problems?

▮ The methods in question involve substituting a difference equation for the differential equation, and for the case $y' = Ay$ it is a difference equation that is linear with constant coefficients. Its solution is, therefore, a combination of terms of the form r_i^k with the r_i the roots of the characteristic equation. One of these roots will be $r_1 = 1 + Ah$, apart from terms of higher degree in h, and r_1^k will then be close to $e^{Ahk} = e^{Ax}$ when h is small. This is the solution we want, the one that converges to the differential solution. Other components, corresponding to the other r_i, are called parasitic solutions. They are the price paid for the lower truncation error that methods such as Milne and Adams bring.

If the parasitic terms are dominated by the r_1 term, then their contribution will be negligible and the relative error will remain acceptable. If, on the other hand, a parasitic solution becomes dominant, it will ruin the computation. In Problem 11.84, for the modified Euler method, the relevant difference equation had only the root

$$r_1 = \frac{1 + Ah/2}{1 - Ah/2} = 1 + Ah + O(h^2)$$

There were no parasitic solutions. In Problem 11.86, the Milne method offered us

$$r_1 = 1 + Ah \qquad r_2 = -1 + \tfrac{1}{3}Ah$$

up to the terms in h^2. For $A > 0$ it is r_1 that dominates, but for $A < 0$ it is r_2 that takes over and the desired solution is buried. In Problem 11.89, apart from the usual $r_1 = 1 + Ah$, we found two parasitic solution terms, both of size about Ah. Both are dominated by the r_1 term, whether A is positive or negative. The Adams method means stable computing in either case.

We are drawn to the conclusion that to avoid a computational blowup any parasitic term should be dominated by the principal term, that is, we want

$$|r_i| \le r_1$$

for $i \ne 1$. Any method for which these conditions are violated is called unstable. In fact, it is best if the inequalities are satisfied by a wide margin.

11.93 Apply the second-order Runge–Kutta method

$$y_{k+1} = y_k + hf\left(x_k + \tfrac{1}{2}h, \, y_k + \tfrac{1}{2}hf(x_k, y_k)\right)$$

to $y' = Ay$. What does this reveal about the stability of this formula?

$\blacksquare$ Substituting Ay for $f(x, y)$ brings

$$y_{k+1} = \left(1 + Ah + \tfrac{1}{2}A^2h^2\right)y_k$$

making

$$y_k = \left(1 + Ah + \tfrac{1}{2}A^2h^2\right)^k$$

which is close to the true solution $y_k = e^{kh} = e^{x_k}$ if Ah is small. But how small should Ah be? Figure 11-13 provides a view of the quadratic $r = 1 + Ah + \tfrac{1}{2}A^2h^2$. When A is positive, r will be greater than 1, so both r^k and e^{kh} will be increasing. The qualitative behavior of r^k is, therefore, correct. But when A is negative, we want a decreasing solution, and this will occur only if Ah is between -2 and 0. Below this interval the approximate solution r^k will be increasing and will bear no resemblance whatsoever to e^{kh}. Here there are no parasitic solutions, since Runge–Kutta methods do not reach back beyond y_k to do their work. The blowup of relative error has a different origin, in the nature of the root r_1 itself.

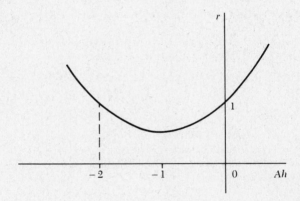

Fig. 11-13

11.94 Apply the fourth-order Runge–Kutta formulas of Problem 11.41 to $y' = Ay$. For what range of Ah values is it stable?

$\blacksquare$ With a little care we find

$$y_{k+1} = \left(1 + Ah + \tfrac{1}{2}A^2h^2 + \tfrac{1}{6}A^3h^3 + \tfrac{1}{24}A^4h^4\right)y_k$$

in which the approximation to e^{Ah} is prominent. Denoting it by r, our approximate solution is again $y_k = r^k$. A plot of r against Ah appears as Fig. 11-14 and, as with the second-order method, suggests that for positive A the

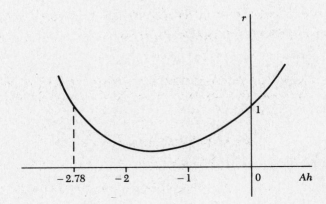

Fig. 11-14

true and the approximate solutions will have the same character, both increasing steadily. But for negative A, just as in the preceding problem, there is a lower bound below which the r^k values will not follow the decreasing trend of the true solution. Here that bound is near -2.78. For Ah smaller than this, we find r greater than 1 and an exploding computation.

11.95   Are the theoretical predictions of the preceding problem supported by actual runs of RK4 with the equation $y' = Ay$?

▮ The test comes with negative A. Choosing $A = -20$ we have a true solution that decays very rapidly and probably needs a rather small h to track it well. Suppose, however, we ask only for the computed solution to decay, avoiding a blowup. With $h = .1$ this does occur, the computed $y(2)$ being $3 \cdot 10^{-10}$. (The true value has exponent -18.) For h up to $.135$ the computed solution still decays, though more and more slowly, but at $h = .14$ it begins to grow while $h = .15$ triggers an impressive blowup. This fits reasonably well with the theory, since $Ah = (-20)(.14) = -2.8$ is just beyond the critical -2.78.

11.96   How can an analysis based on the equation $y' = Ay$ tell us anything useful about the general problem $y' = f(x, y)$?

▮ There are certainly no guarantees, but the general equation is too difficult for such analysis so the issue is really one of doing what is possible. One link that can be established between the two problems is the identification of our constant A with the partial derivative f_y, evaluated originally in the vicinity of the initial point (x_0, y_0), and later at other regions of the plane to which the solution has penetrated. If f_y changes sign along the way, we would expect the stability of Milne's method to react quickly and that of Runge–Kutta methods to show some sensitivity as well.

11.97   Apply the fourth-order Runge–Kutta method to the nonlinear equation $y' = -100xy^2$ with $y(0) = 2$. The exact solution is $y = 2/(1 + 100x^2)$. Test the stability for different step sizes.

▮ Since $f_y = -200xy = -400x/(1 + 100x^2)$, which is zero initially but climbs quickly to -20 at $x = .1$, we recall the stability condition

$$-2.78 \le Ah = -20h$$

and decide to test h values around $.14$. With $h = .10$ the computed solution decays nicely to $.0197$ at $x = 1$ and to $.0050$ at $x = 2$. With $h = .12$, a similar descent is observed. But with $h = .13$, three steps bring us to the very unsatisfactory -29.11, followed by overflow. This definite blowup speaks well for efforts to transfer our linear stability criteria to the nonlinear scene.

11.98   What can be done to control roundoff error?

▮ In a long solution process, roundoff can become a serious factor. If double precision arithmetic is available, it should probably be used, in spite of the additional expense. It may be the only recourse. There is an intermediate step that may be helpful if the use of higher precision throughout the entire computation is deemed too time consuming. To illustrate, many of our formulas for solving differential equations amount to

$$y_{k+1} = y_k + h \, \Delta y_k$$

with the term Δy_k small compared with y_k itself. To perform the addition on the right, this small correction term has to be shifted (to line up the binary points) and this is where many roundoffs occur. To avoid them, the y_k are stored in double precision and this addition is done in double precision. The job of computing Δy_k, usually the heaviest work, is still done in single precision because this term is expected to be small anyway. In this way double precision is used only where it is needed most.

11.99   Show that the midpoint method is unstable.

▮ The solution Y_k generated by this method satisfies

$$Y_{k+1} = Y_{k-1} + 2hf(x_k, Y_k)$$

while in Problem 11.67 we found the true solution to satisfy

$$y_{k+1} = y_{k-1} + 2hf(x_k, y_k) + \tfrac{1}{3}h^3 y^{(3)}$$

Taking the special case $f(x, y) = Ay$ and subtracting,

$$d_{k+1} = d_{k-1} + 2hA d_k$$

with the truncation error ignored in order to focus once again on the long-range effect of a single error d_0. The characteristic equation $r^2 - 2hAr - 1 = 0$ has roots

$$r = hA \pm \sqrt{h^2 A^2 + 1} = hA \pm 1 + O(h^2)$$

which for small hA are near e^{hA} and $-e^{-hA}$. The solution is then

$$d_k = c_1 (1 + Ah)^k + c_2 (-1)^k (1 - Ah)^k \approx c_1 e^{Ahk} + c_2 (-1)^k e^{-Ahk}$$

Setting $k = 0$ finds $d_0 = c_1 + c_2$, and dividing through by y_k shows the relative error to be

$$r_k \approx c_1 + (d_0 - c_1)(-1)^k e^{-2Ahk}$$

For positive Ah this is a decaying oscillation, but for negative Ah it is an unbounded one. The midpoint method is unstable in this case.

11.100 Make a few runs with the midpoint method, using $f(x, y) = -y$ and the initial condition $y(0) = 1$. How well do the computations support the predictions of the preceding problem?

$\blacksquare$ Trying $h = .1$ first, the relative error at $x = 1$ was of size .002, but climbed to 1.6 at $x = 5$ and to 36000 at $x = 10$. Reducing h to .05 delayed the blowup only slightly, the relative error at $x = 5$ now 1.6 and at $x = 10$ a still healthy 5000. One further reduction, to $h = .01$, produced $y(5)$ to almost six decimal places, for a relative error of only .0004. But at $x = 8$ there were no correct digits while at $x = 10$ the relative error was poised for takeoff at the value 6.

11.101 The results in Table 11.19 were obtained by applying the midpoint method to the equation $y' = -xy^2$ with $y(0) = 2$. The interval $h = .1$ was used but only values for $x = .5(.5)5$ are printed. This equation is not linear, but calculate the relative error of each value and discover the rapidly increasing oscillation forecast by the analysis of the previous linear problem.

$\blacksquare$ The corresponding relative errors are

.003   .004 $-.002$ .012 $-.039$ .100 $-.225$ .519 $-.795$ 1.93

and since this seems to be only the edge of the blowup, a few more entries may be of interest. Continuing the computation, errors for $x = 6, 7, 8$ proved to be 4.06, 5.70, 432, followed by overflow.

TABLE 11.19

x_k	Computed y_k	Exact y_k	x_k	Computed y_k	Exact y_k
.5	1.5958	1.6000	3.0	.1799	.2000
1.0	.9962	1.0000	3.5	.1850	.1509
1.5	.6167	.6154	4.0	.0566	.1176
2.0	.3950	.4000	4.5	.1689	.0941
2.5	.2865	.2759	5.0	$-.0713$	.0769

11.102 Show that the formula

$$y_{k+1} \approx y_k + \tfrac{1}{2} h (y'_{k+1} + y'_k) + \tfrac{1}{12} h^2 (-y''_{k+1} + y''_k)$$

has truncation error $h^5 y^{(5)}(\xi)/720$, while the similar predictor

$$y_{k+1} \approx y_k + \tfrac{1}{2} h (-y'_k + 3y'_{k-1}) + \tfrac{1}{12} h^2 (17 y''_k + 7 y''_{k-1})$$

has truncation error $31 h^5 y^{(5)}(\xi)/6!$. These formulas use values of the second derivative to reduce truncation error.

■ The familiar Taylor series argument will be followed. Three expansions are needed for the corrector:

$$y_{k+1} - y_k = hy_k' + \tfrac{1}{2}h^2 y_k'' + \tfrac{1}{6}h^3 y_k^{(3)} + \tfrac{1}{24}h^4 y_k^{(4)} + \tfrac{1}{120}h^5 y_k^{(5)} + \cdots$$

$$\tfrac{1}{2}h(y_k' + y_{k+1}') = hy_k' + \tfrac{1}{2}h^2 y_k'' + \tfrac{1}{4}h^3 y_k^{(3)} + \tfrac{1}{12}h^4 y_k^{(4)} + \tfrac{1}{48}h^5 y_k^{(5)} + \cdots$$

$$\tfrac{1}{12}h^2(y_k'' - y_{k+1}'') = -\tfrac{1}{12}h^3 y_k^{(3)} - \tfrac{1}{24}h^4 y_k^{(4)} - \tfrac{1}{72}h^5 y_k^{(5)} + \cdots$$

According to the formula, the first of these three should be the sum of the other two. Doing the arithmetic one finds a discrepancy in the fifth degree term, with coefficient $1/720$ as stated. For the predictor the first of the preceding three expansions is still needed, plus these two:

$$\tfrac{1}{2}h(-y_k' + 3y_{k-1}') = hy_k' - \tfrac{3}{2}h^2 y_k'' + \tfrac{3}{4}h^3 y_k^{(3)} - \tfrac{1}{4}h^4 y_k^{(4)} + \tfrac{1}{16}h^5 y_k^{(5)} + \cdots$$

$$\tfrac{1}{12}h^2(17y_k'' + 7y_{k-1}'') = 2h^2 y_k'' - \tfrac{7}{12}h^3 y_k^{(3)} + \tfrac{7}{24}h^4 y_k^{(4)} - \tfrac{7}{72}h^5 y_k^{(5)} + \cdots$$

The sum of this pair should again be $y_{k+1} - y_k$; and comparisons again find the discrepancy to be in the fifth degree term, with coefficient $31/720$ as given.

11.103 Investigate the stability of the corrector given in the preceding problem.

■ For the case $y' = f(x, y) = Ay$ we have $y'' = A^2 y$, leading to

$$Y_{k+1} = Y_k + \tfrac{1}{2}h(AY_{k+1} + AY_k) + \tfrac{1}{12}h^2(-A^2 Y_{k+1} + A^2 Y_k)$$

as the corrector. Neglecting the truncation error term we assume an initial error $d_0 = y(x_0) - Y_0$ and have, for d_k, the equation

$$d_{k+1} = d_k + \tfrac{1}{2}h(Ad_{k+1} + Ad_k) + \tfrac{1}{12}h^2(-A^2 d_{k+1} + A^2 d_k)$$

or

$$d_{k+1} = \frac{1 + \tfrac{1}{2}hA + \tfrac{1}{12}h^2 A^2}{1 - \tfrac{1}{2}hA + \tfrac{1}{12}h^2 A^2} d_k = R d_k$$

This is a first-order equation with obvious solution $d_k = R^k d_0$. As h tends to zero it will approach $e^{Akh}d_0$, which means that the error d_0 will propagate as the true solution does. The relative error d_k / e^{Akh} will be constant at the value d_0. Of course, we do not really swallow this whole, but it does seem that no one error will ruin the computation and the method appears to be stable.

11.104 Test the stability prediction of the preceding problem with actual runs using $f(x, y) = Ay$.

■ With $A = 1$ and $h = .1$, a run from $x = 0$ to $x = 20$ found the relative error stirring only a point or two either side of .00002, and then climbing to .00007 when the computation ran on to $x = 30$. At this point the true solution is of order 10^{13} and the computed value was still correct to five figures. With $A = -1$ and the same h, the run to $x = 30$ found the relative error following a ragged, linear trend also ending at .00007. Here the true solution value is of order 10^{-13}.

11.105 Apply the same method to $y' = -xy^2$ with $y(0) = 2$. How well does the approximation follow the decaying solution $y = 2/(x^2 + 1)$?

■ The additional starting value that is needed was found using RK4. A run up to $x = 20$ found the relative error climbing slowly to .000008 and then slipping backward to .000003. Not bad.

11.11 ADAPTIVE METHODS: VARIABLE STEP SIZE

11.106 How can the idea of adaptive integration, introduced in Problem 8.78, be extended to treat differential equations?

■ Suppose the goal is to solve $y' = f(x, y)$ approximately from an initial point $x = a$ to a terminal point $x = b$, arriving with an error no greater than e. Assume that the error will accumulate linearly, so that over a step of length h we can allow an error of size $eh/(b - a)$. This is precisely the idea of adaptive integration used earlier. Let T be an estimate of the truncation error made in taking the step of length h. Then if T does not exceed $eh/(b - a)$, this step is accepted and we move on to the next. Otherwise, the step size h is reduced (to $.5h$ or a suitable alternative) and the step repeated. With a convergent method the requirements will eventually be met, provided the step size h does not become so small that roundoff becomes the dominant error source.

If the Milne predictor–corrector method is being used, then Problem 11.49 provides the needed truncation error estimate $(P - C)/29$ and the condition for acceptance is

$$|P - C| \le \frac{29eh}{b - a}$$

which is easily computed from ingredients already in hand. If the Adams method is being used, then Problem 11.56 leads to the similar condition of acceptance,

$$|P - C| \le \frac{\frac{270}{19}eh}{b - a}$$

In either case, rejection will require reactivating the supplementary start-up procedure.

11.107 To make Runge–Kutta methods adaptive, a practical way of estimating local truncation error is needed. Develop such an estimate, one that does not involve the higher derivatives of $y(x)$.

▌ The now familiar idea of comparing errors for step sizes h and $2h$ will be used. Take the classical fourth-order method and make a step of size $2h$ from the current position x_k. The local error is about

$$T_{2h} = C(2h)^5 = 32Ch^5$$

Now cover the same interval in two steps of size h. The combined error is about

$$2T_h = 2Ch^5$$

leading to these two estimates of the true value y_{k+2}:

$$y_{k+2} = A_{2h} + 32Ch^5 = A_h + 2Ch^5$$

The subscripts $2h$ and h indicate the step sizes used in getting the two approximations. Subtraction now yields the value of C and the error estimate

$$T_h = Ch^5 = \frac{A_h - A_{2h}}{30}$$

which may be doubled for the full forward run. This estimate assumes that Ch^5 is an appropriate error measure and that C (with the higher derivatives imbedded) does not change much over the interval.

11.108 Use the error estimate of the preceding problem to make the Runge–Kutta method adaptive.

▌ For the interval (a, b) let the allowable error be e. For this to be distributed proportionately, we ask that between x_k and x_{k+2} the local error not exceed $2eh/(b - a)$. If $2T_h$ as just estimated does not exceed this, that is, if

$$|A_h - A_{2h}| \le \frac{30eh}{b - a}$$

the value A_h can be accepted at x_{k+2} and one moves on. Otherwise a smaller step size h^* is needed such that the new truncation error T_{h^*} will be suitable. Returning to basics, we assume

$$T_h = Ch^5 \qquad T_{h^*} = Ch^{*5} = \frac{T_h h^{*5}}{h^5}$$

with the latter not exceed $h^*e/(b - a)$ in magnitude. Putting the pieces together, the new step size is determined:

$$h^* = \left[\frac{eh^5}{(b - a)T_h} \right]^{1/4}$$

In view of the various assumptions made in deriving this formula, it is suggested that it not be pushed to the limit. An insurance factor of .8 is usually introduced. Moreover, if h is already quite small, and T_h small with it, the computation of h^* may even cause an overflow. The formula should be used with discretion.

11.109 Which methods are better for adaptive computing, the predictor–corrector pairs or Runge–Kutta?

■ Predictor–corrector methods have the advantage that ingredients for estimating local error are already in hand when needed. With Runge–Kutta a separate application of the formulas must be made, as just outlined. This almost doubles the number of times that $f(x, y)$ has to be evaluated, and since this is where the major computing effort is involved, running time may be almost doubled. On the other hand, and as said before, whenever the step size is changed it will be necessary to assist a predictor–corrector method in making a restart. This means extra programming, and if frequent changes are anticipated, it may be just as well to use Runge–Kutta throughout.

11.110 Try varying the step in the classical Runge–Kutta method as it solves the problem

$$y' = -xy^2 \qquad y(0) = 2$$

for which we have the exact solution $y = 2/(1 + x^2)$.

■ The solution starts with a relatively sharp downward turn, then gradually levels off and becomes rather flat. So we anticipate the need for a small step size at the start and a gradual relaxation as things move along. It is interesting to watch these expectations develop in a run to $x = 27$.

x	.15	1	2	3	4	9	12	17	27
h	.07	.05	.1	.2	.3	.9	1.4	2.7	4.3

11.111 What are variable order methods?

■ Varying the order of the formulas used in integrating a differential equation is another way of trying to achieve a given level of accuracy with a minimum of computing. Starting with a low-order formula to make the process self-starting and a small step size to keep it accurate, both are adjusted as computation proceeds. The idea is to find an optimal order and step size for the current step. A variety of professional programs is available for doing this, all somewhat complex, but the underlying strategy is similar to that in Problems 11.106 and 11.108.

11.112 An object falling toward the earth progresses, under Newtonian theory with only the gravitational attraction of the earth taken into consideration, according to the equation

$$\frac{dy}{dt} = -\sqrt{2gR^2} \sqrt{\frac{H - y}{Hy}}$$

where y is the distance from the earth's center, $g = 32$, $R = 4000$ (5280) and H is the initial distance from the earth's center. The exact solution can be shown to be

$$t = \frac{H^{3/2}}{8y} \left[\sqrt{\frac{y}{H} - \left(\frac{y}{H}\right)^2} + \frac{1}{2} \arccos\left(\frac{2y}{H} - 1\right) \right]$$

if the initial speed is zero. But apply an adaptive method to the differential equation itself with $y(0) = H = 240000$ (5280). At what time does $y = R$? This can be interpreted as the time needed for the moon to fall to earth if it were stopped in its course and the earth remained stationary.

■ A convenient choice of units can be helpful in astronomical problems, and here time in hours and distance in earth radii will be used. The differential equation becomes

$$\frac{dz}{dx} = -\sqrt{\frac{2g}{R}} \sqrt{\frac{60 - z}{60z}}$$

with z the new measure of distance and x the new measure of time. In these units it will be found that $\sqrt{2g/R}$ works out to $72/\sqrt{132}$.

There is a slight problem with beginning the computation at the top of the fall. The constant function $z = 60$ satisfies the differential equation and any numerical method will simply repeat this constant output. The moon will refuse to fall. One way out is to run the fall in reverse, starting at $z(0) = 1$ and using negative values for h. We then ask, for what value of x will z reach 60? Using an adaptive method seems justified since the moon's speed will be zero at the top but enormous at the crash. With RK4 and a policy of doubling h whenever the local error was 100 times smaller than the usual tolerance, the following (partial) output was generated:

Up to $x = .6$	1.0	2.6	4.2	7.4	20.2	33.0
$z = 3.49$	4.69	8.40	11.32	16.08	29.30	38.36
$-h = .1$	.2	.4	.8	1.6	3.2	6.4

The larger (negative) h values are consistent with the two decimal place accuracy requested. The adaptive method has proven its worth up to this point. Unfortunately, the continual doubling of h soon led to a z value exceeding 60 and the program rebelled. A more conservative finish eventually produced $z(110) = 59.89$ and $z(115) = 59.99$ bringing the moon back to about 40 mi from its assumed proper position.

The example offers adaptive methods a good chance to show what they can do. It also shows that it may not always be wise to give them a completely free rein.

11.113   A raindrop of mass m has speed v after falling for time t. Assume air resistance proportional to the square of speed. Then

$$\frac{dv}{dt} = 32 - \frac{cv^2}{m}$$

is the equation of motion. It can be solved by elementary methods to discover that the speed will approach a limiting value, and for $c/m = 2$ that limit is 4. Confirm this by numerical integration of the equation. Is an adaptive method convenient?

▮   As the speed approaches its limiting value and changes less rapidly, it seems that a larger step size may meet the prescribed accuracy requirement. Again choosing RK4, with $h = .01$ initially, and a raindrop that starts from rest, here are a few numbers:

Up to time and Speed	.02 .63	.38 3.98	.62 4.00
$h =$	.01	.02	.04

The method was on the brink of proposing $h = .08$, but the limiting speed (to two places) was already in hand. It hardly seems that for this accuracy an adaptive method was necessary.

11.114   One end of a rope of length 1 is dragged along the x axis. The path of a weight attached to the other end is determined by the differential equation (see Fig. 11-15)

$$y' = -y/\sqrt{1 - y^2}$$

which too may be solved by elementary means. Apply an adaptive numerical method.

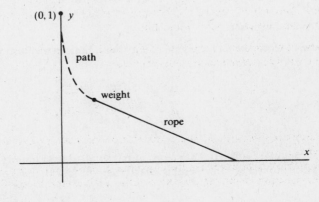

Fig. 11-15

▮ The problem is interesting at both ends. The initial slope of the weight's path is infinite and its ultimate approach to the x axis is asymptotic. The question is how to begin. Suppose we view x as a function of y, with $x(1) = 0$ and

$$\frac{dx}{dy} = -\frac{\sqrt{1-y^2}}{y}$$

Starting with $h = -.01$ and seeking three correct places, some of the results obtained are

y Down to	.98	.96	.92	.84	.68	.36	.20	.12	.04	.02
x up to	.003	.008	.022	.065	.202	.748	1.313	1.817	2.912	3.605
$-h =$	.01	.02	.04	.08	.16	.08	.04	.02	.01	.0025

The step size h is small initially, grows and then shrinks. The asymptotic finish would probably require even smaller step sizes if the computation were continued.

11.115 What does an adaptive method have to contribute to the integration of $y' = y^2/(1 - xy)$, the differential equation of the nose curve already treated in Problems 11.32 and 11.24?

▮ Here the difficulty is the approach to a vertical slope, if we view y as a function of x. Asking for three place accuracy and starting with $h = .1$, the step size was cut several times, reaching $h = .01$ at $x = .34$, then .001 at $x = .366$ and finally .0001 at $x = .3676$. This agrees nicely with the earlier efforts, the true position of infinite slope being $x = 1/e = .3679$ to four places.

11.116 A shot is fired upward against air resistance of cv^2. Assume the equation of motion to be

$$\frac{dv}{dt} = -32 - \frac{cv^2}{m}$$

with $c/m = 2$ and $v(0) = 1$. At what time will the shot reach its maximum height? The exact solution is $\frac{1}{8}\arctan(\frac{1}{4})$ or .03062 to five places, but solve by a numerical method. Is an adaptive method called for?

▮ Experience with the falling raindrop suggests that an adaptive method will not have much to offer, but giving it a try anyway we encounter the numbers

$$\begin{array}{lll} \text{up to time .01 and speed .666,} & h = .005 \\ \qquad\quad .03 \qquad\qquad\quad .020, & \quad .01 \end{array}$$

Unfortunately, at this point the method proposed increasing h to .02, which far overshot the position of zero speed. In fact, even $h = .001$ did the same. The adaptive method eventually found the five place answer, but an appeal to an uncomplicated RK2 also found it without difficulty.

11.117 How does an adaptive method handle the impulse first treated in Problem 11.36?

▮ The adaptive version of RK4 with doubling and halving of the step size as deemed suitable, and asking for three place accuracy, found h increasing from an initial .01 up to .04, then dropping again to .01 in the vicinity of the impulse and finally starting another climb which reached .16 at about $x = 1.6$, where the run was terminated.

11.12 STIFF EQUATIONS

11.118 What is a stiff differential equation?

▮ The term is usually associated with a system of equations, but can be illustrated in principle at a simpler level. Take the equation

$$y' = -100y + 99e^{-x}$$

which has the solution

$$y = e^{-x} - e^{-100x}$$

satisfying the initial condition $y(0) = 0$. Both terms of this solution tend to zero, but the point is, the second decays much faster than the first. At $x = .1$, the term is already zero to four decimal places. It is truly a transient term compared with the first, which could almost be called the "steady state." Systems in which different components operate on quite different time scales are called stiff systems and offer more than normal resistance to numerical solution.

11.119 In view of the rapid decay of the preceding transient term, one might expect a step size of $h = .1$ to generate values of the remaining term e^{-x}. What does the classic Runge–Kutta method actually produce?

▎ Much as in Problem 11.97, we have $f_y = -100$ and associate this with the A of our stability criterion, which becomes

$$-2.78 \leq Ah = -100h$$

and suggests that we keep the step size h less than .0278. This is something of a surprise because it seems to imply that the transient term, negligible in size after $x = .1$, can still influence the computation in an important, underground way. Putting theory to the test, a run was made with $h = .03$. The predicted blowup did occur, values of y quickly descending to the vicinity of -10^{14}. But using $h = .025$ led to a successful run, producing .04980 at $x = 3$. This is just one unit high in the fifth place.

11.120 Develop the Gear formula

$$\nabla y_{n+1} + \tfrac{1}{2} \nabla^2 y_{n+1} + \tfrac{1}{3} \nabla^3 y_{n+1} = h y'_{n+1}$$

where ∇ is the backward difference operator. Show that it is equivalent to

$$y_{n+1} = \frac{18}{11} y_n - \frac{9}{11} y_{n-1} + \frac{2}{11} y_{n-2} + \frac{6h}{11} y'_{n+1}$$

where

$$y'_{n+1} = f(x_{n+1}, y_{n+1})$$

▎ Starting with the Newton backward formula

$$p_k = y_{n+1} + k \nabla y_{n+1} + \frac{k(k+1)}{2} \nabla^2 y_{n+1} + \frac{k(k+1)(k+2)}{6} \nabla^3 y_{n+1}$$

(see Problem 4.15) in which $x - x_{n+1} = kh$ and p_k is a polynomial of degree 3 in k collocating with y at $k = 0, -1, -2, -3$, we differentiate and set $k = 0$

$$\frac{dp}{dx}\Big|_{k=0} = \frac{dp}{dk} \frac{1}{h}\Big|_{k=0} = \frac{1}{h}\left(\nabla y_{n+1} + \frac{1}{2} \nabla^2 y_{n+1} + \frac{1}{3} \nabla^3 y_{n+1}\right)$$

Adopting this as an approximation to y'_{n+1}, we already have the first Gear formula. The second follows easily by replacing the backward differences with their equivalents in terms of the y_i.

These formulas can also be found by the method of undetermined coefficients, requiring exactness for polynomials of degree up to 3. Corresponding formulas of higher order are available by extension. For example, if the Newton formula is extended back to $k = -4$, by introducing the fourth difference term, then $\tfrac{1}{4} \nabla^4 y_{n+1}$ is added to the left side above.

11.121 Why are the formulas of Gear preferred for solving stiff equations?

▎ They prove to be stable for considerably larger values of h than our other formulas. Take once again the equation of Problem 11.118. We have found the Runge–Kutta method unstable for $h = .03$. In contrast, the Gear formula now reduces to

$$y_{n+1} = \frac{18 y_n - 9 y_{n-1} + 2 y_{n-2} + 594 h e^{-(x_n + h)}}{11 + 600h}$$

upon inserting y' from the equation and then solving for y_{n+1}. With $h = .1$, this generated (using three correct starting values)

x	2	4	6
y	.135336	.018316	.002479

the first of which is one unit high in the final place. Even $h = .5$ can be considered a modest success:

x	2	4	6
y	.1350	.01833	.002480

The larger h brings more truncation error but there is no cause to complain about the stability.

11.122 The Gear formulas are usually nonlinear in y_{n+1}. Develop the Newton iteration as it applies to the extraction of this unknown.

▮ In the above example $f(x, y)$ was linear in y, permitting a direct solution for y_{n+1}. Generally, however, we must view the Gear formula as

$$F(y) = y - \frac{6h}{11} f(x_{n+1}, y) - S = 0$$

where y_{n+1} has been abbreviated to y and S stands for the sum of three terms not involving y_{n+1}. Newton's iteration is then

$$y^{(k+1)} = y^{(k)} - \frac{F(y^{(k)})}{F'(y^{(k)})}$$

where

$$F'(y) = 1 - \frac{6h}{11} f_y(x_{n+1}, y)$$

11.13 PROBLEMS OF HIGHER ORDER: THE BASIC MODEL

11.123 What is the basic model for higher-order differential problems?

▮ It is a system of first-order equations such as

$$y_i'(x) = f_i(x, y_1, \ldots, y_n) \qquad i = 1, \ldots, n$$

for determining the n functions $y_i(x)$, given the initial values $y_i(x_0)$. This problem arises in a wide variety of applications. That it is a direct generalization of the initial value problem for a single first-order equation is made especially plain by writing it in the vector form

$$Y'(x) = F(x, Y) \qquad Y(x_0) \text{ given}$$

where Y and F have components y_i and f_i, respectively.

Methods for solving a single first-order equation can easily be extended to handle such systems. Series, Runge–Kutta and predictor–corrector methods will be explored in other problems.

11.124 Show that a second-order differential equation may be replaced by a system of two first-order equations.

▮ Let the second-order equation be $y'' = f(x, y, y')$. Then introducing $p = y'$ we have at once $y' = p$, $p' = f(x, y, p)$. As a result of this standard procedure a second-order equation may be treated by system methods if this seems desirable.

11.125 Show that the general nth-order equation

$$y^{(n)} = f(x, y, y', y^{(2)}, \ldots, y^{(n-1)})$$

may also be replaced by a system of first-order equations.

▮ For convenience we assign $y(x)$ the alias $y_1(x)$ and introduce the additional functions $y_2(x), \ldots, y_n(x)$ by

$$y_1' = y_2 \qquad y_2' = y_3 \qquad \cdots \qquad y_{n-1}' = y_n$$

Then the original nth-order equation becomes

$$y_n' = f(x, y_1, y_2, \ldots, y_n)$$

These n equations are of first order and may be solved by system methods.

11.126 Replace the following equations for the motion of a particle in three dimensions,

$$x'' = f_1(t, x, y, z, x', y', z') \qquad y'' = f_2(t, x, y, z, x', y', z') \qquad z'' = f_3(t, x, y, z, x', y', z')$$

by an equivalent system of first-order equations.

▌ Let $x' = u$, $y' = v$ and $z' = w$ be the velocity components. Then

$$u' = f_1(t, x, y, z, u, v, w) \qquad v' = f_2(t, x, y, z, u, v, w) \qquad w' = f_3(t, x, y, z, u, v, w)$$

These six equations are the required first-order system. Other systems of higher-order equations may be treated in the same way.

11.14 THE TAYLOR SERIES METHOD

11.127 Illustrate the Taylor series procedure for simultaneous equations by solving the system

$$x' = -x - y$$
$$y' = x - y$$

for the two functions $x(t)$ and $y(t)$ satisfying initial conditions $x(0) = 1$ and $y(0) = 0$.

▌ We substitute directly into the two series

$$x(t) = x(0) + tx'(0) + \tfrac{1}{2}t^2 x''(0) + \cdots$$
$$y(t) = y(0) + ty'(0) + \tfrac{1}{2}t^2 y''(0) + \cdots$$

obtaining the needed ingredients from the given system. First $x'(0) = -1$ and $y'(0) = 1$. Then from $x'' = -x' - y'$ and $y'' = x' - y'$ come from $x''(0) = 0$, $y''(0) = -2$. Higher derivatives follow in the same way. The series begin as follows:

$$x(t) = 1 - t + \tfrac{1}{3}t^3 - \tfrac{1}{6}t^4 + \cdots$$
$$y(t) = t - t^2 + \tfrac{1}{3}t^3 + \cdots$$

The given system is not only linear but also has constant coefficients. Writing it in the form

$$X'(t) = AX(t)$$

with

$$X = \begin{pmatrix} x \\ y \end{pmatrix} \quad \text{and} \quad A = \begin{bmatrix} -1 & -1 \\ 1 & -1 \end{bmatrix}$$

the exact solution can be found by trying

$$X = e^{\lambda t} \begin{pmatrix} a \\ b \end{pmatrix}$$

Substituting into the system leads to an eigenvalue problem for the matrix A. For the present A we have

$$(-1 - \lambda)a - \qquad b = 0$$
$$a + (-1 - \lambda)b = 0$$

yielding $\lambda = -1 \pm i$ and after a slight effort

$$x(t) = e^{-t} \cos t \qquad y(t) = e^{-t} \sin t$$

The Taylor series begun above are, of course, the series for these functions. The process as illustrated is easily extended to larger systems of equations.

11.128 Solve the Airy equation $y'' = xy$ in terms of Taylor series.

▌ This brief equation has quite a number of applications. Its Taylor series solution at the point $x = 0$ being $\sum_0^\infty a_k x^k$, we have

$$\sum_0^\infty a_k k(k-1) x^{k-2} = \sum_0^\infty a_k x^{k+1}$$

or, after a shift in the index on the right,

$$\sum_0^\infty a_k k(k-1) x^{k-2} = \sum_3^\infty a_{k-3} x^{k-2}$$

Equating coefficients gives

$$a_k k(k-1) = \begin{cases} 0 & \text{for } k = 0, 1, 2 \\ a_{k-3} & \text{for } k = 3, 4, 5, \dots \end{cases}$$

the upper part of which forces $a_2 = 0$ but leaves a_0 and a_1 arbitrary. The lower part leads in stages to

$$a_{3k} = \frac{a_0}{3k(3k-1)(3k-3)(3k-4) \cdots 5 \cdot 3 \cdot 2}$$

which with a slight effort can be condensed to

$$a_{3k} = \frac{a_0 \Gamma\left(\frac{2}{3}\right)}{3^k k! 3^k \Gamma\left(k + \frac{2}{3}\right)}$$

Similarly

$$a_{3k+1} = \frac{a_1}{(3k+1)(3k)(3k-2)(3k-3) \cdots 6 \cdot 4 \cdot 3}$$

$$= \frac{a_1 \Gamma\left(\frac{4}{3}\right)}{3^k k! 3^k \Gamma\left(k + \frac{4}{3}\right)}$$

and, of course, since $a_2 = 0$, all the $a_{3k+2} = 0$ also. The family of solutions of the Airy equation can now be written in terms of the two constants $a_0 = y(0)$ and $a_1 = y'(0)$:

$$y = a_0 \Gamma\left(\frac{2}{3}\right) \sum_0^\infty \frac{x^{3k}}{9^k k! \Gamma\left(k + \frac{2}{3}\right)} + a_1 \Gamma\left(\frac{4}{3}\right) \sum_0^\infty \frac{x^{3k+1}}{9^k k! \Gamma\left(k + \frac{4}{3}\right)}$$

Two particular solutions are usually defined as

$$y(x) = \text{Ai}(x) \quad \text{has} \quad a_0 = 1/3^{2/3} \Gamma\left(\tfrac{2}{3}\right) \quad a_1 = -1/3^{4/3} \Gamma\left(\tfrac{4}{3}\right)$$
$$y(x) = \text{Bi}(x) \quad \text{has} \quad a_0 = 1/3^{1/6} \Gamma\left(\tfrac{2}{3}\right) \quad a_1 = 1/3^{5/6} \Gamma\left(\tfrac{4}{3}\right)$$

The function $\text{Ai}(x)$ tends to zero and $\text{Bi}(x)$ becomes infinite as x approaches ∞, and both are oscillatory for negative x.

11.129 Discuss the convergence of the Airy series.

▌ A general theorem states that such series will converge in any circle having no singularities of the coefficient functions in the differential equation. Here there are no such singularities in the finite complex plane, so the series should converge for all x. Examining the ratio

$$\frac{a_{k+2}}{a_k} = \frac{x^3 \Gamma\left(k + \frac{2}{3}\right)}{9(k+1) \Gamma\left(k + \frac{5}{3}\right)} = \frac{x^3}{9(k+1)\left(k + \frac{5}{3}\right)}$$

confirms this for the series in hand, since this ratio approaches zero as k becomes infinite. Needless to say, the value of x will have a lot to say about the speed of convergence. Here are a few computational results for the series multiplying a_0. Six correct digits were requested.

x	1	2	3	4	5	6	8	10	20
Terms Needed	4	6	8	11	13	15	21	24	51
Max Term	1	$\frac{4}{3}$	11	68	535	1300	10^5	10^7	10^{24}

The last three maximum terms are given as orders of magnitude. The final Airy value was about 10^{25}. The convergence is obviously very fast, but the numbers involved are large and it is useful to find each term as a multiple of its predecessor rather than facing an almost certain overflow while computing each term from scratch. The convergence of the a_1 series is similar.

11.130   Find Taylor series solutions of the Legendre equation

$$(1 - x^2)\, y'' - 2xy' + n(n+1)\, y = 0$$

in the vicinity of $x = 0$.

▌ We know that there is a polynomial solution of this equation for integer n and the series method will produce it. Since two independent solutions exist, we seek the other as well. Letting $y = \sum_0^\infty a_k x^k$, and substituting into the equation,

$$(1 - x^2)\sum_0^\infty k(k-1)a_k x^{k-2} - 2x\sum_0^\infty ka_k x^{k-1} + n(n+1)\sum_0^\infty a_k x^k = 0$$

or

$$\sum_0^\infty k(k-1)a_k x^{k-2} + \sum_0^\infty \left[n(n+1) - k(k+1) \right] a_k x^k = 0$$

and with a shift of index in the first sum,

$$\sum_0^\infty \left[(k+2)(k+1)a_{k+2} + \left[n(n+1) - k(k+1) \right] a_k \right] x^k = 0$$

Since all coefficients must be zero, we have the recursion

$$a_{k+2} = -\frac{(n-k)(n+k+1)}{(k+2)(k+1)} a_k \qquad k = 0, 1, 2, \ldots$$

Primary interest is in the case $n = 0$ or a positive integer, when it is clear that for $k = n$, a_{k+2} will be zero; so will a_{k+4}, a_{k+6} and so on. Suppose, for example, that $n = 2$. Then $a_4, a_6, \ldots$ are all zero. If we also choose $a_1 = 0$, then $a_3, a_5, \ldots$ are also zero and with $a_2 = -3a_0$ we have the polynomial

$$y = a_0(1 - 3x^2)$$

The further choice $a = -\frac{3}{2}$ brings us to the polynomial $P_2(x)$. For other positive integer values of n the corresponding polynomial $P_n(x)$ can be similarly found.

The other independent solution is also available from the recursion. For example, with $n = 2$ the recursion becomes

$$a_{k+2} = -\frac{(2-k)(k+3)}{(k+2)(k+1)} a_k \qquad k = 0, 1, 2, \ldots$$

This time let $a_0 = 0$, forcing $a_k = 0$ for all even k. With a_1 arbitrary we then have

$$a_3 = -\tfrac{2}{3}a_1 \qquad a_5 = -\tfrac{1}{5}a_1 \qquad a_7 = -\tfrac{4}{35}a_1 \qquad \cdots$$

making

$$y = a_1\left(x - \tfrac{2}{3}x^3 - \tfrac{1}{5}x^5 - \tfrac{4}{35}x^7 - \cdots \right)$$

11.131 Discuss the convergence of the series developed in the preceding problem.

▮ The ratio test, since alternate terms are zero, asks for the limit of $x^2 a_{k+2}/a_k$ for large k, and this is easily seen to be x^2, proving the series convergent for x between -1 and 1. This is also discernible from the differential equation, since in solving for y'' we encounter a coefficient having poles at both of these extremes.

11.132 How fast does the series converge?

▮ Since the ratio of terms eventually approaches x^2, the series begins to resemble a geometric series having that ratio. For $x^2 = \frac{1}{2}$ or $x \approx .7$, this means the error will be the size of the last term used. Here are some computational results, asking for seven place accuracy but using single precision, with consequent potential for some roundoff error:

x	.2	.5	.7	.8	.9
Terms Used	4	10	18	29	57
y	.1946012	.4093316	.4230922	.3473192	.1486817

11.133 Apply the series method to the differential equation of the Chebyshev polynomials,

$$(1 - x^2) y'' - xy' + n^2 y = 0$$

▮ Things develop very much as with the Legendre equation. Let $y = \sum_0^\infty a_k x^k$, making $-xy' = \sum_0^\infty - a_k k x^k$, $y'' = \sum_2^\infty a_k (k-1) x^{k-2}$ and $-x^2 y'' = \sum_0^\infty - a_k k(k-1) x^k$. The third of these is subjected to an index change, converting it to $\sum_0^\infty a_{k+2}(k+2)(k+1) x^k$. With all the powers of x reconciled, we substitute into the equation and soon have

$$\sum_0^\infty \left[(n^2 - k^2) a_k + (k+2)(k+1) a_{k+2} \right] x^k = 0$$

which is satisfied by the recursion

$$a_{k+2} = \frac{k^2 - n^2}{(k+2)(k+1)} a_k \qquad k = 0, 1, 2, \ldots$$

Again it becomes clear that polynomial solutions can be arranged. For $n = 2$, as an example, a_4 will be zero, forcing $a_6, a_8, \ldots$ to be zero also. Choosing the arbitrary a_1 to be zero forces all other coefficients with odd index to join it. We are left with

$$y = a_0 + a_2 x^2 = a_0(1 - 2x^2)$$

reducing to $T_2(x)$ when $a_0 = -1$. For other positive integer values of n, the corresponding polynomials $T_n(x)$ are available.

The second independent solution of the differential equation can also be found by the recursion. Continuing with $n = 2$, this reduces to

$$a_{k+2} = \frac{k-2}{k+1} a_k \qquad k = 0, 1, 2, \ldots$$

and, choosing $a_0 = 0$, leads to the series

$$y = a_1 \left(x - \tfrac{1}{2} x^3 - \tfrac{1}{8} x^5 - \tfrac{1}{16} x^7 - \cdots \right)$$

11.134 Discuss the convergence of the series just produced.

▮ The situation is almost identical with that for the Legendre equation. The function $1/(1 - x^2)$ has poles at 1 and -1, so the Taylor series constructed at $x = 0$ will converge out this far at least. The ratio of consecutive terms being $x^2(k^2 - n^2)/(k+2)(k+1)$, with limit x^2, confirms this. Computations with the $n = 2$ series produce results very similar to those found in Problem 11.132.

x	.2	.5	.7	.8	.9
Terms Used	4	10	17	25	47

Again seven place accuracy was sought. Roundoff error may have upset this goal slightly.

11.135 Show that the series method fails for the equation $x^2 y'' = y$.

▌ From $y = \sum_0^\infty a_k x^k$ we find $x^2 y'' = \sum_0^\infty a_k k(k-1) x^k$ and must then require $a_k = a_k k(k-1)$. This can only be satisfied if $a_k = 0$ for all k. The series method has produced the trivial solution $y = 0$ identically, but fails to find the more interesting stuff. The reason is that the remaining solutions are not analytic in the neighborhood of $x = 0$, and so do not have Taylor series there. A modification of the usual power series proves to be the answer.

11.15 MODIFIED POWER SERIES

11.136 What are ordinary points and regular singular points of this differential equation?

$$y'' + a(x) y' + b(x) y = 0$$

▌ It is convenient to take the point to be $x = 0$, since $x = x_0$ can be reduced to this by the shift $t = x - x_0$. An ordinary point is one at which both $a(x)$ and $b(x)$ are analytic functions. Solutions in Taylor series form will exist in the neighborhood of such points. A regular singular point occurs if at least one of the two coefficient functions is not analytic, but both $xa(x)$ and $x^2 b(x)$ are. That is, $a(x)$ must not be worse than $1/x$, nor $b(x)$ worse than $1/x^2$. Any other singularity is called irregular. The point at infinity can also be included by letting $t = 1/x$ and applying our criteria to the modified differential equation.

11.137 Find solutions of $2x^2 y'' + 7x(x+1) y' - 3y = 0$ in the neighborhood of the regular singular point $x = 0$.

▌ The method of Frobenius for doing this is presented in most textbooks of differential equations, but one of two examples here will be useful. It seeks solutions in the form

$$y = x^p \sum_0^\infty a_k x^k = \sum_0^\infty a_k x^{p+k}$$

with the exponent p not necessarily an integer. Differentiating,

$$y' = \sum_0^\infty a_k (p+k) x^{p+k-1} \qquad y'' = \sum_0^\infty a_k (p+k)(p+k-1) x^{p+k-2}$$

from which we derive

$$2x^2 y'' = \sum_0^\infty 2 a_k (p+k)(p+k-1) x^{p+k}$$

$$7x^2 y' = \sum_0^\infty 7 a_k (p+k) x^{p+k-1} = \sum_1^\infty 7 a_{k-1} (p+k-1) x^{p+k}$$

$$7xy' = \sum_0^\infty 7 a_k (p+k) x^{p+k}$$

$$-3y = \sum_0^\infty -3 a_k x^{p+k}$$

to be added together and set to zero:

$$a_0 [2p(p-1) + 7p - 3] x^p + \sum_1^\infty \left\{ [2(p+k)(p+k-1) + 7(p+k) - 3] a_k + 7(p+k-1) a_{k-1} \right\} x^{p+k} = 0$$

As usual, the coefficient of each power of x must be zero. There is no point in letting a_0 be zero since this simply shifts the p value, so we have

$$2p^2 + 5p - 3 = 0$$

to determine p. This is called the indicial equation, and in this case makes p equal to either -3 or $\frac{1}{2}$. Taking the -3 first, the other coefficients will vanish if

$$a_k = \frac{-7(k-4)}{k(2k-7)} a_{k-1}$$

a recursion which leads to the finite sum

$$y_1(x) = a_0 x^{-3} \left(1 - \tfrac{21}{5}x + \tfrac{49}{5}x^2 - \tfrac{343}{15}x^3\right)$$

Then taking $p = \tfrac{1}{2}$, the corresponding recursion is

$$a_k = \frac{-7(2k-1)}{2k(2k+7)} a_{k-1}$$

and leads to an infinite series.

$$y_2(x) = a_0 x^{1/2} \left(1 - \tfrac{7}{18}x + \tfrac{147}{792}x^2 + \cdots\right)$$

The complete solution family is a blend of these two solutions.

11.138 Discuss the convergence of the series just found.

❚ Theory states that such series will converge in a region of the complex plane containing no other singular points of the coefficient functions in the differential equation. In this case there are no other finite singular points, so convergence should occur for any x. The ratio of consecutive terms supports this, being of order x/k and so tending to zero for any x.

A few computations were made seeking four place accuracy, with the results

x	2	4	6	8	10
Terms Used	17	33	51	68	80
Max Term		50	10^4	10^6	10^9
y	.8561	.9364	.9671	.7831	-101.9

The term of maximum size occurs near $k = 3x$, and after this the theorem on alternating series guarantees an error no greater than the first omitted term. But for increasing x the terms become enormous before starting their decrease, and the computation takes on a volatile look.

11.139 Apply the Frobenius method to the Bessel equation of order n,

$$x^2 y'' + xy' + (x^2 - n^2) y = 0$$

where n is constant.

❚ The point $x = 0$ is a regular singular point, so we try for a solution in the form $y = x^p \sum_0^\infty a_k x^k$. Substituting into the differential equation,

$$\sum_0^\infty (p+k)(p+k-1) a_k x^{p+k} + \sum_0^\infty (p+k) a_k x^{p+k} + \sum_0^\infty (x^2 - n^2) a_k x^{p+k} = 0$$

or

$$\sum_0^\infty \left[(p+k)^2 - n^2\right] a_k x^{p+k} + \sum_0^\infty a_k x^{p+k+2} = 0$$

Adjusting the index in the second series this becomes

$$\sum_0^\infty \left[(p+k)^2 - n^2\right] a_k x^{p+k} + \sum_2^\infty a_{k-2} x^{p+k} = 0$$

and the powers of x are synchronized. All coefficients now vanish if

$$(p^2 - n^2) a_0 = 0$$
$$\left[(p+1)^2 - n^2\right] a_1 = 0$$
$$\left[(p+k)^2 - n^2\right] a_k + a_{k-2} = 0 \qquad k = 2, 3, 4, \ldots$$

Since $a_0 = 0$ is pointless, we find the indicial equation

$$p^2 - n^2 = 0 \quad \text{or} \quad p = \pm n$$

Take first $p = n$, which we assume to be nonnegative. It follows that a_1 must be zero, with the recursion

$$a_k = -\frac{1}{2kn + k^2} a_{k-2} \qquad k = 2, 3, 4, \ldots$$

determining the other coefficients. Since $a_1 = 0$, all other odd coefficients are also zero. Repeated application of the recursion finds the even coefficients to be

$$a_{2k} = (-1)^k \frac{1}{k! 4^k (n+k)(n+k-1) \cdots (n+1)} a_0$$

The choice $a_0 = 1/2^n \Gamma(n+1)$ then allows the series to be written as

$$J_n(x) = \sum_0^\infty \frac{(-1)^k}{k! \Gamma(n+k+1)} \left(\frac{x}{2}\right)^{n+2k}$$

which defines the Bessel function of first kind of order n.

If n is not an integer, the root $p = -n$ of the indicial equation leads to a second solution

$$J_{-n}(x) = \sum_0^\infty \frac{(-1)^k}{k! \Gamma(-n+k+1)} \left(\frac{x}{2}\right)^{-n+2k}$$

and we have the entire solution family. For integral n the second solution will be found by another method. The values $n = 0, 1$ are of particular interest. The gamma function can then be expressed as a factorial, leading to

$$J_0(x) = \sum_0^\infty \frac{(-1)^k}{(k!)^2} \left(\frac{x}{2}\right)^{2k} \qquad J_1(x) = \sum_0^\infty \frac{(-1)^k}{k!(k+1)!} \left(\frac{x}{2}\right)^{2k+1}$$

of which more is to be seen in the next problem.

11.140 How well do the Bessel series converge?

▌ That they converge is clear, both from the fact that the coefficients in the differential equation have no other singular points in the finite plane and from the ratio of consecutive terms which behaves like x^2/k^2 for large k. This is excellent convergence. For example, to compute $J_0(x)$ to seven places takes only 19 terms at $x = 10$ and only 34 terms at $x = 20$. The maximum term in the latter computation was about 7500000 but convergence to the seven digits sought occurred shortly thereafter.

11.141 Find the first few zeros of $J_0(x)$.

▌ The excellent convergence permits a direct approach. The first zero was located between 2 and 3, then between 2.4 and 2.5 and three steps later at 2.4048. Fancier methods exist and will be used in a later chapter, but the zeros at 5.5201, 8.6537 and 11.7915 proved to be other easy targets.

11.142 Locate the first few maximum and minimum points of $J_0(x)$.

▌ They can be found by the direct approach, simply computing the function and gradually narrowing the bounds on an extreme point. As an alternative we first compute the derivative of $J_0(x)$:

$$J_0'(x) = \sum_1^\infty \frac{(-1)^k x^{2k-1}}{k!(k-1)! 2^{2k-1}} = -\sum_0^\infty \frac{(-1)^k}{(k+1)! k!} \left(\frac{x}{2}\right)^{2k+1} = -J_1(x)$$

So the extreme points of J_0 can be found by a familiar theorem. We seek instead the zeros of J_1. There is a slight advantage in this switch. A crossing of the x axis is apt to be sharper than a rounded top, and on the computer screen one has only to look for the change in sign rather than to scan several decimal places to detect the turn. In any event, the zeros of J_1 are easily found, because of the rapid convergence, to be at $x = 3.8317$, 7.0156 and 10.1735 for starters. A graph of these two functions over the interval $(0, 11)$ appears as Fig. 11-16.

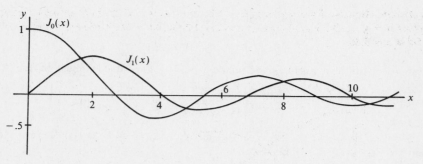

Fig. 11-16

11.143 Find a second independent solution of the Bessel equation of order zero.

❚ One procedure begins from the recursion, which for $n = 0$ is $a_k = -a_{k-2}/(p+k)^2$. Although the value of p is known to be zero this substitution is delayed. Successive applications now manage

$$a_{2k} = \frac{(-1)^k}{(p+2k)^2(p+2k-2)^2 \cdots (p+2)^2}$$

if we take a_0 to be 1. Note that these coefficients now depend on p and may be called $a_{2k}(p)$. If we form the function

$$y(x, p) = x^p \sum_0^\infty a_{2k}(p)x^{2k}$$

and substitute it into the Bessel equation we find, not 0, since we have delayed equating p with 0, but p^2x^p. In fact, for $p = 0$ we also find that $y(x, 0) = J_0(x)$ and, of course, it does reduce p^2x^p to zero as a solution must. However, not only is p^2x^p zero for $p = 0$, so is its derivative relative to p. This suggests that $\partial y(x, p)/\partial p$ may also be a solution and this does prove to be the case. Computing this derivative,

$$\frac{\partial y(x, p)}{\partial p} = x^p \left[\sum_0^\infty \frac{\partial a_{2k}(p)}{\partial p} x^{2k} + (\log x) \sum_0^\infty a_{2k}(p)x^{2k} \right]$$

and then finally setting $p = 0$,

$$\left. \frac{\partial y(x, p)}{\partial p} \right|_{p=0} = \sum_0^\infty \left. \frac{\partial a_{2k}(p)}{\partial p} \right|_{p=0} x^{2k} + (\log x) J_0(x)$$

The derivative of $a_{2k}(p)$ can be computed directly and, after a bit of algebra, is

$$\left. \frac{\partial a_{2k}(p)}{\partial p} \right|_{p=0} = \frac{(-1)^{k+1}}{2^{2k}(k!)^2} \left(1 + \frac{1}{2} + \cdots + \frac{1}{k} \right)$$

All the ingredients are now in hand and this second solution of the zero-order equation is denoted

$$N_0(x) = J_0(x)\log x + \sum_1^\infty \frac{(-1)^{k+1}}{(k!)^2} \left(1 + \frac{1}{2} + \cdots + \frac{1}{k} \right) \left(\frac{x}{2} \right)^{2k}$$

after Neumann.

11.144 How well does the new series converge?

❚ We know from Problem 9.82 that the harmonic sum involved here is, for large k, close to $\log k$ plus Euler's constant. The ratio of one term to the next is, therefore, essentially the same as for the J_0 series. This means excellent convergence. The presence of the logarithm factor does mean that near $x = 0$ this solution grows very large negatively. Its values for a few integer arguments were found easily from the series, using at most 16 terms. A plot appears in Fig. 11-17. The first maximum and minimum were also found directly to be about .83 at $x = 2.12$ and $-.54$ at $x = 5.36$.

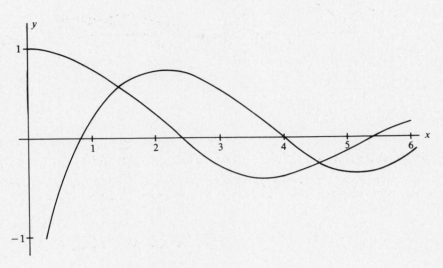

Fig. 11-17

11.45   What is the behavior of solutions to the Legendre equation at infinity?

▮   The change of variable $t = 1/x$ replaces the familiar

$$(1 - x^2) y'' - 2xy' + n(n + 1) y = 0$$

by

$$(t^4 - t^2) \frac{d^2y}{dt^2} + 2t^3 \frac{dy}{dt} + n(n + 1) y = 0$$

for which $t = 0$ is a regular singular point. The Frobenius method applies and with $y(t) = \sum_0^\infty a_k t^{p+k}$ we are led to

$$[-p(p - 1) + n(n + 1)] a_0 t^p + [-p(p + 1) + n(n + 1)] a_1 t^{p+1}$$

$$+ \sum_2^\infty \{[-(p + k)(p + k - 1) + n(n + 1)] a_k + (p + k - 2)(p + k - 1) a_{k-2}\} t^{p+k} = 0$$

The indicial equation $p(p - 1) = n(n + 1)$ is soon seen to have the usual roots $p = -n$ and $p = n + 1$. This leaves a_0 arbitrary as usual but forces $a_1 = 0$. The recursion

$$a_k = \frac{(p + k - 2)(p + k - 1)}{(p + k)(p + k - 1) - n(n + 1)} a_{k-2}$$

then determines the remaining coefficients. For $p = -n$ this is the series that develops:

$$y_1 = a_0 t^{-n} \left[1 - \frac{n(n - 1)}{2(2n - 1)} t^2 + \frac{n(n - 1)(n - 2)(n - 3)}{2 \cdot 4(2n - 1)(2n - 3)} t^4 - \cdots \right]$$

$$= a_0 x^n \left[1 - \frac{n(n - 1)}{2(2n - 1)} x^{-2} + \frac{n(n - 1)(n - 2)(n - 3)}{2 \cdot 4(2n - 1)(2n - 3)} x^{-4} - \cdots \right]$$

If n is a nonnegative integer the series terminates and we have once again the Legendre polynomials, assuming a proper choice of a_0 for each n. This shows only that at infinity we can expect a polynomial to behave like a polynomial.

Taking $p = n + 1$ and $a_0 = 1$ leads to the series

$$y_2 = t^{n+1}\left[1 + \frac{(n+1)(n+2)}{2(2n+3)}t^2 + \frac{(n+1)(n+2)(n+3)(n+4)}{2\cdot 4(2n+3)(2n+5)}t^4 + \cdots\right]$$

$$= \frac{1}{x^{n+1}}\left[1 + \frac{(n+1)(n+2)}{2(2n+3)}\frac{1}{x^2} + \frac{(n+1)(n+2)(n+3)(n+4)}{2\cdot 4(2n+3)(2n+5)}\frac{1}{x^4} + \cdots\right]$$

The family of solutions can now be expressed as a combination of y_1 and y_2, except for some half integer values of n, which bring in zero denominators. For these the method of Problem 11.143 would have to be reactivated.

11.146 How well does the series of the preceding problem converge?

I From the differential equation for $y(t)$ we see that, apart from the singular point at $t = 0$ which has our attention, there are others at ± 1. Convergence is thus guaranteed for $|t| < 1$, and so for $|x| < 1$. The ratio of successive terms is

$$\frac{a_k}{a_{k-2}x^2} = \frac{(n+k-1)(n+k)}{k(2n+k+1)x^2}$$

and approaches $1/x^2$ for large k, again proving convergence for the same values of x. It also points out that the series is basically geometric from that point onward, with ratio $1/x^2$. Computations verify this. Here are a few numbers, computed using $n = 0$:

x	20	10	2	1.5	1.2	1.1	1.01
Terms	2	3	8	15	28	48	struggle
y	.050042	.100335	.549306	.804719	1.19895	1.52226	—

Note that the climb to infinity at $x = 1$ seems to be rather slow starting.

11.147 How do solutions of the Legendre equation behave in the vicinity of $x = 1$?

I Presumably they tend to infinity there. To find out how, the variable change $t = x - 1$ is helpful. It converts the Legendre equation to

$$t(2+t)y'' + 2(1+t)y' - n(n+1)y = 0$$

where the primes now mean derivatives relative to t. So the point $x = 1$ turns out to be a regular singular point and the Frobenius method applies. Substituting $y = t^p\sum_0^\infty a_k t^k$ and doing a bit of work we come eventually to

$$-2a_0 p^2 t^{p-1} + \sum_0^\infty \left\{ -2(k+p+1)^2 a_{k+1} - [(k+p)(k+p+1) - n(n+1)]a_k \right\} t^{p+k} = 0$$

with the indicial equation simply $p^2 = 0$. Certainly this is not the route toward infinity. With $p = 0$ the recursion becomes

$$a_{k+1} = -\frac{k(k+1) - n(n+1)}{2(k+1)^2}a_k$$

and coefficients beginning with $k = n$ with all be zero. We have once again found the Legendre polynomials, which will slide past the singular point without noticing it.

To find the missing solution, the method of Problem 11.143 is available. For simplicity take the case $n = 0$. The recursion then reduces to

$$a_{k+1} = -\frac{k+p}{2(k+p+1)}a_k$$

which, applied successively, makes

$$a_k(p) = \frac{(-1)^k p}{2^k(k+p)} \qquad k = 1, 2, 3, \ldots$$

if we choose $a_0 = 1$. In passing we note that $p = 0$ makes all of these coefficients zero and the solution is $P_0(x) = 1$. But to stay on track, the function

$$y(t, p) = t^p \sum_0^\infty a_k(p) t^k$$

when substituted into the differential equation, leaves a residue of $-2p^2 t^{p-1}$, since the recursion wipes out all other terms. So because of the presence of p^2, not only is $y(t, 0)$ a solution but so is $\partial y(t, p)/\partial p$ for $p = 0$. (There is a matter of interchanging the order of certain differentiations involved here, but it has been proved that this is valid.) Taking the indicated derivative

$$\frac{\partial y(t, p)}{\partial p} = t^p \sum_0^\infty \frac{\partial a_k}{\partial p} t^k + t^p (\log t) \sum_0^\infty a_k(p) t^k$$

and computing $\partial a_k/\partial p = (-1)^k k/2^k (k + p)^2$, we then find

$$y = \frac{\partial y(t, p)}{\partial p} \bigg|_{p=0} = \sum_1^\infty \frac{(-1)^k}{k \cdot 2^k} t^k + \log t = \sum_1^\infty \frac{(-1)^k (x-1)^k}{k \cdot 2^k} + \log(x - 1)$$

which is the solution sought.

The Legendre equation for $n = 0$ can, of course, be solved by elementary methods, the solution in closed form being

$$y = c_1 P_0(x) + c_2 \log \frac{x - 1}{x + 1}$$

for $x > 1$. The series just obtained represents $-\log(x + 1)$ in the neighborhood of $x = 1$, except for a constant term.

11.148 How well does the series of the preceding problem converge?

▮ The logarithm series is familiar and this one converges up to $x = 3$. (Slight adjustments are needed in our presentation if x is to be less than 1.) The closer x is to 1, the fewer terms are needed. Here are a few results:

x	1.001	1.01	1.1	1.2	1.5	2
y	-6.90821	-4.61016	-2.35138	-1.70475	$-.916291$	$-.405465$
Terms	2	3	5	6	11	16

At $x = 3$ it would take a thousand terms to get even three decimal place accuracy. Near the singular point the logarithm term does the job by itself.

11.149 Use the representations and data provided in the preceding four problems to solve the Legendre $(n = 0)$ equation with initial values $y(1.5) = 2$ and $y(2) = 1$.

▮ The only point of interest is the possible interplay between representations. If we call the series of Problem 11.145, $Q_\infty(x)$ and that of Problem 11.147, $Q_1(x)$, indicating the central point of each expansion, then our solution can be represented in either of these ways,

$$y = a_1 + a_2 Q_\infty(x) = b_1 + b_2 Q_1(x)$$

the constant solution $P_0(x) = 1$ accounting for the first term in each case. The coefficients are found from the initial conditions in the usual way, necessary values of the Q functions having been provided in Problems 146 and 148:

$$a_1 + .549306 a_2 = 1 \qquad b_1 - .405465 b_2 = 1$$
$$a_1 + .804719 a_2 = 2 \qquad b_1 - .916291 b_2 = 2$$

Solving and substituting back we have

$$y = -1.15066 + 3.91523 Q_\infty(x) = .206256 - 1.95761 Q_1(x)$$

and may appear to have worked twice as hard as necessary. In fact for some values of x we have, but to compute $y(1.01)$ or $y(10)$ it is convenient to have a choice, and in each case there is little doubt about which representation to choose. A few numbers are offered in Table 11.20.

TABLE 11.20

x	y via $Q_1(x)$	y via $Q_\infty(x)$
1.001	13.7298	struggle
1.01	9.23115	struggle
1.1	4.80933	4.80934
1.2	3.54350	3.54350
1.5	2	2
2	1	1
10	hopeless	−.757825
20	hopeless	−.954735

11.16 IRREGULAR SINGULAR POINTS

11.150 The point $x = 0$ is an irregular singular point of the equation

$$x^3 y'' = y$$

Show that the Frobenius method produces only the trivial solution $y = 0$.

I The singularity at $x = 0$ is not regular because $1/x^3$ is not made analytic when multiplied by x^2. It becomes infinite too rapidly. Writing the Frobenius series out for a change,

$$y = a_0 x^p + a_1 x^{p+1} + a_2 x^{p+2} + \cdots$$
$$x^3 y'' = a_0 p(p-1) x^{p+1} + a_1(p+1) p x^{p+2} + a_2(p+2)(p+1) x^{p+3} + \cdots$$

and comparing coefficients

$$a_0 = 0 \qquad a_1 = a_0 p(p-1) \qquad a_2 = a_1(p+1)p \qquad \cdots$$

forcing all the a_k to be zero.

11.151 How do solutions of $x^3 y'' = y$ behave near $x = 0$?

I The behavior of solutions near such points has been approached by means of asymptotic analysis. The substitution $y = e^u$ has often proved useful. It makes $y' = e^u u'$ and $y'' = e^u(u'' + u'^2)$, so the given equation becomes

$$u'' + (u')^2 = 1/x^3$$

It is often found that u'' is small compared with $(u')^2$, near the singularity. If this term is dropped, the very simple first-order equation

$$u' = \pm x^{-3/2}$$

appears. Integrating now brings

$$u = \pm 2x^{-1/2} + C(x)$$

where $C(x)$ is not assumed to be a constant, but instead a term that is small compared with $x^{-1/2}$. The steps just taken are based to some extent on optimism, and some measure of after-the-fact testing may be wise, but what they suggest is that the behavior of solutions near $x = 0$ is "like" $x^{-1/2}$.

11.152 How can a more detailed estimate of solution behavior be made?

I The behavior of $C(x)$ can be investigated. To do this we substitute back into the equation for u. Taking the plus sign first,

$$u = 2x^{-1/2} + C \qquad u' = -x^{-3/2} + C' \qquad u'' = \tfrac{3}{2} x^{-5/2} + C''$$

leading to

$$\tfrac{3}{2}x^{-5/2} + C'' + x^{-3} - 2x^{-3/2}C' + (C')^2 - x^{-3} = 0$$

But it has been assumed that C is small compared with $2x^{-1/2}$, and will now be further assumed that the relationship can be carried over to their derivatives. That is, C' and C'' are small compared with $x^{-3/2}$ and $x^{-5/2}$. The C'' and $(C')^2$ terms can then be neglected and the equation is reduced to $\tfrac{3}{2}x^{-5/2} = 2x^{-3/2}C'$ or

$$C' = \tfrac{3}{4}x^{-1}$$

But then $C = \tfrac{3}{4}\log x + D$ where D is assumed small compared with the logarithm. At this point we have

$$u = 2x^{-1/2} + \tfrac{3}{4}\log x + D$$

Another similar step estimates D to be constant, clearly smaller than $\log x$ and leading to

$$y = c_1 x^{3/4} e^{2/\sqrt{x}}$$

in which the equality sign is being used to indicate an (unproven) asymptotic relationship. Perhaps the thing to observe is that as x nears zero, y becomes infinite, the exploding exponential being too much for the weaker power of x.

11.153 What is the other independent "solution" of the equation studied in the preceding problem?

▌ Consider the remaining option for $u(x)$, namely $-2x^{-1/2} + C(x)$. The details are now similar but the finish is somewhat different. Since $u' = x^{-3/2} + C'$ and $u'' = -\tfrac{3}{2}x^{-5/2} + C''$ it follows that

$$u'' + (u')^2 - \frac{1}{x^3} = -\frac{3}{2}x^{-5/2} + C'' + 2x^{-3/2}C' + (C')^2 = 0$$

since the x^{-3} terms again cancel. Again assuming that $(C')^2$ is small compared with C'', which in turn is small compared with $x^{-5/2}$, this equation is replaced by

$$-\tfrac{3}{2}x^{-5/2} + 2x^{-3/2}C' = 0 \quad \text{or} \quad C' = \tfrac{3}{4}x^{-1}$$

which yields $C = \tfrac{3}{4}\log x + D$. Another round shows D to be constant and so $u = -2x^{-1/2} + \tfrac{3}{4}\log x + D$. We now have our second "solution":

$$y = e^u = c_2 x^{3/4} e^{-2/\sqrt{x}}$$

This one approaches zero with x, since the powerful exponential is now heading in that direction too.

11.154 Test the asymptotic results just found by applying a Taylor series method to the equation $x^3 y'' = y$, with initial conditions $y(1) = 0$ and $y'(1) = -1$. Compare the numerical output with that of the asymptotic formulas.

▌ The Taylor approach is straighforward, with the change of argument $t = x - 1$ probably a convenience. Substituting $y = \sum_0^\infty a_k t^k$ into $(1 + t)^3 y'' = y$ and multiplying out the cubic leads, with a bit of work and with $a_0 = y(0)$, $a_1 = y'(0)$, to the coefficients

$$a_2 = -\tfrac{1}{2}a_0 \qquad a_3 = \tfrac{1}{6}(a_1 - 6a_2) \qquad a_4 = \tfrac{1}{12}(-5a_2 - 18a_3)$$

and for $k > 2$

$$a_{k+2} = \frac{1}{(k+2)(k+1)}\left[a_k - (k-1)(k-2)a_{k-1} - 3k(k-1)a_k - 3(k+1)ka_{k+1}\right]$$

The series readily produces six place values of y down to $x = .3$ or thereabouts, using up to 40 terms, and four place values a bit further without strain. But below $x = .1$ the computation labors. In Table 11.21 the ratio of the computed y to $x^{3/4}e^{2/\sqrt{x}}$ is given. It appears to converge to a constant near .074, and this is now taken to be the constant c_1 of our first asymptotic result. (The second plays no role this time, since it has zero limiting value and our solution is growing.) Values of $.074x^{3/4}e^{2/\sqrt{x}}$ are also provided in the table. At $x = .4$ the error is still almost 20%, but at $x = .1$ there is agreement to two places. At $x = .05$ the series coefficients were oscillating badly and an overflow occurred, but the asymptotic formula was now on its own ground, providing 59.8. As a final flourish, it also predicted $y(.01)$ to be 1132259, but it is unlikely that all of these digits are correct.

TABLE 11.21

x	Series	Series$/x^{3/4}e^{2/\sqrt{x}}$	$.074x^{3/4}e^{2/\sqrt{x}}$
1	0	0	—
.8	.2019	.0255	—
.6	.4229	.0469	—
.4	.7430	.0625	.877
.2	1.8664	.0713	1.932
.1	7.32	.0738	7.32
.09	9.52	.0738	9.52
.05	—	—	59.8
.01	—	—	1132259

11.155 Show that the Airy equation $y'' = xy$ has an irregular singular point at infinity.

▮ Letting $t = 1/x$, so that $y' = t^2(dy/dt)$ and $y'' = t^4(d^2y/dt^2) + 2t^3(dy/dt)$, the equation

$$\frac{d^2y}{dt^2} + 2t^{-1}\frac{dy}{dt} - t^{-5}y = 0$$

is encountered. The last coefficient is not made analytic when multiplied by t^2, so the singular point at $t = 0$ is not regular.

11.156 Use the substitution $y = e^u$ to study the behavior of solutions of the Airy equation for large x.

▮ Since $y' = e^u u'$ and $y'' = e^u(u'' + (u')^2)$ the given equation is replaced by

$$u'' + (u')^2 = x$$

As a first step we assume that u'' is small compared with $(u')^2$ and have the simpler $u' = \pm\sqrt{x}$, the solution of which is

$$u = \pm\tfrac{2}{3}x^{3/2} + C(x)$$

where we assume that $C(x)$ is small compared with $x^{3/2}$.

To determine $C(x)$ we now substitute what we have into the differential equation for u, taking the minus sign first. Since

$$u' = -x^{1/2} + C' \quad \text{and} \quad u'' = -\tfrac{1}{2}x^{-1/2} + C''$$

we soon have

$$-\tfrac{1}{2}x^{-1/2} + C'' + x - 2x^{1/2}C' + (C')^2 = x$$

But C'' has been assumed small compared with $(C')^2$, which in turn is small compared with $x^{1/2}C'$. (Compare the derivatives of C and $x^{3/2}$.) Eliminating both of these terms leaves

$$-\tfrac{1}{2}x^{-1/2} - 2x^{1/2}C' = 0$$

which makes $C' = -1/4x$. Integrating, $C = -\tfrac{1}{4}\log x + D$, where it is assumed that D is small compared with the logarithm. At this point we have $u(x) = -\tfrac{2}{3}x^{3/2} - \tfrac{1}{4}\log x + D$. We will find that D is asymptotically constant (and arbitrary). Choosing the constant as $1/2\sqrt{\pi}$, we have the solution.

$$y(x) = \left(1/2\sqrt{\pi}\right)x^{-1/4}e^{-(2/3)x^{3/2}}$$

Once again the equality sign is to be interpreted as an asymptotic relationship. As given, $y(x)$ represents the solution Ai(x) first encountered in Problem 11.129.

If the second option, $u = \frac{2}{3}x^{3/2} + C(x)$, is pursued, the analysis differs only by an occasional sign change and the result only in the sign of the power of e, which will be positive.

11.157 Compare the performances of the asymptotic result just found and the series for Ai(x) in Problem 11.128.

▮ Table 11.22 provides a few numbers, only enough digits being included to make the comparisons. The series is to be trusted for small x and the asymptotic formula for large. It seems that the latter has the better range, at least qualitatively. One problem for the series method is that it involves computing two series and subtracting the results. Another local series in the neighborhood of $x = .3$ might be useful. (See also the next few problems.)

TABLE 11.22

x	.1	.5	1	2	3	5	10
Series	.33	.23	.13	.024	$-.039$	—	—
Asymptotic	.49	.26	.14	.036	.0067	.00011	$1.1 \cdot 10^{-10}$

11.158 Derive an asymptotic differential equation for the unknown D.

▮ Renaming it $\log v(x)$ we have

$$u(x) = -\frac{2}{3}x^{3/2} - \frac{1}{4}\log x + \log v(x)$$

$$u'(x) = -x^{1/2} - \frac{1}{4x} + \frac{v'}{v}$$

$$u''(x) = -\frac{1}{2}x^{-1/2} + \frac{1}{4x^2} + \frac{vv'' - (v')^2}{v^2}$$

and again substituting into $u'' + (u')^2 = x$,

$$-\frac{1}{2}x^{-1/2} + \frac{1}{4x^2} + \frac{vv'' - (v')^2}{v^2} + x + \frac{1}{16x^2} + \frac{(v')^2}{v} + \frac{1}{2}x^{-1/2} - 2x^{1/2}\frac{v'}{v} - \frac{v'}{2xv} = x$$

which we proceed to simplify. With the usual assumptions, only the terms

$$\frac{5}{16x^2} + \frac{v''}{v} - 2x^{1/2}\left(\frac{v'}{v}\right) - \frac{v'}{2xv} = 0$$

or

$$x^2 v'' - \left(2x^{5/2} + \frac{1}{2}x\right)v' + \frac{5}{16}v = 0$$

need to be retained. This equation holds the key to the behavior of $v(x)$.

11.159 Solve the equation for $v(x)$ asymptotically.

▮ Assume a representation

$$v = 1 + a_1 x^p + a_2 x^{2p} + a_3 x^{3p} + \cdots$$

leading to

$$v' = pa_1 x^{p-1} + 2pa_2 x^{2p-1} + 3pa_3 x^{3p-1} + \cdots$$

$$x^2 v'' = p(p-1)a_1 x^p + 2p(2p-1)a_2 x^{2p} + 3p(3p-1)a_3 x^{3p} + \cdots$$

and then

$$-2x^{5/2}v' = -2pa_1x^{p+3/2} - 4pa_2x^{2p+3/2} - 6pa_3x^{3p+3/2} + \cdots$$

$$-\tfrac{1}{2}xv' = -\tfrac{1}{2}pa_1x^p - pa_2x^{2p} - \tfrac{3}{2}pa_3x^{3p} + \cdots$$

Substituting into the differential equation, we have

$$\left(\tfrac{5}{16} - 2pa_1x^{p+3/2}\right) + \left(p^2 - \tfrac{3}{2}p + \tfrac{5}{16}\right)a_1x^p + \cdots = 0$$

the terms ordered by decreasing powers, p being assumed negative. To make v small in comparison with $\log x$ and to arrange the leading constant of 1, we choose $p = -\tfrac{3}{2}$ and $a_1 = -\tfrac{5}{48}$. This takes care of the first term listed. It may now be convenient to rewrite the original series with $t = x^p$:

$$v = 1 + a_1t + a_2t^2 + a_3t^3 + \cdots$$

It follows that

$$x^2v'' = a_1p(p-1)t + a_22p(2p-1)t^2 + a_33p(3p-1)t^3 + \cdots$$

$$-2x^{5/2}v' = -2a_1p - 2a_2(2p)t - 2a_3(3p)t^2 - 2a_4(4p)t^3 + \cdots$$

$$-\tfrac{1}{2}xv' = -\tfrac{1}{2}a_1pt - \tfrac{1}{2}a_2(2p)t^2 - \tfrac{1}{2}a_3(3p)t^3 + \cdots$$

and repeating the substitution into the differential equation we find the coefficient of t^k to be

$$a_k\left(\tfrac{5}{16} - \tfrac{3}{2}kp + k^2p^2\right) - 2(k+1)pa_{k+1}$$

which is equated to zero to produce the recursion

$$a_{k+1} = \frac{(6k+5)(6k+1)}{-48(k+1)}a_k \qquad k = 1, 2, \ldots$$

The recursion also happens to be valid for $k = 0$. The series for $v(x)$ is now in hand,

$$v = 1 - \frac{5}{48}x^{-3/2} + \frac{385}{4608}x^{-3} + \cdots$$

Unfortunately it diverges for all values of x.

11.160 Compare values of Ai(x) computed from the Taylor series at $x = 0$ and the asymptotic series

$$y = \frac{1}{2\sqrt{\pi}}x^{-1/4}e^{-(2/3)x^{3/2}}\left(1 - \frac{5}{48}x^{-3/2} - \frac{385}{4608}x^{-3} + \cdots\right)$$

How much does the series improve the estimates of Ai(x) found in Problem 11.157?

▮ The earlier values were obtained using only the leading term of this series, the 1. Table 11.23 shows the new values, more decimal places appearing because of the greater accuracy. For each value of x, the series computation

TABLE 11.23

x	# Terms	Min Term	Asymptotic	Taylor
.1	1		.49	.33
.5	1	.3	.26	.23
1	2	.08	.130	.132
2	4	.0046	.035	.024
3	5	.00015	.00659	minus
4	9	.000003	.00095156	
5	13	$3 \cdot 10^{-11}$	.0001083444	

stopped just short of the absolutely smallest term, making optimal use of the asymptotic series. Values of the Taylor series are repeated for comparisons, and it will be noted that even as low as $x = 1$ the asymptotic result has an error of only 2%. The $x = 4$ value is correct to five significant digits and that for $x = 5$ to the capacity of the single precision operation.

11.161 The Bessel function $J_0(x)$ has been shown to have this asymptotic development for large positive x. Let

$$u(x) = \sum_0^\infty (-1)^k c_{2k} x^{-2k} \quad \text{and} \quad v(x) = \sum_0^\infty (-1)^k c_{2k+1} x^{-2k-1}$$

for large x, where $c_0 = 1$ and for $k = 1, 2, \ldots,$

$$c_k = \frac{(-1)^k \cdot 1^2 \cdot 3^2 \cdots (2k-1)^2}{k! \cdot 8^k}$$

Then

$$J_0(x) = \sqrt{\frac{2}{\pi x}} \left[u(x) \cos\left(x - \frac{\pi}{4}\right) - v(x) \sin\left(x - \frac{\pi}{4}\right) \right]$$

Compare the performance of this series with that of the Taylor series for $J_0(x)$ found in Problem 11.139.

I Table 11.24 provides the numbers. The Taylor series is a very strong competitor here, out as far as $x = 10$ at least. This is a bit surprising in view of the large size and alternating signs of its terms. Somewhat beyond this point, however, the baton must be passed to the asymptotic series.

TABLE 11.24

x	Asymptotic	True	Taylor
1	.748	.7652	.765198
2	.226	.2239	.223891
3	−.2591	−.2601	−.260052
5	−.17772	−.1776	−.177597
10	−.24593	−.2459	−.24589
20	.167025		struggle

11.162 A table of the zeros $J_0(x)$ has the final two listings 18.0711 and 21.2116. Use the asymptotic formula of the preceding problem to verify them and then to find the next zero.

I This simple exercise will serve both to demonstrate the accuracy of the formula and to double-check the table. In fact, the values

$$J_0(18.0710) = -.000014 \qquad J_0(21.2116) = .000008$$
$$J_0(18.0711) = .000005 \qquad J_0(21.2117) = -.000010$$

are produced almost effortlessly. It appears that the table is correct. As for the next zero, it is soon located between 24 and 25, and after a patient search the values found are

$$J_0(24.3524) = -.000012 \qquad J_0(24.3525) = .000004$$

On the evidence, we give the nod to the higher figure.

11.163 What is the behavior of $J_0(x)$ for large x?

I The function $u(x)$ approaches 1, while $v(x)$ behaves like $1/8x$. It seems that the term $\cos(x - \pi/4)/\sqrt{x}$ is dominant. This is a damped oscillation. How well does this dominant term locate the zeros computed in the preceding problem? The zeros of the cosine occur at $\pi/4$ plus the odd multiples of $\pi/2$, or

$$18.064 \qquad 21.206 \qquad 24.347$$

for the proper odd multiples. These are off by a few points in the last place, for a relative error of about .05%.

11.164 Show that the change of variable $y = z/\sqrt{x}$ converts Bessel's equation to

$$z'' + \left(1 - \frac{n^2 - \frac{1}{4}}{x^2}\right)z = 0$$

For large x this resembles $z'' + z = 0$, which suggests that z may be asymptotically like $\sin x$ or $\cos x$.

▮ We find

$$y' = x^{-1/2}z' - \tfrac{1}{2}x^{-3/2}z \qquad y'' = x^{-1/2}z'' - x^{-3/2}z' + \tfrac{3}{4}x^{-5/2}z$$

and substituting into Bessel's equations easily find the required result.

11.17 METHODS FOR SYSTEMS

11.165 Write out the Runge–Kutta formulas for two simultaneous first-order equations using the classical fourth-order set.

▮ Let the given equations be

$$y' = f_1(x, y, p) \qquad p' = f_2(x, y, p)$$

with initial conditions $y(x_0) = y_0$ and $p(x_0) = p_0$. The formulas

$$k_1 = hf_1(x_n, y_n, p_n) \qquad\qquad k_3 = hf_1(x_n + \tfrac{1}{2}h, y_n + \tfrac{1}{2}k_2, p_n = \tfrac{1}{2}l_2)$$

$$l_1 = hf_2(x_n, y_n, p_n) \qquad\qquad l_3 = hf_2(x_n + \tfrac{1}{2}h, y_n + \tfrac{1}{2}k_2, p_n + \tfrac{1}{2}l_2)$$

$$k_2 = hf_1(x_n + \tfrac{1}{2}h, y_n + \tfrac{1}{2}k_1, p_n + \tfrac{1}{2}l_1) \qquad k_4 = hf_1(x_n + hy_n + k_3, p_n + l_3)$$

$$l_2 = hf_2(x_n + \tfrac{1}{2}h, y_n + \tfrac{1}{2}k_1, p_n + \tfrac{1}{2}l_1) \qquad l_4 = hf_2(x_n + h, y_n + k_3, p_n + l_3)$$

$$y_{n+1} = y_n + \tfrac{1}{6}(k_1 + 2k_2 + 2k_3 + k_4)$$

$$p_{n+1} = p_n + \tfrac{1}{6}(l_1 + 2l_2 + 2l_3 + l_4)$$

may be shown to duplicate the Taylor series for both functions up through terms of order 4. The details are identical to those for a single equation and will be omitted. For more than two simultaneous equations, say n, the extension of Runge–Kutta method parallels the preceding method, with n sets of formulas instead of two.

11.166 Write out the Adams-type predictor–corrector formula for the simultaneous equations of the preceding problem.

▮ Assume that four starting values of each function are available, say y_0, y_1, y_2, y_3 and p_0, p_1, p_2, p_3. Then the predictor formulas

$$y_{k+1} \approx y_k + \tfrac{1}{24}h(55y'_k - 59y'_{k-1} + 37y'_{k-2} - 9y'_{k-3})$$

$$p_{k+1} \approx p_k + \tfrac{1}{24}h(55p'_k - 59p'_{k-1} + 37p'_{k-2} - 9p'_{k-3})$$

may be applied with

$$y'_k = f_1(x_k, y_k, p_k) \qquad p'_k = f_2(x_k, y_k, p_k)$$

The results may be used to prime the corrector formulas

$$y_{k+1} \approx y_k + \tfrac{1}{24}h(9y'_{k+1} + 19y'_k - 5y'_{k-1} + y'_{k-2})$$

$$p_{k+1} \approx p_k + \tfrac{1}{24}h(9p'_{k+1} + 19p'_k - 5p'_{k-1} + p'_{k-2})$$

which are then iterated until consecutive outputs agree to a specified tolerance. The process hardly differs from that for a single equation. Extension to more equations or to other predictor–corrector combinations is similar. One may even use different formulas for y and p separately, but this seems fancy.

11.167 Compute the solution of van der Pol's equation

$$y'' - (.1)(1 - y^2)y' + y = 0$$

with initial values $y(0) = 1$ and $y'(0) = 0$ up to the fifth zero of $y'(t)$. Use the Runge–Kutta formulas for two first-order equations.

❚ An equivalent first-order system is

$$y' = p = f_1(t, y, p)$$
$$p' = -y + (.1)(1 - y^2) p = f_2(t, y, p)$$

The Runge–Kutta formulas for this system are

$$k_1 = hp_n \qquad l_1 = h\left[-y_n + (.1)(1 - y_n^2) p_n\right]$$

$$k_2 = h\left(p_n + \tfrac{1}{2}l_1\right) \qquad l_2 = h\left\{-\left(y_n + \tfrac{1}{2}k_1\right) + (.1)\left[1 - \left(y_n + \tfrac{1}{2}k_1\right)^2\right]\left(p_n + \tfrac{1}{2}l_1\right)\right\}$$

$$k_3 = h\left(p_n + \tfrac{1}{2}l_2\right) \qquad l_3 = h\left\{-\left(y_n + \tfrac{1}{2}k_2\right) + (.1)\left[1 - \left(y_n + \tfrac{1}{2}k_2\right)^2\right]\left(p_n + \tfrac{1}{2}l_2\right)\right\}$$

$$k_4 = h\left(p_n + l_3\right) \qquad l_4 = h\left\{-\left(y_n + k_3\right) + (.1)\left[1 - \left(y_n + k_3\right)^2\right]\left(p_n + l_3\right)\right\}$$

and

$$y_{n+1} = y_n + \tfrac{1}{6}\left(k_1 + 2k_2 + 2k_3 + k_4\right) \qquad p_{n+1} = p_n + \tfrac{1}{6}\left(l_1 + 2l_2 + 2l_3 + l_4\right)$$

Choosing $h = .2$, computations produce the following results to three places:

$$k_1 = (.2)(0) = 0 \qquad l_1 = (.2)\left[-1 + (.1)(1 - 1)(0)\right] = -.2$$

$$k_2 = (.2)(-.1) = -.02 \qquad l_2 = (.2)\left[-1 + (.1)(1 - 1)(-.1)\right] = -.2$$

$$k_3 \simeq (.2)(-.1) = -.02 \qquad l_3 = (.2)\left[-.99 + (.1)(.02)(-.1)\right] \simeq -.198$$

$$k_4 \simeq (.2)(-.198) \simeq -.04 \qquad l_4 = (.2)\left[-(.98) + (.1)(.04)(-.198)\right] \simeq -.196$$

These values now combine into

$$y_1 \simeq 1 + \tfrac{1}{6}(-.04 - .04 - .04) = .98$$

$$p_1 \simeq 0 + \tfrac{1}{6}(-.2 - .4 - .396 - .196) \simeq -.199$$

The second step now follows with $n = 1$, and the computation is continued in this way. Results up to $t = 12.5$ when the curve has completed two cycles are shown in Fig. 11-18, in which y and p serve as coordinates. This "phase

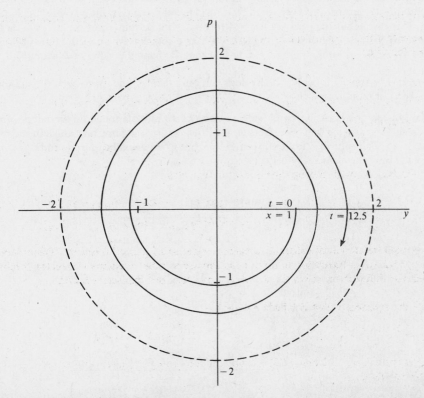

Fig. 11-18

plane" is often used in the study of oscillatory systems. Our oscillation (shown solid) will approach the periodic oscillation (shown dashed) as x becomes infinite. This is proved in the theory of nonlinear oscillations.

11.168 The equations

$$x'(t) = -\frac{2x}{\sqrt{x^2 + y^2}} \qquad y'(t) = 1 - \frac{2y}{\sqrt{x^2 + y^2}}$$

describe the path of a duck attempting to swim across a river by aiming steady at the target position T. The speed of the river is 1 and the duck's speed is 2. The duck starts at S, so that $x(0) = 1$ and $y(0) = 0$ (See Fig. 11-19). Apply the Runge–Kutta formulas for two simultaneous equations to compute the duck's path. Compare with the exact trajectory $y = \frac{1}{2}(x^{1/2} - x^{3/2})$. How long does it take the duck to reach the target?

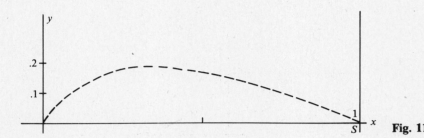

Fig. 11-19

▮ The assumption is that this duck has not learned to handle set and drift of river current, and so is swept downstream to some extent during the earlier part of the crossing. The problem is not without numerical interest, as these numbers, found using $h = .01$ in the classic Runge–Kutta fourth-order method, suggest:

t	0	.2	.4	.6	.8	1
x	1	.60	.24	.02	.005	.005
y	0	.15	.19	.01	.008	.008

The unit of length being a river width, our little friend has been set almost 20% of this before getting the upper webbed foot, but when does the journey end? Logic proposes that the last of the run finds the duck putting its speed of 2 directly against the river speed of 1, for a net of 1. That is, the y value should change just as much as the t value. This just about happens from $t = .4$ to $t = .6$, but certainly not thereafter. The last two entries being challenged by logic, the figure $t = .6$ suggests itself as a rough estimate of the time of crossing. A bit of calculus applied to the exact solution claims $t = \frac{2}{3}$ to be exact. Perhaps the value of h should be reduced?

11.169 Rework the river crossing problem if the duck can only manage the river speed of 1. (Delete the numerator $2s$ in the equations of motion.)

▮ The exact solution is easily found and proves to be a parabola touching the target shore at $y = .5$. Numerical results using $h = .005$ (the reduction from $h = .01$ inspired by the uncertain finish in the preceding problem) appear as Table 11.25. The approach to $(0, .5)$ seems clear, but when does the journey end? Anyone desperate to know the answer will enjoy the elementary calculus required to discover it.

TABLE 11.25

t	x	y
0	1	0
.5	.53	.36
1.0	.22	.48
1.4	.10	.495
1.8	.045	.4990
2.2	.020	.49980
2.4	.014	.49991
2.6	.009	.49996

11.170 A dog, out in a field, sees his master walking along the road and runs toward him. Assuming that the dog always aims directly at his master and that the road is straight, the equation governing the dog's path is (see Fig. 11-20)

$$xy'' = c\sqrt{1 + (y')^2}$$

with c the ratio of the man's speed to the dog's. A well-known line of attack leads to the exact solution

$$y = \frac{1}{2}\left(\frac{x^{1+c}}{1+c} - \frac{x^{1-c}}{1-c}\right) + \frac{c}{1-c^2}$$

for $c < 1$. As x approaches zero, the dog catches his master at position $y = c/(1 - c^2)$. Solve this problem by an approximate method for the case $c = \frac{1}{2}$. The chase should end at $y = \frac{2}{3}$.

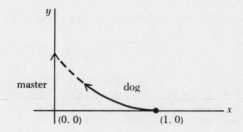

Fig. 11-20

❚ The second-order equation is first replaced by the system

$$y' = p$$

$$p' = \frac{c\sqrt{1 + p^2}}{x}$$

and the initial conditions by $y(1) = 0$ and $p(1) = 0$. The Runge–Kutta formulas of Problem 11.165 can again be used, this time with a negative h. The only difficulty here is that as x nears zero, the slope p grows very large. An adaptive method, and h decreasing in size, seems to be indicated. A primitive strategy was attempted, with $h = -.1$ down to $x = .1$, then $h = -.01$ down to $x = .01$ and so on. The results appear as Table 11.26. The last two x entries appear to contain roundoff error. Values of p are not listed but rose to nearly 1000 in size.

TABLE 11.26

x	y
.1	.3608
.01	.5669
.001	.6350
.0001	.6567
.00001	.6636
.0000006	.6659
$-.0000003$	.6668

11.171 The equations

$$r'' = \frac{9}{r^3} - \frac{2}{r^2} \qquad \theta' = \frac{3}{r^2}$$

in which primes refer to differentiation relative to time t, describe the Newtonian orbit of a particle in an inverse square gravitational field, after suitable choices of some physical constants. If $t = 0$ at the position of minimum r

(Fig. 11-21) and

$$r(0) = 3 \qquad \theta(0) = 0 \qquad r'(0) = 0$$

then the orbit proves to be the ellipse $r = 9/(2 + \cos\theta)$. Use one of our approximation methods and compare with this exact result.

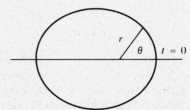

Fig. 11-21

▮ The application is quite straightforward. The familiar reduction to a first-order system comes first,

$$r' = p \qquad p' = \frac{9}{r^3} - \frac{2}{r^2} \qquad \theta' = \frac{3}{r^2}$$

followed by the programming of three sets of Runge–Kutta formulas, still following the model of Problem 11.165. Integration continued until the angle θ exceeded 2π. A selected fragment of the output is provided as Table 11.27 (step size $h = .1$ was used) and it clearly has the desired orbital quality. As a further check, theory offers the period $T = 12\pi\sqrt{3}$, or about 65.3, and this fits in very nicely.

TABLE 11.27

t	r	θ	p
0	3.00	.00	.00
6	4.37	1.51	.33
7	4.71	1.66	.33
32	9.00	3.12	.01
33	9.00	3.15	−.004
59	4.47	4.73	−.33
65	3.00	6.18	−.03
66	3.03	6.52	.08

11.18 PERTURBATION SERIES

11.172 The equation $x''(t) + x + \mu x^3 = 0$ has a family of periodic solutions if $\mu = 0$. This suggests attempting a power series development of the form

$$x(t) = x_0(t) + \mu x_1(t) + \mu^2 x_2(t) + \cdots$$

in the search for a periodic solution for small values of μ. Show that this procedure is unsuccessful.

▮ To be definite, suppose we add the initial conditions

$$x(0) = A \qquad x'(0) = 0$$

with A to be determined. Substituting the series into the differential equation and equating coefficients of the powers of μ leads to a sequence of simpler equations for the determination of the functions $x_i(t)$:

$$x_0'' + x_0 = 0 \qquad x_1'' + x_1 = -x_0^3 \qquad \cdots$$

The initial conditions translate into

$$x_0(0) = A, x_0'(0) = 0 \qquad x_i(0) = 0, x_i'(0) = 0 \qquad i = 1, 2, 3, \ldots$$

Solving our equations successively we find first

$$x_0 = A \cos t$$

and then, since $x_0^3 = \frac{3}{4}A^3 \cos t + \frac{1}{4}A^3 \cos 3t$,

$$x_1 = -\frac{3}{8}A^3 t \sin t - \frac{1}{32}A^3(\cos t - \cos 3t)$$

But x_1 is not periodic, and it seems unwise to continue a process that generates nonperiodic approximations to an anticipated periodic solution, particularly when an alternative is available. (See the next problem.)

11.173 Approximate the periodic solutions of the equation of the preceding problem by the perturbation method.

    I Let $\tau = \omega t$. The equation becomes

$$\omega^2 \ddot{x}(\tau) + x(\tau) + \mu x^3(\tau) = 0$$

the dots meaning derivatives relative to τ. Introduce the power series

$$x(\tau) = x_0(\tau) + \mu x_1(\tau) + \mu^2 x_2(\tau) + \cdots$$
$$\omega = \omega_0 + \mu \omega_1 + \mu^2 \omega_2 + \cdots$$

Substituting and equating the coefficients of the powers of μ, we have the system

$$\omega_0^2 \ddot{x}_0 + x_0 = 0$$
$$\omega_0^2 \ddot{x}_1 + x_1 = -2\omega_0 \omega_1 x_0'' - x_0^3$$
$$\omega_0^2 \ddot{x}_2 + x_2 = -(2\omega_0 \omega_2 + \omega_1^2) x_0'' - 2\omega_0 \omega_1 x_1'' - 3x_0^2 x_1$$
$$\vdots$$

The initial conditions are the same as before, and in addition we have

$$x_i(\tau + 2\pi) = x_i(\tau)$$

since the idea is to find a solution of period $2\pi/\omega$ in the argument t. Solving the first equation, we find $x_0 = A \cos \tau$ and $\omega_0 = 1$, which convert the second equation to

$$\ddot{x}_1 + x_1 = (2\omega_1 - \frac{3}{4}A^2)A \cos \tau - \frac{1}{4}A^3 \cos 3\tau \quad \text{since } (\cos t)^3 = \frac{1}{4}\cos 3t + \frac{3}{4}\cos t$$

Unless the coefficient of $\cos \tau$ is made zero, this equation will lead to nonperiodic terms. Accordingly we choose $\omega_1 = 3A^2/8$ and soon obtain

$$x_1 = \frac{1}{32}A^3(-\cos \tau + \cos 3\tau)$$

Similar handling of the third equation then leads to

$$x(t) = \left(A - \frac{1}{32}\mu A^3 + \frac{23}{1024}\mu^2 A^5\right)\cos \omega t + \left(\frac{1}{32}\mu A^3 - \frac{3}{128}\mu^2 A^5\right)\cos 3\omega t + \frac{1}{1024}\mu^2 A^5 \cos 5\omega t + \cdots$$
$$\omega = 1 + \frac{3}{8}\mu A^2 - \frac{21}{256}\mu^2 A^4 + \cdots$$

and more terms are computable if desired. Notice that the frequency ω is related to the amplitude A, unlike the situation for the linear case $\mu = 0$.

11.174 Apply the perturbation method to van der Pol's equation

$$y'' - \mu(1 - y^2)y' + y = 0$$

    I It is known that for $\mu \neq 0$ one periodic solution exists. To find it let $\tau = \omega t$, converting the equation to

$$\omega^2 \ddot{y} - \mu\omega(1 - y^2)\dot{y} + y = 0$$

Again introduce the series

$$y(\tau) = y_0(\tau) + \mu y_1(\tau) + \mu^2 y_2(\tau) + \cdots$$
$$\omega = \omega_0 + \mu \omega_1 + \mu^2 \omega_2 + \cdots$$

and substitute into the differential equation. Equating coefficients brings

$$\omega_0^2 \ddot{y}_0 + y_0 = 0$$
$$\omega_0^2 \ddot{y}_1 + y_1 = -2\omega_0\omega_1\ddot{y}_0 + \omega_0\left(1 - y_0^2\right)\dot{y}_0$$
$$\omega_0^2 \ddot{y}_2 + y_2 = -\left(2\omega_0\omega_2 + \omega_1^2\right)\ddot{y}_0 - 2\omega_0\omega_1\ddot{y}_1 + \omega_1\left(1 - y_0^2\right)\dot{y}_0 - 2\omega_0 y_0 y_1\dot{y}_0 + \omega_0\left(1 - y_0^2\right)y_1$$
$$\vdots$$

The periodicity condition requires $y_i(\tau + 2\pi) = y_i(\tau)$, and we can also set the initial condition $\dot{y}(0) = 0$, or $\dot{y}_i(0) = 0$, which amounts to choosing $\tau = 0$ when y is at its maximum or minimum value. Using these conditions the first equation yields

$$y_0 = A_0 \cos \tau \qquad \omega_0 = 1$$

with A still arbitrary. Substituting into the second equation,

$$\ddot{y}_1 + y_1 = 2\omega_1 A_0 \cos \tau + A_0\left(\tfrac{1}{4}A_0^2 - 1\right)\sin \tau + \tfrac{1}{4}A_0^3 \sin 3\tau$$

To avoid nonperiodic "resonance" terms in the solution, we must have

$$\omega_1 A_0 = 0 \qquad A_0\left(\tfrac{1}{4}A_0^2 - 1\right) = 0$$

The choice $A_0 = 0$ would lead nowhere, since y_1 would simply assume the role of y_0. Accordingly we choose $\omega_1 = 0$ and $A_0 = 2$. This leads us to

$$y_0 = 2 \cos \tau \qquad y_1 = A_1 \cos \tau + B_1 \sin \tau - \tfrac{1}{4}\sin 3\tau$$

The condition $\dot{y}_1(0) = 0$ forces $B_1 = \tfrac{3}{4}$, and A_1 will be determined in the next step. Substitution into the third equation next brings

$$\ddot{y}_2 + y_2 = \left(4\omega_2 + \tfrac{1}{4}\right)\cos \tau + 2A_1 \sin \tau - \tfrac{3}{2}\cos 3\tau + 3A_1 \sin 3\tau + \tfrac{5}{4}\cos 5\tau$$

and we choose $\omega_2 = -\tfrac{1}{16}$ and $A_1 = 0$ to remove the resonance terms in $\cos \tau$ and $\sin \tau$. Solving for y_2, we get

$$y_2 = A_2 \cos \tau + B_2 \sin \tau + \tfrac{3}{16}\cos 3\tau - \tfrac{5}{96}\cos 5\tau$$

The condition $\dot{y}_2(0) = 0$ forces $B_2 = 0$. The next step would produce $A_2 = -\tfrac{1}{8}$, and so the solution series are

$$y(t) = \left(2 - \tfrac{1}{8}\mu^2\right)\cos \omega t + \tfrac{3}{4}\mu \sin \omega t + \tfrac{3}{16}\mu^2 \cos 3\omega t - \tfrac{1}{4}\mu \sin 3\omega t - \tfrac{5}{96}\mu^2 \cos 5\omega t + \cdots$$
$$\omega = 1 - \tfrac{1}{16}\mu^2 + \cdots$$

with more terms available if desired. This $y(t)$ and its accompanying $y'(t) = p(t)$ correspond to the dotted curve in Fig. 11-18. It is not, of course, a true circle, but is very close to one. The period is $2\pi/\omega \approx 6.32$, which is greater than 2π and not far from the 6.40 computed for the growing oscillation of Problem 11.167.

11.175 Apply the perturbation method to Duffing's equation

$$x''(t) + x = \mu\left(-ax - bx^3 - cx' + F\cos t\right)$$

obtaining a solution with the period 2π of the "forcing term" $F\cos t$.

▮ Though the period of the solution is known in this case, it pays to be open-minded about the phase. In other words, if we let $t = \tau + p$, then $\dot{x}(0) = 0$ can again be required since it will serve to determine the phase p. The series

$$x(\tau) = x_0(\tau) + \mu x_1(\tau) + \mu^2 x_2(\tau) + \cdots$$
$$p = p_0 + \mu p_1 + \mu^2 p_2 + \cdots$$

can now be substituted into the differential equation, with results

$$\ddot{x}_0 + x_0 = 0$$
$$\ddot{x}_1 + x_1 = -ax_0 - bx_0^3 - c\dot{x}_0 + F\cos(\tau + p_0)$$
$$\ddot{x}_2 + x_2 = -ax_1 - 3bx_0^2 x_1 - c\dot{x}_1 - Fp_1 \sin(\tau + p_0)$$
$$\vdots$$

Using $\dot{x}_0(0) = 0$, we at one find $x_0 = A_0 \cos \tau$. Substituting,

$$\ddot{x}_1 + x_1 = -\left(aA_0 + \tfrac{3}{4}bA_0^3 - F\cos p_0\right)\cos \tau + \left(cA_0 - F\sin p_0\right)\sin \tau - \tfrac{1}{4}bA_0^3 \cos 3\tau$$

The periodicity condition $x_1(\tau + 2\pi) = x_1(\tau)$ again requires that the terms in $\cos \tau$ and $\sin \tau$ be absent. Accordingly,

$$aA_0 + \tfrac{3}{4}bA_0^3 - F\cos p_0 = 0 \qquad cA_0 - F\sin p_0 = 0$$

from which we find $\sin p_0 = cA_0/F$ and the equation

$$c^2A_0^2 + \left(aA_0 + \tfrac{3}{4}bA_0^3\right)^2 = 1$$

for A_0. Solving for $x_1(\tau)$ and using $\dot{x}_1(0) = 0$ then brings

$$x_1 = A_1 \cos \tau + \tfrac{1}{32}bA_0^3 \cos 3\tau$$

The third equation is next treated in the now familiar way and determines

$$A_1 = -\frac{3b^2A_0^5}{128\left(a + \tfrac{9}{4}bA_0^2 + c\tan p_0\right)} \qquad p_1 = -\frac{3cb^2A_0^5}{128\left(a + \tfrac{9}{4}bA_0^2 + c\tan p_0\right)F\cos p_0}$$

before going on to the x_2 term. The solution is

$$x(t) = \left(A_0 + \mu A_1\right)\cos\left(t - p_0 - \mu p_1\right) + \tfrac{1}{32}\mu bA_0^3 \cos 3\left(t - p_0 - \mu p_1\right) + \cdots$$

11.19 A SECOND-ORDER EQUATION WITHOUT y'

11.176 Derive Numerov's formula

$$y_{k+1} = 2y_k - y_{k-1} + \tfrac{1}{12}h^2\left(F_{k+1} + 10F_k + F_{k-1}\right) + R$$

for solving $y'' = F(x, y)$.

$\blacksquare$ Notice that in this case y' does not appear explicitly in the differential equation and that the preceding formula exploits this fact by also omitting y' terms. We proceed by the method of undetermined coefficients,

$$y_{k+1} + Ay_k + By_{k-1} + h^2\left(CF_{k+1} + DF_k + EF_{k-1}\right) + R$$

By Taylor's formula,

$$y_{k+1} = y_k + hy_k' + \frac{1}{2}h^2y_k^{(2)} + \frac{1}{6}h^3y_k^{(3)} + \frac{1}{24}h^4y_k^{(4)} + \frac{1}{120}h^5y_k^{(5)} + \frac{1}{6!}h^6y^{(6)}$$

$$y_{k-1} = y_k - hy_k' + \frac{1}{2}h^2y_k^{(2)} - \frac{1}{6}h^3y_k^{(3)} + \frac{1}{24}h^4y_k^{(4)} - \frac{1}{120}h^5y_k^{(5)} + \frac{1}{6!}h^6y^{(6)}$$

$$h^2F_{k+1} = h^2y_k^{(2)} + h^3y_k^{(3)} + \frac{1}{2}h^4y_k^{(4)} + \frac{1}{6}h^5y_k^{(5)} + \frac{1}{24}h^6y^{(6)}$$

$$h^2F_{k-1} = h^2y_k^{(2)} - h^3y_k^{(3)} + \frac{1}{2}h^4y_k^{(4)} - \frac{1}{6}h^5y_k^{(5)} + \frac{1}{24}h^6y^{(6)}$$

and matching powers of h through the fourth on both sides of Numerov's formula,

$$1 = A + B \qquad 1 = -B \qquad \tfrac{1}{2} = \tfrac{1}{2}B + C + D + E \qquad \tfrac{1}{6} = -\tfrac{1}{6}B + C - E \qquad \tfrac{1}{24} = \tfrac{1}{24}B + \tfrac{1}{2}C + \tfrac{1}{2}E$$

These may be solved for $A = 2$, $B = -1$, $C = \tfrac{1}{12}$, $D = \tfrac{5}{6}$ and $E = \tfrac{1}{12}$. The fifth powers of h also match voluntarily. If we pretend that all factors designated as $y^{(6)}$ are the same, we also obtain the error estimate $R = -h^6y^{(6)}/240$.

11.177 Apply Numerov's formula to the simple equation $y'' = y$ with initial conditions $y(0) = 1$ and $y'(0) = -1$.

$\blacksquare$ The exact solution function is clearly $y(x) = e^{-x}$. However, to illustrate Numerov's method we proceed as with a problem of unknown solution. Two starting values are needed. The first is $y(0) = y_0 = 1$. The second may be found by series expansion of $y(x)$. Using the differential equation to produce higher derivatives, we easily find, with $h = .5$ for a simple if crude approximation,

$$y_1 = y(.5) \simeq 1 - .5 + .125 - .0208 + .0026 - .0003 = .6065$$

Since y_{k+1} occurs on both sides, our main formula has the nature of a corrector. To prime it we ignore the F_{k+1} term on the first round and use

$$y_{k+1} \simeq 2y_k - y_{k-1} + \tfrac{1}{12}h^2(10F_k + F_{k-1})$$

as a predictor. With $k = 1$, for example,

$$y_2 \simeq 1.2130 - 1 + \tfrac{1}{48}(6.065 + 1) \simeq .3602$$

Now applying the complete formula,

$$y_2 \simeq .2130 + \tfrac{1}{48}(.3602 + 6.065 + 1) \simeq .3677$$

Reapplying the complete formula,

$$y_2 \simeq .2130 + \tfrac{1}{48}(.3677 + 6.065 + 1) \simeq .3678$$

Another cycle again produces .3678, so we stop. The correct value is $e^{-1} \simeq .36788$, so our y_2, is close. The process now moves to the computation of y_3, beginning with the predictor, but the path is clear and our illustration may stop here. For an accurate solution truncation error must be diminished, by decreasing h, and roundoff error reduced, by carrying more than four places.

11.20 A STIFF SYSTEM

11.178 Apply the Gear method of Problem 11.120 to the stiff system

$$y' = p$$
$$p' = -100y - 101p$$

with initial conditions $y(0) = 1$ and $p(0) = -1$. This system is equivalent to the second-order equation

$$y'' + 101y' + 100y = 0$$

with $y = 1$ and $y' = -1$ initially. The exact solution is $y(x) = e^{-x}$.

▌ Runge–Kutta methods could handle this system, but the classic fourth-order set would require a step size less than .0278 for a stable computation. Writing out the Gear formula for both y and p we have

$$y_{n+1} = \frac{1}{11}(18y_n - 9y_{n-1} + 2y_{n-2}) + \frac{6h}{11}p_{n+1}$$

$$p_{n+1} = \frac{1}{11}(18p_n - 9p_{n-1} + 2p_{n-2}) + \frac{6h}{11}(-100y_{n+1} - 101p_{n+1})$$

which can be rewritten as a linear system for y_{n+1} and p_{n+1}:

$$y_{n+1} - \frac{6h}{11}p_{n+1} = \frac{1}{11}(18y_n - 9y_{n-1} + 2y_{n-2})$$

$$\frac{600h}{11}y_{n+1} + \left(1 + \frac{606h}{11}\right)p_{n+1} = \frac{1}{11}(18p_n - 9p_{n-1} + 2p_{n-2})$$

Since the system is linear, there is no need to use the Newton iteration for its solution. Results for two choices of step size h appear below, both much larger than what is needed for Runge–Kutta. The true values are also listed for comparison.

x	$y = e^{-x}$	$h = .1$	$h = .2$
2	.1353	.1354	.1359
4	.01832	.01833	.0185
6	.002479	.002483	.00251
8	.0003355	.0003362	.000342
10	.0000454	.0000455	.0000465

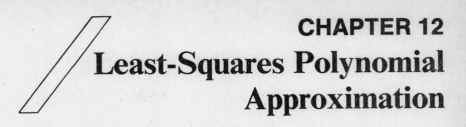

CHAPTER 12
Least-Squares Polynomial Approximation

12.1 What is the principle of least squares?

▌ It is the idea of choosing an approximation to a given function in a way that minimizes (in some sense) the squares of the errors. There are several variations, depending on the set of arguments involved and the error measure to be used. The idea was developed by Gauss during his many years as director of surveying in Germany and was useful in the reduction of field data, as described briefly in Problem 2.54. It has been claimed that, given a set of experimentally produced points such as appear in Fig. 12-1(a), the human eye and hand (assisted by a transparent ruler) will instinctively choose to smooth the data by drawing in a line that minimizes the error squares. No large scale experiment to this effect has, however, been reported, and if one were conducted with the very small data set in Fig. 12-1(b), asking for the best horizontal line, the dashed suggestion would surely get some votes, though logic (see Problem 12.11) argues strongly for the competitor shown solid.

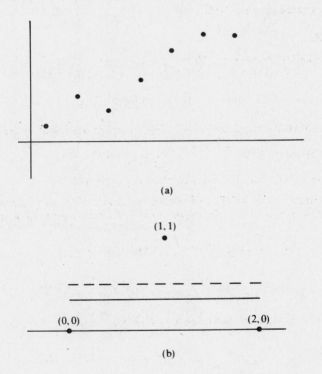

(a)

(1, 1)

(0, 0)   (2, 0)

(b)

Fig. 12-1

12.2 Restate the principle of least squares in the language of norms.

▌ Suppose we want to approximate a data function y by using a polynomial p. Which polynomial would be the best choice? For the present chapter the answer will be, the one that minimizes

$$\|y - p\|^2$$

for an appropriate definition of the norm. For example, if the data function is the discrete

$$y(x_i) = y_i \qquad i = 0, 1, 2, \ldots, N$$

and the polynomial is $p(x)$, then the norm $\sqrt{S}$, where

$$S = \sum_{0}^{N} (y_i - p_i)^2$$

is a likely choice, where $p_i = p(x_i)$. If $y(x)$ is defined for all x arguments in (a, b), then one appropriate norm is $\sqrt{I}$, where

$$I = \int_a^b (y - p)^2 \, dx$$

In general one looks for as simple a polynomial as possible; the chance that S or I will be zero does not exist. The problem is to minimize it.

12.1 DISCRETE DATA: THE LEAST-SQUARES LINE

12.3   Find the straight line $p(x) = Mx + B$ for which $\sum_{i=0}^{N}(y_i - Mx_i - B)^2$ is a minimum, the data (x_i, y_i) being given.

▌ Calling the sum S, we follow a standard minimum-finding course and set derivatives to zero:

$$\frac{\partial S}{\partial B} = -2 \sum_{i=0}^{N} 1 \cdot (y_i - Mx_i - B) = 0 \qquad \frac{\partial S}{\partial M} = -2 \sum_{i=0}^{N} x_i \cdot (y_i - Mx_i - B) = 0$$

Rewriting we have

$$(N+1) B + \left(\sum x_i \right) M = \sum y_i \qquad \left(\sum x_i \right) B + \left(\sum x_i^2 \right) M = \sum x_i y_i$$

which are the "normal equations." Introducing the symbols

$$s_0 = N + 1 \qquad s_1 = \sum x_i \qquad s_2 = \sum x_i^2 \qquad t_0 = \sum y_i \qquad t_1 = \sum x_i y_i$$

these equations may be solved in the form

$$M = \frac{s_0 t_1 - s_1 t_0}{s_0 s_2 - s_1^2} \qquad B = \frac{s_2 t_0 - s_1 t_1}{s_0 s_2 - s_1^2}$$

To show that $s_0 s_2 - s_1^2 \neq 0$, we may first notice that squaring and adding terms such as $(x_0 - x_1)^2$ leads to

$$0 < \sum_{i<j} (x_i - x_j)^2 = N \cdot \sum x_i^2 - 2 \sum_{i<j} x_i x_j$$

But also

$$\left(\sum x_i \right)^2 = \sum x_i^2 + 2 \sum_{i<j} x_i x_j$$

so that $s_0 s_2 - s_1^2$ becomes

$$(N+1) \sum x_i^2 - \left(\sum x_i \right)^2 = N \cdot \sum x_i^2 - 2 \sum_{i<j} x_i x_j > 0$$

Here we have assumed that the x_i are not all the same, which is surely reasonable. The last inequality also helps to prove that the M and B chosen actually produce a minimum. Calculating second derivatives, we find

$$\frac{\partial^2 S}{\partial B^2} = 2 s_0 \qquad \frac{\partial^2 S}{\partial M^2} = 2 s_2 \qquad \frac{\partial^2 S}{\partial B \, \partial M} = 2 s_1$$

Since the first two are positive and since

$$(2 s_1)^2 - 2(N+1)(2 s_2) = 4 \left(s_1^2 - s_0 s_2 \right) < 0$$

the second derivative test for a minimum of a function of two arguments B and M is satisfied. The fact that the first derivatives can vanish together only once, shows that our minimum is an absolute minimum.

12.4 The average scores reported by golfers of various handicaps on a difficult par-3 hole are

Handicap	6	8	10	12	14	16	18	20	22	24
Average	3.8	3.7	4.0	3.9	4.3	4.2	4.2	4.4	4.5	4.5

Find the least-squares linear function for this data by the formulas of Problem 12.3

▮ Let h represent handicap and $x = (h - 6)/2$. Then the x_i are the integers $0, \ldots, 9$. Let y represent average score. Then $s_0 = 10$, $s_1 = 45$, $s_2 = 285$, $t_0 = 41.5$, $t_1 = 194.1$ and so

$$M = \frac{(10)(194.1) - (45)(41.5)}{(10)(285) - (45)^2} \simeq .089 \qquad B = \frac{(285)(41.5) - (45)(194.1)}{(10)(285) - (45)^2} \simeq 3.76$$

This makes $y \simeq p(x)$ where $p(x) = .09x + 3.76 \simeq .045h + 3.49$.

12.5 Use the least-squares line of the previous problem to smooth the reported data.

▮ The effort to smooth data proceeds on the assumption that the reported data contain inaccuracies of a size to warrant correction. In this case the data seem to fall roughly along a straight line, but there are large fluctuations, due perhaps to the natural fluctuations in a golfer's game (see Fig. 12-2). The least-squares line may be assumed to be a better representation of the true relationship between the handicap and the average scores than the original data are. It yields the following smoothed values

Handicap	6	8	10	12	14	16	18	20	22	24
Smoothed y	3.76	3.85	3.94	4.03	4.12	4.21	4.30	4.39	4.48	4.57

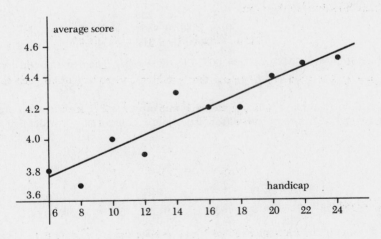

Fig. 12-2

12.6 Estimate the rate at which the average score increases per unit handicap.

▮ From the least-squares line of Problem 12.4 we obtain the estimate .045 stroke per unit handicap.

12.7 Obtain a formula of the type $P(x) = Ae^{Mx}$ from the following data:

x_i	1	2	3	4
P_i	7	11	17	27

▮ Let $y = \log P$, $B = \log A$. Then taking logarithms, $\log P = \log A + Mx$ which is equivalent to $y(x) = Mx + B$.

We now decide to make this the least-squares line for the (x_i, y_i) data points.

x_i	1	2	3	4
y_i	1.95	2.40	2.83	3.30

Since $s_0 = 4$, $s_1 = 10$, $s_2 = 30$, $t_0 = 10.48$ and $t_1 = 28.44$, the formulas of Problem 12.3 make $M \simeq .45$ and $B \simeq 1.5$. The resulting formula is $P = 4.48e^{.45x}$.

It should be noted that in this procedure we do not minimize $\Sigma[P(x_i) - P_i]^2$, but instead choose the simpler task of minimizing $\Sigma[y(x_i) - y_i]^2$. This is a very common decision in such problems.

12.8 The average scores reported by golfers of various handicaps on par-4 hole were

Handicap	6	8	10	12	14	16	18	20	22	24
Average	4.6	4.8	4.6	4.9	5.0	5.4	5.1	5.5	5.6	6.0

Find the least-squares line for this data and use it to smooth the data. How much does the average score really go up with handicap?

▌ The routine of Problem 12.3 finds the line $y = 4.07 + .072x$, which answers the last question first. Scores increase about .07 strokes per unit handicap. The smoothed values are

Smoothed	4.5	4.6	4.8	4.9	5.1	5.2	5.4	5.5	5.6	5.8

with roundoff to blame for the occasional jumps. The changes are minimal, but the irregularities in the original data were clearly flukes and they have been removed.

12.9 Would a parabola do a better job than a straight line at smoothing the same golf data?

▌ The question raises an important point, though a bit prematurely. It will be shown in Problem 12.18 that the coefficients a, b and c of the least-squares parabola $p(x) = a + bx + cx$ are determined by the system

$$10a + 150b + 2580c = 51.5$$
$$150a + 2580b + 48600c = 796.2$$
$$2580a + 48600b + 971088c = 14019.6$$

with solution $a = 4.56$, $b = -.00488$ and $c = .00256$ to three digits. The new smoothed values are

4.6	4.7	4.8	4.9	5.0	5.1	5.3	5.5	5.7	5.9

and show their quadratic parentage in the accelerating rate of increase. The differences are slight and it seems that the golf data hardly require this extra level of sophistication. Problem 12.59 provides a mechanism for making such decisions.

12.10 For the data set

x	2.2	2.7	3.5	4.1
P	65	60	53	50

let $y = \log P$ and compute the least-squares line $y = ax + b$ for the (x, y) data pairs. This makes $P = ce^{ax}$, with $c = e^b$. Then begin again, seeking a formula of the type $P = dx^n$. This time $\log P = \log d + n \log x$. Again find a least-squares line. How much smoothing do the two approximations achieve?

▌ For the first effort we find $y = -.14x + 4.48$ and then $P = 87.9e^{-.14x}$. For the second, $\log P = 4.52 - .43 \log x$ making $P = 91.4x^{-.43}$. As for smoothed values, the first effort manages 64.6, 60.2, 53.8 and 49.5 and the second 65.1, 59.6, 53.3 and 49.8. There seems little reason for preferring one set over the other.

12.11 What does the method of least-squares reduce to when the approximating polynomial is a constant?

▌ With $p(x) = b$ the sum of squared errors becomes

$$S = \sum_0^N (y_i - b)^2$$

and setting the derivative relative to b to zero, we rediscover the grandfather of all least-squares theorems. The best value for b is the average value of y.

12.12 Consider the special case in which $p(x) = mx$. That is, one knows in advance that the line should pass through $(0, 0)$.

▌ The sum to be minimized is

$$S = \sum_0^N (y_i - mx_i)^2$$

and setting the derivative relative to m to zero quickly finds

$$m = \sum x_i y_i \Big/ \sum x_i^2$$

For an example in which the computations can be done mentally, what is the least-squares line through $(0, 0)$ and accommodating the two points $(1, 1)$ and $(2, 1)$? With $\sum x_i y_i = 3$ and $\sum x_i^2 = 5$ the answer is $p(x) = \frac{3}{5}x$.

12.13 The linear relationship $p(x) = mx$ can also be written $p(x)/x = m$, so that this ratio will be constant. Apply the method of Problem 12.11 to determine m. Why is the result different from that found in the preceding problem?

▌ From this new point of view, m will be the average value of y_i/x_i, or $\sum_0^N (y_i/x_i)/(N+1)$. This is hardly the same as the value just found. For the mini example offered in Problem 12.12 we now have

$$m = \tfrac{1}{2}\left(1 + \tfrac{1}{2}\right) = \tfrac{3}{4}$$

The apparent discrepancy is explained by the fact that the two approaches minimize different sums of squared errors. Each line is the better line from its own point of view.

12.14 Recent world records for mens' weight lifting in the snatch and jerk follow.

W (kg)	52	56	60	67.5	75	82.5	90	100	110
L (kg)	252.5	277.5	302.5	345	360	400	415	430	427.5

The higher the weight class, the greater the lift, apart from an anomaly at the upper end. An explanation of this rise has been proposed, that lift depends on the size of muscle cross sections. If body weight W is taken as a three-dimensional measure of body size, then $W^{1/3}$ is a one-dimensional and $W^{2/3}$ a two-dimensional measure. Accordingly, it is proposed that lift should be proportional to $W^{2/3}$,

$$L = cW^{2/3}$$

Determine the constant c by least squares.

▌ Begin by taking logarithms,

$$\log L = \log c + \tfrac{2}{3}\log W$$

Then using $y = \log L$ and $x = \log W$,

$$y = a + \tfrac{2}{3}x$$

We seek a line for which the slope is known to be $\frac{2}{3}$ but the constant is open. The sum to be minimized is

$$S = \sum_0^N \left(y_i - a - \tfrac{2}{3}x_i\right)^2$$

and setting the derivative relative to a to zero,

$$\sum_0^N \left(y_i - a - \tfrac{2}{3}x_i\right) = 0 \quad \text{or} \quad (N+1)a = \sum y_i - \tfrac{2}{3}\sum x_i$$

where $y_i = \log L_i$ and $x_i = \log W_i$.

For the data provided, this least-squares method produced the constant $a = 2.985$, which translates to $c = 19.78$ and

$$L = 19.78 W^{2/3}$$

12.15 Use the result of the preceding problem to determine which of the nine world records is the best performance.

▌ Computing smoothed values by the formula $L = 19.78W^{2/3}$, we have these (actual, theoretical) pairs, rounded to the nearest kilograms. Differences are

Actual	252	278	302	345	360	400	415	430	428
Theory	276	290	303	328	352	375	397	426	454
Diff.	−24	−12	−1	17	8	25	18	4	−26

The 400 kg lift seems to be outstanding.

12.16 For still another variation of the least-squares line problem, suppose it is required to find parallel lines each having its own data set. The requirement of parallelism means that each line cannot be found independently. Consider the case of two such lines, L_1 and L_2, with data points (x_i, y_i) for the first and (x_j, y_j) for the other.

▌ The sum to be minimized is

$$S = \sum_0^N (y_i - mx_i - a)^2 + \sum_0^N (y_j - mx_j - b)^2$$

the first sum referring to line L_1 and the second to L_2. Setting derivatives relative to m, a and b to zero,

$$\partial S/\partial m = 2\sum (y_i - mx_i - a)x_i + 2\sum (y_j - mx_j - b)x_j = 0$$
$$\partial S/\partial a = 2\sum (y_i - mx_i - a) = 0$$
$$\partial S/\partial b = 2\sum (y_j - mx_j - b) = 0$$

which can be rewritten in the form

$$\left(\sum x_i^2 + \sum x_j^2\right)m + \left(\sum x_i\right)a + \left(\sum x_j\right)b = \sum x_i y_i + \sum x_j y_j$$
$$\left(\sum x_i\right)m + (N+1)a + \quad = \sum y_i$$
$$\left(\sum x_j\right)m \quad\quad (N+1)b = \sum y_j$$

if we assume there are $N+1$ points in each data set. Solving this linear system yields the needed parameter values.

12.17 Run the procedure of the preceding problem for these small data sets:

L_1	$(0, -.2)$	$(1, .4)$	$(2, .2)$	$(3, .6)$
L_2	$(0, 1.2)$	$(1, 1.6)$	$(2, 1.8)$	$(3, 2)$

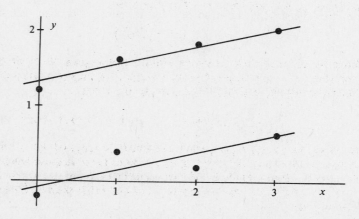

Fig. 12-3

▮ Doing the sums we have the system of equations

$$14m + 3a + 3b = 6.9$$
$$3m + 2a = .5$$
$$3m + 2b = 3.3$$

with solution $m = .24$, $a = -.11$ and $b = 1.29$. The eight data points and parallel lines are shown in Fig. 12-3.

12.2 DISCRETE DATA: THE LEAST-SQUARES POLYNOMIAL

12.18 Generalizing Problem 12.3, find the polynomial $p(x) = a_0 + a_1 x + \cdots + a_m x^m$ for which $S = \sum_{i=0}^{N}(y_i - a_0 - a_1 x_i - \cdots - a_m x_i^m)^2$ is a minimum, the data (x_i, y_i) being given, and $m < N$.

▮ We proceed as in the simpler case of the straight line. Setting the derivatives relative to $a_0, a_1, \ldots, a_m$ to zero produces the $m + 1$ equations

$$\frac{\partial S}{\partial a_k} = -2 \sum_{i=0}^{N} x_i^k (y_i - a_0 - a_1 x_i - \cdots - a_m x_i^m) = 0$$

where $k = 0, \ldots, m$. Introducing the symbols $s_k = \sum_{i=0}^{N} x_i^k$, $t_k = \sum_{i=0}^{N} y_i x_i^k$, these equations may be rewritten as

$$s_0 a_0 + s_1 a_1 + \cdots + s_m a_m = t_0$$
$$s_1 a_0 + s_2 a_1 + \cdots + s_{m-1} a_m = t_1$$
$$\vdots$$
$$s_m a_0 + s_{m+1} a_1 + \cdots + s_{2m} a_m = t_m$$

and are called normal equations. Solving for the coefficients a_i, we obtain the least-squares polynomial. We will show that there is just one solution and that it does minimize S. For smaller integers m, these normal equations may be solved without difficulty. For larger m the system is badly ill-conditioned and an alternative procedure will be suggested.

12.19 Show how the least-squares idea, as just presented in Problem 12.18 and earlier in Problem 12.3, may be generalized to arbitrary vector spaces. What is the relationship with orthogonal projection?

▮ This more general approach will also serve as a model for other variations of the least-squares idea to be presented later in this chapter and focuses attention on the common features that all these variations share. First recall that in Euclidean plane geometry, given a point y and a line S, the point on S closest to y is the unique point p such that $\overline{py}$ is orthogonal to S, p being the *orthogonal projection* point of y onto S. Similarly in Euclidean solid geometry, given a point y and a plane S, the point on S closest to y is the unique point p such that $\overline{py}$ is orthogonal to all vectors in S. Again p is the orthogonal projection of y. This idea is now extended to a more general vector space.

We are given a vector y in a vector space E and are to find a vector p in a given subspace S such that

$$\|y - p\| < \|y - q\|$$

where q is any other vector in S and the *norm* of a vector v is

$$\|v\| = \sqrt{(v, v)}$$

parentheses denoting the scalar product associated with the vector space. We begin by showing that there is a unique vector p for which $y - p$ is orthogonal to every vector in S. This p is called the *orthogonal projection* of y. Let $e_0, \ldots, e_m$ be an orthogonal basis for S and consider the vector

$$p = (y, e_0) e_0 + (y, e_1) e_1 + \cdots + (y, e_m) e_m$$

Direct calculation shows that $(p, e_k) = (y, e_k)$ and therefore $(p - y, e_k) = 0$ for $k = 0, \ldots, m$. It then follows that $(p - y, q) = 0$ for any q in S, simply by expressing q in terms of the orthogonal basis. If another vector p' also had this property $(p' - y, q) = 0$, then it would follow that for any q in S, $(p - p', q) = 0$. Since $p - p'$ is itself in S, this forces $(p - p', p - p') = 0$, which by required properties of any scalar product implies $p = p'$. The orthogonal projection p is thus unique.

But now, if q is a vector other than p in S,

$$\|y - q\|^2 = \|(y - p) + (p - q)\|^2$$
$$= \|y - p\|^2 + \|p - q\|^2 + 2(y - p, p - q)$$

Since the last term is zero, $p - q$ being in S, we deduce that $\|y - p\| < \|y - q\|$ as required.

12.20 If $u_0, u_1, \ldots, u_m$ is an arbitrary basis for S, determine the vector p of the preceding problem in terms of the u_k.

▮ We must have $(y - p, u_k) = 0$ or $(p, u_k) = (y, u_k)$ for $k = 0, \ldots, m$. Since p has the unique representation $p = a_0 u_0 + a_1 u_1 + \cdots + a_m u_m$, substitution leads directly to

$$(u_0, u_k) a_0 + (u_1, u_k) a_1 + \cdots + (u_m, u_k) a_m = (y, u_k)$$

for $k = 0, \ldots, m$. These are the *normal equations* for the given problem and are to be solved for the coefficients $a_0, \ldots, a_m$. A unique solution is guaranteed by the previous problem. Note that in the special case where the $u_0, u_1, \ldots, u_m$ are orthonormal, these normal equations reduce to $a_i = (y, u_i)$ as in the proof given in Problem 12.19.

Note also the following important corollary. If y itself is represented in terms of an orthogonal basis in E which includes $u_0, \ldots, u_m$, say

$$y = a_0 u_0 + a_1 u_1 + \cdots + a_m u_m + a_{m+1} u_{m+1} + \cdots$$

then the orthogonal projection p, which is the least-squares approximation is available by simple *truncation* of the representation after the $a_m u_m$ term:

$$p = a_0 u_0 + a_1 u_1 + \cdots + a_m u_m$$

12.21 How is the specific case treated in Problem 12.18 related to the generalization given in Problems 12.19 and 12.20?

▮ The following identifications must be made:

E: The space of discrete real-valued functions on the set of arguments $x_0, \ldots, x_N$
S: The subset of E involving polynomials of degree m or less
y: The data function having values $y_0, \ldots, y_N$
(v_1, v_2): The scalar product $\sum_{i=0}^{N} v_1(x_i) v_2(x_i)$
$\|v\|^2$: The norm $\sum_{i=0}^{N} [v(x_i)]^2$
u_k: The function with values x_i^k
p: The polynomial with values $p_i = a_0 + a_1 x_i + \cdots + a_m x_i^m$
$\|y - p\|^2$: The sum $S = \sum_{i=0}^{N} (y_i - p_i)^2$
(y, u_k): $t_k = \sum_{i=0}^{N} y_i x_i^k$
(u_j, u_k): $s_{j+k} = \sum_{i=0}^{N} x_i^{j+k}$

With these identifications we also learn that the polynomial p of Problem 12.18 is unique and actually does provide the minimum sum. The general result of Problems 12.19 and 12.20 establishes this.

12.22 Determine the least-squares quadratic function for the data of Problem 12.4.

▮ The sums s_0, s_1, s_2, t_0 and t_1 have already been computed. We also need $s_3 = 2025$, $s_4 = 15,333$ and $t_2 = 1292.9$, which allow the normal equations to be written

$$10a_0 + 45a_1 + 285a_2 = 41.5 \qquad 45a_0 + 285a_1 + 2025a_2 = 194.1 \qquad 285a_0 + 2025a_1 + 15333a_2 = 1248$$

After some labor these yield $a_0 = 3.73$, $a_1 = .11$ and $a_2 = -.0023$ so that our quadratic function is $p(x) = 3.73 + .11x - .0023x^2$.

12.23 Apply the quadratic function of the preceding problem to smooth the reported data.

▮ Assuming that the data should have been values of our quadratic function, we obtain the values

Handicap	6	8	10	12	14	16	18	20	22	24
Smoothed y	3.73	3.84	3.94	4.04	4.13	4.22	4.31	4.39	4.46	4.53

These hardly differ from the predictions of the straight line hypothesis, and the parabola corresponding to our quadratic function would not differ noticeably from the straight line of Fig. 12-2. The fact that a_2 is so small already shows that the quadratic hypothesis may be unnecessary in the golfing problem.

12.24 The five points $(0, .02)$, $(.25, .10)$, $(.5, .20)$, $(.75, .50)$ and $(1, 1.05)$ correspond to points on the parabola $y = x^2$ but with a random error (drawn from a hat) introduced into the y coordinate. Fit a least-squares parabola to this data. Compare its values with the data values given.

▮ The normal equations are found by the procedure of Problem 12.18 and lead to $p = .045 - .319x + 1.303x^2$. Values needed for comparison appear in Table 12.1, and for the most part it is the smoother p values that are closer to true. The example is a small one, but it supports the underlying idea that when the data "want" to be smoothed, they probably will be.

TABLE 12.1

x	0	.25	.50	.75	1.00
Data	.02	.10	.20	.50	1.05
p	.04	.05	.21	.54	1.03
True	.00	.06	.25	.56	1.00

12.25 Consider the special case in which the polynomial $p(x)$ is simply Ax^2. (It is known in advance that the data function corresponds to a function of this family.) Modify the least-squares method to find the optimal A and apply it to the data of the preceding problem.

▮ The sum to be minimized is now

$$S = \sum_0^N \left(y_i - Ax_i^2 \right)^2$$

and setting the derivative relative to A to zero leads quickly to

$$A = \sum x_i^2 y_i \Big/ \sum x_i^4$$

For the data of Problem 12.24 this proves to be $A = 1.003$ and the corresponding smoothed values are identical to two decimal places with the true values. As is often the case, inside information has paid off.

12.26 Consider the special case in which $p(x) = a + bx^2$, it being known that the first degree term should be absent. Develop the method of least squares for determining a and b.

▮ The sum to be minimized is

$$S = \sum_0^N \left(y_i - a - bx_i^2 \right)^2$$

and setting the two appropriate derivatives to zero,

$$\sum \left(y_i - a - bx_i^2 \right) = 0 \qquad \sum \left(y_i - a - bx_i^2 \right) x_i^2 = 0$$

we have the linear system for determining a and b:

$$(N + 1) a + \left(\sum x_i^2 \right) b = \sum y_i$$
$$\left(\sum x_i^2 \right) a + \left(\sum x_i^4 \right) b = \sum x_i^2 y_i$$

12.27 Given the six data points $(0, .98)$, $(.1, 1.01)$, $(.2, .99)$, $(.3, .88)$, $(.4, .85)$ and $(.5, .77)$, which are slightly flawed values of the function $y(x) = 1 - x^2$, find the least-squares quadratic and use it to predict the crossing of the x axis near $x = 1$. How well does it do this?

▮ By the preceding problem, the coefficients a and b of the quadratic polynomial $p(x) = a + bx^2$ are determined by

$$6a + .55b = 5.48 \qquad .55a + .0979b = .4574$$

with solution $a = 1.000$ and $b = -.946$ to three places. The polynomial $p(x) = 1 - .946x^2$ crosses the x axis at $x = 1.03$. Errors of $\sim 2\%$ in the data have led to a similar (3) percent error in the prediction.

12.28 The six points

$$(0, .40) \quad (.5, .35) \quad (1, .24) \quad (1.5, .13) \quad (2, .054) \quad (2.5, .018)$$

are slightly flawed elements of the function $y = (1/\sqrt{2\pi})e^{-x^2/2}$. Find the best approximation of the form $y = ae^{-bx^2}$ using the least-squares principle

▮ Introducing logarithms,

$$\log y = \log a - bx^2 \quad \text{or} \quad p = c - bx^2$$

and the procedure of Problem 12.26 applies. The system

$$6c - 13.75b = -12.3696 \quad 13.75c - 61.1875b = -43.0638$$

determines $c = -.925$, $b = .496$ and then $a = e^c = .396$, so that

$$y = .396e^{-.496x^2}$$

is our approximation. The three place value of $1/\sqrt{2\pi}$ is .399.

12.3 SMOOTHING AND DIFFERENTIATION

12.29 Derive the formula for a least-squares parabola for five points (x_i, y_i) where $i = k - 2, k - 1, k, k + 1, k + 2$.

▮ Let the parabola be $p(t) = a_0 + a_1 t + a_2 t^2$ where $t = (x - x_k)/h$, the arguments x_i being assumed equally spaced at interval h. The five points now have arguments $t = -2, -1, 0, 1, 2$. For this symmetric arrangement the normal equations simplify to

$$\begin{aligned}
5a_0 \quad + 10a_2 &= \sum y_i \\
10a_1 \quad &= \sum t_i y_i \\
10a_0 \quad + 34a_2 &= \sum t_i^2 y_i
\end{aligned}$$

and are easily solved. We find first

$$\begin{aligned}
70a_0 &= 34\sum y_i - 10\sum t_i^2 y_i \\
&= -6y_{k-2} + 24y_{k-1} + 34y_k + 24y_{k+1} - 6y_{k+2} \\
&= 70y_k - 6(y_{k-2} - 4y_{k-1} + 6y_k - 4y_{k+1} + y_{k+2})
\end{aligned}$$

from which

$$a_0 = y_k - \tfrac{3}{35}\delta^4 y_k$$

Substituting back we also obtain

$$a_2 = \tfrac{1}{14}(2y_{k-2} - y_{k-1} - 2y_k - y_{k+1} + 2y_{k+2})$$

and directly from the middle equation

$$a_1 = \tfrac{1}{10}(-2y_{k-2} - y_{k-1} + y_{k+1} + 2y_{k+2})$$

12.30 With $y(x_k)$ representing the exact value of which y_k is an approximation, derive the smoothing formula $y(x_k) \simeq y_k - \tfrac{3}{35}\delta^4 y_k$.

▮ The least-squares parabola for the five points (x_{k-2}, y_{k-2}) to (x_{k+2}, y_{k+2}) is

$$p(x) = a_0 + a_1 t + a_2 t^2$$

At the center argument $t = 0$ this becomes $p(x_k) = a_0 = y_k - \tfrac{3}{35}\delta^4 y_k$ by Problem 12.29. Using this formula amounts to accepting the value of p on the parabola as better than the data value y_k.

12.31 The square roots of the integers from 1 to 10 were rounded to two decimal places and a random error of $-.05, -.04, \ldots, .05$ added to each (determined by drawing cards from a pack of 11 cards so labeled). The results form the top row of Table 12.2. Smooth these values using the formula of the preceding problem.

▌ Differences through the fourth also appear in Table 12.2, as well as $\frac{3}{35}\delta^4 y$. Finally the bottom row contains the smoothed values.

TABLE 12.2

x_k	1	2	3	4	5	6	7	8	9	10
y_k	1.04	1.37	1.70	2.00	2.26	2.42	2.70	2.78	3.00	3.14
δy		33	33	30	26	16	28	8	22	14
$\delta^2 y$		0		-3	-4	-10	12	-20	14	-8
$\delta^3 y$			-3	-1	-6	22	-32	34	-22	
$\delta^4 y$			2	-5	28	-54	66	-56		
$\frac{3}{35}\delta^4 y$			0	0	-5		6	-5		
$p(x_k)$			1.70	2.00	2.24	2.47	2.64	2.83		

12.32 The smoothing formula of Problem 12.30 requires two data values on each side of x_k for producing the smoothed value $p(x_k)$. It cannot therefore be applied to the two first and last entries of a data table. Derive the formulas

$$y(x_0) \simeq y_0 + \tfrac{1}{5}\Delta^3 y_0 + \tfrac{3}{35}\Delta^4 y_0 \qquad y(x_{N-1}) \simeq y_{N-1} + \tfrac{2}{5}\nabla^3 y_N - \tfrac{1}{7}\nabla^4 y_N$$

$$y(x_1) \simeq y_1 - \tfrac{2}{5}\Delta^3 y_0 - \tfrac{1}{7}\Delta^4 y_0 \qquad y(x_N) \simeq y_N - \tfrac{1}{5}\nabla^3 y_N + \tfrac{3}{35}\nabla^4 y_N$$

for smoothing end values.

▌ If we let $t = (x - x_2)/h$, then the quadratic function of Problem 12.29 is the least-squares quadratic for the first five points. We shall use the values of this function at x_0 and x_1 as smoothed values of y. First

$$p(x_0) = a_0 - 2a_1 + 4a_2$$

and inserting our expressions for the a_i, with k replaced by 2,

$$p(x_0) = \tfrac{1}{70}(62 y_0 + 18 y_1 - 6 y_2 - 10 y_3 + 6 y_4)$$
$$= y_0 + \tfrac{1}{70}[(-14 y_0 + 42 y_1 - 42 y_2 + 14 y_3) + (6 y_0 - 24 y_1 + 36 y_2 - 24 y_3 + 6 y_4)]$$

which reduce to the preceding formula for $y(x_0)$. For $p(x_1)$ we have

$$p(x_1) = a_0 - a_1 + a_2$$

and insertion of our expressions for the a_i again leads to the required formula. At the other end of our data supply the change of argument $t = (x - x_{N-2})/h$ applies, the details being similar.

12.33 Apply the formulas of the preceding problem to complete the smoothing of the y values in Table 12.2.

▌ We find these changes to two places:

$$y(x_0) \simeq 1.04 + \tfrac{1}{5}(-.03) + \tfrac{3}{35}(.02) \simeq 1.03 \qquad y(x_{N-1}) \simeq 3.00 + \tfrac{2}{5}(-.22) - \tfrac{1}{7}(-.56) \simeq 2.99$$

$$y(x_1) \simeq 1.37 - \tfrac{2}{5}(-.03) - \tfrac{1}{7}(.02) \simeq 1.38 \qquad y(x_N) \simeq 3.14 - \tfrac{1}{5}(-.22) + \tfrac{3}{35}(-.56) \simeq 3.14$$

12.34 Compute the rms error of both the original data and the smoothed values.

▌ The root-mean-square error of a set of approximations A_i corresponding to exact values T_i is defined by

$$\text{rms error} = \left[\sum_{i=0}^{N} \frac{(T_i - A_i)^2}{N} \right]^{1/2}$$

In this example we have the values

T_i	1.00	1.41	1.73	2.00	2.24	2.45	2.65	2.83	3.00	3.16
y_i	1.04	1.37	1.70	2.00	2.26	2.42	2.70	2.78	3.00	3.14
$p(x_i)$	1.03	1.38	1.70	2.00	2.24	2.47	2.64	2.83	2.99	3.14

The exact roots are given to two places. By the preceding formula,

$$\text{rms error of } y_i \simeq \left(\frac{.0108}{10}\right)^{1/2} \simeq .033$$

$$\text{rms error of } p(x_i) \simeq \left(\frac{.0037}{10}\right)^{1/2} \simeq .019$$

so that the error is less by nearly half. The improvement over the center portion is greater. If the two values at each end are ignored we find rms errors of .035 and .015, respectively, for a reduction of more than half. The Formula of Problem 12.30 appears more effective than those of Problem 12.32.

12.35 Use the five point parabola to obtain the formula

$$y'(x_k) \simeq \frac{1}{10h}(-2y_{k-2} - y_{k-1} + y_{k+1} + 2y_{k+2})$$

for approximate differentiation.

▌ With the symbols of Problem 12.30 we shall use $y'(x_k)$, which is the derivative of our five point parabola, as an approximation to the exact derivative at x_k. This again amounts to assuming that our data values y_i are approximate values of an exact but unknown function, but that the five point parabola will be a better approximation, especially in the vicinity of the center point. On the parabola

$$p = a_0 + a_1 t + a_2 t^2$$

and according to plan, we calculate $p'(t)$ at $t = 0$ to be a_1. To convert this to a derivative relative to x involves merely division by h, and so, recovering the value a_1 found in Problem 12.29 and taking $p'(x)$ as an approximation to $y'(x)$, we come to the required formula.

12.36 Apply the preceding formula to estimate $y'(x)$ from the y_k values given in Table 12.2.

▌ At $x_2 = 3$ we find

$$y'(3) \simeq \tfrac{1}{10}(-2.08 - 1.37 + 2.00 + 4.52) = .307$$

and at $x_3 = 4$,

$$y'(4) \simeq \tfrac{1}{10}(-2.74 - 1.70 + 2.26 + 4.84) = .266$$

The other entries in the top row shown are found in the same way. The second row was computed using the approximation

$$y'(x_k) \simeq \frac{1}{12h}(y_{k-2} - 8y_{k-1} + 8y_{k+1} - y_{k+2})$$

found earlier from Stirling's five point collocation polynomial. Notice the superiority of the present formula. Errors in data were found earlier to be considerably magnified by approximate differentiation formulas. Preliminary smoothing can lead to better results, by reducing such data errors.

$y'(x)$ by least squares	.31	.27	.24	.20	.18	.17
$y'(x)$ by collocation	.31	.29	.20	.23	.18	.14
Correct $y'(x)$	.29	.25	.22	.20	.19	.18

12.37   The formula of Problem 12.35 does not apply near the ends of the data supply. Use a four point parabola at each end to obtain the formulas

$$y'(x_0) \simeq \frac{1}{20h}(-21y_0 + 13y_1 + 17y_2 - 9y_3)$$

$$y'(x_1) \simeq \frac{1}{20h}(-11y_0 + 3y_1 + 7y_2 + y_3)$$

$$y'(x_{N-1}) \simeq \frac{1}{20h}(11y_N - 3y_{N-1} - 7y_{N-2} - y_{N-3})$$

$$y'(x_N) \simeq \frac{1}{20h}(21y_N - 13y_{N-1} - 17y_{N-2} + 9y_{N-3})$$

▮   Four points will be used rather than five, with the thought that a fifth point may be rather far from the position x_0 or x_N where a derivative is required. Depending on the size of h, the smoothness of the data and perhaps other factors, one could use formulas based on five points or more. Proceeding to the four point parabola we let $t = (x - x_1)/h$ so that the first four points have arguments $t = -1, 0, 1, 2$. The normal equations become

$$4a_0 + 2a_1 + 6a_2 = y_0 + y_1 + y_2 + y_3 \qquad 2a_0 + 6a_1 + 8a_2 = -y_0 + y_2 + 2y_3$$

$$6a_0 + 8a_1 + 18a_2 = y_0 + y_2 + 4y_3$$

and may be solved for

$$20a_0 = 3y_0 + 11y_1 + 9y_2 - 3y_3 \qquad 20a_1 = -11y_0 + 3y_1 + 7y_2 + y_3 \qquad 4a_2 = y_0 - y_1 - y_2 + y_3$$

With these and $y'(x_0) = (a_1 - 2a_2)/h$, $y'(x_1) = a_1/h$ the required results follow. Details at the other end of the data supply are almost identical.

12.38   Apply the formulas of the preceding problem to the data of Table 12.2.

▮   We find

$$y'(1) \simeq \tfrac{1}{20}[-21(1.04) + 13(1.37) + 17(1.70) - 9(2.00)] \simeq .35$$

$$y'(2) \simeq \tfrac{1}{20}[-11(1.04) + 3(1.37) + 7(1.70) + 2.00] \simeq .33$$

Similarly $y'(9) \simeq .16$ and $y'(10) \simeq .19$. The correct values are .50, .35, .17 and .16. The poor results obtained at the endpoints are further evidence of the difficulties of numerical differentiation. Newton's original formula

$$y'(x_0) \simeq \Delta y_0 - \tfrac{1}{2}\Delta^2 y_0 + \tfrac{1}{3}\Delta^3 y_0 - \tfrac{1}{4}\Delta^4 y_0 + \cdots$$

produces from this data the value .32, which is worse than our .35. At the other extreme the corresponding backward difference formula manages .25, which is much worse than our .19.

12.39   Apply the formulas for approximate derivatives a second time to estimate $y''(x)$, using the data of Table 12.2.

▮   We have already obtained estimates of the first derivative, of roughly two place accuracy. They are

x	1	2	3	4	5	6	7	8	9	10
$y'(x)$	.35	.33	.31	.27	.24	.20	.18	.17	.16	.19

Now applying the same formula to the $y'(x)$ values will produce estimates of $y''(x)$. For example, at $x = 5$,

$$y''(5) \simeq \tfrac{1}{10}[-2(.31) - (.27) + (.20) + 2(.18)] \simeq -.033$$

which is half again as large as the correct $-.022$. Complete results from our formulas and correct values are

$-y''$ (computed)	.011	.021	.028	.033	.033	.026	.019	.004	.012	-0.32
$-y''$ (correct)	.250	.088	.048	.031	.022	.017	.013	.011	.009	.008

Near the center we have an occasional ray of hope, but at the ends the disaster is evident.

12.40   Develop the smoothing formula

$$y(x_k) = y_k - \tfrac{1}{21}\left(9\,\delta^4 y_k + 2\,\delta^6 y_k\right)$$

which is based on the least-squares parabola for seven points.

▋ The argument used in Problem 12.29 to develop the five point parabola can be imitated. With $t = (x - x_k)/h$ we look for a parabola $p(t) = a_0 + a_1 t + a_2 t^2$ on the set $t = -3, -2, \ldots, 2, 3$. Mental calculations easily produce the set of normal equations

$$
\begin{aligned}
7a_0 \quad\; + 28a_2 &= \sum y_i \\
28a_1 \qquad\quad &= \sum t_i y_i \\
28a_0 \quad + 196a_2 &= \sum t_i^2 y_i
\end{aligned}
$$

Again we find a_1 isolated, while the other pair yield

$$21a_0 = -2y_{k-3} + 3y_{k-2} + 6y_{k-1} + 7y_k + 6y_{k+1} + 3y_{k+2} - 2y_{k+3}$$

Following in the footsteps of our mathematical ancestors we now compute

$$
\begin{aligned}
9\,\delta^4 y_k + 2\,\delta^6 y_k &= 9\left(y_{k-2} - 4y_{k-1} + 6y_k - 4y_{k+1} + y_{k+2}\right) \\
&\quad + 2\left(y_{k-3} - 6y_{k-2} + 15y_{k-1} - 20y_k + 15y_{k+1} - 6y_{k+2} + y_{k+3}\right) \\
&= 2y_{k-3} - 3y_{k-2} - 6y_{k-1} + 14y_k - 6y_{k+1} - 3y_{k+2} + 2y_{k+3}
\end{aligned}
$$

and discover a strong resemblance to $21a_0$. Matching them up now leads quickly to

$$a_0 = y_k - \tfrac{1}{21}\left(9\,\delta^4 y_k + 2\,\delta^6 y_k\right)$$

and since we plan to adopt $p(0) = a_0$ as our smoothed replacement for y_k, this is the required formula.

12.41   Apply the seven point smoothing formula to the data of Table 12.2 and compare with the earlier results from the five point formula.

▋ A row of sixth differences may be added to the table:

$$40 \qquad -115 \qquad 202 \qquad -242$$

Then the formula yields

$$
\begin{aligned}
y(4) &\simeq 2.00 - \tfrac{3}{7}(-.05) - \tfrac{2}{21}(.40) \simeq 1.98 \\
y(5) &\simeq 2.26 - \tfrac{3}{7}(.28) - \tfrac{2}{21}(-1.15) \simeq 2.25
\end{aligned}
$$

and similarly $y(6) \simeq 2.46$ and $y(7) \simeq 2.65$. These are a slight improvement over the results from the five point formula, except for $y(4)$ which is slightly worse.

12.42   The following are values of $y(x) = x^2$ with random errors of from $-.10$ to $.10$ added. (Errors were obtained by drawing cards from an ordinary pack with face cards removed, black meaning plus and red meaning minus.) The correct values T_i are also included:

x_i	1.0	1.1	1.2	1.3	1.4	1.5	1.6	1.7	1.8	1.9	2.0
y_i	.98	1.23	1.40	1.72	1.86	2.17	2.55	2.82	3.28	3.54	3.92
T_i	1.00	1.21	1.44	1.69	1.96	2.25	2.56	2.89	3.24	3.61	4.00

Apply the five and seven point smoothing formulas and compare the results.

▋ It will be enough to limit attention to the center five pairs for which both formulas can be applied. Here the numbers:

True	1.69	1.96	2.25	2.56	2.89
Five point	1.66	1.90	2.18	2.51	2.88
Seven point	1.65	1.91	2.19	2.51	2.87

Things seem to be just about dead even. It is a disappointment that the values from the seven point formula are not superior.

12.43 Use the seven point parabola to derive the approximate differentiation formula

$$y'(x) = \frac{1}{28h}(-3y_{k-3} - 2y_{k-2} - y_{k-1} + y_{k+1} + 2y_{k+2} + 3y_{k+3})$$

Then apply it, together wih the five point formula

$$y'(x) = \frac{1}{10h}(-2y_{k-2} - y_{k-1} + y_{k+1} + 2y_{k+2})$$

to the data of Problem 12.42.

∎ The quadratic $p(t) = a_0 + a_1 t + a_2 t^2$ has $p'(0) = a_1$. This is to be our estimate of $y'(0)$ and the system of Problem 12.40 gives it up instantly, it being necessary only to insert the $1/h$ needed to yield a derivative relative to x.

Applied to the center five points of the suggested data, the two formulas produce the numbers

x	1.3	1.4	1.5	1.6	1.7
5 point	2.34	2.75	2.89	3.49	3.47
7 point	2.52	2.69	3.05	3.20	3.45

the true values being $2x$. The given values of y were correct to two places, but this is not quite true of these derivatives. It is worth noting that in this example the true function is really a parabola, so there is no "truncation" error, so to speak. If no random errors had been introduced and no roundoffs made in the computations, we would have exact results.

12.44 Find a single least-squares parabola for the same data. How does it compare with $y = x^2$?

∎ The normal equations prove to be

$$\begin{aligned}
11a_0 + 16.5a_1 + 25.85a_2 &= 25.47 \\
16.5a_0 + 25.85a_1 + 42.075a_2 &= 41.452 \\
25.85a_0 + 42.075a_1 + 70.73331a_2 &= 69.68359
\end{aligned}$$

and lead to the parabola

$$p(x) = .095 - .128x + 1.027x^2$$

with a bit of rounding. It does bear some resemblance to x^2 and a sampling of its values is slightly reassuring, the error rising from 0 to 4 points in the second decimal place as the interval is crossed, with the approximation running low.

12.45 How well does the parabola of the preceding problem represent the derivative $y'(x) = 2x$? Does it do as well as the local five and seven point formulas, which use a different parabola at each data point?

∎ The plan now is to use $p'(x) = -.128 + 2.054x$ for the entire interval. Its values range from 8 to 2 points low in the second decimal place, improving as they go. For the center five points, where the other formulas were applied, the values are

$$2.54 \quad 2.75 \quad 2.95 \quad 3.16 \quad 3.36$$

They are no worse than those produced by the seven point formula and definitely better than the output of the five.

12.46 Estimate $y''(x)$ from the parabola $p(x)$.

∎ Obviously we get the constant 2.054 instead of the correct 2, for a relative error of under 3%. If we also apply the five point formula to the $y'(x)$ estimates obtained earlier,

$$2.34, 2.75, 2.89, 3.49, 3.47$$

just enough points to estimate $y'(1.5)$, we find it generating the value 3. In this case the global parabola is a clear winner.

12.47 The following are values of sin x with random errors of $-.10$ to $.10$ added. Find the least-squares parabola and use it to compute smoothed values. Compare with the performance of the five and seven point local formulas.

x	0	.2	.4	.6	.8	1.0	1.2	1.4	1.6
$\sin x$	$-.09$	.13	.44	.57	.64	.82	.97	.98	1.04

■ The normal equations prove to be

$$9a_0 + \quad 7.2a_1 + \quad 8.16a_2 = 5.5$$
$$7.2a_0 + \quad 8.16a_1 + \quad 10.368a_2 = 6.076$$
$$8.16a_0 + 10.368a_1 + 14.0352a_2 = 7.4904$$

and lead to the parabola $p = -.0823 + 1.28x - .361x^2$. A few values of p, together with corresponding values found by the five and seven point formulas, are listed in Table 12.3. At the few points where all three are active, the global parabola seems to have a slight edge.

TABLE 12.3

x	**0**	**.6**	**.8**	**1.0**	**1.6**
Five point		.57	.67	.82	
Seven point		.55	.71	.81	
Global	$-.08$	.55	.71	.83	1.04
True	0	.565	.717	.841	.999

12.48 Compare derivatives obtained by the three methods used in the preceding problem.

■ Here are the numbers:

x	Five pt.	Seven pt.	Global	True
.6	.79	.85	.84	.825
.8	.66	.69	.70	.697
1.0	.58	.53	.55	.540

Once again the global parabola has come closest. Some of the data errors in this example were more than 10% of true value and may have caused local troubles for the local methods.

12.49 A simple and ancient smoothing procedure that still finds use is the method of moving averages. In this method each value y_k is replaced by the average of itself and certain neighbors. For example, if two neighbors on each side are used, the formula

$$p_k = \tfrac{1}{5}(y_{k-2} + y_{k-1} + y_k + y_{k+1} + y_{k+2})$$

applies, while only one neighbor to a side brings

$$p_k = \tfrac{1}{3}(y_{k-1} + y_k + y_{k+1})$$

instead. For what type of data values y_k will such an operation change nothing?

■ Taking the simpler three point average first, what is asked is that $p_k = \tfrac{1}{3}(y_{k-1} + y_k + y_{k+1}) = y_k$. This can be rewritten as

$$y_{k+1} - 2y_k + y_{k-1} = 0$$

a familiar difference equation, with solution $y_k = a + bk$. The answer may have been guessed; the data points must lie on a line. Taking moving averages of such data will change nothing.

Proceeding to the broader average, the corresponding difference equation is

$$y_{k+2} + y_{k+1} - 4y_k + y_{k-1} + y_{k-2} = 0$$

which has constant coefficients and so can be solved through its characteristic equation. We find

$$y_k = a + bk + cr_1^k + dr_2^k$$

where the r_i satisfy $r^2 + 3r + 1 = 0$. The linear part was anticipated, but the rest is a surprise.

12.50 Why is the preceding result of interest?

▮ Suppose the given data happen to be exact. Then we would not want a smoothing method to alter it. Straight line data will not be altered, but consider the parabolic values

$$0 \quad 1 \quad 4 \quad 9 \quad 16$$

The three point moving average changes the inside three to

$$\tfrac{5}{3} \qquad \tfrac{14}{3} \qquad \tfrac{29}{3}$$

each of which is too big by $\tfrac{2}{3}$. Since

$$\tfrac{1}{3}\left[(k+1)^2 + k^2 + (k-1)^2\right] = k^2 + \tfrac{2}{3}$$

the result is general for this parabolic function. It seems that to smooth nearly parabolic data by moving averages. It would be necessary to reduce output by $\tfrac{2}{3}$.

12.51 How do three point moving averages affect cubic data?

▮ Since $(k+1)^3 + k^3 + (k-1)^3 = 3k^3 + 6k$, the "smoothed" value is $k^3 + 2k$. For example, the three center values of the exact data function

$$0 \quad 1 \quad 8 \quad 27 \quad 64$$

would be replaced by $\tfrac{9}{3}$, $\tfrac{36}{3}$ and $\tfrac{99}{3}$, each of which is $2k$ too big.

12.52 Assume that a linear data supply is contaminated by a single error of size 1. What does smoothing by the method of moving averages do to this error?

▮ The error function has the character

$$0 \quad 0 \quad 0 \quad 0 \quad 1 \quad 0 \quad 0 \quad 0 \quad 0$$

One application of the three point formula manages

$$0 \quad 0 \quad \tfrac{1}{3} \quad \tfrac{1}{3} \quad \tfrac{1}{3} \quad 0 \quad 0$$

and a second brings

$$\tfrac{1}{9} \quad \tfrac{2}{9} \quad \tfrac{3}{9} \quad \tfrac{2}{9} \quad \tfrac{1}{9}$$

and so on. The total error remains the same, but it is gradually spread over a wider and wider range, just as if the error were a bump in the road and the smoothing process a steamroller.

12.53 The following data supply is almost linear, contaminated by the addition of random errors. How well can smoothing by the method of moving averages identify the line?

x	0	1	2	3	4	5	6	7	8	9	10
y	1.08	1.26	1.70	1.92	2.12	2.48	2.70	3.14	3.40	3.76	3.94

■ No effort has been made to be secretive. The true line can probably be guessed, but the question is whether moving averages can identify it. One application of the simpler formula produces

| 1.35 | 1.63 | 1.91 | 2.17 | 2.43 | 2.77 | 3.08 | 3.43 | 3.70 |

followed by

| | 1.63 | 1.90 | 2.17 | 2.46 | 2.76 | 3.09 | 3.40 |

and then

| | | 1.90 | 2.18 | 2.46 | 2.77 | 3.08 |

and perhaps

| | | | 2.18 | 2.47 | 2.77 |

or even

2.47

at which point enthusiasm must be reined in. The apparent slope is $(2.77 - 2.18)/2 = .295$ leading to the line

$$y = .295(x - 5) + 2.47 = .295x + .995$$

which is not too far from the uncontaminated $y = .3x + 1$.

12.4 ORTHOGONAL POLYNOMIALS: DISCRETE CASE

12.54 For large N and m the set of normal equations may be badly ill-conditioned. To see this show that for equally spaced x_i from 0 to 1 the matrix of coefficients is approximately

$$\begin{bmatrix} 1 & \dfrac{1}{2} & \dfrac{1}{3} & \cdots & \dfrac{1}{m+1} \\ \dfrac{1}{2} & \dfrac{1}{3} & \dfrac{1}{4} & \cdots & \dfrac{1}{m+2} \\ & & & \cdots & \\ \dfrac{1}{m+1} & \dfrac{1}{m+2} & \dfrac{1}{m+3} & \cdots & \dfrac{1}{2m+1} \end{bmatrix}$$

if a factor of N is deleted from each term. This matrix is the Hilbert matrix of order $m + 1$.

■ For large N the area under $y(x) = x^k$ between 0 and 1 will be approximately the sum of N rectangular areas (see Fig. 12-4). Since the exact area is given by an integral, we have

$$\frac{1}{N} \sum_{i=0}^{N} x_i^k \simeq \int_0^1 x^k \, dx = \frac{1}{k+1}$$

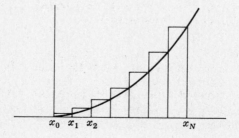

Fig. 12-4

Thus $s_k \simeq N/(k + 1)$, and deleting the N we have at once the Hilbert matrix. This matrix will later be shown to be extremely troublesome for large N.

12.55 How can the Hilbert matrices be avoided?

❚ The preceding problem shows that the normal equations that arise with the basis $1, x, \ldots, x^m$ and equally spaced arguments involve an approximately Hilbert matrix, which is troublesome. It is computationally more efficient to find an orthogonal basis so that the corresponding normal equations become trivial. A convenient orthogonal basis is constructed in the next problem. It is interesting to note that in developing this basis we will deal directly with the Hilbert matrix itself, not with approximations to it, and that the system of equations encountered will be solved exactly, thus avoiding pitfalls of computing with ill-conditioned systems.

12.56 Construct a set of polynomials $P_{m,N}(t)$ of degrees $m = 0, 1, 2, \ldots$ such that

$$\sum_{t=0}^{N} P_{m,N}(t) P_{n,N}(t) = 0 \quad \text{for } m > n$$

Such polynomials are called orthogonal over the set of arguments t.

❚ Let the polynomial be

$$P_{m,N}(t) = 1 + c_1 t + c_2 t^{(2)} + \cdots + c_m t^{(m)}$$

where $t^{(i)}$ is the factorial $t(t-1) \cdots (t-i+1)$. We first make the polynomial orthogonal to $(t+s)^{(s)}$ for $s = 0, 1, \ldots, m-1$, which means that we require

$$\sum_{t=0}^{N} (t+s)^{(s)} P_{m,N}(t) = 0$$

Since

$$(t+s)^{(s)} P_{m,N}(t) = (t+s)^{(s)} + c_1 (t+s)^{(s+1)} + \cdots + c_m (t+s)^{(s+m)}$$

summing over the arguments t and using Problem 3.83 brings

$$\sum_{t=0}^{N} (t+s)^{(s)} P_{m,N}(t) = \frac{(N+s+1)^{(s+1)}}{s+1} + c_1 \frac{(N+s+1)^{(s+2)}}{s+2} + \cdots + c_m \frac{(N+s+1)^{(s+m+1)}}{s+m+1}$$

which is to be zero. Removing the factor $(N+s+1)^{(s+1)}$, the sum becomes

$$\frac{1}{s+1} + \frac{Nc_1}{s+2} + \frac{N^{(2)} c_2}{s+3} + \cdots + \frac{N^{(m)} c_m}{s+m+1} = 0$$

and setting $N^{(i)} c_i = a_i$ this simplifies to

$$\frac{1}{s+1} + \frac{a_1}{s+2} + \frac{a_2}{s+3} + \cdots + \frac{a_m}{s+m} = 0$$

for $s = 0, 1, \ldots, m-1$. The Hilbert matrix again appears in this set of equations, but solving the system exactly will still lead us to a useful algorithm. If the last sum were merged into a single quotient it would take the form $Q(s)/(s+m+1)^{(m+1)}$ with $Q(s)$ a polynomial of degree at most m. Since $Q(s)$ must be zero at the m arguments $s = 0, 1, \ldots, m-1$, we must have $Q(s) = Cs^{(m)}$ where C is independent of s. To determine C we multiply both the sum and the equivalent quotient by $(s+1)$ and have

$$1 + (s+1) \left(\frac{a_1}{s+2} + \cdots + \frac{a_m}{s+m+1} \right) = \frac{Cs^{(m)}}{(s+2) \cdots (s+m+1)}$$

which must be true for all s except zeros of denominators. Setting $s = -,1$ we see that $C = m!/[(-1)(-2) \cdots (-m)] = (-1)^m$. We now have

$$\frac{1}{s+1} + \frac{a_1}{s+2} + \cdots + \frac{a_m}{s+m+1} = \frac{(-1)^m s^{(m)}}{(s+m+1)^{(m+1)}}$$

The device that produced C now produces the a_i. Multiply by $(s+m+1)^{(m+1)}$ and then set $s = -i-1$ to find for

$i = 1, \ldots, m,$

$$(-1)^i i! (m-i)! a_i = (-1)^m (-i-1)^{(m)} = (m+i)^{(m)}$$

and then solve for

$$a_i = (-1)^i \frac{(m+i)^{(m)}}{(m-i)! i!} = (-1)^i \binom{m}{i} \binom{m+i}{i}$$

Recalling that $a_i = c_i N^{(i)}$, the required polynomials may be written as

$$P_{m,N}(t) = \sum_{i=0}^{m} (-1)^i \binom{m}{i} \binom{m+i}{i} \frac{t^{(i)}}{N^{(i)}}$$

What we have proved is that each $P_{m,N}(t)$ is orthogonal to the functions

$$1 \qquad t+1 \qquad (t+2)(t+1) \qquad \cdots \qquad (t+m-1)^{(m-1)}$$

but in Problem 3.91 we saw that the powers $1, t, t^2, \ldots, t^{m-1}$ may be expressed as combinations of these, so that $P_{m,N}(t)$ is orthogonal to each of these powers as well. Finally, since $P_{n,N}(t)$ is a combination of these powers we find $P_{m,N}(t)$ and $P_{n,N}(t)$ to be themselves orthogonal. The first five of these polynomials are

$$P_{0,N} = 1$$

$$P_{1,N} = 1 - \frac{2t}{N}$$

$$P_{2,N} = 1 - \frac{6t}{N} + \frac{6t(t-1)}{N(N-1)}$$

$$P_{3,N} = 1 - \frac{12t}{N} + \frac{30t(t-1)}{N(N-1)} - \frac{20t(t-1)(t-2)}{N(N-1)(N-2)}$$

$$P_{4,N} = 1 - \frac{20t}{N} + \frac{90t(t-1)}{N(N-1)} - \frac{140t(t-1)(t-2)}{N(N-1)(N-2)} + \frac{70t(t-1)(t-2)(t-3)}{N(N-1)(N-2)(N-3)}$$

12.57 Determine the coefficients a_k so that

$$p(x) = a_0 P_{0,N}(t) + a_1 P_{1,N}(t) + \cdots + a_m P_{m,N}(t)$$

[with $t = (x - x_0)/h$] will be the least-squares polynomial of degree m for the data (x_i, y_i), $t = 0, 1, \ldots, N$.

∥ We are to minimize

$$S = \sum_{t=0}^{N} [y_t - a_0 P_{0,N}(t) - \cdots - a_m P_{m,N}(t)]^2$$

Setting derivatives relative to the a_k equal to zero, we have

$$\frac{\partial S}{\partial a_k} = -2 \sum_{t=0}^{N} [y_t - a_0 P_{0,N}(t) - \cdots - a_m P_{m,N}(t)] P_{k,N}(t) = 0$$

for $k = 0, 1, \ldots, m$. But by the orthogonality property most terms here are zero, only two contributing:

$$\sum_{t=0}^{N} [y_t - a_k P_{k,N}(t)] P_{k,N}(t) = 0$$

Solving for a_k, we find

$$a_k = \frac{\sum_{t=0}^{N} y_t P_{k,N}(t)}{\sum_{t=0}^{N} P_{k,N}^2(t)}$$

This is one advantage of the orthogonal functions. The coefficients a_k are uncoupled, each appearing in a single normal equation. Substituting the a_k into $p(x)$, we have the least-squares polynomial.

The same result follows directly from the general theorem of Problems 2.19 and 2.20. Identifying E, S, y, (v_1, v_2) and $\|v\|$ exactly as before, we now take $u_k = P_{k,N}(t)$ so that the orthogonal projection is still $p = a_0 u_0 + \cdots + a_m u_m$. The kth normal equation is $(u_k, u_k) a_k = (y, u_k)$ and leads to the expression for a_k already found.

Our general theory now also guarantees that we have actually minimized S and that our $p(x)$ is the unique solution. An argument using second derivatives could also establish this but is now not necessary.

12.58 Show that the minimum value of S takes the form $\sum_{t=0}^{N} y_t^2 - \sum_{k=0}^{m} W_k a_k^2$ where $W_k = \sum_{t=0}^{N} P_{k,N}^2(t)$.

▮ Expansion of the sum brings

$$S = \sum_{t=0}^{N} y_t^2 - 2 \sum_{t=0}^{N} y_t \sum_{k=0}^{m} a_k P_{k,N}(t) + \sum_{t=0}^{N} \sum_{k=0}^{m} a_j a_k P_{j,N}(t) P_{k,N}(t)$$

The second term on the right equals $-2\sum_{k=0}^{m} a_k (W_k a_k) = -2\sum_{k=0}^{m} W_k a_k^2$. The last term vanishes by the orthogonality except when $j = k$, in which case it becomes $\sum_{k=0}^{m} W_k a_k^2$. Putting the pieces back together,

$$S_{\min} = \sum_{i=0}^{N} y_t^2 - \sum_{k=0}^{m} W_k a_k^2$$

Notice what happens to the minimum of S as the degree m of the approximating polynomial is increased. Since S is nonnegative, the first sum in $S_{\min}$ clearly dominates the second. But the second increases with m, steadily diminishing the error. When $m = n$ we know by our earlier work that a collocation polynomial exists, equal to y_t at each argument $t = 0, 1, \ldots, N$. This reduces S to zero.

12.59 Apply the orthogonal functions algorithm to find a least-squares polynomial of degree 3 for the following data:

x_i	0	1	2	3	4	5	6	7	8	9	10
y_i	1.22	1.41	1.38	1.42	1.48	1.58	1.84	1.79	2.03	2.04	2.17

x_i	11	12	13	14	15	16	17	18	19	20
y_i	2.36	2.30	2.57	2.52	2.85	2.93	3.03	3.07	3.31	3.48

▮ The coefficients a_j are computed directly by the formula of the preceding problem. For hand computing, tables of the W_k and $P_{k,N}(t)$ exist and should be used. Although we have "inside information" that degree 3 is called for,

TABLE 12.4

x	Given Data	1	2	3	4	5	Correct Results
0	1.22	1.12	1.231	1.243	1.27	1.27	1.250
1	1.41	1.23	1.308	1.313	1.31	1.31	1.327
2	1.38	1.34	1.389	1.388	1.37	1.38	1.406
3	1.42	1.45	1.473	1.469	1.45	1.45	1.489
4	1.48	1.56	1.561	1.554	1.54	1.54	1.575
5	1.58	1.67	1.652	1.645	1.63	1.63	1.663
6	1.84	1.78	1.747	1.740	1.74	1.73	1.756
7	1.79	1.89	1.845	1.839	1.84	1.84	1.852
8	2.03	2.01	1.947	1.943	1.95	1.95	1.951
9	2.04	2.12	2.053	2.051	2.07	2.07	2.054
10	2.17	2.23	2.162	2.162	2.18	2.18	2.160
11	2.36	2.34	2.275	2.277	2.29	2.29	2.270
12	2.30	2.45	2.391	2.395	2.41	2.41	2.383
13	2.57	2.56	2.511	2.517	2.52	2.52	2.500
14	2.52	2.67	2.635	2.642	2.64	2.64	2.621
15	2.85	2.78	2.762	2.769	2.76	2.76	2.746
16	2.93	2.89	2.892	2.899	2.88	2.88	2.875
17	3.03	3.00	3.027	3.031	3.01	3.01	3.008
18	3.07	3.12	3.164	3.165	3.15	3.15	3.144
19	3.31	3.23	3.306	3.301	3.30	3.30	3.285
20	3.48	3.34	3.451	3.439	3.47	3.47	3.430

it is instructive to go slightly further. Up through $m = 5$ we find $a_0 = 2.2276$, $a_1 = -1.1099$, $a_2 = .1133$, $a_3 = .0119$, $a_4 = .0283$ and $a_5 = -.0038$, and with $x = t$,

$$p(x) = 2.2276 - 1.1099P_{1,20} + .1133P_{2,20} + .0119P_{3,20} + .0283P_{4,20} - .0038P_{5,20}$$

By the nature of orthogonal function expansions we obtain least-squares approximations of various degrees by truncation of this result. The values of such polynomials from degree 1 to degree 5 are given in Table 12.4 along with the original data. The final column lists the values of $y(x) = (x + 50)^3/10^5$ from which the data were obtained by adding random errors of size up to .10. Our goal has been to recover this cubic, eliminating as much error as we can by least-squares smoothing. Without prior knowledge that a cubic polynomial was our target, there would be some difficulty in choosing our approximation. Fortunately the results do not disagree violently after the linear approximation. A computation of the rms error shows that the quadratic has, in this case, outperformed the cubic approximation.

Degree	1	2	3	4	5	Raw Data
rms	.060	.014	.016	.023	.023	.069

12.60 Plot the first few polynomials $P_{m,10}(t)$.

▌ The plots for $m = 0$ to 4 appear in Fig. 12-5 and are reasonably normal polynomials of these degrees.

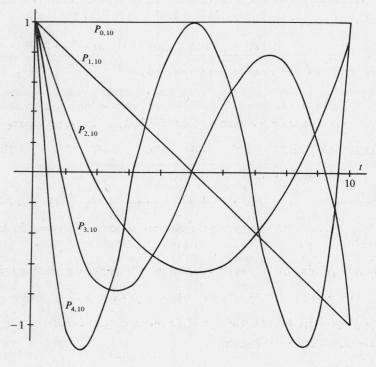

Fig. 12-5

12.61 Apply the method of orthogonal polynomials to the golf data of Problem 12.4. Does it support the hypothesis that these scores are essentially linear?

▌ The first few coefficients generated were

$$4.15 \qquad -.401 \qquad -.023 \qquad .019 \qquad .022$$

and the hypothesis already looks good. The first of these is the mean value of the scores, the second is the linear term ($P_{1,10}$ has a negative slope) and the rest seem harmless. For more evidence the rms errors can be examined.

They prove to be

$$.27 \quad .09 \quad .09 \quad .09 \quad .09$$

so including nonlinear terms brings no further improvement. In the preceding problem there was some uncertainty about where to truncate. Here there is none at all.

12.62 The following are the squares of $1, 1.1, \dots, 1.9, 2$ to two decimal places, with no added errors. The data are reasonably smooth and one would expect the orthogonal polynomial method to have no difficulty in discovering its quadratic nature. Does it pass this minor test?

▮ Computation of the coefficients finds them to be

$$2.35 \quad -1.5 \quad .15 \quad -.0000006 \quad -.000003$$

and things are certainly looking good. The approximation

$$p(t) = 2.35 - 1.5P_{1,10}(t) + .15P_{2,10}(t)$$

is clearly indicated and it does reproduce the data. What of the rms errors? They prove to be

$$.95 \quad .09 \quad .0007 \quad .0007 \quad .0007$$

definitely marking the chop after the quadratic term.

12.63 The following are values of e^x with random errors of from $-.10$ to $.10$ added. Use orthogonal polynomials to find a least-squares approximation. Choose a degree that seems appropriate and smooth the data. Is there improvement?

▮ These coefficients were found:

$$1.749 \quad -.892 \quad .143 \quad -.0491 \quad -.0285$$

The rms errors for approximations of degrees 0 to 4 were

$$.57 \quad .10 \quad .06 \quad .05 \quad .04$$

and where to chop is once again not quite so clear. Retaining all five terms leading to the smoothed values

$$.92 \quad 1.13 \quad 1.27 \quad 1.38 \quad 1.49 \quad 1.63 \quad 1.80 \quad 2.02 \quad 2.27 \quad 2.54 \quad 2.80$$

which may be compared with the true

$$1.00 \quad 1.11 \quad 1.22 \quad 1.35 \quad 1.49 \quad 1.65 \quad 1.82 \quad 2.01 \quad 2.23 \quad 2.46 \quad 2.82$$

as well as with the original data. Improvement is a bit spotty but does exist and, in any case, the new values are definitely smoother.

12.64 In Problem 12.31 these contaminated square root values were subjected to the five point parabola for smoothing:

$$1.04 \quad 1.37 \quad 1.70 \quad 2.00 \quad 2.26 \quad 2.42 \quad 2.70 \quad 2.78 \quad 3.00 \quad 3.14$$

Use orthogonal polynomials to show that no real improvement can be expected from higher-degree polynomials.

▮ As usual the coefficients must come first:

$$2.241 \quad -1.036 \quad -.158 \quad -.019 \quad .008$$

It is not instantly clear where to chop, but the rms errors of the five polynomials are

$$.67 \quad .10 \quad .02 \quad .02 \quad .02$$

and seem to answer the question. Taking the (global) quadratic we end up with these smoothed values, not much different from the five point output:

$$1.04 \quad 1.38 \quad 1.70 \quad 1.99 \quad 2.24 \quad 2.46 \quad 2.65 \quad 2.82 \quad 2.98 \quad 3.15$$

12.65 On a more recreational note, what do our orthogonal polynomials make of this oscillating function?

$$0, 1, 0, 1, 0, 1, 0, 1, 0, 1, 0$$

▌ The first five coefficients are

$$.454545 \quad 0 \quad -.1748 \quad 0 \quad -.1678$$

and the associated fourth degree polynomial has the "smoothed" values

$$.11 \quad .55 \quad .63 \quad .55 \quad .45 \quad .40 \quad .45 \quad .55 \quad .63 \quad .55 \quad .11$$

Are these preferable to the original? Not if the true function were a pure oscillation, sampled too sparsely. What do the rms errors have to say? They are

$$.50 \quad .50 \quad .49 \quad .49 \quad .47$$

and their message is that we have made little progress. The data apparently do not want to be smoothed.

12.66 What is the Gram–Schmidt process for generating orthogonal polynomials?

▌ It is an alternative to the procedure of Problem 12.56. Let $u_0(x), u_1(x), u_2(x), \ldots$ be a set of linearly independent functions. In fact, for simplicity we consider the special case for which $u_k(x) = x^k$, the generalization being direct. A set of functions $p_k(x)$ orthogonal on a discrete set of arguments x_i can then be constructed as follows. Take $p_0(x) = 1$. Next

$$p_1(x) = x - cp_0(x)$$

where c is to be determined by the orthogonality requirement

$$\sum p_0(x) p_1(x) = 0$$

the sum being over the arguments x_i. This means that

$$\sum x p_0(x) - c \sum p_0^2(x) = 0 \qquad c = \sum x_i / n$$

with n the number of the x_i. So we have

$$p_1(x) = x - \bar{x}$$

where $\bar{x}$ is the mean of the x_i. Next take

$$p_2(x) = x^2 - c_0 p_0(x) - c_1 p_1(x)$$

and require the two orthogonality conditions

$$\sum p_2(x) p_0(x) = \sum x^2 p_0(x) - c_0 \sum p_0(x) p_0(x) - c_1 \sum p_0(x) p_1(x) = 0$$
$$\sum p_2(x) p_1(x) = \sum x^2 p_1(x) - c_0 \sum p_0(x) p_1(x) - c_1 \sum p_1(x) p_1(x) = 0$$

or, in view of prior orthogonalities,

$$c_0 = \sum x^2 / n \qquad c_1 = \sum x^2 p_1(x) / \sum p_1^2(x)$$

The process can easily be continued as long as functions $u_k(x)$ are available, with

$$p_k(x) = x^k - \sum_{0}^{k-1} c_i p_i(x)$$

and the coefficients determined by

$$c_i = \sum x^k p_i(x) / \sum p_i^2(x)$$

It is useful to visualize the Gram–Schmidt process as one that removes from each $u_k(x)$ the components "parallel to" the earlier and mutually orthogonal $p_i(x)$, for $i = 0, 1, \ldots, k - 1$. With all of these components gone, the remaining $p_k(x)$ will be orthogonal to all its predecessors. The idea is an extension of a simpler vector process.

12.5 CONTINUOUS DATA: THE LEAST-SQUARES POLYNOMIAL

12.67 Determine the coefficients a_i so that

$$I = \int_{-1}^{1} \left[y(x) - a_0 P_0(x) - a_1 P_1(x) - \cdots - a_m P_m(x) \right]^2 dx$$

will be a minimum, the function $P_k(x)$ being the kth Legendre polynomial.

■ Here it is not a sum of squares that is to be minimized but an integral, and the data are no longer discrete values y_i but a function $y(x)$ of the continuous argument x. The use of the Legendre polynomials is very convenient. As in the previous section it will reduce the normal equations, which determine the a_k, to a very simple set, and since any polynomial can be expressed as a combination of Legendre polynomials, we are actually solving the problem of least-squares polynomial approximation for continuous data. Setting the usual derivatives to zero, we have

$$\frac{\partial I}{\partial a_k} = -2 \int_{-1}^{1} \left[y(x) - a_0 P_0(x) - \cdots - a_m P_m(x) \right] P_k(x) \, dx = 0$$

for $k = 0, 1, \ldots, m$. By the orthogonality of these polynomials, these equations simplify at once to

$$\int_{-1}^{1} \left[y(x) - a_k P_k(x) \right] P_k(x) \, dx = 0$$

Each equation involves only one of the a_k so that

$$a_k = \frac{\int_{-1}^{1} y(x) P_k(x) \, dx}{\int_{-1}^{1} P_k^2(x) \, dx} = \frac{2k+1}{2} \int_{-1}^{1} y(x) P_k(x) \, dx$$

Here again it is true that our problem is a special case of Problems 12.19 and 12.20, with these identifications:

$\quad E$: The space of real-valued functions on $-1 \leq x \leq 1$
$\quad S$: Polynomials of degree m or less
$\quad y$: The data function $y(x)$
(v_1, v_2): The scalar product $\int_{-1}^{1} v_1(x) v_2(x) \, dx$
$\quad \|v\|$: The norm $\int_{-1}^{1} [v(x)]^2 \, dx$
$\quad u_k$: $P_k(x)$
$\quad p$: $a_0 P_0(x) + \cdots + a_m P_m(x)$
$\quad a_k$: $(y, u_k)/(u_k, u_k)$

These problems therefore guarantee that our solution $p(x)$ is unique and does minimize the integral I.

12.68 Find the least-squares approximation to $y(t) = t^2$ on the interval $(0, 1)$ by a straight line.

■ Here we are approximating a parabolic arc by a line segment. First let $t = (x + 1)/2$ to obtain the interval $(-1, 1)$ in the argument x. This makes $y = (x + 1)^2/4$. Since $P_0(x) = 1$ and $P_1(x) = x$, the coefficients a_0 and a_1 are

$$a_0 = \frac{1}{2} \int_{-1}^{1} \frac{1}{4}(x+1)^2 \, dx = \frac{1}{3} \qquad a_1 = \frac{3}{2} \int_{-1}^{1} \frac{1}{4}(x+1)^2 x \, dx = \frac{1}{2}$$

and the least-squares line is $y = \frac{1}{3} P_0(x) + \frac{1}{2} P_1(x) = \frac{1}{3} + \frac{1}{2} x = t - \frac{1}{6}$.

Both the parabolic arc and the line are shown in Fig. 12-6. The difference between y values on the line and the parabola is $t^2 - t + \frac{1}{6}$, and this takes extreme values at $t = 0, \frac{1}{2}$ and 1 of amounts $\frac{1}{6}, -\frac{1}{12}$ and $\frac{1}{6}$. The error made in

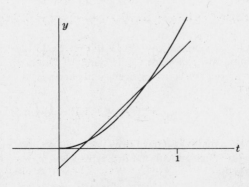

Fig. 12-6

substituting the line for the parabola is therefore slightly greater at the ends than at the center of the interval. This error can be expressed as

$$\tfrac{1}{4}(x+1)^2 - \tfrac{1}{3}P_0(x) - \tfrac{1}{2}P_1(x) = \tfrac{1}{6}P_2(x)$$

and the shape of $P_2(x)$ corroborates this error behavior.

12.69 Find the least-squares approximation to $y(t) = \sin t$ on the interval $(0, \pi)$ by a parabola.

▮ Let $t = \pi(x+1)/2$ to obtain the interval $(-1, 1)$ in the argument x. Then $y = \sin[\pi(x+1)/2]$. The coefficients are

$$a_0 = \frac{1}{2} \int_{-1}^{1} \sin\left[\frac{\pi(x+1)}{2}\right] dx = \frac{2}{\pi}$$

$$a_1 = \frac{3}{2} \int_{-1}^{1} \sin\left[\frac{\pi(x+1)}{2}\right] x \, dx = 0$$

$$a_2 = \frac{5}{2} \int_{-1}^{1} \sin\left[\frac{\pi(x+1)}{2}\right] \frac{1}{2}(3x^2 - 1) \, dx = \frac{10}{\pi}\left(1 - \frac{12}{\pi^2}\right)$$

so that the parabola is

$$y = \frac{2}{\pi} + \frac{10}{\pi}\left(1 - \frac{12}{\pi^2}\right) \frac{1}{2}(3x^2 - 1) = \frac{2}{\pi} + \frac{10}{\pi}\left(1 - \frac{12}{\pi^2}\right)\left[\frac{6}{\pi^2}\left(t - \frac{\pi}{2}\right)^2 - \frac{1}{2}\right]$$

The parabola and sine curve are shown in Fig. 12-7 with slight distortions to better emphasize the over and under nature of the approximation.

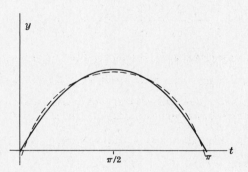

Fig. 12-7

12.70 What are the "shifted Legendre polynomials"?

▮ These result from a change of argument that converts the interval $(-1, 1)$ into $(0, 1)$. Let $t = (1 - x)/2$ to effect this change. The familiar Legendre polynomials in the argument x then become

$$P_0 = 1 \qquad\qquad P_2 = \tfrac{1}{2}(3x^2 - 1) = 1 - 6t + 6t^2$$
$$P_1 = x = 1 - 2t \qquad P_3 = \tfrac{1}{2}(5x^3 - 3x) = 1 - 12t + 30t^2 - 20t^3$$

and so on. These polynomials are orthogonal over $(0, 1)$ and we could have used them as the basis of our least-squares analysis of continuous data in place of the standard Legendre polynomials. With this change of argument the integrals involved in our formulas for coefficients become

$$\int_0^1 [P_n(t)]^2 \, dt = \frac{1}{2n + 1} \qquad a_k = (2k + 1)\int_0^1 y(t) P_k(t) \, dt$$

The argument change $t = (x + 1)/2$ might also have been used, altering the sign of each odd-degree polynomial, but the device used leads to a close analogy with the orthogonal polynomials for the discrete case developed in Problem 12.56.

12.71 Suppose that an experiment produces the curve shown in Fig. 12-8. It is known or suspected that the curve should be a straight line. Show that the least-squares line is approximately given by $y = .21t + .11$, which is shown dashed in the diagram.

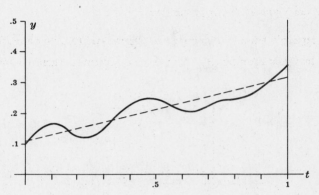

Fig. 12-8

❚ Instead of reducing the interval to $(-1, 1)$ we work directly with the argument t and the shifted Legendre polynomials. Two coefficients are needed,

$$a_0 = \int_0^1 y(t)\, dt \qquad a_1 = 3\int_0^1 y(t)(1 - 2t)\, dt$$

Since $y(t)$ is not available in analytic form, these integrals must be evaluated by approximate methods. Reading from the diagram, we may estimate y values as follows:

t	0	.1	.2	.3	.4	.5	.6	.7	.8	.9	1.0
y	.10	.17	.13	.15	.23	.25	.21	.22	.25	.29	.36

Applying Simpson's rule now makes $a_0 \simeq .214$ and $a_1 \simeq -.105$. The resulting line is

$$y = .214 - .105(1 - 2t) = .21t + .11$$

and this appears in Fig. 12-8. An alternative treatment of this problem could involve applying the methods for discrete data to the y values read from the diagram.

12.72 Find the least-squares line for $y(x) = x^2$ on the interval $(-1, 1)$.

❚ The two coefficients are

$$a_0 = \tfrac{1}{2}\int_{-1}^1 x^2\, dx = \tfrac{1}{3} \qquad a_1 = \tfrac{3}{2}\int_{-1}^1 x^3\, dx = 0$$

so the line is the horizontal $y(x) = \tfrac{1}{3}$. Perhaps the one point worth noting is that the "best" line in the least-squares sense is not $y = \tfrac{1}{2}$, which makes maximum errors of $\tfrac{1}{2}$ at the ends and in the middle. The larger errors near the ends here occur over a relatively short range and so their influence is limited.

12.73 Find the least-squares line for $y(x) = x^3$ on the interval $(-1, 1)$.

❚ This time the two coefficients are

$$a_0 = \tfrac{1}{2}\int_{-1}^1 x^3\, dx = 0 \qquad a_1 = \tfrac{3}{2}\int_{-1}^1 x^4\, dx = \tfrac{3}{5}$$

so the line is $y(x) = \tfrac{3}{5}x$. It misses the cubic by $\tfrac{2}{5}$ at the two endpoints, while a bit of calculus shows that the largest error inside the interval is just under $\tfrac{1}{5}$ in size.

Asking for the least-squares quadratic approximation to this cubic leads to the computation of a third coefficient.

$$a_2 = \tfrac{5}{2}\int_{-1}^1 x^3\left(\tfrac{3}{2}x^2 - \tfrac{1}{2}\right) dx = 0$$

So the quadratic degenerates to the line already in hand.

Although no one is likely to ask for the least-squares cubic, there may be some minor satisfaction in computing

$$a_3 = \tfrac{7}{2}\int_{-1}^{1} x^3\left(\tfrac{5}{2}x^3 - \tfrac{3}{2}x\right) dx = \tfrac{2}{5}$$

and reconstructing

$$y(x) = \tfrac{3}{5}x + \tfrac{2}{5}\left(\tfrac{5}{2}x^3 - \tfrac{3}{2}x\right) = x^3$$

12.74 Find the least-squares line for $y(x) = e^x$ on $(-1, 1)$.

▮ The coefficients are

$$a_0 = \tfrac{1}{2}\int_{-1}^{1} e^x \, dx = \tfrac{1}{2}(e - e^{-1}) \qquad a_1 = \tfrac{3}{2}\int_{-1}^{1} x e^x \, dx = 3e^{-1}$$

making the line

$$y(x) = \tfrac{1}{2}(e - e^{-1}) + 3e^{-1}x$$

with errors of .30, −.18, .43 at $x = -1, 0, 1$. The line follows an under-over-under path relative to the exponential function, which is what intuition and experience expect, but the error sizes are more obscure and governed by the least-squares principle.

12.75 Find approximately the least-squares parabola for the function in Fig. 12-9, evaluating the integrals by Simpson's rule. This curve should be imagined to be an experimental result which theory claims ought to have been a parabola.

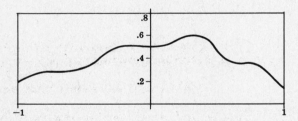

Fig. 12-9

▮ The suggested approach involves first estimating some values of the function. With no microscope readily available, suppose we settle for

x	−1	−.8	−.6	−.4	−.2	0	.2	.4	.6	.8	1
y	.2	.3	.3	.4	.5	.5	.56	.59	.38	.35	.16

Simpson's rule then manages

$$a_0 = \tfrac{1}{2}\int_{-1}^{1} y \, dx = .41$$

$$a_1 = \tfrac{3}{2}\int_{-1}^{1} yx \, dx = .054$$

$$a_2 = \tfrac{5}{2}\int_{-1}^{1} y P_2(x) \, dx = -.225$$

giving us the parabola $p(x) = .41 + .054x - .225(\tfrac{3}{2}x^2 - \tfrac{1}{2})$. A look at the smoothed values produced by this parabola,

| .13 | .26 | .37 | .45 | .50 | .52 | .52 | .49 | .43 | .35 | .24 |

finds numerous adjustments, particularly in the vicinity of the bump at $x = .4$.

12.76 The function of Fig. 12-9 having been approximated by a discrete function, the method of Problem 12.59 can be applied. What does it have to offer?

▮ The first few coefficients prove to be

$$.385 \qquad -.031 \qquad -.207 \qquad .059 \qquad .009$$

with the corresponding rms errors for the first five polynomial approximations:

$$.14 \qquad .13 \qquad .06 \qquad .04 \qquad .04$$

The quadratic seems a reasonable place to chop, so to this extent the data agree with theory. Trying a higher-degree polynomial would probably produce values more faithful to the given data, but presumably they would be farther from the true parabola.

12.77 Use the Gram–Schmidt algorithm to produce the first few members of an orthogonal family $p_0(x)$, $p_1(x)$, $p_2(x), \ldots$ from the set of powers $1, x, x^2, \ldots$. What are these orthogonal polynomials?

▮ Let $p_0(x) = 1$. The algorithm then continues with

$$p_1(x) = x - c$$

and the requirement that p_0 and p_1 be orthogonal. We choose the definition

$$\int_{-1}^{1} p_i p_j \, dx = 0$$

for orthogonality, which here becomes $\frac{1}{2}x^2 - cx|_{-1}^{1} = 0$ or $c = 0$, identifying $p_1(x) = x$. The next step begins

$$p_2(x) = x^2 - c_0 p_0(x) - c_1 p_1(x)$$

and asks that p_2 be orthogonal to both p_0 and p_1. First

$$\int_{-1}^{1} p_2 p_0 \, dx = 0 = \frac{1}{3}x^3 \Big|_{-1}^{1} - 2 \int_{-1}^{1} p_0^2 \, dx = \frac{2}{3} - 2c_0$$

determines $c_0 = \frac{1}{3}$, after which

$$\int_{-1}^{1} p_2 p_1 \, dx = 0 = \frac{1}{4}x^4 \Big|_{-1}^{1} - c_1 \int_{-1}^{1} p_1^2 \, dx = -\frac{2}{3}c_1$$

which leaves no doubt that $c_1 = 0$. So $p_2(x) = x^2 - \frac{1}{3}$.

The continuation is clear, but one more step may not be amiss. Let

$$p_3(x) = x^3 - c_0 p_0(x) - c_1 p_1(x) - c_2 p_2(x)$$

with p_3 to be orthogonal to its three predecessors. First

$$\int_{-1}^{1} p_3 p_0 \, dx = 0 = \frac{1}{4}x^4 \Big|_{-1}^{1} - c_0 \int_{-1}^{1} p_0^2 \, dx = -2c_0$$

forcing $c_0 = 0$. Then

$$\int_{-1}^{1} p_3 p_1 \, dx = 0 = \frac{1}{5}x^5 \Big|_{-1}^{1} - c_1 \int_{-1}^{1} p_1^2 \, dx = \frac{2}{5} - \frac{2}{3}c_1$$

and $c_1 = \frac{3}{5}$. Finally

$$\int_{-1}^{1} p_3 p_2 \, dx = 0 = \left(\frac{1}{6}x^6 - \frac{1}{3}\frac{1}{4}x^4 \right)\Big|_{-1}^{1} - c_2 \int_{-1}^{1} p_2^2 \, dx$$

and c_2 must be zero. Putting the pieces together

$$p_3(x) = x^3 - \frac{3}{5}x$$

The polynomials $p_k(x)$ have no doubt been recognized as Legendre polynomials, apart from a multiplicative constant that could be attached to each without destroying the orthogonality.

12.6 CONTINUOUS DATA: A GENERALIZED TREATMENT

12.78 Develop the least-square polynomial in terms of a set of orthogonal polynomials on the interval (a, b) with nonnegative weight function $w(x)$.

▌ The details are very similar to those of earlier derivations. We are to minimize

$$I = \int_a^b w(x)[y(x) - a_0 Q_0(x) - \cdots - a_m Q_m(x)]^2 \, dx$$

by choice of the coefficients a_k, where the functions $Q_k(x)$ satisfy the orthogonality condition

$$\int_a^b w(x) Q_j(x) Q_k(x) \, dx = 0$$

for $j \neq k$. Without stopping for the duplicate argument involving derivatives, we appeal at once to Problems 12.19 and 12.20 with the scalar product

$$(v_1, v_2) = \int_a^b w(x) v_1(x) v_2(x) \, dx$$

and other obvious identifications, and find

$$a_k = \frac{\int_a^b w(x) y(x) Q_k(x) \, dx}{\int_a^b w(x) Q_k^2(x) \, dx}$$

With these a_k the least-squares polynomial is $p(x) = a_0 Q_0(x) + \cdots + a_m Q_m(x)$.

12.79 What is the importance of the fact that a_k does not depend on m?

▌ This means that the degree of the approximation polynomial does not have to be chosen at the start of a computation. The a_k may be computed successively and the decision of how many terms to use can be based on the magnitudes of the computed a_k. In nonorthogonal developments a change of degree will usually require that all coefficients be recomputed.

12.80 Show that the minimum value of I can be expressed in the form

$$\int_a^b w(x) y^2(x) \, dx - \sum_{k=0}^m W_k a_k^2 \quad \text{where} \quad W_k = \int_a^b w(x) Q_k^2(x) \, dx$$

▌ Explicitly writing out the integral makes

$$I = \int_a^b w(x) y^2(x) \, dx - 2 \sum_{k=0}^m \int_a^b w(x) y(x) a_k Q_k(x) \, dx + \sum_{j,\, k=0}^m \int_a^b w(x) a_j a_k Q_j(x) Q_k(x) \, dx$$

The second term on the right equals $-2\sum_{k=0}^m a_k(W_k a_k) = -2\sum_{k=0}^m W_k a_k^2$. The last term vanishes by the orthogonality except when $j = k$, in which case it becomes $\sum_{k=0}^m W_k a_k^2$. Putting the pieces back together, $I_{\min} = \int_a^b w(x) y^2(x) \, dx - \sum_{k=0}^m W_k a_k^2$.

12.81 Prove Bessel's inequality, $\sum_{k=0}^m W_k a_k^2 \leq \int_a^b w(x) y^2(x) \, dx$.

▌ Assuming $w(x) \geq 0$, it follows that $I \geq 0$ so that Bessel's inequality is an immediate consequence of the preceding problem.

12.82 Prove the series $\sum_{k=0}^\infty W_k a_k^2$ to be convergent.

▌ It is a series of positive terms with partial sums bounded above by the integral in Bessel's inequality. This guarantees convergence. Of course, it is assumed all along that the integrals appearing in our analysis exist, in other words that we are dealing with functions that are integrable on the interval (a, b).

12.83 Is it true that as m tends to infinity, the value of $I_{\min}$ tends to zero?

▮ With the families of orthogonal functions ordinarily used, the answer is yes. The process is called convergence in the mean and the set of orthogonal functions is called complete. The details of proof are more extensive than will be attempted here.

12.84 Use the Gram–Schmidt algorithm to produce the first few members of an orthogonal family $L_0(x), L_1(x), L_2(x), \ldots$ from the usual ingredients $1, x, x^2, \ldots$ and with orthogonality meaning

$$\int_0^\infty e^{-x} L_i(x) L_j(x) \, dx = 0$$

▮ Using the letter L is a clear indicator of the outcome to be expected, more or less. Let $L_0(x) = 1$ and $L_1(x) = x - c$. For orthogonality we need

$$\int_0^\infty e^{-x} L_1(x) L_0(x) \, dx = 0 = \int_0^\infty e^{-x}(x - c) \, dx = 1 - c$$

so $c = 1$ and $L_1(x) = x - 1$, which differs only in sign from our earlier definition of this Laguerre polynomial. Next

$$L_2(x) = x^2 - c_0 L_0(x) - c_1 L_1(x)$$

and orthogonality requires

$$\int_0^\infty e^{-x} L_2 L_0 \, dx = 0 = \int_0^\infty e^{-x}(x^2 - c_0) \, dx = 2 - c_0$$

$$\int_0^\infty e^{-x} L_2 L_1 \, dx = 0 = \int_0^\infty e^{-x}(x^2 - c_1 L_1) L_1 \, dx = 6 - 2 - c_1$$

with the prior orthogonality of L_0 and L_1 helpful. This makes

$$L_2(x) = x^2 - 4x + 2$$

and this time there is no problem with the sign. Continuation is routine and eased by recalling that $\int_0^\infty e^{-x} x^p \, dx = p!$.

12.85 Use the Gram–Schmidt algorithm to produce some of the orthogonal polynomials $H_0(x), H_1(x), \ldots$ with orthogonality defined by

$$\int_{-\infty}^\infty e^{-x^2} H_i(x) H_j(x) \, dx = 0$$

▮ As usual we start with the powers of x and subtract away the components parallel to earlier selections. Let $H_0(x) = 1$ and $H_1(x) = x - c$. Then we ask that

$$\int_{-\infty}^\infty e^{-x^2} H_1 H_0 \, dx = 0 = \int_{-\infty}^\infty e^{-x^2}(x - c) \, dx = -c\sqrt{\pi}$$

forcing $c = 0$ and making $H_1(x) = x$. [Whenever the weight function $w(x)$ is an even function on $(-1, 1)$ this will happen.] Next

$$H_2(x) = x^2 - c_0 H_0(x) - c_1 H_1(x)$$

and we require

$$\int_{-\infty}^\infty e^{-x^2} H_2 H_0 \, dx = 0 = \int_{-\infty}^\infty e^{-x^2}(x^2 - c_0) \, dx = \tfrac{1}{2}\sqrt{\pi} - c_0\sqrt{\pi}$$

$$\int_{-\infty}^\infty e^{-x^2} H_2 H_1 \, dx = 0 = \int_{-\infty}^\infty e^{-x^2}(x^2 - c_1 H_1) H_1 \, dx = -\tfrac{1}{2} c_1 \sqrt{\pi}$$

which together make $H_2(x) = x^2 - \tfrac{1}{2}$. Apart from constant factors we have begun the sequence of Hermite polynomials.

12.86 Use the Gram–Schmidt algorithm to produce polynomials orthogonal relative to the weight function of Fig. 12-10.

$$w(x) = \begin{cases} 1 + x & x < 0 \\ 1 - x & x > 0 \end{cases}$$

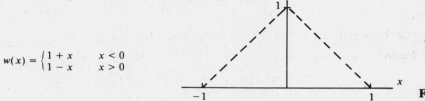

Fig. 12-10

▌ Let $S_0(x) = 1$ and, noting that $w(x)$ is even, $S_1(x) = x$. Next

$$S_2(x) = x^2 - c_0 - c_1 x$$

and by the required orthogonality

$$\int_{-1}^{1} w S_2 S_0 \, dx = 0 = \int_{-1}^{0} (1 + x)(x^2 - c_0) \, dx + \int_{0}^{1} (1 - x)(x^2 - c_0) \, dx = \tfrac{1}{6} - c_0$$

$$\int_{-1}^{1} w S_2 S_1 \, dx = 0 = \int_{-1}^{0} (1 + x)(x^2 - c_1 x) x \, dx + \int_{0}^{1} (1 - x)(x^2 - c_1 x) x \, dx = -\tfrac{1}{6} c_1$$

leading to $S_2(x) = x^2 - \tfrac{1}{6}$.

12.7 APPROXIMATION WITH CHEBYSHEV POLYNOMIALS

12.87 The Chebyshev polynomials are defined for $1 \le x \le 1$ by $T_n(x) = \cos(n \arccos x)$. Find the first few such polynomials directly from this definition.

▌ For $n = 0$ and 1 we have at once $T_0(x) = 1$ and $T_1(x) = x$. Let $A = \arccos x$. Then

$$T_2(x) = \cos 2A = 2\cos^2 A - 1 = 2x^2 - 1$$

$$T_3(x) = \cos 3A = 4\cos^3 A - 3\cos A = 4x^3 - 3x \quad \text{etc.}$$

12.88 Prove the recursion relation $T_{n+1}(x) = 2x T_n(x) - T_{n-1}(x)$.

▌ The trigonometric relationship $\cos(n + 1)A + \cos(n - 1)A = 2\cos A \cos nA$ translates directly into $T_{n+1}(x) + T_{n-1}(x) = 2x T_n(x)$.

12.89 Use the recursion to produce the next few Chebyshev polynomials.

▌ Beginning with $n = 3$,

$$T_4(x) = 2x(4x^3 - 3x) - (2x^2 - 1) = 8x^4 - 8x^2 + 1$$

$$T_5(x) = 2x(8x^4 - 8x^2 + 1) - (4x^3 - 3x) = 16x^5 - 20x^3 + 5x$$

$$T_6(x) = 2x(16x^5 - 20x^3 + 5x) - (8x^4 - 8x^2 + 1) = 32x^6 - 48x^4 + 18x^2 - 1$$

$$T_7(x) = 2x(32x^6 - 48x^4 + 18x^2 - 1) - (16x^5 - 20x^3 + 5x) = 64x^7 - 112x^5 + 56x^3 - 7x \quad \text{etc.}$$

12.90 Prove the orthogonality property

$$\int_{-1}^{1} \frac{T_m(x) T_n(x)}{\sqrt{1 - x^2}} \, dx = \begin{cases} 0 & m \neq n \\ \pi/2 & m = n \neq 0 \\ \pi & m = n = 0 \end{cases}$$

▌ Let $x = \cos A$ as before. The preceding integral becomes

$$\int_{0}^{\pi} (\cos mA)(\cos nA) \, dA = \left[\frac{\sin(m + n)A}{2(m + n)} + \frac{\sin(m - n)A}{2(m - n)} \right]_{0}^{\pi} = 0$$

for $m \neq n$. If $m = n = 0$, the result π is immediate. If $m = n \neq 0$, the integral is

$$\int_0^\pi \cos^2 nA \, dA = \left[\frac{1}{2} \left(\frac{\sin nA \cos nA}{n} + A \right) \right]_0^\pi = \frac{\pi}{2}$$

12.91 Express the powers of x in terms of Chebyshev polynomials.

▮ We find

$$
\begin{aligned}
1 &= T_0 & x^4 &= \tfrac{1}{8}(3T_0 + 4T_2 + T_4) \\
x &= T_1 & x^4 &= \tfrac{1}{16}(10T_1 + 5T_3 + T_5) \\
x^2 &= \tfrac{1}{2}(T_0 + T_2) & x^6 &= \tfrac{1}{32}(10T_0 + 15T_2 + 6T_4 + T_6) \\
x^3 &= \tfrac{1}{4}(3T_1 + T_3) & x^7 &= \tfrac{1}{64}(35T_1 + 21T_3 + 7T_5 + T_7)
\end{aligned}
$$

and so on. Clearly the process may be continued to any power.

12.92 Find the least-squares polynomial that minimizes the integral

$$\int_{-1}^1 \frac{1}{\sqrt{1 - x^2}} \left[y(x) - a_0 T_0(x) - \cdots - a_m T_m(x) \right]^2 dx$$

▮ By results of the previous section the coefficients a_k are

$$a_k = \frac{\int_{-1}^1 w(x) y(x) T_k(x) \, dx}{\int_{-1}^1 w(x) T_k^2(x) \, dx} = \frac{2}{\pi} \int_{-1}^1 \frac{y(x) T_k(x)}{\sqrt{1 - x^2}} \, dx$$

except for a_0, which is

$$a_0 = \frac{1}{\pi} \int_{-1}^1 \frac{y(x)}{\sqrt{1 - x^2}} \, dx$$

The least-squares polynomial is $a_0 T_0(x) + \cdots + a_m T_m(x)$.

12.93 Show that $T_n(x)$ has n zeros inside the interval $(-1, 1)$ and none outside. What is the *equal ripple* property?

▮ Since $T_n(x) = \cos n\theta$, with $x = \cos\theta$ and $-1 \le x \le 1$, we may require $0 \le \theta \le \pi$ without loss. Actually this makes the relationship between θ and x more precise. Clearly $T_n(x) = 0$ for $\theta = (2i + 1)\pi/2n$, or

$$x_i = \cos[(2i + 1)\pi/2n] \qquad i = 0, 1, \ldots, n - 1$$

These are n distinct arguments between -1 and 1. Since $T_n(x)$ has only n zeros, there can be none outside the interval $(-1, 1)$. Being equal to a cosine in the interval $(-1, 1)$, the polynomial $T_n(x)$ cannot exceed 1 in magnitude there. It reaches this maximum size at $n + 1$ arguments, including the endpoints:

$$T_n(x) = (-1)^i \quad \text{at} \quad x = \cos i\pi/n \qquad i = 0, 1, \ldots, n$$

This oscillation between extreme values of equal magnitude is known as the equal ripple property. This property is illustrated in Fig. 12-11, which shows $T_2(x)$, $T_3(x)$, $T_4(x)$ and $T_5(x)$.

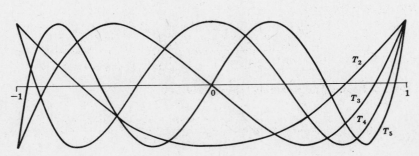

Fig. 12-11

12.94 In what way does the equal ripple property make the least-squares approximation

$$y(x) \simeq a_0 T_0(x) + \cdots + a_m T_m(x)$$

superior to similar approximations using other polynomials in place of the $T_k(x)$?

▌ Suppose we assume that, for the $y(x)$ concerned, the series obtained by letting m tend to infinity converges to $y(x)$, and also that it converges quickly enough so that

$$y(x) - a_0 T_0(x) - \cdots - a_m T_m(x) \simeq a_{m+1} T_{m+1}(x)$$

In other words, the error made in truncating the series is essentially the first omitted term. Since $T_{m+1}(x)$ has the equal ripple property, the error of our approximation will fluctuate between a_{m+1} and $-a_{m+1}$ across the entire interval $(-1, 1)$. The error will not be essentially greater over one part of the interval compared with another. This error uniformity may be viewed as a reward for accepting the unpleasant weighting factor $1/\sqrt{1 - x^2}$ in the integrals.

12.95 Find the least-squares line for $y(t) = t^2$ over the interval $(0, 1)$ using the weight function $1/\sqrt{1 - x^2}$.

▌ The change of argument $t = (x + 1)/2$ converts the interval to $(-1, 1)$ in the argument x and makes $y = \frac{1}{4}(x^2 + 2x + 1)$. If we note first the elementary result

$$\int_{-1}^{1} \frac{x^p}{\sqrt{1 - x^2}}\, dx = \int_0^{\pi} (\cos A)^{\nu}\, dA = \begin{cases} \pi & p = 0 \\ 0 & p = 1 \\ \pi/2 & p = 2 \\ 0 & p = 3 \end{cases}$$

then the coefficient a_0 becomes (see Problem 12.92) $a_0 = \frac{1}{4}(\frac{1}{2} + 0 + 1) = \frac{3}{8}$, and since $y(x)T_1(x)$ is $\frac{1}{4}(x^3 + 2x^2 + x)$, we have $a_1 = \frac{1}{4}(0 + 2 + 0) = \frac{1}{2}$. The least-squares polynomial is therefore

$$\tfrac{3}{8}T_0(x) + \tfrac{1}{2}T_1(x) = \tfrac{3}{8} + \tfrac{1}{2}x$$

There is a second and much briefer path to this result. Using the results in Problem 12.91,

$$y(x) = \tfrac{1}{4}\left(\tfrac{1}{2}T_0 + \tfrac{1}{2}T_2 + 2T_1 + T_0\right) = \tfrac{3}{8}T_0 + \tfrac{1}{2}T_1 + \tfrac{1}{8}T_2$$

Truncating this after the linear terms, we have at once the result just found. Moreover we see that the error is, in the case of this quadratic $y(x)$, precisely the equal ripple function $T_2(x)/8$. This is, of course, a consequence of the series of Chebyshev polynomials terminating with this term. For most functions the error will only be approximately the first omitted term, and therefore only approximately an equal ripple error. Comparing the extreme errors here $(\frac{1}{8} - \frac{1}{8}, \frac{1}{8})$ with those in Problem 12.68, which were $(\frac{1}{6}, -\frac{1}{12}, \frac{1}{6})$, we see that the present approximation sacrifices some accuracy in the center for improved accuracy at the extremes plus the equal ripple feature. Both lines are shown in Fig. 12-12.

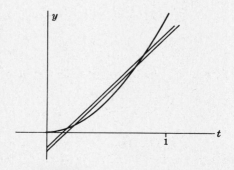

Fig. 12-12

12.96 Find a cubic approximation in terms of Chebyshev polynomials for $y(x) = \sin x$.

❚ The integrals that must be computed to obtain the coefficients of the least-squares polynomial with weight function $w(x) = 1/\sqrt{1 - x^2}$ are too complicated in this case. Instead we will illustrate the process of *economization of polynomials*. Beginning with

$$\sin x \simeq x - \tfrac{1}{6}x^3 + \tfrac{1}{120}x^5$$

we replace the powers of x by their equivalents in terms of Chebyshev polynomials, using Problem 12.91:

$$\sin x \simeq T_1 - \tfrac{1}{24}(3T_1 + T_3) + \tfrac{1}{1920}(10T_1 + 5T_3 + T_5) = \tfrac{169}{192}T_1 - \tfrac{5}{128}T_3 + \tfrac{1}{1920}T_5$$

The coefficients here are not exactly the a_k of Problem 12.92 since higher powers of x from the sine series would make further contributions to the T_1, T_3 and T_5 terms. But those contributions would be relatively small, particularly for the early T_k terms. For example, the x^5 term has altered the T_1 term by less than 1% and the x^7 term would alter it by less than .01%. In contrast the x^5 term has altered the T_3 term by about 6%, though x^7 will contribute only about .02% more. This suggests that truncating our expansion will give us a close approximation to the least-squares cubic. Accordingly we take for our approximation

$$\sin x \simeq \tfrac{169}{192}T_1 - \tfrac{5}{128}T_3 \simeq .9974x - .1562x^3$$

The accuracy of this approximation may be estimated by noting that we have made two "truncation errors," first by using only three terms of the power series for $\sin x$ and second in dropping T_5. Both affect the fourth decimal place. Naturally, greater accuracy is available if we seek a least-squares polynomial of higher degree, but even the one we have has accuracy comparable to that of the fifth degree Taylor polynomial with which we began. The errors of our present cubic and the Taylor cubic, obtained by dropping the x^5 term, are compared in Fig. 12-13. The Taylor cubic is superior near zero but the almost-equal-error property of the (almost) least-squares polynomial is evident and should be compared with $T_5(x)$.

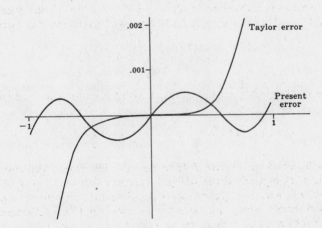

Fig. 12-13

12.97 Prove that for m and n less than or equal to N,

$$\sum_{i=0}^{N-1} T_m(x_i) T_n(x_i) = \begin{cases} 0 & m \neq n \\ N/2 & m = n \neq 0 \\ N & m = n = 0 \end{cases}$$

where $x_i = \cos A_i = \cos[(2i + 1)\pi/2N]$, $i = 0, 1, \dots, N - 1$.

❚ From the trigonometric definition of the Chebyshev polynomials, we find directly

$$\sum_{i=0}^{N-1} T_m(x_i) T_n(x_i) = \sum_{i=0}^{N-1} \cos mA_i \cos nA_i = \tfrac{1}{2} \sum_{i=0}^{N-1} \left[\cos(m + n)A_i + \cos(m - n)A_i \right]$$

Since $\cos ai = (\tfrac{1}{2}\sin\tfrac{1}{2}a)[\Delta \sin a(i - \tfrac{1}{2})]$ both cosine sums may be telescoped. It is simpler, however, to note that except when $m + n$ or $m - n$ is zero each sum vanishes by symmetry, the angles A_i being equally spaced between 0 and π. This already proves the result for $m \neq n$. If $m = n \neq 0$ the second sum contributes $N/2$, while if $m = n = 0$ both sums together total N. It should be noticed that the Chebyshev polynomials are orthogonal under summation as well as under integration. It is often a substantial advantage, since sums are far easier to compute than integrals of complicated functions, particularly when the factor $\sqrt{1 - x^2}$ appears in the latter but not in the former.

12.98 What choice of coefficients a_k will minimize

$$\sum_{x_i} [y(x_i) - a_0 T_0(x_i) - \cdots - a_m T_m(x_i)]^2$$

where the x_i are the arguments of the preceding problem?

∎ With proper identifications it follows directly from Problems 12.19 and 12.20 that the orthogonal projection $p = a_0 T_0 + \cdots + a_m T_m$ determined by

$$a_k = \frac{\sum_i y(x_i) T_k(x_i)}{\sum_i [T_k(x_i)]^2}$$

provides the minimum. Using Problem 12.97 the coefficients are

$$a_0 = \frac{1}{N} \sum_i y(x_i) \qquad a_k = \frac{2}{N} \sum_i y(x_i) T_k(x_i) \qquad k = 1, \ldots, m$$

For $m = N - 1$ we have the collocation polynomial for the N points $(x_i, y(x_i))$ and the minimum sum is zero.

12.99 Find the least-squares line for $y(t) = t^2$ over $(0,1)$ by the method of Problem 12.98.

∎ We have already found a line that minimizes the integral of Problem 12.92. To minimize the sum of Problem 12.98, choose $t = (x + 1)/2$ as before. Suppose we use only two points, so that $N = 2$. These points will have to be $x_0 = \cos \pi/4 = 1/\sqrt{2}$ and $x_1 = \cos 3\pi/4 = -1\sqrt{2}$. Then

$$a_0 = \tfrac{1}{2} \left[\tfrac{1}{8}(3 + 2\sqrt{2}) + \tfrac{1}{8}(3 - 2\sqrt{2}) \right] = \tfrac{3}{8}$$
$$a_1 = \tfrac{1}{8}(3 + 2\sqrt{2})(1/\sqrt{2}) + \tfrac{1}{8}(3 - 2\sqrt{2})(-1/\sqrt{2}) = \tfrac{1}{2}$$

and the line is given by $p(x) = \tfrac{3}{8}T_0 + \tfrac{1}{2}T_1 = \tfrac{3}{8} + \tfrac{1}{2}x$. This is the same line as before, and using a larger N would reproduce it again. The explanation of this is simply that y itself can be represented in the form $y = a_0 T_0 + a_1 T_1 + a_2 T_2$ and, since the T_k are orthogonal relative to both integration and summation, the least-squares line in either sense is also available by *truncation*. (See the last paragraph of Problem 12.20.)

12.100 Find least-squares lines for $y(x) = x^3$ over $(-1, 1)$ by minimizing the sum of Problem 12.98.

∎ In this problem the line we get will depend somewhat on the number of points we use. First take $N = 2$, which means that we use $x_0 = -x_1 = 1/\sqrt{2}$ as before. Then

$$a_0 = \tfrac{1}{2}(x_0^3 + x_1^3) = 0 \qquad a_1 = x_0^4 + x_1^4 = \tfrac{1}{2}$$

Choosing $N = 3$ we find $x_0 = \sqrt{3}/2$, $x_1 = 0$ and $x_2 = -\sqrt{3}/2$. This makes

$$a_0 = \tfrac{1}{3}(x_0^3 + x_1^3 + x_2^3) = 0 \qquad a_1 = \tfrac{2}{3}(x_0^4 + x_1^4 + x_2^4) = \tfrac{3}{4}$$

Taking the general case of N points, we have $x_i = \cos A_i$ and

$$a_0 = \frac{1}{N} \sum_{i=0}^{N-1} \cos^3 A_i = 0$$

by the symmetry of the A_i in the first and second quadrants. Also,

$$a_1 = \frac{2}{N} \sum_{i=0}^{N-1} \cos^4 A_i = \frac{2}{N} \sum_{i=0}^{N-1} \left(\frac{3}{8} + \frac{1}{2}\cos 2A_i + \frac{1}{8}\cos 4A_i \right)$$

Since the A_i are the angles $\pi/2N, 3\pi/2N, \ldots, (2N-1)\pi/2N$, the doubled angles are $\pi/N, 3\pi/N, \ldots, (2N-1)\pi/N$ and these are symmetrically spaced around the entire circle. The sum of the $\cos 2A_i$ is therefore zero. Except when $N = 2$, the sum of the $\cos 4A_i$ will also be zero so that $a_1 = \tfrac{3}{4}$ for $N > 2$. For N tending to infinity we thus have trivial convergence to the line $p(x) = 3T_1/4 = 3x/4$.

If we adopt the minimal integral approach, then we find

$$a_0 = \frac{1}{\pi} \int_{-1}^{1} \frac{x^3}{\sqrt{1-x^2}} \, dx = 0 \qquad a_1 = \frac{2}{\pi} \int_{-1}^{1} \frac{x^4}{\sqrt{1-x^2}} \, dx = \frac{3}{4}$$

which leads to the same line.

The present example may serve as further elementary illustration of the Problem 12.98 algorithm, but the result is more easily found and understood by noting that $y = x^3 = \frac{3}{4} T_1 + \frac{1}{4} T_3$ and once again appealing to the corollary in Problem 12.20 to obtain $3T_1/4$ or $3x/4$ by truncation. The truncation process fails for $N = 2$ since then the polynomials T_0, T_1, T_2, T_3 are not orthogonal. (See Problem 12.97.)

12.101 Find least-squares lines for $y(x) = |x|$ over $(-1, 1)$ by minimizing the sum of Problem 12.98.

▮ With $N = 2$ we quickly find $a_0 = 1/\sqrt{2}$ and $a_1 = 0$. With $N = 3$ the results $a_0 = 1/\sqrt{3}$ and $a_1 = 0$ are just as easy. For arbitrary N,

$$a_0 = \frac{1}{N} \sum_{i=0}^{N-1} |\cos A_i| = \frac{2}{N} \sum_{i=0}^{I} \cos A_i$$

where I is $(N-3)/2$ for odd N and $(N-2)/2$ for even N. This trigonometric sum may be evaluated by telescoping or otherwise, with the result

$$a_0 = \frac{\sin[\pi(I+1)/N]}{N \sin(\pi/2N)}$$

It is a further consequence of symmetry that $a_1 = 0$ for all N. For N tending to infinity it now follows that

$$\lim a_0 = \lim \frac{1}{N \sin \pi/2N} = \frac{2}{\pi}$$

As more and more points are used, the limiting line is approached. Turning to the minimum integral approach, we of course anticipate this same line. The computation produces

$$a_0 = \frac{1}{\pi} \int_{-1}^{1} \frac{|x|}{\sqrt{1-x^2}} \, dx = \frac{2}{\pi}$$

$$a_1 = \frac{2}{\pi} \int_{-1}^{1} \frac{x|x|}{\sqrt{1-x^2}} \, dx = 0$$

and so we are not disappointed. The limiting line is the solid line in Fig. 12.14. For more details and a continuation to higher-degree polynomials, see Problem 12.119.

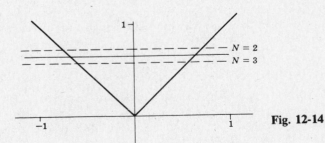

Fig. 12-14

12.102 Apply the method of the previous problems to the experimentally produced curve of Fig. 12-8.

▮ For such a function, of unknown analytic character, any of our methods must involve discretization at some point. We have already chosen one discrete set of values of the function for use in Simpson's rule, thus maintaining at least in spirit the idea of minimizing an integral. We could have used the same equidistant set of arguments and minimized a sum. With the idea of obtaining a more nearly equal ripple error, however, we now choose the arguments $x_i = \cos A_i = 2t_i - 1$ instead. With 11 points, the number used earlier, the arguments, $x_i = \cos A_i = \cos[(2i+1)\pi/22]$ and corresponding t_i as well as y_i values read from the curve are

x_i	.99	.91	.75	.54	.28	.00	$-.28$	$-.54$	$-.75$	$-.91$	$-.99$
t_i	1.00	.96	.88	.77	.64	.50	.36	.23	.12	.04	.00
y_i	.36	.33	.28	.24	21	.25	.20	.12	.17	.13	.10

The coefficients become

$$a_0 = \tfrac{1}{11}\sum y_i \simeq .22 \qquad a_1 = \tfrac{2}{11}\sum x_i y_i \simeq .11$$

making the line $p(x) = .22 + .11x = .22t + .11$, which is almost indistinguishable from the earlier result. The data inaccuracies have not warranted the extra sophistication.

12.103 Show that the Chebyshev series for arcsin x is

$$\arcsin x = \frac{4}{\pi}\left(T_1 + \tfrac{1}{9}T_3 + \tfrac{1}{25}T_5 + \tfrac{1}{49}T_7 + \cdots\right)$$

by evaluating the coefficient integrals directly. Truncate after T_3 to obtain the least-squares cubic for this function. Compute the actual error of this cubic and compare with the first omitted term (the T_5 term). Notice the (almost) equal ripple behavior of the error.

∎ With $T_k(x) = \cos(kA)$ and $\cos A = x$, it follows that arcsin x will be $\pi/2 - A$. The coefficients can now be evaluated. First

$$a_0 = \frac{1}{\pi}\int_\pi^0 \frac{\pi/2 - A}{\sin A}(-\sin A)\,dA = \frac{1}{\pi}\left(\frac{\pi}{2}A - \frac{1}{2}A^2\right)\Big|_0^\pi = 0$$

a consequence of the fact that arcsin x is an odd function. Then

$$a_k = \frac{2}{\pi}\int_\pi^0 \frac{(\pi/2 - A)\cos(kA)}{\sin A}(-\sin A)\,dA = -\frac{2}{\pi}\int_0^\pi A\cos(kA)\,dA$$

since the cosine integrates to zero, after which

$$a_k = \frac{-2}{\pi k^2}\left[\cos(kA) + kA\sin(kA)\right]\Big|_0^\pi = \frac{2}{\pi k^2}[1 - \cos(\pi k)]$$

for $k = 1, 2, 3, \ldots$. The series begins

$$\arcsin x = \frac{4}{\pi}T_1 + 0\cdot T_2 + \frac{4}{9\pi}T_3 + \cdots$$

and continues as given.

The least-squares cubic is found by truncation and consists of the two nonzero terms just exhibited. While the values of the first omitted term are close to $T_5(x)/20$, and so swing between $-.05$ and $.05$, the error of our cubic swings as

$$-.16 \qquad .06 \qquad -.03 \qquad .03 \qquad -.06 \qquad .16$$

Too much has been truncated to get a true equal error behavior.

12.104 Find the least-squares line for $y(x) = x^2$ on $(-1, 1)$ with weight function $w(x) = 1/\sqrt{1 - x^2}$. Compare it with the line found in Problem 12.72.

∎ Since $x^2 = \frac{1}{2}(T_0 + T_2)$, we get the least-squares line simply by truncating to $\frac{1}{2}T_0 = \frac{1}{2}$. The error made is $\frac{1}{2}T_2$, which swings from $\frac{1}{2}$ to $-\frac{1}{2}$ and back to $\frac{1}{2}$. In the earlier problem we found the line $y = \frac{1}{3}$ instead. An alternative to the procedure used there is to express x^2 in terms of Legendre polynomials

$$x^2 = \tfrac{1}{3}P_0 + \tfrac{2}{3}P_2$$

and truncate the P_2 term. The error made is, of course, $\frac{2}{3}P_2$ and does not have the equal error property.

12.105 Find the least-squares parabola for $y(x) = x^3$ on the interval $(-1, 1)$ using weight function $w(x) = 1/\sqrt{1 - x^2}$. Compare it with the one found in Problem 12.73.

▮ Since $x^3 = \frac{3}{4}T_1 + \frac{1}{4}T_3$, we get the least-squares parabola by truncating the T_3 term and keeping $\frac{3}{4}T_1$. The error made is $\frac{1}{4}T_3$, which swings between $-\frac{1}{4}$ and $\frac{1}{4}$. In the earlier problem we found a different degenerate parabola and can find it again by writing

$$x^3 = \tfrac{3}{5}P_1 + \tfrac{2}{5}P_3$$

and truncating the P_3 term. However, it is the new approximation that has the equal error property.

12.106 Represent $y(x) = e^{-x}$ by terms of its power series through x^7, the maximum error being in the fifth decimal place for x in $(-1, 1)$. Rearrange the sum into a combination of Chebyshev polynomials and determine how many terms could be dropped without affecting the fourth decimal place.

▮ The required terms of the Taylor series are

$$1 - x + \tfrac{1}{2}x^2 - \tfrac{1}{6}x^3 + \tfrac{1}{24}x^4 - \tfrac{1}{120}x^5 + \tfrac{1}{720}x^6 - \tfrac{1}{5040}x^7$$

and rearrange, using the representations of x^p listed in Problem 12.91, as

$$\left(1 + \frac{1}{4} + \frac{1}{64} + \frac{1}{72 \cdot 32}\right)T_0 - \left(1 + \frac{1}{8} + \frac{1}{192} + \frac{35}{64 \cdot 5040}\right)T_1 + \left(\frac{1}{4} + \frac{1}{48} + \frac{1}{48 \cdot 32}\right)T_2$$

$$- \left(\frac{1}{24} + \frac{1}{384} + \frac{21}{64 \cdot 5040}\right)T_3 + \left(\frac{1}{192} + \frac{1}{32 \cdot 120}\right)T_4$$

$$- \left(\frac{1}{16 \cdot 120} + \frac{7}{64 \cdot 5040}\right)T_5 + \frac{1}{32 \cdot 720}T_6 - \frac{1}{64 \cdot 5040}T_7$$

$$= 1.26606 T_0 - 1.13032 T_1 + .27148 T_2 - .044336 T_3$$

$$+ .005469 T_4 - .000543 T_5 + .000043 T_6 - .000003 T_7$$

Since the Chebyshev polynomials do not exceed 1 in magnitude, it seems that the last two terms can be truncated without doing much damage to the fourth decimal place. The result would be a fifth-degree polynomial approximation to the exponential. It is worth noting that the fifth-degree Taylor polynomial does not give this same accuracy over the entire interval $(-1, 1)$.

12.107 Show that for $y(x) = T_n(x) = \cos(n \arccos x) = \cos(nA)$ it follows that $y'(x) = (n \sin nA)/\sin A$. Then show that

$$(1 - x^2)y'' - xy' + n^2 y = 0$$

which is the classical differential equation of the Chebyshev polynomials.

▮ Differentiating relative to A and then multiplying by the derivative of A relative to x, which is $-1/\sin A$, we have

$$y'(x) = (-n \sin nA)(-1/\sin A) = (n \sin nA)/\sin A$$

as predicted. Differentiating once more

$$y''(x) = \frac{n^2 \sin A \cos(nA) - n \cos A \sin(nA)}{\sin^2 A} \cdot \frac{-1}{\sin A}$$

Putting the pieces together and using $1 - x^2 = \sin^2 A$,

$$(1 - x^2)y'' - xy' + n^2 y = \frac{n \sin(nA)\cos A - n^2 \sin A \cos(nA)}{\sin A}$$

$$+ \frac{n^2 \sin A \cos(nA) - n \sin(nA)\cos A}{\sin A}$$

which is clearly zero.

12.108 Show that $S_n(x) = \sin(n \arccos x)$ also satisfies the differential equation of the preceding problem.

▌ The proof is very similar. First $y' = -n\cos(nA)/\sin A$ and then

$$y''(x) = \frac{n^2 \sin A \sin(nA) + n\cos(nA)\cos A}{\sin^2 A} \cdot \frac{-1}{\sin A}$$

Observing that

$$-xy' + n^2 y = \frac{n\cos(nA)\cos A + n^2 \sin A \sin(nA)}{\sin A}$$

and again recalling $1 - x^2 = \sin^2 A$, the last two expressions will be seen to add to zero.

12.109 Let $U_n(x) = S_n(x)/\sqrt{1 - x^2}$ and prove the recursion

$$U_{n+1}(x) = 2xU_n(x) - U_{n-1}(x)$$

▌ Referring to the preceding problem, we find

$$U_{n+1}(x) = \frac{\sin(n+1)A}{\sin A} \qquad U_{n-1}(x) = \frac{\sin(n-1)A}{\sin A}$$

so adding the two and doing a bit of trigonometry,

$$U_{n+1}(x) + U_{n-1}(x) = \frac{2\sin(nA)\cos A}{\sin A} = 2xU_n(x)$$

as stated. Since by the definition $U_0(x) = 0$ and $U_1(x) = 1$, we can use this recursion to write out a few of the polynomials:

$$U_2 = 2x$$
$$U_3 = 4x^2 - 1$$
$$U_4 = 8x^3 - 2x - 2x = 8x^3 - 4x$$
$$U_5 = 16x^4 - 8x^2 - 4x^2 + 1 = 16x^4 - 12x^2 + 1$$
$$U_6 = 32x^5 - 24x^3 + 2x - 8x^3 + 4x = 32x^5 - 32x^3 + 6x$$
$$U_7 = 64x^6 - 64x^4 + 12x^2 - 16x^4 + 12x^2 - 1$$
$$= 64x^6 - 80x^4 + 24x^2 - 1$$

12.110 Show that $T_{m+n}(x) + T_{m-n}(x) = 2T_m(x)T_n(x)$ and then put $m = n$ to obtain $T_{2n}(x) = 2T_n^2(x) - 1$. Use this recursion to find T_2, T_4 and T_8 from T_1.

▌ A direct trigonometric proof is convenient:

$$T_{m+n} = \cos(m+n)A = \cos(mA)\cos(nA) - \sin(mA)\sin(nA)$$
$$T_{m-n} = \cos(m-n)A = \cos(mA)\cos(nA) + \sin(mA)\sin(nA)$$

Adding these,

$$T_{m+n} + T_{m-n} = 2\cos(mA)\cos(nA) = 2T_m T_n$$

For $m = n$ this becomes $T_{2n} + T_0 = 2T_n^2$ as expected.
Starting from $T_1 = x$ this recursion produces $T_2 = 2x^2 - 1$ and then

$$T_4 = 2(2x^2 - 1)^2 - 1 = 8x^4 - 8x^2 + 1$$
$$T_8 = 2(8x^4 - 8x^2 + 1)^2 - 1 = 128x^8 - 256x^6 + 160x^4 - 32x^2 + 1$$

and so on.

12.111 Prove $(1/n)T_n' = 1/(n-2)T_{n-2}' + 2T_{n-1}$ and then deduce

$$T_{2n}' = 2(2n)(T_{2n-1} + T_{2n-3} + \cdots + T_1)$$
$$T_{2n+1}' = 2(2n+1)(T_{2n} + T_{2n-2} + \cdots + T_2) + (2n+1)T_0$$

$\blacksquare$ Since $T_n'(x) = n\sin(nA)/\sin A$ was computed in Problem 12.107, we have $T_n'/n = \sin(nA)\sin A$. Then

$$\frac{T_n'}{n} - \frac{T_{n-2}'}{n-2} = \frac{\sin(nA) - \sin(n-2)A}{\sin A} = \frac{2\sin\frac{1}{2}(2A)\cos\frac{1}{2}(2nA - 2A)}{\sin A}$$
$$= 2\cos(n-1)A = 2T_{n-1}$$

and the recursion is in hand. Rewriting it as

$$T_n' = 2nT_{n-1} + \frac{n}{n-2}T_{n-2}'$$

it generates in succession,

$$T_3' = 6T_2 + 3T_0,\, T_4' = 8T_3 + 2T_2' = 8T_3 + 8T_1$$
$$T_5' = 2(5)T_4 + \tfrac{5}{3}T_3' = 2(5)T_4 + \tfrac{5}{3}\left[2(3)T_2 + 3T_0\right]$$
$$= 2(5)(T_4 + T_2) + 5T_0$$
$$T_6' = 2(6)T_5 + \tfrac{3}{2}T_4' = 2(6)T_5 + \tfrac{3}{2}\left[2(4)T_3 + \tfrac{4}{2}T_2'\right]$$
$$= 2(6)T_5 + 2(6)T_3 + 2(6)T_1$$

and so on.

12.112 Show that $T_{2n} = 2x(T_{2n-1} - T_{2n-3} + \cdots \pm T_1) \mp T_0$.

$\blacksquare$ For $n = 1$ it claims that $T_2 = 2xT_1 - T_0$, which we know to be true. For $n = 2$ it becomes

$$T_4 = 2x(T_3 - T_1) + T_0 = 2x(4x^3 - 4x) + 1 = 8x^4 - 8x^2 + 1$$

which is also true. Assuming it true for $n = k$,

$$T_{2k} = 2x(T_{2k-1} - T_{2k-3} + \cdots \pm T_1) \mp T_0$$

we apply the basic recursion of these polynomials in the form $T_{2k+2} = 2xT_{2k+1} - T_{2k}$ obtaining

$$T_{2k+2} = 2xT_{2k+1} - 2x(T_{2k-1} - T_{2k-3} + \cdots \mp T_1) \pm T_0$$

and the induction is complete.

12.113 Also prove that $T_{2n+1} = 2x(T_{2n} - T_{2n-2} + T_{2n-4} - \cdots \pm T_2) \mp xT_0$.

$\blacksquare$ For $n = 1$ it claims that $T_3 = 2xT_2 - xT_0$, which is easily checked. Assume it true for $n = k - 1$:

$$T_{2k-1} = 2x(T_{2k-2} - T_{2k-4} + \cdots \pm T_2) \mp xT_0$$

Applying the basic recursion in the form $T_{2k+1} = 2xT_{2k} - T_{2k-1}$, we have

$$T_{2k+1} = 2xT_{2k} - 2x(T_{2k-2} - T_{2k-4} + \cdots \pm T_2) \mp xT_0$$

which is our theorem for $n = k$ and completes the induction.

12.114 Economize the polynomial $x - \frac{1}{2}x^2 + \frac{1}{3}x^3 - \frac{1}{4}x^4 + \frac{1}{5}x^5$, which is the fifth-degree Taylor polynomial for $\log(1 + x)$, to second degree by rearrangement into a combination of Chebyshev polynomials. Show that the result has about the same accuracy as the fourth-degree Taylor polynomial.

$\blacksquare$ The rearrangement produces

$$T_1 - \tfrac{1}{4}(T_0 + T_2) + \tfrac{1}{12}(3T_1 + T_3) - \tfrac{1}{32}(3T_0 + 4T_2 + T_4) + \tfrac{1}{80}(10T_1 + 5T_3 + T_5)$$
$$= -\tfrac{11}{32}T_0 + \tfrac{11}{8}T_1 - \tfrac{3}{8}T_2 + \tfrac{7}{48}T_3 - \tfrac{1}{32}T_4 + \tfrac{1}{80}T_5$$

so extracting the first three terms we have

$$-\tfrac{11}{32} + \tfrac{11}{8}x - \tfrac{3}{8}(2x^2 - 1) = \tfrac{1}{32} + \tfrac{11}{8}x - \tfrac{3}{4}x^2$$

as the result of economization. Its error will not exceed the sum of the truncated coefficients, $\tfrac{7}{48} + \tfrac{1}{32} + \tfrac{1}{80}$, which is just under .2. The error of the fourth-degree Taylor polynomial is not greater than $\tfrac{1}{5}$, by the theorem on alternating series, so we have about the same minimum accuracy over the interval $(-1, 1)$ where the series converges. The level of accuracy is, of course, useless, but the underlying idea of economization is illustrated.

12.115　Economize the polynomial $1 + x + \tfrac{1}{2}x^2 + \tfrac{1}{6}x^3 + \tfrac{1}{24}x^4$ to a linear function. Compare this with the second-degree part of the given polynomial, considering both as approximations to e^x.

　∎　Rearranged as a combination of Chebyshev polynomials, the given function is

$$T_0 + T_1 + \tfrac{1}{4}(T_0 + T_2) + \tfrac{1}{24}(3T_1 + T_3) + \tfrac{1}{192}(3T_0 + 4T_2 + T_4)$$
$$= \tfrac{81}{64}T_0 + \tfrac{9}{8}T_1 + \tfrac{13}{48}T_2 + \tfrac{1}{24}T_3 + \tfrac{1}{192}T_4$$

and the linear portion is apparent. The maximum error of this approximation is the sum of the other coefficients and works out to about .3. The first three terms as given, $1 + x + \tfrac{1}{2}x^2$, viewed as a quadratic approximation to e^x, have a maximum error on the interval $(-1, 1)$ of $e/6$ or about .45. Needless to say, the Taylor polynomial is excellent near $x = 0$, but the idea of economization is to produce a more uniform error. Incidentally, the error of the original fifth-degree polynomial does not exceed $e/120$ on the same interval and so is not a factor for present purposes.

12.116　The change of argument $x = 2t - 1$ converts the interval $(-1, 1)$ to $(0, 1)$ in terms of t. How does it modify the Chebyshev polynomials and how does it affect the basic recursion?

　∎　We find

$$T_0^*(t) = 1 \qquad T_1^*(t) = 2t - 1 \qquad T_2^*(t) = 2(2t - 1)^2 - 1 = 8t^2 - 8t + 1$$
$$T_3^*(t) = 4(2t - 1)^3 - 3(2t - 1) = 32t^3 - 48t^2 + 18t - 1$$

and so on. The recursion $T_{n+1}(x) = 2xT_n(x) - T_{n-1}(x)$ translates immediately:

$$T_{n+1}^*(t) = 2(2t - 1)T_n^*(t) - T_{n-1}^*(t)$$

12.117　It is clear that $\int T_0(x)\,dx = T_1(x)$ and $\int T_1(x)\,dx = \tfrac{1}{4}T_2(x)$. Find an indefinite integral of $T_n(x)$ for $n > 1$.

　∎　Rewriting the result of Problem 12.111 as

$$2T_n = \frac{1}{n+1}T_{n+1}' - \frac{1}{n-1}T_{n-1}'$$

the integral is almost in hand. We find

$$\int T_n(x)\,dx = \frac{1}{2}\left[\frac{1}{n+1}T_{n+1} - \frac{1}{n-1}T_{n-1}\right]$$

12.118　Use the method of Problem 12.98 to find a least-squares parabola for $y(x) = x^3$ on $(-1, 1)$. Choose $N = 3$. Why would the same results be found for larger N?

　∎　The three points involved were found in Problem 12.100 to be 0 and $\pm\tfrac{1}{2}\sqrt{3}$. The coefficients $a_0 = 0$ and $a_1 = \tfrac{3}{4}$ will be just as before. As for a_2, we can compute directly

$$a_2 = \frac{\Sigma x_i^3(2x_i^2 - 1)}{\Sigma(2x_i^2 - 1)} = \frac{(3\sqrt{3}/8)(\tfrac{1}{2}) + (-3\sqrt{3}/8)(\tfrac{1}{2})}{\tfrac{1}{4} + \tfrac{1}{4}} = 0$$

or simply note that T_2 is an even function while x^3 is odd, so the numerator sum for symmetric arguments will have to be zero. The quadratic is once again a degenerate, the line $y = \tfrac{3}{4}x$ found earlier in Problem 12.100. For larger N the representation

$$x^3 = \tfrac{3}{4}T_1 + \tfrac{1}{4}T_3$$

can simply be truncated, since T_1 and T_3 are orthogonal over such point sets.

12.119 Apply the method of Problem 12.98 to the function $y(x) = |x|$ on the interval $(-1, 1)$ for arbitrary N. Evaluate the coefficients directly and determine their behavior as N becomes infinite.

▮ Take the case N odd, only minor adjustments being needed to accommodate N even. The angles $A_i = (2i + 1)\pi/2N$ will then run from $\theta = \pi/2N$, in steps of size 2θ, up to $\pi - \theta$, with the angle $\pi/2$ occurring for $i = (N-1)/2$. Noting that the function $y(x)$ is even, we see that coefficients of odd $T_k(x)$ will all be zero and that for the rest it will be possible to limit the summations to first quadrant angles only, doubling the resulting sums. This means stopping at $i = (N-3)/2$ (see Fig. 12-15).

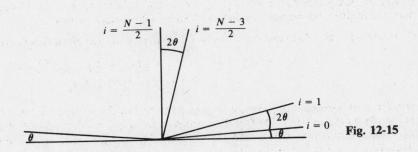

Fig. 12-15

After this preliminary we now find

$$a_0 = \frac{1}{N} \sum |\cos A_i| = \frac{2}{N} \sum_{i=0}^{i=(N-3)/2} \cos A_i$$

$$= \frac{2}{N} [\cos \theta + \cos(3\theta) + \cdots + \cos(N-2)\theta]$$

But a finite integral is available in the form

$$\Delta^{-1} \cos(2i+1)\theta = \sin(2i\theta)/2\sin\theta$$

so the sum is evaluated as

$$a_0 = \frac{2}{N} \frac{\sin(2i\theta)}{2\sin\theta} \bigg|_{i=0}^{i=(N-1)/2} = \frac{\sin(N-1)\theta}{N\sin\theta}$$

As N becomes infinite this approaches $2/\pi$, since $N\theta = \pi/2$.

Turning to the a_k for even k, with $T_k(x) = \cos(kA)$ and using the symmetry,

$$a_k = \frac{4}{N}[\cos\theta\cos(k\theta) + \cdots + \cos(N-2)\theta\cos k(N-2)\theta]$$

$$= \frac{2}{N} \sum_{i=0}^{i=(N-3)/2} [\cos(k+1)(2i+1)\theta + \cos(k-1)(2i+1)\theta]$$

by the cosine product identity. We can now use the preceding finite integral again, replacing θ by $(k+1)\theta$ and $(k-1)\theta$ in turn:

$$a_k = \frac{2}{N}\left[\frac{\sin 2i(k+1)\theta}{2\sin(k+1)\theta} + \frac{\sin 2i(k-1)\theta}{2\sin(k-1)\theta}\right]_{i=0}^{i=(N-1)/2}$$

$$= \frac{1}{N}\left[\frac{\sin(N-1)(k+1)\theta}{\sin(k+1)\theta} + \frac{\sin(N-1)(k-1)\theta}{\sin(k-1)\theta}\right]$$

and the coefficients are in hand. For a least-squares parabola we put $k = 2$ and have

$$a_2 = \frac{1}{N}\left[\frac{\sin(N-1)3\theta}{\sin 3\theta} + \frac{\sin(N-1)\theta}{\sin\theta}\right]$$

leading to $a_0 T_0 + a_2 T_2$.

As N becomes infinite the a_k will approach

$$\frac{\sin \pi/2(k+1)}{\pi/2(k+1)} + \frac{\sin \pi/2(k-1)}{\pi/2(k-1)}$$

because $\theta = \pi/2N$.

12.120 Also apply the minimum integral method of Problem 12.92 to the function $y(x) = |x|$. Are the coefficients the limiting values of those just found?

▮ We now have

$$a_0 = \frac{1}{\pi} \int_{-1}^{1} \frac{y(x)}{\sqrt{1-x^2}} \, dx = \frac{1}{\pi} \int_0^\pi |\cos A| \, dA$$

$$= \frac{2}{\pi} \int_0^{\pi/2} \cos A \, dA = \frac{2}{\pi}$$

and one limiting value has been reproduced. Then

$$a_k = \frac{2}{\pi} \int_{-1}^{1} \frac{y(x) T_k(x)}{\sqrt{1-x^2}} \, dx = \frac{2}{\pi} \int_0^\pi |\cos A| \cos(kA) \, dA$$

$$= \frac{4}{\pi} \int_0^{\pi/2} \cos A \cos(kA) \, dA = \frac{2}{\pi} \int_0^{\pi/2} [\cos(k+1)A + \cos(k-1)A] \, dA$$

$$= \frac{2}{\pi} \left[\frac{\sin(k+1)A}{k+1} + \frac{\sin(k-1)A}{k-1} \right]_{A=0}^{A=\pi/2}$$

$$= \frac{\sin(k+1)\pi/2}{(k+1)\pi/2} + \frac{\sin(k-1)\pi/2}{(k-1)\pi/2}$$

and the other limiting values have been reproduced.

12.121 Given the luxury of choosing the points at which experimental data are to be collected, the possibility of selecting $x_i = \cos A_i$ as in the preceding problems may be worth considering. The benefits of almost equal error approximation are then obtained without having to face integrals involving $1/\sqrt{1-x^2}$. Suppose this thought had led to the collection of the following data [actually computed from $y(x) = \sqrt{1+x}$ for $A_i = (2i+1)\pi/2N$ and $N = 11$]. Find second- and fourth-degree least-squares approximations by the indicated method. Compare with the correct values.

▮ This time there is no symmetry and the sums run from $\theta = \pi/22$ to $\pi - \theta$. The data points $x_i = \cos(2i+1)\theta$ take values from .99 to $-.99$, some of them listed in Table 12.5. The first five coefficients were computed, providing the terms

$$.901T_0 + .599T_1 - .118T_2 + .0498T_3 - .00269T_4$$

The sums of the first three and of all five are given, along with the correct values to two places, in the table. Also provided for comparison are values of the Taylor polynomials of degrees 2 and 4 in x. The latter appears very slightly better over most of the interval, but falls down badly at the left end.

TABLE 12.5

x	y	Least Square 3 Terms	Least Square 5 Terms	Taylor 3 Terms	Taylor 5 Terms
.99	1.41	1.38	1.40	1.37	1.40
.76	1.33	1.34	1.34	1.31	1.33
.28	1.13	1.17	1.12	1.13	1.13
−.28	.85	.83	.86	.85	.85
−.76	.49	.43	.48	.55	.51
−.99	.10	.19	.13	.38	.28

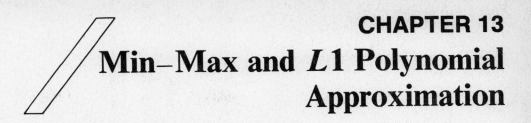

CHAPTER 13
Min–Max and $L1$ Polynomial Approximation

13.1 What is the central idea of min–max approximation?

▌ A given function $y(x)$ is approximated by the member $p(x)$ of a given set of functions, say polynomials of degree n or less, for which the maximum absolute error $|y(x) - p(x)|$ over the data set involved is a minimum. For example, if the data are discrete (x_i, y_i) pairs and the errors are denoted $h_i = y_i - p(x_i)$, with H the largest absolute h_i, then it is H that is to be minimized. Min–max approximations are also known as Chebyshev approximations since it was Chebyshev who proved the first existence and uniqueness theorems in this area. By its definition a min–max approximation guarantees an absolute error no larger than H over the entire data set. For this reason it is also called a uniform approximation. In the language of norms, the central idea of min–max approximation is to minimize the infinity norm

$$\| y(x) - p(x) \|_\infty$$

13.2 What is the equal error property?

▌ Examples have already been offered, in the approximation of polynomials of degree n by polynomials of degree $n - 1$. In Problem 12.95 we found a least-squares line for a quadratic polynomial, with error $\frac{1}{8} T_2(x)$, which "ripples" back and forth between equal extremes of $\pm \frac{1}{8}$ which it reaches three times over the interval $(-1, 1)$. In Problem 12.105 a least-squares (degenerate) quadratic approximation to x^3 was found, with error $\frac{1}{4} T_3$, which reaches its extremes of $\pm \frac{1}{4}$ four times on $(-1, 1)$. The equal error property is the identifying feature of min–max approximations. In the case of nth-degree polynomials, the maximum absolute error will be reached at least $n + 2$ times over the active interval.

13.1 DISCRETE DATA: THE MIN–MAX LINE

13.3 Show that for any three points (x_i, Y_i) with the arguments x_i distinct, there is exactly one straight line that misses all three points by equal amounts and with alternating signs. This is the *equal error* line or Chebyshev line.

▌ Let $y(x) = Mx + B$ represent an arbitrary line and let $h_i = y(x_i) - Y_i = y_i - Y_i$ be the "errors" at the three data points. An easy calculation shows that, since $y_i = Mx_i + B$, for any straight line at all

$$(x_3 - x_2) y_1 - (x_3 - x_1) y_2 + (x_2 - x_1) y_3 = 0$$

Defining $\beta_1 = x_3 - x_2$, $\beta_2 = x_3 - x_1$ and $\beta_3 = x_2 - x_1$, the preceding equation becomes

$$\beta_1 y_1 - \beta_2 y_2 + \beta_3 y_3 = 0$$

We may take it that $x_1 < x_2 < x_3$ so that the three βs are positive numbers. We are to prove that there is one line for which

$$h_1 = h \qquad h_2 = -h \qquad h_3 = h$$

making the three errors of equal size and alternating sign. (This is what will be meant by an equal error line.) Now, if a line having this property does exist, then

$$y_1 = Y_1 + h \qquad y_2 = Y_2 - h \qquad y_3 = Y_3 + h$$

and substituting above,

$$\beta_1(Y_1 + h) - \beta_2(Y_2 - h) + \beta_3(Y_3 + h) = 0$$

Solving for h,

$$h = -\frac{\beta_1 Y_1 - \beta_2 Y_2 + \beta_3 Y_3}{\beta_1 + \beta_2 + \beta_3}$$

This already proves that at most one equal error line can exist and that it must pass through the three points $(x_1, Y_1 + h)$, $(x_2, Y_2 - h)$ and $(x_3, Y_3 + h)$ for the value h just computed. Though normally one asks a line to pass through only two designated points, it is easy to see that in this special case the three points do fall on a line. The slopes of $P_1 P_2$ and $P_2 P_3$ (where P_1, P_2, P_3 are the three points taken from left to right) are

$$\frac{Y_2 - Y_1 - 2h}{x_2 - x_1} \quad \text{and} \quad \frac{Y_3 - Y_2 + 2h}{x_3 - x_2}$$

and using our earlier equations these are easily proved to be the same. So there is exactly one equal error, or Chebyshev, line.

13.4 Find the equal error line for the data points $(0,0)$, $(1,0)$ and $(2,1)$.

∥ First we find $\beta_1 = 2 - 1 = 1$, $\beta_2 = 2 - 0 = 2$ and $\beta_3 = 1 - 0 = 1$, and then compute

$$h = -\frac{(1)(0) - (2)(0) + (1)(1)}{1 + 2 + 1} = -\frac{1}{4}$$

The line passes through $(0, -\frac{1}{4})$, $(1, \frac{1}{4})$ and $(2, \frac{3}{4})$ and so has the equation $y(x) = \frac{1}{2}x - \frac{1}{4}$. The line and points appear in Fig. 13-1.

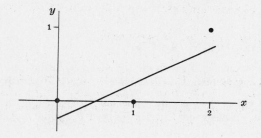

Fig. 13-1

13.5 Show that the equal error line is also the min–max line for the three points (x_i, Y_i).

∥ The errors of the equal error line are $h, -h, h$. Let h_1, h_2, h_3 be the errors for any other line. Also let H be the largest of $|h_1|, |h_2|, |h_3|$. Then using our earlier formulas,

$$h = -\frac{\beta_1 Y_1 - \beta_2 Y_2 + \beta_3 Y_3}{\beta_1 + \beta_2 + \beta_3} = -\frac{\beta_1(y_1 - h_1) - \beta_2(y_2 - h_2) + \beta_3(y_3 - h_3)}{\beta_1 + \beta_2 + \beta_3}$$

where y_1, y_2, y_3 here refer to the "any other line." This rearranges to

$$h = -\frac{(\beta_1 y_1 - \beta_2 y_2 + \beta_3 y_3) - (\beta_1 h_1 - \beta_2 h_2 + \beta_3 h_3)}{\beta_1 + \beta_2 + \beta_3}$$

and the first term being zero we have a relationship between the h of the equal error line and the h_1, h_2, h_3 of the other line,

$$h = \frac{\beta_1 h_1 - \beta_2 h_2 + \beta_3 h_3}{\beta_1 + \beta_2 + \beta_3}$$

Since the βs are positive, the right side of this equation will surely be increased if we replace h_1, h_2, h_3 by $H, -H, H$, respectively. Thus $|h| \leq H$ and the maximum error size of the Chebyshev line, which is $|h|$, comes out no greater than that of any other line.

13.6 Show that no other line can have the same maximum error as the Chebyshev line, so that the min–max line is unique.

▌ Suppose equality holds in our last result, $|h| = H$. This means that the substitution of $H, -H, H$ that produced this result has not actually increased the size of $\beta_1 h_1 - \beta_2 h_2 + \beta_3 h_3$. But this can be true only if h_1, h_2, h_3 themselves are all of equal size H and alternating sign, and these are the features that led us to the three points through which the Chebyshev line passes. Surely these are not two straight lines through these three points. This proves that the equality $|h| = H$ identifies the Chebyshev line. We have now proved that the equal error line and the min–max line for three points are the same.

13.7 Illustrate the *exchange method* by applying it to the following data:

x_i	0	1	2	6	7
Y_i	0	0	1	2	3

▌ We will prove shortly that there exists a unique min–max line for N points. The proof uses the exchange method, which is also an excellent algorithm for computing this line, and so this method will first be illustrated. It involves four steps.

Step 1. Choose any three of the data points. (A set of three data points will be called a triple. This step simply selects an initial triple. It will be changed in Step 4.)

Step 2. Find the Chebyshev line for this triple. The value h for this line will of course be computed in the process.

Step 3. Compute the errors at all data points for the Chebyshev line just found. Call the largest of these h_i values (in absolute value) H. If $|h| = H$ the search is over. The Chebyshev line for the triple in hand is the min–max line for the entire set of N points. (We shall prove this shortly.) If $|h| < H$ proceed to Step 4.

Step 4. This is the exchange step. Choose a new triple as follows. Add to the old triple a data point at which the greatest error size H occurs. Then discard one of the former points, in such a way that the remaining three have errors of alternating sign. (A moment's practice will show that this is always possible.) Return, with the new triple, to Steps 2 and 3.

To illustrate, suppose we choose for the initial triple

$$(0,0) \qquad (1,0) \qquad (2,1)$$

consisting of the first three points. This is the triple of Problem 13.4, for which we have already found the Chebyshev line to be $y = \frac{1}{2}x - \frac{1}{4}$ with $h = -\frac{1}{4}$. This completes Steps 1 and 2. Proceeding to Step 3 we find the errors at all five data points to be $-\frac{1}{4}, \frac{1}{4}, -\frac{1}{4}, \frac{3}{4}, \frac{1}{4}$. This makes $H = h_4 = \frac{3}{4}$. This Chebyshev line is an equal error line on its own triple but it misses the fourth data point by a larger amount. (See the dashed line in Fig. 13-2.)

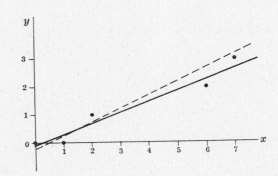

Fig. 13-2

Moving therefore to Step 4 we now include the fourth point and eliminate the first to obtain the new triple

$$(1,0) \qquad (2,1) \qquad (6,2)$$

on which the errors of the old Chebyshev line do have the required alternation of sign $(\frac{1}{4}, -\frac{1}{4}, \frac{3}{4})$. With this triple we return to Step 2 and find a new Chebyshev line. The computation begins with

$$\beta_1 = 6 - 2 = 4 \qquad \beta_2 = 6 - 1 = 5 \qquad \beta_3 = 2 - 1 = 1$$
$$h = -\frac{(4)(0) - (5)(1) + (1)(2)}{4 + 5 + 1} = \frac{3}{10}$$

so that the line must pass through the three points $(1, \frac{3}{10})$, $(2, \frac{7}{10})$ and $(6, \frac{23}{10})$. This line is found to be $y = \frac{2}{5}x - \frac{1}{10}$. Repeating Step 3 we find the five errors $-\frac{1}{10}, \frac{3}{10}, -\frac{3}{10}, \frac{3}{10}, -\frac{3}{10}$, and since $H = 3/10 = |h|$, the job is done.

The Chebyshev line for the new triple is the min–max line for the entire point set. Its maximum error is $\frac{3}{10}$. The new line is shown solid in Fig. 13-2. Notice that the $|h|$ value of our new line $(\frac{3}{10})$ is larger than that of the first line $(\frac{1}{4})$. But over the entire point set the maximum error has been reduced from $\frac{3}{4}$ to $\frac{3}{10}$ and it is the min–max error. This will now be proved for the general case.

13.8   Prove that the condition $|h| = H$ in Step 3 of the exchange method will be satisfied eventually, so that the method will stop. (Conceivably we could be making exchanges forever.)

▮ Recall that after any particular exchange the old Chebyshev line has errors of size $|h|$, $|h|$, H on the new triple. Also recall that $|h| < H$ (or we would have stopped) and that the three errors alternate in sign. The Chebyshev line for this new triple is then found. Call its errors on this new triple h^*, $-h^*$, h^*. Returning to the formula for h in Problem 13.5, with the old Chebyshev line playing the role of "any other line," we have

$$h^* = \frac{\beta_1 h_1 - \beta_2 h_2 + \beta_3 h_3}{\beta_1 + \beta_2 + \beta_3}$$

where h_1, h_2, h_3 are the numbers h, h, H with alternating sign. Because of this alternation of sign all three terms in the numerator of this fraction have the same sign, so that

$$|h^*| = \frac{\beta_1 |h| + \beta_2 |h| + \beta_3 H}{\beta_1 + \beta_2 + \beta_3}$$

if we assume that the error H is at the third point, just to be specific. (It really makes no difference in which position it goes.) In any event, $|h^*| > |h|$ because $H > |h|$. The new Chebyshev line has a greater error size on its triple than the old one had on its triple. This result now gives excellent service. If it comes as a surprise, look at it this way: The old line gave excellent service ($h = \frac{1}{4}$ in our example) on its own triple, but poor service ($H = \frac{3}{4}$) elsewhere. The new line gave good service ($h = \frac{3}{10}$) on its own triple and just as good service on the other points also.

We can now prove that the exchange method must come to a stop sometime, for there are only so many triples and no triple is ever chosen twice, since as just proved the h values increase steadily. At some stage the condition $|h| = H$ will be satisfied.

13.9   Prove that the last Chebyshev line computed in the exchange method is the min–max line for the entire set of N points.

▮ Let h be the equal error value of the last Chebyshev line on its own triple. Then the maximum error size on the entire point set is $H = |h|$, or we would have proceeded by another exchange to still another triple and another line. Let $h_1, h_2, \ldots, h_N$ be the errors for any other line. Then $|h| < \max |h_i|$ where h_i is restricted to the three points of the last triple, because no line outperforms a Chebyshev line on its own triple. But then certainly $|h| < \max |h_i|$ for h_i unrestricted, for including the rest of the N points can only make the right side even bigger. Thus $H = |h| < \max |h_i|$ and the maximum error of the last Chebyshev line is the smallest maximum error of all. In summary, the min–max line for the set of N points is an equal error line on a properly chosen triple.

13.10   Apply the exchange method to find the min–max line for the following data:

x_i	0	1	2	3	4	5	6	7	8	9	10	11	12	13	14	15
Y_i	0	1	1	2	1	3	2	2	3	5	3	4	5	4	5	6

x_i	16	17	18	19	20	21	22	23	24	25	26	27	28	29	30
Y_i	6	5	7	6	8	7	7	8	7	9	11	10	12	11	13

▮ The number of available triples is $C(31, 3) = 4495$, so that finding the correct one might seem comparable to needle-hunting in haystacks. However, the exchange method wastes very little time on inconsequential triples. Beginning with the very poor triple at $x = (0, 1, 2)$, only three exchanges are necessary to produce the min–max line $y(x) = .38x - .29$, which has coefficients rounded off to two places. The successive triples with h and H values were

Triple at $x =$	$(0, 1, 2)$	$(0, 1, 24)$	$(1, 24, 30)$	$(9, 24, 30)$
h	.250	.354	-1.759	-1.857
H	5.250	3.896	2.448	1.857

Note that in this example no unwanted point is ever brought into the triple. Three points are needed; three exchanges suffice. Note also the steady increase of $|h|$, as forecast. The 31 points, the min–max line and the final triple (dashed vertical lines show the equal errors) appear in Fig. 13-3.

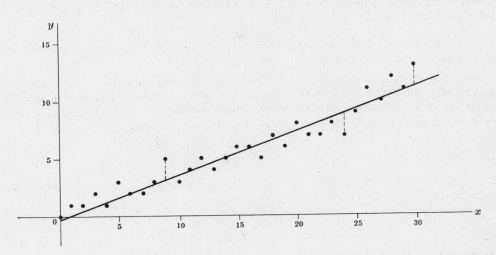

Fig. 13-3

13.11 Find the least-squares line for the data of Problem 13.4 and compare it with the Chebyshev line $y = \frac{1}{2}x - \frac{1}{4}$ found there. Verify in particular that the Chebyshev line does have the smaller maximum error and the least-squares line the smaller sum of errors squared.

■ The sums $s_0 = 3$, $s_1 = 3$, $s_2 = 5$, $t_0 = 1$ and $t_1 = 2$ can be computed mentally and lead to the slope $m = \frac{1}{2}$ and intercept $b = -\frac{1}{6}$. The least-squares line is $y = \frac{1}{2}x - \frac{1}{6}$. Its errors at the three data points are $\frac{1}{6}, -\frac{1}{3}, \frac{1}{6}$ and are to be compared with the $\frac{1}{4}, -\frac{1}{4}, \frac{1}{4}$ of the min–max line. Because of the $\frac{1}{3}$ the Chebyshev line does have the smaller maximum error. The sums of squares being $\frac{1}{6}$ and $\frac{3}{16}$, respectively, the least-squares line also lives up to its credentials.

13.12 Apply the exchange algorithm to the average golf scores of Problem 12.4, producing the min–max line. Use this line to compute smoothed average scores. Do they differ very much from those found earlier by a least-squares line?

■ Starting with an uninspired initial triple consisting of the first three points $x = 6, 8, 10$ we find the developments

Triple	h	Slope	$H = \max|h_i|$	x_i
6, 8, 10	$-.1$	.05	.2	14
6, 8, 14	$-.11$	.0625	.3125	24
8, 14, 24	.15	.05	.15	8

It has taken only two exchanges to achieve the ultimate equality of h and H, the point $x = 8$ proving to be not entirely uninspired. The final column shows the point at which maximum error occurred. Since the min–max line takes the value $Y_1 + h$ at the first point of the triple, here $3.7 + .15 = 3.85$, the line is

$$y = .05(\text{hcp} - 8) + 3.85 = .05\text{hcp} + 3.45$$

The least-squares line (lsq) found earlier was $y = .045\text{hcp} + 3.49$. Here are the smoothed values from each of these lines:

x	6	8	10	12	14	16	18	20	22	24
lsq	3.76	3.85	3.94	4.03	4.12	4.21	4.30	4.39	4.48	4.57
mm	3.75	3.85	3.95	4.05	4.15	4.25	4.35	4.45	4.55	4.65

Since hundredths of a stroke are hardly of interest in golf, the two sets are just about equivalent. Their prediction of a score of 3.5 strokes for a scratch player (hcp $= 0$) on this par-3 hole suggests that it is not an easy one.

13.13 Apply the exchange method to the data of Problem 12.7, obtaining the min–max line and then the corresponding exponential function $P(x) = Ae^{mx}$. Compare with the result found earlier.

▌ The needed logarithms were introduced in the second table of the earlier problem and serve as input to the exchange algorithm. With so few points active, one expects a short computation and here is the report:

Triple	h	m	H	x
1, 2, 3	.005	.44	.025	4
2, 3, 4	−.01	.45	.01	

One exchange has done it, the h and H being of equal size. The value corresponding to the first point of the triple ($x = 2$) being $Y_1 + h = 2.40 - .01 = 2.39$, our min–max line is

$$y = .45(x - 2) + 2.39 = .45x + 1.49$$

and is to be compared with the least-squares line found earlier,

$$y = .45x + 1.5$$

Clearly there is little difference, the earlier exponential

$$P(x) = 4.48e^{.45x}$$

having its coefficient replaced by 4.44 for a change of less than 1%.

13.14 In Problem 12.10 the power function $P(x) = 91.4x^{-.43}$ was found for the data

x	2.2	2.7	3.5	4.1
P	65	60	53	50

by replacing an assumed $P = dx^n$ with $\log P = \log d + n \log x$ and then applying the least-squares method. The line found was $\log P = 4.52 - .43 \log x$, which does exponentiate to the result just reproduced. Now find the min–max line for the same data and compare.

▌ Here is the output of the exchange method:

Triple	h	m	H	x
1, 2, 3	.005	−.440	.0063	4
2, 3, 4	−.0054	−.436	.0054	

One exchange was enough. The min–max line is

$$\log P = -.436(\log x - \log 2.7) + 4.089 = -.436 \log x + 4.52$$

so again there is not much difference from the least-squares result.

13.15 In Problem 12.10 a second least-squares result was found, the exponential $P(x) = 87.9e^{-.14x}$. With $y = \log P$, it was the line $y = -.14x + 4.48$ that was produced by the least-squares method, with P following by exponentiation. Now find the corresponding min–max line for this data.

▌ This time two exchanges were needed:

Triple	h	m	H	x
1, 2, 3	−.0008	−.157	.037	4
1, 2, 4	−.0055	−.138	.019	3
1, 3, 4	−.0123	−.138	.0123	

The min–max line is

$$y = -.138(x - 2.2) + 4.162 = -.14x + 4.47$$

and is almost identical with the least-squares line.

13.16 How well does a function of the type $P(x) = Ae^{Mx}$ represent the following data if a min–max approach is used?

x	1	2	3	4
P	60	30	20	15

❚ With $y = \log P$ we look for the line $y = \log A + Mx$. The exchange method runs through the steps

Triple	h	m	H	x
1,2,3	−.072	−.549	.33	4
1,2,4	−.1155	−.462	.1155	

leading to the line

$$y = -.462(x - 1) + 3.98 = -.462x + 4.44$$

Its values at the data arguments are 53, 34, 21 and 13, and the suitability of the exponential function is a matter of judgment.

13.17 Integers from 0 to 10 were drawn from a table of random numbers:

7 0 8 10 7 10 2 8 9 4 7 3 7 6 10 1 3 7 6 3

What is the constant function that best approximates this set in the min–max sense?

❚ The problem is to minimize the maximum error and it takes only a moment to see that the number 5 does this. It is midway between the extreme values of 0 and 10, making the maximum error 5. Any other number will clearly cause a greater error somewhere along the list. A general result follows, that the best min–max constant approximation to a set of numbers is midway between the extreme members of the set. The least-squares approximation is, of course, the average value of the set.

13.18 In a recreational spirit, apply the exchange algorithm to the random data of the preceding problem. Is a constant function produced?

❚ Even short experience with random processes suggests that we not expect so neat a result. Instead we may wonder how small the calculated slope will be. From the usual initial triple at 1, 2, 3 only two exchanges were needed to get the answer

Triple	h	m	H	x
1,2,3	−3.75	.5	9.75	16
2,3,16	3.96	.07	5.89	4
2,4,16	4.93	.07	4.93	

a slope of .07. In fact, a result of zero might have raised one or two statistical eyebrows and suspicions about the randomness of the input numbers.

13.19 Show that if the arguments x_i are not distinct, then the min–max line may not be uniquely determined.

❚ An example is all that is required. Consider the three point set of Fig. 13-4, consisting of $(0,0)$, $(0,1)$ and $(1,0)$. The line $y = \frac{1}{2}$ misses all three data points by half a unit. This performance cannot be bettered, since the two points $(0,0)$ and $(0,1)$ are a unit apart, so $y = \frac{1}{2}$ is a min–max line. But then so is $y = \frac{1}{2} - x$ and, in fact, so are all the lines between them, all of which have $H = \frac{1}{2}$.

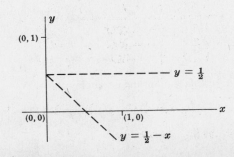

Fig. 13-4

13.2 DISCRETE DATA: THE MIN–MAX POLYNOMIAL

13.20   Extend the exchange method to find the min–max parabola for the data

x_i	-2	-1	0	1	2
y_i	2	1	0	1	2

❚ The data are of course drawn from the function $y = |x|$, but this simple function will serve to illustrate how all the essential ideas of the exchange method carry over from the straight line problems just treated to the discovery of a min–max polynomial. The proofs of the existence, uniqueness and equal error properties of such a polynomial are extensions of our proofs for the min–max line and will not be given. The algorithm now begins with the choice of an "initial quadruple" and we take the first four points at $x = -2, -1, 0, 1$. For this quadruple we seek an equal error parabola, say

$$p_1(x) = a + bx + cx^2$$

This means that we require $p(x_i) - y_i = \pm h$ alternately, or

$$
\begin{aligned}
a - 2b + 4c - 2 &= h \\
a - b + c - 1 &= -h \\
a &= h \\
a + b + c - 1 &= -h
\end{aligned}
$$

Solving these four equations, we find $a = \frac{1}{4}$, $b = 0$, $c = \frac{1}{2}$ and $h = \frac{1}{4}$ so that $p_1(x) = \frac{1}{4} + \frac{1}{2}x^2$. This completes the equivalent of Steps 1 and 2, and we turn to Step 3 and compute the errors of our parabola at all five data points. They are $\frac{1}{4}, -\frac{1}{4}, \frac{1}{4}, -\frac{1}{4}, \frac{1}{4}$ so that the maximum error on the entire set ($H = \frac{1}{4}$) equals the maximum on our quadruple ($|h| = \frac{1}{4}$). The algorithm is ended and our first parabola is the min–max parabola. It is shown in Fig. 13-5.

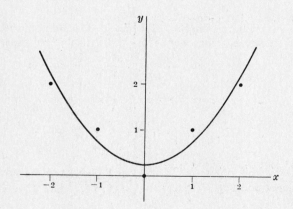

Fig. 13-5

13.21   Summarize the procedure for finding a min–max parabola for given discrete data.

❚ An initial quadruple (x_1, x_2, x_3, x_4) is chosen and the system

$$
\begin{aligned}
a + bx_1 + cx_1^2 - y_1 &= h \\
a + bx_2 + cx_2^2 - y_2 &= -h \\
a + bx_3 + cx_3^2 - y_3 &= h \\
a + bx_4 + cx_4^2 - y_4 &= -h
\end{aligned}
$$

is solved for h and the three coefficients of the parabola,

$$p = a + bx + cx^2$$

Then errors at all the data points x_i are calculated,

$$e_i = a + bx_i + cx_i^2 - y_i$$

and the absolutely largest located. This identifies $H = \max|e_i|$. If $|h| = H$, the algorithm is done and the current parabola is min–max. Otherwise a new quadruple is formed by including an x_i with error H and deleting one former member, in such a way that the new quadruple has alternating error signs (for the current parabola). With the new quadruple, a new linear system is formed and the process repeated.

13.22 Carry through the procedure just summarized for the seven points $y = |x|$ and $x = -3(1)3$.

  ∎ This adds a new point at each end of the data supply of Problem 13.20. Suppose we choose the same initial quadruple. Then we again have the equal error parabola $p_1(x)$ of that problem. It has errors of size $\frac{7}{4}$ at the new data points, so $H = \frac{7}{4}$ while $|h|$ is only $\frac{1}{4}$. Accordingly we introduce one of the new points, say $x = 3$, into the new quadruple and abandon $x = -2$. On the new quadruple the old parabola has the errors $-\frac{1}{4}, \frac{1}{4}, -\frac{1}{4}, \frac{7}{4}$, which do alternate in sign. Having made the exchange, a new equal error parabola

$$p_2(x) = a + bx + cx^2$$

must be found. The new system of equations is

$$
\begin{aligned}
a - b + c - h &= 1 \\
a + h &= 0 \\
a + b + c - h &= 1 \\
a + 3b + 9c + h &= 3
\end{aligned}
$$

with solution $h = -\frac{1}{3}$ and $a, b, c = \frac{1}{3}, 0, \frac{1}{3}$. The new parabola is $p_2(x) = \frac{1}{3}(1 + x^2)$. Its errors at the seven data points are all of size $\frac{1}{3}$ so that $|h| = H$ and the algorithm stops. The fact that all errors are of the same size is a bonus, not characteristic of min–max polynomials generally, as the straight line problems solved earlier show.

13.23 Find the equal error parabola for the four points $(0,0)$, $(\pi/6, \frac{1}{2})$, $(\pi/3, \sqrt{3}/2)$ and $(\pi/2, 1)$ of the function $y = \sin x$.

  ∎ Here is the linear system that must be solved.

$$
\begin{aligned}
a \phantom{+ \frac{\pi}{6}b + \frac{\pi^2}{36}c - \frac{1}{2}} &= h \\
a + \frac{\pi}{6}b + \frac{\pi^2}{36}c - \frac{1}{2} &= -h \\
a + \frac{\pi}{3}b + \frac{\pi^2}{9}c - \frac{\sqrt{3}}{2} &= h \\
a + \frac{\pi}{2}b + \frac{\pi^2}{4}c - 1 &= -h
\end{aligned}
$$

Pen and paper is sufficient here and manages

$$h = a = (5 - 3\sqrt{3})\big/16 \qquad b = 3\big(\sqrt{3} - \tfrac{1}{2}\big)\big/\pi \qquad c = 9(1 - \sqrt{3})\big/2\pi^2$$

with the parabola, of course, $p(x) = a + bx + cx^2$. The maximum error h is near .01. The values of $p(x)$ at the four data points are

$$-.012 \qquad .51 \qquad .85 \qquad 1.01$$

and the alternation of error sign is conspicuous, the parabola taking an under-over-under-over path relative to the sine curve.

13.24 Find the min–max parabola for the five points $y = x^3$ and $x = 0(\frac{1}{4})1$.

  ∎ Taking as initial quadruple the first four arguments, $0, \frac{1}{4}, \frac{1}{2}, \frac{3}{4}$, two exchanges were made. Here are the key elements:

Quadruple	h	Errors	H
$0, \frac{1}{4}, \frac{1}{2}, \frac{3}{4}$	.0117	$h, -h, h, -h, -.175$	.175
$0, \frac{1}{4}, \frac{1}{2}, 1$	.0208	$h, -h, h, .052, -h$	.052
$0, \frac{1}{4}, \frac{3}{4}, 1$	.03125	$h, -h, 0, h, -h$	h

The third and last parabola was

$$p(x) = .03125 - .5625x + 1.5x^2 = \tfrac{1}{32} - \tfrac{9}{16}x + \tfrac{3}{2}x^2$$

with maximum error $h = \tfrac{1}{32}$.

13.25 Use the exchange method to find the min−max parabola for the seven points $y = \cos x$ and $x = 0(\pi/12)\pi/2$. What is the maximum error of this parabola and how does it compare with the Taylor parabola $1 - \tfrac{1}{2}x^2$?

▎ The temporary variable $t = 12x/\pi$ is slightly helpful and is used in listing the quadruples in Table 13.1.

TABLE 13.1

Quadruple	h	Errors	H
0, 1, 2, 3	.001	$h, -h, h, -h, -.02, -.06, -.15$	.150
0, 1, 2, 6	.004	$h, -h, h, .02, .034, .031, -h$	.034
0, 1, 4, 6	.011	$h, -h, -.0137, -.003, h, .0139, -h$	.014
0, 1, 5, 6	.012	$h, -h, -.016, -.005, .009, h, -h$	.016
0, 2, 5, 6	.013	$h, -.010, -h, -.002, .011, h, -h$	h

There are 35 quadruples available in a set of seven arguments, so this trail of 5 is still a shortcut. The final parabola was

$$p(x) = 1.0130 - .03445t - .02276t^2$$

with t still equal to $12x/\pi$. Its maximum error is $h = .013$ over the data set. Comparisons with the Taylor polynomial of degree 2 may be made in Table 13.2.

TABLE 13.2

t	0	1	2	3	4	5	6
min−max	1.01	.956	.853	.705	.511	.272	−.013
Taylor	1	.966	.863	.692	.452	.143	−.234
True	1	.966	.866	.707	.500	.259	0

13.26 Extend the exchange method to find a min−max cubic polynomial for data points (x_i, y_i).

▎ The extension involves setting up the linear system

$$a + bx_1 + cx_1^2 + dx_1^3 - y_1 = h$$
$$a + bx_2 + cx_2^2 + dx_2^3 - y_2 = -h$$
$$a + bx_3 + cx_3^2 + dx_3^3 - y_3 = h$$
$$a + bx_4 + cx_4^2 + dx_4^3 - y_4 = -h$$
$$a + bx_5 + cx_5^2 + dx_5^3 - y_5 = h$$

for the equal error h and coefficients a, b, c, d, in which the set x_1, x_2, x_3, x_4, x_5 is a selected quintuple. The errors of the cubic

$$p = a + bx + cx^2 + dx^3$$

are then found at all data points and their largest absolute value H located. If $|h| = H$ the algorithm stops, with the cubic in hand the min−max cubic. Otherwise an exchange is made in the now familiar way, bringing a point of error size H into a new quintuple and abandoning one former member, in such a way that errors of the old cubic on the new quintuple alternate in sign. The new quintuple then leads to a new linear system, new h, new cubic and so on.

13.27 Apply the exchange method to find the min–max cubic polynomial for the seven points $y = \sin x$ and $x = 0(\pi/12)\pi/2$. What is the maximum error of this cubic? Compare its accuracy with that of the Taylor cubic $x - \frac{1}{6}x^3$.

▮ Using the variable $t = 12x/\pi$ here and there for convenience, the initial quintuple $t = 0, 1, 2, 3, 4$ was chosen. Of the 21 quintuples available in a set of seven arguments, the method selected three others before reaching its goal. The record of progress appears in Table 13.3, the final cubic being

$$p = -.001343 + .2683t - .004813t^2 - .002022t^3$$

with equal error $h = -.0013$. Errors in the third decimal place are to be expected. A comparison of this and the Taylor cubic is also offered in the table. This time the min–max result proves superior even as early as $t = 2$.

TABLE 13.3

Quintuple	h	Errors	H
0, 1, 2, 3, 4	−0001	$h, -h, h, -h, h, -.005, -.024$	.024
0, 1, 2, 3, 6	−.0004	$h, -h, h, -h, .003, .005, h$	.005
0, 1, 2, 5, 6	−.0010	$h, -h, h, -.002, -.001, -h, h$	.002
0, 1, 3, 5, 6	−.0013	$h, -h, -.0001, h, -.0004, -h, h$	h

t	0	1	2	3	4	5	6
min–max	−.0013	.2602	.4999	.7058	.8656	.9673	.9987
Taylor	0	.2588	.4997	.7047	.8558	.9352	.9248
true	0	.2588	.5000	.7071	.8660	.9659	1

13.28 The exchange method for finding min–max polynomials has been developed for degrees 1 to 3 and the continuation is clear. Can it also be used for degree 0, that is, a constant?

▮ The question does have a slight entertainment value so let us make the most of it. Take this data function, for example:

x	0	1	2	3
y	0	4	6	2

The best initial "double" is probably conspicuous enough, but for present purposes suppose we choose $x = 0, 1$. The min–max constant is then $p = 2$, since it has the equal error property, with $h = 2$, on this double. But it makes the errors $h, -h, -4, 0$ overall, so $H = 4$ and an exchange is called for. The new double is $x = 0, 2$ (preserving the sign alternation) and leading to $p = 3$, $h = 3$ and the new errors $h, -1, -h, 1$. Now $H = h$ and the run is over.

Recalling Problem 13.17, the pair $x = 0, 2$ was clearly indicated for initial double since it would bring the extreme values of y into action at once. It is also apparent that at most two exchanges should be needed whatever the initial double, one to bring in each extreme value.

13.3 CONTINUOUS DATA: THE WEIERSTRASS THEOREM

13.29 Prove that $\sum_{k=0}^{n} p_{nk}^{(x)}(k - nx) = 0$, where $p_{nk}^{(x)} = \binom{n}{k} x^k (1 - x)^{n-k}$.

▮ The binomial theorem for integers n and k,

$$(p + q)^n = \sum_{k=0}^{n} \binom{n}{k} p^k q^{n-k}$$

is an identity in p and q. Differentiating relative to p brings

$$n(p + q)^{n-1} = \sum_{k=0}^{n} \binom{n}{k} kp^{k-1} q^{n-k}$$

Multiplying by p and then setting $p = x$ and $q = 1 - x$, this becomes $nx = \sum_{k=0}^{n} kp_{nk}^{(x)}$. Using the same p and q in the binomial theorem itself shows that $1 = \sum p_{nk}^{(x)}$ and so finally

$$\sum_{k=0}^{n} p_{nk}^{(x)}(k - nx) = nx - nx = 0$$

13.30 Prove also that $\sum_{k=0}^{n} p_{nk}^{(x)}(k - nx)^2 = nx(1 - x)$.

▌ A second differentiation relative to p brings

$$n(n-1)(p+q)^{n-2} = \sum_{k=0}^{n} \binom{n}{k} k(k-1) p^{k-2} q^{n-k}$$

Multiplying by p^2 and then setting $p = x$ and $q = 1 - x$, this becomes

$$n(n-1)x^2 = \sum_{k=0}^{n} k(k-1) p_{nk}^{(x)}$$

from which we find

$$\sum_{k=0}^{n} k^2 p_{nk}^{(x)} = n(n-1)x^2 + \sum_{k=0}^{n} k p_{nk}^{(x)} = n(n-1)x^2 + nx$$

Finally we compute

$$\sum_{k=0}^{n} p_{nk}^{(x)}(k - nx)^2 = \sum k^2 p_{nk}^{(x)} - 2nx \sum k p_{nk}^{(x)} + n^2 x^2 \sum p_{nk}^{(x)}$$

$$= n(n-1)x^2 + nx - 2nx(nx) + n^2 x^2 = nx(1 - x)$$

13.31 Prove that if $d > 0$ and $0 \leq x \leq 1$, then

$$\sum' p_{nk}^{(x)} \leq \frac{x(1-x)}{nd^2}$$

where $\sum'$ is the sum over those integers k for which $|(k/n) - x| \geq d$. (This is a special case of the famous Chebyshev inequality.)

▌ Breaking the sum of the preceding problem into two parts,

$$nx(1-x) = \sum' p_{nk}^{(x)}(k - nx)^2 + \sum'' p_{nk}^{(x)}(k - nx)^2$$

where $\sum''$ includes those integers k omitted in $\sum'$. But then

$$nx(1-x) \geq \sum' p_{nk}^{(x)}(k - nx)^2$$

$$\geq \sum' p_{nk}^{(x)} n^2 d^2$$

the first of these steps being possible since $\sum''$ is nonnegative and the second because in $\sum'$ we find $|k - nx| \geq nd$. Dividing through by $n^2 d^2$, we have the required result.

13.32 Derive these estimates for $\sum'$ and $\sum''$:

$$\sum' p_{nk}^{(x)} \leq \frac{1}{4nd^2} \qquad \sum'' p_{nk}^{(x)} \geq 1 - \frac{1}{4nd^2}$$

▌ The function $x(1 - x)$ takes its maximum at $x = \frac{1}{2}$ and so $0 \leq x(1 - x) \leq \frac{1}{4}$ for $0 \leq x \leq 1$. The result for $\sum'$ is thus an immediate consequence of the preceding problem. But then $\sum'' = 1 - \sum' \geq 1 - (1/4nd^2)$.

13.33 Prove that if $f(x)$ is continuous for $0 \leq x \leq 1$, then $\lim \sum_{k=0}^{n} p_{nk}^{(x)} f(k/n) = f(x)$ uniformly as n tends to infinity.

▌ This will prove the Weierstrass theorem, by exhibiting a sequence of polynomials

$$B_n(x) = \sum_{k=0}^{n} p_{nk}^{(x)} f\left(\frac{k}{n}\right)$$

which converges uniformly to $f(x)$. These polynomials are called the Bernstein polynomials for $f(x)$. The proof begins with the choice of an arbitrary positive number ϵ. Then for $|x' - x| < d$,

$$|f(x') - f(x)| < \frac{\epsilon}{2}$$

and d is independent of x by the uniform continuity of $f(x)$. Then with M denoting the maximum of $|f(x)|$, we have

$$|B_n(x) - f(x)| = \left| \sum p_{nk}^{(x)} \left[f\left(\frac{k}{n}\right) - f(x) \right] \right|$$

$$\leq \sum' p_{nk}^{(x)} \left| f\left(\frac{k}{n}\right) - f(x) \right| + \sum'' p_{nk}^{(x)} \left| f\left(\frac{k}{n}\right) - f(x) \right|$$

$$\leq 2M \sum' p_{nk}^{(x)} + \frac{1}{2} \epsilon \sum'' p_{nk}^{(x)}$$

with k/n in the $\sum''$ part playing the role of x'. The definition of $\sum''$ guarantees $|x' - x| < d$. Then

$$|B_n(x) - f(x)| \leq \left(\frac{2M}{4nd^2}\right) + \frac{1}{2}\epsilon$$

$$\leq \frac{1}{2}\epsilon + \frac{1}{2}\epsilon = \epsilon$$

for n sufficiently large. This is the required result. An interval other than $(0,1)$ can be accommodated by a simple change of argument.

13.34 Show that in the case of $f(x) = x^2$, $B_n(x) = x^2 + x(1-x)/n$ so that Bernstein polynomials are not the best approximations of given degree to $f(x)$. [Surely the best quadratic approximation to $f(x) = x^2$ is x^2 itself.]

❚ Since the sum $\sum k^2 p_{nk}^{(x)}$ was found in Problem 13.30,

$$B_n(x) = \sum_{k=0}^{n} p_{nk}^{(x)} f\left(\frac{k}{n}\right) = \sum_{k=0}^{n} \frac{p_{nk}^{(x)} k^2}{n^2} = \frac{1}{n^2}\left[n(n-1)x^2 + nx \right] = x^2 + \frac{x(1-x)}{n}$$

as required. The uniform convergence for n tending to infinity is apparent, but clearly $B_n(x)$ does not duplicate x^2.

13.35 What are the Bernstein polynomials $B_n(x)$ for the cases $f(x) = 1$ and $f(x) = x$?

❚ For the first,

$$B_n(x) = \sum_{k=0}^{n} p_{nk}^{(x)}(1) = 1$$

while for the second, borrowing a result from Problem 13.29,

$$B_n(x) = \sum_{k=0}^{n} p_{nk}^{(x)}\left(\frac{k}{n}\right) = x$$

so in both cases there is instant convergence.

13.36 Find the Bernstein polynomials for $f(x) = x^3$.

❚ First we recall from Problem 13.30 the result of differentiating the binomial theorem twice with respect to p:

$$n(n-1)(p+q)^{n-2} = \sum_{k=0}^{n} \binom{n}{k} k(k-1) p^{k-2} q^{n-k}$$

Take one more derivative,

$$n(n-1)(n-2)(p+q)^{n-3} = \sum_{k=0}^{n} \binom{n}{k} k(k-1)(k-2) p^{k-3} q^{n-k}$$

multiply by p^3 and then let $p = x$ and $q = 1 - x$:

$$n(n-1)(n-2)x^3 = \sum_{k=0}^{n} k(k-1)(k-2)\binom{n}{k}x^k(1-x)^{n-k}$$

$$= \sum_{k=0}^{n} (k^3 - 3k^2 + 2k)p_{nk}^{(x)}$$

This rearranges to

$$\sum_{k=0}^{n} k^3 p_{nk}^{(x)} = n(n-1)(n-2)x^3 + 3\sum_{k=0}^{n} k^2 p_{nk}^{(x)} - 2\sum_{k=0}^{n} kp_{nk}^{(x)}$$

in which the remaining sums on the right can be replaced by their values as found in Problem 13.30:

$$\sum_{k=0}^{n} k^3 p_{nk} = n(n-1)(n-2)x^3 + 3[n(n-1)x^2 + nx] - 2nx$$

Finally

$$B_n(x) = \frac{n(n-1)(n-2)}{n^3}x^3 + \frac{3(n-1)}{n^2}x^2 + \frac{1}{n^2}x$$

which does have limit x^3 as n becomes infinite.

13.4 CONTINUOUS DATA: THE CHEBYSHEV THEORY

13.37 Prove that if $y(x)$ is continuous for $a \le x \le b$, then there is a polynomial $P(x)$ of degree n or less such that $\max|P(x) - y(x)|$ on the interval (a, b) is a minimum. In other words, no other polynomial of this type produces a smaller maximum.

▌ Let $p(x) = a_0 + a_1 x + \cdots + a_n x^n$ be any polynomial of degree n or less. Then

$$M(\bar{a}) = \max|p(x) - y(x)|$$

depends on the polynomial $p(x)$ chosen, that is, it depends on the coefficient set $(a_0, a_1, \ldots, a_n)$, which we shall call $\bar{a}$ as indicated. Since $M(\bar{a})$ is a continuous function of $\bar{a}$ and nonnegative, it has a greatest lower bound. Call this bound L. What has to be proved is that for some particular coefficient set A, the coefficients of $P(x)$, the lower bound L is actually attained, that is, $M(A) = L$. By way of contrast, the function $f(t) = 1/t$ for positive t has greatest lower bound zero, but there is no argument t for which $f(t)$ actually attains this bound. The infinite range of t is of course the factor that allows this situation to occur. In our problem the coefficient set $\bar{a}$ also has unlimited range, but we now show that $M(A) = L$ nevertheless. To begin, let $a_i = Cb_i$ for $i = 0, 1, \ldots, n$ in such a way that $\sum b_i^2 = 1$. We may also write $\bar{a} = C\bar{b}$. Consider a second function

$$m(\bar{b}) = \max|b_0 + b_1 x + \cdots + b_n x^n|$$

where max refers as usual to the maximum of the polynomial on the interval (a, b). This is a continuous function on the unit sphere $\sum b_i^2 = 1$. On such a set (closed and bounded) a continuous function does assume its minimum value. Call this minimum μ. Plainly $\mu \ge 0$. But the zero value is impossible since only $p(x) = 0$ can produce this minimum and the condition on the b_i temporarily excludes this polynomial. Thus $\mu > 0$. But then

$$m(\bar{a}) = \max|a_0 + a_1 x + \cdots + a_n x^n| = \max|p(x)| = Cm(\bar{b}) \ge C\mu$$

Now returning to $M(\bar{a}) = \max|p(x) - y(x)|$ and using the fact that the absolute value of a difference exceeds the difference of absolute values, we find

$$M(\bar{a}) \ge m(\bar{a}) - \max|y(x)|$$

$$\ge C\mu - \max|y(x)|$$

If we choose $C > (L + 1 + \max|y(x)|)/\mu = R$, then at once $M(\bar{a}) \ge L + 1$. Recalling that L is the greatest lower bound of $M(\bar{a})$, we see that $M(\bar{a})$ is relatively large for $C > R$ and that its greatest lower bound under the constraint $C \le R$ will be this same number L. But this constraint is equivalent to $\sum a_i^2 \le R$, so that now it is again a matter of a continuous function $M(\bar{a})$ on a closed and bounded set (a solid sphere, or ball). On such a set the greatest lower bound is actually assumed, say at $\bar{a} = A$. Thus $M(A)$ is L and $P(x)$ is a min–max polynomial.

13.38 Let $P(x)$ be a min–max polynomial approximation to $y(x)$ on the interval (a, b), among all polynomials of degree n or less. Let $E = \max|y(x) - P(x)|$ and assume $y(x)$ is not itself a polynomial of degree n or less, so that $E > 0$. Show that there must be at least one argument for which $y(x) - P(x) = E$ and similarly for $-E$. [We continue to assume $y(x)$ continuous.]

▌ Since $y(x) - P(x)$ is continuous for $a \le x \le b$, it must attain either $\pm E$ somewhere. We are to prove that it must achieve both. Suppose that it did not equal E anywhere in (a, b). Then

$$\max[y(x) - P(x)] = E - d$$

where d is positive, and so

$$-E \le y(x) - P(x) \le E - d$$

But this can be written as

$$-E + \tfrac{1}{2}d \le y(x) - [P(x) - \tfrac{1}{2}d] \le E - \tfrac{1}{2}d$$

which flatly claims that $P(x) - \tfrac{1}{2}d$ approximates $y(x)$ with a maximum error of $E - \tfrac{1}{2}d$. This contradicts the original assumption that $P(x)$ itself is a min–max polynomial, with maximum error of E. Thus $y(x) - P(x)$ must equal E somewhere in (a, b). A very similar proof shows it must also equal $-E$. Figure 13-6 illustrates the simple idea of this proof. The error $y(x) - P(x)$ for the min–max polynomial cannot behave as shown solid, because raising the curve by $\tfrac{1}{2}d$ then brings a new error curve (shown dashed) with a smaller maximum absolute value of $E - \tfrac{1}{2}d$, and this is a contradiction.

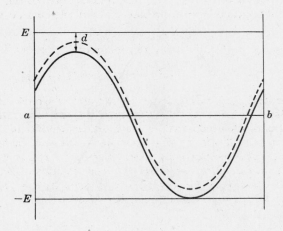

Fig. 13-6

13.39 Continuing the previous problem, show that, for $n = 1$, approximation by linear polynomials, there must be a third point at which the error $|y(x) - P(x)|$ of a min–max $P(x)$ assumes its maximum value E.

▌ Let $y(x) - P(x) = E(x)$ and divide (a, b) into subintervals small enough so that for x_1, x_2 within any subinterval,

$$|E(x_1) - E(x_2)| \le \tfrac{1}{2}E$$

Since $E(x)$ is continuous for $a \le x \le b$, this can surely be done. In one subinterval, call it I_1, we know the error reaches E, say at $x = x_+$. It follows that throughout this subinterval,

$$|E(x) - E(x_+)| = |E(x) - E| \le \tfrac{1}{2}E$$

making $E(x) \ge \tfrac{1}{2}E$. Similarly, in one subinterval, call it I_2, we find $E(x_-) = -E$, and therefore $|E(x)| \le -\tfrac{1}{2}E$. These two subintervals cannot therefore be adjacent and so we can choose a point u_1 between them. Suppose that I_1 is to the left of I_2. (The argument is almost identical for the reverse situation.) Then $u_1 - x$ has the same sign as $E(x)$ in each of the two subintervals discussed. Let $R = \max|u_1 - x|$ in (a, b).

Now suppose that there is no third point at which the error is $\pm E$. Then in all but the two subintervals just discussed we must have

$$\max |E(x)| < E$$

and since there are finitely many subintervals,

$$\max\left[\max |E(x)|\right] = E^* < E$$

Naturally $E^* \geq \frac{1}{2}E$ since these subintervals extend to the endpoints of I_1 and I_2 where $|E(x)| \geq \frac{1}{2}E$. Consider the following alteration of $P(x)$, still a linear polynomial:

$$P^*(x) = P(x) + \epsilon(u_1 - x)$$

If we choose ϵ small enough so that $\epsilon R < E - E^* \leq \frac{1}{2}E$, then $P^*(x)$ becomes a better approximation than $P(x)$ for

$$|y(x) - P^*(x)| = |E(x) - \epsilon(u_1 - x)|$$

so that in I_1 the error is reduced but is still positive while in I_2 it is increased but remains negative; in both subintervals the error size has been reduced. Elsewhere, though the error size may grow, it cannot exceed $E^* + \epsilon R < E$ and so $P^*(x)$ has a smaller maximum error than $P(x)$. This contradiction shows that a third point with error $\pm E$ must exist. Figure 13-7 illustrates the simple idea behind this proof. The error curve $E(x)$ cannot behave like the solid curve (only two $\pm E$ points) because adding the linear correction term $\epsilon(u_1 - x)$ to $P(x)$ then diminishes the error by this same amount, leading to a new error curve (shown dashed) with smaller maximum absolute value.

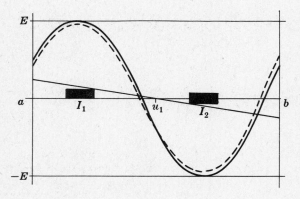

Fig. 13-7

13.40 Show that for the $P(x)$ of the previous problem there must be three points at which errors of size E and with alternating sign occur.

❚ The proof of the previous problem is already sufficient. If, for example, the signs were $+, +, -$, then choosing u_1 between the adjacent $+$ and $-$ our $P^*(x)$ is again better than $P(x)$. The pattern $+, -, -$ is covered by exactly the same remark. Only the alternation of signs can avoid the contradiction.

13.41 Show that in the general case of the min–max polynomial of degree n or less, there must be $n + 2$ points of maximum error size with alternating sign.

❚ The proof is illustrated by treating the case $n = 2$. Let $P(x)$ be a min–max polynomial of degree 2 or less. By Problem 13.38 it must have at least two points of maximum error. The argument of Problems 13.39 and 13.40, with $P(x)$ now quadratic instead of linear but with no other changes, then shows that a third such point must exist and signs must alternate, say $+, -, +$ just to be definite. Now suppose that no fourth position of maximum error occurs. We repeat the argument of Problem 13.39, choosing two points u_1 and u_2 between the subintervals I_1, I_2 and I_3 in which the errors $\pm E$ occur, and using the correction term $\epsilon(u_1 - x)(u_2 - x)$, which agrees in sign with $E(x)$ in these subintervals. No other changes are necessary. The quadratic $P^*(x)$ will have a smaller maximum error than $P(x)$ and this contradiction proves that the fourth $\pm E$ point must exist. The alternation of sign is established by the same argument used in Problem 13.40, and the extension to higher values of n is entirely similar.

13.42   Prove that there is just one min–max polynomial for each n.

▮   Suppose there were two, $P_1(x)$ and $P_2(x)$. Then

$$-E \le y(x) - P_1(x) \le E \qquad -E \le y(x) - P_2(x) \le E$$

Let $P_3(x) = \frac{1}{2}(P_1 + P_2)$. Then

$$-E \le y(x) - P_3(x) \le E$$

and P_3 is also a min–max polynomial. By Problem 13.41 there must be a sequence of $n + 2$ points at which $y(x) - P_3(x)$ is alternately $\pm E$. Let $P_3(x_+) = E$. Then at x_+ we have $y - P_3 = E$ or

$$(y - P_1) + (y - P_2) = 2E$$

Since neither term on the left can exceed E, each must equal E. Thus $P_1(x_+) = P_2(x_+)$. Similarly $P_1(x_-) = P_2(x_-)$. The polynomials P_1 and P_2 therefore coincide at the $n + 2$ points and so are identical. This proves the uniqueness of the min–max polynomial for each n.

13.43   Prove that a polynomial $p(x)$ of degree n or less, for which the error $y(x) - p(x)$ takes alternate extreme values of $\pm e$ on a set of $n + 2$ points, must be the min–max polynomial.

▮   This will show that only the min–max polynomial can have this equal error feature, and it is useful in finding and identifying such polynomials. We have

$$\max|y(x) - p(x)| = e \ge E = \max|y(x) - P(x)|$$

$P(x)$ being the unique min–max polynomial. Suppose $e > E$. Then since

$$P - p = (y - p) + (P - y)$$

we see that, at the $n + 2$ extreme points of $y - p$, the quantities $P - p$ and $y - p$ have the same sign. (The first term on the right equals e at these points and so dominates the second.) But the sign of $y - p$ alternates on this set, so the sign of $P - p$ does likewise. This is $n + 1$ alternations in all and means $n + 1$ zeros for $P - p$. Since $P - p$ is of degree n or less it must be identically zero, making $p = P$ and $E = e$. This contradicts our supposition of $e > E$ and leaves us with the only alternative, namely $e = E$. The polynomial $p(x)$ is thus the (unique) min–max polynomial $P(x)$.

13.5   CONTINUOUS DATA: EXAMPLES OF MIN–MAX POLYNOMIALS

13.44   Show that on the interval $(-1, 1)$ the min–max polynomial of degree n or less for $y(x) = x^{n+1}$ can be found by expressing x^{n+1} as a sum of Chebyshev polynomials and dropping the $T_{n+1}(x)$ term.

▮   Let

$$x^{n+1} = a_0 T_0(x) + \cdots + a_n T_n(x) + a_{n+1} T_{n+1}(x) = p(x) + a_{n+1} T_{n+1}(x)$$

Then the error is

$$E(x) = x^{n+1} - p(x) = a_{n+1} T_{n+1}(x)$$

and we see that this error has alternate extremes of $\pm a_{n+1}$ at the $n + 2$ points where $T_{n+1} = \pm 1$. These points are $x_k = \cos[k\pi/(n+1)]$, with $k = 0, 1, \ldots, n+1$. Comparing coefficients of x^{n+1} on both sides above, we also find that $a_{n+1} = 2^{-n}$. [The leading coefficient of $T_{n+1}(x)$ is 2^n.] The result of Problem 13.43 now applies and shows that $p(x)$ is the min–max polynomial, with $E = 2^{-n}$. As illustrations the sums in Problem 12.91 may be truncated to obtain

$$n = 1 \qquad x^2 \simeq \frac{1}{2} T_0 \qquad\qquad \text{Error} = \frac{T_2}{2}$$

$$n = 2 \qquad x^3 \simeq \frac{3}{4} T_1 \qquad\qquad \text{Error} = \frac{T_3}{4}$$

$$n = 3 \qquad x^4 \simeq \frac{1}{8}(3T_0 + 4T_2) \qquad \text{Error} = \frac{T_4}{8}$$

$$n = 4 \qquad x^5 \simeq \frac{1}{16}(10T_1 + 5T_3) \qquad \text{Error} = \frac{T_5}{16}$$

and so on. Note that in each case the min–max polynomial (of degree n or less) is actually of degree $n - 1$.

13.45 Show that in any series of Chebyshev polynomials $\sum_{i=0}^{\infty} a_i T_i(x)$, each partial sum S_n is the min–max polynomial of degree n or less for the next sum S_{n+1}. [The interval is again taken to be $(-1,1)$.]

■ Just as in the previous problem, but with $y(x) = S_{n+1}(x)$ and $p(x) = S_n(x)$, we have

$$E(x) = S_{n+1}(x) - S_n(x) = a_{n+1} T_{n+1}(x)$$

The result of Problem 13.43 again applies. Note also, however, that $S_{n-1}(x)$ may not be the min–max polynomial of degree $n-1$ or less, since $a_n T_n + a_{n+1} T_{n+1}$ is not necessarily an equal ripple function. (It was in the previous problem, however, since a_n was zero.)

13.46 Use the result of Problem 13.44 to economize the polynomial $y(x) = x - \frac{1}{6}x^3 + \frac{1}{120}x^5$ to a cubic polynomial, for the interval $(-1,1)$.

■ This was actually accomplished in Problem 12.96, but we may now view the result in a new light. Since

$$x - \frac{1}{6}x^3 + \frac{1}{120}x^5 = \frac{169}{192}T_1 - \frac{5}{128}T_3 + \frac{1}{1920}T_5$$

the truncation of the T_5 term leaves us with the min–max polynomial of degree 4 or less for $y(x)$, namely

$$P(x) = \frac{169}{192}x - \frac{5}{128}(4x^3 - 3x)$$

This is still only approximately the min–max polynomial of the same degree for $\sin x$. Further truncation of the T_3 term would not produce a min–max polynomial for $y(x)$, not exactly anyway.

13.47 Find the min–max polynomial of degree 1 or less, on the interval (a, b), for a function $y(x)$ with $y''(x) > 0$.

■ Let the polynomial be $P(x) = Mx + B$. We must find three points $x_1 < x_2 < x_3$ in (a, b) for which $E(x) = y(x) - P(x)$ attains its extreme values with alternate signs. This puts x_2 in the interior of (a, b) and requires $E'(x_2)$ to be zero, or $y'(x_2) = M$. Since $y'' > 0$, y' is strictly increasing and can equal M only once, which means that x_2 can be the only interior extreme point. Thus $x_1 = a$ and $x_3 = b$. Finally, by the equal ripple property,

$$y(a) - P(a) = -[y(x_2) - P(x_2)] = y(b) - P(b)$$

Solving, we have

$$M = \frac{y(b) - y(a)}{b - a} \qquad B = \frac{y(a) + y(x_2)}{2} - \frac{(a + x_2)[y(b) - y(a)]}{2(b - a)}$$

with x_2 determined by $y'(x_2) = [y(b) - y(a)]/(b - a)$.

13.48 Apply the previous problem to $y(x) = -\sin x$ on the interval $(0, \pi/2)$.

■ We find $M = -2/\pi$ first. Then from $y'(x_2) = M$, $x_2 = \arccos(2/\pi)$. Finally,

$$B = -\frac{1}{2}\sqrt{1 - \frac{4}{\pi^2}} + \frac{1}{\pi}\arccos\frac{2}{\pi}$$

and from $P(x) = Mx + B$ we find

$$\sin x \simeq \frac{2x}{\pi} + \frac{1}{2}\sqrt{1 - \frac{4}{\pi^2}} + \frac{1}{\pi}\arccos\frac{2}{\pi}$$

the approximation being the min–max line.

13.49 Show that $P(x) = x^2 + \frac{1}{8}$ is the min–max *cubic* (or less) approximation to $y(x) = |x|$ over the interval $(-1,1)$.

■ The error is $E(x) = |x| - x^2 - \frac{1}{8}$ and takes the extreme values $-\frac{1}{8}, \frac{1}{8}, -\frac{1}{8}, \frac{1}{8}, -\frac{1}{8}$ at $x = -1, -\frac{1}{2}, 0, \frac{1}{2}, 1$. These alternating errors of maximal size $E = \frac{1}{8}$ at $n + 2 = 5$ points guarantee (by Problem 13.43) that $P(x)$ is the min–max polynomial of degree $n = 3$ or less.

13.50 Find the min–max polynomial of degree 5 or less for $y(x) = x^6$ on the interval $(-1, 1)$. What is the error?

▌ The procedure is now routine. From Problem 12.91,

$$x^6 = \tfrac{1}{32}(10T_0 + 15T_2 + 6T_4 + T_6)$$

so chopping off the T_6 term, and thereby making an equal ripple error, we have the approximation

$$\tfrac{1}{32}\left[10 + 15(2x^2 - 1) + 6(8x^4 - 8x^2 + 1)\right] = \tfrac{1}{32}[1 - 18x^2 + 48x^4]$$

For an interesting comparison, see Problem 13.24 where almost the same approximation was found for $y = x^3$ on a discrete set.

For an encore, the same routine takes

$$x^7 = \tfrac{1}{64}(35T_1 + 21T_3 + 7T_5 + T_7)$$

and chops the final term to achieve the approximation

$$\tfrac{1}{64}\left[35x + 21(4x^3 - 3x) + 7(16x^5 - 20x^3 + 5x)\right] = \tfrac{7}{64}[x - 8x^3 + 16x^5]$$

leaving the equal ripple error $\tfrac{1}{64}T_7$.

13.51 The function $T_0 + T_1 + T_2 + T_3$ has the min–max approximation $T_0 + T_1 + T_2$, since the truncated term has just the proper error behavior. Show, however, that $T_0 + T_1$ is not the min–max line for the given function.

▌ The error of $T_0 + T_1$, considered as an approximation to the original function, is $T_2 + T_3$. All that is needed is to show that this is not an equal ripple. Calculus would surely oblige with impeccable logic, but with computer in hand it is convincing enough to run off a string of values. This error term begins at 0, rises to about .7, falls to -1.6 and then climbs back up to 2. It does ripple, but not equally. Of course, if we think of the pair $T_0 + T_1$ as an approximation to $T_0 + T_1 + T_2$, then we do have the equal ripple error T_2 and a min–max approximation.

13.52 Apply the method of Problem 13.47 to the function $y(x) = -\cos x$ on $(0, \pi/2)$.

▌ First we find

$$M = \frac{y(b) - y(a)}{b - a} = \frac{2}{\pi}$$

Then x_2 is needed and from

$$y'(x_2) = \sin x_2 = M = \frac{2}{\pi}$$

it follows that $x_2 = \arcsin(2/\pi)$. This leads to

$$B = \frac{1}{2}[y(a) + y(x_2) - (a + x_2)M] = \frac{1}{2}\left(-1 - \sqrt{1 - \frac{4}{\pi^2}}\right) - \frac{1}{\pi}\arcsin\left(\frac{2}{\pi}\right)$$

which works out to roughly -1.10. The min–max line is $y = Mx + B$.

13.53 Apply the same method to $y(x) = x^n$ on $(0, 1)$.

▌ The slope M will be 1, so x_2 is found from $y'(x_2) = nx_2^{n-1} = 1$, making it

$$x_2 = (1/n)^{1/(n-1)}$$

after which $B = \frac{1}{2}(0 + x_2^n - x_2)$ and we have the min–max line

$$p = x + \tfrac{1}{2}(x_2^n - x_2)$$

For $n = 2$ it reduces to $x - \tfrac{1}{8}$ while for $n = 3$ it is $x - 1/3\sqrt{3}$. As n becomes infinite, x_2 has limit 1 and x_2^n limit 0. The min–max line approaches $y = x - \tfrac{1}{2}$, which still has the equal error property, as Fig. 13-8 suggests.

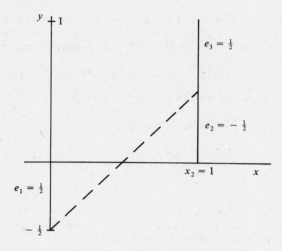

Fig. 13-8

13.54 The function $y(x) = |x|$ is concave upward, but its second derivative is hardly positive. Can the method of Problem 13.47 be used with this function, say on $(-1, 1)$?

❚ Giving it a try, we at once find $M = 0$ and are faced with the impossible $y'(x_2) = 0$. Cheating just a bit, we choose the tempting $x_2 = 0$ and then find B to be $\frac{1}{2}$. The line $y = \frac{1}{2}$ is then in our hands and is definitely min–max, with three maximum errors of size $\frac{1}{2}$ and alternating sign. So yes, the method can be used (or abused) with this function.

13.55 Use the function $y(x) = e^x$ on the interval $(-1, 1)$ to illustrate the *exchange method* for finding a min–max line.

❚ The method of Problem 13.47 would produce the min–max line, but for a simple first illustration, we momentarily ignore that method and proceed by exchange, imitating the procedure of Problem 13.7. Since we are after a line, we need $n + 2 = 3$ points of maximum error $\pm E$. Try $x = -1, 0, 1$ for an initial triple. The corresponding values of $y(x)$ are about .368, 1 and 2.718. The equal error line for this triple is easily found to be

$$p_1(x) = 1.175x + 1.272$$

with errors $h = \pm .272$ on the triple. Off the triple, a computation of the error at intervals of .1 discovers a maximum error of size $H = .286$ (and negative) at $x = .2$. Accordingly we form a new triple, exchanging the old argument $x = 0$ for the new $x = .2$. This retains the alternation of error signs called for in Step 4 of the exchange method as presented earlier and which we are now imitating. On the new triple $y(x)$ takes the values .368, 1.221 and 2.718 approximately. The equal error line is found to be

$$p_2(x) = 1.175x + 1.264$$

with errors $h = \pm .278$ on the triple. Off the triple, anticipating maximum errors near $x = .2$, we check this neighborhood at intervals of .01 and find an error of .279 at $x = .16$. Since we are carrying only three places, this is the best we can expect. A shift to the triple $x = -1, .16, 1$ would actually reproduce $p_2(x)$.

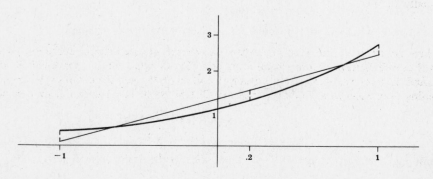

Fig. 13-9

Let us now see what the method of Problem 13.47 manages. With $a = -1$ and $b = 1$ it at once produces $M = (2.718 - .368)/2 = 1.175$. Then the equation $y'(x_2) = e^{x_2} = 1.175$ leads to $x_2 \simeq .16$, after which the result $B = 1.264$ is direct. The line is shown in Fig. 13-9, with the vertical scale compressed.

13.56 Use the exchange method to find the min–max quadratic polynomial for $y(x) = e^x$ over $(-1, 1)$.

▌ Recalling that truncation of a series of Chebyshev polynomials often leads to nearly equal ripple errors resembling the first omitted term, we take as our initial quadruple the four extreme points of $T_3(x)$, which are $x = \pm 1, \pm \frac{1}{2}$. The parabola that misses the four points

x	-1	$-\frac{1}{2}$	$\frac{1}{2}$	1
e^x	.3679	.6065	1.6487	2.7183

alternately by $\pm h$ proves to have its maximum error at $x = .56$. The new quadruple $(-1, -.5, .56, 1)$ then leads to a second parabola with maximum error at $x = -.44$. The next quadruple is $(-1, -.44, .56, 1)$ and proves to be our last. Its equal ripple parabola is, to five decimal places,

$$p(x) = .55404x^2 + 1.13018x + .98904$$

and its maximum error both inside and outside the quadruple is $H = .04502$.

13.57 Use the exchange method to find the min–max line for $y(x) = \cos x$ on the interval $(0, \pi/2)$.

▌ With a continuous data supply we have plenty of triples. The choice $x = 0, .75, \pi/2$ seems impartial and leads to $h = .1046$ and $m = -.6366$. This slope is, of course, $-2/\pi$, since the endpoints of the triple are what determine it. The next task is to scan the interval $(0, \pi/2)$ for error behavior. Calculating at steps of size .1 the largest error is found to be $-.1059$ at $x = .7$. A slight adjustment is indicated for the triple, and we replace .75 by .7, leading to $h = .1052$ and the same m as before. Repeating the scan of errors we find none larger than h. The question at this point is how exhaustive a search to make. Should we scan again at intervals of .01? The answer depends on the level of accuracy desired and the likelihood of improvement. The decision here is to settle for what we have,

$$y = -.6366x + 1.105$$

since one straight line cannot represent the cosine function very well over this range.

13.58 Is there an optimal choice of initial triple?

▌ Only the ultimate triple is optimal, but there is this simple argument. If the given function $y(x)$ were quadratic, then the min–max line would be available by truncating the $T_2(x)$ term in the representation

$$y(x) = aT_0(x) + bT_1(x) + cT_2(x)$$

placing the error extremes at those of $T_2(x)$, namely $x = -1, 0, 1$. A variable change is needed to shift the standard interval $(-1, 1)$ to whatever interval (a, b) is active, but the indicated choice of initial triple is clear, equal spacing. In the preceding example this advice was more or less heeded and the initial triple was almost the last.

13.59 Find the min–max line for $y(x) = \sqrt{x}$ over $(1, 9)$.

▌ The initial triple $x = 1, 4, 9$ almost shouts to be used and is so attractive that ignoring the expert advice just received is to be forgiven, and 4 is fairly close to the midpoint 5 anyway. So proceeding, we find $h = \frac{1}{8}$, $m = \frac{1}{4}$ and eventually the line

$$y = \tfrac{1}{4}x + \tfrac{7}{8}$$

Checking the errors we find none larger than h anywhere, so our choice of initial triple was not only attractive, but optimal as well. Is there a more general result here? Trying again, on $(9, 25)$ with initial triple $x = 9, 16, 25$, we once more find the min–max line without exchanges, and on $(9, 49)$ with the initial triple $x = 9, 25, 49$ it is the same story.

Passing from experiment to theory, the basic system for the equal error line on a triple

$$a + bx_1 - h = y_1$$
$$a + bx_2 + h = y_2$$
$$a + bx_3 - h = y_3$$

at once requires its slope to be $b = (y_3 - y_1)/(x_3 - x_1)$, that is, the slope of the segment joining the endpoints. The error of the resulting line is $E = a + bx - y(x)$. To find its interior maximum we set $b - y'(x)$ to zero, and have the mean-value point where tangent and chord are parallel. But for the square root curve, with x_1 and x_3 the endpoints of the triple, this is

$$\frac{1}{2\sqrt{x}} = \frac{\sqrt{x_3} - \sqrt{x_1}}{x_3 - x_1} \quad \text{or} \quad \frac{1}{2y} = \frac{y_3 - y_1}{y_3^2 - y_1^2}$$

which can quickly be reduced to $y = \frac{1}{2}(y_1 + y_3)$. So if the y values are in arithmetic progression, then their squares make an optimal initial triple for this function.

13.60 Use the exchange method to find the min–max parabola for $y(x) = \cos x$ to over $(0, \pi/2)$.

∎ The min–max line was found in Problem 13.57 and had an error $h = .1$. How much better can a parabola do? Choosing the initial quadruple $x = 0, .5, 1, \pi/2$ led to this chain of events:

Quadruple	h	Max error
$0, .5, 1, \pi/2$	.012	.017 at $x = 1.2$
$0, .5, 1.2, \pi/2$	.0137	$-.0140$ at $x = .4$
$0, .4, 1.2, \pi/2$	.0138	on the quadruple

The error is down by an order of magnitude. The final parabola was $p = 1.014 - .1339x - .3312x^2$. How close is the final quadruple to the set of extreme points of $T_3(x)$, recommended by an extension of the argument used in Problem 13.58 as initial quadruple? The extremes of $4x^3 - 3x$ occur for $x = \pm 1, \pm \frac{1}{2}$ and translate to the interval $(0, \pi/2)$ as $0, \pi/8, 3\pi/8, \pi/2$. To three places the two interior arguments are .393 and 1.178, not too far from our .4 and 1.2. Using this professional starting quadruple would have led to the equal error $h = .0137$ and slightly larger errors at both $x = .4$ and $x = 1.2$.

13.61 Find a polynomial of minimum degree that approximates the cosine function over $(0, \pi/2)$ with maximum error error .005.

∎ The parabola of the preceding problem has not quite made this grade, so we move to a cubic. The extreme points of $T_4(x)$ are at $x = 0, \pm 1, \pm 1/\sqrt{2}$, which translate to $(0, \pi/2)$ as $0, .24, .79, 1.34$ and $\pi/2$ more or less. Choosing this as initial quintuple leads to $h = -.00136$, so apparently we have the accuracy requested. It is hard to resist the temptation to test the error across the interval, and computations show that errors slightly larger than h do occur. Minor adjustments eventually settle on the quintuple $0, .22, .76, 1.32$ and $\pi/2$, with $h = -.001367$ and the cubic

$$p = .9996 + .0296x - .6009x^2 + .1125x^3$$

13.62 What is the minimum degree of a polynomial approximation to e^x on $(0, 1)$ with maximum error .005 or less?

∎ Try a quadratic first. The extremes of $T_3(x)$ are at $x = \pm 1$ and $\pm \frac{1}{2}$, which translate to our $(0, 1)$ interval as $0, \frac{1}{4}, \frac{3}{4}, 1$. With this as initial quadruple, the equal error $h = .009$ is found with no significantly larger errors off the quadruple. Quadratic approximation is not good enough. Turning to cubics, the extreme points of $T_4(x)$ are still at $0, \pm 1$ and $\pm 1/\sqrt{2}$, which now translate to $0, .15, .50, .85, 1$. For this initial quintuple the error h turns out to be $-.00054$, almost an order of magnitude better than the specification. Checking the errors at intervals of .01 locates a few slightly larger than h, leading to minor adjustments, a final quintuple of $0, .15, .51, .86, 1$ and the cubic

$$p = .9995 + 1.0166x + .4217x^2 + .2800x^3$$

13.63 The Taylor series for $\log(1 + x)$ converges so slowly that hundreds of terms would be needed for reasonable accuracy on the interval $(0, 1)$. How well does the min–max cubic perform?

∎ Taking the extremes of $T_4(x)$ as initial quintuple, translated as before to $0, .15, .5, .85, 1$ for the $(0, 1)$ interval, we find the equal error $h = .00043$ on the quintuple. Searching elsewhere we find slightly larger errors near .13, .47 and .84. Making these adjustments to the quintuple causes only an insignificant change in h, to .00044, with no larger errors elsewhere. The min–max cubic proves to be

$$p = .00044 + .98350x - .40004x^2 + .10969x^3$$

From this cubic we get three place accuracy.

13.64   Continue the preceding problem to find the min–max polynomial of degree 4 for $\log(1 + x)$.

▪   The extreme values of $T_5(x)$, ± 1 of course, occur at $0, 1$ and where $x^2 = (3 \pm \sqrt{5})/8$. A bit of arithmetic then recommends the initial sextuple $0, .10, .35, .65, .90, 1$ for our $(0, 1)$ interval. The modifications required in the linear system that determines h and the coefficients are fairly obvious and lead to $h = .00006$. Scanning the error spectrum across the interval, we find a few very slightly larger ones near $.32$ and $.63$. Making these adjustments to the sextuple leaves h the same to five places, with no larger errors elsewhere. The min–max quartic is

$$p = .00006 + .99652x - .46772x^2 + .22072x^3 - .05649x^4$$

and gives almost four place accuracy.

The sixth degree

$$p(x) = .999902x - .497875x^2 + .317650x^3 - .193761x^4 + .085569x^5 - .018339x^6$$

has been used for this function. An error scan will find a maximum less than two units in the sixth decimal place.

13.65   The sum $1 - x + x^2 - x^3 + x^4 - x^5 + x^6$ is the beginning of the series for $1/(1 + x)$, a binomial series. Find a minimum degree polynomial that approximates it with error no greater than $.005$.

▪   Trying degree 3 we find $h = -.02$, which is not good enough. For degree 4 we make the usual start, the extremes of $T_5(x)$ providing the initial sextuple. The error h now proves to be $.0039$, which is satisfactory. The usual error scan finds none larger than h, so we accept the accompanying polynomial

$$p = 1.0040 - 1.1768x + 2.2327x^2 - 3.7509x^3 + 2.6870x^4$$

13.66   Prove that the min–max polynomial approximation to $y(x) = 0$, among all polynomials of degree n with leading coefficient 1, is $2^{1-n}T_n(x)$. The interval of approximation is $(-1, 1)$.

▪   The proof of this historic theorem runs as follows. Let

$$p(x) = x^n + a_1 x^{n-1} + \cdots + a_n$$

be any polynomial of the type described. Since $T_n(x) = \cos(n \arccos x)$, we have

$$\max \left| 2^{1-n}T_n(x) \right| = 2^{1-n}$$

Notice that this polynomial takes its extreme values of $\pm 2^{1-n}$ alternately at the arguments $x_k = \cos k\pi/n$, where $k = 0, 1, \ldots, n$. Suppose that some polynomial $p(x)$ were such that

$$\max \left| p(x) \right| < 2^{1-n}$$

and let

$$P(x) = p(x) - 2^{1-n}T_n(x)$$

Then $P(x)$ is of degree $n - 1$ or less and it does not vanish identically since this would require $\max|p(x)| = 2^{1-n}$. Consider the values $P(x_k)$. Since $p(x)$ is dominated by $2^{1-n}T_n(x)$ at these points, we see that the $P(x_k)$ have alternating signs. Being continuous, $P(x)$ must therefore have n zeros between the consecutive x_k. But this is impossible for a polynomial of degree $n - 1$ or less which does not vanish identically. This proves that $\max|p(x)| \geq 2^{1-n}$.

13.67   Compare approximations to $y(x) = x$ by a constant c on $(0, 1)$ using various L_p norms.

▪   Suppose we take L_1 first. Then the area between the line $y = x$ and the horizontal $y = c$ is to be minimized. This consists of two triangles with combined area $\frac{1}{2}(1 - c)^2 + \frac{1}{2}c^2$ and is least when $c = \frac{1}{2}$. The minimum value is $\frac{1}{4}$.

For other norms L_p we have, denoting the norm N,

$$N^p = \int_0^1 (x - c)^p \, dx = \frac{(1 - c)^{p+1} - (-c)^{p+1}}{p + 1} = \frac{(1 - c)^{p+1} + c^{p+1}}{p + 1}$$

where it is assumed that p is even. This is a minimum when $c = \frac{1}{2}$, as the first derivative will testify, and the minimum value is

$$N^p = \frac{1}{2^p(p+1)} \quad \text{or} \quad N = \frac{1}{2}(p+1)^{-1/p}$$

which approaches $\frac{1}{2}$ as p becomes infinite. The min–max, or L_∞ norm, is clearly optimal when $c = \frac{1}{2}$, with extreme errors of size $\frac{1}{2}$ at $x = 0$ and 1. The L_p minima thus approach the L_∞ minimum as p increases, as anticipated.

13.68   Show that the min–max norm determines the optimal value $c = \frac{1}{2}$ more sharply than any of the others.

▮   A plot of some of the curves

$$N = \left[\frac{(1-c)^{p+1} + c^{p+1}}{p+1} \right]^{1/p}$$

appears in Fig. 13-10. For small p values the curves are rather flat-bottomed, the value of N changing little in the neighborhood of the minimum. In contrast, the min–max norm points to $c = \frac{1}{2}$ like an arrowhead. Confirmation by the second derivative is at hand. Differentiating N^p brings $pN^{p-1}N'(c) = -(1-c)^p + c^p$, or

$$N'(c) = \frac{-(1-c)^p + c^p}{pN^{p-1}}$$

In taking the second derivative it is useful to note that at the point of interest, $c = \frac{1}{2}$, the numerator is zero. The familiar rule for derivative of a quotient thus manages

$$N''\left(\frac{1}{2}\right) = \left. \frac{p(1-c)^{p-1} + pc^{p-1}}{pN^{p-1}} \right|_{c=1/2} = \frac{1}{2^{p-2}N^{p-1}}$$

and inserting the norm

$$N''\left(\frac{1}{2}\right) = \frac{2^{p-1}(p+1)^{(p-1)/p}}{2^{p-2}} = 2(p+1)^{(p-1)/p}$$

so that this derivative becomes infinite with p. The minimum thus becomes increasingly sharp.

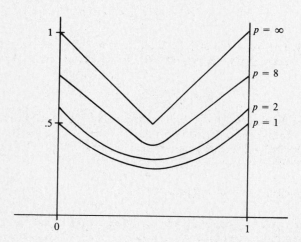

Fig. 13-10

13.69   Find optimal lines of the form $y = mx$ for the data points $(0,0)$, $(1,0)$, $(2,0)$ and $(3,1)$ using various norms.

▮   A glance at Fig. 13-11 will make it clear that only values of m from 0 to $\frac{1}{3}$ need be considered. At the four data points the errors are 0 (forced) and $m, 2m, 1 - 3m$. In the L_1 norm we sum these errors and find a total of 1 for any m under consideration. In other words, all lines $y = mx$ with m between 0 and $\frac{1}{3}$ are optimal, with minimum norm of 1.

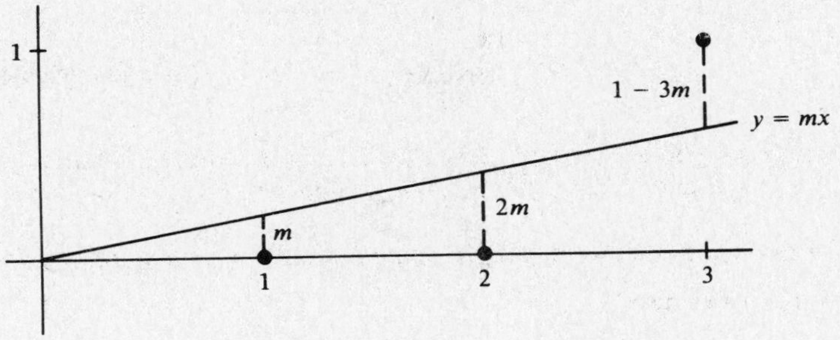

Fig. 13-11

Turning to the L_p norm, again denoted N, we have

$$N^p = m^p + (2m)^p + (1 - 3m)^p = (1 + 2^p)m^p + (1 - 3m)^p$$

and to find its minimum compute

$$pN^{p-1}N'(m) = p(1 + 2^p)m^{p-1} - 3p(1 - 3m)^{p-1}$$

$$N'(m) = \frac{(1 + 2^p)m^{p-1} - 3(1 - 3m)^{p-1}}{N^{p-1}}$$

which is zero when

$$m = \frac{a}{1 + 3a} \quad \text{with} \quad a = \left[\frac{3}{1 + 2^p}\right]^{1/(p-1)}$$

as a bit of algebra shows. Substituting back, we then discover the minimum value itself:

$$N_{\min} = \left[\frac{(1 + 2^p)a^p + 1}{(1 + 3a)^p}\right]^{1/p}$$

For example, taking $p = 2$ leads to $a = \frac{3}{5}$ and $m = \frac{3}{14}$, and then $N = \sqrt{46}/14$, roughly .48.

What happens as p becomes infinite? Noting that

$$(p - 1)\log a = \log 3 - \log(1 + 2^p)$$

which for large p is essentially $-p \log 2$, we find that a has the limit $\frac{1}{2}$. This makes $\lim m = \frac{1}{5}$ and also leads to the limit $\frac{2}{5}$ for $N_{\min}$. Is this also the optimal error for the norm L_∞, under which we must minimize the maximum error? Since

$$\max(m, 2m, 1 - 3m) = \max(2m, 1 - 3m)$$

for the range of m needing to be considered, experience suggests equal errors at the two active points, or $2m = 1 - 3m$. This does mean $m = \frac{1}{5}$ and equal errors of size $\frac{2}{5}$ with alternating signs on the "double" $x = 2, 3$. It is in fact the min–max line and has maximum error of $\frac{2}{5}$. Notice that in this problem, as p ranges from 1 to ∞, the estimate of optimal error decreases from 1 to $\frac{2}{5}$.

13.70 Show that, as in Problem 13.68, the min–max norm determines its optimal value of m more sharply than any of the other norms do.

▌ Figure 13-12 provides a plot of $N(m)$ for various values of p. The "curve" for $p = 1$ is truly flat-bottomed and the gradual modulation into the $p = \infty$ arrowhead is evident. It is probably not worth observing that the arrowhead is asymmetric this time, since a downshift in m causes the error at $x = 3$ to grow at a rate of $3m$, while an upshift causes the error at $x = 2$ to grow at a rate of $2m$.

The second derivative confirms this gradual sharpening of the optimum. Returning to

$$N'(m) = \frac{(1 + 2^p)m^{p-1} - 3(1 - 3m)^{p-1}}{N^{p-1}}$$

and again noting that the numerator will be zero for the optimal m, differentiation brings

$$N''(m)|_{\text{opt}} = \frac{(1+2^p)(p-1)m^{p-2} + 9(p-1)(1-3m)^{p-2}}{N^{p-1}}\bigg|_{\text{opt}}$$

both terms of which will be found to behave like p as p becomes infinite.

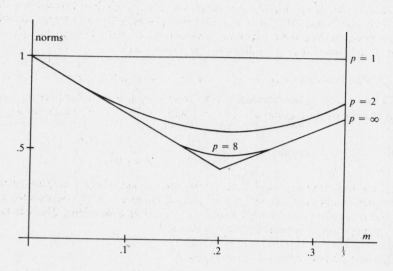

Fig. 13-12

13.71 Show that for the data points $(0,0)$, $(1,1)$, $(2,0)$ and $(3,1)$ the min–max and L_1 lines of type $y = mx$ are the same. Does this imply that the lines for other L_p norms will conform?

❚ It takes only a moment to decide that the line will have to pass under-over-under the last three points. Any other line will definitely have larger errors. The three active errors are thus $1 - m$, $2m$ and $1 - 3m$. With L_1 their sum $2 - 2m$ must be minimal, so we make m as large as we can, which is $m = \frac{1}{3}$. The minimum norm is $\frac{4}{3}$. Turning to L_∞, we seek to minimize $\max(1 - m, 2m, 1 - 3m)$. To arrange equal errors (opposite signs) on the $x = 1, 2$ double, we set $1 - m$ equal to $2m$ and find $m = \frac{1}{3}$. The error on this double is $\frac{2}{3}$ and off the double it is 0, familiar characteristics of a min–max approximation. Even intuitively one can judge that any small rotation of this line will cause a larger error somewhere. So the two norms do lead to the same line.

As for the L_2 norm, now we are to minimize

$$(1-m)^2 + 4m^2 + (1-3m)^2$$

and find in the usual way that m should be $\frac{2}{7}$. So it seems that the answer to the question posed is, no.

13.72 For the data points $(0,0)$, $(1,1)$, $(2,0)$ and $(3,0)$ determine the L_1, L_2 and L_∞ optimal lines of type $y = mx$ mentally.

❚ The slopes are 0, $\frac{1}{14}$ and $\frac{1}{4}$ in that order.

13.6 THE $L1$ NORM

13.73 Why is the problem of finding the best approximation to a function $y(x)$ according to the $L1$ norm usually exchanged for that of arranging collocation at certain selected points?

❚ For one thing, $L1$ norms are not all that easy to minimize. The following reasoning leads to a substitute process that often comes close. Taking the somewhat general approximating function

$$f(x) = a_1\phi_1(x) + a_2\phi_2(x) + \cdots + a_n\phi_n(x)$$

with the $L1$ error measured by the norm

$$\int_a^b |y(x) - f(x)|\, dx$$

and optimistically differentiating relative to the coefficient a_i (even though the derivative fails to exist at any point where the integrand is zero) we have

$$\int_a^b \text{sign}[\,y(x) - f(x)\,]\phi_i(x)\,dx = 0$$

since the derivative of $|x|$ relative to x is either 1 or -1. It is not easy to implement this equation fully, but it does suggest that what really matters is the sign of the error. This focuses attention on where the sign changes, that is, points where $y(x)$ and $f(x)$ collocate. For some sets of $\phi_i(x)$, more or less optimal choices of these points of collocation have been found. They are called canonical points and good *L1* approximations, often optimal, can be found by interpolating the given function at them. For polynomial approximation on $(-1,1)$ with $\phi_i(x) = x^{i-1}$, a set of points is $x_i = \cos i\pi/(n+1)$, called the Chebyshev points.

13.74 Apply the preceding process to $y(x) = e^x$ over $(-1,1)$, seeking a line $p = a + bx$ that minimizes the *L1* norm.

▌ Here $\phi_1(x) = 1$ and $\phi_2(x) = x$, with $n = 2$. The Chebyshev points are $\cos \pi/3$ and $\cos 2\pi/3$, or $\pm \frac{1}{2}$. Collocation at these points requires $a + \frac{1}{2}b = e^{1/2}$ and $a - \frac{1}{2}b = e^{-1/2}$, which make

$$a = \tfrac{1}{2}(e^{1/2} + e^{-1/2}) = 1.1276$$

$$b = e^{1/2} - e^{-1/2} = 1.0422$$

The value of the norm for these coefficients can be found by any of our quadrature methods and proves to be .2681. Parallel integrations will show that shifting the value of either a or b will increase the norm. That is, the method of canonical points seems to have actually found the optimal approximation. This has been proved to be the case whenever no other contact is made by the two curves.

13.75 Also find the best *L1* line over $(-1,1)$ for $y(x) = x^3$.

▌ Symmetry suggests that the line will have to pass through the origin. If we leave it to the canonical points, which remain $\pm \frac{1}{2}$, the collocation equations $a + b/2 = \frac{1}{8}$ and $a - b/2 = -\frac{1}{8}$ will force it through there, since they make $a = 0$ and $b = \frac{1}{4}$. The line $p = x/4$ seems to be indicated.

For this example, the *L1* integral can be evaluated exactly. The intersection points of $y(x) = x^3$ and $p(x) = bx$ for any likely value of b will be at $x = \pm \sqrt{b}$. By symmetry we can compute the integral over $(0,1)$ and double, so that the two subintervals $(0, \sqrt{b})$ and $(\sqrt{b}, 1)$ are sufficient. The result

$$\int_{-1}^1 |y(x) - p(x)|\,dx = b^2 - b + \tfrac{1}{2}$$

follows by elementary calculus. The minimum value is clearly $\frac{1}{2}$, disagreeing with the canonical points. For $b = \frac{1}{2}$ the *L1* error is $\frac{1}{4}$, while for $b = \frac{1}{4}$ it is $\frac{5}{16}$. The discrepancy is modest and the blame rests with the changing curvature of the cubic curve.

13.76 Apply the canonical point method to $y(x) = 1 - x^4$ over $(-1,1)$, seeking an approximation of the form $p(x) = ax^2 + bx + c$.

▌ The canonical points are now $x_i = \cos \pi i/4$, or $0, \pm 1/\sqrt{2}$. Requiring $y(x) = p(x)$ at these points leads to $c = 1$ and

$$\frac{3}{4} = 1 + \frac{\sqrt{2}}{2}b + \frac{1}{2}a \qquad \frac{3}{4} = 1 - \frac{\sqrt{2}}{2}b + \frac{1}{2}a$$

from which come $b = 0$ and $a = -\frac{1}{2}$. The corresponding polynomial is $p(x) = 1 - \frac{1}{2}x^2$. This is a slight surprise, since it misses the two endpoints by half a unit. Is it the *L1* solution? To find out we use numerical quadrature methods on the error integral, since finding the points of intersection needed to manage a direct evaluation seems pointless. The value for $a, b, c = -\frac{1}{2}, 0, 1$ proves to be .1657. Retaining $b = 0$ for the symmetry, informal experimenting with a and c soon finds lower errors near $a = -.65$ and $c = 1.04$, the minimum being about .135 and the optimal

$$p(x) = -.65x^2 + 1.04$$

The value of c means a slight overshoot at top center. What of the choice $a = -1$, with $p(x) = 1 - x^2$ and end values 0 to match those of the given function? The error for this quadratic approximation turns out to be .27, just about double the minimum error.

CHAPTER 14
Approximation by Rational Functions

14.1 THE COLLOCATION RATIONAL FUNCTION

14.1 Find the rational function $y(x) = 1/(a + bx)$ given that $y(1) = 1$ and $y(3) = \frac{1}{2}$.

▮ Substitution requires $a + b = 1$ and $a + 3b = 2$, which force $a = b = \frac{1}{2}$. The required function is $y(x) = 2/(1 + x)$. This simple problem illustrates the fact that finding a rational function by collocation is equivalent to solving a set of *linear* equations for the unknown coefficients.

14.2 Also find rational functions $y_2(x) = Mx + B$ and $y_3(x) = c + d/x$ which have $y(1) = 1$ and $y(3) = \frac{1}{2}$.

▮ The linear function $y_2(x) = (5 - x)/4$ may be found by inspection. For the other we need to satisfy the coefficient equations $c + d = 1$ and $3c + d = \frac{1}{2}$ and this means that $c = \frac{1}{4}$ and $d = \frac{3}{4}$, making $y_3(x) = (x + 3)/4x$. We now have three rational functions that pass through the three given points. Certainly there are others, but in a sense these are the simplest. At $x = 2$ the three functions offer us the interpolated values $\frac{2}{3}$, $\frac{3}{4}$ and $\frac{5}{8}$. Inside the interval $(1, 3)$ all three resemble each other to some extent. Outside they differ violently (see Fig. 14-1). The diversity of rational functions exceeds that of polynomials and it is very helpful to have knowledge of the type of rational function required.

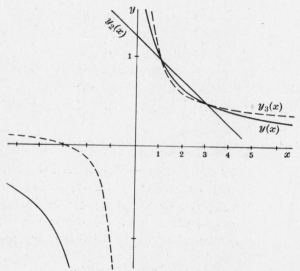

Fig. 14-1

14.3 Suppose it is known that $y(x)$ is of the form $y(x) = (a + bx^2)/(c + dx^2)$. Determine $y(x)$ by the requirements $y(0) = 1$, $y(1) = \frac{2}{3}$ and $y(2) = \frac{5}{9}$.

▮ Substitution brings the linear system

$$a = c \qquad a + b = \tfrac{2}{3}(c + d) \qquad a + 4b = \tfrac{5}{9}(c + 4d)$$

Since only the ratio of the two polynomials is involved one coefficient may be taken to be 1, unless it later proves to be 0. Try $d = 1$. Then one discovers that $a = b = c = \frac{1}{2}$ and $y(x) = (1 + x^2)/(1 + 2x^2)$. Note that the rational function $y_2(x) = 10(10 + 6x - x^2)$ also includes these three points and so does $y_3(x) = (x + 3)/[3(x + 1)]$.

14.2 CONTINUED FRACTIONS AND RECIPROCAL DIFFERENCES

14.4 Evaluate the continued fraction

$$y = 1 + \cfrac{x}{-3 + \cfrac{x-1}{-\frac{2}{3}}} \quad \text{at } x = 0, 1 \text{ and } 2$$

∎ Direct computation shows $y(0) = 1$, $y(1) = \frac{2}{3}$ and $y(2) = \frac{5}{9}$. These are again the values of the previous problem. The point here is that the structure of a continued fraction of this sort makes these values equal to the successive "convergents" of the fraction, that is, the parts obtained by truncating the fraction before the x and $x-1$ terms and, of course, at the end. One finds easily that the fraction also rearranges into our $y_3(x)$.

14.5 Develop the connection between rational functions and continued fractions in the case

$$y(x) = \frac{a_0 + a_1 x + a_2 x^2}{b_0 + b_1 x + b_2 x^2}$$

∎ We follow another historical path. Let the five data points (x_i, y_i) for $i = 1, \ldots, 5$ be given. For collocation at these points,

$$a_0 - b_0 y + a_1 x - b_1 xy + a_2 x^2 - b_2 x^2 y = 0$$

for each x_i, y_i pair. The determinant equation

$$\begin{vmatrix} 1 & y & x & xy & x^2 & x^2 y \\ 1 & y_1 & x_1 & x_1 y_1 & x_1^2 & x_1^2 y_1 \\ 1 & y_2 & x_2 & x_2 y_2 & x_2^2 & x_2^2 y_2 \\ 1 & y_3 & x_3 & x_3 y_3 & x_3^2 & x_3^2 y_3 \\ 1 & y_4 & x_4 & x_4 y_4 & x_4^2 & x_4^2 y_4 \\ 1 & y_5 & x_5 & x_5 y_5 & x_5^2 & x_5^2 y_5 \end{vmatrix} = 0$$

clearly has the required features. The second row is now reduced to $1, 0, 0, 0, 0, 0$ by these operations:

> multiply column 1 by y_i and subtract from column 2
> multiply column 3 by y_1 and subtract from column 4
> multiply column 5 by y_1 and subtract from column 6
> multiply column 3 by x_1 and subtract from column 5
> multiply column 1 by x_1 and subtract from column 3

At this point the determinant has been replaced by the substitute

$$\begin{vmatrix} 1 & y - y_1 & x - x_1 & x(y - y_1) & x(x - x_1) & x^2(y - y_1) \\ 1 & 0 & 0 & 0 & 0 & 0 \\ 1 & y_2 - y_1 & x_2 - x_1 & x_2(y_2 - y_1) & x_2(x_2 - x_1) & x_2^2(y_2 - y_1) \\ 1 & y_3 - y_1 & x_3 - x_1 & x_3(y_3 - y_1) & x_3(x_3 - x_1) & x_3^2(y_3 - y_1) \\ 1 & y_4 - y_1 & x_4 - x_1 & x_4(y_4 - y_1) & x_4(x_4 - x_1) & x_4^2(y_4 - y_1) \\ 1 & y_5 - y_1 & x_5 - x_1 & x_5(y_5 - y_1) & x_5(x_5 - x_1) & x_5^2(y_5 - y_1) \end{vmatrix}$$

Expand this determinant by its second row and then

> divide row 1 by $y - y_1$
> divide row i by $y_i - y_1$, for $i = 2, 3, 4, 5$

Introducing the symbol $\rho_1(xx_1) = (x - x_1)/(y - y_1)$, the equation may now be written as

$$\begin{vmatrix} 1 & \rho_1(xx_1) & x & x\rho_1(xx_1) & x^2 \\ 1 & \rho_1(x_2x_1) & x_2 & x_2\rho_1(x_2x_1) & x_2^2 \\ 1 & \rho_1(x_3x_1) & x_3 & x_3\rho_1(x_3x_1) & x_3^2 \\ 1 & \rho_1(x_4x_1) & x_4 & x_4\rho_1(x_4x_1) & x_4^2 \\ 1 & \rho_1(x_5x_1) & x_5 & x_5\rho_1(x_5x_1) & x_5^2 \end{vmatrix} = 0$$

The operation is now repeated, to make the second row $1,0,0,0,0$:

multiply column 1 by $\rho_1(x_2x_1)$ and subtract from column 2
multiply column 3 by $\rho_1(x_2x_1)$ and subtract from column 4
multiply column 3 by x_2 and subtract from column 5
multiply column 1 by x_2 and subtract from column 3

The determinant then has the form

$$\begin{vmatrix} 1 & \rho_1(xx_1) - \rho_1(x_2x_1) & x - x_2 & x[\rho_1(xx_1) - \rho_1(x_2x_1)] & x(x - x_2) \\ 1 & 0 & 0 & 0 & 0 \\ 1 & \rho_1(x_3x_1) - \rho_1(x_2x_1) & x_3 - x_2 & x[\rho_1(x_3x_1) - \rho_1(x_2x_1)] & x_3(x_3 - x_2) \\ 1 & \rho_1(x_4x_1) - \rho_1(x_2x_1) & x_4 - x_2 & x[\rho_1(x_4x_1) - \rho_1(x_2x_1)] & x_4(x_4 - x_2) \\ 1 & \rho_1(x_5 - x_1) - \rho_1(x_2x_1) & x_5 - x_2 & x[\rho_1(x_5x_1) - \rho_1(x_2x_1)] & x_5(x_5 - x_2) \end{vmatrix}$$

Expand by the second row and then

divide row 1 by $\rho_1(xx_1) - \rho_1(x_2x_1)$
divide row i by $\rho_1(x_{i+1}x_1) - \rho_1(x_2x_1)$, for $i = 2, 3, 4$

An additional step is traditional at this point in order to assure a symmetry property of the ρ quantities to be defined:

multiply column 1 by y_1 and add to column 2
multiply column 3 by y_1 and add to column 4

Introducing the symbol $\rho_2(xx_1x_2) = (x - x_2)/(\rho_1(xx_1) - \rho_1(x_2x_1)) + y_1$, the equation has now been reduced to

$$\begin{vmatrix} 1 & \rho_2(xx_1x_2) & x & x\rho_2(xx_1x_2) \\ 1 & \rho_2(x_3x_1x_2) & x_3 & x_3\rho_2(x_3x_1x_2) \\ 1 & \rho_2(x_4x_1x_2) & x_4 & x_4\rho_2(x_4x_1x_2) \\ 1 & \rho_2(x_5x_1x_2) & x_5 & x_5\rho_2(x_5x_1x_2) \end{vmatrix} = 0$$

Another similar reduction produces

$$\begin{vmatrix} 1 & \rho_3(xx_1x_2x_3) & x \\ 1 & \rho_3(x_4x_1x_2x_3) & x_4 \\ 1 & \rho_3(x_5x_1x_2x_3) & x_5 \end{vmatrix} = 0$$

where

$$\rho_3(xx_1x_2x_3) = \frac{x - x_3}{\rho_2(xx_1x_2) - \rho_2(x_3x_1x_2)} + \rho_1(x_1x_2)$$

Finally, the last reduction manages

$$\begin{vmatrix} 1 & \rho_4(xx_1x_2x_3x_4) \\ 1 & \rho_4(x_5x_1x_2x_3x_4) \end{vmatrix} = 0$$

where

$$\rho_4(xx_1x_2x_3x_4) = \frac{x - x_4}{\rho_3(xx_1x_2x_3) - \rho_3(x_4x_2x_3x_1)} + \rho_2(x_1x_2x_3)$$

We deduce that $\rho_4(xx_1x_2x_3x_4) = \rho_4(x_5x_1x_2x_3x_4)$. The various ρ_is just introduced are called *reciprocal differences* of order i, and the equality of these fourth-order reciprocal differences is equivalent to the determinant equation with which we began and that identifies the rational function we are seeking.

The definitions of reciprocal differences now lead in a natural way to a continued fraction. We find successively

$$y = y_1 + \frac{x - x_1}{\rho_1(xx_1)} = y_1 + \cfrac{x - x_1}{\rho_1(x_2x_1) + \cfrac{x - x_2}{\rho_2(xx_1x_2) - y_1}}$$

$$= y_1 + \cfrac{x - x_1}{\rho_1(x_2x_1) + \cfrac{x - x_2}{\rho_2(x_3x_1x_2) - y_1 + \cfrac{x - x_3}{\rho_3(xx_1x_2x_3) - \rho_1(x_1x_2)}}}$$

$$= y_1 + \cfrac{x - x_1}{\rho_1(x_2x_1) + \cfrac{x - x_2}{\rho_2(x_3x_1x_2) - y_1 + \cfrac{x - x_3}{\rho_3(x_4x_1x_2x_3) - \rho_1(x_1x_2) + \cfrac{x - x_4}{\rho_4(x_5x_1x_2x_3x_4) - \rho_2(x_1x_2x_3)}}}}$$

where, in the last denominator, the equality of certain fourth differences, which was the culmination of our extensive determinant reduction, has finally been used. This is what makes the preceding continued fraction the required rational function. (Behind all these computations there has been the assumption that the data points do actually belong to such a rational function and that the algebraic procedure will not break down at some point. See the problems for exceptional examples.)

14.6 Prove that reciprocal differences are symmetric.

❚ For first-order differences it is at once clear that $\rho_1(x_1x_2) = \rho_1(x_2x_1)$. For second-order differences one verifies first that

$$\frac{x_3 - x_2}{\dfrac{x_3 - x_1}{y_3 - y_1} - \dfrac{x_2 - x_1}{y_2 - y_1}} + y_1 = \frac{x_3 - x_1}{\dfrac{x_3 - x_2}{y_3 - y_2} - \dfrac{x_1 - x_2}{y_1 - y_2}} + y_2 = \frac{x_2 - x_1}{\dfrac{x_2 - x_3}{y_2 - y_3} - \dfrac{x_1 - x_3}{y_1 - y_3}} + y_3$$

from which it follows that in $\rho_2(x_1x_2x_3)$ the x_i may be permuted in any way. For higher-order differences the proof is similar.

14.7 Apply reciprocal differences to recover the function $y(x) = 1/(1 + x^2)$ from the x, y data in the first two columns of Table 14.1.

TABLE 14.1

x	y				
0	1				
		-2			
①	$\frac{1}{2}$		-1		
		$-\frac{10}{3}$		0	
2	$\frac{1}{5}$		$\left(-\frac{1}{10}\right)$		0
		$\left(-10\right)$		40	
3	$\frac{1}{10}$		$\left(-\frac{1}{25}\right)$		0
		$-\frac{170}{7}$		140	
④	$\frac{1}{17}$		$-\frac{1}{46}$		
		$-\frac{442}{9}$			
5	$\frac{1}{26}$				

❚ Various reciprocal differences also appear in this table. For example, the entry 40 is obtained from the looped entries as follows:

$$\rho_3(x_2 x_3 x_4 x_5) = \frac{4-1}{\left(-\frac{1}{25}\right) - \left(-\frac{1}{10}\right)} + (-10) = 40$$

$$= \frac{x_5 - x_2}{\rho_2(x_3 x_4 x_5) - \rho_2(x_2 x_3 x_4)} + \rho_1(x_3 x_4)$$

From the definition given in Problem 14.5 this third difference should be

$$\rho_3(x_2 x_3 x_4 x_5) = \frac{x_2 - x_5}{\rho_2(x_2 x_3 x_4) - \rho_2(x_5 x_3 x_4)} + \rho_1(x_3 x_4)$$

but by the symmetry property this is the same as what we have. The other differences are found in the same way.
The continued fraction is constructed from the top diagonal

$$y = 1 + \cfrac{x-0}{-2 + \cfrac{x-1}{-1-1+\cfrac{x-2}{0-(-2)+\cfrac{x-3}{0-(-1)}}}}$$

and easily rearranges to the original $y(x) = 1/(1 + x^2)$. This test case merely illustrates the continued fractions algorithm.

By substituting successively the arguments $x = 0, 1, 2, 3, 4$ into this continued fraction it is easy to see that as the fraction becomes longer it absorbs the (x, y) data pairs one by one. This further implies that truncating the fraction will produce a rational collocation function for an initial segment of the data. The same remarks hold for the case of Problem 14.5. It should also be pointed out that the zeros in the last column of the table cause the fraction to terminate without an $x - x_4$ term, but that the fraction in hand absorbs the (x_5, x_5) data pair anyway.

14.8 Use a rational approximation to interpolate for tan 1.565 from the data provided in Table 14.2.

❚ The table also includes reciprocal differences through fourth order.
The interpolation then proceeds as

$$\tan 1.565 \simeq 24.498 + \cfrac{1.565 - 1.53}{.0012558 + \cfrac{1.565 - 1.54}{-24.531 + \cfrac{1.565 - 1.55}{2.7266 + \cfrac{1.565 - 1.56}{-.3837}}}}$$

which works out to 172.552. This result is almost perfect, which is remarkable considering how terribly close we are to the pole of the tangent function at $x = \pi/2$. Newton's backward formula, using the same data, produces the

TABLE 14.2

x	$\tan x$				
1.53	24.498				
		.0012558			
1.54	32.461		−.033		
		.0006403		2.7279	
1.55	48.078		−.022		−.4167
		.0002245		1.7145	
1.56	92.631		−.0045		
		.0000086			
1.57	1255.8				

value 433, so it is easy to see that our rational approximation is an easy winner. It is interesting to notice the results obtained by stopping at the earlier differences, truncating the fraction at its successive "convergents." Those results are

$$52.37 \qquad 172.36 \qquad 172.552$$

so that stopping at third and fourth differences we find identical values. This convergence is reassuring, suggesting implicitly that more data pairs and continuation of the fraction are unnecessary and that even the final data pair has served only as a check or safeguard.

14.9 It is possible that more than one rational function of the form in Problem 14.5 may include the given points. Which one will the continued fraction algorithm produce?

▌ As the continued fraction grows it represents successively functions of the forms

$$a_0 + a_1 x \qquad \frac{a_0 + a_1 x}{b_0 + b_1 x} \qquad \frac{a_0 + a_1 x + a_2 x^2}{b_0 + b_1 x} \qquad \frac{a_0 + a_1 x + a_2 x^2}{b_0 + b_1 x + b_2 x^2} \qquad \cdots$$

Our algorithm chooses the simplest form (left to right) consistent with the data.

14.10 Given that $y(x)$ has a simple pole at $x = 0$ and is of the form used in Problem 14.5 determine it from these (x, y) points: $(1, 30), (2, 10), (3, 5), (4, 3)$.

▌ Such a function may be sought directly starting with

$$y(x) = \frac{1 + a_1 x + a_2 x^2}{b_1 x + b_2 x^2}$$

It may also be found by this slight variation of the continued fractions algorithm. The table of reciprocal differences

x	y				
1	30				
		$-\frac{1}{20}$			
2	10		$-\frac{10}{3}$		
		$-\frac{1}{5}$		$\frac{8}{5}$	
3	5		$-\frac{5}{3}$		0
		$-\frac{1}{2}$		1	
4	3		-3		
		0			
0	∞				

leads to the continued fraction

$$y = 30 + \cfrac{x - 1}{-\cfrac{1}{20} + \cfrac{x - 2}{-\cfrac{100}{3} + \cfrac{x - 3}{\cfrac{33}{20} + \cfrac{x - 4}{\frac{10}{3}}}}}$$

which collapses to $y(x) = 60/[x(x + 1)]$.

14.11 Find directly, as in Problem 14.1, a function $y(x) = 1/(a + bx)$ such that $y(1) = 3$ and $y(3) = 1$. Does the continued fractions method produce this rational function?

▌ The two conditions require

$$a + b = \tfrac{1}{3} \qquad a + 3b = 1$$

which are satisfied by $a = 0$ and $b = \tfrac{1}{3}$. The rational function of this type is thus $y(x) = 3/x$. The continued fractions algorithm sets up the little difference table below left, and then the fraction to its right, which reduces to

$y(x) = 4 - x$ instead of to $3/x$:

$$\begin{array}{cc} 1 & 3 \\ & \quad -1 \qquad y = 3 + \dfrac{x-1}{-1} \\ 3 & 1 \end{array}$$

14.12 Find a rational function $y(x) = 1/(a + bx + cx^2)$ including the points $(0,1)$, $(1, \frac{1}{2})$ and $(10, \frac{1}{4})$. Does our method of continued fractions produce this function?

▮ The three conditions require

$$a = 1 \qquad a + b + c = 2 \qquad a + 10b + 100c = 4$$

from which come $b = 97/90$ and $c = -70/90$. The function is

$$y(x) = 90/(90 + 97x - 7x^2)$$

The continued fractions method constructs the table below left and then the fraction to its right.

$$\begin{array}{cc} 0 & 1 \\ & \qquad -2 \\ 1 & \frac{1}{2} \qquad\qquad \frac{7}{34} \qquad y = 1 + \dfrac{x - 0}{-2 + \dfrac{x-1}{\frac{7}{34} - 1}} = \dfrac{7x + 20}{34x + 20} \\ & \qquad -36 \\ 10 & \frac{1}{4} \end{array}$$

14.13 Use the continued fractions method to find a rational function having the values

x	0	1	2	3	4
y	-1	0	$\frac{3}{5}$	$\frac{4}{5}$	$\frac{15}{17}$

▮ The difference table and continued fraction appear as Table 14.3. The fraction reduces to $y(x) = (x^2 - 1)/(x^2 + 1)$.

TABLE 14.3

$$\begin{array}{llllll}
0 & -1 \\
 & & 1 \\
1 & 0 & & 3 \\
 & & 5/3 & & 0 \\
2 & 3/5 & & 6/5 & & 1 \qquad y = -1 + \dfrac{x-0}{1 + \dfrac{x-1}{4 + \dfrac{x-2}{-1 + \dfrac{x-3}{-2}}}} \\
 & & 5 & & -20 \\
3 & 4/5 & & 27/25 \\
 & & 85/7 \\
4 & 5/17
\end{array}$$

14.14 Use the continued fractions method to produce a rational function having the values

x	0	1	9	19
y	0	$\frac{1}{2}$	8.1	18.05

▮ The data foretell some amount of numerical nuisance, and the computer is always ready to assist, but the entries in Table 14.4 are exact. The fact that a very simple result was in store may also have been foretold by the nature of the data.

TABLE 14.4

$$\begin{array}{llllll}
0 & 0 \\
 & & 2 \\
1 & 1/2 & & -9 \\
 & & 20/19 & & 1 \qquad y = 0 + \dfrac{x-0}{2 + \dfrac{x-1}{-9 + \dfrac{x-9}{-1}}} = \dfrac{x^2}{1 + x} \\
 & & & -370 \\
9 & 8.1 & & 200/199 \\
 & & 200/199 \\
19 & 18.05
\end{array}$$

14.15   Find a rational function with the values

x	0	1	$+\infty$
y	$\frac{1}{2}$	$\frac{2}{3}$	1

▮ For a direct approach the function $(a + x)/(b + x)$ almost nominates itself, the limiting value of 1 being built in. The remaining two conditions require

$$a/b = 1/2 \qquad (a+1)(b+1) = 2/3$$

which are satisfied by $a = 1$, $b = 2$. The function is then $y(x) = (1 + x)/(2 + x)$.

The continued fractions method can also produce this. Table 14.5 is developed with the intention of pushing B to infinity while b goes to zero. The entry Q will then have limit 1, and the continued fraction can be constructed.

TABLE 14.5

0	$\dfrac{1}{2}$				
		6			
1	$\dfrac{2}{3}$		$Q = \dfrac{B}{\dfrac{B-1}{\frac{1}{3}-b} - 6} + \frac{2}{3}$	$y = \dfrac{1}{2} + \dfrac{x-0}{6 + \dfrac{x-1}{1-\frac{1}{2}}}$	
		$\dfrac{B-1}{\frac{1}{3}-b}$			
B	$1-b$				

14.16   Find a rational function meeting the limited specifications

x	-1	1	0
y	-1	1	∞

▮ The following difference table leads to the continued fraction at its right. Reduction to $y = 1/x$ is instant.

$$
\begin{array}{cccc}
-1 & -1 & & \\
 & & 1 & \\
1 & 1 & & 0 \qquad y = -1 + \dfrac{x+1}{1 + \dfrac{x-1}{1}} \\
 & & 0 & \\
0 & \infty & &
\end{array}
$$

The use of a temporary entry B with eventual target of infinity has been submerged.

14.17   Find a rational function with the values

x	0	1	2	4	∞
y	-2	$\pm\infty$	2	$\frac{6}{5}$	1

▮ Returning to the direct method, with

$$y(x) = \frac{x^2 + ax + b}{x^2 + cx + d}$$

guaranteeing the limiting value 1 at infinity, we are led to the equations

$$b = -2d \qquad 4 + 2a + b = 2(4 + 2c + d)$$
$$c + d = -1 \qquad 16 + 4a + b = -(16 + 4c + d)$$

with solution $a, b, c, d = 0, 2, 0, -1$. The required function is

$$y(x) = (x^2 + 2)/(x^2 - 1)$$

14.18   What sort of function includes this data? Interpolate for its values at .5 and 1.5.

x	0	± 1	± 2
y	$\frac{1}{2}$	1	$-\frac{1}{2}$

∥ There are surely millions of suitable functions, including this unique polynomial of degree 4:

$$p(x) = \tfrac{1}{2} + \tfrac{3}{4}x^2 - \tfrac{1}{4}x^4$$

which offers $\frac{43}{64}$ at $x = \pm\frac{1}{2}$ and $\frac{59}{64}$ at $x = \pm\frac{3}{2}$.

What do continued fractions have to say? Table 14.6 has the details, the continued fraction reducing after a bit of a struggle to $y(x) = 1/(2 - x^2)$. At $x = \pm.5$ this interpolates to $\frac{4}{7}$, while at $x = 1.5$ its value is -4, this point being close to the pole at $x = \sqrt{2}$. Interpolations here make no sense unless we have inside information about the nature of the background function.

Realizing that an even function was in hand, the assumption $y(x) = (x^2 + b)/(cx^2 + d)$ would seem a fair start on a direct search, having divided through by the upstairs coefficient of x^2 for simplicity. Unfortunately, as we now know, that coefficient proves to be zero and this start fails. A modified start would quickly find the earlier result.

TABLE 14.6

-2	$-\frac{1}{2}$			
		$\frac{2}{3}$		
-1	1	$\frac{1}{4}$		
		-2	2	
0	$\frac{1}{2}$	1	0	
		2	-2	
1	1	$\frac{1}{4}$		
		$-\frac{2}{3}$		
2	$-\frac{1}{2}$			

$$y = -\frac{1}{2} + \cfrac{x+2}{\frac{2}{3} + \cfrac{x+1}{\frac{3}{4} + \cfrac{x}{\frac{4}{3} + \cfrac{x-1}{-\frac{1}{4}}}}}$$

14.3 MIN–MAX RATIONAL FUNCTIONS

14.19   How can a rational function $R(x) = 1/(a + bx)$, that misses the three points (x_1, y_1), (x_2, y_2) and (x_3, y_3) alternately by $\pm h$ be found?

∥ The three conditions

$$y_i - \frac{1}{a + bx_i} = h, -h, h \quad \text{for } i = 1, 2, 3$$

can be rewritten as

$$a(y_1 - h) + b(y_1 - h)x_1 - 1 = 0$$
$$a(y_2 + h) + b(y_2 + h)x_2 - 1 = 0$$
$$a(y_3 - h) + b(y_3 - h)x_3 - 1 = 0$$

Eliminating a and b, we find that h is determined by the quadratic equation

$$\begin{vmatrix} y_1 - h & (y_1 - h)x_1 & -1 \\ y_2 + h & (y_2 + h)x_2 & -1 \\ y_3 - h & (y_3 - h)x_3 & -1 \end{vmatrix} = 0$$

Choosing the root with smaller absolute value, we substitute back and obtain a and b. (It is not hard to show that real roots will always exist.)

14.20 Apply the routine of Problem 14.19 to the points $(0, .83)$, $(1, 1.06)$ and $(2, 1.25)$.

▌ A bit of algebra is involved but the quadratic equation for h proves to be

$$Ah^2 + Bh + C = 0$$

with

$$A = 2(x_1 - x_3)$$

$$B = 2(x_3 y_3 - x_1 y_1) + (x_2 y_1 + x_1 y_2) - (x_3 y_2 + x_2 y_3) + (x_3 y_1 - x_1 y_3)$$

$$C = x_1 y_1(y_3 - y_2) + x_2 y_2(y_1 - y_3) + x_3 y_3(y_2 - y_1)$$

after which the coefficients are determined from

$$d = (y_1 - h)(y_2 + h)(x_2 - x_1)$$

$$a = [(y_2 + h)x_2 - (y_1 - h)x_1]/d$$

$$b = (y_1 - y_2 - 2h)/d$$

For the example in hand, the equation for h is

$$4h^2 - 4.12h - .130 = 0$$

with smaller root $h = -.03$. The first two equations for a and b are then simply $.86a - 1 = 0$ and $1.03a + 1.03b - 1 = 0$ which readily yield $a = 1.16$, $b = -.19$.

14.21 Extending the previous problem, apply an exchange method to find a min–max rational function of the form $R = 1/(a + bx)$ for these points: $(0, .83)$, $(1, 1.06)$, $(2, 1.25)$, $(4, 4.15)$.

▌ Our problem will be a close parallel to earlier exchange methods. Let the triple of the previous problem serve as initial triple. The equal error rational function for this triple was found to be $R_1(x) = 1/(1.16 - .19x)$. At the four data points its errors may be computed to be $-.03, .03, -.03, 1.65$ and we see that $R_1(x)$ is very poor at $x = 4$. For a new triple we choose the last three points, to retain alternating error signs. The new quadratic equation is

$$6h^2 - 21.24h + 1.47 = 0$$

making $h = .07$. The new equations for a and b are

$$a + b = 1.010 \qquad a + 2b = .758 \qquad a + 4b = .245$$

making $a \simeq 1.265$ and $b \simeq -.255$. The errors at the four data points are now $.04, .07, -.07, .07$, and since no error exceeds the $.07$ of our present triple we stop, accepting

$$R_2(x) = \frac{1}{1.265 - .255x}$$

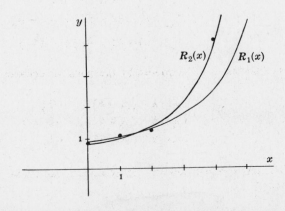

Fig. 14-2

as the min–max approximation. This is the typical development of an exchange algorithm. Our result is of course accurate only to a point, but the data themselves are given to only two places, so a greater struggle seems unwarranted. It is interesting to notice that the computation is quite sensitive. Rounding the third digit 5s in our $R_2(x)$, for instance, can change $R_2(4)$ by almost half a unit. This sensitivity is due to the pole near $x = 5$. Both $R_1(x)$ and $R_2(x)$ are shown in Fig. 14-2.

14.22   The data points of the preceding problem were chosen by adding random "noise" of up to 5% to values of $y(x) = 4/(5 - x)$. Use $R_2(x)$ to compute smoothed values and compare with the correct values and the original data.

▌ The required values are as follows, with entries at $x = 3$ added:

Original "noisy" data	.83	1.06	1.25	—	4.15
Values of $R_2(x)$	.79	.99	1.32	2.00	4.08
Correct values of $y(x)$	.80	1.00	1.33	2.00	4.00

Only the error at $x = 4$ is sizable and this has been reduced by almost half. The influence of the pole at $x = 5$ is evident. Approximation by means of polynomials would be far less successful.

14.23   What is the min–max rational function $R(x) = 1/(a + bx)$ for $y(x) = x^2 - 1$ on $(-1, 1)$?

▌ The function $y(x)$ is a parabolic segment, symmetric over the given interval. This suggests a symmetric $R(x)$, which only happens for $b = 0$. The constant $R = -\frac{1}{2}$ passes under-over-under this parabolic segment with equal errors and alternating signs. Elsewhere it makes only smaller errors, and proves to be the best approximation.

14.24   Apply the exchange method to find the min–max approximation of type $R(x) = 1/(a + bx)$ for this data:

x	0	1	2	3	4	5
y	.38	.30	.16	.20	.12	.10

Use $R(x)$ to smooth the data. The parent function of this data was $y(x) = 1/(x + 3)$, with random errors added. Does smoothing bring improvement?

▌ Choosing the uninspired initial triple $x = 0, 1, 2$ led to the following brief run:

Triple	h	Errors
0, 1, 2	−.035	$h, -h, h, .046, -.007, -.009$
1, 2, 3	.039	$.001, h, -h, h, -.015, -.016$

The first triple was almost satisfactory. The coefficients for the second were $a = 2.64$, $b = 1.19$, and the function

$$R(x) = \frac{1}{2.64 + 1.19x}$$

produced the smoothed values that follow. Original data and correct values are attached for comparisons. There does seem to be good improvement.

Data	.38	.30	.16	.20	.12	.10
$R(x)$	.38	.26	.20	.16	.14	.12
True	.33	.25	.20	.17	.14	.12

14.25   Use the exchange method to find the min–max approximation of the form $R(x) = 1/(a + bx)$ to $y(x) = e^x$ on the interval $(0, 1)$.

▌ Here is the record of progress:

Triple	h	Max Error	at $x =$
0, .5, 1	−.085	.11	.7
0, .7, 1	−.0977	.0978	.69

The run was terminated, with $R(x) = 1/(.91 - .56x)$. Even this best $R(x)$ makes errors of roughly .1. Its pole is near 1.6.

14.26 Proceed as in the preceding problem but with $y(x) = \log x$ and the interval $(1, e)$.

▮ Taking the initial triple $x = 1, 2, e$ we find the equal error $h = -.24$ with no larger errors elsewhere. The best $R(x)$ is $1/(6.11 - 1.95x)$ more or less. It passes over-under-over the logarithm curve on its way to a pole near $x = 3.1$.

14.27 Apply the exchange method to this data, seeking the min–max function of type $R(x) = 1/(a + bx)$. The parent function of the data is of this type, but small random errors have been introduced to partially obscure it. Where is the pole?

x	0	1	2	3	4	5	6	7	8	9
y	.10	.11	.12	.13	.18	.21	.23	.31	.53	1.04

▮ The triple $x = 0, 5, 9$ began a slightly ragged run, reported in Table 14.7, with the final triple at $x = 7, 8, 9$. The rational function produced was $R(x) = 1/(10.2 - 1.03x)$, with a pole at $x = 9.9$. As may be suspected, the true function was $1/(10 - x)$. A comparison of raw, smoothed and true values also appears in the table.

TABLE 14.7

Triple	h	Max Error	at $x =$
$0, 5, 9$	$-.003$	$-.035$	7
$0, 5, 7$	$-.007$	.32	9
$5, 7, 9$	.016	.035	8
$5, 7, 8$	.019	$-.115$	9
$7, 8, 9$	$-.023$	h	

Data	.10	.11	.12	.13	.18	.21	.23	.31	.53	1.04
Smoothed	.10	.11	.12	.14	.16	.20	.25	.33	.51	1.06
True	.10	.11	.12	.14	.17	.20	.25	.33	.50	1

14.28 Is there a unique function of type $R(x) = 1/(a + bx)$ passing through any two given points?

▮ Writing the collocation equations as

$$a + bx_1 = 1/y_1 \qquad a + bx_2 = 1/y_2$$

shows at once that there is just one a, b pair for given points (x_1, y_1) and (x_2, y_2), provided neither y_i is zero and the two x_i are distinct. Given exact data, say $(0, 1/10)$ and $(1, 1/9)$, the function $R(x) = 1/(10 - x)$ of the problem could have been found perfectly by the direct method. The min–max approach would also produce it given one more exact point, say $(2, 1/8)$, since the basic equations of Problem 14.19 are then satisfied with $h = 0$.

14.29 Develop equations for determining the min–max function of type $R(x) = (a + bx)/(c + x)$ for a given quadruple of points (x_i, y_i). [It is assumed that the denominator coefficient of x (say d) will not be zero, reducing $R(x)$ to a linear function, so that d can be normalized at 1.]

▮ The governing equations are, for $i = 1$ to 4,

$$y_i - \frac{a + bx_i}{c + x_i} = \pm h$$

the signs of the h terms alternating. Writing out the set,

$$a + bx_1 - c(y_1 - h) - x_1(y_1 - h) = 0$$
$$a + bx_2 - c(y_2 + h) - x_2(y_2 + h) = 0$$
$$a + bx_3 - c(y_3 - h) - x_3(y_3 - h) = 0$$
$$a + bx_4 - c(y_4 + h) - x_4(y_4 + h) = 0$$

we seem to have a homogeneous, linear system in a, b, c, d with a nontrivial solution (remember the normalizing $d = 1$ which would otherwise have appeared in column four). The determinant of the coefficients would then have to be zero:

$$\begin{vmatrix} 1 & x_1 & y_1 - h & x_1(y_1 - h) \\ 1 & x_2 & y_2 + h & x_2(y_2 + h) \\ 1 & x_3 & y_3 - h & x_3(y_3 - h) \\ 1 & x_4 & y_4 + h & x_4(y_4 + h) \end{vmatrix} = 0$$

Of course, the equal error h is also unknown, which makes this system like the earlier one of Problem 14.19 nonlinear. But as earlier, the determinant equation is quadratic and determines h. The values of a, b, c are then accessible.

14.30 Given the data function

x	0	1	2	3	4	5
y	0	1	2	4	6	10

and the inside information that the first, second and fourth y values are exact, a collocation rational function of the type $R(x) = (a + bx)/(c + x)$ can be found directly. Find it.

▌ The three collocations dictate

$$a = 0 \qquad a + b = c + 1 \qquad a + 3b = 4c + 12$$

which quickly give up $b = -8$ and $c = -9$. The function is

$$R(x) = 8x/(9 - x)$$

with a pole at $x = 9$. The sixth y value also proves to be exact.

14.31 Without inside information, use the method of Problem 14.29 to find a min–max approximation to the data function.

▌ Suppose we choose the initial quadruple $x = 0, 1, 2, 3$. It is a bit of a nuisance to set up the quadratic equation for determining h, but there seems to be no obviously better alternative. A few moments spent in organizing the various ingredients will pay off, since it is a step that will surely have to be repeated. In this case the quadratic proves to be

$$8h^2 + 12h + 1 = 0$$

with roots at $-.0886$ and -1.41 more or less. The smaller one is taken, since we prefer small errors. With this h the basic set of four equations becomes redundant, its determinant being zero, so we choose three, say the last three. They are

$$a + b - .9114c = .9114$$
$$a + 2b - 2.0886c = 4.1772$$
$$a + 3b - 3.9114c = 11.7342$$

and determine $a = -.5877$, $b = -4.559$ and $c = -6.647$. It is time to find the errors made by the corresponding $R(x)$ at the other data points, and the largest turns out to be a hefty -1.4 at (hardly a surprise, in view of the preceding problem) $x = 5$. The new quadruple is $x = 1, 2, 3, 5$ and the search continues as follows.

Quadruple	h	Max Error	at $x =$
1, 2, 3, 5	.142	$-.32$	4
1, 2, 3, 4	.163	.39	5
2, 3, 4, 5	$-.196$	h	

The final coefficients a, b, c make

$$R(x) = \frac{-1.4 + 7.999x}{8.93 - x}$$

which clearly has its good points and its bad. The zero of the function is called at $x = -.2$ and the pole at 8.93.

14.32   Determine the zero and the pole of the function represented by this data, using the smoothing method of Problem 14.29:

x	0	2	4	6	8
y	.09	.34	.67	1.7	4.6

❚   The experience of the preceding problem suggests choosing an initial quadruple with bias toward the larger y values. Taking this advice led to the brief run

Quadruple	h	Max Error	at $x =$
2, 4, 6, 8	.057	.067	0
0, 4, 6, 8	.059	h	

Including $x = 0$ in the quadruple did bring a slight reduction in the maximum error. The rational function produced was

$$R(x) = \frac{.287 + .973x}{9.73 - x}$$

and has a zero at $x = -.3$ and a pole at 9.73. The true parent function was $y(x) = (1 + x)/(10 - x)$, but random errors were added. These random errors seem to have affected the zero more seriously than the pole.

14.33   Here are some three place values of the tangent function:

x	0	.3	.6	.9	1.2	1.5
y	0	.309	.684	1.260	2.572	14.101

The pole at $\pi/2$ is looming. How well does the min–max approximation to this data determine it?

❚   Since the question concerns the pole, it seems best to begin with the quadruple $x = .6, .9, 1.2, 1.5$. It leads to the quadratic

$$1.44h^2 + 8.592h + .2087 = 0$$

with smaller root $h = -.024$. The three equations

$$a + .9b - 1.236c = 1.1124$$
$$a + 1.2b - 2.596c = 3.1152$$
$$a + 1.5b - 14.077c = 21.1155$$

then determine a, b, c and the rational function

$$R(x) = (.4 + .49x)/(1.58 - x)$$

with pole at 1.58. The correct value to two places is, of course 1.57. At $x = 0$ this $R(x)$ offers us .25 instead of zero, but we have focused on the other end of the given data interval. Checking errors at the other data points, it is no surprise to find the largest at $x = 0$, its value $-.25$ definitely larger than h. The run to the overall min–max $R(x)$ was then completed:

Quadruple	h	Max	at
0, .9, 1.2, 1.5	$-.085$	.093	.6
0, .6, 1.2, 1.5	$-.089$	h	

On the given discrete set, no error exceeds the equal error h, so at $x = 0$ there has been considerable improvement. What news of the pole? The final rational function was

$$R(x) = \frac{.14 + .75x}{1.59 - x}$$

so we have lost just a bit of ground there. The input data were correct to three digits, but the pole has not been pinpointed to quite the same precision.

14.34 Another common function has these values, rounded to two places,

x	0	.2	.4	.6	.8
y	0	.22	.51	.92	1.61

with a singularity approaching at the right. Approximate it by a min–max function of type $R(x) = (a + bx)/(c + x)$. Where does $R(x)$ predict the singularity to occur?

▮ Using the quadruple at the right, $x = .2, .4, .6, .8$, we find an equal error of $h = -.004$ and then

$$R(x) = (.0273 + 1.241x)/(1.435 - x)$$

with singularity at $x = 1.435$. Seeking the overall min–max $R(x)$ leads us to the quadruple $x = 0, .4, .6, .8$ and

$$R(x) = (.0093 + 1.313x)/(1.461 - x)$$

with little change in the forecast. Trying the collocation $R(x)$ for $x = .4, .6, .8$ we find the denominator element $(1.4 - x)$. In view of the fact that the parent function is $-\log(1 - x)$ with singularity at $x = 1$, none of these predictions is exciting. The data are too far removed from the event.

14.35 In these various min–max efforts we have ignored the larger root of the quadratic equation determining h. What does it represent?

▮ There are two curves of the $R(x)$ family that have the equal error feature for a given quadruple. The smaller error is surely what we want. Curves of this family are, of course, hyperbolas and we end up on the appropriate branch of one of them. The larger h value is useless to us except perhaps as entertainment. If we return to the quadruple $x = .2, .4, .6, .8$ of the preceding problem, with h determined by (coefficients abbreviated)

$$.64h^2 + .3h + .0012 = 0$$

we find the larger root $h = -.46$. The corresponding $R(x)$ is

$$R(x) = (-.4 + x)/(-.46 + x)$$

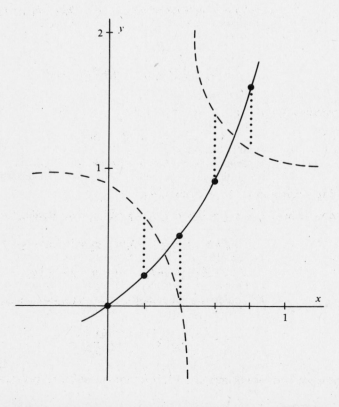

Fig. 14-3

and is plotted in Fig. 14-3. The corresponding curve for the smaller $h = -.0041$ is also sketched, and appears to pass through each point of the quadruple. Notice that both branches of the new hyperbola are involved in its over-under-over-under journey.

14.4 PADÉ APPROXIMATIONS

14.36 Derive conditions on the coefficients such that the Padé rational function

$$R_{mn}(x) = \frac{P_m(x)}{Q_n(x)}$$

with

$$P_m(x) = a_0 + a_1 x + a_2 x^2 + \cdots + a_m x^m$$

$$Q_n(x) = 1 + b_1 x + b_2 x^2 + \cdots + b_n x^n$$

will satisfy

$$R_{mn}^{(k)}(0) = y^{(k)}(0) \qquad k = 0, 1, \ldots, N$$

for $N = m + n$, assuming that $y(x)$ has the series representation

$$y(x) = c_0 + c_1 x + c_2 x^2 + \cdots$$

▮ We have

$$y(x) - R_{mn}(x) = \frac{\left(\sum_0^\infty c_i x^i\right)\left(\sum_0^n b_i x^i\right) - \sum_0^m a_i x^i}{\sum_0^n b_i x^i}$$

and will have achieved the required goal if the numerator on the right has no terms of lower degree than x^{N+1}. For this we need

$$a_0 = b_0 c_0 \qquad a_1 = b_0 c_1 + b_1 c_0 \qquad a_2 = b_0 c_2 + b_1 c_1 + b_2 c_0$$

and in general

$$a_j = \sum_{i=0}^j b_i c_{j-i} \qquad j = 0, 1, \ldots, N$$

subject to the constraints $b_0 = 1$ and

$$a_i = 0 \quad \text{if } i > m$$

$$b_i = 0 \quad \text{if } i > n$$

14.37 Apply the preceding problem to $y(x) = e^x$ with $m = n = 2$.

▮ For this function we have $c_0 = 1, c_1 = 1, c_2 = \frac{1}{2}, c_3 = \frac{1}{6}$ and $c_4 = \frac{1}{24}$, leading to the equations

$$a_0 = 1 \qquad a_1 = 1 + b_1 \qquad a_2 = \frac{1}{2} + b_1 + b_2$$

$$0 = \frac{1}{6} + \frac{1}{2}b_1 + b_2 \qquad 0 = \frac{1}{24} + \frac{1}{6}b_1 + \frac{1}{2}b_2$$

Their solution is $a_0 = 1$, $a_1 = \frac{1}{2}$, $a_2 = \frac{1}{12}$, $b_1 = -\frac{1}{2}$ and $b_2 = \frac{1}{12}$. Substituting back we have finally

$$R_{22}(x) = \frac{12 + 6x + x^2}{12 - 6x + x^2}$$

for the Padé approximation. On the interval $(-1, 1)$ its absolute error ranges from zero at the center to .004 at

$x = 1$. It is interesting to note that the approximation reflects a basic property of the exponential function, namely that replacing x by $-x$ produces the reciprocal.

14.38 For $y(x) = e^x$ it is clear that

$$R_{40} = 1 + x + \tfrac{1}{2}x^2 + \tfrac{1}{6}x^3 + \tfrac{1}{24}x^4$$

but use the method of Problem 14.36 to find $R_{04}(x)$.

▮ The appropriate equations include $a_0 = 1$ and then the triangular system

$$0 = 1 + b_1$$
$$0 = \tfrac{1}{2} + b_1 + b_2$$
$$0 = \tfrac{1}{6} + \tfrac{1}{2}b_1 + b_2 + b_3$$
$$0 = \tfrac{1}{24} + \tfrac{1}{6}b_1 + \tfrac{1}{2}b_2 + b_3 + b_4$$

leading to the approximation

$$R_{04}(x) = \frac{1}{1 - x + \tfrac{1}{2}x^2 - \tfrac{1}{6}x^3 + \tfrac{1}{24}x^4}$$

of which the denominator is a five term approximation to the reciprocal of $y(x)$. Presumably this could have been anticipated.

Over $(-1, 1)$ R_{04} is closer to e^x on the left half and farther from it on the right, relative to R_{40}. It is inferior all the way to R_{22} and this is generally true of Padé approximations. Those with m and n equal or nearly equal are the most accurate.

14.39 Find the Padé approximation $R_{31}(x)$ for $y(x) = e^x$.

▮ Since $c_0 = 1$, $c_1 = 1$, $c_2 = \tfrac{1}{2}$, $c_3 = \tfrac{1}{6}$ and $c_4 = \tfrac{1}{24}$, we have $a_0 = b_0 c_0 = 1$ and then

$$a_1 = 1 + b_1 \qquad a_2 = \tfrac{1}{2} + b_1$$
$$a_3 = \tfrac{1}{6} + \tfrac{1}{2}b_1 \qquad a_4 = 0 = \tfrac{1}{24} + \tfrac{1}{6}b_1$$

which determine $a_1 = \tfrac{3}{4}$, $a_2 = \tfrac{1}{4}$, $a_3 = \tfrac{1}{24}$ and $b_1 = -\tfrac{1}{4}$. The approximation is

$$R_{31}(x) = \frac{1 + \tfrac{3}{4}x + \tfrac{1}{4}x^2 + \tfrac{1}{24}x^3}{1 - \tfrac{1}{4}x} = \frac{24 + 18x + 6x^2 + x^3}{24 - 6x}$$

14.40 Also find the $R_{13}(x)$ approximation to e^x.

▮ After $a_0 = 1$ the needed equations are

$$a_1 = 1 + b_1 \qquad\qquad 0 = \tfrac{1}{2} + b_1 + b_2$$
$$0 = \tfrac{1}{6} + \tfrac{1}{2}b_1 + b_2 + b_3 \qquad 0 = \tfrac{1}{24} + \tfrac{1}{6}b_1 + \tfrac{1}{2}b_2 + b_3$$

the left side zeros being the a_i for $i > m$. The system has the solution $a_1 = \tfrac{1}{4}$, $b_1 = -\tfrac{3}{4}$, $b_2 = \tfrac{1}{4}$ and $b_3 = -\tfrac{1}{24}$, which makes

$$R_{13}(x) = \frac{1 + \tfrac{1}{4}x}{1 - \tfrac{3}{4}x + \tfrac{1}{4}x^2 - \tfrac{1}{24}x^3} = \frac{24 + 6x}{24 - 18x + 6x^2 - x^3}$$

14.41 Compare the errors of the five Padé approximations to e^x that are now in hand. Use the interval $(-1, 1)$.

▮ Error curves are shown in Fig. 14-4. All are nearest to zero around $x = 0$ since this is the point where the fit was made, derivatives through the fourth agreeing with those of e^x. The Taylor polynomial R_{40} is weak at both ends, while R_{13} and R_{04} perform poorly near $x = 1$, the latter simply exploding. Best overall marks go to R_{31} and R_{22}.

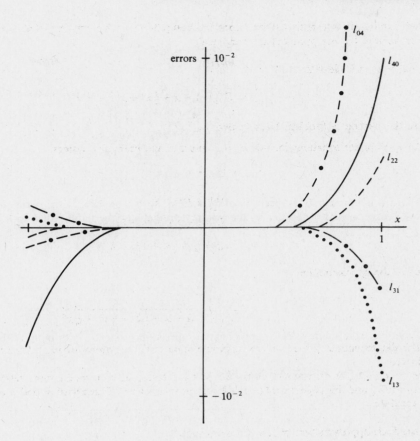

Fig. 14-4

Here are the extreme values (figures given are in the fifth decimal place):

	e_{04}	e_{13}	e_{22}	e_{31}	e_{40}
$x = -1$	-135	53	-54	121	-712
$x = 1$	5160	-900	400	-394	995

14.42 Find the R_{22} approximation to $y(x) = \log(1 + x)$.

$\blacksquare$ The series

$$\log(1 + x) = x - \tfrac{1}{2}x^2 + \tfrac{1}{3}x^3 - \tfrac{1}{4}x^4 + \tfrac{1}{5}x^5 - \cdots$$

identifies the various coefficients c_i and, truncated after the x^4 term, is R_{40}. For R_{22} we begin with $a_0 = c_0 = 0$ and then

$$a_1 = 1 \qquad a_2 = -\tfrac{1}{2} + b_1 \qquad a_3 = 0 = \tfrac{1}{3} - \tfrac{1}{2}b_1 + b_2$$

$$a_4 = 0 = -\tfrac{1}{4} + \tfrac{1}{3}b_1 - \tfrac{1}{2}b_2$$

which yield $b_1 = 1$, $b_2 = \tfrac{1}{6}$ and $a_2 = \tfrac{1}{2}$. So

$$R_{22}(x) = \frac{x + \tfrac{1}{2}x^2}{1 + x + \tfrac{1}{6}x^2} = \frac{6x + 3x^2}{6 + 6x + x^2}$$

which can be checked by direct long division, recovering the R_{40} portion of the given series.

In passing it may be noted that $R_{04} = 0$ identically, since we have already found a_0 to be zero, and with $m = 0$ all other a_i are required to be zero.

14.43 Also find the R_{33} Padé approximation to $\log(1 + x)$.

 ❚ With a_0 still zero, the remaining equations for a_i, b_i take the shape

$$a_1 = 1$$
$$a_2 = -\tfrac{1}{2} + b_1$$
$$a_3 = \tfrac{1}{3} - \tfrac{1}{2}b_1 + b_2$$
$$0 = -\tfrac{1}{4} + \tfrac{1}{3}b_1 - \tfrac{1}{2}b_2 + b_3$$
$$0 = \tfrac{1}{5} - \tfrac{1}{4}b_1 + \tfrac{1}{3}b_2 - \tfrac{1}{2}b_3$$
$$0 = -\tfrac{1}{6} + \tfrac{1}{5}b_1 - \tfrac{1}{4}b_2 + \tfrac{1}{3}b_3$$

In this slightly larger version the structure of the system shows more clearly, the c_i running in reverse order through each row, broken toward the bottom by the requirement that all b_i be zero beyond $i = n$. Solving the system we find

$$a_1 = a_2 = 1 \qquad a_3 = \tfrac{11}{60} \qquad b_1 = \tfrac{3}{2} \qquad b_2 = \tfrac{3}{5} \qquad b_3 = \tfrac{1}{20}$$

after which

$$R_{33}(x) = \frac{x + x^2 + \tfrac{11}{60}x^3}{1 + \tfrac{3}{2}x + \tfrac{3}{5}x^2 + \tfrac{1}{20}x^3}$$

14.44 How do the errors of the two Padé approximations just found compare with those of the Taylor polynomials of degrees 4 and 6?

 ❚ Here are a few numbers for R_{40} and R_{22}:

x	$-.8$	$-.6$	$-.4$	0	.4	.6	.8	1
e_{40}	$-.220$	$-.032$	$-.003$	0	.002	.010	.040	.109
e_{22}	$-.044$	$-.003$	$-.000$	0	.000	.000	.000	.001

There seems little doubt as to the winner. As for R_{33}, it seems fair to compare it with R_{60} and the numbers appear below. A few entries are given only as orders of magnitude, but there is no question which is the better performer.

x	$-.8$	$-.6$	$-.4$	$-.2$	0	.2	.4	.6	.8	1
e_{33}	$-.107$	$-.009$	$-.0004$	10^{-6}	10^{-8}	10^{-6}	.0002	.003	.018	.076
e_{60}	$-.007$	$-.0002$	10^{-6}	10^{-8}	10^{-8}	10^{-8}	10^{-7}	10^{-6}	10^{-5}	10^{-5}

14.5 MISCELLANEOUS PROBLEMS

14.45 Find a rational function that includes the points

x	-2	-1	0	1	2
y	$-\infty$	0	3	8	∞

TABLE 14.8

-1	0			
		$\tfrac{1}{3}$		
0	3		-12	
		$\tfrac{1}{5}$		$\tfrac{1}{2}$
1	8		-2		0
		0		0
2	∞		∞	
		0		
-2	$-\infty$			

$$R(x) = 0 + \cfrac{x + 1}{\cfrac{1}{3} + \cfrac{x - 0}{-12 + \cfrac{x - 1}{\cfrac{1}{6} + \cfrac{x - 2}{12}}}}$$

▮ A bit of optimism pays off here. The two poles suggest the denominator factor $x^2 - 4$, while the zero of $y(x)$ recommends a factor $x + 1$ for the numerator. This leaves only two points to be accommodated. Trying

$$R(x) = A(x+1)/(x^2 - 4)$$

the point $(0, 3)$ forces $A = -12$. But does this satisfy the point $(1, 8)$? It does, and optimism is justified.

The method of continued fractions also works, in spite of the two poles. Table 14.8 provides the details.

14.46 Find a rational function that includes the points

x	-1	0	1	2	3
y	∞	4	2	4	7

▮ The method of continued fractions again obliges, with the same casual treatment of infinity employed in the preceding problem. Table 14.9 gives the details.

TABLE 14.9

0	4		
		$-\frac{1}{2}$	
1	2		4
		$\frac{1}{2}$	$\frac{1}{4}$
2	4		-8
		$\frac{1}{3}$	$\frac{1}{4}$
3	7		16
		0	
-1	∞		

$$R(x) = 4 + \cfrac{x-0}{-\cfrac{1}{2} + \cfrac{x-1}{0 + \cfrac{x-2}{\cfrac{3}{4} + \cfrac{x-3}{\infty}}}}$$

$$= \frac{4(x^2 - x + 1)}{x + 1}$$

14.47 Find a rational function that includes the points

x	-2	-1	0	1	2	3
y	$\frac{4}{3}$	2	2	$\frac{4}{3}$	$\frac{8}{7}$	$\frac{14}{13}$

▮ The fractional values, with each numerator one greater than its denominator, are enough to raise the eyebrows. A heavy user of quadratic functions may also notice something else. Here are the numerator values and a few finite differences:

$$
\begin{array}{ccccccccccc}
4 & & 2 & & 2 & & 4 & & 8 & & 14 \\
& -2 & & 0 & & 2 & & 4 & & 6 &
\end{array}
$$

The next row hardly needs printing. Newton's formula quickly arranges

$$N(x) = 4 - 2(x + 2) + \tfrac{2}{2}(x + 1)(x + 2) = x^2 + x + 2$$

and we have this rational function

$$R(x) = \frac{x^2 + x + 2}{x^2 + x + 1}$$

Needless to say, there are others available, including one polynomial of degree 5 or less.

14.48 Interpolate for $y(1.5)$ in the following table, using a rational approximation function.

x	1	2	3	4
y	57.298677	28.653706	19.107321	14.335588

▮ First we try a function of the type $R(x) = 1/(a + bx)$. Substituting coordinates of the first two points leads quickly to $a = 5.32344 \cdot 10^{-6}$ and $b = .01744709$. The interpolated value is then $y(1.5) = 38.203$. The parent function here is $\csc x$, with x given in degrees, so the correct value is $\csc 1.5 = 38.201547$ to six places. Our effort has gotten only two decimal places right.

How would a continued fraction do? Following the routine set up in Problem 14.5, with the computer developing the reciprocal difference table and fraction, we come to

$$R(x) = 57.29868 + \cfrac{x - 1}{-.03491014 + \cfrac{x - 2}{-57.28122 + \cfrac{x - 3}{227.9604}}}$$

which manages $R(1.5) = 38.20179$, so we are now off in the fourth place. For still greater accuracy, the data supply would have to be supplemented.

14.49 Find a rational function, in the form of a cubic polynomial over a quadratic, including the points

x	0	1	2	3	4	5
y	12	0	-4	-6	6	4

▮ Reciprocal differences and a continued fraction seem to be indicated, but a direct approach is also available. The form

$$y = \frac{a_0 + a_1 x + a_2 x^2 + a_3 x^3}{1 + b_1 x + b_2 x^2}$$

can be rewritten as

$$a_0 + a_1 x + a_2 x^2 + a_3 x^3 - b_1 xy - b_2 x^2 y = y$$

which, upon substitution of the various x, y pairs, leads to the system of equations

$$
\begin{aligned}
a_0 &&&&&&&&&&&&= 12 \\
a_0 &+ a_1 &+ a_2 &+ a_3 &&&&&&= 0 \\
a_0 &+ 2a_1 &+ 4a_2 &+ 8a_3 &+ 8b_1 &+ 16b_2 &= -4 \\
a_0 &+ 3a_1 &+ 9a_2 &+ 27a_3 &+ 18b_1 &+ 54b_2 &= -6 \\
a_0 &+ 4a_1 &+ 16a_2 &+ 64a_3 &- 24b_1 &- 96b_2 &= 6 \\
a_0 &+ 5a_1 &+ 25a_2 &+ 125a_3 &- 20b_1 &- 100b_2 &= 4
\end{aligned}
$$

Appealing to the computer, we learn the identity of the various coefficients and have our rational function,

$$\frac{12 - 17.7x + 6.407144x^2 - .7071429x^3}{1 + .1714278x - .1214283x^2}$$

14.50 Apply the continued fractions method to the data of the preceding problem.

▮ In the order presented, the data leads to Table 14.10, the next reciprocal difference being infinite. The attempt to construct the function, at the right of the table, fails.

TABLE 14.10

0	12		
		$-\frac{1}{12}$	
1	0		-12
		$-\frac{1}{4}$	
2	-4		-12
		$-\frac{1}{2}$	
3	-6		

$$y = 12 + \cfrac{x}{-\cfrac{1}{12} + \cfrac{x - 1}{-24 + \cfrac{x - 2}{\infty}}}$$

$$= 12(1 - x)/(1 + x)$$

A slight rearrangement permits the process to run its course, a bit of persistence being required if exact fractional values are necessary. Table 14.11 gives the details. Comparisons will show that the result in the preceding problem is in agreement, roundoff errors being forgiven.

TABLE 4.11

1	0				
	$-\frac{1}{4}$				
2	-4	-12			
	$-\frac{1}{2}$	$-\frac{2}{11}$			
3	-6	$-\frac{18}{7}$	$\frac{15}{7}$		
	$\frac{1}{12}$	$\frac{2}{3}$	$\frac{17}{99}$		
4	6	$\frac{18}{7}$	$\frac{204}{49}$		
	$-\frac{1}{2}$	$-\frac{23}{39}$			
5	4	36			
	$-\frac{5}{8}$				
0	12				

$$y = 0 + \cfrac{x-1}{-\cfrac{1}{4} + \cfrac{x-2}{-12 + \cfrac{x-3}{\cfrac{3}{44} + \cfrac{x-4}{\cfrac{99}{7} + \cfrac{x-5}{\cfrac{35}{99}}}}}} = \frac{1680 - 2478x + 897x^2 - 99x^3}{140 + 24x - 17x^2}$$

14.51 Why are rational functions so commonly used in the approximation of other discrete and continuous functions?

❚ Rational functions are the most general functions that can be computed directly, involving only the four elementary operations or arithmetic. They offer more flexibility than polynomials, which are, of course, included. Experience suggests that for the same amount of computing, rational functions produce smaller maximum errors than polynomials.

CHAPTER 15
Trigonometric Approximation

15.1 Why are trigonometric functions so often used as approximations both to discrete data and to functions defined over an interval (a, b)?

▌ The sine and cosine functions share many of the desirable features of polynomials. They are easy to compute by rapidly convergent series over a basic interval and by reduction properties elsewhere. Their successive derivatives and integrals are again sines and cosines. They have orthogonality properties and, of course, periodicity, which polynomials do not have. This last feature makes them a natural for use with periodic functions, but if interest in a nonperiodic function is limited to some finite interval, it can be extended (even if only in the mind) beyond that interval in a periodic way. This is often convenient.

15.1 TRIGONOMETRIC SUMS BY COLLOCATION

15.2 Prove the orthogonality conditions

$$\sum_{x=0}^{N} \sin\frac{2\pi}{N+1}jx \sin\frac{2\pi}{N+1}kx = \begin{cases} 0 & \text{if } j \neq k \text{ or } j = k = 0 \\ (N+1)/2 & \text{if } j = k \neq 0 \end{cases}$$

$$\sum_{x=0}^{N} \sin\frac{2\pi}{N+1}jx \cos\frac{2\pi}{N+1}kx = 0$$

$$\sum_{x=0}^{N} \cos\frac{2\pi}{N+1}jx \cos\frac{2\pi}{N+1}kx = \begin{cases} 0 & \text{if } j \neq k \\ (N+1)/2 & \text{if } j = k \neq 0 \\ N+1 & \text{if } j = k = 0 \end{cases}$$

for $j + k \leq N$.

▌ The proofs are by elementary trigonometry. As an example,

$$\sin\frac{2\pi}{N+1}jx \sin\frac{2\pi}{N+1}kx = \frac{1}{2}\left[\cos\frac{2\pi}{N+1}(j-k)x - \cos\frac{2\pi}{N+1}(j+k)x\right]$$

and each cosine sums to zero since the angles involved are symmetrically spaced between 0 and 2π, except when $j = k \neq 0$, in which case the first sum of cosines is $(N+1)/2$. The other two parts are proved in similar fashion.

15.3 For collocation at an odd number of arguments $x = 0, 1, \ldots, N = 2L$, the trigonometric sum may take the form

$$\frac{1}{2}a_0 + \sum_{k=1}^{L}\left(a_k \cos\frac{2\pi}{2L+1}kx + b_k \sin\frac{2\pi}{2L+1}kx\right)$$

Use Problem 15.2 to determine the coefficients a_k and b_k.

▌ To obtain a_j multiply by $\cos[2\pi/(2L+1)]jx$ and sum. We find

$$a_j = \frac{2}{2L+1}\sum_{x=0}^{2L} y(x)\cos\frac{2\pi}{2L+1}jx \qquad j = 0, 1, \ldots, L$$

since all other terms on the right are zero. The factor $\frac{1}{2}$ in $y(x)$ makes this result true also for $j = 0$. To obtain b_j we multiply $y(x)$ by $\sin[2\pi/(2L+1)]jx$ and sum, getting

$$b_j = \frac{2}{2L+1}\sum_{x=0}^{2L} y(x)\sin\frac{2\pi}{2L+1}jx \qquad j = 1, 2, \ldots, L$$

Thus only one such expression can represent a given $y(x)$, the coefficients being uniquely determined by the values of $y(x)$ at $x = 0, 1, \ldots, 2L$. Notice that this function will have period $N + 1$.

15.4 Verify that, with the coefficients of Problem 15.3, the trigonometric sum does equal $y(x)$ for $x = 0, 1, \ldots, 2L$. This will prove the existence of a unique sum of this type which collocates with $y(x)$ for these arguments.

▮ Calling the sum $T(x)$ for the moment and letting x^* be any one of the $2L + 1$ arguments, substitution of our formulas for the coefficients leads to

$$T(x^*) = \frac{2}{2L+1} \sum_{x=0}^{2L} y(x) \left[\frac{1}{2} + \sum_{k=1}^{L} \left(\cos\frac{2\pi}{2L+1} kx \cos\frac{2\pi}{2L+1} kx^* + \sin\frac{2\pi}{2L+1} kx \sin\frac{2\pi}{2L+1} kx^* \right) \right]$$

$$= \frac{2}{2L+1} \sum_{x=0}^{2L} y(x) \left[\frac{1}{2} + \sum_{k=1}^{L} \cos\frac{2\pi}{2L+1} k(x - x^*) \right]$$

in which the order of summation has been altered. The last sum is now written as

$$\sum_{k=1}^{L} \cos\frac{2\pi}{2L+1} k(x - x^*) = \frac{1}{2} \sum_{k=1}^{L} \cos\frac{2\pi}{2L+1} k(x - x^*) + \frac{1}{2} \sum_{k=L+1}^{2L} \cos\frac{2\pi}{2L+1} k(x - x^*)$$

which is possible because of the symmetry property

$$\cos\frac{2\pi}{2L+1} k(x - x^*) = \cos\frac{2\pi}{2L+1}(2L + 1 - k)(x - x^*)$$

of the cosine function. Filling in the $k = 0$ term, we now find

$$T(x^*) = \frac{1}{2L+1} \sum_{x=0}^{2L} y(x) \left[\sum_{k=0}^{2L} \cos\frac{2\pi}{2L+1} k(x - x^*) \right]$$

But the term in brackets is zero by the orthogonality conditions unless $x = x^*$, when it becomes $2L + 1$. Thus $T(x^*) = y(x^*)$, which was to be proved.

15.5 Suppose $y(x)$ is known to have the period 3. Find a trigonometric sum that includes the following data points and use it to interpolate for $y(\frac{1}{2})$ and $y(\frac{3}{2})$.

x	0	1	2
y	0	1	1

▮ Using the formulas of Problem 15.3, we find

$$a_0 = \frac{2}{3}(0 + 1 + 1) = \frac{4}{3} \qquad a_1 = \frac{2}{3}\left(\cos\frac{2\pi}{3} + \cos\frac{4\pi}{3} \right) = -\frac{2}{3}$$

$$b_1 = \frac{2}{3}\left(\sin\frac{2\pi}{3} + \sin\frac{4\pi}{3} \right) = 0$$

The trigonometric sum

$$y(x) = \frac{1}{2} \cdot \frac{4}{3} - \frac{2}{3} \cos\frac{2\pi}{3} x = \frac{2}{3}\left(1 - \cos\frac{2\pi}{3} x \right)$$

then reproduces the three data points and goes on to complete a period with $y(3) = 0$. It also manages $y(\frac{1}{2}) = \frac{1}{3}$ and $y(\frac{3}{2}) = \frac{4}{3}$.

15.6   Apply the same method to the data

x	0	1	2	3	4	5
y	0	1	2	2	1	0

▮ We can ignore the last point, take $N = 2L = 4$ and assume a period of 5. Then

$$a_0 = \frac{2}{5}(0 + 1 + 2 + 2 + 1) = \frac{12}{5}$$

$$a_1 = \frac{2}{5}\left[\cos\left(\frac{2\pi}{5}\right) + 2\cos\left(\frac{4\pi}{5}\right) + 2\cos\left(\frac{6\pi}{5}\right) + \cos\left(\frac{8\pi}{5}\right)\right] = -1.04721$$

$$a_2 = \frac{2}{5}\left[\cos\left(\frac{4\pi}{5}\right) + 2\cos\left(\frac{8\pi}{5}\right) + 2\cos\left(\frac{12\pi}{5}\right) + \cos\left(\frac{16\pi}{5}\right)\right] = -.152786$$

with $b_1 = b_2 = 0$. The data thus have the representation

$$y = 1.2 + a_1 \cos\left(\frac{2\pi}{5}x\right) + a_2 \cos\left(\frac{4\pi}{5}x\right)$$

Beyond $x = 5$ the y values are repeated, $y(6)$ being 1 and so on.

15.7   Find coefficients such that for an even number of arguments $x = 0, 1, \ldots, N = 2L - 1$, the sum

$$y(x) = \frac{1}{2}a_0 + \sum_{k=1}^{L-1}\left(a_k \cos\frac{\pi}{L}kx + b_k \sin\frac{\pi}{L}kx\right) + \frac{1}{2}a_L \cos\pi x$$

takes prescribed y values.

▮ The orthogonality properties now read

$$\sum_{x=0}^{2L-1} \sin\frac{\pi}{L}jx \sin\frac{\pi}{L}kx = \begin{cases} 0 & j \neq k \text{ or } j = k = 0 \\ L & j = k \neq 0 \end{cases}$$

$$\sum_{x=0}^{2L-1} \sin\frac{\pi}{L}jx \sin\frac{\pi}{L}kx = 0$$

$$\sum_{x=0}^{2L-1} \cos\frac{\pi}{L}jx \cos\frac{\pi}{L}kx = \begin{cases} 0 & j \neq k \\ L & j = k \neq 0 \text{ or } L \\ 2L & j = k = 0 \text{ or } L \end{cases}$$

as a bit of trigonometry will demonstrate. Multiplying the given $y(x)$ by $\cos(\pi/L)jx$ and summing,

$$\sum_{x=0}^{2L-1} y(x)\cos\frac{\pi}{L}jx = a_j \cdot \begin{cases} L & \text{if } j \neq 0 \text{ or } L \\ 2L & \text{if } j = 0 \text{ or } L \end{cases}$$

the formula

$$a_j = \frac{1}{L}\sum_{x=0}^{2L-1} y(x)\cos\frac{\pi}{L}jx \qquad j = 0, \ldots, L$$

then covering all cases because of the two $\frac{1}{2}$ factors in $y(x)$ as given. Similarly, multiplying $y(x)$ by $\sin(\pi/L)jx$ and summing brings

$$b_j = \frac{1}{L}\sum_{x=0}^{2L-1} y(x)\sin\frac{\pi}{L}jx \qquad j = 1, \ldots, L - 1$$

15.8   Apply the preceding problem to the data function

x	0	1	2	3	4
y	0	1	2	1	0

▮ Depending on periodicity to cover the last point we take $N = 3$, which makes $L = 2$. The coefficients are

$$a_0 = \frac{1}{2}(0 + 1 + 2 + 1) = 2$$

$$a_1 = \frac{1}{2}\left(\cos\frac{\pi}{2} + 2\cos\frac{2\pi}{2} + \cos\frac{3\pi}{2}\right) = -1$$

$$a_2 = \frac{1}{2}(\cos\pi + 2\cos 2\pi + \cos 3\pi) = 0$$

with $b_1 = 0$. So $y(x)$ is simply $1 - \cos(\pi/2)x$.

15.9 The following data differ in only one position from that of the preceding problem. In what respect does the representation of $y(x)$ differ?

x	0	1	2	3	4
y	0	1	2	2	0

▮ Again choosing $N = 3$ and $L = 2$, we are led to the coefficients

$$a_0 = \frac{1}{2}(0 + 1 + 2 + 2) = \frac{5}{2}$$

$$a_1 = \frac{1}{2}\left(\cos\frac{\pi}{2} + 2\cos\frac{2\pi}{2} + 2\cos\frac{3\pi}{2}\right) = -1$$

$$a_2 = \frac{1}{2}\left(\cos\frac{2\pi}{2} + 2\cos\frac{4\pi}{2} + 2\cos\frac{6\pi}{2}\right) = -\frac{1}{2}$$

$$b_1 = \frac{1}{2}\left(\sin\frac{\pi}{2} + 2\sin\frac{2\pi}{2} + 2\sin\frac{3\pi}{2}\right) = -\frac{1}{2}$$

and the representation

$$y(x) = \frac{5}{4} - \cos\frac{\pi}{2}x - \frac{1}{2}\sin\frac{\pi}{2}x - \frac{1}{4}\cos\pi x$$

Why has a sine term finally proved useful? All the earlier data were symmetric about the center of the period. If repeated to the left of $x = 0$, the extended function would be symmetric about that point, which cosines are but sines are not. In this example the symmetry is lost.

15.10 What happens if the procedure of Problem 15.7 is applied to the data of Problem 15.5, repeated here for convenience?

x	0	1	2	3
y	0	1	1	0

▮ The example serves only to clarify the application of these two procedures. The one suggested requires an even number of active data points, here $N = 4$. That is, the fourth point is not now a passive end to the period. The coefficients are

$$a_0 = \frac{1}{2}(2) = 1$$

$$a_1 = \frac{1}{2}\left(\cos\frac{\pi}{2} + \cos\frac{2\pi}{2}\right) = -\frac{1}{2}$$

$$a_2 = \frac{1}{2}(\cos\pi + \cos 2\pi) = 0$$

$$b_1 = \frac{1}{2}\left(\sin\frac{\pi}{2} + \sin\frac{2\pi}{2}\right) = \frac{1}{2}$$

and the representation

$$y(x) = \frac{1}{2} - \frac{1}{2}\cos\frac{\pi}{2}x + \frac{1}{2}\sin\frac{\pi}{2}x$$

which differs from that of Problem 15.5. Both will reproduce the four points provided, but with different continuations. The earlier representation produces the (period 3) sequence

$$0 \quad 1 \quad 1 \quad 0 \quad 1 \quad 1 \quad 0 \quad \cdots$$

with the second managing the (period 4)

$$0 \quad 1 \quad 1 \quad 0 \quad 0 \quad 1 \quad 1 \quad 0 \quad 0 \quad \cdots$$

The latter is not symmetric about the period center and so needs help from the sine function.

15.2 TRIGONOMETRIC SUMS BY LEAST SQUARES: DISCRETE DATA

15.11 Determine the coefficients A_k and B_k so that the sum of squares

$$S = \sum_{x=0}^{2L} [y(x) - T_m(x)]^2 = \text{minimum}$$

where $T_m(x)$ is the trigonometric sum

$$T_m(x) = \frac{1}{2}A_0 + \sum_{k=1}^{M} \left(A_k \cos\frac{2\pi}{2L+1}kx + B_k \sin\frac{2\pi}{2L+1}kx \right)$$

and $M < L$.

▌ Since by Problem 15.4 we have

$$y(x) = \frac{1}{2}a_0 + \sum_{k=1}^{L} \left(a_k \cos\frac{2\pi}{2L+1}kx + b_k \sin\frac{2\pi}{2L+1}kx \right)$$

the difference is

$$y(x) - T_m(x) = \frac{1}{2}(a_0 - A_0) + \sum_{k=1}^{M} \left[(a_k - A_k)\cos\frac{2\pi}{2L+1}kx + (b_k - B_k)\sin\frac{2\pi}{2L+1}kx \right]$$
$$+ \sum_{k=M+1}^{L} \left[a_k \cos\frac{2\pi}{2L+1}kx + b_k \sin\frac{2\pi}{2L+1}kx \right]$$

Squaring, summing over the arguments x and using the orthogonality conditions,

$$S = \sum_{x=0}^{2L} [y(x) - T_m(x)]^2 = \frac{2L+1}{4}(a_0 - A_0)^2 + \frac{2L+1}{2}\sum_{k=1}^{M} \left[(a_k - A_k)^2 + (b_k - B_k)^2 \right]$$
$$+ \frac{2L+1}{2} \sum_{k=M+1}^{L} \left(a_k^2 + b_k^2 \right)$$

Only the first two terms depend on the A_k and B_k, and since these terms are nonnegative the minimum sum can be achieved in only one way, by making these terms zero. Thus for a minimum,

$$A_k = a_k \qquad B_k = b_k$$

and we have the important result that truncation of the collocation sum $T(x)$ at $k = M$ produces the least-squares trigonometric sum $T_M(x)$. (This is actually another special case of the general result found in Problem 12.20.) We also find

$$S_{\min} = \frac{2L+1}{2} \sum_{k=M+1}^{L} \left(a_k^2 + b_k^2 \right)$$

Since an almost identical computation shows that

$$\sum_{x=0}^{2L} [y(x)]^2 = \sum_{x=0}^{2L} [T(x)]^2 = \frac{2L+1}{4} a_0^2 + \frac{2L+1}{2} \sum_{k=1}^{L} (a_k^2 + b_k^2)$$

this may also be expressed in the form

$$S_{\min} = \sum_{x=0}^{2L} [y(x)^2] - \frac{2L+1}{4} a_0^2 - \frac{2L+1}{2} \sum_{k=1}^{M} (a_k^2 + b_k^2)$$

As M increases this sum steadily decreases, reaching zero for $M = L$, since then the least-squares and collocation sums are identical.

15.12 Find least-squares approximations from our results in Problems 15.5 and 15.6.

❚ There is not much to work with here, but the one cosine term can be truncated from the $y(x)$ in Problem 15.5, leaving $T_0 = \frac{2}{3}$. In the other problem we get $T_1(x) = 1.2 + a_1 \cos((2\pi/5)x)$ by dropping the a_2 term and then $T_0 = 1.2$ by dropping the remaining cosine. Each of these three trigonometric sums will minimize the sum of squared errors among the appropriate field of competitors. For example, to find the least-squares constant C for the data of Problem 15.5 we would minimize

$$C^2 + (C-1)^2 + (C-1)^2$$

and a quick mental calculation produces $C = \frac{2}{3}$. The constant 1.2 of Problem 15.6 pops out just as easily, but continuing on to a direct verification of our $T_1(x)$ would take something more than a mental effort.

15.13 Imitate the reasoning in Problem 15.11 to obtain a similar result for the case of an even number of x arguments.

❚ The modifications needed are routine, such as getting the correct limits on the summations and not forgetting the stray end term. We are to minimize

$$S = \sum_{x=0}^{2L-1} [y(x) - T_M(x)]^2$$

where $T_M(x)$ is the trigonometric sum

$$T_M(x) = \frac{1}{2} A_0 + \sum_{k=1}^{M} \left(A_k \cos \frac{\pi}{L} kx + B_k \sin \frac{\pi}{L} kx \right)$$

and $M < L$. But from Problem 15.7,

$$y(x) = \frac{1}{2} a_0 + \sum_{k=1}^{L-1} \left(a_k \cos \frac{\pi}{L} kx + b_k \sin \frac{\pi}{L} kx \right) + \frac{1}{2} a_L \cos \pi x$$

so taking the difference

$$y(x) - T_M(x) = \frac{1}{2}(a_0 - A_0) + \sum_{k=1}^{M} \left[(a_k - A_k) \cos \frac{\pi}{L} kx + (b_k - B_k) \sin \frac{\pi}{L} kx \right]$$

$$+ \sum_{k=M+1}^{L-1} \left[a_k \cos \frac{\pi}{L} kx + b_k \sin \frac{\pi}{L} kx \right]$$

Squaring, summing over the arguments x and making heavy use of the orthogonality conditions,

$$S = \frac{1}{4}(a_0 - A_0)^2 2L + L \sum_{k=1}^{M} \left[(a_k - A_k)^2 + (b_k - B_k)^2 \right] + L \sum_{k=M+1}^{L-1} (a_k^2 + b_k^2) + \frac{1}{4} a_L^2 2L$$

The only terms involving A_k and B_k are nonnegative, so the minimum of S is arranged as before by choosing $A_k = a_k$ and $B_k = b_k$. Again the least-squares sum T_M is available by truncation of the collocation sum at $k = M$.

15.14 Apply the result just found to find least-squares approximations to the data of Problems 15.8 to 15.10.

▮ The three constant approximations are 1, $\frac{5}{4}$ and $\frac{1}{2}$ in that order, all verifiable by mental arithmetic. In the first and last of these problems only this truncation to T_0 was possible, but in the middle one the less severe cut to

$$T_1(x) = \frac{5}{4} - \cos\frac{\pi}{2}x - \frac{1}{2}\sin\frac{\pi}{2}x$$

can also be made.

15.3 ODD OR EVEN PERIODIC FUNCTIONS

15.15 Suppose $y(x)$ has the period $P = 2L$, that is, $y(x + P) = y(x)$ for all x. Show that the formulas for a_j and b_j in Problem 15.7 may be written as

$$a_j = \frac{2}{P} \sum_{x=-L+1}^{L} y(x)\cos\frac{2\pi}{P}jx \quad j = 0,1,\ldots,L$$

$$b_j = \frac{2}{P} \sum_{x=-L+1}^{L} y(x)\sin\frac{2\pi}{P}jx \quad j = 1,\ldots,L-1$$

▮ Since the sine and cosine also have period P, it makes no difference whether the arguments $x = 0,\ldots,2L-1$ or the arguments $-L+1,\ldots,L$ are used. Any such set of P consecutive arguments will lead to the same coefficients.

15.16 Suppose $y(x)$ has the period $P = 2L$ and is also an odd function, that is, $y(-x) = -y(x)$. Prove that

$$a_j = 0 \qquad b_j = \frac{4}{P}\sum_{x=1}^{L-1} y(x)\sin\frac{2\pi}{P}jx$$

▮ By periodicity, $y(0) = y(P) = y(-P)$. But since $y(x)$ is an odd function, $y(-P) = -y(P)$ also. This implies $y(0) = 0$. In the same way we find $y(L) = y(-L) = -y(L) = 0$. Then in the sum for a_j each remaining term at positive x cancels its mate at negative x, so that all a_j will be 0. In the sum for b_j the terms for x and $-x$ are identical and so we find b_j by doubling the sum over positive x.

15.17 Find a trigonometric sum $T(x)$ for the function of Problem 15.10, assuming it extended to an odd function of period $P = 6$.

▮ By the previous problem all $a_j = 0$, and since $L = 3$,

$$b_1 = \frac{2}{3}\left(\sin\frac{\pi}{3} + \sin\frac{2\pi}{3}\right) = \frac{2}{\sqrt{3}} \qquad b_2 = \frac{2}{3}\left(\sin\frac{2\pi}{3} + \sin\frac{4\pi}{3}\right) = 0$$

making $T(x) = (2/\sqrt{3})\sin(\pi x/3)$.

15.18 Find a sum of sines to represent this extension of the data in Problem 15.8.

x	0	1	2	3	4	5	6	7	8
y	0	1	2	1	0	-1	-2	-1	0

▮ Since the target is a sum of sines, we are to view the data as one period of an odd function and use the coefficient formulas in Problem 15.16:

$$b_1 = \frac{1}{2}\left(\sin\frac{\pi}{4} + 2\sin\frac{2\pi}{4} + \sin\frac{3\pi}{4}\right) = 1 + \frac{\sqrt{2}}{2}$$

$$b_2 = \frac{1}{2}\left(\sin\frac{2\pi}{4} + 2\sin\frac{4\pi}{4} + \sin\frac{6\pi}{4}\right) = 0$$

$$b_3 = \frac{1}{2}\left(\sin\frac{3\pi}{4} + 2\sin\frac{6\pi}{4} + \sin\frac{9\pi}{4}\right) = \frac{\sqrt{2}}{2} - 1$$

The representation is then

$$y(x) = \left(1 + \frac{\sqrt{2}}{2}\right)\sin\frac{\pi}{4}x + \left(\frac{\sqrt{2}}{2} - 1\right)\sin\frac{3\pi}{4}x$$

15.19   If $y(x)$ has the period $P = 2L$ and is an even function, that is, $y(-x) = y(x)$, show that the formulas of Problem 15.15 become

$$a_j = \frac{2}{P}\left[y(0) + y(L)\cos j\pi\right] + \frac{4}{P}\sum_{x=1}^{L-1} y(x)\cos\frac{2\pi}{P}jx \qquad j = 0,1,\ldots,L$$
$$b_j = 0$$

❚   The terms for $\pm x$ in the formula for b_j cancel in pairs. In the a_j formula the terms for $x = 0$ and $x = L$ may be separated as above, after which the remaining terms come in matching pairs for $\pm x$.

15.20   Find a trigonometric sum to represent the data

x	-3	-2	-1	0	1	2	3
y	0	2	1	0	1	2	0

considering it as one period of an even function.

❚   The preceding problem applies with $P = 2L = 6$. We can if we wish continue the data stream to $y = 2,1,0$ for $x = 4,5,6$. The coefficients are then

$$a_0 = \frac{2}{3}(1 + 2) = 2$$
$$a_1 = \frac{2}{3}\left(\cos\frac{\pi}{3} + 2\cos\frac{2\pi}{3}\right) = -\frac{1}{3}$$
$$a_2 = \frac{2}{3}\left(\cos\frac{2\pi}{3} + 2\cos\frac{4\pi}{3}\right) = -1$$
$$a_3 = \frac{2}{3}\left(\cos\frac{3\pi}{3} + 2\cos\frac{6\pi}{3}\right) = \frac{2}{3}$$

which, substituted into the formula of Problem 15.7, produce the required representation.

$$y(x) = 1 - \frac{1}{3}\cos\frac{\pi}{3}x - \cos\frac{2\pi}{3}x + \frac{1}{3}\cos\pi x$$

15.21   Extend the following data to an even function of period $P = 8$ and then find a representation as a sum of cosines.

x	0	1	2	3	4
y	0	1	0	2	0

❚   The extension involves the y sequence $0,2,0,1$ for $x = -4$ to -1 or the sequence $2,0,1,0$ for $x = 5$ to 8. Choosing the latter for our computations we have, since $P = 2L = 8$,

$$a_0 = \frac{1}{2}(1 + 0 + 2) = \frac{3}{2}$$
$$a_1 = \frac{1}{2}\left(\cos\frac{\pi}{4} + 2\cos\frac{3\pi}{4}\right) = -\frac{\sqrt{2}}{4}$$
$$a_2 = \frac{1}{2}\left(\cos\frac{2\pi}{4} + 2\cos\frac{6\pi}{4}\right) = 0$$
$$a_3 = \frac{1}{2}\left(\cos\frac{3\pi}{4} + 2\cos\frac{9\pi}{4}\right) = \frac{\sqrt{2}}{4}$$
$$a_4 = \frac{1}{2}\left(\cos\frac{4\pi}{4} + 2\cos\frac{12\pi}{4}\right) = -\frac{3}{2}$$

so that

$$y(x) = \frac{3}{4} - \frac{\sqrt{2}}{4}\cos\frac{\pi}{4}x + \frac{\sqrt{2}}{4}\cos\frac{3\pi}{4}x - \frac{3}{4}\cos\pi x$$

which can easily be verified at the data points. The truncation to the least-squares constant $\frac{3}{4}$ can also be checked, mentally of course.

15.4 CONTINUOUS DATA: THE FOURIER SERIES

15.22 Prove the orthogonality conditions

$$\int_0^{2\pi} \sin jt \sin kt\, dt = \begin{cases} 0 & \text{if } j \neq k \\ \pi & \text{if } j = k \neq 0 \end{cases}$$

$$\int_0^{2\pi} \sin jt \cos kt\, dt = 0$$

$$\int_0^{2\pi} \cos jt \cos kt\, dt = \begin{cases} 0 & \text{if } j \neq k \\ \pi & \text{if } j = k \neq 0 \\ 2\pi & \text{if } j = k = 0 \end{cases}$$

where $j, k = 0, 1, \ldots$ to infinity.

∥ The proofs are elementary calculus. For example,

$$\sin jt \sin kt = \tfrac{1}{2}\left[\cos(j-k)t - \cos(j+k)t\right]$$

and each cosine integrates to zero since the interval of integration is a period of the cosine, except when $j = k \neq 0$, in which case the first integral becomes $\frac{1}{2}(2\pi)$. The other two parts are proved in similar fashion.

15.23 Derive the coefficient formulas

$$\alpha_j = \frac{1}{\pi}\int_0^{2\pi} y(t)\cos jt\, dt \qquad \beta_j = \frac{1}{\pi}\int_0^{2\pi} y(t)\sin jt\, dt$$

of the Fourier series

$$y(t) = \tfrac{1}{2}\alpha_0 + \sum_{k=1}^{\infty}\left(\alpha_k \cos kt + \beta_k \sin kt\right)$$

These are called the Fourier coefficients. As a matter of fact all such coefficients in sums or series of orthogonal functions are frequently called Fourier coefficients.

∥ The proof follows a familiar path. Multiply $y(t)$ by $\cos jt$ and integrate over $(0, 2\pi)$. All terms but one on the right are zero and the formula for α_j emerges. The factor $\frac{1}{2}$ in the α_0 term makes the result true also for $j = 0$. To obtain β_j we multiply by $\sin jt$ and integrate. Here we are assuming that the series will converge to $y(t)$ and that term by term integration is valid. This is proved, under very mild assumptions about the smoothness of $y(t)$, in the theory of Fourier series. Clearly $y(t)$ must also have the period 2π.

15.24 Obtain the Fourier series for $y(t) = |t|$, $-\pi \leq t \leq \pi$.

∥ Let $y(t)$ be extended to an even function of period 2π. (See solid curve in Fig. 15-1.) The limits of integration in our coefficient formulas may be shifted to $(-\pi, \pi)$ and we see that all $\beta_j = 0$. Also $\alpha_0 = \pi$ and for $j > 0$

$$\alpha_j = \frac{2}{\pi}\int_0^{\pi} t \cos jt\, dt = \frac{2(\cos j\pi - 1)}{\pi j^2}$$

Thus

$$y(t) = \frac{\pi}{2} - \frac{4}{\pi}\left(\cos t + \frac{\cos 3t}{3^2} + \frac{\cos 5t}{5^2} + \cdots\right)$$

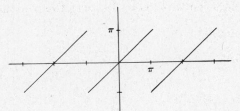

Fig. 15-1

15.25 Obtain the Fourier series for $y(t) = t$, $-\pi < t < \pi$.

▌ We extend $y(t)$ to an odd function of period 2π. (See Fig. 15-2.) Again shifting to limits $(-\pi, \pi)$ we find all $\alpha_j = 0$ and

$$\beta_j = \frac{2}{\pi} \int_0^\pi t \sin jt \, dt = \frac{2(-1)^{j-1}}{j}$$

Thus

$$y(t) = 2\left(\sin t - \frac{\sin 2t}{2} + \frac{\sin 3t}{3} - \frac{\sin 4t}{4} + \cdots \right)$$

Notice that the cosine series of Problem 15.24 converges more rapidly than the sine series. This is related to the fact that the $y(t)$ of that problem is continuous, while this one is not. The smoother $y(t)$ is, the more rapid the convergence. Notice also that at the points of discontinuity our sine series converges to zero, which is the average of the left and right extreme values (π and $-\pi$) of $y(t)$.

Fig. 15-2

15.26 Find the Fourier series for

$$y(t) = \begin{cases} t(\pi - t) & 0 \leq t \leq \pi \\ t(\pi + t) & -\pi \leq t \leq 0 \end{cases}$$

▌ Extending the function to an odd function of period 2π, we have the result shown in Fig. 15-3. Notice that this function has no corners. At $t = 0$ its derivative is π from both sides, while both $y'(\pi)$ and $y'(-\pi)$ are $-\pi$ so that

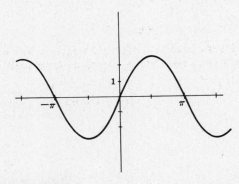

Fig. 15-3

even the extended periodic function has no corners. This extra smoothness will affect the Fourier coefficients. Using limits $(-\pi, \pi)$ we again find all $\alpha_j = 0$ and

$$\beta_j = \frac{2}{\pi}\int_0^\pi t(\pi - t)\sin jt\, dt = \frac{2}{\pi}\int_0^\pi \frac{\pi - 2t}{j}\cos jt\, dt = \frac{4}{\pi j^2}\int_0^\pi \sin jt\, dt = \frac{4(1 - \cos j\pi)}{\pi j^3}$$

The series is therefore

$$y(t) = \frac{8}{\pi}\left(\sin t + \frac{\sin 3t}{3^3} + \frac{\sin 5t}{5^3} + \cdots\right)$$

The coefficients diminish as reciprocal cubes, which makes for very satisfactory convergence. The extra smoothness of the function has proved useful.

15.27 Show that for the Bernoulli function

$$F_n(x) = B_n(x) \qquad 0 < x < 1 \qquad F_n(x \pm m) = F_n(x) \qquad m \text{ an integer}$$

$B_n(x)$ being a Bernoulli polynomial, the Fourier series is

$$F_n(x) = (-1)^{(n/2)+1} n!\left[\frac{2}{(2\pi)^n}\right]\sum_{k=1}^\infty \frac{\cos 2\pi kx}{k^n}$$

when n is even and

$$F_n(x) = (-1)^{(n+1)/2} n!\left[\frac{2}{(2\pi)^n}\right]\sum_{k=1}^\infty \frac{\sin 2\pi kx}{k^n}$$

when n is odd. This result was used in Problem 9.66 of the chapter on sums and series.

Since $B_1(x) = x - \frac{1}{2}$, the series for $F_1(x)$ may be found directly from the coefficient formulas to be

$$F_1(x) = -\frac{1}{\pi}\left(\frac{\sin 2\pi x}{1} + \frac{\sin 4\pi x}{2} + \frac{\sin 6\pi x}{3} + \cdots\right)$$

Integrating, and recalling that

$$B_n'(x) = nB_{n-1}(x) \qquad \int_0^1 B_n(x)\, dx = 0 \quad \text{for } n > 0$$

we soon find

$$F_2(x) = \frac{2 \cdot 2!}{(2\pi)^2}\left(\frac{\cos 2\pi x}{1} + \frac{\cos 4\pi x}{2^2} + \frac{\cos 6\pi x}{3^2} + \cdots\right)$$

The next integration makes

$$F_3(x) = \frac{2 \cdot 3!}{(2\pi)^3}\left(\frac{\sin 2\pi x}{1} + \frac{\sin 4\pi x}{2^3} + \frac{\sin 6\pi x}{3^3} + \cdots\right)$$

and an induction may be used to complete a formal proof. (Here it is useful to know that integration of a Fourier series term by term always produces the Fourier series of the integrated function. The analogous statement for differentiation is not generally true. For details see a theoretical treatment of Fourier series.

15.28 How are the collocation coefficients of Problem 15.7, or of Problem 15.3, related to the Fourier coefficients of Problem 15.23?

❚ There are many ways of making the comparisons. One of the most interesting is to notice that in Problem 15.7, assuming $y(x)$ to have the period $P = 2L$, we may rewrite a_j as

$$a_j = \frac{1}{L}\left[\frac{1}{2}y(0) + \frac{1}{2}y(2L) + \sum_{x=1}^{2L-1} y(x)\cos\frac{\pi}{L}jx\right]$$

and this is the trapezoidal rule approximation to the Fourier coefficient

$$\alpha_j = \frac{1}{\pi}\int_0^{2\pi} y(t)\cos jt\, dt = \frac{1}{L}\int_0^{2L} y(x)\cos\frac{\pi}{L}jx\, dx$$

Similar results hold for b_j and β_j and for the coefficients in Problem 15.3. Since the trapezoidal rule converges to the integral for L becoming infinite, we see that the collocation coefficients converge on the Fourier coefficients. (Here we may fix the period at 2π for convenience.)

15.29 Show that the Fourier series for $y(x) = |\sin x|$, the "fully rectified" sine wave, is

$$y(x) = \frac{4}{\pi}\left(\frac{1}{2} - \frac{\cos 2x}{1\cdot 3} - \frac{\cos 4x}{3\cdot 5} - \frac{\cos 6x}{5\cdot 7} - \cdots\right)$$

$\blacksquare$ Since $y(x)$ is an even function we need only the cosine coefficients. They are

$$a_j = \frac{1}{\pi}\int_0^{2\pi}|\sin x|\cos jx\, dx = \frac{2}{\pi}\int_0^{\pi}\sin x \cos jx\, dx$$

$$= \frac{2}{\pi}\int_0^{\pi}\frac{1}{2}\left[\sin(j+1)x - \sin(j-1)x\right]dx$$

$$= \frac{1}{\pi}\left[\frac{\cos(j+1)x}{-(j+1)} + \frac{\cos(j-1)x}{j-1}\right]_0^{\pi}$$

$$= \frac{1}{\pi}\left[\frac{\cos(j+1)\pi}{-(j+1)} + \frac{\cos(j-1)\pi}{j-1} + \frac{1}{j+1} - \frac{1}{j-1}\right]$$

For j odd this works out to zero, while for j even it reduces to

$$a_j = -4/\pi(j-1)(j+1)$$

leading to the series as given.

15.30 Use the series just found, with $x = 0$ and with $x = \pi/2$, to evaluate

$$\frac{1}{1\cdot 3} + \frac{1}{3\cdot 5} + \frac{1}{5\cdot 7} + \cdots$$

and

$$\frac{1}{1\cdot 3} - \frac{1}{3\cdot 5} + \frac{1}{5\cdot 7} - \cdots$$

$\blacksquare$ Since $y(0) = 0$ and all the cosines become 1, the first sum is easily seen to be $\frac{1}{2}$, a result also available by telescoping. Since $y(\pi/2) = 1$ and the cosines are alternately $-1, 1, -1, \ldots$ the second sum proves to be $\pi/4 - 1/2$.

15.31 Show that the Fourier series for $y(x) = x^2$ for x between $-\pi$ and π, and of period 2π, is

$$y(x) = \frac{\pi^2}{3} - 4\sum_{k=1}^{\infty}\frac{(-1)^{k-1}\cos kx}{k^2}$$

Use the result to evaluate the series $\sum_{k=1}^{\infty}(-1)^{k-1}/k^2$ and $\sum_{k=1}^{\infty}1/k^2$.

$\blacksquare$ Once again $y(x)$ is an even function. The cosine coefficients are, introducing $u = jx$,

$$a_j = \frac{2}{\pi}\int_0^{\pi}x^2\cos jx\, dx = \frac{2}{\pi j^3}\int_0^{\pi}u^2\cos u\, dx$$

$$= \frac{2}{\pi j^3}\left[2u\cos u + (u^2 - 2)\sin u\right]_0^{j\pi}$$

$$= \frac{2}{\pi j^3}\left[2\pi j\cos j\pi + (\pi^2 - 2)\sin j\pi\right]$$

$$= \begin{cases} \dfrac{4}{j^2} & \text{for } j \text{ even but not } 0 \\[2mm] -\dfrac{4}{j^2} & \text{for } j \text{ odd} \end{cases}$$

For $j = 0$ a mental integration manages $2\pi^2/3$ and we have the series as stated.

When $x = 0$ the function $y(x) = 0$ and all the cosines are 1, which gives the first of the two series requiring evaluation the value $\pi^2/12$. When $x = \pi$ we have $y(x) = \pi^2$ and $\cos kx = (-1)^k$, giving the second series the familiar value $\pi^2/6$.

15.32 Use the series in Problems 15.24 to 15.26 to evaluate

$$\sum_{k=1}^{\infty} \frac{1}{(2k-1)^2} \qquad \sum_{k=1}^{\infty} \frac{(-1)^{k+1}}{2k-1} \qquad \sum_{k=1}^{\infty} \frac{(-1)^{k+1}}{(2k-1)^3}$$

▌ From Problem 15.24 with $t = 0$ the first series quickly proves to be $\pi^2/8$. The same result comes by letting $t = \pi$. Putting $t = \pi/2$ in Problem 15.25 allows the Leibnitz series for $\pi/4$ to be read directly. Finally, setting $t = \pi/2$ in Problem 15.26 we find the third series above equal to $\pi^3/32$.

15.33 It seems that Fourier series can be exploited endlessly to evaluate series, simply by introducing various values for the argument. Give other illustrations.

▌ The only cloud in this otherwise rosy sky is the fact that most values of the argument lead to series with uninteresting coefficient patterns. Here are two more examples having at least some recognizable structure. Putting $t = \pi/4$ in Problem 15.24,

$$\frac{\pi}{4} = \frac{\pi}{2} - \frac{4}{\pi}\frac{\sqrt{2}}{2}\left(1 - \frac{1}{3^2} - \frac{1}{5^2} + \frac{1}{7^2} + \frac{1}{9^2} - \cdots\right) \qquad 1 - \frac{1}{9} - \frac{1}{25} + \frac{1}{36} + \frac{1}{49} - \cdots = \frac{\pi^2}{8\sqrt{2}}$$

and putting $t = \pi/4$ in the series of Problem 15.26,

$$\frac{\pi}{4}\cdot\frac{3\pi}{4} = \frac{8}{\pi}\frac{\sqrt{2}}{2}\left(1 + \frac{1}{3^3} - \frac{1}{5^3} - \frac{1}{7^3} + \cdots\right) \qquad 1 + \frac{1}{3^3} - \frac{1}{5^3} - \frac{1}{7^3} + \cdots = \frac{3\pi^3}{4^3\sqrt{2}}$$

In both examples the double plus, double minus pattern is due to the sign behavior of the trigonometric functions. Many similar examples exist.

15.5 LEAST SQUARES: CONTINUOUS DATA

15.34 Determine the coefficients A_k and B_k so that the integral

$$I = \int_0^{2\pi} [y(t) - T_M(t)]^2 \, dt$$

will be a minimum where $T_M(t) = \frac{1}{2}A_0 + \sum_{k=1}^{M}(A_k \cos kt + B_k \sin kt)$.

▌ More or less as in Problem 15.11, we first find

$$y(t) - T_M(t) = \frac{1}{2}(\alpha_0 - A_0) + \sum_{k=1}^{M}\left[(\alpha_k - A_k)\cos kt + (\beta_k - B_k)\sin kt\right] + \sum_{k=M+1}^{\infty}(\alpha_k \cos kt + \beta_k \sin kt)$$

and then square, integrate and use the orthogonality conditions to get

$$I = \frac{\pi}{2}(\alpha_0 - A)^2 + \pi\sum_{k=1}^{M}\left[(\alpha_k - A_k)^2 + (\beta_k - B_k)^2\right] + \pi\sum_{k=M+1}^{\infty}(\alpha_k^2 + \beta_k^2)$$

For a minimum we choose all $A_k = \alpha_k$ and $B_k = \beta_k$ so that

$$I_{\min} = \pi\sum_{k=M+1}^{\infty}(\alpha_k^2 + \beta_k^2)$$

Again we have the important result that truncation of the Fourier series at $k = M$ produces the least-squares sum $T_M(t)$. (Once again this is a special case of Problem 12.20.) The minimum integral may be rewritten as

$$I_{\min} = \int_0^{2\pi}[y(t)]^2 \, dt - \frac{1}{2}\pi\alpha_0^2 - \pi\sum_{k=1}^{M}(\alpha_k^2 + \beta_k^2)$$

As M increases, this diminishes, and it is proved in the theory of Fourier series that $I_{\min}$ tends to zero for M becoming infinite. This is called convergence in the mean.

15.35 Find the least-squares sum with $M = 1$ for the function $y(t)$ of Problem 15.24.

    ❚ Truncation brings $T_1(t) = \pi/2 - (4/\pi)\cos t$. This function is shown dashed in Fig. 15-1. Notice that it smooths the corners of $y(t)$.

15.36 How do the two term and four term least-squares approximations to the series of Problem 15.29 compare with $y(x)$?

    ❚ Figure 15-4 shows the rectified sine wave solid and the two term approximation dotted. The corners have been rounded and there is some overshoot at the top. The four term approximation is shown only near the corners, since elsewhere its graph on this scale is indistinguishable from $y(x)$.

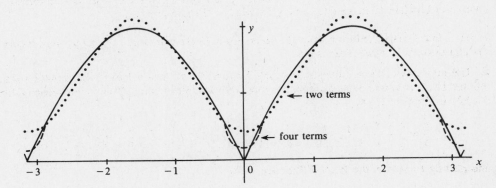

Fig. 15-4

15.6 SMOOTHING BY FOURIER ANALYSIS

15.37 What is the basis of the Fourier analysis method for smoothing data?

    ❚ If we think of given numerical data as consisting of the true values of a function with random errors superposed, the true functions being relatively smooth and the superposed errors quite unsmooth, then the examples in Problems 15.24 to 15.26 suggest a way of partially separating functions from error. Since the true function is smooth, its Fourier coefficients will decrease quickly. But the unsmoothness of the error suggests that its Fourier coefficients may decrease very slowly, if at all. The combined series will consist almost entirely of error, therefore, beyond a certain place. If we simply truncate the series at the right place, then we are discarding mostly error. There will still be error contributions in the terms retained. Since truncation produces a least-squares approximation, we may also view this method as least-squares smoothing.

15.38 Apply the method of the previous problem to the following data:

x	0	1	2	3	4	5	6	7	8	9	10
y	0	4.3	8.5	10.5	16.0	19.0	21.1	24.9	25.9	26.3	27.8

x	11	12	13	14	15	16	17	18	19	20
y	30.0	30.4	30.6	26.8	25.7	21.8	18.4	12.7	7.1	0

    ❚ Assuming the function to be truly zero at both ends, we may suppose it extended to an odd function of period $P = 40$. Such a function will even have a continuous first derivative, which helps to speed convergence of Fourier series. Using the formulas of Problem 15.16, we now compute the b_j.

j	1	2	3	4	5	6	7	8	9	10
b_j	30.04	-3.58	1.35	$-.13$	$-.14$	$-.43$	.46	.24	$-.19$	.04

j	11	12	13	14	15	16	17	18	19	20
b_j	.34	.19	.20	$-.12$	$-.36$	$-.18$	$-.05$	$-.37$	.27	

The rapid decrease is apparent and we may take all b_j beyond the first three or four to be largely error effects. If four terms are used, we have the trigonometric sum

$$T(x) = 30.04 \sin\frac{\pi x}{20} - 3.58 \sin\frac{2\pi x}{20} + 1.35 \sin\frac{3\pi x}{20} - .13 \sin\frac{4\pi x}{20}$$

The values of this sum may be compared with the original data, which were actually values of $y(x) = x(400 - x^2)/100$ contaminated by artificially introduced random errors (see Table 15.1). The rms error of the given data was 1.06 and of the smoothed data .80.

TABLE 15.1

x	Given	Correct	Smoothed	x	Given	Correct	Smoothed
1	4.3	4.0	4.1	11	30.0	30.7	29.5
2	8.5	7.9	8.1	12	30.4	30.7	29.8
3	10.5	11.7	11.9	13	30.6	30.0	29.3
4	16.0	15.6	15.5	14	26.8	28.6	28.0
5	19.0	18.7	18.6	15	25.7	26.2	25.8
6	21.1	22.7	21.4	16	21.8	23.0	22.4
7	24.9	24.6	23.8	17	18.4	18.9	18.0
8	25.9	26.9	25.8	18	12.7	13.7	12.6
9	26.3	28.7	27.4	19	7.1	7.4	6.5
10	27.8	30.0	28.7	20			

15.39 Approximate the derivative $y'(x) = (400 - 3x^2)/100$ of the function in the preceding problem on the basis of the same given data.

▎ First we shall apply the formula

$$y'(x) \simeq \tfrac{1}{10}[-2y(x-2) - y(x-1) + y(x-1) + 2y(x+2)]$$

derived earlier from the least-squares parabola for the five arguments $x-2, \ldots, x+2$. With similar formulas for the four end arguments, the results form the second column of Table 15.2. Using this local least-squares parabola already amounts to local smoothing of the original x, y data. We now attempt further overall smoothing by the Fourier method. Since the derivative of an odd function is even, the formula of Problem 15.19 is appropriate:

$$a_j = \frac{1}{20}[y'(0) + y'(20)\cos j\pi] + \frac{1}{10}\sum_{x=1}^{19} y'(x)\cos\frac{\pi}{20}jx$$

These coefficients may be computed to be

j	0	1	2	3	4	5	6	7	8	9	10
a_j	0	4.81	−1.05	.71	−.05	.05	−.20	.33	.15	.00	.06

j	11	12	13	14	15	16	17	18	19	20
a_j	.06	.06	−.03	.11	.06	.14	−.04	.16	−.09	.10

TABLE 15.2

x	Local	Fourier	Correct	x	Local	Fourier	Correct
0	5.3	4.4	4.0	11	1.1	.5	.4
1	4.1	4.4	4.0	12	−.1	−.1	−.3
2	3.8	4.1	3.9	13	−1.2	−.9	−1.1
3	3.7	3.8	3.7	14	−2.2	−1.8	−1.9
4	3.4	3.4	3.5	15	−2.9	−2.9	−2.8
5	3.4	3.0	3.2	16	−3.6	−4.0	−3.7
6	2.6	2.5	2.9	17	−4.6	−5.0	−4.7
7	1.9	2.1	2.5	18	−5.5	−5.8	−5.7
8	1.5	1.8	2.1	19	−7.1	−6.4	−6.8
9	1.2	1.4	1.6	20	−6.4	−6.6	−8.0
10	1.3	1.0	1.0				

Again the sharp drop is noticeable. Neglecting all terms beyond $j = 4$, we have

$$y'(x) \simeq 4.81 \cos \frac{\pi x}{20} - 1.05 \cos \frac{2\pi x}{20} + .71 \cos \frac{3\pi x}{20} - .05 \cos \frac{4\pi x}{20}$$

Computing this for $x = 0, \ldots, 20$ produces the third column of Table 15.2. The last column gives the correct values. The rms error in column 2, after local smoothing by a least-squares parabola, is .54, while the rms error in column 3, after additional Fourier smoothing, is .39.

15.40 Apply Fourier smoothing to the following data, assuming that the end values are actually zero and extending the function to an odd function of period $P = 2L = 40$. Compare smoothed values with the correct $y(x) = x(1 - x)$ for $x = 0(.05)1$. The data were obtained by adding random errors of up to 20%.

.00	.06	.10	.11	.14	.22	.22	.27	.28	.21	.22	.27	.21	.20	.19
									.21	.19	.12	.08	.04	.00

▌ The formulas of Problem 15.16 produce the b_i coefficients.

.258	.008	.015	−.012	−.007	.007	.002	.017	.001	−.009
	.011	.008	−.003	−.012	−.005	.007	−.000	.001	.005

Except for the first, they show the more or less chaotic behavior of noise. Setting up the corresponding one, two, three and four term trigonometric sums, and computing four sets of smoothed values, we find their rms errors to be .007, .009, .007 and .011, so there seems to be little use in retaining more than one:

$$T(x) = .258 \sin \pi x$$

The results of smoothing by this one term appear in Table 15.3, along with correct two place values. In spite of the bumpy data there is now no error greater than .01.

TABLE 15.3

T	y	T	y	T	y	T	y
.04	.05	.21	.21	.25	.25	.15	.16
.08	.09	.23	.23	.25	.24	.12	.13
.12	.13	.25	.24	.23	.23	.08	.09
.15	.16	.25	.25	.21	.21	.04	.05
.18	.19	.26	.25	.18	.19		

15.7 COMPLEX FORMS

15.41 Prove the following orthogonality property of the functions e^{ijx} and e^{ikx} for j and k integers. The overbar denotes a complex conjugate:

$$\int_0^{2\pi} \overline{e^{ijx}} e^{ikx} \, dx = \begin{cases} 0 & \text{if } k \neq j \\ 2\pi & \text{if } k = j \end{cases}$$

▌ The proof is elementary, the integral reducing at once to

$$\int_0^{2\pi} e^{i(k-j)x} \, dx = \frac{1}{i(k-j)} e^{i(k-j)x} \Big|_0^{2\pi}$$

for $k \neq j$. But this is equal to 1 at both limits, hence zero. For $k = j$, the left side above is clearly 2π.

15.42 Derive the formula for Fourier coefficients in complex form.

▌ The proof takes a familiar path. The Fourier series is

$$f(x) = \sum_{j=-\infty}^{\infty} f_j e^{ijx}$$

Multiplying by e^{ikx} and integrating brings

$$\int_0^{2\pi} f(x) e^{ikx} \, dx = \sum_{j=-\infty}^{\infty} f_j e^{ikx} e^{ijx} \, dx$$

and since all terms on the right vanish by orthogonality except the one for $j = k$, the required result is found,

$$f_k = \frac{1}{2\pi} \int_0^{2\pi} f(x) e^{-ikx} \, dx$$

15.43 Show that the two sums $\sum_{j=-l}^{l} c_j e^{ijx}$ and $a_0/2 + \sum_{j=1}^{l} (a_j \cos jx + b_j \sin jx)$ are equivalent by reconciling the various coefficients.

▌ Using Euler's theorem $e^{i\theta} = \cos\theta + i\sin\theta$, the complex form can be written

$$\sum_{j=-l}^{l} c_j e^{ijx} = \sum_{j=-l}^{l} c_j (\cos jx + i \sin jx)$$

$$= c_0 + \sum_{j=1}^{l} \left[(c_j + c_{-j}) \cos jx + i(c_j - c_{-j}) \sin jx \right]$$

where in the second step we group terms for j and $-j$. This shows that

$$a_0 = 2c_0 \qquad a_j = c_j + c_{-j} \qquad b_j = i(c_j - c_{-j})$$

which can be inverted to read

$$c_0 = \tfrac{1}{2} a_0 \qquad c_j = \tfrac{1}{2}(a_j - ib_j) \qquad c_{-j} = \tfrac{1}{2}(a_j + ib_j)$$

15.44 Show that the functions e^{ijx_n} and e^{ikx_n} are orthogonal in the following sense:

$$\sum_{n=0}^{N-1} \overline{e^{ijx_n}} e^{ikx_n} = \begin{cases} N & \text{if } k = j \\ 0 & \text{if } k \neq j \end{cases}$$

Here as before, $x_n = 2\pi n/N$.

▌ We will find a geometric sum with ratio $r = e^{i(k-j)2\pi/N}$:

$$\sum_{n=0}^{N-1} \overline{e^{ijx_n}} e^{ikx_n} = \sum_{n=0}^{N-1} e^{i(k-j)x_n} = e^{i(k-j)x_0} \left(1 + r + r^2 + \cdots + r^{N-1} \right)$$

For $j = k$ we have $r = 1$ and the sum is N. Otherwise the sum of the powers of r is $(1 - r^N)/(1 - r)$ by a familiar formula. But r^N is $e^{2\pi i(k-j)}$, which is 1, making the numerator zero and establishing the orthogonality.

15.45 Show that if $N = 2l + 1$, then the trigonometric sum

$$\sum_{j=-l}^{l} d_j e^{ijx}$$

must have coefficients $d_j = f_j^*$ if it is to collocate with the function $f(x)$ at $x_n = 2\pi n/N$.

▌ Assume that collocation occurs, multiply by $\overline{e^{ikx_n}}$ and sum:

$$\sum_{n=0}^{N-1} f(x_n) \overline{e^{ikx_n}} = \sum_{n=0}^{N-1} \overline{e^{ikx_n}} \sum_{j=-l}^{l} d_j e^{ijx_n} = \sum_{j=-l}^{l} d_j \sum_{n=0}^{N-1} \overline{e^{ikx_n}} e^{ijx_n}$$

Again all terms on the right are zero except one, for $j = k$, and we have

$$\sum_{n=0}^{N-1} f(x_n) e^{ikx_n} = d_k(N) = f_k^* N$$

15.46 How are the coefficients $f_j{}^*$ related to discrete Fourier transforms?

▮ Let V be the vector with components $f(x_0), \ldots, f(x_{N-1})$. For $N = 2l + 1$ this makes V $(2l + 1)$-dimensional, as is the vector of coefficients $f_j{}^*$ for the trigonometric sum

$$\sum_{j=-l}^{l} f_j{}^* e^{ijx}$$

in which

$$f_j{}^* = \frac{1}{N} \sum_{n=0}^{N-1} f(x_n) e^{-ijx_n}$$

for $j = -l$ to $j = l$. Comparing with

$$v_j^T = \sum_{n=0}^{N-1} v_n \omega_N^{jn} = \sum_{n=0}^{N-1} f(x_n) e^{-ijx_n}$$

where $x_n = 2\pi n/N$ and $j = 0$ to $j = N - 1$, the match is conspicuous. We do have one problem: the ranges of validity do not coincide. But we may deduce that where the ranges overlap, from $j = 0$ to $j = l$,

$$v_j^T = N f_j{}^* \qquad j = 0, \ldots, l$$

Now we observe that

$$v_{j+N}^T = \sum_{n=0}^{N-1} f(x_n) e^{-i(j+N)x_n} = \sum_{n=0}^{N-1} f(x_n) e^{-ijx_n}$$

for $j + N = 0, \ldots, N - 1$ or $j = -1, \ldots, -N$. Once again we have a match, this time for $j = -1$ to $j = -l$:

$$v_{j+N} = N f_j{}^* \qquad j = -l, \ldots, -1$$

Apart from the factor $1/N$ the components v_j^T do, therefore, match the coefficients $f_j{}^*$, though in a slightly scrambled order. Taking the v_j^T in their natural order v_0^T to v_{2l}^T it is easy to verify that the order of the coefficients will be

$$f_0{}^*, \ldots, f_l{}^* \qquad f_{-l}^*, \ldots, f_{-1}^*$$

15.47 Work through the details of the preceding problem for the simple example $V = (1, 0, -1)$.

▮ Here $N = 3$ and $l = 1$:

$$3 f_j{}^* = \sum_{n=0}^{2} f(x_n) e^{-ijx_n} = \sum_{n=0}^{2} f(x_n) \omega_3^{jn} = 1 - \omega_3^{2j}$$

This makes

$$3 f_{-1}^* = 1 - \omega_3 \qquad 3 f_0{}^* = 0 \qquad 3 f_1{}^* = 1 - \omega_3^2$$

and we have the three coefficients directly. Turning to the transform,

$$v_j^T = \sum_{n=0}^{2} f(x_n) \omega_3^{jn} = 1 - \omega_3^{2j}$$

we find

$$v_0^T = 0 \qquad v_1^T = 1 - \omega_3^2 \qquad v_3^T = 1 - \omega_3$$

and the correspondence discovered in Problem 15.46 is confirmed.

15.48 What is the central idea behind the fast Fourier transform (FFT)?

▮ When N is the product of integers, the numbers $f_j{}^*$ prove to be closely interdependent. This interdependence can be exploited to substantially reduce the amount of computing required to generate these numbers.

15.49   Develop the FFT for the simplest case, when N is the product of two integers t_1 and t_2.

▋ Let $j = j_1 + t_1 j_2$ and $n = n_2 + t_2 n_1$. Then for j_1, $n_1 = 0$ to $t_1 - 1$, and j_2, $n_2 = 0$ to $t_2 - 1$ both j and n run their required ranges 0 to $N - 1$. Now

$$\omega^{(j_1 + t_1 j_2)(n_2 + t_2 n_1)} = \omega_N^{j_1 t_2 n_1 + j_1 n_2 + t_1 j_2 n_2}$$

since $t_1 t_2 = N$ and $\omega_N^N = 1$. The transform can then be written as a double sum

$$v_j^T = \sum_{n_2=0}^{t_2-1} \sum_{n_1=0}^{t_1-1} v_n \omega_N^{j_1 t_2 n_1} \omega_N^{j_1 n_2 + t_1 j_2 n_2}$$

This can also be arranged in a two-step algorithm:

$$F_1(j_1, n_2) = \sum_{n_1=0}^{t_1-1} v_n \omega_N^{j_1 t_2 n_1}$$

$$v_j^T = F_2(j_1, j_2) = \sum_{n_2=0}^{t_2-1} F_1(j_1, n_2) \omega_N^{j_1 n_2 + t_1 j_2 n_2}$$

15.50   What is the gain in computing efficiency if the FFT of Problem 15.49 is used? In other words, just how fast is the fast Fourier transform?

▋ To compute F_1 there are t_1 terms to be processed; to compute F_2 there are t_2. The total is $t_1 + t_2$. This must be done for each (j_1, n_2) and (j_1, j_2) pair, or N pairs. The final count is, thus, $N(t_1 + t_2)$ terms processed. The original form of the transform

$$v_j^T = \sum_{n=0}^{N-1} v_n \omega_N^{jn}$$

processed N terms for each j, a total of N^2 terms. The gain in efficiency, if measured by this standard, is thus

$$\frac{t_1 + t_2}{N}$$

and depends very much on N. For a small data set, say $N = 12 = 3 \times 4$, the FFT will need about $\frac{7}{12}$ the computing time of a direct approach. This is hardly significant but points out the direction of things to come.

15.51   Run the FFT of Problem 15.49 for the vector

n	0	1	2	3	4	5	
v_n	0	1	1	0	-1	-1	0 $\cdots$

▋ The small scale of the problem, $N = 6$, makes it easy to see all the detail. Here $N = t_1 t_2 = 2 \times 3$ so we first find the F_1 values from

$$F_1(j_1, n_2) = \sum_{n_1=0}^{1} v_n \omega_6^{3 j_1 n_1} \qquad n = n_2 + 3 n_1$$

and they prove to be, with $\omega = \omega_6$,

$$F_1(0,0) = v_0 + v_3 = 0 \qquad F_1(1,0) = v_0 - v_3 = 0$$
$$F_1(0,1) = v_1 + v_4 = 0 \qquad F_1(1,1) = v_1 - v_4 = 2$$
$$F_1(0,2) = v_2 + v_5 = 0 \qquad F_1(1,2) = v_2 - v_5 = 2$$

Then

$$v_j^T = F_2(j_1, j_2) = \sum_{n_2=0}^{2} F_1(j_1, n_2) \omega^{j_1 n_2 + t_1 j_2 n_2}$$

leading to, since $j = j_1 + 2j_2$,

$$v_0^T = F_2(0,0) = v_0 + v_1 + v_2 + v_3 + v_4 + v_5 = 0$$

$$v_1^T = F_2(1,0) = F_1(1,0) + F_1(1,1)\,\omega + F_1(1,2)\,\omega^2 = 2\omega + 2\omega^2 = 2\sqrt{3}\,i$$

$$v_2^T = F_2(0,1) = F_1(0,0) + F_1(0,1)\,\omega^2 + F_2(0,2)\,\omega^4 = 0$$

$$v_3^T = F_2(1,1) = F_1(1,0) + F_1(1,1)\,\omega^3 + F_1(1,2)\,\omega^6 = 0$$

and similarly

$$v_4^T = F_2(0,2) = 0$$

$$v_5^T = F_2(1,2) = -2\sqrt{3}\,i$$

Note that Nt_1 terms were involved in computing the F_1 values and Nt_2 terms in getting F_2, a total of $12 + 18 = 30$ terms. The direct computation would have used 36 and would confirm the results just found. Also note the order of processing j_1, j_2 pairs. In programming language, the j_2 loop is external to the j_1 loop.

15.52 Extend the FFT of Problem 15.49 to the case $N = t_1 t_2 t_3$.

$\blacksquare$ The details will suggest the way to generalization for still longer products. Let

$$j = j_1 + t_1 j_2 + t_1 t_2 j_3 \qquad n = n_3 + t_3 n_2 + t_3 t_2 n_1$$

and observe that of the nine possible power terms in

$$\omega_N^{(j_1 + t_1 j_2 + t_1 t_2 j_3)(n_3 + t_3 n_2 + t_3 t_2 n_1)}$$

three will contain the product $t_1 t_2 t_3$ and may be neglected since $\omega_N^N = 1$. The remaining six may be grouped in the transform

$$v_j^T = \sum_{n_3=0}^{t_3-1} \left[\sum_{n_2=0}^{t_2-1} \left(\sum_{n_1=0}^{t_1-1} v_n \omega_N^{j t_3 t_2 n_1} \right) \omega_N^{(j_1 + t_1 j_2) t_3 n_2} \right] \omega_N^{(j_1 + t_1 j_2 + t_1 t_2 j_3) n_3}$$

with n_1 appearing only in the inner sum and n_2 not appearing in the outer. As before, this triple sum can be expressed as an algorithm, this time having three steps:

$$F_1(j_1, n_2, n_3) = \sum_{n_1=0}^{t_1-1} v_n \omega_N^{j t_3 t_2 n_1}$$

$$F_2(j_1, j_2, n_3) = \sum_{n_2=0}^{t_2-1} F_1(j_1, n_2, n_3)\, \omega_N^{(j_1 + t_1 j_2) t_3 n_2}$$

$$v_j^T = F_3(j_1, j_2, j_3) = \sum_{n_3=0}^{t_3-1} F_2(j_1, j_2, n_3)\, \omega_N^{(j_1 + t_1 j_2 + t_1 t_2 j_3) n_3}$$

This is the required FFT.

15.53 Estimate the saving in computing time if this algorithm is used.

$\blacksquare$ At each of the three steps the number of triples, such as (j_1, n_2, n_3), that must be processed is $t_1 t_2 t_3 = N$. In the sums we find the number of terms to be t_1, t_2, t_3 in turn. This makes a total of $N(t_1 + t_2 + t_3)$ terms altogether. The transform as defined still uses N^2 terms, so the efficiency of the FFT may be estimated as

$$\frac{t_1 + t_2 + t_3}{N}$$

If, for instance, $N = 1000 = 10 \times 10 \times 10$, then only 3% of the original 1000000 terms are needed.

15.54   Run the FFT algorithm of Problem 15.52 manually for the input vector

n	0	1	2	3	4	5	6	7
v_n	1	$1+i$	i	$i-1$	-1	$-1-i$	$-i$	$1-i$

▌ We have $N = 8 = 2 \times 2 \times 2$, making $j = j_1 + 2j_2 + 4j_3$ and $n = n_3 + 2n_2 + 4n_1$. The formula for F_1 is then

$$F_1(j_1, n_2, n_3) = \sum_{n_1=0}^{1} v_n \omega_8^{4j_1 n_1}$$

and we have

$$F_1(0,0,0) = v_0 + v_4 = 0 \qquad F_1(1,0,0) = v_0 + v_4\omega^4 = 2$$
$$F_1(0,0,1) = v_1 + v_5 = 0 \qquad F_1(1,0,1) = v_1 + v_5\omega^4 = 2 + 2i$$
$$F_1(0,1,0) = v_2 + v_6 = 0 \qquad F_1(1,1,0) = v_2 + v_6\omega^4 = 2i$$
$$F_1(0,1,1) = v_3 + v_7 = 0 \qquad F_1(1,1,1) = v_3 + v_7\omega^4 = 2i - 2$$

with ω_8 abbreviated to ω. Notice the $Nt_1 = 8 \times 2$ terms used. Next we use

$$F_2(j_1, j_2, n_3) = \sum_{n_2=0}^{1} F_1(j_1, n_2, n_3)\,\omega^{2(j_1 + 2j_2)n_2}$$

to compute

$$F_2(0,0,0) = 0 \qquad F_2(1,0,0) = F_1(1,0,0) + F_1(1,1,0)\,\omega^2 = 4$$
$$F_2(0,0,1) = 0 \qquad F_2(1,0,1) = F_1(1,0,1) + F_1(1,1,1)\,\omega^2 = 4 + 4i$$
$$F_2(0,1,0) = 0 \qquad F_2(1,1,0) = F_1(1,0,0) + F_1(1,1,0)\,\omega^6 = 0$$
$$F_2(0,1,1) = 0 \qquad F_2(1,1,1) = F_1(1,0,1) + F_1(1,1,1)\,\omega^6 = 0$$

and finally

$$v_j^T = F_3(j_1, j_2, j_3) = \sum_{n_3=0}^{1} F_2(j_1, j_2, n_3)\,\omega^{jn_3}$$

to get the transform

$$v_0^T = F_3(0,0,0) = F_2(0,0,0) + F_2(0,0,1) = 0$$
$$v_1^T = F_3(1,0,0) = F_2(1,0,0) + F_2(1,0,1)\,\omega = 4 + 4\sqrt{2}$$
$$v_2^T = F_3(0,1,0) = F_2(0,1,0) + F_2(0,1,1)\,\omega^2 = 0$$
$$v_3^T = F_3(1,1,0) = F_2(1,1,0) + F_2(1,1,1)\,\omega^3 = 0$$
$$v_4^T = F_3(0,0,1) = F_2(0,0,0) + F_2(0,0,1)\,\omega^4 = 0$$
$$v_5^T = F_3(1,0,1) = F_2(1,0,0) + F_2(1,0,1)\,\omega^5 = 4 - 4\sqrt{2}$$
$$v_6^T = F_3(0,1,1) = F_2(0,1,0) + F_2(0,1,1)\,\omega^6 = 0$$
$$v_7^T = F_3(1,1,1) = F_2(1,1,0) + F_2(1,1,1)\,\omega^7 = 0$$

A total of $N(t_1 + t_2 + t_3) = 48$ terms has been processed, only a slight saving from $N^2 = 64$ because of the problem's small scale.

15.55   The inverse discrete transform may be defined by

$$u_k^{-T} = \frac{1}{N} \sum_{j=0}^{N-1} u_j \omega^{-jk} = \frac{1}{N} \sum_{j=0}^{N-1} u_j e^{ikx_j}$$

Show that this definition does give an inverse relationship by inserting $u_j = v_j^T$ and discovering that $u_k^{-T} = v_k$. That is, the components of the original vector V are regained.

$\blacksquare$ It may be useful to first rewrite the result of Problem 15.48 using

$$\omega^{jn} = e^{-ijx_n}$$

to obtain

$$\sum_{n=0}^{N-1} \omega^{jn}\omega^{-kn} = \begin{cases} N & \text{if } k=j \\ 0 & \text{if } k \neq j \end{cases}$$

for j, k in the interval $(0, N-1)$. Now

$$\frac{1}{N}\sum_{j=0}^{N-1} v_j^T \omega^{-jk} = \frac{1}{N}\sum_{j=0}^{N-1}\sum_{n=0}^{N-1} v_n \omega^{jn}\omega^{-jk} = \frac{1}{N}\sum_{n=0}^{N-1} v_n \sum_{j=0}^{N-1} \omega^{(n-k)j}$$

and the last sum being zero, unless n takes the value k, we soon have the anticipated v_k.

15.56 Invert the transform found in Problem 15.54.

$\blacksquare$ The FFT could be used, but in view of the large number of zero components this is a good chance to proceed directly:

$$8u_0^{-T} = \sum_{j=0}^{7} v_j^T = 8 \qquad u_0^{-T} = 1 = v_0$$

$$8u_1^{-T} = \sum_{j=0}^{7} v_j^T \omega^{-j} = \left(4 + 4\sqrt{2}\right)\omega^{-1} + \left(4 - 4\sqrt{2}\right)\omega^{-5}$$

$$= 8(1+i) \qquad u_1^{-T} = 1 + i = v_2$$

$$8u_2^{-T} = \sum_{j=0}^{7} v_j^T \omega^{-2j} = \left(4 + 4\sqrt{2}\right)\omega^{-2} + \left(4 - 4\sqrt{2}\right)\omega^{-10}$$

$$= 8i \qquad u_2^{-T} = i = v_3$$

$$\vdots$$

The remaining components may be verified.

15.8 THE $L1$ NORM

15.57 Recalling the discussion of the $L1$ norm in Problem 13.73, which suggested replacing the search for an optimal $L1$ approximation by an interpolation problem, find the best function of the form $t(x) = b \sin x$ for $y(x) = x(\pi - x)$ over $(0, \pi)$.

$\blacksquare$ Canonical points for approximations of the form

$$t(x) = a_1 \sin x + a_2 \sin 2x + \cdots + a_n \sin nx$$

are $x_i = i\pi/(n+1)$. That is, these have been found to produce optimal or near optimal solutions by interpolation. In this case we have $n = 1$, so $x_1 = \pi/2$ is our only canonical point. For collocation we need $(\pi/2)^2 = b$, and the indicated approximation is $t(x) = (\pi^2/4)\sin x$.

Is this optimal? Relying on quadrature methods, the error of this $t(x)$ is found to be .233 to three places. Recomputing with a variety of other b values, lower errors are soon found with a minimum near 2.53 ($\pi^2/4$ is just below 2.47). The method of canonical points has gotten us close, with

$$t(x) = 2.53 \sin x$$

a slight improvement.

15.58 Find the best $L1$ approximation of form

$$t(x) = a_1 \sin x + a_2 \sin 2x + a_3 \sin 3x$$

for $y(x) = x(\pi - x)$.

❙ Now $n = 3$ and the canonical points are $x_i = \pi/4$, $2\pi/4$ and $3\pi/4$. This makes the collocation equations

$$\frac{\pi}{4} \cdot \frac{3\pi}{4} = \frac{\sqrt{2}}{2} a_1 + a_2 + \frac{\sqrt{2}}{2} a_3$$

$$\frac{\pi}{2} \cdot \frac{\pi}{2} = a_1 \qquad - \quad a_3$$

$$\frac{3\pi}{4} \cdot \frac{\pi}{4} = \frac{\sqrt{2}}{2} a_1 - a_2 + \frac{\sqrt{2}}{2} a_3$$

with solution

$$a_2 = 0 \qquad a_1 = \frac{\pi^2}{8}\left(1 + \frac{3}{2\sqrt{2}}\right) = 2.542 \qquad a_3 = \frac{\pi^2}{8}\left(-1 + \frac{3}{2\sqrt{2}}\right) = .07484$$

The $L1$ error for these coefficients, found by quadrature, is near .0477, and an informal search of the neighborhood finds that improvement lies in the direction of increasing a_3. The optimum is close to $a_3 = .09$, with changes in a_1 apparently making things worse instead of better. Our $L1$ solution is

$$t(x) = 2.542 \sin x + .09 \sin 3x$$

The middle term disappears ($a_2 = 0$) since it is not symmetric about $x = \pi/2$ as $y(x)$ is.

CHAPTER 16
Roots of Equations

16.1 THE ITERATIVE METHOD

16.1 Prove that if r is a root of $f(x) = 0$ and if this equation is rewritten in the form $x = F(x)$ in such a way that $|F'(x)| \leq L < 1$ in an interval I centered at $x = r$, then the sequence $x_n = F(x_{n-1})$ with x_0 arbitrary but in the interval I has $\lim x_n = r$.

▮ First we find

$$|F(x) - F(y)| = |F'(\xi)(x - y)| \leq L|x - y|$$

provided both x and y are close to r. Actually it is this Lipschitz condition rather than the more restrictive condition on $F'(x)$ that we need. Now

$$|x_n - r| = |F(x_{n-1}) - F(r)| \leq L|x_{n-1} - r|$$

so that, since $L < 1$, each approximation is at least as good as its predecessor. This guarantees that all our approximations are in the interval I, so that nothing interrupts the algorithm. Applying the last inequality n times, we have

$$|x_n - r| \leq L^n|x_0 - r|$$

and since $L < 1$, $\lim x_n = r$.

The convergence is illustrated in Fig. 16-1. Note that choosing $F(x_{n-1})$ as the next x_n amounts to following one of the horizontal line segments over to the line $y = x$. Notice also that in Fig. 16-2 the case $|F'(x)| > 1$ leads to divergence.

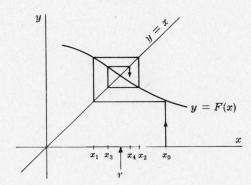

Fig. 16-1

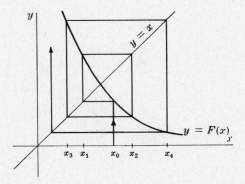

Fig. 16-2

16.2 In the year 1225 Leonardo of Pisa studied the equation

$$f(x) = x^3 + 2x^2 + 10x - 20 = 0$$

and produced $x = 1.368808107$. Nobody knows by what method Leonardo found this value but it is a remarkable result for his time. Apply the method of Problem 16.1 to obtain this result.

▌ The equation can be put into the form $x = F(x)$ in many ways. We take $x = F(x) = 20/(x^2 + 2x + 10)$, which suggests the iteration

$$x_n = \frac{20}{x_{n-1}^2 + 2x_{n-1} + 10}$$

With $x_0 = 1$ we find $x_1 = \frac{20}{13} \simeq 1.538461538$. Continuing the iteration produces the sequence of Table 16.1. Sure enough, on the twenty-fourth round Leonardo's value appears.

TABLE 16.1

n	x_n	n	x_n
1	1.538461538	13	1.368817874
2	1.295019157	14	1.368803773
3	1.401825309	15	1.368810031
4	1.354209390	16	1.368807254
5	1.375298092	17	1.368808486
6	1.365929788	18	1.368807940
7	1.370086003	19	1.368808181
8	1.368241023	20	1.368808075
9	1.369059812	21	1.368808122
10	1.368696397	22	1.368808101
11	1.368857688	23	1.368808110
12	1.368786102	24	1.368808107

16.3 Why is the convergence of the algorithm of the previous problem so slow?

▌ The rate of convergence may be estimated from the relation

$$e_n = r - x_n = F(r) - F(x_{n-1}) = F'(\xi)(r - x_{n-1}) = F'(\xi)e_{n-1}$$

which compares the nth error e_n with the preceding error. As n increases we may take $F'(r)$ as an approximation to $F'(\xi)$, assuming the existence of this derivative. Then $e_n \simeq F'(r)e_{n-1}$. In our example,

$$F'(r) = -\frac{40(r+1)}{(r^2 + 2r + 10)^2} \simeq -.44$$

making each error about $-.44$ times the one before it. This suggests that two or three iterations will be required for each new correct decimal place and this is what the algorithm has actually achieved.

16.4 Apply the idea of extrapolation to the limit to accelerate the previous algorithm.

▌ The idea may be used whenever information about the character of the error in an algorithm is available. Here we have the approximation $e_n \simeq F'(r)e_{n-1}$. Without knowledge of $F'(r)$ we may still write

$$r - x_{n+1} \simeq F'(r)(r - x_n)$$
$$r - x_{n+2} \simeq F'(r)(r - x_{n+1})$$

Dividing we find

$$\frac{r - x_{n+1}}{r - x_{n+2}} \simeq \frac{r - x_n}{r - x_{n+1}}$$

and solving for the root

$$r \simeq x_{n+2} - \frac{(x_{n+2} - x_{n+1})^2}{x_{n+2} - 2x_{n+1} + x_n} = x_{n+2} - \frac{(\Delta x_{n+1})^2}{\Delta^2 x_n}$$

This is often called the *Aitken* Δ^2 *process*.

16.5   Apply extrapolation to the limit to the computation of Problem 16.2.

▮ Using x_{10}, x_{11} and x_{12}, the formula produces

$$r \simeq 1.368786102 - \frac{(.000071586)^2}{-.000232877} \simeq 1.368808107$$

which is once again Leonardo's value. With this extrapolation, only half the iterations are needed. Using it earlier might have made still further economies by stimulating the convergence.

16.6   Apply the iteration method to the equation $x = e^{-x}$ to find a root near $x = .5$. Also apply the Aitken Δ^2 process. How much faster is it?

▮ Taking .5 as the initial approximation, it is found that x_{10} and x_{11} agree to three places at .567, the approximations again being alternately high and low, and so giving continual bounds on the root. The Aitken process produced three correct places from x_1, x_2 and x_3, and managed seven places from x_8, x_9 and x_{10}. Only at x_{25} was the original iteration able to achieve this same figure of .5671433.

16.7   Rewrite the equation $x^3 = x^2 + x + 1$ in the form $x = F(x)$ and apply the iteration method to find a positive root.

▮ Even a casual look at the equation suggests a root between 1 and 2. As for rearrangements, there seem to be several options, but one such as $x = f(x) = x^3 - x^2 - 1$ would be useless since $F'(x) = 3x^2 - 2x > 1$ in the active area. In contrast, the rearrangement

$$x = 1 + 1/x + 1/x^2$$

will have small derivatives for such x. Choosing this form and the initial approximation $x_0 = 2$, we find x_{13} and x_{14} agreeing to three decimal places at 1.839. The Aitken acceleration process squeezes the same value out of x_3, x_4, x_5 and gets the six place 1.839287 from x_9, x_{10}, x_{11}. Why is convergence just a bit slower here than in the preceding example? The derivative at the root turns out to be near $-.62$ here, while for the exponential function it was $-.567$. The difference may be small but its effect is observable.

16.8   Will iterations generated by this method always oscillate back and forth across the true root?

▮ No, as Fig. 16-3 shows, the convergence can be one-sided. A trivial example is offered by

$$x = \tfrac{1}{2}x$$

with one-sided convergence to zero for any x_0. There can also be one-sided divergence, as Fig. 16-4 shows.

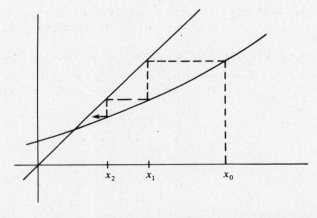

Fig. 16-3

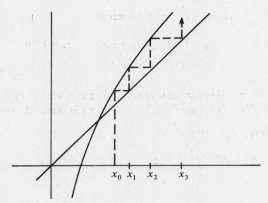

Fig. 16-4

16.9   What happens when the equation $x^2 = 2$ is rearranged as $x = 2/x$ and an iterative attempt is made to produce the square root of 2?

▮   From x_0 one creates $x_1 = 2/x_0$, then $x_2 = 2/x_1 = x_0$ with obvious continuation. A diagram of the computation shows the cycle as a rectangular track around the root. To make a short story long, the derivative of $F(x)$ in this example is less than 1 on one side of the root and greater than 1 on the other side, resulting in alternate efforts at convergence and divergence. No such fortunate balancing of forces occurs with the attempt

$$x = Q/x^2$$

to find the cube root of Q. here $F'(x) = -2Q/x^3 = -2$ at the root, which means divergence on either side.

16.10   Using extrapolation to the limit systematically after each three iterations is what is known as *Steffensen's method*. Apply this to Leonardo's equation.

▮   The first three approximations x_0, x_1 and x_2 may be borrowed from Problem 16.2. Aitken's formula is now used to produce x_3:

$$x_3 = x_2 - \frac{(x_2 - x_1)^2}{x_2 - 2x_1 + x_0} = 1.370813882$$

The original iteration is now resumed as in Problem 16.2 to produce x_4 and x_5:

$$x_4 = F(x_3) = 1.367918090 \qquad x_5 = F(x_4) = 1.369203162$$

Aitken's formula then yields x_6:

$$x_6 = x_5 - \frac{(x_5 - x_4)^2}{x_5 - 2x_4 + x_3} = 1.368808169$$

The next cycle brings the iterates

$$x_7 = 1.368808080 \qquad x_8 = 1.368808120$$

from which Aitken's formula manages $x_9 = 1.368808108$.

16.11   Show that other rearrangements of Leonardo's equation may not produce convergent sequences.

▮   As an example we may take $x = (20 - 2x^2 - x^3)/10$, which suggests the iteration

$$x_n = \frac{20 - 2x_{n-1}^2 - x_{n-1}^3}{10}$$

Again starting with $x_0 = 1$, we are led to the sequence

$$x_1 \simeq 1.70 \qquad x_3 \simeq 1.75 \qquad x_5 \simeq 1.79 \qquad x_7 \simeq 1.83$$

$$x_2 \simeq .93 \qquad x_4 \simeq .85 \qquad x_6 \simeq .79 \qquad x_8 \simeq .72$$

and so on. It seems clear that alternate approximations are headed in opposite directions. Comparing with Problem 16.1 we find that here $F'(r) = (-4r - 3r^2)/10 < -1$, confirming the computational evidence.

16.2 THE NEWTON METHOD

16.12 Derive the Newton iterative formula

$$x_n = x_{n-1} - \frac{f(x_{n-1})}{f'(x_{n-1})}$$

for solving $f(r) = 0$.

▌ Beginning with Taylor's formula,

$$f(r) = f(x_{n-1}) + (r - x_{n-1})f'(x_{n-1}) + \tfrac{1}{2}(r - x_{n-1})^2 f''(\xi)$$

we retain the linear part, recall that $f(r) = 0$ and define x_n by putting it in place of the remaining r to obtain

$$0 = f(x_{n-1}) + (x_n - x_{n-1})f'(x_{n-1})$$

which rearranges at once into

$$r \simeq x_n = x_{n-1} - \frac{f(x_{n-1})}{f'(x_{n-1})}$$

16.13 What is the geometric interpretation of Newton's formula?

▌ It amounts to using the tangent line to $y = f(x)$ at x_{n-1} in place of the curve. In Fig. 16-5 it can be seen that this leads to

$$\frac{f(x_{n-1}) - 0}{x_{n-1} - x_n} = f'(x_{n-1})$$

which is once again Newton's formula. Similar steps follow, as suggested by the arrow.

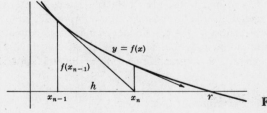

Fig. 16-5

16.14 Apply Newton's formula to Leonardo's equation.

▌ With $f(x) = x^3 + 2x^2 + 10x - 20$ we find $f'(x) = 3x^2 + 4x + 10$, and the iterative formula becomes

$$x_n = x_{n-1} - \frac{x_{n-1}^3 + 2x_{n-1}^2 + 10x_{n-1} - 20}{3x_{n-1}^2 + 4x_{n-1} + 10}$$

Once more choosing $x_0 = 1$, we obtain the results in Table 16.2.

TABLE 16.2

n	1	2	3	4
x_n	1.411764706	1.369336471	1.368808189	1.368808108

The speed of convergence is remarkable. In four iterations we have essentially Leonardo's value. In fact, computation shows that

$$f(1.368808107) \simeq -.000000016$$
$$f(1.368808108) \simeq -.000000005$$

which suggests that the Newton result is the winner by a nose.

16.15 Explain the rapid convergence of Newton's iteration by showing that the convergence is "quadratic."

❚ Recalling the equations of Problem 16.12 that led to the Newton formula,

$$f(r) = f(x_{n-1}) + (r - x_{n-1})f'(x_{n-1}) + \tfrac{1}{2}(r - x_{n-1})^2 f''(\xi)$$
$$0 = f(x_{n-1}) + (x_n - x_{n-1})f'(x_{n-1})$$

we subtract to obtain

$$0 = (r - x_n)f'(x_{n-1}) + \tfrac{1}{2}(r - x_{n-1})^2 f''(\xi)$$

or, letting $e_n = r - x_n$,

$$0 = e_n f'(x_{n-1}) + \tfrac{1}{2}e_{n-1}^2 f''(\xi)$$

Assuming convergence, we replace both x_{n-1} and ξ by the root r and have

$$e_n \simeq -\frac{f''(r)}{2f'(r)} e_{n-1}^2$$

Each error is therefore roughly proportional to the square of the previous error. This means that the number of correct decimal places roughly doubles with each approximation and is what is called quadratic convergence. It may be compared with the slower, linear convergence in Problem 16.3, where each error was roughly proportional to the previous error. Since the error of our present x_3 is about .00000008 and $[f''(r)]/[2f'(r)]$ is about .3, we see that if we had been able to carry more decimal places in our computation the error of x_4 might have been about two units in the fifteenth place. This superb speed suggests that the Newton algorithm deserves a reasonably accurate first approximation to trigger it and that its natural role is the conversion of such a reasonable approximation into an excellent one. In fact, other algorithms to be presented are better suited than Newton's for the "global" problem of obtaining first approximations to all the roots. Such methods usually converge very slowly, however, and it seems only natural to use them only as a source of reasonable first approximations, the Newton method then providing the polish. Such procedures are very popular and will be mentioned again as we proceed. It may also be noted that occasionally, given an inadequate first approximation, the Newton algorithm will converge at quadratic speed, but not to the root expected. Recalling the tangent line geometry behind the algorithm, it is easy to diagram a curve for which this happens, simply putting the first approximation near a maximum or minimum point.

16.16 Show that the formula for determining square roots,

$$x_n = \frac{1}{2}\left(x_{n-1} + \frac{Q}{x_{n-1}}\right)$$

is a special case of Newton's iteration.

❚ With $f(x) = x^2 - Q$, it is clear that making $f(x) = 0$ amounts to finding a square root of Q. Since $f'(x) = 2x$, the Newton formula becomes

$$x_n = x_{n-1} - \frac{x_{n-1}^2 - Q}{2x_{n-1}} = \frac{1}{2}\left(x_{n-1} + \frac{Q}{x_{n-1}}\right)$$

16.17   Apply the square root iteration with $Q = 2$.

▮ Choosing $x_0 = 1$, we find the results in Table 16.3. Notice once again the quadratic nature of the convergence. Each result has roughly twice as many correct digits as the one before it. Figure 16-6 illustrates the action. Since the first approximation was on the concave side of $y = x^2 - 2$, the next is on the other side of the root. After this the sequence is monotone, remaining on the convex side of the curve as tangent lines usually do.

TABLE 16.3

n	x_n
1	1.5
2	1.416666667
3	1.414215686
4	1.414213562
5	1.414213562

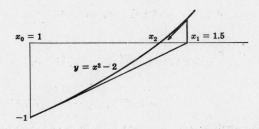

Fig. 16-6

16.18   Apply Newton's method to find the root of $x = e^{-x}$.

▮ With $f(x) = x - e^{-x}$, the formula is

$$x_n = x_{n-1} - \frac{x_{n-1} - e^{-x_{n-1}}}{1 + e^{-x_{n-1}}}$$

and we have only to choose x_0. The choice 1 leads to the brief sequence

$$1 \qquad .5378828 \qquad .5669870 \qquad .5671433$$

the last element of which may be recognized from Problem 16.6. A casual check shows the number of correct digits roughly doubling with each iteration.

There is a measure of entertainment value in Newton's method. Trying an apparently ridiculous starting value, like $x_0 = 100$ or 1000, one finds $x_1 = 0$. A moment's thought would have predicted this, since for large x the function $f(x) = x - e^{-x}$ is almost x. From 0 the usual rapid convergence resumes. The run beginning at $x_0 = -10$ is interesting in its own way, stopping very close to $-9, -8, -7, -6, -5, -4$ on its way to $x_{13} = .5671433$. It is easy to see why the method is believed to be almost fool-proof.

16.19   Find the real root of $f(x) = x^3 - x^2 - x - 1 = 0$ by Newton's method.

▮ The iteration can be written

$$x_{n+1} = x_n - \frac{x_n^3 - x_n^2 - x_n - 1}{3x_n^2 - 2x_n - 1}$$

and with $x_0 = 2$ generates the sequence of approximations

$$2 \qquad 1.857143 \qquad 1.839545 \qquad 1.839287$$

The last agrees with the work of Problem 16.7, but reaches it in 3 steps instead of an Aitken assisted 11.

The choice $x_0 = .5$ inspired this merry chase, only a few key points of which are reproduced:

n	0	1	3	7	8	14	15	16	18	20
x_n	.5	$-.8$	-3.4	$-.3$	-10	$-.6$	.04	$-.9$	1.86	1.839287

The oscillations are consistent with a graph of the function $f(x)$ and relevant tangent lines. The choice $x_0 = -10$ also triggered a curious run:

n	0	6	7	8	9	10	13	17	19
x	-10	$-.6$	.09	$-.9$	$-.35$	12	4	1.845	1.839287

In contrast to these oscillations, the choice $x_0 = -100$ produced a monotone increasing sequence reaching its goal at x_{16}. It goes without saying that $x_0 = 1$ led to division by zero, so it is just possible for the method to fail, but note the following histories of success, x_0 matched with the first correct x_n:

$$1.1, x_8 \qquad 1.01, x_{13} \qquad 1.001, x_{19} \qquad .99, x_{22} \qquad .9, x_{61}$$

The last of these runs oscillated nine times between positive and negative approximations, but found its goal.

16.20   Find the square root of 3 by Newton's method.

❚   The iteration $x_{n+1} = \frac{1}{2}(x_n + 3/x_n)$ had no difficulty:

$$1 \qquad 2 \qquad 1.75 \qquad 1.732143 \qquad 1.732051$$

Starting with $x_0 = .1$ provoked a leap to 15.05 but the root was found in eight steps. A start at $x_0 = 100$ required nine steps, and an unintended start $x_0 = -1$ led quickly to the negative root.

16.21   Derive the iteration $x_n = x_{n-1} - (x_{n-1}^p - Q)/px_{n-1}^{p-1}$ for finding a pth root of Q.

❚   With $f(x) = x^p - Q$ and $f'(x) = px^{p-1}$, the result is at once a special case of Newton's method.

16.22   Apply the preceding problem to find a cube root of 2.

❚   With $Q = 2$ and $p = 3$, the iteration simplifies to $x_n = \frac{2}{3}(x_{n-1} + 1/x_{n-1}^2)$. Choosing $x_0 = 1$, we find $x_1 = \frac{4}{3}$ and then

$$x_2 = 1.263888889 \qquad x_3 = 1.259933493 \qquad x_4 = 1.259921049 \qquad x_5 = 1.259921049$$

The quadratic convergence is conspicuous.

16.23   Find the seventh root of 2 and the seventy-seventh.

❚   The iteration of Problem 16.21 obliges, first with $Q = 2$ and $p = 7$. Choosing $x_0 = 1$, the approximation $x_4 = 1.104089$ with all later x_n concurring. Switching to $p = 77$, once again it is x_4 that first hits the mark, with 1.009043. As an encore, the power $p = 777$ was offered, and again it was x_4 that established the eventual consensus, with 1.000893. As an absolutely final encore, the 7777th root of 2 was sought, and at last there was a new winner: $x_3 = 1.000089$.

16.24   Apply Newton's method to $f(x) = 1/x - Q$, developing an iteration to produce the reciprocal of Q without division. Then put $Q = e$ and try both $x_0 = .3$ and 1.

❚   The derivation is routine:

$$x_n = x_{n-1} - \frac{1/x_{n-1} - Q}{-1/x_{n-1}^2} = x_{n-1}(2 - Qx_{n-1})$$

Putting $Q = e = 2.718282$ and trying $x_0 = .3$, we find x_4 to be .3678794, which is correct to seven places. However, any $x_0 > .8$ or there about will cause overflow, since it brings a shift to the negative branch of the hyperbola involved.

16.3 INTERPOLATION METHODS

16.25 This ancient method uses two previous approximations and constructs the next approximation by making a linear interpolation between them. Derive the *regula falsi* (see Fig. 16-7)

$$c = a - \frac{(a-b)f(a)}{f(a)-f(b)}$$

▐ The linear function

$$y = f(a) + \frac{f(a)-f(b)}{a-b}(x-a)$$

clearly has $y = f(x)$ at a and b. It vanishes at the argument c given in the *regula falsi*. This zero serves as our next approximation to the root of $f(x) = 0$, so effectively we have replaced the curve $y = f(x)$ by a linear collocation polynomial in the neighborhood of the root. It will also be noticed in Fig. 16-7 that the two given approximations a and b are on opposite sides of the exact root. Thus $f(a)$ and $f(b)$ have opposite signs. This opposition of signs is assumed when using *regula falsi*. Accordingly, having found c, to reapply *regula falsi* we use this c as either the new a or the new b, whichever choice preserves the opposition of signs. In Fig. 16-7, c would become the new a. In this way a sequence of approximations $x_0, x_1, x_2, \ldots$ may be generated, x_0 and x_1 being the original a and b.

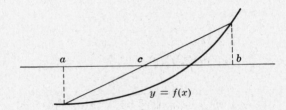

Fig. 16-7

16.26 Apply *regula falsi* to Leonardo's equation.

▐ Choosing $x_0 = 1$ and $x_1 = 1.5$, the formula produces

$$x_2 = 1.5 - \frac{.5(2.875)}{9.875} \simeq 1.35 \qquad x_3 = 1.35 - \frac{(-.15)(-.3946)}{-3.2696} \simeq 1.368$$

and so on. The rate of convergence can be shown to be better than the rate in Problem 16.2 but not so good as that of Newton's method.

16.27 Apply *regula falsi* to $x - e^{-x} = 0$.

▐ Starting with $a, b = 0, 1$, seven steps were needed to produce the familiar .5671433. Because of the local geometry the b value was continually replaced by c, the left end of the interval being steady at 0. Curiously, the seemingly poorer start $a, b = 1, 2$ got there faster, by first bringing a down to .4871417 and then allowing b this steady retreat:

$$.58 \qquad .5674 \qquad .5671468 \qquad .5671433$$

16.28 Find the real root of $x^3 - x^2 - x - 1 = 0$ by *regula falsi*.

▐ Starting with $a, b = 1, 2$, eight steps were needed to reach 1.839287, with the a value making all the changes:

$$1 \qquad 1.67 \qquad 1.82 \qquad 1.836 \qquad 1.8389 \qquad 1.839246 \qquad 1.839282 \qquad 1.839286$$

16.29 A natural next step is to use a quadratic interpolation polynomial rather than a linear one. Assuming three approximations x_0, x_1, x_2 are in hand, derive a formula for a new approximation x_3, which is a root of such a quadratic.

▐ It is not hard to verify that the quadratic through the three points (x_0, y_0), (x_1, y_1) and (x_2, y_2), where $y = f(x)$, can be written as

$$p(x) = \frac{x_1 - x_0}{x_2 - x_0}(Ah^2 + Bh + C)$$

where $h = x - x_2$ and A, B, C are

$$A = \frac{(x_1 - x_0)y_2 + (x_0 - x_2)y_1 + (x_2 - x_1)y_0}{(x_2 - x_1)(x_1 - x_0)^2}$$

$$B = \frac{(x_1 - x_0)(2x_2 - x_1 - x_0)y_2 - (x_2 - x_0)^2 y_1 + (x_2 - x_1)^2 y_0}{(x_2 - x_1)(x_1 - x_0)^2}$$

$$C = \frac{x_2 - x_0}{x_1 - x_0} y_2$$

Solving $p(x) = 0$ for h we find

$$h = -\frac{2C}{B \pm \sqrt{B^2 - 4AC}}$$

this form of the quadratic formula being chosen to avoid loss of significant digits during subtraction. Here the sign that makes the denominator larger in absolute value should be chosen. Then

$$x_3 = x_2 + h$$

becomes the next approximation and the process may be repeated with all subscripts advanced by 1.

The method just described is what is known as *Muller's method* and has been found to converge to both real and complex roots. For the latter it is necessary, of course, to run the algorithm in complex arithmetic, but even with real roots, complex arithmetic is the wiser choice since traces of imaginary parts occasionally enter.

16.30 Apply Muller's method to $x - e^{-x} = 0$.

▐ Starting with this triple of x arguments $(0, 1, 2)$ progress is recorded in Table 16.4, the coefficients of the quadratic $Ah^2 + Bh + C$ being given (only roughly and just to follow their general behavior) along with c and the new x argument, which always replaces the first member of the triple. The variations in A and B can be slightly obscure, but if h is to become small, then C is likely to move toward zero. Three steps were needed to reach the root.

TABLE 16.4

	A, B, C		h	New x
$-.619$	2.645	1.264	$-.4338998$	.5661002
$-.033$	.208	$-.0002$	.0010414	.5671416
$-.246$	1.563	$-.000003$	.0000017	.5671433
18.038	1.589	$6 \cdot 10^{-8}$	$-4 \cdot 10^{-8}$	same
239,389				

16.31 Apply Muller's method to Leonardo's equation.

▐ The initial triple $x = (0, 1, 2)$ led to what is recorded in Table 16.5, the correct value (single precision this time) arriving on the third step.

In programming Muller's method it is best to avoid computing the larger h value, since in late stages this will have almost a zero denominator. (For $c = 0$, one of the denominator entries $B \pm \sqrt{B^2 - 4AC}$ would vanish.)

TABLE 16.5

	A, B, C		h	New x
10	56	32	$-.6459341$	1.354066
2.25	7.49	$-.110$	.0145814	1.368647
6.57	20.6	$-.0033$	.0001607	1.368808
5.94	21.3	$-.000002$	.0000001	same

16.32 Use Muller's method to solve $x^7 = 7$ and $x^{11} = 11$.

❚ Keeping the starting triple $x = (0, 1, 2)$, which made the first quadratic $126h^2 + 380h + 242 = 0$, six steps were needed in reaching 1.320469. The last quadratic had the b coefficient 36.77152 and the denominator square root 36.77151.

With $x^{11} - 11 = 0$ eight steps were needed to find the root 1.243575 with a final useful quadratic of $400h^2 + 100h - .00005$ more or less.

16.4 BERNOULLI'S METHOD

16.33 Prove that if the polynomial of degree n,

$$p(x) = a_0 x^n + a_1 x^{n-1} + \cdots + a_n$$

has a single dominant zero, say r_1, then it may be found by computing a solution sequence for the difference equation of order n,

$$a_0 x_k + a_1 x_{k-1} + \cdots + a_n x_{k-n} = 0$$

and taking $\lim(x_{k+1}/x_k)$.

❚ This difference equation has $p(x) = 0$ for its characteristic equation and its solution can therefore be written as

$$x_k = c_1 r_1^k + c_2 r_2^k + \cdots + c_n r_n^k$$

If we choose initial values so that $c_1 \neq 0$, then

$$\frac{x_{k+1}}{x_k} = r_1 \frac{1 + (c_2/c_1)(r_2/r_1)^{k+1} + \cdots + (c_n/c_1)(r_n/r_1)^{k+1}}{1 + (c_2/c_1)(r_2/r_1)^k + \cdots + (c_n/c_1)(r_n/r_1)^k}$$

and since r_1 is the dominant root,

$$\lim \frac{r_i}{r_1} = 0 \qquad i = 2, 3, \ldots, n$$

making $\lim(x_{k+1}/x_k) = r_1$ as claimed. It can be shown using complex variable theory that the initial values $x_{-n+1} = \cdots = x_{-1} = 0$ and $x_0 = 1$ will guarantee $c_1 \neq 0$.

16.34 Apply the Bernoulli method to the equation $x^4 - 5x^3 + 9x^2 - 7x + 2 = 0$.

❚ The associated difference equation is

$$x_k - 5x_{k-1} + 9x_{k-2} - 7x_{k-3} + 2x_{k-4} = 0$$

and if we take the initial values $x_{-3} = x_{-2} = x_{-1} = 0$ and $x_0 = 1$, then the succeeding x_k are given in Table 16.6. The ratio x_{k+1}/x_k is also given. The convergence to $r = 2$ is slow, the rate of convergence of Bernoulli's method being linear. Frequently the method is used to generate a good starting approximation for Newton's or Steffensen's iteration, both of which are quadratic.

TABLE 16.6

k	x_k	x_{k+1}/x_k	k	x_k	x_{k+1}/x_k
1	5	3.2000	9	4017	2.0164
2	16	2.6250	10	8100	2.0096
3	42	2.3571	11	16278	2.0056
4	99	2.2121	12	32647	2.0032
5	219	2.1279	13	65399	2.0018
6	466	2.0773	14	130918	2.0010
7	968	2.0465	16	261972	2.0006
8	1981	2.0278	16	524097	

16.35 Modify the Bernoulli method for the case in which a pair of complex conjugate roots are dominant.

▮ Let r_1 and r_2 be complex conjugate roots. Then $|r_i| < |r_1|$ for $i = 3, \ldots, n$, since the r_1, r_2 pair is dominant. Using real starting values, the solution of the difference equation may be written as

$$x_k = c_1 r_1^k + c_2 r_2^k + \cdots + c_n r_n^k$$

where c_1 and c_2 are also complex conjugate. Let $r_1 = re^{i\phi} = \bar{r}_2$, $c_1 = ae^{i\theta} = \bar{c}_2$ with $r > 0$, $a > 0$, and $0 < \phi < \pi$ so that r_1 is the root in the upper half plane. Then

$$x_k = 2ar^k \cos(k\phi + \theta) + c_3 r_3^k + \cdots + c_n r_n^k$$

$$= 2ar^k \left[\cos(k\phi + \theta) + \frac{c_3}{2a}\left(\frac{r_3}{r}\right)^k + \cdots + \frac{c_n}{2a}\left(\frac{r_n}{r}\right)^k \right]$$

All terms except the first have limit zero, and so for large k, $x_k \simeq 2ar^k \cos(k\phi + \theta)$. We now use this result to determine r and ϕ. First we observe that

$$x_{k+1} - 2r\cos\phi\, x_k + r^2 x_{k-1} \simeq 0$$

as may be seen by substituting for x_k from the previous equation and using the identities for cosines of sums and differences. Reducing the subscripts, we also have

$$x_k - 2r\cos\phi\, x_{k-1} + r^2 x_{k-2} \simeq 0$$

Now solving these two simultaneously,

$$r^2 \simeq \frac{x_k^2 - x_{k+1}x_{k-1}}{x_{k-1}^2 - x_k x_{k-2}} \qquad -2r\cos\phi \simeq \frac{x_{k+1}x_{k-2} - x_{k-1}x_k}{x_{k-1}^2 - x_k x_{k-2}}$$

The necessary ingredients for determining r_1 and r_2 are now in hand.

16.36 Apply Bernoulli's method to Leonardo's equation.

▮ The associated difference equation is $x_k = -2x_{k-1} - 10x_{k-2} + 20x_{k-3}$ and the solution sequence for initial values $x_{-2} = x_{-1} = 0$ and $x_0 = 1$ appears in Table 16.7. Some approximations to r^2 and $-2r\cos\phi$ also appear. The fluctuating $\pm$ signs are an indication that dominant complex roots are present. This may be seen by recalling the form of the x_k as given in Problem 16.35, namely $x_k \simeq 2ar^k \cos(k\phi + \theta)$. As k increases, the value of the cosine will vary between ± 1 in a somewhat irregular way that depends on the value of ϕ.

From the last approximations we find

$$r\cos\phi \simeq -1.6844 \qquad r\sin\phi = \pm\sqrt{r^2 - (r\cos\phi)^2} \simeq \pm 3.4313$$

making the dominant pair of roots $r_1 r_2 \simeq -1.6844 \pm 3.4313i$. Since Leonardo's equation is cubic, these roots could also be found by using the real root found earlier to reduce to a quadratic equation. The Bernoulli method was not really needed in this case. The results found may be checked by computing the sum (-2) and product (20) of all the roots.

TABLE 16.7

k	x_k	k	x_k	r^2	$-2r\cos\phi$
1	-2	7	-2608	14.6026	3.3642
2	-6	8	-32464	14.6076	3.3696
3	52	9	147488	14.6135	3.3692
4	-84	10	-22496	14.6110	3.3686
5	-472	11	-2079168	14.6110	3.3688
6	2,824	12	7333056		

16.37 Apply the Bernoulli method to the Fibonacci equation.

The equation $x^2 = x + 1$ leads to the difference equation

$$x_{k+1} = x_k + x_{k-1}$$

and we take the usual initial conditions $x_{-1} = 0$ and $x_0 = 1$. The Fibonacci numbers $1, 2, 3, 5, 8, \ldots$ are, of course, generated accompanied by the ratios $1, 2, \frac{3}{2}, \frac{5}{3}$ and so on. With the eventual appearance of $x_{16} = 1597$ and $x_{17} = 2584$, the corresponding ratios agreed to six places upon 1.618034, which is the dominant root. The exact roots are still $(1 \pm \sqrt{5})/2$.

16.38 Apply the Bernoulli method to the Fibonacci-like equation

$$x^3 = x^2 + x + 1$$

The application is routine, using the difference equation $x_{k+1} = x_k + x_{k-1} + x_{k-2}$ and initial values $0, 0, 1$ for $k = -2, -1, 0$. The sequence of x_k values continues $1, 2, 4, 7, 13, 24$ with companion ratios $1, 2, 2, 1.75, 1.857, 1.846$. The later entries $x_{15} = 5768$ and $x_{16} = 10609$ brought identical ratios of 1.839286.

16.39 What happens if the Bernoulli method is attempted with the equation $x^3 - x^2 - 4x + 4 = 0$ which has the roots 2, -2 and 1?

The sequence generated begins $1, 1, 5, 5, 21, 21, 85, 85, \ldots$. Alternate ratios are either 1 or approaching 4. Curiously it is the smaller root that comes through, plus an extraneous value. The exact solution of the difference equation is

$$x_k = 2^k + \tfrac{1}{3}(-2)^k - \tfrac{1}{3}$$

as may be verified, or derived by the methods of Chapter 10. For large k the end term is negligible, making

$$\frac{x_{k+1}}{x_k} = \frac{2^{k+1} + \frac{1}{3}(-2)^{k+1}}{2^k + \frac{1}{3}(-2)^k} = \frac{2 + \frac{2}{3}(-1)^{k+1}}{1 + \frac{1}{3}(-1)^k} = \begin{cases} 1 & k \text{ even} \\ 4 & k \text{ odd} \end{cases}$$

confirming the computational outlook. There is not just one dominant root in this example, as Problem 16.33 supposes, but a pair.

However, the equation $x^3 - 5x^2 + 8x - 4 = 0$ which has the roots 2, 2 and 1 caused no theoretical difficulty, only a practical one. The sequence x_{k+1}/x_k did converge toward 2, but painfully slowly. The run was stopped at $k = 100$, with $x_k = 3 \cdot 10^{31}$ more or less and ratio 2.02. The exact solution of the difference equation is $x_k = k \cdot 2^{k+1} + 1$ and, ignoring the 1, one sees that the ratio x_{k+1}/x_k behaves like $2(k+1)/k$ for large k.

16.40 Apply Bernoulli's method to find the dominant pair of complex roots of $4x^4 + 4x^3 + 3x^2 - x - 1 = 0$.

The difference equation

$$x_{k+1} = -x_k - \tfrac{3}{4}x_{k-1} + \tfrac{1}{4}x_{k-2} + \tfrac{1}{4}x_{k-3}$$

produces values whose signs seem to fluctuate chaotically, which suggests dominant complex roots. Table 16.8 does not list these but instead provides some of the $r\cos\phi$ and $r\sin\phi$ values instead, the real and imaginary parts of the approximate roots. The last few entries are certainly suggestive and it is no heavy task to verify that $(-1 \pm i\sqrt{3})/2$ are exact roots. The quadratic $x^2 + x + 1$ is thus a factor of the given fourth degree polynomial and it soon follows that $4x^2 - 1$ is another. Bernoulli's method has converged to the dominant roots.

TABLE 16.8

k	$r\cos\phi$	$r\sin\phi$
5	$-.5$	.707
10	$-.498$	.876
15	$-.5001$	.86613
20	$-.499992$	.866019
25	$-.4999999$	.8660254

16.41 Run the Bernoulli algorithm just a few steps by hand (a mental run probably being too much to ask) for $x^2 - 2x + 2 = 0$.

▮ The difference equation is

$$x_{k+1} = 2x_k - 2x_{k-1} = 2(x_k - x_{k-1})$$

and with $x_{-1} = 0$ and $x_0 = 1$ for starters we soon have the sequence

$$0, 1, 2, 2, 0, -4, -8, -8, 0, 16, 32, 32, 0, \cdots$$

exhibiting an attractive pattern that makes it hard to stop. The sign behavior is far from chaotic but knowing there are complex roots ahead we seize the first quadruple $0, 1, 2, 2$ and compute

$$r^2 = \frac{4 - 2 \cdot 1}{1 - 2 \cdot 0} = 2 \qquad -2r\cos\phi = \frac{2 \cdot 0 - 1 \cdot 2}{1 - 2 \cdot 0} = -2$$

and then $r\cos\phi = 1$ and $r\sin\phi = \sqrt{2-1} = 1$. The anticipated root $1 + i$ is found on the first try. Its conjugate $1 - i$ automatically qualifies. Any other quadruple will reproduce the same numbers.

16.5 DEFLATION

16.42 Use the simple equation $x^4 - 10x^3 + 35x^2 - 50x + 24 = 0$ to illustrate the idea of deflation.

▮ The dominant root of this equation is exactly 4. Applying the factor theorem we remove the factor $x - 4$ by division:

1	−10	35	−50	24	|4
	4	−24	44	−24	
1	−6	11	−6	0	

The quotient is the cubic $x^3 - 6x^2 + 11x - 6$ and we say that the original quartic polynomial has been deflated to this cubic. The dominant root of the cubic is exactly 3. Removing this factor,

1	−6	11	−6	|3
	3	−9	6	
1	−3	2	0	

we achieve a second deflation, to the quadratic $x^3 - 3x + 2$, which may then be solved for the remaining roots 2 and 1. Alternatively, the quadratic may be deflated to the linear function $x - 1$. The idea of deflation is that, one root having been found, the original equation may be exchanged for one of lower degree. Theoretically, a method for finding the dominant root of an equation, such as Bernoulli's method, could be used to find all the roots one after another, by successive deflations which remove each dominant root as it is found, and assuming no two roots are of equal size. Actually there are error problems that limit the use of this procedure, as the next problem suggests.

16.43 Show that if the dominant root is not known exactly, then the method of deflation may yield the next root with still less accuracy and suggest a procedure for obtaining this second root to the same accuracy as the first.

▮ Suppose, for simplicity, that the dominant root of the previous equation has been found correct to only two places to be 4.005. Deflation brings

1	−10	35	−50	24	|4.005
	4.005	−24.01	44.015	−23.97	
1	−5.995	10.99	−5.985	.03	

and the cubic $x^3 - 5.995x^2 + 10.99x - 5.985$. The dominant zero of this cubic (correct to two places) is 2.98. As far as the original quartic equation is concerned, this is incorrect in the last place. The natural procedure at this point is to use the 2.98 as the initial approximation for a Newton iteration, which would rapidly produce a root of the original equation correct to two places. A second deflation could then be made. In practice it is found that the smaller "roots" require substantial correction and that for polynomials of even moderate degree the result obtained by deflation may not be good enough to guarantee convergence of the Newton iteration to the desired root. Similar remarks hold when complex conjugate roots $a \pm bi$ are removed through division by the quadratic factor $x^2 - 2ax + a^2 + b^2$.

16.6 THE QUOTIENT-DIFFERENCE ALGORITHM

16.44 What is the quotient-difference scheme?

    ▌ Given a polynomial $a_0 x^n + a_1 x^{n-1} + \cdots + a_n$ and the associated difference equation

$$a_0 x_k + a_1 x_{k-1} + \cdots + a_n x_{k-n} = 0$$

consider the solution sequence for which $x_{-n+1} = \cdots = x_{-1} = 0$ and $x_0 = 1$. Let $q_k^1 = x_{k+1}/x_k$ and $d_k^0 = 0$. Then define

$$q_k^{j+1} = \left(\frac{d_{k+1}^j}{d_k^j} \right) q_{k+1}^j \qquad d_k^j = q_{k+1}^j - q_k^j + d_{k+1}^{j-1}$$

where $j = 1, 2, \ldots, n-1$ and $k = 0, 1, 2, \ldots$. These various quotients (q) and differences (d) may be displayed as in Table 16.9. The definitions are easily remembered by observing the rhombus-shaped parts of the table. In a rhombus centered in a (q) column the sum of the SW pair equals the sum of the NE pair. In a rhombus centered in a (d) column the corresponding products are equal. These are the *rhombus rules*.

TABLE 16.9

q_0^1			
0	d_0^1		
q_1^1		q_0^2	
0	d_1^1		d_0^2	
q_2^1		q_1^2		q_0^3
0	d_2^1		d_1^2		d_0^3
q_3^1		q_2^2		q_1^3		q_0^4
0	d_3^1		d_2^2		d_1^3
q_4^1		q_3^2		q_2^3		q_1^4
0	d_4^1		d_3^2		d_2^3
q_5^1		q_4^2		q_3^3		q_2^4
$\vdots$		$\vdots$		$\vdots$		$\vdots$

16.45 Compute the quotient-difference scheme for the polynomial $x^2 - x - 1$ associated with the Fibonacci sequence.

    ▌ The results appear in Table 16.10.

TABLE 16.10

k	x_k	d_k^0	q_k^1	d_k^1	q_k^2	d_k^2
0	1	0				
			1.0000			
1	1	0		1.0000		
			2.0000		−1.0000	
2	2	0		−.5000		−.0001
			1.5000		−.5001	
3	3	0		.1667		−.0001
			1.6667		−.6669	
4	5	0		−.0667		.0005
			1.6000		−.5997	
5	8	0		.0250		.0007
			1.6250		−.6240	
6	13	0		−.0096		−.0082
			1.6154		−.6226	
7	21	0		.0037		
			1.6190			
8	34	0				

16.46 What is the first convergence theorem associated with the quotient-difference scheme?

▮ Suppose no two zeros of the given polynomial have the same absolute value. Then

$$\lim q_k^j = r_j \qquad j = 1, 2, \ldots, n$$

for k tending to infinity, where $r_1, r_2, \ldots, r_n$ are in the order of diminishing absolute value. For $j = 1$ this is Bernoulli's result for the dominant root. For the other values of j the proof requires complex function theory and will be omitted. It has also been assumed that none of the denominators involved in the scheme is zero. The convergence of the qs to the roots implies the convergence of the ds to zero. This may be seen as follows. By the first of the defining equations of Problem 16.44,

$$\frac{d_{k+1}^j}{d_k^j} = \frac{q_k^{j+1}}{q_{k+1}^j} \to \frac{r_{j+1}}{r_j} < 1$$

The d_k^j therefore converge geometrically to zero. The beginning of this convergence, in the present problem, is evident already in Table 16.10, except in the last column, which will be discussed shortly. In this table the (q) columns should, by the convergence theorem, be approaching the roots $(1 \pm \sqrt{5})/2$, which are approximately 1.61803 and $-.61803$. Clearly we are closer to the first than to the second.

16.47 How can a quotient-difference scheme produce a pair of complex conjugate roots?

▮ The presence of such roots may be indicated by (d) columns that do not converge to zero. Suppose the column of d_k^j entries does not. Then one forms the polynomial

$$p_j = x^2 - A_j x + B_j$$

where for k tending to infinity,

$$A_j = \lim \left(q_{k+1}^j + q_k^{j+1} \right) \qquad B_j = \lim q_k^j q_k^{j+1}$$

The polynomial will have the roots r_j and r_{j+1}, which will be complex conjugates. Essentially, a quadratic factor of the original polynomial will have been found. Here we have assumed that the columns of d_k^{j-1} and d_k^{j+1} entries do converge to zero. If they do not, then more than two roots have equal absolute value and a more complicated procedure is needed. The details, and also the proofs of convergence claims just made, are given in *National Bureau of Standards Applied Mathematics Series*, vol. 49.

16.48 What is the row-by-row method of generating a quotient-difference scheme and what are its advantages?

▮ The column-by-column method first introduced in Problem 16.44 is very sensitive to roundoff error. This is the explanation of the fact that the final column of Table 16.10 is not converging to zero as a (d) column should but instead shows the typical start of an error explosion. The following row-by-row method is less sensitive to error. Fictitious entries are supplied to fill out the top two rows of a quotient-difference scheme as follows, starting with the d_k^0 column and ending with d_k^n. Both of these boundary columns are to consist of zeros for all values of k. This amounts to forcing proper behavior of these boundary differences in an effort to control roundoff error effects:

$$
\begin{array}{ccccc}
-a_1/a_0 & & 0 & 0 & 0 \\
0 & a_2/a_1 & a_3/a_2 & a_4/a_3 & 0
\end{array}
$$

The rhombus rules are then applied, filling each new row in its turn. It can be shown that the same scheme found in Problem 16.44 will be developed by this method, assuming no errors in either procedure. In the presence of error the row-by-row method is more stable. Note that in this method it is not necessary to compute the x_k.

16.49 Apply the row-by-row method to the polynomial of the Fibonacci sequence $x^2 - x - 1$.

▮ The top rows are filled as suggested in the previous problem. The others are computed by the rhombus rules. Table 16.11 exhibits the results. The improved behavior in the last (q) column is apparent.

TABLE 16.11

k	d	q	d	q	d
		1		0	
1	0		1		0
		2		-1	
2	0		-.5000		0
		1.5000		-.5000	
3	0		.1667		0
		1.6667		-.6667	
4	0		-.0667		0
		1.6000		-.6000	
5	0		.0250		0
		1.6250		-.6250	
6	0		-.0096		0
		1.6154		-.6154	
7	0		.0037		0
		1.6191		-.6191	
8	0				0

16.50 Apply the quotient-difference algorithm to find all the roots of

$$x^4 - 10x^3 + 35x^2 - 50x + 24 = 0$$

▌ The roots of this equation are exactly 1, 2, 3, and 4. No advance information about the roots is, however, required by this algorithm, so the equation serves as a simple test case. The quotient-difference scheme, generated by the method of Problem 16.48, appears as Table 16.12. Clearly the convergence is slow, but the expected pattern is emerging. The (d) columns seem headed for zero and the (q) columns for 4, 3, 2, 1 in that order. Probably it would be wise to switch at this point to Newton's method, which very quickly converts reasonable first approximations such as we now have into accurate results. The quotient-difference algorithm is often used for exactly this purpose, to prime the Newton iteration.

TABLE 16.12

k	d	q	d	q	d	q	d	q	d
		10		0		0		0	
1	0		-3.5000		-1.4286		-.4800		0
		6.5000		2.0714		.9486		.4800	
2	0		-1.1154		-.6542		-.2429		0
		5.3846		2.5326		1.3599		.7229	
3	0		-.5246		-.3513		-.1291		0
		4.8600		2.7059		1.5821		.8520	
4	0		-.2921		-.2054		-.0695		0
		4.5679		2.7926		1.7180		.9215	
5	0		-.1786		-.1264		-.0373		0
		4.3893		2.8448		1.8071		.9588	
6	0		-.1158		-.0803		-.0198		0
		4.2735		2.8803		1.8676		.9786	
7	0		-.0780		-.0521		-.0104		0
		4.1955		2.9062		1.9093		.9890	
8	0		-.0540		-.0342		-.0054		0
		4.1415		2.9260		1.9381		.9944	

16.51 Apply the quotient-difference algorithm to Leonardo's equation.

▌ Again using the row-by-row method, we generate the scheme displayed in Table 16.13.
The convergence being slow, suppose we stop here. The second (d) column hardly seems headed for zero, suggesting that r_1 and r_2 are complex, as we already know anyway. The next (d) column does appear to be tending

TABLE 16.13

k	d	q	d	q	d	q	d
		-2		0		0	
1	0		5		-2		0
		3		-7		2	
2	0		-11.6667		$.5714$		0
		-8.6667		5.2381		1.4286	
3	0		7.0513		$.1558$		0
		-1.6154		-1.6574		1.2728	
4	0		7.2346		$-.1196$		0
		5.6192		-9.0116		1.3924	
5	0		-11.6022		$.0185$		0
		-5.9830		2.6091		1.3739	
6	0		5.0596		$.0097$		0
		$-.9234$		-2.4408		1.3642	

to zero, suggesting a real root which we know to be near 1.369. The Newton method would quickly produce an accurate root from the initial estimate of 1.3642 we now have here. Returning to the complex pair, we apply the procedure of Problem 16.47. From the first two (q) columns we compute

$$5.6192 - 9.0116 \simeq -3.3924 \qquad (-1.6154)(-9.0116) \simeq 14.5573$$
$$-5.9830 + 2.6091 \simeq -3.3739 \qquad (5.6192)(2.6091) \simeq 14.6611$$
$$-.9234 - 2.4408 \simeq -3.3642 \qquad (-5.9830)(-2.4408) \simeq 14.6033$$

so that $A_1 \simeq -3.3642$ and $B_1 \simeq 14.6033$. The complex roots are therefore approximately given by $x^2 + 3.3642x + 14.6033 = 0$, which makes them $r_1, r_2 \simeq -1.682 \pm 3.431i$.

Newton's method using complex arithmetic could be used to improve these values, but an alternative procedure known as Bairstow's method will be presented shortly. Once again in this problem we have used the quotient-difference algorithm to provide respectable estimates of all the roots. A method that can do this should not be expected to converge rapidly, and the switch to a quadratically convergent algorithm at some appropriate point is a natural step.

16.7 STURM SEQUENCES

16.52 Define a Sturm sequence.

■ A sequence of functions $f_0(x), f_1(x), \ldots, f_n(x)$ which satisfy on an interval (a, b) of the real line the conditions:

(1) Each $f_i(x)$ is continuous.
(2) The sign of $f_n(x)$ is constant.
(3) If $f_i(r) = 0$, then $f_{i-1}(r)$ and $f_{i+1}(r) \neq 0$.
(4) If $f_i(r) = 0$, then $f_{i-1}(r)$ and $f_{i+1}(r)$ have opposite signs.
(5) If $f_0(r) = 0$, then for h sufficiently small

$$\operatorname{sign}\frac{f_0(r-h)}{f_1(r-h)} = -1 \qquad \operatorname{sign}\frac{f_0(r+h)}{f_1(r+h)} = 1$$

is called a Sturm sequence.

16.53 Prove that the number of roots of the function $f_0(x)$ on the interval (a, b) is the difference between the number of changes of sign in the sequences $f_0(a), f_1(a), \ldots, f_n(a)$ and $f_0(b), f_1(b), \ldots, f_n(b)$.

■ As x increases from a to b the number of sign changes in the Sturm sequence can only be affected by one or more of the functions having a zero, since all are continuous. Actually only a zero of $f_0(x)$ can affect it. Suppose $f_i(r) = 0$ with $i \neq 0, n$. Then by properties 1, 3 and 4 the following sign patterns are possible for small h:

	f_{i-1}	f_i	f_{i+1}			f_{i-1}	f_i	f_{i+1}
$r-h$	$+$	$\pm$	$-$	or	$r-h$	$-$	$\pm$	$+$
r	$+$	0	$-$		r	$-$	0	$+$
$r+h$	$+$	$\pm$	$-$		$r+h$	$-$	$\pm$	$+$

In all cases there is one sign change, so that moving across such a root does not affect the number of sign changes. By condition 2 the function $f_n(x)$ cannot have a zero, so we come finally to $f_0(x)$. By condition 5 we lose one sign change, between f_0 and f_1, as we move across the root r. This proves the theorem. One sees that the five conditions have been designed with this root-counting feature in mind.

16.54 If $f_0(x)$ is a polynomial of degree n with no multiple roots, how can a Sturm sequence for enumerating its roots be constructed?

▌ Let $f_1(x) = f_0'(x)$ and then apply the Euclidean algorithm to construct the rest of the sequence as follows:

$$f_0(x) = f_1(x) L_1(x) - f_2(x)$$
$$f_1(x) = f_2(x) L_2(x) - f_3(x)$$
$$\vdots$$
$$f_{n-2}(x) = f_{n-1}(x) L_{n-1}(x) - f_n(x)$$

where $f_i(x)$ is of degree $n - i$ and the $L_i(x)$ are linear.

The sequence $f_0(x), f_1(x), \ldots, f_n(x)$ will be a Sturm sequence. To prove this we note first that all $f_i(x)$ are continuous, since f_0 and f_1 surely are. Condition 2 follows since f_n is a constant. Two consecutive $f_i(x)$ cannot vanish simultaneously since then all would vanish including f_0 and f_1 and this would imply a multiple root. This proves condition 3. Condition 4 is a direct consequence of our defining equations and 5 is satisfied since $f_1 = f_0'$.

If the method were applied to a polynomial having multiple roots, then the simultaneous vanishing of all the $f_i(x)$ would give evidence of them. Deflation of the polynomial to remove multiplicities allows the method to be applied to find the simple roots.

16.55 Apply the method of Sturm sequences to locate all real roots of

$$x^4 - 2.4x^3 + 1.03x^2 + .6x - .32 = 0$$

▌ Denoting this polynomial $f_0(x)$, we first compute its derivative. Since we are concerned only with the signs of the various $f_i(x)$, it is often convenient to use a positive multiplier to normalize the leading coefficient. Accordingly we multiply $f_0'(x)$ by $\frac{1}{4}$ and take

$$f_1(x) = x^3 - 1.8x^2 + .515x + .15$$

The next step is to divide f_0 by f_1. One finds the linear quotient $L_1(x) = x - .6$, which is of no immediate interest, and a remainder of $-.565x^2 + .759x - .23$. A common error at this point is to forget that we want the *negative* of this remainder. Also normalizing, we have

$$f_2(x) = x^2 - 1.3434x + .4071$$

Dividing f_1 by f_2 brings a linear quotient $L_2(x) = x - .4566$ and a remainder whose negative, after normalizing, is

$$f_3(x) = x - .6645$$

Finally, dividing f_2 by f_3 we find the remainder to be $-.0440$. Taking the negative and normalizing, we may choose

$$f_4(x) = 1$$

We now have our Sturm sequence and are ready to search out the roots. It is a simple matter to confirm the signs displayed in Table 16.14. They show that there is one root in the interval $(-1, 0)$, one in $(1, 2)$, and two roots in $(0, 1)$.

TABLE 16.14

	f_0	f_1	f_2	f_3	f_4	Changes
$-\infty$	+	−	+	−	+	4
-1	+	−	+	−	+	4
0	−	+	+	−	+	3
1	−	−	+	+	+	1
2	+	+	+	+	+	0
∞	+	+	+	+	+	0

16.56 For polynomials of higher degree, with coefficients that are not quite simple integers, manual operation of the Euclidean algorithm as used in the Sturm method becomes tedious. Develop the formulas needed for machine computation of the negative remainder.

▮ Let the dividend be

$$p(x) = a_n x^n + a_{n-1} x^{n-1} + \cdots + a_0$$

and the divisor

$$d(x) = b_{n-1} x^{n-1} + b_{n-2} x^{n-2} + \cdots + b_0$$

The algorithm represents $p(x)$ as $q(x)\,d(x) - r(x)$ where

$$q(x) = q_1 x + q_0 \qquad r(x) = c_{n-2} x^{n-2} + c_{n-1} x^{n-1} + \cdots + c_0$$

and the coefficients c_i are our main interest. A patient effort, writing out the steps of the long division process, will now be rewarded as follows.

$$q_1 = \frac{a_n}{b_{n-1}} \qquad q_0 = \frac{a_{n-1}}{b_{n-1}} - \frac{a_n b_{n-2}}{b_{n-1}^2}$$

$$c_{n-j} = -a_{n-j} + q_1 b_{n-j-1} + q_0 b_0 \qquad j = 2 \text{ to } n$$

with the artifice $b_{-1} = 0$ making a separate formula for c_0 unnecessary. This algorithm can be used to produce each of the needed Sturm functions in its turn.

16.57 Use Sturm's method to show that the equation

$$f_0(x) = 288x^5 - 720x^4 + 694x^3 - 321x^2 + 71x - 6 = 0$$

has five closely packed real roots. Then pinpoint these roots in some way.

▮ The Sturm functions begin with $f_0(x)$ and

$$f_1(x) = f_0'(x) = 1440x^4 - 2880x^3 + 2082x^2 - 642x + 71$$

after which the Euclidean algorithm generates f_2 to f_5. In an abbreviated form these are

$$f_2(x) = 10.4x^3 - 15.6x^2 + 7.4x - 1.1$$

$$f_3(x) = 22.6x^2 - 22.6x + 5.2$$

$$f_4(x) = .17x - .085$$

$$f_5(x) = .5$$

Table 16.15 offers an edited version of computations based on these Sturm functions. The first thing to note is that there are five sign changes at $-\infty$ and none at ∞, so there will be five real roots somewhere. The remaining rows of

TABLE 16.15

x	f_0	f_1	f_2	f_3	f_4	f_5	Count
$-\infty$	−	+	−	+	−	+	5
.2	−	+	−	+	−	+	5
.3	+	−	−	+	−	+	4
.4	−	−	+	−	−	+	3
.5	0	+	+	−	+	−	3
.6	+	−	−	−	+	+	2
.7	−	−	+	+	+	+	1
.8	+	+	+	+	+	+	0
∞	+	+	+	+	+	+	0

the table then show them to fall within the intervals

$$(.2, .3) \qquad (.3, .4) \qquad (.5, .5) \qquad (.6, .7) \qquad (.7, .8)$$

the zero for $f_0(.5)$ indicating a bull's-eye. Speculation as to the exact whereabouts of the other four roots is permissible and will very likely be successful. Those preferring more solid logic may test suspicions in the original polynomial.

16.58 Use a Sturm sequence to show that

$$f_0(x) = 36x^6 + 36x^5 + 23x^4 - 13x^3 - 12x^2 + x + 1 = 0$$

has only four real roots and to locate these four. Then apply Newton's method to pinpoint them.

▌ First we find

$$f_1(x) = 216x^5 + 180x^4 + 92x^3 - 39x^2 - 24x + 1$$

and then, appealing to the Euclidean algorithm,

$$f_2(x) = -2.67x^4 + 9.06x^3 + 6.92x^2 - 1.5x - .97$$
$$f_3(x) = -3750x^3 - 2209x^2 + 617x + 332$$
$$f_4(x) = -.228x^2 - .0091x + .033$$
$$f_5(x) = -162x - 37.6$$
$$f_6(x) = -.02$$

all in an abbreviated form. Table 16.16 provides information on sign changes. From $-\infty$ to ∞ the count drops from 5 to 1, showing that only four real roots exist. These four are then located in the intervals $(-.6, .4)$, $(-.4, -.2)$, $(.2, .4)$ and $(.4, .6)$ after which Newton's method strongly recommends $\pm \frac{1}{3}, \pm \frac{1}{2}$.

TABLE 16.16

x	f_0	f_1	f_2	f_3	f_4	f_5	f_6	Count
$-\infty$	+	−	−	+	−	+	−	5
$-.6$	+	−	+	−	−	+	−	5
$-.4$	−	+	+	−	−	+	−	4
$-.2$	+	+	−	+	+	−	−	3
$.2$	+	−	−	+	+	−	−	3
$.4$	−	−	+	−	−		−	2
$.6$	+	+	+	−	−	−	−	1
∞	+	+	−	−	−	−	−	1

16.59 What does the Sturm method have to offer for this equation?

$$x^5 + 1492x^4 - 1776x^3 + 1066x^2 - 1620x + 2000 = 0$$

▌ For an equation that emerges from some field of application there is a fair chance that special knowledge of the field will suggest the rough location of its roots. Here we have a somewhat frivolous set of coefficients and no special knowledge, except perhaps that the sum of the roots is -1492 and the product -2000. Giving the Sturm method a chance we find, after a humbling reconnaissance of much of the real line, this information:

x	f_0	f_1	f_2	f_3	f_4	f_5	Count
$-\infty$	−	+	−	−	+	+	3
-2000	−	+	−	−	+	+	3
-1000	+	−	−	−	+	+	2
∞	+	+	+	−	−	+	2

There seems to be just one real root, in the interval shown. Can it possibly be -1492? A little more help from the Sturm sequence places it between -1493 and -1494. One Newton iteration then pinpoints the root at -1493.19.

16.8 GRAEFFE'S ROOT-SQUARING METHOD

16.60 Develop Graeffe's method for the case of a cubic polynomial.

■ This method deserves mention both for its historical interest and for the attractive idea involved. Let

$$p(x) = a_0 + a_1 x + a_2 x^2 + a_3 x^3 = a_3 (x - r_1)(x - r_2)(x - r_3)$$

with r_1, r_2, r_3 to be determined. Note that

$$(-1)^3 p(-x) = a_3 (x + r_1)(x + r_2)(x + r_3)$$

so that with $t = x^2$,

$$f_1(t) = (-1)^3 p(x) p(-x) = a_3^2 (x^2 - r_1^2)(x^2 - r_2^2)(x^2 - r_3^2)$$
$$= a_3^2 (t - r_1^2)(t - r_2^2)(t - r_3^2)$$

will have roots that are squares of the originals. To find the coefficients of $f_1(t)$ we can simply do the multiplications,

$$f_1(t) = -(a_0 + a_1 x + a_2 x^2 + a_3 x^3)(a_0 - a_1 x + a_2 x^2 - a_3 x^3)$$
$$= a_3^2 t^3 - (a_2^2 - 2a_1 a_3) t^2 + (a_1^2 - 2a_0 a_2) t - a_0^2$$

with all the odd powers of x vanishing.

The idea of Graeffe's method is to apply the preceding process a number of times, obtaining polynomials f_i of the same degree as the original, but with roots that are the 2^i powers of the original roots. (The coefficients a_k should carry superscripts i to match f_i but they are omitted for simplicity.) Under certain circumstances this process will separate the roots more and more widely. At each step, or occasionally, the famous theorem that relates roots and coefficients can then be called on. For example, the coefficient of x^2 (in a cubic) is minus the sum of all roots, and with the roots widely separated, this sum will be essentially the dominant root, say r_1. Similarly, the coefficient of x is $r_1 r_2 + r_1 r_3 + r_2 r_3$ here dominated by (say) $r_1 r_2$. With r_1 in hand, r_2 can be found. Finally, the constant term being $-r_1 r_2 r_3$, the last root is gathered in. It will be remembered that what we now have are the 2^i powers of the original roots, so the job is finished by elevating them to power 2^{-i}.

16.61 Apply the Graeffe method to the cubic equation

$$x^3 - 15x^2 + 23x + 231 = 0$$

■ Table 16.17 shows both the coefficients of the successive polynomials (abbreviated) and the root estimates obtained from each. The coefficients became very large and eventually led to overflow, but the estimates have converged sufficiently to guess the original magnitudes of 11, 7 and 3. The proper sign of each of these must be determined by referring to the original cubic, since minus signs will have been lost at the first squaring. It will be found that the 3 should be negative.

TABLE 16.17

i			Coefficients		Original Roots (est)		
1	1	-179	7459	-53361	13.4	6.5	2.7
2	1	-17123	$3.65 \cdot 10^7$	$-2.85 \cdot 10^9$	11.4	6.8	2.97
3	1	$-2.20 \cdot 10^8$	$1.24 \cdot 10^{15}$	$-8.11 \cdot 10^{18}$	11.04	6.98	2.9996
4	1	$-4.60 \cdot 10^{16}$	$1.53 \cdot 10^{30}$	$-6.57 \cdot 10^{37}$	11.0005	6.9997	3.0000

16.62 Extend the Graeffe method to fourth degree polynomials.

■ The basics are the same. The function $f_1(t) = (-1)^4 p(x) p(-x)$ will have roots that are squares of those of $p(x)$ itself. The coefficients of $f_1(t)$ are found directly:

$$p(x) p(-x) = (a_0 + a_1 x + a_2 x^2 + a_3 x^3 + a_4 x^4)(a_0 - a_1 x + a_2 x^2 - a_3 x^3 + a_4 x^4)$$
$$= a_4 t^4 - (a_3^2 - 2a_2 a_4) t^3 + (a_2^2 - 2a_1 a_3 + 2a_0 a_4) t^2 - (a_1^2 - 2a_0 a_2) t + a_0^2$$

with all odd powers of x vanishing.

16.63 Apply the root-squaring method to the fourth degree equation

$$x^4 - 16x^3 + 86x^2 - 176x + 105 = 0$$

❚ Table 16.18 has the details. Again the coefficients became large, producing overflow at $i = 5$. The root estimates are not as satisfactory as in the last example, $i = 4$ doing no better than $i = 3$. But Newton's method can be used to finish up and would confirm what the present evidence leads one to hope, that the exact roots are $7, 5, 3, 1$.

TABLE 16.18

i			Coefficients				Original Roots (est)		
1	1	-84	1936	-12916	11025	9.2	4.8	2.6	.9
2	1	-3184	$1.6 \cdot 10^6$	$-1.2 \cdot 10^8$	$1.2 \cdot 10^8$	7.5	4.7	2.98	.99
3	1	$-7 \cdot 10^6$	$1.7 \cdot 10^{12}$	$-1.5 \cdot 10^{16}$	$1.5 \cdot 10^{16}$	7.2	4.7	3.1	.998

16.9 ADDITIONAL PROBLEMS

16.64 Show that the second degree term can be removed from the general cubic equation

$$x^3 + ax^2 + bx + c = 0$$

by a translation of the variable.

❚ Let $x = y - h$ and substitute.

$$(y - h)^3 + a(y - h)^2 + b(y - h) + c = y^3 - 3y^2h + 3yh^2 - h^3 + a(y^2 - 2yh + h^2) + b(y - h) + c$$

The coefficient of y^2 is $a - 3h$. Choosing $h = a/3$, the remaining terms provide the new equation:

$$y^3 + \left(b - \tfrac{1}{3}a^2\right)y + \left(\tfrac{2}{27}a^3 - \tfrac{1}{3}ab\right) + c = 0$$

16.65 In 1545 the Italian mathematician Cardano published this formula for solving the cubic equation $x^3 + bx + c = 0$:

$$x = \left[-\frac{c}{2} + \sqrt{\left(\frac{c}{2}\right)^2 + \left(\frac{b}{3}\right)^3}\right]^{1/3} - \left[\frac{c}{2} + \sqrt{\left(\frac{c}{2}\right)^2 + \left(\frac{b}{3}\right)^3}\right]^{1/3}$$

(Note the absent second degree term.) Apply it to the equation

$$x^3 + 3x - 4 = 0$$

❚ With $b = 3$ and $c = -4$ we soon have

$$x = \left(2 + \sqrt{5}\right)^{1/3} - \left(-2 + \sqrt{5}\right)^{1/3}$$

which the computer number-crunches to $1.618034 - .618034$ or 1.000000, and, of course, 1 is an exact root, the other two being complex. By properly pairing up the three available values of each $\frac{1}{3}$ power, all three roots of the equation can be found. The real values given above will be recognized. In fact, the Fibonacci roots $(1 \pm \sqrt{5})/2$ cube to $2 \pm \sqrt{5}$.

16.66 Apply Cardano's formula to the equation $x^3 - 15x - 4 = 0$.

❚ This time it leads to

$$x = \left(2 + \sqrt{-121}\right)^{1/3} - \left(-2 + \sqrt{-121}\right)^{1/3}$$

Since the equation has the three real roots 4 and $-2 \pm \sqrt{3}$, one may enjoy the pleasures of complex arithmetic in reducing Cardano's result.

16.67 Observing that 1 is an exact root of $x^3 - 2x^2 - 5x + 6 = 0$, find the other two by deflation.

❚ This should only be done as a mental exercise. The quadratic has simple factors. Roots appear at the finish of the next problem.

16.68   Find the largest root of

$$p(x) = x^4 - 2.0379x^3 - 15.4245x^2 + 15.6696x + 35.4936 = 0$$

▮   Simple measures can be very reassuring. A numerical scan of the polynomial draws the picture

x	-4	-3	-1	3	5
$p(x)$	112	-14	7	-30	98

Since a quartic is limited to four sign changes, we see all its ups and downs and approximate roots. The largest is between 3 and 5. Starting from 4, Newton's method soon manages 4.327506. (3 and -2)

16.69   Find two roots near $x = 1$ of $p(x) = 2x^4 + 16x^3 + x^2 - 74x + 56 = 0$.

▮   A scan can do no harm. Here are a few numbers:

x	-8	-6	-1	0	1	2
p	712	-328	117	56	1	72

Certainly the vicinity of $x = 1$ deserves more careful study. As a courtesy we may offer *regula falsi* a chance. From the initial interval $(1, 2)$ it came to 1.115 in 15 steps. Making a new start with $(1.11, 1.5)$ it reached 1.121 in 15 more steps, but with doubt remaining about the second decimal place. It was easy to get the impression that *regula falsi* was overmatched.

Switching to Newton, the start $x = 1$ led in five steps to 1.1213, with the corresponding value of p reported as zero. A start from the *regula falsi* 1.121 ended up at 1.121284 and also reported p as zero. The equation seems easy to satisfy to six decimal places, but what is the best estimate of this root and what of the nearby root? Observing that the derivative value at 1.1213 is negative and rechecking the earlier scan, we realize that the companion root must lie to the right. Priming Newton's method with 1.5, to be fairly certain of being above the root, we find 1.123098 in 11 steps, with $p = 0$. Then starting from 1.13 we also find 1.123092, with $p = 0$.

Turning to the method of last resort, we compute $p(x)$ on a fine argument mesh. It changes sign between 1.12131 and 1.12132, determining one root to about five places. Moving to the right, another sign change occurs between 1.12309 and 1.12313, with the three intermediate arguments $x = 1.12310$ and 1.12312 all giving p the value zero. Further progress, if needed, would be possible by computing p to double precision.

16.70   Find any real roots of the cubic equation $x^3 - x - 4 = 0$.

▮   Since the equation is already in Cardano's form, we must see where his result leads. With $b = -1$ and $c = -4$ the roots are represented by

$$x = \left(2 + \sqrt{4 - 1/27}\right)^{1/3} - \left(-2 + \sqrt{4 - 1/27}\right)^{1/3}$$

which computes in real arithmetic to 1.79632. Two complex values are also available with persistence.

Without Cardano's formula we have numerous options. A scan of the cubic finds a local maximum near $(-.3, -3.7)$, which explains the complex roots, and a local minimum near $(.3, -4.3)$. The curve then climbs to the zero level and beyond. Bernoulli's method found the zero to be near 1.7962 after 70 iterations, after which the next two decimal places oscillated. Newton's method converted the initial approximation $x = 2$ into 1.79632 in just three steps.

16.71   Find two real roots of $f(x) = x^{1.8632} - 5.2171x + 2.1167 = 0$. Also find the minimum value of the function for positive x.

▮   The function involved is "almost quadratic" in the sense that the exponent is roughly 2. We anticipate then a drop from the level 2.1167 at $x = 0$ to a minimum point, followed by a climb to infinity. The second derivative $f''(x) = 1.8632(.8632)x^{-.1368}$ being positive for positive x confirms this general shape. Scanning the function we find one zero falling between 0 and 1, the other between 6 and 7. Newton's method pinpoints both:

0	.41	.4484	.4488	.4488
6	6.29	6.2732	6.2732	

Suppose we seek the minimum value in the classic way. The first derivative is $f'(x) = 1.8632x^{.8632} - 5.2171$. The scan suggests a minimum near $x = 3$, and the Newton method finds 3.2964, with the corresponding function value -5.8507.

16.72 Find a root near $x = 2$ of $x = 2 \sin x$.

▌ Instead of using a scan to get the general picture, suppose we fall back on the geometric device of plotting both sides on the same graph. Figure 16-8 shows that $y = x$ and $y = 2 \sin x$ will cross only once for positive x, somewhat to the right of $\pi/2$. Since both functions are odd, there will be a second crossing in quadrant three. So there are three real roots, including 0. The positive one responds quickly to the Newton iteration. Starting with $x = 2$, it reaches 1.895494 in three steps.

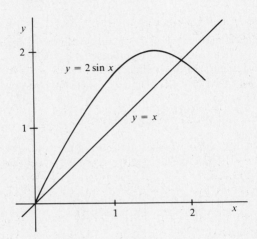

Fig. 16-8

16.73 Find a root near $x = 2$ of $x^4 - x = 10$.

▌ Even a visual inspection suggests a root between 1 and 2, and closer to 2. This is a good equation with which to reactivate the iterative method. Rewriting it as

$$x = F(x) = (x + 10)^{1/4}$$

and choosing 2 as the starter, we find the successive approximations

$$1.86121 \qquad 1.855805 \qquad 1.855593 \qquad 1.855585$$

the last value then repeating. Why has convergence been so good? Computing $F'(r)$, where r is the root, we find it to be near .04. For convergence it need only be less than 1. Here it is rather close to zero.

16.74 Find a root near 1.5 of $x = .5 + \sin x$.

▌ Since the equation is already in the form $x = F(x)$, the derivative of the sine will not exceed 1 and the vicinity of the root suggests a small cosine value, we have another chance to exercise the iterative method. Starting at $x = 1.5$, it finds a good approximation in just four steps:

$$1.497495 \qquad 1.497315 \qquad 1.497301 \qquad 1.497300$$

Again, why has convergence been so good? The value of $F'(r)$ at the root proves to be near .07, so each error should be less than a tenth the preceding error (once we are somewhere near the root). A quick check of errors

$$.003 \qquad .0002 \qquad .000015 \qquad .000001$$

confirms this prediction.

16.75 Find the smallest positive root of

$$1 - x + \frac{x^2}{(2!)^2} - \frac{x^3}{(3!)^2} + \frac{x^4}{(4!)^2} - \cdots = 0$$

▌ The series converges rapidly and has terms of alternating sign if x is positive. This makes it relatively easy to estimate the truncation error, and ten terms will give seven place values at least up to $x = 2$. Scanning the series values over this range we find them changing from plus to minus between $x = 1.4$ and 1.5. Zooming in a bit, the interval is narrowed to 1.44, 1.45, with end values of .002507634 and $-.001812261$ for the series. It is tempting to continue in this way, zooming in even closer, but probably more efficient to switch to one of our more professional devices. Giving Newton's method a little more rest, *regula falsi* chooses the three new points

$$(1.445805, -4 \cdot 10^{-6}) \quad (1.445797, -7 \cdot 10^{-8}) \quad (1.445796, 4 \cdot 10^{-9})$$

the series values being abbreviated. The root seems to be rather well determined.

16.76   Find the three smallest positive roots of $e^{-x} = \sin x$.

Figure 16-9 is offered as an alternative to a numerical scan of the function $e^{-x} - \sin x$. There will be infinitely many roots, all of them positive, and the asymptotic decay of the exponential function places all but the first one close to the integer multiples of pi. Still keeping Newton's method on the shelf, we try the quadratic interpolation device of Muller. Offering it the starting triple $x = 0, 1, 2$ (just to see which root it chooses) we find it generating the sequence of approximations

$$2.93 \quad 3.06 \quad 3.095 \quad 3.096364 \quad 3.096364$$

So the second root is in hand, and is less than 2% short of pi. Restarting, with the triple $x = .4, .5, .6$ and the first root our target, only two quadratic interpolates are needed. The new approximations are

$$.5885290 \quad .5885327$$

with the second satisfying the equation to seven places. The final quadratic was

$$-.06432731h^2 + .1590986h - .00000059$$

with discriminant .1590981, so as usual we have drawn close to a zero denominator in the equation determining h. Finally, using the initial triple $x = 6.2, 6.3, 6.4$ led to

$$6.285025 \quad 6.285049$$

and repetition of the last value, which differs from 2π in the third decimal place.

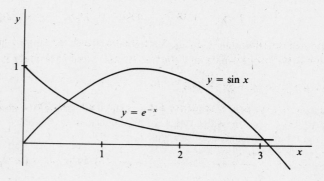

Fig. 16-9

16.77   Find all the roots of $2x^3 - 13x^2 - 22x + 3 = 0$.

[]   A scan at integer spacing discovers three real roots, in the intervals $(-2, -1)$, $(0, 1)$ and $(7, 8)$. Newton's method would locate them easily, but suppose we run a different course. With the difference equation

$$x_{k+1} = 6.5x_k + 11x_{k-1} - 1.5x_{k-2}$$

and $x_{-2} = x_{-1} = 0$, $x_0 = 1$, we are ready for the Bernoulli algorithm. It generates a sequence of x_k beginning with $x_1 = 6.5$, $x_2 = 53.25$ and going on to an x_{12} close to 10^8. The ratios of consecutive x_k show oscillating convergence,

$$6.5 \quad 8.2 \quad 7.8 \quad 7.88 \quad 7.871 \quad 7.8734 \quad 7.87291 \quad 7.872998 \quad 7.872980$$

and then 7.872983, which is repeated. As expected, the method has found the dominant root.

Next we deflate the cubic to a quadratic. Calling our root r, the familiar routine

1	-6.5	-11	1.5
	r	$r^2 - 6.5r$	$r^3 - 6.5r^2 - 11r$
1	$r - 6.5$	$r^2 - 6.5r - 11$	$r^3 - 6.5r^2 - 11r + 1.5$

is easily programmed. The lower right entry proves to be zero to seven places, confirming our Bernoulli result, and the remaining entries on this line form the target of deflation,

$$x^2 + 1.372983x - .1905289 = 0$$

The roots .127019 and -1.500002 follow, and if substituted into the original cubic would yield $-3 \cdot 10^{-5}$. This is not bad, but not as good as the seven place zero created by the dominant root. Something has been lost along the itinerary.

As a final stage, we offer all three values in hand as input to Newton's method. It confirms the first and produces

$$.1270166 \qquad -1.500000$$

from the other two. Both of these last values satisfy the cubic equation to seven places.

16.78 What is meant by purification of roots?

▮ This was just illustrated. Purification is the process of refining approximations, obtained through deflation and perhaps other roundabout methods, by taking them back to the original problem for testing. Newton's method is the popular purifier of roots.

16.79 Find a root near $x = 1.5$ of $x^6 = x^4 + x^3 + 1$.

▮ Spurning roundabout journeys, we let Newton's method work it out directly. Given $x = 1.5$ it manages

$$1.42 \quad 1.4045 \quad 1.403605 \quad 1.403602$$

followed by repeats. The equation is satisfied with an error of $2 \cdot 10^{-7}$.

How does the Muller method do? Given the initial triple $x = 1, 1.5, 2$ it generates

$$1.45 \quad 1.38 \quad 1.407 \quad 1.403615 \quad 1.403602$$

involving just one more iteration than Newton. Both methods have performed very well. Newton of course has the help of the derivative, while Muller brings a third point into each step to get a better interpolation.

16.80 Find all the roots of $f(x) = x^4 - 5x^3 - 12x^2 + 76x - 79 = 0$.

▮ Partly for the exercise, the Sturm sequence for this equation was found and sign changes counted, with the results

x	$-\infty$	-4	-3	1	2	3	4	5	∞
Count	4	4	3	3	2	1	1	0	0

There are roots in the intervals $(-4, -3)$, $(1, 2)$, $(2, 3)$ and $(4, 5)$. Newton's method found them as shown in Table 16.19.

TABLE 16.19

Start	Finish	Steps	$f(x)$
-4	-3.996909	4	$-2 \cdot 10^{-4}$
1.5	1.768387	5	10^{-6}
2.5	2.240988	4	$8 \cdot 10^{-6}$
5	4.987534	2	10^{-6}

16.81 Why is the value of $f(x)$ in Table 16.19 greater at the negative root than at the others?

I The curve is steeper there, so a slight error in estimating the root will carry $f(x)$ farther away from zero. The derivative values at the four roots, taken from left to right, are near -320, 9, -8 and 80. Surrounding the negative root we have

$$f(-3.996909) = -1.8 \cdot 10^{-4} \qquad f(-3.996910) = 2.1 \cdot 10^{-4}$$

approximately. Newton's method has chosen the closer of the two best options available to it in single precision arithmetic. The shift of 10^{-6} in the argument has brought a shift 400 times as large in $f(x)$, more or less consistent with the derivative value. The root has chosen to lie about midway between these single precision gridpoints.

In contrast, near the second root we find the values

$$f(1.768387) = 10^{-6} \qquad f(1.768388) = 8 \cdot 10^{-6}$$

The first argument was selected, but even at the other the error in $f(x)$ is small because the derivative is only of size 9, and near the third root we find

$$f(2.240987) = 1.5 \cdot 10^{-5} \qquad f(2.240988) = 7.6 \cdot 10^{-6} \qquad f(2.240989) = -1.5 \cdot 10^{-5}$$

so the root appears to be near the center argument, with the other function values consistent with a derivative of size 8. The fourth root also falls near a single precision gridpoint, so $f(x)$ has a comfortably small value there in spite of the large $f'(x)$.

In Problem 16.69 we found several single precision gridpoints making $f(x)$ essentially a flat zero. The derivative near that root was only about $\frac{1}{8}$. In a sense, it is easier to bring $f(x)$ close to zero when the derivative near the root is small, but not so easy to pinpoint the root itself. When $f'(x)$ is large, the facts are reversed.

16.82 What is meant by a polyalgorithm for root finding?

I This word has been applied to library routines. Faced with a nonlinear equation, one can always try scanning the interval of interest informally, then apply Newton's method as seems fit. If the output is scrutinized carefully, the probability is probably high that no very serious blunder will have been made. Library routines are designed to solve both easy and difficult problems without the user's personal intervention. Like the preceding simple plan they consist of stages and checkpoints, perhaps beginning with a scan or preliminary search, then iterations and tests of convergence, "consideration" of the preliminary results, followup of likely looking opportunities, rejection of the opposite and so on. They have to be cautious, but at the same time must not give up all interest in speed. Such algorithms, which combine a number of the methods presented, are called polyalgorithms. It may be that any such procedure can be fooled by a sadistic placement of roots, but professional libraries included some excellent ones.

16.10 SYSTEMS OF EQUATIONS: NEWTON'S METHOD

16.83 Derive the formulas for solving $f(x, y) = 0$, $g(x, y) = 0$,

$$x_n = x_{n-1} + h_{n-1}$$
$$y_n = y_{n-1} + k_{n-1}$$

where h and k satisfy

$$f_x(x_{n-1}, y_{n-1})h_{n-1} + f_y(x_{n-1}, y_{n-1})k_{n-1} = -f(x_{n-1}, y_{n-1})$$
$$g_x(x_{n-1}, y_{n-1})h_{n-1} + g_y(x_{n-1}, y_{n-1})k_{n-1} = -g(x_{n-1}, y_{n-1})$$

These formulas are known as the Newton method for solving two simultaneous equations.

I Approximate f and g by the linear parts of their Taylor series for the neighborhood of (x_{n-1}, y_{n-1}):

$$f(x, y) \simeq f(x_{n-1}, y_{n-1}) + (x - x_{n-1})f_x(x_{n-1}, y_{n-1}) + (y - y_{n-1})f_y(x_{n-1}, y_{n-1})$$
$$g(x, y) \simeq g(x_{n-1}, y_{n-1}) + (x - x_{n-1})g_x(x_{n-1}, y_{n-1}) + (y - y_{n-1})g_y(x_{n-1}, y_{n-1})$$

This assumes that the derivatives involved exist. With (x, y) denoting an exact solution, both left sides vanish. Defining $x = x_n$ and $y = y_n$ as the numbers that make the right sides vanish, we have at once that equations required. This idea of replacing a Taylor series by its linear part is what led to the Newton method for solving a single equation in Problem 16.12.

16.84 Find the intersection points of the circle $x^2 + y^2 = 2$ with the hyperbola $x^2 - y^2 = 1$.

    ❚ This particular problem can easily be solved by elimination. Addition brings $2x^2 = 3$ and $x \simeq \pm 1.2247$. Subtraction brings $2y^2 = 1$ and $y = \pm .7071$. Knowing the correct intersections makes the problem a simple test case for Newton's method. Take $x_0 = 1$ and $y_0 = 1$. The formulas for determining h and k are

$$2x_{n-1}h_{n-1} + 2y_{n-1}k_{n-1} = 2 - x_{n-1}^2 - y_{n-1}^2$$
$$2x_{n-1}h_{n-1} - 2y_{n-1}k_{n-1} = 1 - x_{n-1}^2 + y_{n-1}^2$$

and with $n = 1$ become $2h_0 + 2k_0 = 0$ and $2h_0 - 2k_0 = 1$. Then $h_0 = -k_0 = \frac{1}{4}$, making

$$x_1 = x_0 + h_0 = 1.25 \qquad y_1 = y_0 + k_0 = .75$$

The next iteration brings $2.5h_1 + 1.5k_1 = -.125$ and $2.5h_1 - 1.5k_1 = 0$ making $h_1 = -.025$, $k_1 = -.04167$ and

$$x_2 = x_1 + h_1 = 1.2250 \qquad y_2 = y_1 + k_1 = .7083$$

A third iteration manages $2.45h_2 + 1.4167k_2 = -.0024$ and $2.45h_2 - 1.4167k_2 = .0011$ making $h_2 = -.0003$, $k_2 = -.0012$ and

$$x_3 = x_2 + h_2 = 1.2247 \qquad y_3 = y_2 + k_2 = .7071$$

The convergence to the correct results is evident. It can be proved that for sufficiently good initial approximations the convergence of Newton's method is quadratic. The idea of the method can easily be extended to any number of simultaneous equations.

16.85 Other iterative methods may also be generalized for simultaneous equations. For example, if our basic equations $f(x, y) = 0$, $g(x, y) = 0$ are rewritten as

$$x = F(x, y) \qquad y = G(x, y)$$

then under suitable assumptions on F and G, the iteration

$$x_n = F(x_{n-1}, y_{n-1}) \qquad y_n = G(x_{n-1}, y_{n-1})$$

will converge for sufficiently accurate initial approximations. Apply this method to the equations $x = \sin(x + y)$, $y = \cos(x - y)$.

    ❚ These equations are already in the required form. Starting with the uninspired initial approximations $x_0 = y_0 = 0$, we obtain the results given below. Convergence for such poor starting approximations is by no means the rule. Often one must labor long to find a convergent rearrangement of given equations and good first approximations.

n	0	1	2	3	4	5	6	7
x_n	0	0	.84	.984	.932	.936	.935	.935
y_n	0	1	.55	.958	1.000	.998	.998	.998

16.86 Find a solution of the system

$$x = \sin x \cosh y \qquad y = \cos x \sinh y$$

near $x = 7$, $y = 3$.

    ❚ The equations are in the form of Problem 16.85 but this is no help since the iterations rapidly diverge. Rewriting as

$$f = x - \sin x \cosh y = 0 \qquad g = y - \cos x \sinh y = 0$$

and computing the needed derivatives, the linear system for h and k is arranged:

$$(1 - \cos x \cosh y)h - (\sin x \sinh y)k = -x + \sin x \cosh y$$
$$(\sin x \sinh y)h + (1 - \cos x \cosh y)k = -y + \cos x \sinh y$$

The good initial approximations $(x, y) = (7, 3)$ then lead to the improvements

x	7.37	7.51	7.497643	7.497646
y	2.68	2.765	2.768607	2.768678

The quadratic convergence is apparent.

There is often some entertainment value in offering Newton's method a variety of starting values. With systems, the supply is of course much richer. Neighbor points of $(7, 3)$ within a unit up or down, left or right, all led to the preceding solution point, with the possible exception of $(6, 2)$, which produced a tiresome oscillation and was abandoned. Starting from $x = 5$, $y = 3, 4, 5$ led (surprisingly) to the symmetric solution with y negative, but $(6, 1)$ then spurned this new solution and led to still another at $(13.9, -3.35221)$. An assortment of other starts then chose among the three solutions in hand, or became troublesome, and the search for entertainment was terminated.

16.87 Solve the system $x^4 + y^4 - 67 = 0$, $x^3 - 3xy^2 + 35 = 0$ in the vicinity of $x = 2$, $y = 3$.

▮ From $(2, 3)$ Newton's method quickly runs the course

x	1.77	1.83	1.86	1.880	1.8835	1.883645
y	2.79	2.74	2.72	2.717	2.7160	2.715948

with some remaining oscillation in the final digit 5. From $(1, 2)$ it finds this digit to be a 4, with no oscillations. The given equations are satisfied by these nubmers with residue of only a few points in the sixth decimal place.

16.88 Change the starting point for the preceding problem and observe developments.

▮ The starts $(1, 2)$, $(1, 3)$ and $(2, 3)$ led quickly to the solution just found. Completing the circumnavigation with the start $(2, 2)$ and anticipating the same result, brought rapid convergence to

$$(1.94242, 2.695166)$$

instead. Figure 16-10 shows the results of further explorations of the scene. The two solutions are shown, Ⓐ being the original and Ⓑ the new. Various starts are labeled by the end to which they lead. It seems that Ⓐ is the preferred target except from the east and northeast and except from curious islands near $(0, 2)$ and $(1.5, 1)$. There is also a troublesome start at $(1, 1.5)$, which was not pursued.

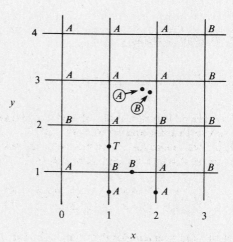

Fig. 16-10

16.89 Apply the iterative method to find a solution of

$$x = .7 \sin x + .2 \cos y \qquad y = .7 \cos x - .2 \sin y$$

near $(.5, .5)$.

❚ The equations are already in the form $x = F(x, y)$, $y = G(x, y)$. Substituting .5 for each argument on the right, we get the new estimates $x = .516$, $y = .510$ to three places. Continuing in this way, the successive estimates will be found to converge, but at the rather deliberate rate we have come to expect of the method. Table 16.20 shows every other iterate. If the last are accepted as correct to at least five places, then a scan of the two lists finds between two and three entries (four to six iterations) needed to gain one decimal place.

TABLE 16.20

x	y	x	y
.516	.510	.52642	.50795
.523	.509	.52648	.50793
.5249	.5084	.526506	.507924
.5259	.5081	.526516	.507922
.5263	.5080	.526520	.507920

16.90 Develop a theoretical estimate of convergence speed for the iterative method and apply it to the preceding problem.

❚ We have the equations

$$x_{n+1} = F(x_n, y_n) \qquad y_{n+1} = G(x_n, y_n)$$

for the successive approximations and

$$x = F(x, y) \qquad y = G(x, y)$$

for the true solution. The error in x_{n+1} can be estimated as

$$e_{n+1} = x - x_{n+1} = F(x, y) - F(x_n, y_n) = F_x(x - x_n) + F_y(y - y_n)$$

if higher power terms of the Taylor series are ignored. This makes

$$e_{n+1} = F_x e_n + F_y f_n$$

where f_n is the error in y_n. Similarly

$$f_{n+1} = G_x e_n + G_y f_n$$

and the pair can be written in vector language:

$$V_{n+1} = \begin{pmatrix} e_n \\ f_n \end{pmatrix} = \begin{pmatrix} F_x & F_y \\ G_x & G_y \end{pmatrix} \begin{pmatrix} e_n \\ f_n \end{pmatrix} = J_n V_n$$

The subscript on the Jacobian matrix indicates that it should be evaluated at (x_n, y_n). But then $\|V_{n+1}\| \le \|J_n\| \cdot \|V_n\|$. So it is the norm of the Jacobian that controls convergence speed.

For the application, suppose we choose the maximum norm and use the initial Jacobian matrix. Recalling that the maximum norm is the largest absolute row sum (Problem 2.61),

$$\|J_0\| = \left\| \begin{matrix} .7\cos x & -.2\sin y \\ -.7\sin x & -.2\cos y \end{matrix} \right\| = \left\| \begin{matrix} .62 & -.10 \\ -.34 & -.18 \end{matrix} \right\| = .72$$

more or less. This measures the improvement per iteration, and since $.72^n$ will be near .1 for $n = 7$, suggests about seven iterates to gain a decimal place. We did slightly better.

16.91 Also apply the Newton method to the system of Problem 16.89.

❚ Now we want the equations in the form

$$x - .7\sin x - .2\cos y = 0 \qquad y - .7\cos x + .2\sin y = 0$$

which makes the system for the h and k at each step

$$(1 - .7\cos x)h + (.2\sin y)k = -x + .7\sin x + .2\cos y$$

$$(.7\sin x)h + (1 + .2\cos y)k = -y + .7\cos x - .2\sin y$$

Starting at $(.5, .5)$ these equations produce

x	.5268	.5265235	.5265226
y	.5080	.5079194	.5079197

followed by repeats or occasional shifts of one unit in the final place. The original equations are satisfied by these numbers to the order 10^{-8}, compared with 10^{-6} and 10^{-7} for the results in Problem 16.89. Measured in this way, Newton's method has done 10 to 100 times better while using three instead of 20 iterations.

16.92   Find a solution of $x = x^2 + y^2$, $y = x^2 - y^2$ near $(.8, .4)$.

▮ Either unknown can, of course, be eliminated to obtain a one-dimensional problem, but there is no difficulty in proceeding directly. Even the shift to

$$x = \sqrt{y + y^2} \qquad y = \sqrt{x - x^2}$$

works, the iterative method converting $(.8, .4)$ into $(.7755, .4172)$ in 10 steps and into $(.7720, .4195)$ in 10 more, with the x_n and y_n plodding patiently in opposite directions toward their goals. Newton's method does no plodding. Remembering to rearrange the equation as

$$f = x^2 + y^2 - x = 0 \qquad g = x^2 - y^2 - y = 0$$

we soon have this output:

x	.8	.7729	.771846	.7718445
y	.4	.4203	.419644	.4196434

The final value of f was reported as flat zero, and that of g as 10^{-8}.

16.93   What is the modified Newton method?

▮ One modification retains the original values of f_x, f_y, g_x, g_y for several iterations, instead of updating them every time. This cuts down the amount of computing that has to be done. It is especially useful with higher-order systems, where the size of the coefficient matrix becomes more of a burden.

Applying the idea to the system of the preceding problem, we keep the original coefficients for the entire run, with the results

x	.8	.772	.77190	.771848	.7718447	.7718446	.7718445
y	.4	.4203	.4196417	.4196447	.4196434		

Both sequences have reached the same goals as before, the second doing better than the first. The x sequence is hardly demonstrating quadratic convergence toward its finish, but even so has managed with only six steps.

16.11 QUADRATIC FACTORS: BAIRSTOW'S METHOD

16.94   Develop a recursion for the coefficients b_k in

$$q(x) = b_0 x^{n-2} + \cdots + b_{n-2} \qquad r(x) = b_{n-1}(x - u) + b_n$$

when $q(x)$ and $r(x)$ are defined by

$$p(x) = a_0 x^n + \cdots + a_n = (x^2 - ux - v)q(x) + r(x)$$

▮ Multiplying out on the right and comparing the powers of x, we have

$$b_0 = a_0$$

$$b_1 = a_1 + ub_0$$

$$b_k = a_k + ub_{k-1} + vb_{k-2} \qquad k = 2, \ldots, n$$

If we artificially set $b_{-1} = b_{-2} = 0$, the last recursion holds for $k = 0, 1, \ldots, n$. The b_k depend of course on the numbers u and v.

16.95 How may the recursion of the previous problem be used to calculate $p(x)$ for a complex argument $x = a + bi$? (Assume the a_k are real.)

▮ With $u = 2a$ and $v = -a^2 - b^2$, we have $x^2 - ux - v = 0$ so that

$$p(x) = b_{n-1}(x - 2a) + b_n$$

The advantage of this procedure is that the b_k are found by real arithmetic, so that no complex arithmetic occurs until the final step. In particular, if $b_{n-1} = b_n = 0$, then we have $p(x) = 0$. The complex conjugates $a \pm bi$ are then zeros of $p(x)$.

16.96 Develop Bairstow's method for using the Newton iteration to solve the simultaneous equations $b_{n-1}(u, v) = 0$, $b_n(u, v) = 0$.

▮ To use Newton's iteration, as described in Problem 16.83, we need the partial derivatives of b_{n-1} and b_n relative to u and v. First taking derivatives relative to u and letting $c_k = \partial b_{k+1} / \partial u$, we find $c_{-2} = c_{-1} = 0$, $c_0 = b_0$, $c_1 = b_1 + uc_0$ and then

$$c_k = b_k + uc_{k-1} + vc_{k-2}$$

The last result is actually valid for $k = 0, 1, \ldots, n - 1$. Thus the c_k are computed from the b_k just as the b_k were obtained from the a_k. The two results we need are

$$\frac{\partial b_{n-1}}{\partial u} = c_{n-2} \qquad \frac{\partial b_n}{\partial u} = c_{n-1}$$

Similarly taking derivatives relative to v and letting $d_k = \partial b_{k+2} / \partial v$ we find $d_{-2} = d_{-1} = 0$, then $d_1 = b_1 + ud_0$, after which

$$d_k = b_k + ud_{k-1} + vd_{k-2}$$

The latter holds for $k = 0, 1, \ldots, n - 2$. Since the c_k and d_k therefore satisfy the same recursion with the same initial conditions, we have proved $c_k = d_k$ for $k = 0, 1, \ldots, n - 2$. In particular,

$$\frac{\partial b_{n-1}}{\partial v} = c_{n-3} \qquad \frac{\partial b_n}{\partial v} = c_{n-2}$$

and we are ready for Newton's iteration.

Suppose we have approximate roots $a \pm bi$ of $p(x) = 0$ and the associated quadratic factor $x^2 - ux - v$ of $p(x)$. This means we have approximate roots of $b_{n-1} = b_n = 0$ and are seeking improved approximations $u + h, v + k$. The corrections h and k are determined by

$$c_{n-2}h + c_{n-3}k = -b_{n-1}$$

$$c_{n-1}h + c_{n-2}k = -b_n$$

These are the central equations of Newton's iteration. Solving for h and k,

$$h = \frac{b_n c_{n-3} - b_{n-1} c_{n-2}}{c_{n-2}^2 - c_{n-1} c_{n-3}} \qquad k = \frac{b_{n-1} c_{n-1} - b_n c_{n-2}}{c_{n-2}^2 - c_{n-1} c_{n-3}}$$

16.97   Apply Bairstow's method to determine the complex roots of Leonardo's equation correct to nine places.

▮   We have already found excellent initial approximations by the quotient-difference algorithm (see Problem 16.51): $u_0 \simeq -3.3642$, $v_0 \simeq -14.6033$. Our recursion now produces the following b_k and c_k:

k	0	1	2	3
a_k	1	2	10	-20
b_k	1	-1.3642	$-.01386$	$-.03155$
c_k	1	-4.7284	1.2901	

The formulas of Problem 16.96 then produce $h = -.004608$, $k = -.007930$ making

$$u_1 = u_0 + h = -3.368808 \qquad v_1 = v_0 + k = -14.611230$$

Repeating the process, we next find new b_k and c_k:

k	0	1	2	3
a_k	1	2	10	-20
b_k	1	-1.368808	.000021341	$-.000103380$
c_k	1	-4.73616	1.348910341	

These bring

$$h = -.000000108 \qquad k = -.000021852$$
$$u_2 = -3.368808108 \qquad v_2 = -14.611251852$$

Repeating the cycle once more finds $b_2 = b_3 = h = k = 0$ to nine places. The required roots are now

$$x_1, x_2 = \tfrac{1}{2}u \pm i\sqrt{-v - \tfrac{1}{4}u^2} = -1.684404054 \pm 3.431331350i$$

These may be further checked by computing the sum and product of all three roots and comparing with the coefficients of 2 and 20 in Leonardo's equation.

16.98   Apply Bairstow's method to $x^4 - 3x^3 + 20x^2 + 44x + 54 = 0$ to find a quadratic factor close to $x^2 + 2x + 2$.

▮   From the start $(u, v) = (-2, -2)$, two iterations were enough to arrange convergence to six places.

u	-1.941281	-1.941278
v	-1.950178	-1.953789

Restarting from $(-1, -1)$, $(0, 0)$ and $(3, 3)$ reached the same values in at most six steps. The corresponding factor $x^2 - ux - v$ has zeros at $x = -.97064 \pm 1.00581i$, compared with the $-1 \pm i$ of the starting quadratic. The minor changes can be viewed as purification.

The coefficients b_0, b_1, b_2 produced by the algorithm represent the remaining quadratic factor of the original fourth degree polynomial $x^2 - 4.941278x + 27.63861$. Its zeros are the other pair of roots and prove to be $x = 2.47064 \pm 4.64053i$.

16.99   Find all the roots of $x^4 + 2x^3 + 7x^2 - 11 = 0$ using deflation supported by the Newton and Bairstow iterations.

▮   For a homemade polyalgorithm, suppose we begin with a scan of the neighborhood of zero. The value of this polynomial is found to change sign in the interval $(1, 2)$ and in $(-2, -1)$. Turning the Newton method loose, roots at $x = -1.34107$ and 1.04036 are determined. Noting the sum of these to be $-.30071$ and the product -1.39520, the quadratic factor $x^2 + .30071 \times -1.39520$ is then in hand. The other quadratic factor is then accessible by the Bairstow method, which from this point of view becomes a method of deflation. The b_0, b_1, b_2 generated make that quadratic

$$x^2 + 1.69929x + 7.88420$$

with roots $-.849646 \pm 2.67625i$. The values of b_3 and b_4 were $-7 \cdot 10^{-7}$ and $-9 \cdot 10^{-7}$, so appealing to Problem

16.95 the original function can be evaluated for this final pair. We have

$$p(x) = 7 \cdot 10^{-7}(x - 2a) - 9 \cdot 10^{-7}$$

where $x = a \pm bi$. This works out to $p(x) = (-15 \pm 19i) \cdot 10^{-7}$.

16.100 Split the fourth degree polynomial $x^4 + 5x^3 + 3x^2 - 5x - 9$ into quadratic factors.

▌ Assuming this implied all complex roots, Bernoulli's method was exercised to find the dominant pair. What it produced was $x = -4.10186$, in about 10 iterations, so at least two of the roots are real. A scan located the other in the interval $(1, 2)$ and Newton's method then honed both roots into -4.101858 and 1.198905. The sum and product of these are -2.902953 and -4.917738, providing one of the required quadratic factors

$$x^2 + 2.902953x - 4.917738$$

Turning to Bairstow's method, with $u = -2.902953$ and $v = 4.917738$ we find it reproducing this input quadratic factor without change and giving us the quotient quadratic

$$x^2 + 2.097046x + 1.830110$$

The remainder term was $(.95x - 3.8) \cdot 10^{-6}$. Out of curiosity, this second quadratic was then used as input to Bairstow's method to see how well the first might be recovered. The result was a change of the terminal 3 in the middle coefficient to 4. The remainder term was $(4.8x - 9.5) \cdot 10^{-7}$.

CHAPTER 17
Linear Systems

17.1 What kinds of method are available for the solution of a linear system $Ax = b$?

❚ This may well be the principal problem of numerical analysis, and the extensive set of algorithms available for finding x when A and b are given indicates that the apparently elementary character of the problem is deceptive. There are numerous pitfalls.

Gaussian elimination is one of the oldest algorithms. It involves replacing equations by combinations of equations in such a way that a triangular system is obtained:

$$u_{11}x_1 + u_{12}x_2 + \cdots + u_{1n}x_n = c_1$$
$$u_{22}x_2 + \cdots + u_{2n}x_n = c_2$$
$$\vdots$$
$$u_{nn}x_n = c_n$$

After this, the components of x are easily found, one after the other, by the process of back-substitution. The last equation determines x_n, which is then substituted into the next-to-last equation to get x_{n-1} and so on.

Factorization methods convert A into products such as LU or LDU, where L is zero above the main diagonal, U is zero below it and D has only diagonal elements different from zero. The matrix L is called lower triangular and U is upper triangular. The methods of Gauss, Doolittle, Crout, Cholesky and others all produce factorizations. With A factored, the solution is easily accessible. Since

$$Ax = LUx = L(Ux) = Ly = b$$

we first solve $Ly = b$ for y and then $Ux = y$ for x. The first of the triangular systems responds to forward-substitution and the second to back-substitution.

Iterative methods generate sequences of successive approximations to x. The classic of this type is the Gauss–Seidel method, which reshapes $Ax = b$ into

$$x_1 = \cdots$$
$$x_2 = \cdots$$
$$\vdots$$
$$x_n = \cdots$$

by solving the ith equation for x_i. Initial approximations then allow each component to be corrected in its turn and when the cycle is complete to begin another cycle. Numerous variations have been devised. Iterative methods are especially convenient for sparse matrices A, in which many elements are zero.

17.1 GAUSSIAN ELIMINATION

17.2 Solve by Gaussian elimination:

$$x_1 + \tfrac{1}{2}x_2 + \tfrac{1}{3}x_3 = 1$$

$$\tfrac{1}{2}x_1 + \tfrac{1}{3}x_2 + \tfrac{1}{4}x_3 = 0$$

$$\tfrac{1}{3}x_1 + \tfrac{1}{4}x_2 + \tfrac{1}{5}x_3 = 0$$

❚ We begin by seeking the absolutely largest coefficient in column 1. Here it is in the top place. If this were not so, an interchange of rows would be made to arrange it. This largest element is called the *first pivot*. Now define

$$l_{21} = \frac{a_{21}}{a_{11}} = \frac{1}{2} \qquad l_{31} = \frac{a_{31}}{a_{11}} = \frac{1}{3}$$

and reduce the two lower coefficients in column 1 to zero in a familiar way, subtracting from the ith equation the product of l_{i1} by the first. Here is the result:

$$x_1 + \tfrac{1}{2}x_2 + \tfrac{1}{3}x_3 = 1$$

$$\tfrac{1}{12}x_2 + \tfrac{1}{12}x_3 = -\tfrac{1}{2}$$

$$\tfrac{1}{12}x_2 + \tfrac{4}{45}x_3 = -\tfrac{1}{3}$$

This is the first modified system. The same procedure is now applied to the smaller system consisting of the two lower equations. Again the absolutely largest coefficient is already at the top of the leading column, so no interchange of rows is needed. We find

$$l_{32} = \frac{a_{32}^{(1)}}{a_{22}^{(1)}} = 1$$

and so subtract from the third equation the product of l_{32} and the second equation. [The superscript (1) refers to the first modified system.] We then have

$$x_1 + \tfrac{1}{2}x_2 + \tfrac{1}{3}x_3 = 1$$

$$\tfrac{1}{12}x_2 + \tfrac{1}{12}x_3 = -\tfrac{1}{2}$$

$$\tfrac{1}{180}x_3 = \tfrac{1}{6}$$

and the triangular system is evident. The solution process is then completed by back-substitution, which finds the components x_i from the bottom up and in reverse order:

$$x_3 = 30 \qquad x_2 = -36 \qquad x_1 = 9$$

17.3 Why is pivoting important?

❚ Consider this extreme example:

$$10^{-5}x_1 + x_2 = 1$$

$$x_1 + x_2 = 2$$

The very small coefficient makes it clear that the solution ought to be quite close to $x_1 = x_2 = 1$. Suppose we solve without pivoting and with the assumption that only four decimal places can be carried. Exact subtraction would yield the equation

$$(1 - 10^5)x_2 = 2 - 10^5$$

but with the restriction on decimal places we must settle for

$$10^5 x_2 = 10^5$$

which still presents us with $x_2 = 1$. However, continuing the back-substitution we then face

$$10^{-5}x_1 + 1 = 1$$

making $x_1 = 0$ instead of the anticipated 1.

But now interchange the two equations, bringing the largest coefficient of column 1 in to the pivot position:

$$x_1 + x_2 = 2$$

$$10^{-5}x_1 + x_2 = 1$$

Exact subtraction would now bring

$$(1 - 10^{-5})x_2 = 1 - 2(10^{-5})$$

which the same restrictions would round to $x_2 = 1$. This time the back-substitution manages

$$x_1 + 1 = 2$$

and $x_1 = 1$. Pivoting has made the difference between nonsense and a perfect result. Experience with many less dramatic systems has shown that pivoting is an important part of the elimination algorithm. The technique described is called *partial pivoting*, since the search for largest coefficient is limited to the immediate column. The value of a broader search, into other columns, and leading to column interchanges, is a matter of debate.

The example in hand may be used to illustrate a further point. Multiply the first equation by 10^5 to obtain

$$x_1 + 10^5 x_2 = 10^5$$
$$x_1 + \quad x_2 = 2$$

and make pivoting unnecessary. The usual subtraction manages

$$(1 - 10^5) x_2 = 2 - 10^5$$

when done exactly, but becomes

$$-10^5 x_2 = -10^5$$

after rounding. So $x_2 = 1$. But then

$$x_1 = 10^5 - 10^5 = 0$$

and we have the earlier "solution." The point is, even pivoting may not help when very large coefficients occur elsewhere. One way out of the difficulty might be to interchange columns, but an alternative is to *normalize* each equation, making the absolutely largest coefficient in each about the same. A popular way to do this is dividing each equation by its coefficient of greatest size. The "norm" of each equation will then be 1. In our example we would, of course, return to the original system. The lesson appears to be that the combination of normalization and partial pivoting has a good chance of yielding a good result.

17.4 Summarize the Gauss algorithm for the general n by n linear system.

▮ Suppose that k steps of the type described in Problem 17.2 have been made, bringing the system to the form

$$u_{11} x_1 + u_{12} x_2 + \cdots + u_{1k} x_k + u_{1,k+1} x_{k+1} + \cdots + u_{1n} x_n \quad = b_1'$$
$$u_{22} x_2 + \cdots + u_{2k} x_k + u_{2,k+1} x_{k+1} + \cdots + u_{2n} x_n \quad = b_2'$$
$$\vdots$$
$$u_{kk} x_k + u_{k,k+1} x_{k+1} + \cdots + u_{kn} x_n \quad = b_k'$$
$$a_{k+1,k+1}^{(k)} x_{k+1} + \cdots + a_{k+1,n}^{(k)} x_n = b_{k+1}^{(k)}$$
$$\vdots$$
$$a_{n,k+1}^{(k)} x_{k+1} + \cdots + a_{nn}^{(k)} x_n \quad = b_n^{(k)}$$

The top k equations are in their final form, with $u_{11}, \ldots, u_{kk}$ the first k pivots. In the remaining $n - k$ equations the coefficients bear the superscript (k) of this modified system. We next seek the $(k + 1)$th pivot among the coefficients of x_{k+1} in the lower $n - k$ equations. It will be the absolutely largest and its equation will be interchanged with equation $k + 1$. With this new pivot in place, now called $u_{k+1,k+1}$, a new set of multipliers is found

$$l_{i,k+1} = \frac{a_{i,k+1}^{(k)}}{u_{k+1,k+1}} \qquad i = k + 2, \ldots, n$$

and zeros are arranged under the new pivot by subtracting equations. Coefficient changes are governed by

$$a_{ij}^{(k+1)} = a_{ij}^{(k)} - l_{i,k+1} a_{k+1,j}^{(k)} \qquad k = 0, \ldots, n - 2$$
$$j = k + 2, \ldots, n$$
$$b_i^{(k+1)} = b_i^{(k)} - l_{i,k+1} b_k^{(k)} \qquad i = k + 2, \ldots, n$$

with $k = 0$ referring to the original system. The back-substitution part of the algorithm is represented by

$$x_i = \frac{1}{u_{ii}} \left(b_i' - \sum_{j=i+1}^{n} u_{ij} x_j \right) \qquad i = n, \ldots, 1$$

17.5 What is the Gauss–Jordan variation?

∎ Here zeros are generated both below and above each pivot, by further subtractions. The final matrix is thus diagonal rather than triangular and back-substitution is eliminated. The idea is attractive, but it involves more computing than the original algorithm and so is little used.

17.6 Estimate the amount of computing needed to carry out the Gauss algorithm for an n by n system.

∎ Consider the reduction of the coefficient matrix A to triangular form. This is where the lion's share of the effort occurs. At the first step, $(n-1)^2$ modified coefficients are obtained. We further limit our attention to a count of such coefficients. In successive steps this number is reduced and the grand total will be

$$(n-1)^2 + (n-2)^2 + \cdots + 1$$

coefficients. By a well known result of algebra this is equal to $(2n^3 - 3n^2 + n)/6$, from which the principal term $n^3/3$ is extracted as a simple measure of the computation's size. If $n = 100$, this number runs to six figures.

17.7 Apply Gaussian elimination to this system, assuming that a computer capable of carrying only two floating-point digits is to do the calculations:

$$x_1 + .67x_2 + .33x_3 = 2$$
$$.45x_1 + \quad x_2 + .55x_3 = 2$$
$$.67x_1 + .33x_2 + \quad x_3 = 2$$

∎ With $l_{21} = .45$ and $l_{31} = .67$, the array below left summarizes the first stage of the process and then with $l_{32} = -.17$ the array at the right shows the final triangularization.

```
1    .67   .33 | 2.0      1    .67   .33 | 2.0
0    .70   .40 | 1.1      0    .70   .40 | 1.1
0   -.12   .78 |  .7      0     0    .85 |  .89
```

Back-substitution now begins with

$$x_3 = \frac{.89}{.85} = 1.047$$

if we assume a double precision accumulator, but rounding to 1.0 in any event. Then

$$x_2 = \left(\frac{1}{.7}\right)(1.1 - .4) = 1.0$$
$$x_1 = 2 - .67 - .33 = 1.0$$

and the exact $(1, 1, 1)$ solution has been found in spite of the severe limitations of the computer. This is because we have a very cooperative matrix. (See also Problem 17.57.)

17.8 Use the Gauss algorithm to show that this system is inconsistent except for one value of A:

$$x + y + 2z = 1$$
$$2x + y - 3z = 0$$
$$-3x - y + 8z = A$$

∎ Since we are dealing with integers it will be simpler not to pivot. Two Gaussian steps bring the systems

$$
\begin{array}{ll}
x + y + 2z = 1 \qquad & x + y + 2z = 1 \\
\;\; -y - 7z = -2 & \;\; -y - 7z = -2 \\
\;\; 2y + 14z = A + 3 & \qquad\;\; 0 = A - 1
\end{array}
$$

The facts are clear. There is no solution at all unless $A = 1$, in which case z is arbitrary and y, x follow.

The elimination method may be the simplest way to test the consistency of a system. The old rule of the zero determinant is a poor alternative, requiring many more arithmetic steps when the system is large scale.

17.9 Solve the following system using only four significant digits at each step, first with pivot .002110 and then with .3370, and compare results:

$$.002110x + .08204y = .04313$$
$$.3370x + 12.84y = 6.757$$

| With the first pivot Gaussian elimination leads to the system on the left; with the second, it manages the one at the right:

$$2110x + .08204y = .04313 \qquad .3370x + 12.84y = 6.757$$
$$- .26y = - .131 \qquad .001650y = .0008200$$

The first has the four digit solution $(x, y) = (.8531, .5038)$ while the second offers $(1.116, .4970)$. The correct values are $(1, .5)$ so the larger pivot has brought some slight improvement.

17.10 What is complete pivoting? Apply it to the system of the preceding problem.

| Complete pivoting involves searching throughout the entire coefficient matrix, or the part still active in the elimination process, for the largest coefficient. Both rows and columns are then interchanged to bring this element to the pivot position. If columns are interchanged, it will be necessary to record in some way the new positions of the unknowns.

In the given system the largest coefficient is the 12.84, and pivoting rearranges the system as

$$12.84y + .3370x = 6.757$$
$$.08204y + .002110x = .04313$$

Elimination now produces the new second equation

$$- .000043x = - .00004$$

from which $(x, y) = (.9302, .5018)$ is found. This is again an improvement over former results. In general, complete pivoting has been found to produce better results than partial pivoting, in which only the active column is searched. In practice, the gain in accuracy is usually not considered worth the price of the additional computing time required.

17.11 Solve this system by Gaussian elimination, computing in rational form so that no roundoff errors are introduced, and so getting an exact solution. The coefficient matrix is the Hilbert matrix of order 4 (the Hilbert matrix of order 3 appeared in Problem 17.2):

$$x_1 + \tfrac{1}{2}x_2 + \tfrac{1}{3}x_3 + \tfrac{1}{4}x_4 = 1$$
$$\tfrac{1}{2}x_1 + \tfrac{1}{3}x_2 + \tfrac{1}{4}x_3 + \tfrac{1}{5}x_4 = 0$$
$$\tfrac{1}{3}x_1 + \tfrac{1}{4}x_2 + \tfrac{1}{5}x_3 + \tfrac{1}{6}x_4 = 0$$
$$\tfrac{1}{4}x_1 + \tfrac{1}{5}x_2 + \tfrac{1}{6}x_3 + \tfrac{1}{7}x_4 = 0$$

| The three stages of the elimination process are presented in Table 17.1, variables omitted for simplicity. Back substitution quickly finds $(x_1, x_2, x_3, x_4) = (16, -120, 240, -140)$.

TABLE 17.1

1	1/2	1/3	1/4	1
0	1/12	1/12	3/40	-1/2
0	1/12	4/45	1/12	-1/3
0	3/40	1/12	9/112	-1/4
1	1/2	1/3	1/4	1
0	1/12	1/12	3/40	-1/2
0	0	1/180	1/120	1/6
0	0	1/120	9/700	1/5
1	1/2	1/3	1/4	1
0	1/12	1/12	3/40	-1/2
0	0	1/180	1/120	1/6
0	0	0	1/2800	-1/20

17.12 $\quad$ Now try to solve the system of the preceding problem with a three significant digit restriction on all values. (Hilbert matrices of higher order are very troublesome even when many digits can be carried.)

▮ $\quad$ Table 17.2 parallels Table 17.1. In places only one significant digit is available, due to disappearance in the subtraction of near equals. Just how one does the rounding off has a profound effect on the lower corner entry. The .0025 found here does not bode well, having little resemblance to the exact $1/2800$. If we insist on perservering, back-substitution begins with $x_4 = -8.4$, which thoroughly discourages continuation.

TABLE 17.2

1	.500	.333	.250	1
0	.083	.083	.075	$-.5$
0	.083	.089	.083	$-.333$
0	.075	.083	.081	$-.250$
1	.500	.333	.250	1
0	.083	.083	.075	$-.5$
0	0	.006	.008	.167
0	0	.008	.0132	.202
1	.500	.333	.250	1
0	.083	.083	.075	$-.5$
0	0	.006	.008	.167
0	0	0	.0025	$-.021$

17.13 $\quad$ Lest it be thought that all systems are troublesome, apply the Gauss method to

$$\begin{aligned} w + 2x - 12y + 8z &= 27 \\ 5w + 4x + 7y - 2z &= 4 \\ -3w + 7x + 9y + 5z &= 11 \\ 6w - 12x - 8y + 3z &= 49 \end{aligned}$$

▮ $\quad$ A program with partial pivoting and no scaling, using a seven digit computer, printed the output $3, -2, 1, 5$, which is exact.

17.14 $\quad$ Solve the following system by Gaussian elimination:

$$\begin{aligned} x_1 + x_2 + x_3 + x_4 + x_5 &= 1 \\ x_1 + 2x_2 + 3x_3 + 4x_4 + 5x_5 &= 0 \\ x_1 + 3x_2 + 6x_3 + 10x_4 + 15x_5 &= 0 \\ x_1 + 4x_2 + 10x_3 + 20x_4 + 35x_5 &= 0 \\ x_1 + 5x_2 + 15x_3 + 35x_4 + 70x_5 &= 0 \end{aligned}$$

▮ $\quad$ The same program and computer printed $5, -10, 10, -5, 1$, which also happens to be exact. The "average" system is handled very well by Gaussian elimination.

17.15 $\quad$ Apply Gaussian elimination to $Ax = b$ where

$$A = \begin{pmatrix} 8 & 1 & 6 \\ 3 & 5 & 7 \\ 4 & 9 & 2 \end{pmatrix}$$

and b is in turn $(1, 0, 0)^T$, $(0, 1, 0)^T$ and $(0, 0, 1)^T$.

▮ $\quad$ The magic square matrix causes no trouble and is of course very small scale. For the first b vector the result

$$x^T = (.1472222, -.06111111, -.01944445)$$

was outputted. The repetitions suggest nines, and quick work with pen and paper comes up with the theory $x^T = (53, -22, -7)/360$, which is easily tested. For the other two b vectors, the results

$$(-13, 2, 17)/90 \quad \text{and} \quad (23, 38, -37)/360$$

follow by the same combination of seven place accuracy and inspiration.

17.16 Solve the systems in Fig. 17-1 by Gaussian elimination.

▮ The smaller is satisfied to seven places by

$$x^T = (-.03538461, -.004615384, .003076923, .01076923, .04153846)$$

with inspiration unequal to the task of revealing the rational counterparts. The larger is also satisfied to seven places, by

$$x^T = (-.01953251, .001231608, -.002090649, .0008163261, .003723303, .0004010435, .02116516)$$

and the true rational identities are well concealed. The elimination method is having no difficulty with magic square matrices. (See also Problem 17.37.)

$$\begin{pmatrix} 17 & 24 & 1 & 8 & 15 \\ 23 & 5 & 7 & 14 & 16 \\ 4 & 6 & 13 & 20 & 22 \\ 10 & 12 & 19 & 21 & 3 \\ 11 & 18 & 25 & 2 & 9 \end{pmatrix} \begin{pmatrix} x_1 \\ x_2 \\ x_3 \\ x_4 \\ x_5 \end{pmatrix} = \begin{pmatrix} 0 \\ 0 \\ 1 \\ 0 \\ 0 \end{pmatrix}$$

$$\begin{pmatrix} 30 & 39 & 48 & 1 & 10 & 19 & 28 \\ 38 & 47 & 7 & 9 & 18 & 27 & 29 \\ 46 & 6 & 8 & 17 & 26 & 35 & 37 \\ 5 & 14 & 16 & 25 & 34 & 36 & 45 \\ 13 & 15 & 24 & 33 & 42 & 44 & 4 \\ 21 & 23 & 32 & 41 & 43 & 3 & 12 \\ 22 & 31 & 40 & 49 & 2 & 11 & 20 \end{pmatrix} \begin{pmatrix} x_1 \\ x_2 \\ x_3 \\ x_4 \\ x_5 \\ x_6 \\ x_7 \end{pmatrix} = \begin{pmatrix} 0 \\ 0 \\ 0 \\ 1 \\ 0 \\ 0 \\ 0 \end{pmatrix}$$

Fig. 17-1

17.2 FACTORIZATIONS

17.17 What is the connection between Gaussian elimination and factors of the coefficient matrix?

▮ Form matrices L and U as follows, using results of Problem 17.2:

$$L = \begin{bmatrix} 1 & 0 & 0 \\ l_{21} & 1 & 0 \\ l_{31} & l_{32} & 1 \end{bmatrix} = \begin{bmatrix} 1 & 0 & 0 \\ \frac{1}{2} & 1 & 0 \\ \frac{1}{3} & 1 & 1 \end{bmatrix}$$

$$U = \begin{bmatrix} u_{11} & u_{12} & u_{13} \\ 0 & u_{22} & u_{23} \\ 0 & 0 & u_{33} \end{bmatrix} = \begin{bmatrix} 1 & \frac{1}{2} & \frac{1}{3} \\ 0 & \frac{1}{12} & \frac{1}{12} \\ 0 & 0 & \frac{1}{180} \end{bmatrix}$$

Then

$$LU = \begin{bmatrix} 1 & \frac{1}{2} & \frac{1}{3} \\ \frac{1}{2} & \frac{1}{3} & \frac{1}{4} \\ \frac{1}{3} & \frac{1}{4} & \frac{1}{5} \end{bmatrix} = A$$

For a general proof of this factorization, see the following problem.

17.18 Show that if L is a lower triangular matrix with elements l_{ij} and $l_{ii} = 1$, and if U is an upper triangular matrix with elements u_{ij}, then $LU = A$.

▮ The proof involves some easy exercise with triangular matrices. Returning briefly to the opening example, define

$$S_1 = \begin{bmatrix} 1 & 0 & 0 \\ -\frac{1}{2} & 1 & 0 \\ -\frac{1}{3} & 0 & 1 \end{bmatrix} \quad S_2 = \begin{bmatrix} 1 & 0 & 0 \\ 0 & 1 & 0 \\ 0 & -1 & 1 \end{bmatrix}$$

and observe that the product $S_1 A$ effects Step 1 of the Gauss algorithm, as it applies to the left sides of the equations, while $S_2 S_1 A$ then effects Step 2. This means that

$$S_2 S_1 A = U \qquad A = S_1^{-1} S_2^{-1} U = LU$$

with $L = S_1^{-1} S_2^{-1}$. Also note that

$$S_1^{-1} = \begin{bmatrix} 1 & 0 & 0 \\ \frac{1}{2} & 1 & 0 \\ \frac{1}{3} & 0 & 1 \end{bmatrix} \qquad S_2^{-1} = \begin{bmatrix} 1 & 0 & 0 \\ 0 & 1 & 0 \\ 0 & 1 & 0 \end{bmatrix}$$

so that inversions are achieved by changing the signs of the l_{ij} entries.

For the general problem assume at first that no interchanges will be needed. Define matrices

$$L_i = \begin{bmatrix} 1 & & & & \\ & \ddots & & & \\ & & 1 & & \\ & & -l_{i+1,i} & & \\ & & \vdots & \ddots & \\ & & -l_{n,i} & & 1 \end{bmatrix} \qquad i = 1, \ldots, n-1$$

with all other elements zero. As in the example, each of these effects one step of the elimination process, making

$$L_{n-1} L_{n-2} \cdots L_1 A = U$$

This means that

$$A = L_1^{-1} \cdots L_{n-1}^{-1} U = LU$$

Since the product of lower triangles with diagonal 1s is itself of the same type, we have our factorization. In addition, since each inversion is achieved by changing the signs of the l_{ij} entries, these are readily in hand and may be multiplied to rediscover

$$L = \begin{bmatrix} 1 & 0 & \cdots & & 0 \\ l_{21} & 1 & \cdots & & 0 \\ & & \cdots & & \\ l_{n1} & l_{n2} & \cdots & l_{n,n-1} & 1 \end{bmatrix}$$

Now suppose that some interchanges are to be made. Introduce the interchange matrices

$$I_{ij} = \begin{bmatrix} 1 & & & & & \\ & \ddots & & & & \\ & & 0 & 1 & & \\ & & 1 & 0 & & \\ & & & & \ddots & \\ & & & & & 1 \end{bmatrix} \begin{matrix} \\ \\ \text{row } i \\ \text{row } j \\ \\ \\ \end{matrix}$$

$$\begin{matrix} \text{col} & \text{col} \\ i & j \end{matrix}$$

The product $I_{ij} A$ will have rows i and j of A interchanged, while $A I_{ij}$ has the corresponding columns interchanged. The elimination algorithm now uses a chain of I_{ij} interchanges and L_i operations, leading to the representation

$$L_{n-1} I_{n-1, r_{n-1}} L_{n-2} I_{n-2, r_{n-2}} \cdots L_1 L_{1, r_1} A = U$$

where the r_i are the rows containing the selected pivots. This can be rearranged as

$$(L_{n-1} L_{n-2} \cdots L_1)(I_{n-1, r_{n-1}} \cdots I_{1, r_1}) A = U$$

or

$$L^{-1} PA = U \qquad PA = LU$$

with P the permutation matrix including the $n-1$ interchanges. Assuming A nonsingular, this means that there is a permutation of rows such that PA has an LU factorization. The uniqueness of this factorization will be evident from Problem 17.24.

17.19 Solve the system $Ax = b$ assuming an LU factorization has been done.

▮ We have, since L, U and P are in hand,

$$Ax = LUx = PAx = Pb$$

and letting $y = Ux$, first solve $Ly = Pb$ for y. This is easily done by forward-substitution. Then $Ux = y$ is solved by back-substitution. More specifically, and with p_i denoting an element of Pb, the system $Ly = Pb$ is

$$l_{11} y_1 \qquad\qquad\qquad = p_1$$
$$l_{21} y_1 + l_{22} y_2 \qquad\qquad = p_2$$
$$\vdots$$
$$l_{n1} y_1 + l_{n2} y_2 + \cdots + l_{nn} y_n = p_n$$

with all $l_{ii} = 1$. The solution by forward-substitution is clearly $y_1 = p_1$, $y_2 = p_2 - l_{21} y_1$, or more generally,

$$y_r = p_r - l_{r1} y_1 - \cdots - l_{r-1,\,r-1} y_{r-1}$$

for $r = 1, \ldots, n$. The backward-substitution is then achieved by the formula of Problem 17.4, modified only by the replacement of the vector b' by y:

$$x_i = \left(\frac{1}{u_{ii}}\right)\left(y_i - u_{i,\,i+1} x_{i+1} - \cdots - u_{in} x_n\right)$$

with $i = n, \ldots, 1$. The combination of factoring and forward-backward substitution is particularly useful if the system must be solved for more than one vector b.

17.20 What is a compact algorithm?

▮ When Gaussian elimination was done by hand, many elements of A were copied many times. In a computer this would be equivalent to making liberal use of storage space. With large-scale systems it is advisable to be economical both of storage space and computer time. For this reason, compact algorithms have been devised. For example, as elimination proceeds, the lower triangle of matrix A is replaced by zeros. These storage locations may better be used to record successively the values l_{ij}, for $j < i$. At the end of the run the upper triangle of A will then have been replaced by U and the lower triangle by L without its unit diagonal. There is no need to store all the interchanging matrices I_{ij}. It is enough to define initially a vector v with elements $(1, 2, 3, \ldots, n)$ and at each step to simply interchange the appropriate elements. If, for instance, the first pivot is in row 3, then $(3, 2, 1, 4, \ldots, n)$ records this. It is not necessary to physically interchange the rows, thus saving the time that would have been used for this maneuver. From the final v the permutation matrix P can be constructed, if desired, or v itself used to permute the elements of vector b.

17.21 Apply the procedure of Problem 17.20 to the matrix

$$A = \begin{bmatrix} 0 & 1 & 2 & 3 \\ 3 & 0 & 1 & 2 \\ 2 & 3 & 0 & 1 \\ 1 & 2 & 3 & 0 \end{bmatrix}$$

▮ The essential computations are displayed in Fig. 17-2. In three steps the original matrix is replaced by a four by four array containing all the information needed, except for the vector v, which traces the interchanges.

At this point matrix A has been replaced by a triangular matrix in the LU factorization of PA. The vector v tells us that the triangle will be evident if we look at rows 2, 3, 4, 1 in that order. Indeed the unstarred elements are the factor U. The factor L can also be read by taking the starred elements in the same row order. As for the

$$\begin{bmatrix} 0 & 1 & 2 & 3 \\ 3 & 0 & 1 & 2 \\ 2 & 3 & 0 & 1 \\ 1 & 2 & 3 & 0 \end{bmatrix}$$

The given matrix A

$$v = (1, 2, 3, 4)$$

Identify the first pivot, 3.

$$\begin{bmatrix} 0^* & 1 & 2 & 3 \\ ③ & 0 & 1 & 2 \\ \dfrac{2^*}{3} & 3 & -\dfrac{2}{3} & -\dfrac{1}{3} \\ \dfrac{1^*}{3} & 2 & \dfrac{8}{3} & -\dfrac{2}{3} \end{bmatrix}$$

Bring its row number to the first position in v. $v = (2, 1, 3, 4)$.
Compute and store the l_{i1} (starred).
Compute the nine new entries by subtractions (right of the solid line).

$$\begin{bmatrix} 0^* & \dfrac{1^*}{3} & \dfrac{20}{9} & \dfrac{28}{9} \\ 3 & 0 & 1 & 2 \\ \dfrac{2^*}{3} & ③ & -\dfrac{2}{3} & -\dfrac{1}{3} \\ \dfrac{1^*}{3} & \dfrac{2^*}{3} & \dfrac{28}{9} & -\dfrac{4}{9} \end{bmatrix}$$

Identify the second pivot (column 2 and right of the solid line).
Bring its row number to second position in v (2, 3, 1, 4).
Compute the l_{i2} and store them (starred).
Compute the four new entries.

$$\begin{bmatrix} 0^* & \dfrac{1^*}{3} & \dfrac{5^*}{7} & \dfrac{24}{7} \\ 3 & 0 & 1 & 2 \\ \dfrac{2^*}{3} & 3 & -\dfrac{2}{3} & -\dfrac{1}{3} \\ \dfrac{1^*}{3} & \dfrac{2^*}{3} & \boxed{\dfrac{28}{9}} & -\dfrac{4}{9} \end{bmatrix}$$

Identify the last pivot (column 3 and right of the solid line). Bring its row number to third position in v (2, 3, 4, 1).
Compute the l_{i3} and store them.
Compute the one new entry.

Fig. 17-2

permutation matrix P, it is constructed by placing 1s in columns 2, 3, 4, 1 of an otherwise zero matrix, as follows:

$$P = \begin{bmatrix} 0 & 1 & 0 & 0 \\ 0 & 0 & 1 & 0 \\ 0 & 0 & 0 & 1 \\ 1 & 0 & 0 & 0 \end{bmatrix}$$

One may now calculate

$$PA = LU = \begin{bmatrix} 3 & 0 & 1 & 2 \\ 2 & 3 & 0 & 1 \\ 1 & 2 & 3 & 0 \\ 0 & 1 & 2 & 3 \end{bmatrix}$$

and so verify all steps taken.

17.22 Using the results of the preceding problem and given the vector b with components $(0, 1, 2, 3)$, solve $Ax = b$.

▮ We use the suggestion in Problem 17.19. First either Pb or the vector v rearranges the components of b in the order $(1, 2, 3, 0)$. Although it is not necessary, suppose we display the system $Ly = Pb$ directly:

$$\begin{bmatrix} 1 & 0 & 0 & 0 \\ \dfrac{2}{3} & 1 & 0 & 0 \\ \dfrac{1}{3} & \dfrac{2}{3} & 1 & 0 \\ 0 & \dfrac{1}{3} & \dfrac{5}{7} & 1 \end{bmatrix} \begin{bmatrix} y_1 \\ y_2 \\ y_3 \\ y_4 \end{bmatrix} = \begin{bmatrix} 1 \\ 2 \\ 3 \\ 0 \end{bmatrix}$$

Forward-substitution then manages $y = (1, \frac{4}{3}, \frac{16}{9}, -\frac{12}{7})^T$. Turning to $Ux = y$ we face

$$
\begin{bmatrix}
3 & 0 & 1 & 2 \\
0 & 3 & -\frac{2}{3} & -\frac{1}{3} \\
0 & 0 & \frac{28}{9} & -\frac{4}{9} \\
0 & 0 & 0 & \frac{24}{7}
\end{bmatrix}
\begin{bmatrix}
x_1 \\ x_2 \\ x_3 \\ x_4
\end{bmatrix}
=
\begin{bmatrix}
1 \\ \frac{4}{3} \\ \frac{16}{9} \\ -\frac{12}{7}
\end{bmatrix}
$$

from which comes $x = (\frac{1}{2}, \frac{1}{2}, \frac{1}{2}, -\frac{1}{2})^T$, which may be verified directly in $Ax = b$.

17.23 Prove the fundamental theorem of linear algebra.

▮ We use the Gauss algorithm. If it can be continued to the end, producing a triangular system, then back-substitution will yield the unique solution. If all the b_i are zero, this solution has all zero components. This is already a principal part of the theorem. But suppose the algorithm cannot be continued to the anticipated triangular end. This happens only when at some point all coefficients below a certain level are zero. To be definite, say the algorithm has reached this point:

$$
\begin{aligned}
u_{11}x_1 + \cdots \qquad\qquad\qquad &= b_1' \\
u_{22}x_2 + \cdots \qquad\qquad &= b_2' \\
\vdots\qquad\quad & \\
u_{kk}x_k + \cdots &= b_k' \\
0 &= b_{k+1}^{(k)} \\
\vdots\quad & \\
0 &= b_n^{(k)}
\end{aligned}
$$

Then in the homogeneous case, where all the bs are zero, we may choose x_{k+1} to x_n as we please and then determine the other x_i. But in the general case, unless $b_{k+1}^{(k)}$ to $b_n^{(k)}$ are all zero, no solution is possible. If these bs do happen to be zero, then again we may choose x_{k+1} to x_n freely, after which the other x_i are determined. This is the content of the fundamental theorem.

17.24 Determine the elements of matrices L and U such that $A = LU$ by a direct comparison of corresponding elements.

▮ Assume that no interchanges will be necessary. Then we are to equate corresponding elements from the two sides of

$$
\begin{bmatrix}
1 & 0 & 0 & \cdots & 0 \\
l_{21} & 1 & 0 & \cdots & 0 \\
l_{31} & l_{32} & 1 & \cdots & 0 \\
 & & \cdots & & \\
l_{n1} & l_{n2} & l_{n3} & \cdots & 1
\end{bmatrix}
\begin{bmatrix}
u_{11} & u_{12} & u_{13} & \cdots & u_{1n} \\
0 & u_{22} & u_{23} & \cdots & u_{2n} \\
0 & 0 & u_{33} & \cdots & u_{3n} \\
 & & & \cdots & \\
0 & 0 & 0 & \cdots & u_{nn}
\end{bmatrix}
=
\begin{bmatrix}
a_{11} & \cdots & a_{1n} \\
a_{21} & \cdots & a_{2n} \\
a_{31} & \cdots & a_{3n} \\
 & \cdots & \\
a_{m1} & \cdots & a_{nn}
\end{bmatrix}
$$

which amounts to n^2 equations in the n^2 unknowns l_{ij} and u_{ij}. The determination runs as follows. First multiply the top row of L by all columns of U to get

$$ u_{1j} = a_{1j} \qquad j = 1, \ldots, n $$

Next multiply the rows of L (omit the first) by column 1 of U, finding $l_{i1}u_{11} = a_{i1}$, from which the l_{i1} follow:

$$ l_{i1} = \frac{a_{i1}}{u_{11}} \qquad i = 2, \ldots, n $$

It is next the turn of the second row of L to multiply the columns of U (omit the first). The second row of U is then

$$ u_{2j} = a_{2j} - l_{21}u_{1j} \qquad j = 2, \ldots, n $$

Now multiply the rows of L (omit the first two) by column 2 of U. All elements involved except the l_{i2} are in hand, so we solve for these:

$$l_{i2} = \frac{a_{i2} - l_{i1}u_{12}}{u_{22}} \qquad i = 3, \ldots, n$$

Continuing in this recursive way, we alternately find the rows of U to be

$$u_{rj} = a_{rj} - \sum_{k=1}^{r-1} l_{rk}u_{kj} \qquad j = r, \ldots, n$$

each row followed by the corresponding column of L:

$$l_{ir} = \frac{a_{ir} - \sum_{k=1}^{r-1} l_{ik}u_{kr}}{u_{rr}} \qquad i = r+1, \ldots, n$$

This procedure is called the *Doolittle algorithm.*

17.25 What is the Crout algorithm?

▮ The Crout algorithm also produces a factorization of A, in the form $L'U'$, with U' having the diagonal of 1s and L' the general diagonal. Formulas for the elements of the factors may be found very much as in Problem 17.24, but it is of interest to note that, with D denoting the matrix of diagonal elements of our earlier U and zeros elsewhere,

$$A = LU = L(DD^{-1})U = (LD)(D^{-1}U) = L'U'$$

so the two factorizations are closely related.

17.26 Develop the Choleski method for factoring a real, symmetric, positive definite matrix.

▮ Here we will find factors of the form

$$A = LL^T$$

the T denoting the transpose. The procedure is almost identical with that of Problem 17.24, with symmetry allowing us to consider only the lower triangle of A. The Hilbert matrix of order 3 can once again serve as a small-scale introduction:

$$\begin{bmatrix} l_{11} & 0 & 0 \\ l_{21} & l_{22} & 0 \\ l_{31} & l_{32} & l_{33} \end{bmatrix} \begin{bmatrix} l_{11} & l_{21} & l_{31} \\ 0 & l_{22} & l_{32} \\ 0 & 0 & l_{33} \end{bmatrix} = \begin{bmatrix} 1 & \frac{1}{2} & \frac{1}{3} \\ \frac{1}{2} & \frac{1}{3} & \frac{1}{4} \\ \frac{1}{3} & \frac{1}{4} & \frac{1}{5} \end{bmatrix}$$

The elements of L will be found from top to bottom and left to right:

$$l_{11}l_{11} = 1 \qquad l_{11} = 1$$

$$l_{21}l_{11} = \frac{1}{2} \qquad l_{21} = \frac{1}{2}$$

$$l_{21}^2 + l_{22}^2 = \frac{1}{3} \qquad l_{22} = \frac{1}{\sqrt{12}}$$

$$l_{31}l_{11} = \frac{1}{3} \qquad l_{31} = \frac{1}{3}$$

$$l_{31}l_{21} + l_{32}l_{22} = \frac{1}{4} \qquad l_{32} = \frac{1}{\sqrt{12}}$$

$$l_{31}^2 + l_{32}^2 + l_{33}^2 = \frac{1}{5} \qquad l_{33} = \frac{1}{\sqrt{180}}$$

The computation is again recursive, each line having only one unknown.

Because of the way the algorithm develops, should we now wish to extend our effort to the Hilbert matrix of order 4, it is only necessary to border L with a new bottom row and fourth column:

$$LL^T = \begin{bmatrix} 1 & 0 & 0 & 0 \\ \dfrac{1}{2} & \dfrac{1}{\sqrt{12}} & 0 & 0 \\ \dfrac{1}{3} & \dfrac{1}{\sqrt{12}} & \dfrac{1}{\sqrt{180}} & 0 \\ l_{41} & l_{42} & l_{43} & l_{44} \end{bmatrix} L^T = \begin{bmatrix} 1 & \dfrac{1}{2} & \dfrac{1}{3} & \dfrac{1}{4} \\ \dfrac{1}{2} & \dfrac{1}{3} & \dfrac{1}{4} & \dfrac{1}{5} \\ \dfrac{1}{3} & \dfrac{1}{4} & \dfrac{1}{5} & \dfrac{1}{6} \\ \dfrac{1}{4} & \dfrac{1}{5} & \dfrac{1}{6} & \dfrac{1}{7} \end{bmatrix}$$

We then find

$$l_{41}l_{11} = \frac{1}{4} \qquad l_{41} = \frac{1}{4}$$

$$l_{41}l_{21} + l_{42}l_{22} = \frac{1}{5} \qquad l_{42} = \frac{3\sqrt{3}}{20}$$

and so on to $l_{43} = \sqrt{5}/20$ and $l_{44} = \sqrt{7}/140$.

The algorithm can be summarized in the equations

$$\sum_{j=1}^{i-1} l_{rj}l_{ij} + l_{ri}l_{ii} = a_{ri} \qquad i = 1, \ldots, r-1$$

$$\sum_{j=1}^{r-1} l_{rj}^2 + l_{rr}^2 = a_{rr}$$

to be used for $r = 1, \ldots, n$ in turn.

17.27 Use the Doolittle method to factor the matrix

$$A = \begin{pmatrix} 1 & 2 & 3 \\ 8 & 9 & 4 \\ 7 & 6 & 5 \end{pmatrix}$$

❚ Since the top row of U will agree with that of A, there are only six entries to be determined:

$$LU = \begin{pmatrix} 1 & 0 & 0 \\ l_{21} & 1 & 0 \\ l_{31} & l_{32} & 1 \end{pmatrix} \begin{pmatrix} 1 & 2 & 3 \\ 0 & u_{22} & u_{23} \\ 0 & 0 & u_{33} \end{pmatrix} = A$$

Clearly $l_{21} = 8$ and $l_{31} = 7$. Then $16 + u_{22} = 9$, $24 + u_{23} = 4$ and so on. The results $u_{22} = -7$, $u_{23} = -20$, $l_{32} = 8/7$ and $u_{33} = 48/7$ come in turn.

17.28 Use the factors just found to solve the system $Ax = (1,0,0)^T$.

❚ It is a two step process. First forward-substitution is applied to $Ly = b = (1,0,0)^T$,

$$\begin{pmatrix} 1 & 0 & 0 \\ 8 & 1 & 0 \\ 7 & 8/7 & 1 \end{pmatrix} \begin{pmatrix} y_1 \\ y_2 \\ y_3 \end{pmatrix} = \begin{pmatrix} 1 \\ 0 \\ 0 \end{pmatrix}$$

to obtain $(y_1, y_2, y_3) = (1, -8, 15/7)$. Then back-substitution is applied to $Ux = y$,

$$\begin{pmatrix} 1 & 2 & 3 \\ 0 & -7 & -20 \\ 0 & 0 & 48/7 \end{pmatrix} \begin{pmatrix} x_1 \\ x_2 \\ x_3 \end{pmatrix} = \begin{pmatrix} 1 \\ -8 \\ 15/7 \end{pmatrix}$$

yielding the solution $(x_1, x_2, x_3) = (79/112, 1/4, 15/48)$.

17.29   Factor the matrix A of the preceding two problems by the Gaussian elimination method. Also solve the system $Ax = (1, 0, 0)^T$.

▌ The two steps involved in the elimination are summarized in Fig. 17-3 with the factor U at the bottom. The computation of L from S_1 and S_2 is also shown. To solve the system $Ax = b$ it is converted to $Ux = S_2 S_1 b = (1, -8, 15/7)$ and back-substitution is carried out.

$$A \qquad \begin{matrix} 1 & 2 & 3 \\ 8 & 9 & 4 \\ 7 & 6 & 5 \end{matrix} \qquad S_1 = \begin{pmatrix} 1 & 0 & 0 \\ -8 & 1 & 0 \\ -7 & 0 & 1 \end{pmatrix}$$

$$S_1 A \qquad \begin{matrix} 1 & 2 & 3 \\ 0 & -7 & -20 \\ 0 & -8 & -16 \end{matrix} \qquad S_2 = \begin{pmatrix} 1 & 0 & 0 \\ 0 & 1 & 0 \\ 0 & -\frac{8}{7} & 1 \end{pmatrix}$$

$$U = S_2 S_1 A \qquad \begin{matrix} 1 & 2 & 3 \\ 0 & -7 & -20 \\ 0 & 0 & \frac{48}{7} \end{matrix} \qquad L = S_1^{-1} S_2^{-1} = \begin{pmatrix} 1 & 0 & 0 \\ 8 & 1 & 0 \\ 7 & 0 & 1 \end{pmatrix} \begin{pmatrix} 1 & 0 & 0 \\ 0 & 1 & 0 \\ 0 & \frac{8}{7} & 1 \end{pmatrix} \qquad \textbf{Fig. 17-3}$$

17.30   Convert the LU factorization just found into the form LDU, where D is a diagonal matrix.

▌ The factor L can be left untouched. Since each diagonal element of D will multiply the corresponding row of U, we simply extract such a factor from this row:

$$LU = \begin{pmatrix} 1 & 0 & 0 \\ 8 & 1 & 0 \\ 7 & 8/7 & 1 \end{pmatrix} \begin{pmatrix} 1 & 2 & 3 \\ 0 & -7 & -20 \\ 0 & 0 & 48/7 \end{pmatrix} = L \begin{pmatrix} 1 & 0 & 0 \\ 0 & -7 & 0 \\ 0 & 0 & 48/7 \end{pmatrix} \begin{pmatrix} 1 & 2 & 3 \\ 0 & 1 & 20/7 \\ 0 & 0 & 1 \end{pmatrix} = LDU'$$

17.31   Show that this effort at matrix factorization is unsuccessful:

$$\begin{pmatrix} 1 & 2 & 3 \\ 2 & 4 & 1 \\ -1 & 0 & 2 \end{pmatrix} = \begin{pmatrix} 1 & 0 & 0 \\ 2 & 1 & 0 \\ -1 & l_{32} & 1 \end{pmatrix} \begin{pmatrix} 1 & 2 & 3 \\ 0 & u_{22} & u_{23} \\ 0 & 0 & u_{33} \end{pmatrix}$$

▌ This is a routine Doolittle approach to $A = LU$. A few of the L and U entries are already in place, since they respond to quick mental computations. But now, $4 + u_{22} = 4$ forces u_{22} to be zero, a bad omen, since zeros on the U diagonal mean a singular A, and this A is not singular. Trouble follows at once from the row three by column two product, which suggests that $-2 = 0$. The assumption of an LU factorization has failed.

17.32   What does Gaussian elimination have to say about the matrix of the preceding problem?

▌ If pivoting is ignored at the first step, one immediately encounters the new second row 0   0 -5 and will be forced into pivoting (or some improvisation) on the next. So one may as well accept the inevitable and follow prescribed protocol. Interchanging the first two rows, doing one Gaussian step and then interchanging the last two rows creates a triangular matrix U. In the language of I and S matrices (Problem 17.18),

$$I_{23} S_1 I_{12} A = \begin{pmatrix} 0 & 1 & 0 \\ 1 & 0 & 0 \\ 0 & 0 & 1 \end{pmatrix} \begin{pmatrix} 1 & 0 & 0 \\ -1/2 & 1 & 0 \\ 1/2 & 0 & 1 \end{pmatrix} \begin{pmatrix} 1 & 0 & 0 \\ 0 & 0 & 1 \\ 0 & 1 & 0 \end{pmatrix} = \begin{pmatrix} 2 & 4 & 1 \\ 0 & 2 & 5/2 \\ 0 & 0 & 5/2 \end{pmatrix} = U$$

so we do have a factorization $A = I_{12}^{-1} S_1^{-1} I_{23}^{-1} U$ with U an upper triangle. However, its premultiplier does not turn out to be a lower triangle, as may be verified.

17.33 Do a Cholesky factorization of the Hilbert-like matrix A:

$$A = \begin{pmatrix} 1 & 1/3 & 1/5 \\ 1/3 & 1/5 & 1/7 \\ 1/5 & 1/7 & 1/9 \end{pmatrix} \qquad L = \begin{pmatrix} 1 & 0 & 0 \\ 1/3 & l_{22} & 0 \\ 1/5 & l_{32} & l_{33} \end{pmatrix}$$

▮ Easy mental calculations bring us to the L shown. Since LL^T must equal A, we next find $1/9 + l_{22}^2 = 1/5$ which makes $l_{22} = \sqrt{4/45}$. Similarly, $1/15 + l_{32}l_{22} = 1/7$ arranges $l_{32} = 4\sqrt{5}/35$. Finally, l_{33} proves to be $8/105$.

17.34 Does the Doolittle method lead to the same factorization of the matrix in the preceding problem?

▮ No, it leads instead to the factorization

$$A = LU = \begin{pmatrix} 1 & 0 & 0 \\ 1/3 & 1 & 0 \\ 1/5 & 6/7 & 1 \end{pmatrix} \begin{pmatrix} 1 & 1/3 & 1/5 \\ 0 & 4/45 & 8/3 \cdot 5 \cdot 7 \\ 0 & 0 & (8/3 \cdot 5 \cdot 7)^2 \end{pmatrix}$$

17.3 TWO LARGER-SCALE SYSTEMS

17.35 Consider the random walker of Fig. 17-4, who takes steps to right or left without prejudice until an endpoint is reached. Let x_k be his probability of reaching the left end if currently at position k. If x_0 and x_{n+1} are defined as 1 and 0, then

$$x_k = \tfrac{1}{2}(x_{k-1} + x_{k+1})$$

follows by probability theory at each interior point, that is, for $k = 1$ to n. For $n = 4$ the system looks like

$$\begin{aligned} -2x_1 + x_2 & = -1 \\ x_1 - 2x_2 + x_3 & = 0 \\ x_2 - 2x_3 + x_4 &= 0 \\ x_3 - 2x_4 &= 0 \end{aligned}$$

Use Gaussian elimination to solve such systems for various values of n.

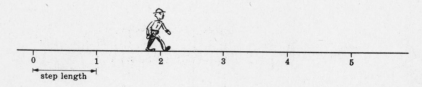

step length

Fig. 17-4

▮ Since the governing equation makes the second differences of the x_k zero, it is clear from the start that x_k is a linear function of k, the boundary values determining it uniquely. This gives us a chance to test the Gaussian output for various sizes of system. Here are a few numbers. Computations were made on a PC using seven decimal digits. Running times are in seconds.

n	10	20	30	50	70	100
Max Error	10^{-7}	10^{-7}	$2 \cdot 10^{-7}$	$3 \cdot 10^{-7}$	$3 \cdot 10^{-7}$	$3 \cdot 10^{-7}$
Running Time	—	1	3	12	28	80

The error is growing slowly, as we must expect with all the internal roundoffs. Running time is also growing, with something like the n^3 spirit suggested by Problem 17.6, the earlier values here being only as accurate as an ordinary wrist watch allows.

17.36 Why might Gaussian elimination not be the ideal method for the system of the preceding problem?

▮ It does seem to be getting the job done, with acceptable accuracy even for a 100 by 100 system. The question that arises is, is there no way to exploit the fact that most of the elements of the coefficient matrix are zero? The Gaussian method succeeds in creating $n - 1$ new zeros below the main diagonal, wiping out this triangle. But it does this at the cost of destroying all the zeros above that diagonal. There is another point. Ten thousand memory locations were needed for the A matrix alone in the final effort. This is a lot of storage space filled with the numbers 0, 1 and -2, each of which has to be retrieved when its active moments arrive. The matrix A is called sparse, meaning that its nonzero elements are far outnumbered. For obvious reasons it is also called triple diagonal. For such matrices, alternative procedures have been developed. (See, for example, Problem 17.38.)

17.37 Apply Gaussian elimination to $Ax = b$ where A is an n by n magic square of the sort illustrated in Problems 17.15 and 17.16, n being odd, and b is a vector of zeros with a central 1.

$\blacksquare$ For n odd, the following rough version of an ancient recipe for magic squares may have been deduced from the earlier examples.

> Proceed northeasterly, except
> upon reaching row 1, drop to row n
> upon reaching column n, go back to column 1
> if blocked, drop down one row.

From this recipe the computer can generate substantial squares. Here are some results of Gaussian elimination:

n	9	21	35	45	65	99
Max Error	$6 \cdot 10^{-8}$	$2 \cdot 10^{-7}$	$6 \cdot 10^{-8}$	$2 \cdot 10^{-7}$	10^{-7}	$5 \cdot 10^{-7}$
Running Time	—	2	5	10	28	95

Here, since the true solutions are unknown, error was measured by the differences (also known as residuals) between left and right sides of the equations. Only a few residuals were close to the reported maximum values, with the central 1 value of b being hit perfectly each time. The running time grows more or less as in Problem 17.35, the absence of zero elements in the given matrix apparently exacting little if any penalty.

17.4 ITERATIVE METHODS

17.38 Illustrate the Gauss–Seidel iteration for solving linear systems using the following well known example. A dog is lost in a square maze of corridors (Fig. 17-5). At each intersection it chooses a direction at random and proceeds to the next intersection, where it again chooses at random and so on. What is the probability that a dog starting at intersection i will eventually emerge on the south side?

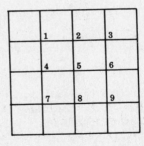

Fig. 17-5

$\blacksquare$ Suppose there are just nine interior intersections, as shown. Let P_1 stand for the probability that a dog starting at intersection 1 will eventually emerge on the south side. Let $P_2, \ldots, P_9$ be similarly defined. Assuming that at each intersection he reaches, a dog is as likely to choose one direction as another, and that having reached any exit his walk is over, probability theory then offers the following nine equations for the P_k:

$$P_1 = \tfrac{1}{4}(0 + 0 + P_2 + P_4) \qquad P_4 = \tfrac{1}{4}(P_1 + 0 + P_5 + P_7) \qquad P_7 = \tfrac{1}{4}(P_4 + 0 + P_8 + 1)$$

$$P_2 = \tfrac{1}{4}(0 + P_1 + P_3 + P_5) \qquad P_5 = \tfrac{1}{4}(P_2 + P_4 + P_6 + P_8) \qquad P_8 = \tfrac{1}{4}(P_5 + P_7 + P_9 + 1)$$

$$P_3 = \tfrac{1}{4}(0 + P_2 + 0 + P_6) \qquad P_6 = \tfrac{1}{4}(P_3 + P_5 + 0 + P_9) \qquad P_9 = \tfrac{1}{4}(P_6 + P_8 + 0 + 1)$$

Leaving the equations in this form, we choose initial approximations to the P_k. It would be possible to make intelligent guesses here, but suppose we choose the uninspired initial values $P_k = 0$ for all k. Taking the equations in the order listed we compute second approximations, one by one. First P_1 comes out zero and so do

$P_2, P_3, \ldots, P_6$. But then we find

$$P_7 = \tfrac{1}{4}\left(0 + 0 + 0 + 1\right) = \tfrac{1}{4} \qquad P_8 = \tfrac{1}{4}\left(0 + \tfrac{1}{4} + 0 + 1\right) = \tfrac{5}{16} \qquad P_9 = \tfrac{1}{4}\left(0 + \tfrac{5}{16} + 0 + 1\right) = \tfrac{21}{64}$$

and the second approximation to each P_k is in hand. Notice that in computing P_8 and P_9, the newest approximations to P_7 and P_8, respectively, have been used. There seems little point in using more antique approximations. This procedure leads to the correct results more rapidly. Succeeding approximations are now found in the same way and the iteration continues until no further changes occur in the required decimal places. Working to three places, the results of Table 17.3 are obtained. Note that P_5 comes out .250, which means that one-fourth of the dogs starting at the center should emerge on the south side. From the symmetry this makes sense. All nine values may be substituted back into the original equations as a further check, to see if the residuals are small.

In this example of the Gauss–Seidel method each of the nine equations comes to us in the form

$$P_i = \cdots$$

and is used to update the approximation to P_i using the most recent values of the other components. It is worth noting that in each equation the unknown on the left side has the dominant coefficient.

TABLE 17.3

Iteration	P_1	P_2	P_3	P_4	P_5	P_6	P_7	P_8	P_9
0	0	0	0	0	0	0	0	0	0
1	0	0	0	0	0	0	.250	.312	.328
2	0	0	0	.062	.078	.082	.328	.394	.328
3	.016	.024	.027	.106	.152	.127	.375	.464	.398
4	.032	.053	.045	.140	.196	.160	.401	.499	.415
5	.048	.072	.058	.161	.223	.174	.415	.513	.422
6	.058	.085	.065	.174	.236	.181	.422	.520	.425
7	.065	.092	.068	.181	.244	.184	.425	.524	.427
8	.068	.095	.070	.184	.247	.186	.427	.525	.428
9	.070	.097	.071	.186	.249	.187	.428	.526	.428
10	.071	.098	.071	.187	.250	.187	.428	.526	.428

17.39 Apply the Gauss–Seidel iteration to the random walk of Problem 17.35, using $n = 9$. How well does it converge to the linear solution?

❚ Starting with all the interior x_k equal to zero, a deliberately weak first approximation, the first iteration manages

$$.5 \qquad .25 \qquad .125 \qquad .0625 \qquad .03125 \qquad \cdots$$

which is at least a start. Thirty iterations were needed to get the first digit correct for all x_k, the wave of accuracy working across the line from x_1 to x_9. After 70 iterations all were correct to three places and after 150 the maximum error was $3 \cdot 10^{-7}$. The last run took < 1 s.

17.40 What happens when the number of points is increased?

❚ Choosing $n = 49$ and the same initial approximations, here are the results at $k = 0, 10, \ldots, 50$:

200 iterations	1	.62	.33	.15	.05	0
500 iterations	1	.75	.52	.32	.15	0
1000 iterations	1	.79	.59	.39	.19	0

The last run took 22 s. The method seems to be struggling.

17.41 Modify the random walk problem so that the probability of a step to the left is $\tfrac{3}{4}$. Set up and solve the appropriate difference equation and so obtain the exact solution.

❚ The difference equation is

$$x_k = \tfrac{3}{4}x_{k-1} + \tfrac{1}{4}x_{k+1}$$

with characteristic equation $r^2 - 4r + 3 = 0$. The roots $1, 3$ lead to the solution family $x_k = c_1 + c_2 \cdot 3^k$. Using the boundary conditions then determines the two coefficients

$$x_k = 1 - (3^k - 1)/(3^{n+1} - 1)$$

For $n = 9$ we find the sample values which will be used to judge the accuracy of approximation efforts:

$$x_1 = .9999661 \qquad x_9 = .6666780$$

17.42 Use the Gauss–Seidel iteration to solve the random walk problem for left, right probabilities of $\frac{3}{4}$, $\frac{1}{4}$ and $n = 9$.

$\blacksquare$ The difference equation in the form $x_k = \frac{3}{4} x_{k-1} + \frac{1}{4} x_{k+1}$ is ready for use. Here is an abbreviated report of the computation. The first iteration manages

$$1 \qquad \tfrac{3}{4} \qquad \tfrac{9}{16} \qquad \tfrac{27}{64} \qquad \cdots$$

and is a giant step in the right direction. The two correct values given above were then duplicated to the following number of places, all runs taking less than a second:

Iterations	10	20	30	40	50
x_1	2	4	6	7	7
x_9	1	3	4	5	6

This is good convergence, assisted by the leftward leaning of the walker.

17.43 Develop the Gauss–Seidel method for a general linear system.

$\blacksquare$ The algorithm is applied most often to systems $Ax = b$ for which the diagonal elements of A are dominant. In any event, one should arrange by row and column interchanges that larger elements fall along the diagonal, to the extent that this is possible. The ith equation of the system is then solved for x_i in terms of the other unknowns. If we use the symbol $x_i^{(k)}$ to represent the kth approximation to x_i, then the algorithm proceeds as in the example:

$$x_1^{(1)} = \frac{b_1 - a_{12} x_2^{(0)} - \cdots - a_{1n} x_n^{(0)}}{a_{11}}$$

$$x_2^{(1)} = \frac{b_2 - a_{21} x_1^{(1)} - a_{23} x_3^{(0)} - \cdots - a_{2n} x_n^{(0)}}{a_{22}}$$

$$x_3^{(1)} = \frac{b_3 - a_{31} x_1^{(1)} - a_{32} x_2^{(1)} - a_{34} x_4^{(0)} - \cdots - a_{3n} x_n^{(0)}}{a_{33}}$$

$$\vdots$$

the superscript (0) denoting an initial approximation. More generally we have for the kth approximation to x_i

$$x_i^{(k)} = \frac{b_i - \sum_{j=1}^{i-1} a_{ij} x_j^{(k)} - \sum_{j=i+1}^{n} a_{ij} x_j^{(k-1)}}{a_{ii}}$$

in which the first sum uses kth approximations to x_j having $j < i$, while the second uses $(k-1)$th approximations to x_j with $j > i$. Here $i = 1, \ldots, n$ and $k = 1, \ldots$.

17.44 Express the Gauss–Seidel algorithm in matrix form.

$\blacksquare$ First the matrix A is split into

$$A = L + D + U$$

where L and U are lower and upper triangles with zero elements on the diagonal. The general formula for Problem 17.43 can then be written as

$$x^{(k)} = D^{-1}(b - Lx^{(k)} - Ux^{(k-1)})$$

which can be solved for $x^{(k)}$. First

$$(I + D^{-1}L)x^{(k)} = D^{-1}b - D^{-1}Ux^{(k-1)}$$

which leads to

$$x^{(k)} = (I + D^{-1}L)^{-1}(D^{-1}b - D^{-1}Ux^{(k-1)})$$

or

$$x^{(k)} = -(D + L)^{-1}Ux^{(k-1)} + (D + L)^{-1}b$$

17.45 What is a stationary matrix iteration?

❚ A matrix iteration of the form

$$x^{(k)} = M_k x^{(k-1)} + C_k b$$

is called stationary if M_k and C_k are independent of k. The iteration then becomes

$$x^{(k)} = Mx^{(k-1)} + Cb$$

The Gauss–Seidel method is stationary, with this M and C:

$$M = -(D + L)^{-1}U \qquad C = (D + L)^{-1}$$

17.46 Discuss the convergence of matrix iterations.

❚ First we ask that the exact solution of $Ax = b$ be a fixed point of the iteration. That is, we substitute $x = A^{-1}b$ for both the input and output approximations in

$$x^{(k)} = M_k x^{(k-1)} + C_k b$$

and have

$$x = A^{-1}b = M_k A^{-1}b + C_k b = M_k x + C_k b$$

This is to hold for all vectors b, so we equate coefficients.

$$A^{-1} = M_k A^{-1} + C_k$$
$$I = M_k + C_k A$$

Now define $e^{(k)}$ as the error of the kth approximation.

$$e^{(k)} = x - x^{(k)}$$

Then

$$e^{(k)} = x - M_k x^{(k-1)} - C_k b$$
$$= M_k(x - x^{(k-1)}) = M_k e^{(k-1)}$$

which shows that it is the matrices M_k that control error behavior. Using this result repeatedly,

$$e^{(k)} = M_k M_{k-1} \cdots M_1 e^{(0)}$$

where $e^{(0)}$ is the initial error. For a stationary iteration this becomes

$$e^{(k)} = M^k e^{(0)}$$

17.47 Prove that the Gauss–Seidel iteration converges for an arbitrary initial vector $x^{(0)}$, if the matrix A is positive definite, symmetric.

❚ Because of the symmetry, $A = L + D + L^T$, which makes

$$M = -(D + L)^{-1}L^T$$

If λ and v are an eigenvalue and eigenvector of M, then

$$(D + L)^{-1} L^T v = -\lambda v$$
$$L^T v = -\lambda (D + L) v$$

Premultiplying by the conjugate transpose of v (denoted v^*)

$$v^* L^T v = -v^* \lambda (D + L) v$$

and then adding $v^*(D + L)v$ to both sides,

$$v^* A v = (1 - \lambda) v^* (D + L) v$$

since $A = L + D + L^T$. But the conjugate transpose of $v^* A v$ is $v^* A v$, so the same must be true for the right side of this last equation. Thus, with $\bar{\lambda}$ denoting the conjugate of λ,

$$(1 - \bar{\lambda}) v^* (D + L)^T v = (1 - \lambda) v^* (D + L) v$$
$$= (1 - \lambda)(v^* D v + v^* L v)$$
$$= (1 - \lambda)(v^* D v - \bar{\lambda} v^* (D + L)^T v)$$

Combining terms,

$$(1 - |\lambda|^2) v^* (D + L)^T v = (1 - \lambda) v^* D v$$

multiplying both sides by $(1 - \bar{\lambda})$ and doing a little algebra we have finally

$$(1 - |\lambda|^2) v^* A v = |1 - \lambda|^2 v^* D v$$

But both $v^* A v$ and $v^* D v$ are nonnegative and λ cannot equal 1 (since this would lead back to $A v = 0$), so

$$|\lambda|^2 < 1$$

placing all eigenvalues within the unit circle and guaranteeing that $\lim M^k = 0$. Thus $e^{(k)}$ has limit zero for any $e^{(0)}$.

17.48 How can an acceleration method be applied to the Gauss–Seidel iteration?

▌ Since $e^{(k)} = M e^{(k-1)}$, we anticipate that errors may diminish in a constant ratio, much as in Problem 16.4. The extrapolation to the limit idea then suggests itself. Here it would take the form

$$x_i = x_i^{(k+2)} - \frac{\Delta x_i^{(k+1)}}{\Delta^2 x_i^{(k)}}$$

for $i = 1, \ldots, n$. The superscripts denote three successive approximations.

For example, using the center column of Table 17.3, in which we know the correct value to be .250, the errors in rows 4 to 8 are 54, 27, 14, 6 and 3 in the third decimal place. This is very close to a steady reduction by one-half. Suppose we try extrapolation to the limit using the three entries below, together with the corresponding differences as given:

$$
\begin{array}{llc}
.196 & & \\
& .027 & \\
.223 & & -.014 \\
& .013 & \\
.236 & &
\end{array}
$$

We find

$$P_5 = .236 - \frac{(.013)^2}{-.014} = .248$$

which is in the right direction if not especially dramatic.

17.49 What are relaxation and overrelaxation methods?

❚ The central idea is to use residuals as indicators of how to correct approximations already in hand. For example, the iteration

$$x^{(k)} = x^{(k-1)} + \left(b - A x^{(k-1)} \right)$$

has the character of a relaxation method. It has been found that giving extra weight to the residual can speed convergence, leading to overrelaxation formulas such as

$$x^{(k)} = x^{(k-1)} + w \left(b - A x^{(k-1)} \right)$$

with $w > 1$. Other variations of the idea have also been used.

17.50 Adapt the overrelaxation method to accelerate the convergence of Gauss–Seidel.

❚ The natural adaptation is

$$x^{(k)} = x^{(k-1)} + w \left[b - L x^{(k)} - (D + U) x^{(k-1)} \right]$$

with $A = L + D + U$ as before. We take $w = 1.2$, $x^{(0)} = 0$ and try once more the problem of the dog in the maze, finding zeros generated as earlier until

$$P_7^{(1)} = P_7^{(0)} + 1.2 \left(.250 + \tfrac{1}{4} P_4^{(1)} - P_7^{(0)} + \tfrac{1}{4} P_8^{(0)} \right) = .300$$

$$P_8^{(1)} = P_8^{(0)} + 1.2 \left(.250 + \tfrac{1}{4} P_5^{(1)} + \tfrac{1}{4} P_7^{(1)} - P_8^{(0)} + \tfrac{1}{4} P_9^{(0)} \right) = .390$$

$$P_9^{(1)} = P_9^{(0)} + 1.2 \left(.250 + \tfrac{1}{4} P_6^{(1)} + \tfrac{1}{4} P_8^{(1)} - P_9^{(0)} \right) = .418$$

Succeeding approximations are found in the same way and are listed in Table 17.4. Notice that about half as many iterations are now needed.

TABLE 17.4

Iteration	P_1	P_2	P_3	P_4	P_5	P_6	P_7	P_8	P_9
0	0	0	0	0	0	0	0	0	0
1	0	0	0	0	0	0	.300	.390	.418
2	0	0	0	.090	.144	.169	.384	.506	.419
3	.028	.052	.066	.149	.234	.182	.420	.520	.427
4	.054	.096	.071	.183	.247	.187	.427	.526	.428
5	.073	.098	.071	.188	.251	.187	.428	.527	.428
6	.071	.098	.071	.187	.250	.187	.428	.526	.428

17.51 Apply relaxation methods to the random walk of Problem 17.35.

❚ For this problem the governing equations are

$$x_k = p x_{k-1} + (1 - p) x_{k+1} \qquad k = 1, \dots, n$$

with p the probability of a leftward step. The relaxation equivalent is, with x_k' denoting the new approximation at point k,

$$x_k' = x_k + w \left[p x_{k-1} - x_k + (1 - p) x_{k+1} \right]$$

which collapses to the Gauss–Seidel formula if $w = 1$.

In Problem 17.39 it was found, using $p = .5$ and $w = 1$, that 30 iterations produced only one correct digit for all x_k and that 150 were needed to get something like seven digits right. With $w = 1.2$, the first digit was correct after 15 iterations and seven were right after 100. Increasing w to 1.5 found 35 iterations managing seven figure accuracy, but $w = 1.7$ was less successful and $w = 2$ a disaster. It seems that overrelaxation is fine in moderation, with $w = 1.2$ a popular choice.

17.52 Does overrelaxation help the struggling computation in Problem 17.40?

▮ Here the walker had positions $k = 0$ to 50 available, so the linear system involved 49 equations. The Gauss–Seidel method labored through 1000 iterations to manage something close to two correct digits. Recklessly choosing $w = 1.5$, we now find that 500 iterations leave a maximum error of 10^{-3}, while 1000 bring results almost correct to six places after a run of 32 s. Overrelaxation has done rather well here.

17.53 Will overrelaxation speed up the computation in Problem 17.42, in which the probability of a leftward step was $p = \frac{3}{4}$? Also consider the case $p = \frac{1}{4}$.

▮ Perservering with $w = 1.5$, it was found that 10 iterations yielded three to four correct digits, but with x_2 slightly larger than 1 and unacceptable as a probability. Carrying on to 20 iterations produced six or seven correct digits all the way. The more conservative $w = 1.2$ reached the same goal almost as fast, which is more than twice as fast as the earlier runs with $w = 1$.

As for $p = \frac{1}{4}$, the exact solution is accessible from its difference equation, as in Problem 17.41. Staying with the smaller $n = 9$ grid and using $w = 1$, we find 20 iterations yielding three to four correct digits. Switching to $w = 1.2$, the same number of iterations manages seven correct digits instead.

17.5 ERRORS AND NORMS

17.54 What is a condition number of a matrix A?

▮ It is a measure of how trustworthy the matrix is in computations. For a given norm, we define the condition number as

$$C(A) = \|A\| \cdot \|A^{-1}\|$$

and observe, using Problem 2.56, that $C(I) = 1$, where I is the identity matrix. Moreover, using Problem 2.55,

$$C(A) = \|A\| \cdot \|A^{-1}\| \geq \|I\| = 1$$

so the identity matrix has the lowest condition number.

17.55 Suppose the vector b of the system $Ax = b$ contains input errors. Estimate the influence of such errors on the solution vector x.

▮ Rewrite the system as

$$Ax_e = b + e$$

and combine with $Ax = b$ to obtain

$$A(x_e - x) = e \qquad x_e - x = A^{-1}e$$

from which it follows that, using Problem 2.55,

$$\|x - x_e\| \leq \|A^{-1}\| \cdot \|e\|$$

To convert this to a relative error estimate, we have, from $Ax = b$,

$$\|A\| \cdot \|x\| \geq \|b\| \qquad \|x\| \geq \frac{\|b\|}{\|A\|}$$

and finally

$$\frac{\|x_e - x\|}{\|x\|} \leq \|A\| \cdot \|A^{-1}\| \frac{\|e\|}{\|b\|} = C(A) \frac{\|e\|}{\|b\|}$$

in which the condition number of A appears.
Similarly from

$$\|e\| \leq \|A\| \cdot \|x_e - x\| \quad \text{and} \quad \|A^{-1}\| \cdot \|b\| \geq \|x\|$$

we find

$$\frac{\|e\|}{C(A)\|b\|} \leq \frac{\|x_e - x\|}{\|x\|}$$

giving us both a lower and an upper bound for the relative error.

17.56 Suppose the matrix A of the system $Ax = b$ contains input errors. Estimate the influence of such errors on the solution vector x.

▌ Write the system as

$$(A + E)x_e = b$$

and combine with $Ax = b$ to obtain

$$A(x_e - x) = -Ex_e$$

leading to

$$\|x_e - x\| \le \|A^{-1}\| \cdot \|E\| \cdot \|x_e\|$$
$$\frac{\|x_e - x\|}{\|x_e\|} \le \|A^{-1}\| \cdot \|A\| \cdot \frac{\|E\|}{\|A\|} = C(A)\frac{\|E\|}{\|A\|}$$

which estimates the error relative to the solution x_e. Here again the condition number of A appears. Here and in the preceding problem it measures how much the input errors are inflated.

An estimate relative to the solution x can also be found. One such estimate is

$$\frac{\|x_e - x\|}{\|x\|} \le \frac{C(A)(\|E\|/\|A\|)}{1 - C(A)(\|E\|/\|A\|)}$$

17.57 Rework the opening example (Problem 17.2) under the assumption that a computer carrying only two floating-point digits is to do the computations.

▌ The system now takes the form

$$1.0x_1 + .50x_2 + .33x_3 = 1.0$$
$$.50x_1 + .33x_2 + .25x_3 = 0$$
$$.33x_1 + .25x_2 + .20x_3 = 0$$

and with $l_{21} = .5$ and $l_{31} = .33$ is soon converted to

$$.08x_2 + .09x_3 = -.50$$
$$.09x_2 + .09x_3 = -.33$$

with the first equation retained as is. Here we may as well complete the triangularization by simply subtracting what we have:

$$.01x_2 = .17$$

Now back-substitution manages $x_2 = 17$, $x_3 = -21$, $x_1 = -.6$ and a "solution" vector $(-.6, 17, -21)$. Comparing with the correct $(9, -36, 30)$ we see no resemblance whatsoever. The point is, the matrix of this system is a junior member of a notorious family, the Hilbert matrices. Coupling this with the severe limitations of our computer has led to a grotesque result.

In Problem 17.73 the inverse matrix will be found to be

$$\begin{bmatrix} 9 & -36 & 30 \\ -36 & 192 & -180 \\ 30 & -180 & 180 \end{bmatrix}$$

in which the large elements should be noted. The maximum norm is $36 + 192 + 180 = 408$, making a condition number of

$$C(A) = \|A\| \cdot \|A^{-1}\| = \tfrac{11}{6}(408) = 748$$

By Problem 17.56 we now have the estimate

$$\frac{\|x_e - x\|}{\|x_e\|} \le 748\left(\frac{.005}{\frac{11}{6}}\right) = 2.04$$

suggesting a relative error of 200%. Clearly the computation was naive. At least four digits are needed.

By way of contrast, recall the cooperative matrix of Problem 17.7, which permitted an exact solution to be found even by a two digit computer. For that matrix the maximum norm is 2 and the inverse also has norm near 2. The condition number is then near 4 and we estimate

$$\frac{\|x_e - x\|}{\|x_e\|} \le 4\left(\frac{.005}{1}\right) = .02$$

or a maximum error of 2%.

17.58   What is the "nearest singular matrix" theorem?

▌ Suppose A is nonsingular and B singular. Then, by the fundamental theorem of linear algebra, there exists a vector $x \ne 0$ satisfying $Bx = 0$. For this x,

$$\|Ax\| = \|Ax - Bx\| = \|(A - B)x\| \le \|A - B\| \cdot \|x\|$$

and since $x = A^{-1}Ax$, we also have $\|x\| \le \|A^{-1}\| \cdot \|Ax\|$. Since A is nonsingular, we cancel the factor $\|Ax\|$ and have

$$\frac{1}{\|A - B\|} \le \|A^{-1}\|$$

which is the required theorem.

Its message is that the size of the inverse matrix of A is at least the reciprocal of the "distance" of A from the nearest singular matrix B. If A is nearly singular, then A^{-1} will have a large norm. If A is normalized, in the sense $\|A\| = 1$, the condition number will also be large.

As a corollary we have the following intuitive result. If B is "close enough" to the nonsingular A, in the sense that $1/\|A - B\|$ is larger than $\|A^{-1}\|$, then B is nonsingular also.

17.59   Use the theorem of Problem 17.58 to estimate the condition of the matrix of this system, presented earlier in Problem 2.39:

$$x_1 + x_2 = 1$$
$$1.1x_1 + x_2 = 2$$

▌ The point is that A^{-1}, required for the condition number, is not always easy to find with accuracy. Though this is not true here, we observe that the matrix of coefficients is close to the singular matrix

$$B = \begin{bmatrix} 1 & 1 \\ 1 & 1 \end{bmatrix}$$

and find, using maximum norms, $\|A\| = 2.1$, $\|A - B\| = .1$, so that

$$\|A^{-1}\| \ge \frac{1}{.1} = 10 \qquad C(A) \ge (2.1)(10) = 21$$

17.60   Estimate the error caused by using $1.01x_2$ in place of x_2 in the second equation in Problem 17.59.

▌ The error matrix is

$$E = \begin{bmatrix} 0 & 0 \\ 0 & .01 \end{bmatrix}$$

with maximum norm .01. Thus

$$\frac{\|x_e - x\|}{\|x_e\|} \le C(A)\frac{\|E\|}{\|A\|} \le 21\left(\frac{.01}{2.1}\right) = .1$$

which is our estimate. For an input error of 1% we anticipate an output error of 10%. This inflation is due to the ill-condition of A, as measured by $C(A)$.

Solving the system directly, we find $x = (10, -9)$ and $x_e = (11, -10)$. This makes $\|x_e - x\| = 1$ and $\|x_e\| = 11$, for a relative error of .09. So the 10% inflation is just about realized.

17.61 The many intermediate computations that are made in solving a linear system make roundoff error an important factor. How can this error be estimated?

▮ Backward error analysis has produced the only real successes in this difficult area. It shows that the cumulative effect of roundoffs can be estimated by considering the substitute system $(A + E)x = b$, where E is a perturbation of A. It then finds bounds for the elements of E. The error in x can then be estimated by the formula of Problem 17.56. The details are far from trivial but have been carried through for most of the solution algorithms.

The somewhat deeper estimate

$$\|E\| \le \left(1.06 \max |u_{ij}|\right)\left(3n^2 + n^3\right)2^{-p}$$

may be easier to apply. For example, if A has order 10 ($n = 10$) and the equivalent of eight decimal places are carried ($2^{-p} = 10^{-8}$) and we make the crude guess of 10 for the first factor, then we find

$$\|E\| \le (1.3) \cdot 10^{-4}$$

suggesting that perhaps half the digits being carried may no longer be significant. The estimate is, of course, conservative, since it ignores the fact that errors often cancel one another to some extent.

17.62 How does the condition of the coefficient matrix A enter into the roundoff error estimation process?

▮ Recalling Problem 17.56, the relative error of the solution is bounded by

$$\frac{\|x_e - x\|}{\|x_e\|} \le C(A)\frac{\|E\|}{\|A\|}$$

where E is now the perturbation of A due to internal roundoffs. For a normalized A, the relative error in x_e is thus the product of two factors, the condition of A and the norm of E.

17.63 If double-precision arithmetic is available, how much does it improve the roundoff situation?

▮ By the formula in Problem 17.61, if the factor 2^{-p} can be reduced from 10^{-8} to 10^{-16}, eight additional decimal figures will be gained, surely a significant improvement. But there is a side effect. A large-scale system uses a lot of computer storage space, even at single precision. Doubling the precision may just burst the seams. There is a compromise, similar to the one described in Problem 11.110, where the motivation was computing time rather than storage space. Instead of doing and storing everything in double precision, limit this higher level of activity to the numerous inner product evaluations which permeate these algorithms. Once calculated, their values can be stored in single precision, making just one roundoff where there might have been n. Only a modest programming effort is needed to incorporate this feature and the reward can be dramatic.

17.64 The *residual* of an approximate solution x_e is defined as the vector

$$r = b - Ax_e$$

and gives the amount by which each equation of the linear system fails to be satisfied. How is the residual related to the error of x_e?

▮ Since $Ax = b$ for the exact solution x, we have

$$r = A(x - x_e) \qquad x - x_e = A^{-1}r$$

and, using Problem 2.55,

$$\frac{\|r\|}{\|A\|} \le \|x - x_e\| \le \|A^{-1}\| \cdot \|r\|$$

From $Ax = b$ we have similarly

$$\|A^{-1}\| \cdot \|b\| \ge \|x\| \ge \frac{\|b\|}{\|A\|}$$

so dividing corresponding elements leads to the required result

$$\frac{1}{C(A)}\frac{\|r\|}{\|b\|} \le \frac{\|x - x_e\|}{\|x\|} \le C(A)\frac{\|r\|}{\|b\|}$$

The relative error of x_e is bounded above and below by multiples of the relative residual, the multipliers involving the condition number of A. If $C(A)$ is near 1, then the relative error is close to the relative residual, which is, of course, readily available. If, however, $C(A)$ is large, there may be good reason to suspect inaccuracy in x_e even though r may be small. In other words, if A is ill-conditioned, the system may be nearly satisfied by an x_e containing large error. On the optimistic side, and looking primarily at the left half of the above inequality, when $C(A)$ is large, even a large residual still allows the error $x - x_e$ to be small, though the probability of this happening is likely to be rather small too.

17.65 What is the method of iterative refinement?

❚ Let $h = x - x_e$ and rewrite the equation $A(x - x_e) = r$ of the preceding problem as

$$Ah = r$$

This system has the same coefficient matrix as the original. If A has been factored, or the steps of Gaussian elimination retained in some way, it is solved with relatively little cost. With h in hand, one computes

$$x = x_e + h$$

and has a new, and presumably better, approximation to the true solution. New residuals may now be calculated and the process repeated as long as seems fruitful. This is the idea of iterative refinement. If double-precision arithmetic is available, this is an excellent opportunity to use it.

17.66 Suppose it has been found that the system

$$1.7x_1 + 2.3x_2 - 1.5x_3 = 2.35$$

$$1.1x_1 + 1.6x_2 - 1.9x_3 = -.94$$

$$2.7x_1 - 2.2x_2 + 1.5x_3 = 2.70$$

has a solution near $(1, 2, 3)$. Apply the method of Problem 17.65 to obtain an improved approximation.

❚ The approximation $(1, 2, 3)$ leaves the residuals $(.55, .46, -.1)$. Using these as the new right sides of our three equations and replacing the x_i on the left by the corrections h_i, we solve to find $(h_1, h_2, h_3) = (.1, .1, -.1)$. Adding these corrections to the given $(1, 2, 3)$ yields $(1.1, 2.1, 2.9)$ for the new estimates of the x_i. Another round duplicates these figures to six places.

17.67 Apply the method of iterative refinement to the system

$$\begin{pmatrix} -2 & 1 & .1 & 0 \\ 1 & -2 & 1 & .1 \\ .1 & 1 & -2 & 1 \\ 0 & .1 & 1 & -2 \end{pmatrix} \begin{pmatrix} x_1 \\ x_2 \\ x_3 \\ x_4 \end{pmatrix} = \begin{pmatrix} -1 \\ 0 \\ 0 \\ 0 \end{pmatrix}$$

❚ The superficial resemblance to the system of the random walk allows an initial guess of $(.8, .6, .4, .2)$ to be made for the solution. The .1 entries momentarily suggest a walker who does an occasional skip, but probability theory does not support this interpretation. Proceeding anyway, we find the residual vector to be $(-.04, -.02, -.08, -.06)$. Solving $Ah = r$ and adding h to the initial x produces the new approximate solution

$$x = (.9219288, .7829077, .6094973, .3438941)$$

which is almost unchanged by another cycle.

Changing each .1 of this coefficient matrix to .2 and using the solution just found as a guide, the initial estimate $(1, .8, .6, .4)$ was tried for this further distortion of the random walk system. The residuals were found to be $(.08, -.08, -.2, .04)$ and led to the new approximation vector

$$(x_1, x_2, x_3, x_4) = (1.172708, 1.146055, .9968016, .6130064)$$

with another cycle changing the terminal 6 to an 8.

17.68 Replace the .1 entries in the matrix A of the preceding problem by zeros, but now distort the right-side vector to $(-1, .1, .1, .1)$. Solve by iterative refinement.

▮ There is of course no difficulty in solving these little systems at once by Gaussian elimination, but proceeding as directed suppose we return to the initial estimate $x = (.8, .6, .4, .2)$ used previously. The companion residuals prove to be near .2 and .3, and the system $Ah = r$ suggests reducing our estimates to

$$x = (.68, .36, .14, .02)$$

with another cycle recommending no changes to seven places, the final residuals being of order 10^{-8}.

Changing the .1 elements to .2, with the solution just found serving as initial approximation, residuals of $(0, .1, .1, .1)$ lead to $x = (.56, .12, -.12, -.16)$ with another cycle making no further changes. The terminal residuals were of order 10^{-8}.

17.6 MATRIX INVERSION

17.69 Extend the Gaussian elimination algorithm to produce the inverse of the coefficient matrix A, that is, the A^{-1} such that $AA^{-1} = I$.

▮ Taking once again the system of Problem 17.2, we simply treat three b vectors simultaneously. The starting point is the array

$$
\begin{array}{cccccc}
1 & \frac{1}{2} & \frac{1}{3} & 1 & 0 & 0 \\
\frac{1}{2} & \frac{1}{3} & \frac{1}{4} & 0 & 1 & 0 \\
\frac{1}{3} & \frac{1}{4} & \frac{1}{5} & 0 & 0 & 1
\end{array}
$$

the left half of which is A and the right half I. The first Gaussian step now leads to the new array

$$
\begin{array}{cccccc}
1 & \frac{1}{2} & \frac{1}{3} & 1 & 0 & 0 \\
0 & \frac{1}{12} & \frac{1}{12} & -\frac{1}{2} & 1 & 0 \\
0 & \frac{1}{12} & \frac{4}{45} & -\frac{1}{3} & 0 & 1
\end{array}
$$

Here the method is modified slightly by reducing the next pivot to 1, a multiplication by 12 performing this service:

$$
\begin{array}{cccccc}
1 & \frac{1}{2} & \frac{1}{3} & 1 & 0 & 0 \\
0 & 1 & 1 & -6 & 12 & 0 \\
0 & 0 & \frac{1}{180} & \frac{1}{6} & -1 & 1
\end{array}
$$

The second step has also been performed to triangularize the system. At this point back-substitution could be used to solve three separate systems, each involving one of the last three column vectors. Instead, however, we extend the second Gaussian step. Continuing with the second row as pivotal row, we subtract half of it from row 1 to create one more zero:

$$
\begin{array}{cccccc}
1 & 0 & -\frac{1}{6} & 4 & -6 & 0 \\
0 & 1 & 1 & -6 & 12 & 0 \\
0 & 0 & \frac{1}{180} & \frac{1}{6} & -1 & 1
\end{array}
$$

The third Gaussian step then follows, after reducing the last pivot to 1. The purpose of this step is to create zeros above the new pivot. The final array then appears:

$$
\begin{array}{cccccc}
1 & 0 & 0 & 9 & -36 & 30 \\
0 & 1 & 0 & -36 & 192 & -180 \\
0 & 0 & 1 & 30 & -180 & 180
\end{array}
$$

Since we have actually solved three linear systems of the form $Ax = b$, with vectors $b^T = (1, 0, 0)$, $(0, 1, 0)$ and $(0, 0, 1)$ in turn, it is clear that the last three columns now contain A^{-1}. The original array was (A, I). The final array is (I, A^{-1}). The same process can be applied to other matrices A, row or column interchanges being made if required. If such interchanges are made, they must be restored at the completion of the algorithm.

17.70 Assuming that the matrix A has been factored as $A = LU$, how can A^{-1} be found from the factors?

$\blacksquare$ Since $A^{-1} = U^{-1}L^{-1}$, the question is one of inverting triangular matrices. Consider L and seek an inverse in the same form:

$$\begin{bmatrix} 1 & 0 & 0 & \cdots & 0 \\ l_{21} & 1 & 0 & \cdots & 0 \\ l_{31} & l_{32} & 1 & \cdots & 0 \\ & & \cdots & & \\ l_{n1} & l_{n2} & l_{n3} & \cdots & 1 \end{bmatrix} \begin{bmatrix} 1 & 0 & 0 & \cdots & 0 \\ c_{21} & 1 & 0 & \cdots & 0 \\ c_{31} & c_{32} & 1 & \cdots & 0 \\ & & \cdots & & \\ c_{n1} & c_{n2} & c_{n3} & \cdots & 1 \end{bmatrix} = LL^{-1} = I$$

The validity of the assumption will be clear as we proceed. Now match the elements of the two sides, much as in the Choleski factorization algorithm, top to bottom and left to right. We find

$$\begin{aligned} l_{21} + c_{21} &= 0 & c_{21} &= -l_{21} \\ l_{31} + l_{32}c_{21} + c_{31} &= 0 & c_{31} &= -(l_{31} + l_{32}c_{21}) \\ l_{32} + c_{32} &= 0 & c_{32} &= -l_{32} \\ l_{41} + l_{42}c_{21} + l_{43}c_{31} + c_{41} &= 0 & c_{41} &= -(l_{41} + l_{42}c_{21} + l_{43}c_{31}) \\ l_{42} + l_{43}c_{32} + c_{42} &= 0 & c_{42} &= -(l_{42} + l_{43}c_{32}) \\ l_{43} + c_{43} &= 0 & c_{43} &= -l_{43} \\ &\vdots \end{aligned}$$

The elements are determined recursively, the general formula being

$$c_{ij} = -\sum_{k=j}^{i-1} l_{ik}c_{kj} \qquad i = 2, \ldots, n, \ j = 1, \ldots, i-1$$

All diagonal elements are 1.

The inversion of U is similar. Assuming the inverse to be an upper triangle, with elements d_{ij}, we proceed from bottom to top and right to left, finding

$$d_{ii} = \frac{1}{u_{ii}} \qquad i = n, \ldots, 1$$

and

$$d_{ij} = \frac{-1}{u_{ii}} \sum_{k=i+1}^{j} u_{ik}d_{kj} \qquad i = n, \ldots, 1, \ j = n, \ldots, i+1$$

17.71 Apply the method of the preceding problem to the matrix of Problem 17.21.

$\blacksquare$ In that problem the factorization

$$PA = LU = \begin{bmatrix} 1 & 0 & 0 & 0 \\ \frac{2}{3} & 1 & 0 & 0 \\ \frac{1}{3} & \frac{2}{3} & 1 & 0 \\ 0 & \frac{1}{3} & \frac{5}{7} & 1 \end{bmatrix} \begin{bmatrix} 3 & 0 & 1 & 2 \\ 0 & 3 & -\frac{2}{3} & -\frac{1}{3} \\ 0 & 0 & \frac{28}{9} & -\frac{4}{9} \\ 0 & 0 & 0 & \frac{24}{7} \end{bmatrix}$$

was made. Applying the preceding recursions, we now have

$$L^{-1} = \frac{1}{63}\begin{bmatrix} 63 & 0 & 0 & 0 \\ -42 & 63 & 0 & 0 \\ 7 & -42 & 63 & 0 \\ 9 & 9 & -45 & 63 \end{bmatrix} \qquad U^{-1} = \frac{1}{168}\begin{bmatrix} 56 & 0 & -18 & -35 \\ 0 & 56 & 12 & 7 \\ 0 & 0 & 54 & 7 \\ 0 & 0 & 0 & 49 \end{bmatrix}$$

from which there comes eventually

$$(PA)^{-1} = U^{-1}L^{-1} = \frac{1}{24}\begin{bmatrix} 7 & 1 & 1 & -5 \\ -5 & 7 & 1 & 1 \\ 1 & -5 & 7 & 1 \\ 1 & 1 & -5 & 7 \end{bmatrix}$$

To produce the ultimate A^{-1}, we use $A^{-1} = (PA)^{-1}P$ and recall that postmultiplication by a permutation matrix P rearranges the columns. Referring back to the earlier problem, it is found that the preceding columns should be taken in the order $4, 1, 2, 3$.

17.72 Derive the formula for making an *exchange step* in a linear system.

I Let the linear system be $Ax = b$, or

$$\sum_{k=1}^{n} a_{ik} x_k = b_i \qquad i = 1, \ldots, n$$

The essential ingredients may be displayed as in this array for $n = 3$:

	x_1	x_2	x_3
b_1	a_{11}	a_{12}	a_{13}
b_2	a_{21}	a_{22}	a_{23}
b_3	a_{31}	a_{32}	a_{33}

We proceed to exchange one of the "dependent" variables (say b_2) with one of the independent variables (say x_3). Solving the second equation for x_3, $x_3 = (b_2 - a_{21}x_1 - a_{22}x_2)/a_{23}$. This requires that the *pivot* coefficient a_{23} not be zero. Substituting the expression for x_3 in the remaining two equations brings

$$b_1 = a_{11}x_1 + a_{12}x_2 + \frac{a_{13}(b_2 - a_{21}x_1 - a_{22}x_2)}{a_{23}}$$

$$b_3 = a_{31}x_1 + a_{32}x_2 + \frac{a_{33}(b_2 - a_{21}x_1 - a_{22}x_2)}{a_{23}}$$

The array for the new system, after the exchange, is

	x_1	x_2	b_2
b_1	$a_{11} - \dfrac{a_{13}a_{21}}{a_{23}}$	$a_{12} - \dfrac{a_{13}a_{22}}{a_{23}}$	$\dfrac{a_{13}}{a_{23}}$
x_3	$-\dfrac{a_{21}}{a_{23}}$	$-\dfrac{a_{22}}{a_{23}}$	$\dfrac{1}{a_{23}}$
b_3	$a_{31} - \dfrac{a_{33}a_{21}}{a_{23}}$	$a_{32} - \dfrac{a_{33}a_{22}}{a_{23}}$	$\dfrac{a_{33}}{a_{23}}$

This may be summarized in four rules:

(1) The pivot coefficient is replaced by its reciprocal.
(2) The rest of the pivot column is divided by the pivot coefficient.
(3) The rest of the pivot row is divided by the pivot coefficient with a change of sign.
(4) Any other coefficient (say a_{lm}) is replaced by $a_{lm} - (a_{lk}a_{im}/a_{ik})$ where a_{ik} is the pivot coefficient.

17.73 Illustrate the *exchange method* for finding the inverse matrix.

I Once again we take the matrix of Problem 17.2:

	x_1	x_2	x_3
b_1	1	$\frac{1}{2}$	$\frac{1}{3}$
b_2	$\frac{1}{2}$	$\frac{1}{3}$	$\frac{1}{4}$
b_3	$\frac{1}{3}$	$\frac{1}{4}$	$\frac{1}{5}$

For error control it is the practice to choose the largest coefficient for the pivot, in this case 1. Exchanging b_1 and x_1, we have this new array:

	b_1	x_2	x_3
x_1	1	$-\frac{1}{2}$	$-\frac{1}{3}$
b_2	$\frac{1}{2}$	$\frac{1}{12}$	$\frac{1}{12}$
b_3	$\frac{1}{3}$	$\frac{1}{12}$	$\frac{4}{45}$

Two similar exchanges of b_3 and x_3, then of b_2 and x_2, lead to the two arrays that follow. In each case the largest coefficient in a b row and an x column is used as pivot.

	b_1	x_2	b_3		b_1	b_2	b_3
x_1	$\frac{9}{4}$	$-\frac{3}{16}$	$-\frac{15}{4}$	x_1	9	-36	30
b_2	$\frac{3}{16}$	$\frac{1}{192}$	$\frac{15}{16}$	x_2	-36	192	-180
x_3	$-\frac{15}{4}$	$-\frac{15}{16}$	$\frac{45}{4}$	x_3	30	-180	180

Since what we have done is to exchange the system $b = Ax$ for the system $x = A^{-1}b$, the last matrix is A^{-1}.

17.74 In the preceding problem all pivots happened to be on the main diagonal of A. When this does not occur the variables get somewhat scrambled and the output matrix must be adjusted. Set up a procedure for doing this.

   ▮ One or two examples may be enough to clarify this point. The following matrix A was converted into B by four exchanges:

$$A = \begin{pmatrix} 1 & 1 & 1 & -1 \\ -1 & -1 & 1 & 1 \\ 1 & -1 & -1 & 1 \\ -1 & 1 & -1 & 1 \end{pmatrix} \qquad B = \tfrac{1}{2}\begin{pmatrix} 1 & 1 & 0 & 0 \\ 1 & 0 & 1 & 0 \\ 1 & 0 & 0 & 1 \\ 1 & 1 & 1 & 1 \end{pmatrix}$$

the pivots being in (row, column) positions $(1,1)$, $(2,3)$, $(3,2)$ and $(4,4)$, making the final system

$(*)$ $\qquad\qquad\qquad B(b_1, b_3, b_2, b_4)^T = (x_1, x_3, x_2, x_4)^T$

What we want is, of course, $A^{-1}(b_1, b_2, b_3, b_4)^T = (x_1, x_2, x_3, x_4)^T$. To achieve this we need to permute the variables. Choose the permutation matrix

$$P = (1324) = \begin{pmatrix} 1 & 0 & 0 & 0 \\ 0 & 0 & 1 & 0 \\ 0 & 1 & 0 & 0 \\ 0 & 0 & 0 & 1 \end{pmatrix}$$

and obtain from Equation $(*)$,

$$PBP^{-1}P(b_1, b_3, b_2, b_4) = P(x_1, x_3, x_2, x_4)$$

in which it will be seen that all the b_i and x_i are in the order desired. We deduce that $A^{-1} = PBP^{-1}$.

   In this example $P^{-1} = P$, so we have only to interchange the second and third rows of B to manage the premultiplication, and then interchange the second and third columns of the result to do the postmultiplication. The result is the required inverse:

$$A^{-1} = \tfrac{1}{2}\begin{pmatrix} 1 & 0 & 1 & 0 \\ 1 & 0 & 0 & 1 \\ 1 & 1 & 0 & 0 \\ 1 & 1 & 1 & 1 \end{pmatrix}$$

(The elements of A were selected by a random binary process.)

17.75 The elements of the matrix A in Fig. 17-6 were also selected at random, from a table of random integers 0 to 9. The companion matrix B is an abbreviated output of the exchange method, which used pivots at $(1,2)$, $(4,1)$, $(2,4)$ and $(3,3)$. Make the row and column permutations needed to get A^{-1}.

$$A = \begin{pmatrix} 1 & 9 & 4 & 9 \\ 0 & 1 & 2 & 9 \\ 8 & 3 & 7 & 1 \\ 9 & 4 & 3 & 5 \end{pmatrix} \qquad B = \begin{pmatrix} .0262 & .1302 & -.0457 & -.1397 \\ .0356 & -.0155 & -.0382 & .1110 \\ -.1734 & .0044 & .1946 & .0703 \\ .1375 & -.0508 & -.0233 & -.0230 \end{pmatrix}$$

▮ A review of the exchange process will discover these general rules.

1 Order the pivots by their second elements and read off the first elements. In this example we get (4132). Set up a permutation matrix P_1 with its 1 elements placed accordingly. In the example we would have the P_1 that follows.

2 Order the pivots by their first elements and read off the second elements. In this example we get (2431). Set up a permutation matrix P_2 accordingly.

$$P_1 = \begin{pmatrix} 0 & 0 & 0 & 1 \\ 1 & 0 & 0 & 0 \\ 0 & 0 & 1 & 0 \\ 0 & 1 & 0 & 0 \end{pmatrix} \qquad P_2 = \begin{pmatrix} 0 & 1 & 0 & 0 \\ 0 & 0 & 0 & 1 \\ 0 & 0 & 1 & 0 \\ 1 & 0 & 0 & 0 \end{pmatrix}$$

3 Find $A^{-1} = P_1 B P_2^{-1}$, where B is the exchange output. In the example this proves to be

$$A^{-1} = \begin{pmatrix} -.0508 & -.0230 & -.0233 & .1375 \\ .1302 & -.1397 & -.0457 & .0262 \\ .0044 & .0703 & .1946 & -.1734 \\ -.0155 & .1110 & -.0382 & .0356 \end{pmatrix}$$

Multiplying A by this A^{-1} (but using the original unabbreviated values) produced the identity matrix except for roundoff errors in the last two decimal places.

17.76 Compare the condition numbers of the two fourth order random matrices just inverted, using the infinity norm.

▮ The condition number of A is $C(A) = \|A\| \cdot \|A^{-1}\|$. Since the infinity norm is so accessible, being the maximum absolute row sum, it is a convenient choice. For the matrix of Problem 17.74, $\|A\|$ is 4 and $\|A^{-1}\|$ is 2, making $C(A) = 8$. For the matrix of Problem 17.75, $\|A\|$ is 23 and $\|A^{-1}\|$ is about .44, making $C(A) = 10$ more or less. The matrices are of comparable condition.

17.77 Compare condition numbers of the two matrices

$$\begin{pmatrix} 1 & 1 & 1 & 1 \\ 1 & 2 & 3 & 4 \\ 1 & 3 & 6 & 10 \\ 1 & 4 & 10 & 20 \end{pmatrix} \qquad \begin{pmatrix} -2 & 1 & 0 & 0 \\ 1 & -2 & 1 & 0 \\ 0 & 1 & -2 & 1 \\ 0 & 0 & 1 & -2 \end{pmatrix}$$

▮ The first is a section of Pascal's triangle and the second a small version of the random walk matrix. Inverses were found by the elimination method and, apart from obvious roundoff error in the seventh place, were

$$\begin{pmatrix} 4 & -6 & 4 & -1 \\ -6 & 14 & -11 & 3 \\ 4 & -11 & 10 & -3 \\ -1 & 3 & -3 & 1 \end{pmatrix} \qquad -\frac{1}{5} \begin{pmatrix} 4 & 3 & 2 & 1 \\ 3 & 6 & 4 & 2 \\ 2 & 4 & 6 & 3 \\ 1 & 2 & 3 & 4 \end{pmatrix}$$

From the absolute row sums we find the two condition numbers

$$35 \cdot 34 = 1190 \qquad 4 \cdot 3 = 12$$

so the second is much better conditioned.

17.78 Find a condition number for Wilson's matrix:

$$W = \begin{pmatrix} 10 & 7 & 8 & 7 \\ 7 & 5 & 6 & 5 \\ 8 & 6 & 10 & 9 \\ 7 & 5 & 9 & 10 \end{pmatrix}$$

▮ This innocent seeming matrix is somewhat troublesome. Using single precision, its inverse was computed to be

$$W^{-1} = \begin{pmatrix} 25.00031 & -41.00052 & 10.00014 & -6.000081 \\ -41.00051 & 68.00085 & -17.00022 & 10.00013 \\ 10.00013 & -17.00022 & 5.000059 & -3.000036 \\ -6.000083 & 10.00014 & -3.000036 & 2.000021 \end{pmatrix}$$

but the product $W \cdot W^{-1}$ then reproduced the identity matrix only to about four figures. At least three digits are lost in this brief computation. The integer parts of this pseudoinverse form the real thing and we use them to find the condition number

$$C(A) = 33 \cdot 136 = 4488$$

17.79 Where does the Hilbert matrix of order 4 stand on the ladder of condition?

▌ As with the Wilson matrix, the computed inverse suffers from roundoff error in the last few places. Extracting the integer parts, we have

$$H = \begin{pmatrix} 1 & \frac{1}{2} & \frac{1}{3} & \frac{1}{4} \\ \frac{1}{2} & \frac{1}{3} & \frac{1}{4} & \frac{1}{4} \\ \frac{1}{3} & \frac{1}{4} & \frac{1}{5} & \frac{1}{6} \\ \frac{1}{4} & \frac{1}{5} & \frac{1}{6} & \frac{1}{7} \end{pmatrix} \qquad H^{-1} = \begin{pmatrix} 16 & -120 & 240 & -140 \\ -120 & 1200 & -2700 & 1680 \\ 240 & -2700 & 6480 & -4200 \\ -140 & 1680 & -4200 & 2800 \end{pmatrix}$$

and a condition number of $C(A) = (25/12)(13620) = 28375$. This puts it squarely at the bottom of the ladder.

17.80 What is the condition of these matrices?

$$P = \begin{pmatrix} 0 & 0 & 1 & 0 \\ 1 & 0 & 0 & 0 \\ 0 & 0 & 0 & 1 \\ 0 & 1 & 0 & 0 \end{pmatrix} \qquad D = \begin{pmatrix} 1 & 0 & 0 & 0 \\ 0 & 2 & 0 & 0 \\ 0 & 0 & 3 & 0 \\ 0 & 0 & 0 & 4 \end{pmatrix} \qquad U = \begin{pmatrix} 1 & 1 & 1 & 1 \\ 0 & 1 & 1 & 1 \\ 0 & 0 & 1 & 1 \\ 0 & 0 & 0 & 1 \end{pmatrix}$$

▌ Taking them from left to right, any permutation matrix has as inverse another permutation matrix. Using the infinity norm, all row sums are 1 and so is $C(P)$. This holds for any order. For the diagonal matrix D, the inverse is also diagonal and has the reciprocals of $1, 2, 3, 4$ along it. So $C(D) = 4$. In general, it will be the ratio of the largest to the smallest absolute diagonal entry, for any order. The upper triangle is easily verified to have for its inverse a main diagonal of 1s and a next higher diagonal of -1s, elsewhere zero. This makes $C(U) = 4 \cdot 2 = 8$. For order n we would have $C(U) = 2n$.

17.81 Compare the condition of these fifth order matrices:

$$\begin{pmatrix} 1 & 1 & 1 & 1 & 1 \\ 1 & 2 & 3 & 4 & 5 \\ 1 & 3 & 6 & 10 & 15 \\ 1 & 4 & 10 & 20 & 35 \\ 1 & 5 & 15 & 35 & 70 \end{pmatrix} \qquad \begin{pmatrix} -2 & 1 & 0 & 0 & 0 \\ 1 & -2 & 1 & 0 & 0 \\ 0 & 1 & -2 & 1 & 0 \\ 0 & 0 & 1 & -2 & 1 \\ 0 & 0 & 0 & 1 & -2 \end{pmatrix}$$

▌ Their origins are clear. The inverses are computed easily:

$$\begin{pmatrix} 5 & -10 & 10 & -5 & 1 \\ -10 & 30 & -35 & 19 & -4 \\ 10 & -35 & 46 & -27 & 6 \\ -5 & 19 & -27 & 17 & -4 \\ 1 & -4 & 6 & -4 & 1 \end{pmatrix} \qquad -\frac{1}{6}\begin{pmatrix} 5 & 4 & 3 & 2 & 1 \\ 4 & 8 & 6 & 4 & 2 \\ 3 & 6 & 9 & 6 & 3 \\ 2 & 4 & 6 & 8 & 4 \\ 1 & 2 & 3 & 4 & 5 \end{pmatrix}$$

The condition numbers are $(126)(124)$ and $(4)(27)$.

17.82 How easily does the Hilbert matrix of order 5 invert?

▌ Computing in single precision, its huskiest row turns out to be

$$-1399 \qquad 26862 \qquad -117519 \qquad 179076 \qquad -88139$$

more or less. The alternation of signs occurs throughout and is always a bad omen. When this pseudoinverse is multiplied by the original, the diagonal elements range from .85 to 1.08, while off-diagonal elements rise to size .4, certainly not very close to the identity. The condition number runs to six figures.

17.83 Derive the formula $A^{-1} = (I + R + R^2 + \cdots)B$ where $R = I - BA$.

▮ The idea here is that B is an approximate inverse of A, so that the residual R has small elements. A few terms of the series involved may therefore be enough to produce a much better approximation to A^{-1}. To derive the formula note first that $(I - R)(I + R + R^2 + \cdots) = I$ provided the matrix series is convergent. Then $I + R + R^2 + \cdots = (I - R)^{-1}$ and so

$$(I + R + R^2 + \cdots)B = (I - R)^{-1}B = (BA)^{-1}B = A^{-1}B^{-1}B$$

which reduces to A^{-1}.

17.84 Apply the formula of the preceding problem to the matrix

$$A = \begin{bmatrix} 1 & 10 & 1 \\ 2 & 0 & 1 \\ 3 & 3 & 2 \end{bmatrix}$$

assuming only a three digit computer is available. Since any computer carries only a limited number of digits, this will again illustrate the power of a method of successive corrections.

▮ First we apply Gaussian elimination to obtain a first approximation to the inverse. The three steps, using the largest pivot available in each case, appear below along with the approximate inverse B, which results from two interchanges of rows, bringing the bottom row to the top.

.1	1	.1	.1	0	0		0	1	.037	.111	0	−.0371
2.0	0	1.0	0	1	0		0	0	−.260	.222	1	−.742
2.7	0	1.7	−.3	0	1		1	0	.630	−.111	0	.371

Step 1   Step 2

0	1	0	.143	.143	−.143		
0	0	1	−.854	−3.85	2.85		
1	0	0	.427	2.43	−1.43		

Step 3

$$\begin{bmatrix} .427 & 2.43 & -1.43 \\ .143 & .143 & -.143 \\ -.854 & -3.85 & 2.85 \end{bmatrix}$$

The Matrix B

Next we easily compute

$$R = I - BA = \begin{bmatrix} .003 & .020 & .003 \\ 0 & -.001 & 0 \\ .004 & -.010 & .004 \end{bmatrix}$$

after which RB, $B + RB$, $R^2B = R(RB)$ and $B + RB + R^2B$ are found in that order. (Notice that because the elements in R^2B are so small, a factor of 10000 has been introduced for simplicity in presentation.)

$$\begin{bmatrix} .001580 & -.001400 & .001400 \\ -.000143 & -.000143 & .000143 \\ -.003140 & -.007110 & .007110 \end{bmatrix}$$

RB

$$\begin{bmatrix} .428579 & 2.428600 & -1.428600 \\ .142857 & .142857 & -.142857 \\ -.857138 & -3.857110 & 2.857110 \end{bmatrix}$$

$B + RB$

$$\begin{bmatrix} -.07540 & -.28400 & .28400 \\ .00143 & .00143 & -.00143 \\ -.04810 & -.32600 & .32600 \end{bmatrix}$$

$10^4 \cdot R(RB)$

$$\begin{bmatrix} .4285715 & 2.4285716 & -1.4285716 \\ .1428571 & .1428571 & -.1428571 \\ -.8571428 & -3.8571426 & 2.8571426 \end{bmatrix}$$

$B + RB + R^2B$

Notice that except in the additive processes, only three significant digits have been carried. Since the exact inverse is

$$A^{-1} = \tfrac{1}{7} \begin{bmatrix} 3 & 17 & -10 \\ 1 & 1 & -1 \\ -6 & -27 & 20 \end{bmatrix}$$

it can be verified that $B + RB + R^2B$ is at fault only in the seventh decimal place. More terms of the series formula would bring still further accuracy. This method can often be used to improve the result of inversion by Gaussian elimination, since that algorithm is far more sensitive to roundoff error accumulation.

17.85   Use the correction method of Problem 17.83 with the random walk matrix of Problem 17.77, experimenting with various distortions of the inverse. Are they corrected?

    I Deleting the minus signs of the bottom row .2 and .4 in the inverse, we find $B + RB$ restoring them perfectly. But the same change applied to the .6 or .8 led to a blowup of these entries. Changing all four $-.8$ values to either $-.75$ or $-.85$, we find five terms of the series correcting these elements to four places but introducing similar errors in all the other elements. Nine terms are enough to restore the entire inverse. A variety of other small distortions were similarly removed.

17.86   Also experiment with the binomial coefficient matrix of Problem 17.77 and its inverse. How well does the series method correct flaws in the inverse?

    I Changing the top row of the inverse to $4.5, -6.5, 3.5, -.5$, we find the two terms $B + RB$ enough to make perfect corrections. Similar changes in the center four elements led to oscillations, but smaller errors were quickly removed. Errors larger than this usually led to blowup or other deterioration.

17.87   Try the same method on the Wilson matrix of Problem 17.78, using the approximate inverse obtained by the exchange method. Does it remove the errors in the last two places?

    I No, these decimal places remain untrustworthy.

17.7 DETERMINANTS

17.88   Determinants are no longer used extensively in the solution of linear systems, but continue to have application in other ways. Direct evaluation of a determinant of order n would require the computation of $n!$ terms, which is prohibitive except for small n. What is the alternative?

    I From the properties of determinants, no step in a Gaussian elimination alters the determinant of the coefficient matrix except normalization and interchanges. If these were not performed, the determinant is available by multiplication of the diagonal elements after triangularization. For the matrix of Problem 17.2 the determinant is, therefore, a quick $(\frac{1}{12})(\frac{1}{180}) = \frac{1}{2160}$. This small value is another indication of the troublesome character of the matrix. Determinants can also be found from the factorization $PA = LU$. Since $A = P^{-1}LU$ we have

$$\det(A) = \det(P^{-1})\det(L)\det(U) = (-1)^{P}\det(U)$$

where p is the number of interchanges represented by the permutation matrix P, or P^{-1}. For the matrix of Problem 17.21,

$$\det(U) = 3(3)\left(\tfrac{28}{9}\right)\left(\tfrac{24}{7}\right) = 96$$

while $\det(P)$ is easily found to be -1. (Or recall that three interchanges were made during factorization, making $p = 3$.) Thus

$$\det(A) = -96$$

17.89   Find the determinant of the binomial coefficient matrix in Problem 17.81.

    I Gaussian elimination triangularizes this matrix using one row interchange and finds the diagonal elements 1, 4, -2, $-\frac{1}{2}$ and $-\frac{1}{4}$. Their product, together with a minus for the interchange, is 1. All the matrices of this type have this same determinant.

17.90   Find the determinant of the random walk matrix of Problem 17.81.

    I No interchanges are needed in the elimination process and the following diagonal elements result:

$$-2, \qquad -\tfrac{3}{2}, \qquad -\tfrac{4}{3}, \qquad -\tfrac{5}{4}, \qquad -\tfrac{6}{5}$$

Their product is -6. The fourth order determinant also appearing in earlier problems has determinant 5, while the still smaller third order version has -4. A more general result is certainly suspected and easy enough to prove. Let D_n be the determinant of order n and expand it by minors of the last column to discover this recursion, from which the general result follows:

$$D_n = -2D_{n-1} - D_{n-2}$$

17.91 What is the determinant of Wilson's matrix?

∎ The Gauss method uses two interchanges to triangularize this matrix and leaves diagonal elements that strongly resemble

$$10, \quad .4, \quad 2.5, \quad .1$$

to four places. These are in fact correct, the remaining places being contaminated by roundoff. The determinant is 1.

17.8 EIGENVALUE PROBLEMS: THE CHARACTERISTIC POLYNOMIAL

17.92 What are eigenvalues and eigenvectors of a matrix A?

∎ A number λ for which the system $Ax = \lambda x$ or $(A - \lambda I)x = 0$ has a nonzero solution vector x is called an eigenvalue of the system. Any corresponding nonzero solution vector x is called an eigenvector. Clearly, if x is an eigenvector then so is Cx for any number C.

17.93 Find the eigenvalues and eigenvectors of the system

$$
\begin{aligned}
(2-\lambda)x_1 - \quad\quad x_2 \quad\quad\quad\quad &= 0 \\
-x_1 + (2-\lambda)x_2 - \quad\quad x_3 &= 0 \\
-x_2 + (2-\lambda)x_3 &= 0
\end{aligned}
$$

which arises in various physical settings, including the vibration of a system of three masses connected by springs.

∎ We illustrate the method of finding the *characteristic polynomial* directly and then obtaining the eigenvalues as roots of this polynomial. The eigenvectors are then found last. The first step is to take linear combinations of equations much as in Gaussian elimination, until only the x_3 column of coefficients involves λ. For example, if E_1, E_2 and E_3 denote the three equations, then $-E_2 + \lambda E_3$ is the equation

$$x_1 - 2x_2 + (1 + 2\lambda - \lambda^2)x_3 = 0$$

Calling this E_4, the combination $E_1 - 2E_2 + \lambda E_4$ becomes

$$4x_1 - 5x_2 + (2 + \lambda + 2\lambda^2 - \lambda^3)x_3 = 0$$

These last two equations together with E_3 now involve λ in only the x_3 coefficients.

The second step of the process is to triangularize this system by the Gauss elimination algorithm or its equivalent. With this small system we may take a few liberties as to pivots, retain

$$
\begin{aligned}
x_1 - 2x_2 + (1 + 2\lambda - \lambda^2)x_3 &= 0 \\
-x_2 + \quad\quad (2-\lambda)x_3 &= 0
\end{aligned}
$$

as our first two equations and soon achieve

$$(4 - 10\lambda + 6\lambda^2 - \lambda^3)x_3 = 0$$

to complete the triangularization. To satisfy the last equation we must avoid making $x_3 = 0$, because this at once forces $x_2 = x_1 = 0$ and we do not have a nonzero solution vector. Accordingly we must require

$$4 - 10\lambda + 6\lambda^2 - \lambda^3 = 0$$

This cubic is the *characteristic polynomial* and the eigenvalues must be its zeros since in no other way can we obtain a nonzero solution vector. By methods of an earlier chapter we find those eigenvalues to be $\lambda_1 = 2 - \sqrt{2}$, $\lambda_2 = 2$, $\lambda_3 = 2 + \sqrt{2}$ in increasing order.

The last step is to find the eigenvectors, but with the system already triangularized this involves no more than back-substitution. Taking λ_1 first and recalling that eigenvectors are determined only to an arbitrary multiplier so that we may choose $x_3 = 1$, we find $x_2 = \sqrt{2}$ and then $x_1 = 1$. The other eigenvectors are found in the same way, using λ_2 and λ_3. The final results are

λ	x_1	x_2	x_3
$2 - \sqrt{2}$	1	$\sqrt{2}$	1
2	-1	0	1
$2 + \sqrt{2}$	1	$-\sqrt{2}$	1

In this case the original system of three equations has three distinct eigenvalues, to each of which there corresponds one independent eigenvector. This is the simplest, but not the only, possible outcome of an eigenvalue problem. It should be noted that the present matrix is both real and symmetric. For a real, symmetric $n \times n$ matrix an important theorem of algebra states that

1 All eigenvalues are real, though perhaps not distinct.
2 n independent eigenvectors always exist.

This is not true of all matrices. It is fortunate that many of the matrix problems that computers currently face are real and symmetric.

17.94 To make the algorithm for direct computation of the characteristic polynomial more clear, apply it to this larger system:

$$
\begin{aligned}
E_1: \quad & (1-\lambda)x_1 + & x_2 + & x_3 + & x_4 = 0 \\
E_2: \quad & x_1 + (2-\lambda)x_2 + & 3x_3 + & 4x_4 = 0 \\
E_3: \quad & x_1 + & 3x_2 + (6-\lambda)x_3 + & 10x_4 = 0 \\
E_4: \quad & x_1 + & 4x_2 + & 10x_3 + (20-\lambda)x_4 = 0
\end{aligned}
$$

▮ Calling these equations E_1, E_2, E_3, E_4, the combination $E_1 + 4E_2 + 10E_3 + \lambda E_4$ is

$$15x_1 + 39x_2 + 73x_3 + (117 + 20\lambda - \lambda^2)x_4 = 0$$

and is our second equation in which all but the x_4 term are free of λ. We at once begin triangularization by subtracting $15E_4$ to obtain

$$E_5: \quad -21x_2 - 77x_3 + (-183 + 35\lambda - \lambda^2)x_4 = 0$$

The combination $-21E_2 - 77E_3 + \lambda E_5$ becomes

$$-98x_1 - 273x_2 - 525x_3 + (-854 - 183\lambda + 35\lambda^2 - \lambda^3)x_4 = 0$$

and is our third equation in which all but the x_4 term are free of λ. The triangularization continues by blending this last equation with E_4 and E_5 to obtain

$$E_6: \quad 392x_3 + (1449 - 1736\lambda + 616\lambda^2 - 21\lambda^3)x_4 = 0$$

Now the combination $392E_3 + \lambda E_6$ is formed,

$$392x_1 + 1176x_2 + 2352x_3 + (3920 + 1449\lambda - 1736\lambda^2 + 616\lambda^3 - 21\lambda^4)x_4 = 0$$

and the triangularization is completed by blending this equation with E_4, E_5 and E_6 to obtain

$$E_7: \quad (1 - 29\lambda + 72\lambda^2 - 29\lambda^3 + \lambda^4)x_4 = 0$$

The system E_4, E_5, E_6, E_7 is now the triangular system we have been aiming for. To avoid the zero solution vector, λ must be a zero of $1 - 29\lambda + 72\lambda^2 - 29\lambda^3 + \lambda^4$, which is the characteristic polynomial.

17.95 The characteristic equation just found has four real roots, since the matrix is real and symmetric. One of these roots is near 2. Refine this estimate and then find the corresponding eigenvector.

▮ Newton's method quickly settles on 2.203446. (It also finds two smaller eigenvalues .453835 and .038016 and the dominant 26.3047.) Since we have a triangular system in hand, it is easy enough to backtrack via E_6 and E_5 to obtain the designated eigenvector:

$$x = (-1.346453, -1.625634, -.9947566, 1)^T$$

As a check, both Ax and λx can be computed. They agree to six figures. The other three eigenvectors respond to the same treatment.

17.96 Illustrate the use of the Cayley–Hamilton theorem for finding the characteristic equation of a matrix.

▌ Writing the equation as

$$f(\lambda) = \lambda^n + c_1 \lambda^{n-1} + \cdots + c_{n-1}\lambda + c_n = 0$$

the Cayley–Hamilton theorem states that the matrix A itself satisfies this equation. That is,

$$f(A) = A^n + c_1 A^{n-1} + \cdots + c_{n-1}A + c_n I = 0$$

where the right side is now the zero matrix. This comes to n^2 equations for the n coefficients c_i so there is substantial redundancy.

Take, for example, the Fibonacci matrix $F = \begin{bmatrix} 1 & 1 \\ 1 & 0 \end{bmatrix}$. Since $F^2 = \begin{bmatrix} 2 & 1 \\ 1 & 1 \end{bmatrix}$, we have

$$\begin{bmatrix} 2 & 1 \\ 1 & 1 \end{bmatrix} + c_1 \begin{bmatrix} 1 & 1 \\ 1 & 0 \end{bmatrix} + c_2 \begin{bmatrix} 1 & 0 \\ 0 & 1 \end{bmatrix} = \begin{bmatrix} 0 & 0 \\ 0 & 0 \end{bmatrix}$$

or

$$2 + c_1 + c_2 = 0$$
$$1 + c_1 = 0 \qquad 1 + c_2 = 0$$

with the second of these repeated. The familiar equation $\lambda^2 = \lambda + 1$ is again in hand. (See Problems 10.47 and 17.146.)

Or consider the permutation matrix P with

$$P = \begin{bmatrix} 0 & 0 & 1 \\ 1 & 0 & 0 \\ 0 & 1 & 0 \end{bmatrix} \qquad P^2 = \begin{bmatrix} 0 & 1 & 0 \\ 0 & 0 & 1 \\ 1 & 0 & 0 \end{bmatrix} \qquad P^3 = \begin{bmatrix} 1 & 0 & 0 \\ 0 & 1 & 0 \\ 0 & 0 & 1 \end{bmatrix}$$

which leads quickly to the set

$$1 + c_3 = 0 \qquad c_1 = 0 \qquad c_2 = 0$$

repeated twice. The characteristic equation is $\lambda^3 - 1 = 0$.

Several devices have been suggested for selecting a suitable subset of the available n^2 equations. One such device calls for computing

$$f(A)v = 0$$

for an appropriate vector v and solving this system.

17.97 Use the Cayley–Hamilton theorem to find the characteristic equation of the familiar matrix A at the left in Fig. 17-7.

$$\begin{pmatrix} -2 & 1 & 0 & 0 \\ 1 & -2 & 1 & 0 \\ 0 & 1 & -2 & 1 \\ 0 & 0 & 1 & -2 \end{pmatrix} \begin{pmatrix} 5 & -4 & 1 & 0 \\ -4 & 6 & -4 & 1 \\ 1 & -4 & 6 & -4 \\ 0 & 1 & -4 & 5 \end{pmatrix}$$

$$\begin{pmatrix} -14 & 14 & -6 & 1 \\ 14 & -20 & 15 & -6 \\ -6 & 15 & -20 & 14 \\ 1 & -6 & 14 & -14 \end{pmatrix} \begin{pmatrix} 42 & -48 & 27 & -8 \\ -48 & 69 & -56 & 27 \\ 27 & -56 & 69 & -48 \\ -8 & 27 & -48 & 42 \end{pmatrix} \quad \textbf{Fig. 17-7}$$

▌ The other matrices in the figure are the needed powers of A. The top rows are sufficient for our purposes, leading to the set of equations

$$\begin{aligned} 42 - 14c_1 + 5c_2 - 2c_3 + c_4 &= 0 \\ -48 + 14c_1 - 4c_2 + c_3 &= 0 \\ 27 - 6c_1 + c_2 &= 0 \\ -8 + c_1 &= 0 \end{aligned}$$

The solution is $(8, 21, 20, 5)$ making the characteristic equation

$$\lambda^4 + 8\lambda^3 + 21\lambda^2 + 20\lambda + 5 = 0$$

The eigenvalues were computed as

$$-3.618035, \qquad -2.618033, \qquad -1.381966, \qquad -.381966$$

from which one may speculate as to their true identities.

17.98 Repeat the preceding problem using the fifth order random walk matrix.

❚ The triple diagonal pattern of the preceding A has to be extended to one more row and column. The top two rows of the needed power matrices are then as shown in Fig. 17-8.

The top rows alone are again enough for our purposes, leading to the equations

$$
\begin{array}{r}
-132 + 42c_1 - 14c_2 + 5c_3 - 2c_4 + c_5 = 0 \\
165 - 48c_1 + 14c_2 - 4c_3 + \ c_4 \qquad = 0 \\
-110 + 27c_1 - \ 6c_2 + \ c_3 \qquad\qquad = 0 \\
44 - \ 8c_1 + \ c_2 \qquad\qquad\qquad = 0 \\
-10 + \ c_1 \qquad\qquad\qquad\qquad = 0
\end{array}
$$

The solution $(c_1, c_2, c_3, c_4, c_5) = (10, 36, 56, 35, -192)$ can be checked by using one of the second row positions to write still another equation. The characteristic equation is

$$\lambda^5 + 10\lambda^4 + 36\lambda^3 + 56\lambda^2 + 35\lambda - 192 = 0$$

$$
A^2 = \begin{pmatrix} 5 & -4 & 1 & 0 & 0 \\ -4 & 6 & -4 & 1 & 0 \\ & & \cdots & & \end{pmatrix}
\qquad
A^3 = \begin{pmatrix} -14 & 14 & -6 & 1 & 0 \\ 14 & -20 & 15 & -6 & 1 \\ & & \cdots & & \end{pmatrix}
$$

$$
A^4 = \begin{pmatrix} 42 & -48 & 27 & -8 & 1 \\ -48 & 69 & -56 & 28 & -8 \\ & & \cdots & & \end{pmatrix}
\qquad
A^5 = \begin{pmatrix} -132 & 165 & -110 & 44 & -10 \\ 165 & -242 & 209 & -120 & 44 \\ & & \cdots & & \end{pmatrix}
\qquad \textbf{Fig. 17-8}
$$

17.99 Find the exact characteristic equation of the Hilbert matrix of order 3.

❚ Suppose we try once more the method of Problem 17.93, starting with

$$
\begin{array}{ll}
E_1: & (1 - \lambda)x_1 + \ \tfrac{1}{2}x_2 + \ \tfrac{1}{3}x_3 = 0 \\
E_2: & \tfrac{1}{2}x_1 + \left(\tfrac{1}{3} - \lambda\right)x_2 + \ \tfrac{1}{4}x_3 = 0 \\
E_3: & \tfrac{1}{3}x_1 + \ \tfrac{1}{4}x_2 + \left(\tfrac{1}{5} - \lambda\right)x_3 = 0
\end{array}
$$

The combination $\tfrac{1}{3}E_1 + \tfrac{1}{4}E_2 + \lambda E_3$ is

$$\tfrac{11}{24}x_1 + \tfrac{1}{4}x_2 + \left(\tfrac{25}{144} + \tfrac{1}{5}\lambda - \lambda^2\right)x_3 = 0$$

and subtracting $11E_3/8$ begins the triangularization.

$$E_4: \qquad -\tfrac{3}{32}x_2 + \left(-\tfrac{73}{720} + \tfrac{63}{40}\lambda - \lambda^2\right)x_3 = 0$$

Now the combination $-\tfrac{3}{32}E_2 + \lambda E_4$ is

$$-\tfrac{3}{64}x_1 - \tfrac{1}{32}x_2 + \left(-\tfrac{3}{128} - \tfrac{73}{720}\lambda + \tfrac{63}{40}\lambda^2 - \lambda^3\right)x_3 = 0$$

and a bit of a struggle completes the triangularization:

$$\tfrac{1}{256}x_2 + \left(\tfrac{3}{640} - \tfrac{697}{2880}\lambda + \tfrac{63}{40}\lambda^2 - \lambda^3\right)x_3 = 0$$

$$\left(\tfrac{1}{2160} - \tfrac{127}{720}\lambda + \tfrac{23}{15}\lambda^2 - \lambda^3\right)x_3 = 0$$

The characteristic equation results from setting this last coefficient to zero.

17.9 GERSCHGORIN'S THEOREM

17.100 Prove Gerschgorin's theorem, which states that every eigenvalue of a matrix A falls within one of the complex circles having centers at a_{ii} and radii

$$R_i = \sum_{j \neq i} |a_{ij}|$$

with $i = 1, \ldots, n$.

▌ Let x_i be the component of largest magnitude of one of the eigenvectors of A. From the ith equation of the system $(A - \lambda I)x = 0$, we have

$$(a_{ii} - \lambda)x_i = -\sum_{j \neq i} a_{ij}x_j$$

$$|a_{ii} - \lambda| \leq \sum_{j \neq i} |a_{ij}| \left| \frac{x_j}{x_i} \right| \leq \sum_{j \neq i} |a_{ij}|$$

which is the theorem.

17.101 What does the Gerschgorin theorem tell us about the eigenvalues of a permutation matrix that has a single 1 in each row and column, with zeros elsewhere?

▌ The circles either have center at 0 with radius 1 or center at 1 with radius 0. All eigenvalues lie within a unit of the origin. For example, the eigenvalues of

$$\begin{bmatrix} 0 & 0 & 1 \\ 1 & 0 & 0 \\ 0 & 1 & 0 \end{bmatrix}$$

and the cube roots of 1. In particular, the eigenvalues of the identity matrix must be within the circle having center at 1 and radius 0.

17.102 The Gerschgorin theorem is especially useful for matrices having a dominant diagonal. Apply it to the matrix

$$\begin{bmatrix} 4 & -1 & -1 & 0 \\ -1 & 4 & -1 & -1 \\ -1 & -1 & 4 & -1 \\ 0 & -1 & -1 & 4 \end{bmatrix}$$

▌ All the eigenvalues must fall inside the circle with center at 4 and radius 3. By the symmetry, they must also be real.

17.103 What does Gerschgorin's theorem have to say about the eigenvalues in Problem 17.94?

▌ The four circles have the (center, radius) pairs

$$(1, 3) \qquad (2, 8) \qquad (6, 14) \qquad (20, 15)$$

They are not too helpful, but the last does suggest a possibly large root, which we found to be near 26.

17.104 Compare the eigenvalues found in Problem 17.97 with the forecast of the Gerschgorin theorem.

▌ The theorem requires them to be no farther than two units from -2. We found the smallest of the four closely packed roots near -3.6 and the largest near $-.4$.
For the corresponding fifth order matrix, the characteristic equation found in Problem 17.98 must have all five roots in this same $(-4, 0)$ interval.

17.105 Can a Hilbert matrix have large eigenvalues?

▌ Not for small order, but the harmonic series does diverge, so for large order n, the Gerschgorin theorem allows the entire real axis out to something like $\log n$.

17.10 THE POWER METHOD

17.106 What is the power method for producing the dominant eigenvalue and eigenvector of a matrix?

▮ Assume that the matrix A is of size $n \times n$ with n independent eigenvectors $V_1, V_2, \ldots, V_n$ and a truly dominant eigenvalue λ_1: $|\lambda_1| > |\lambda_2| \geq \cdots \geq |\lambda_n|$. Then an arbitrary vector V can be expressed as a combination of eigenvectors,

$$V = a_1 V_1 + a_2 V_2 + \cdots + a_n V_n$$

It follows that

$$AV = a_1 AV_1 + a_2 AV_2 + \cdots + a_n AV_n = a_1 \lambda_1 V_1 + a_2 \lambda_2 V_2 + \cdots + a_n \lambda_n V_n$$

Continuing to multiply by A we arrive at

$$A^p V = a_1 \lambda_1^p V_1 + a_2 \lambda_2^p V_2 + \cdots + a_n \lambda_n^p V_n = \lambda_1^p \left[a_1 V_1 + a_2 \left(\frac{\lambda_2}{\lambda_1} \right)^p V_2 + \cdots + a_n \left(\frac{\lambda_n}{\lambda_1} \right)^p V_n \right]$$

provided $a_1 \neq 0$. Since λ_1 is dominant, all terms inside the brackets have limit zero except the first term. If we take the ratio of any corresponding components of $A^{p+1} V$ and $A^p V$, this ratio should therefore have limit λ_1. Moreover, $\lambda_1^{-p} A^p V$ will converge to the eigenvector $a_1 V_1$.

17.107 Apply the power method to find the dominant eigenvalue and eigenvector of the matrix used in Problem 17.93:

$$A = \begin{bmatrix} 2 & -1 & 0 \\ -1 & 2 & -1 \\ 0 & -1 & 2 \end{bmatrix}$$

▮ Choose the initial vector $V = (1, 1, 1)$. Then $AV = (1, 0, 1)$ and $A^2 V = (2, -2, 2)$. It is convenient here to divide by 2 and in the future we continue to divide by some suitable factor to keep the numbers reasonable. In this way we find

$$A^7 V = c(99, -140, 99) \qquad A^8 V = c(338, -478, 338)$$

where c is some factor. The ratios of components are

$$\tfrac{338}{99} \simeq 3.41414 \qquad \tfrac{478}{140} \simeq 3.41429$$

and we are already close to the correct $\lambda_1 = 2 + \sqrt{2} \simeq 3.414214$. Dividing our last output vector by 338, it becomes $(1, -1.41420, 1)$ approximately and this is close to the correct $(1, -\sqrt{2}, 1)$ found in Problem 17.93.

17.108 What is the Rayleigh quotient and how may it be used to find the dominant eigenvalue?

▮ The Rayleigh quotient is $x^T A x / x^T x$, where T denotes the transpose. If $Ax = \lambda x$ this collapses to λ. If $Ax \simeq \lambda x$ then it is conceivable that the Rayleigh quotient is approximately λ. Under certain circumstances the Rayleigh quotients for the successive vectors generated by the power method converge to λ_1. For example, let x be the last output vector of the preceding problem, $(1, -1.41420, 1)$. Then

$$Ax = (3.41420, -4.82840, 3.41420) \qquad x^T A x = 13.65672 \qquad x^T x = 3.99996$$

and the Rayleigh quotient is 3.414214 approximately. This is correct to six decimal places, suggesting that the convergence to λ_1 here is more rapid than for ratios of components.

17.109 Assuming all eigenvalues are real, how may the other extreme eigenvalue be found?

▮ If $Ax = \lambda x$, then $(A - qI)x = (\lambda - q)x$. This means that $\lambda - q$ is an eigenvalue of $A - qI$. By choosing q properly, perhaps $q = \lambda_1$, we make the other extreme eigenvalue dominant and the power method can be applied. For the matrix of Problem 17.108 we may choose $q = 4$ and consider

$$A - 4I = \begin{bmatrix} -2 & -1 & 0 \\ -1 & -2 & -1 \\ 0 & -1 & -2 \end{bmatrix}$$

Again taking $V = (1, 1, 1)$ we soon find the Rayleigh quotient -3.414214 for the vector $(1, 1.41421, 1)$ which is essentially $(A - 4I)^8V$. Adding 4 we have $.585786$, which is the other extreme eigenvalue $2 - \sqrt{2}$ correct to six places. The vector is also close to $(1, \sqrt{2}, 1)$, the correct eigenvector.

17.110 How may the absolutely smallest eigenvalue be found by the power method?

❚ If $Ax = \lambda x$, then $A^{-1}x = \lambda^{-1}x$. This means that the absolutely smallest eigenvalue of A can be found as the reciprocal of the dominant λ of A^{-1}. For the matrix of Problem 17.108 we first find

$$A^{-1} = \tfrac{1}{4} \begin{bmatrix} 3 & 2 & 1 \\ 2 & 4 & 2 \\ 1 & 2 & 3 \end{bmatrix}$$

Again choosing $V = (1, 1, 1)$ but now using A^{-1} instead of A, we soon find the Rayleigh quotient 1.707107 for the vector $(1, 1.41418, 1)$. The reciprocal quotient is $.585786$ so that we again have this eigenvalue and vector already found in Problems 17.93 and 17.109. Finding A^{-1} is ordinarily no simple task, but this method is sometimes the best approach to the absolutely smallest eigenvalue.

17.111 How may a next dominant eigenvalue be found by a suitable choice of starting vector V?

❚ Various algorithms have been proposed, with varying degrees of success. The difficulty is to sidetrack the dominant eigenvalue itself and to keep it sidetracked. Roundoff errors have spoiled several theoretically sound methods by returning the dominant eigenvalue to the main line of the computation and obscuring the next dominant or limiting the accuracy to which this runnerup can be determined. For example, suppose that in the argument of Problem 17.106 it could be arranged that the starting vector V is such that a_1 is zero. Then λ_1 and V_1 never actually appear and if λ_2 dominates the remaining eigenvalues it assumes the role formerly played by λ_1 and the same reasoning proves convergence to λ_2 and V_2. With our matrix of Problem 17.107 this can be nicely illustrated. Being real and symmetric, this matrix has the property that its eigenvectors are orthogonal. (Problem 17.93 allows a quick verification of this.) This means that $V_1^T V = a_1 V_1^T V_1$ so that a_1 will be zero if V is orthogonal to V_1. Suppose we take $V = (-1, 0, 1)$. This is orthogonal to V_1. At once we find $AV = (-2, 0, 2) = 2V$, so that we have the exact $\lambda_2 = 2$ and $V_2 = (-1, 0, 1)$. However, our choice of starting vector here was fortunate.

It is almost entertaining to watch what happens with a reasonable but not so fortunate V, say $V = (0, 1, 1.4142)$, which is also orthogonal to V_1 as required. Then we soon find $A^3 V \simeq 4.8(-1, .04, 1.20)$, which is something like V_2 and from which the Rayleigh quotient yields the satisfactory $\lambda_2 \simeq 1.996$. After this, however, the computation deteriorates and eventually we come to $A^{20}V \simeq c(1, -1.419, 1.007)$, which offers us good approximations once again to λ_1 and V_1. Roundoff errors have brought the dominant eigenvalue back into action. By taking the trouble to alter each vector $A^p V$ slightly, to make it orthogonal to V_1, a better result can be achieved. Other devices also have been attempted using several starting vectors.

17.112 Use the power method to reproduce the dominant eigenvalue 26.3047 found in Problem 17.95 for the fourth order matrix A of binomial coefficients.

❚ Choosing the impartial initial vector $x = (1, 1, 1, 1)^T$, the first step can be handled mentally, producing the vector of row sums $Ax = (4, 10, 20, 35)^T$. The ratios of output to input for the four components give estimates from 4 to 35 for the eigenvalue, but the Rayleigh quotient is already 26.21. Normalizing each output by dividing through by its first element, we find the third power of A producing a Rayleigh quotient of 26.3047. With the eighth power of A, the normalized output vector has converged to the eigenvector

$$(1, 3.342476, 7.61102, 14.35121)$$

Using A^{-1} to find the absolutely smallest eigenvalue brings an interesting result. The Rayleigh quotient corresponding to power 5 is the same 26.3047, meaning that this eigenvalue is its reciprocal, the $.038016$ found in Problem 17.95. The corresponding eigenvector is $(1, -2.34248, 1.92607, -.545576)$. Moreover, the dominant and absolutely smallest eigenvalues of the next larger matrix of this type, shown in Problem 17.81, are also reciprocals, the first being 92.29044.

17.113 Apply the power method to the random walk matrix of Problem 17.97.

❚ The dominant eigenvalue was found to be $\lambda_1 = -3.618035$, but making the usual impartial start with $x = (1, 1, 1, 1)$ we get a surprise. The Rayleigh quotients converge quickly to -2.618034, which is the second largest eigenvalue λ_2. This can mean only one thing, the choice $(1, 1, 1, 1)$ must be orthogonal to an eigenvector belonging to λ_1. The eigenvector produced was

$$(1, -.618026, -.618026, 1)$$

Trying the new starting vector $(0, 1, 0, 0)$ then led very slowly (power 50 needed) to λ_1 and its eigenvector

$$(1, -1.618034, 1.618034, -1)$$

which does prove to be orthogonal to $(1, 1, 1, 1)$. Curiosity having been aroused, the starts $(0, 0, 1, 0)$ and $(0, 0, 0, 1)$ were also tried. Both led to overflow. $(1, 0, 0, 1)$ was also tried and of course led back to λ_2. Just why the dominant eigenvalue chose to remain sidetracked was not explored.

17.114 Apply the power method to Wilson's matrix.

▌ The seventh power of this matrix was sufficient to determine the dominant 30.2887 and its companion vector,

$$(1, .719420, 1.04425, .98554)$$

Only the fifth power of A^{-1} was needed to find the absolutely smallest $1/98.6248$ and $(1, -1.65575, .425897, -.246678)$.

17.115 Apply the power method to the Hilbert matrix of order 3. See also Problem 17.99.

▌ Seventh power is enough to find $\lambda_1 = 1.408319$ and its eigenvector $(1, .5560325, .3909079)$. Six powers of the inverse also determine the small eigenvalue $1/372.1151 = .00268734$ and the companion vector $(1, -5.59103, 5.39460)$.

The third eigenvalue and vector of this matrix respond well to the method of Problem 17.111. From the usual starting vector $(1, 1, 1)$ we first subtract the component parallel to the λ_1 eigenvector V_1. The result $(-.3317137, .2595239, .4794226)$ is then orthogonal to V_1 and the a_1 coefficient in Problem 17.106 is zero. The first power step now finds a Rayleigh quotient of .1223255. Newton's method, applied to the characteristic equation developed in Problem 17.99, finds this eigenvalue to be .1223271, so things are certainly off to a good start. The power step also produces this first approximation to the companion eigenvector,

$$(1, -.9611477, -1.191013)$$

Removing the component parallel to V_1 modifies this slightly to

$$(1.000004, -.9611455, -1.191012)$$

and we are ready for another cycle. After the fourth round of approximations there is little further change, the Newton value of .1223271 being reproduced along with this eigenvector,

$$(1, -.9650062, -1.185513)$$

in which the final two places of the middle element remain a bit unsteady.

17.116 Develop the inverse power method.

▌ This is an extension of the eigenvalue shift used in Problem 17.109. If A has eigenvalues λ_i, then $A - tI$ and $(A - tI)^{-1}$ have eigenvalues $\lambda_i - t$ and $(\lambda_i - t)^{-1}$, respectively. Applying the power method as in Problem 17.106, but using $(A - tI)^{-1}$ in place of A, we have

$$(A - tI)^{-p}V = a_1(\lambda_1 - t)^{-p}V_1 + \cdots + a_n(\lambda_n - t)^{-p}V_n$$

If t is near an eigenvalue λ_k, then the term $a_k(\lambda_k - t)^{-p}V_k$ will dominate the sum, assuming that $a_k \neq 0$ and λ_k is an isolated eigenvalue. The powers being computed will then lead to an eigenvalue of A, because all these matrices have the same eigenvectors. This is the basis of the inverse power method.

An interesting variation of this idea uses a sequence of values t_j. Given an initial approximation to an eigenvector, say $x^{(0)}$, compute successively

$$t_{i+1} = \frac{x^{(i)T}Ax^{(i)}}{x^{(i)T}x^{(i)}} \qquad x^{(i+1)} = c_{i+1}(A - t_{i+1}I)^{-1}x^{(i)}$$

the t_{i+1} being Rayleigh quotient estimates to λ_k and the $x^{(i+1)}$ approximations to V_k. Convergence has been proved under various hypotheses. The factor c_{i+1} is chosen to make $\|x^{(i+1)}\| = 1$ for some norm.

It is not actually necessary to compute the inverse matrix. What is needed is the vector $w^{(i+1)}$ defined by

$$w^{(i+1)} = (A - t_{i+1}I)^{-1}x^{(i)}$$

so it is more economical to get it by solving the system

$$(A - t_{i+1}I)w^{(i+1)} = x^{(i)}$$

for this vector. Then $x^{(i+1)} = c_{i+1}w^{(i+1)}$. As the sequence develops, the matrices $A - t_{i+1}I$ will approach singularity, suggesting that the method may have a perilous character, but with attention to normalization and pivoting, accurate results can be obtained.

17.117 Use the inverse power method to find the dominant eigenvalue 26.3047 of Problem 17.112, together with its vector.

▌ Suppose we casually begin with the vector $(1, 1, 1, 1)$, which is hardly very close to the target eigenvector. On the first cycle of inverse powers we find $t = 17.25$, which is not discouraging. The output vector, maximum element normalized to 1, is

$$x = (-.27, -.06, .34, 1)$$

and not a startling step toward the true eigenvector found in the earlier problem. But improvement is rapid, and three more cycles bring the familiar results.

Going for the smallest eigenpair we choose $(-.5, 1, -1, .25)$ to start, a slightly roughened version of the known eigenvector. In three cycles we have .038016 as before, along with

$$x = (-.426899, 1, -.822236, .232906)$$

which conforms to the present strategy of normalizing, but is just about equivalent to the earlier vector. The point is that all the eigenpairs are accessible by this method, given suitable starts and assuming no near duplicates.

17.118 Find all the eigenpairs of the Hilbert matrix of order 3 by using inverse powers.

▌ With an eye on Problem 17.115, we take $(1, .5, .4)$ as a start. Two cycles prove enough to repeat our earlier eigenvalue and its vector to seven figures. Starting again, with $(1, -5.6, 5.4)$ and normalizing by dividing through by -5.6, two cycles again prove to be enough to duplicate the earlier values, apart from a unit change here and there in the seventh figure. Finally, the start $(-.8, .8, 1)$ leads quickly to the expected .1223271 and its mate.

Here we did, of course, have an almost infinite head start, but the method does not usually need such an advantage.

17.119 What is inverse iteration?

▌ Given an accurate approximation to an eigenvalue of A, inverse iteration is a fast way to obtain the corresponding eigenvector. Let t be an approximation to λ, obtained from the characteristic polynomial or other method that produces eigenvalues only. Then $A - tI$ is near singular, but still has a factorization

$$P(A - tI) = LU \qquad A - tI = P^{-1}LU$$

as in Problem 17.18. Just as in the preceding problem, we begin an iteration with

$$(A - tI)x^{(1)} = P^{-1}LUx^{(1)} = x^{(0)}$$

using an $x^{(0)}$ with a nonzero component in the direction of x, the eigenvector corresponding to λ. The choice $x^{(0)} = P^{-1}L(1, 1, \dots, 1)^T$ has sometimes been suitable, or what is the same thing,

$$Ux^{(1)} = (1, 1, \dots, 1)^T$$

17.120 Apply inverse iteration to the matrix of Problem 17.93, using .586 as an approximation to the eigenvalue $2 - \sqrt{2}$. Since the eigenvector $x = (1, \sqrt{2}, 1)$ has already been found, this will serve as a small-scale illustration of the method's potential.

▌ To start, we need the factors L and U, which prove to be

$$L = \begin{bmatrix} 1 & 0 & 0 \\ -.70721 & 1 & 0 \\ 0 & -1.4148 & 1 \end{bmatrix} \qquad U = \begin{bmatrix} 1.414 & -1 & 0 \\ 0 & .7068 & -1 \\ 0 & 0 & -.0008 \end{bmatrix}$$

In this example $P = I$. The solution of $Ux^{(1)} = (1, 1, \ldots, 1)^T$, found by back-substitution, is $x^{(1)} = (1250, 1767, -1250)^T$, after which

$$LUx^{(2)} = x^{(1)}$$

yields $x^{(2)} = (13, 319, 44, 273, 31, 265)^T$ to five figures. Normalizing then brings the approximate eigenvector $(1, 1.414, .998)^T$.

17.11 REDUCTION TO CANONICAL FORMS

17.121 A basic theorem of linear algebra states that a real symmetric matrix A has only real eigenvalues and that there exists a real orthogonal matrix Q such that $Q^{-1}AQ$ is diagonal. The diagonal elements are then the eigenvalues and the columns of Q are the eigenvectors. Derive the Jacobi formulas for producing this orthogonal matrix Q.

▮ In the Jacobi method Q is obtained as an infinite product of "rotation" matrices of the form

$$Q_1 = \begin{bmatrix} \cos\phi & -\sin\phi \\ \sin\phi & \cos\phi \end{bmatrix}$$

all other elements being identical with those of the unit matrix I. If the four entries shown are in positions (i, i), (i, k), (k, i) and (k, k), then the corresponding elements of $Q_1^{-1}AQ_1$ may easily be computed to be

$$b_{ii} = a_{ii}\cos^2\phi + 2a_{ik}\sin\phi\cos\phi + a_{kk}\sin^2\phi$$

$$b_{ki} = b_{ik} = (a_{kk} - a_{ii})\sin\phi\cos\phi + a_{ik}(\cos^2\phi - \sin^2\phi)$$

$$b_{kk} = a_{ii}\sin^2\phi - 2a_{ik}\sin\phi\cos\phi + a_{kk}\cos^2\phi$$

Choosing ϕ such that $\tan 2\phi = 2a_{ik}/(a_{ii} - a_{kk})$ then makes $b_{ik} = b_{ki} = 0$. Each step of the Jacobi algorithm therefore makes a pair of off-diagonal elements zero. Unfortunately the next step, while it creates a new pair of zeros, introduces nonzero contributions to formerly zero positions. Nevertheless, successive matrices of the form $Q_2^{-1}Q_1^{-1}AQ_1Q_2$ and so on, approach the required diagonal form and $Q = Q_1Q_2 \cdots$.

17.122 Apply Jacobi's method to

$$A = \begin{bmatrix} 2 & -1 & 0 \\ -1 & 2 & -1 \\ 0 & -1 & 2 \end{bmatrix}$$

▮ With $i = 1$ and $k = 2$ we have $\tan 2\phi = -2/0$, which we interpret to mean $2\phi = \pi/2$. Then $\cos\phi = \sin\phi = 1/\sqrt{2}$ and

$$A_1 = Q_1^{-1}AQ_1 = \begin{bmatrix} \dfrac{1}{\sqrt{2}} & \dfrac{1}{\sqrt{2}} & 0 \\ -\dfrac{1}{\sqrt{2}} & \dfrac{1}{\sqrt{2}} & 0 \\ 0 & 0 & 1 \end{bmatrix} \begin{bmatrix} 2 & -1 & 0 \\ -1 & 2 & -1 \\ 0 & -1 & 2 \end{bmatrix} \begin{bmatrix} \dfrac{1}{\sqrt{2}} & -\dfrac{1}{\sqrt{2}} & 0 \\ \dfrac{1}{\sqrt{2}} & \dfrac{1}{\sqrt{2}} & 0 \\ 0 & 0 & 1 \end{bmatrix} = \begin{bmatrix} 1 & 0 & -\dfrac{1}{\sqrt{2}} \\ 0 & 3 & -\dfrac{1}{\sqrt{2}} \\ -\dfrac{1}{\sqrt{2}} & -\dfrac{1}{\sqrt{2}} & 2 \end{bmatrix}$$

Next we take $i = 1$ and $k = 3$ making $\tan 2\phi = -\sqrt{2}/(-1) = \sqrt{2}$. Then $\sin\phi \simeq .45969$, $\cos\phi \simeq .88808$ and we compute

$$A_2 = Q_2^{-1}A_1Q_2 = \begin{bmatrix} .88808 & 0 & .45969 \\ 0 & 1 & 0 \\ -.45969 & 0 & .88808 \end{bmatrix} A_1 \begin{bmatrix} .88808 & 0 & -.45969 \\ 0 & 1 & 0 \\ .45969 & 0 & .88808 \end{bmatrix}$$

$$= \begin{bmatrix} .63398 & -.32505 & 0 \\ -.32505 & 3 & -.62797 \\ 0 & -.62797 & 2.36603 \end{bmatrix}$$

The convergence of the off-diagonal elements toward zero is not startling, but at least the decrease has begun. After nine rotations of this sort we achieve

$$A_9 = \begin{bmatrix} .58578 & .00000 & .00000 \\ .00000 & 2.00000 & .00000 \\ .00000 & .00000 & 3.41421 \end{bmatrix}$$

in which the eigenvalues found earlier have reappeared. We also have

$$Q = Q_1 Q_2 \cdots Q_9 = \begin{bmatrix} .50000 & .70710 & .50000 \\ .70710 & .00000 & -.70710 \\ .50000 & -.70710 & .50000 \end{bmatrix}$$

in which the eigenvectors are also conspicuous.

17.123 Apply Jacobi's method to the binomial coefficient matrix of Problem 17.94 and elsewhere.

▌ An abbreviated version of the first cycle of rotations is provided as Fig. 17.9. Notice the creation of zeros as planned and the "small" contributions that work back into positions once zeroed. Also notice the somewhat familiar entry at lower right.

.4	0	−.7	−1.2	.29	.39	0	.02	.29	.39	−.01	0
0	2.6	3.1	3.9	.39	2.6	3	3.9	.38	2.6	3	3.9
−.7	3.1	6	10	0	3	6.1	10	−.01	3	6.1	10
−1.2	3.9	10	20	.02	3.9	10	20	0	3.9	10	20
.29	.34	.18	0	.29	.34	.18	−.03	.29	.34	.17	.07
.34	.84	0	−1.7	.34	.70	.92	0	.34	.70	.80	.46
.18	0	7.9	10.7	.18	.92	7.9	10.7	.17	.80	1.7	0
0	−1.7	10.7	20	−.03	0	10.7	20.1	.07	.46	0	26.30

Fig. 17-9

After two more such cycles of rotations the diagonal entries were .0380163, .4538346, 2.203445, 26.3047 and no longer changing to seven figures. The first and last of these agree with our earlier values. The product of all Q matrices used, the columns of which are the eigenvectors, appears below. Testing $Ax = \lambda x$, with x a column and λ the appropriate eigenvalue, will find minor discrepancies in the last place. The vectors are orthogonal:

$$Q = \begin{pmatrix} 1 & 1 & 1 & 1 \\ -2.342474 & -.2073395 & 1.20734 & 3.342477 \\ 1.926065 & -.6758844 & .7387961 & 7.611021 \\ -.5455748 & -.3370586 & -.7426899 & 14.35121 \end{pmatrix}$$

17.124 Use Jacobi's method on the Hilbert matrix of order 3.

▌ On the first cycle of rotations it manages eigenvalue approximations of 1.408, .1225 and .0028. Two cycles later these have been fixed at 1.408009, .1224966 and .002827618, with elements off the diagonal of orders 10^{-9} to 10^{-19}. The Q matrix was

$$Q = \begin{pmatrix} 1 & 1 & 1 \\ .5560326 & -.9650047 & -5.591029 \\ .390908 & -1.185512 & 5.394606 \end{pmatrix}$$

and the $Ax = \lambda x$ test finds errors of only 1 unit in the final figure. The orthogonality of eigenvectors can also be verified.

17.125 Apply the Jacobi method to the fourth order Hilbert matrix.

▌ The results of three and four cycles of rotations agree on the eigenvalues 1.500214, .1691412, .0000967155, .00673829 and the eigenvectors in the Q matrix

$$Q = \begin{pmatrix} 1 & 1 & 1 & 1 \\ .5701721 & -.636519 & -11.2599 & -4.140485 \\ .406779 & -.8754507 & 27.10957 & .559358 \\ .318141 & -.8831295 & -17.62589 & 3.562113 \end{pmatrix}$$

The equation $Ax = \lambda x$ is satisfied if one or two units in the final place are forgiven. The orthogonality of eigenvectors can also be verified.

17.126 How well does Jacobi's method do with the Hilbert matrices of orders 5 and 6?

▮ For order 5, four cycles were enough for convergence to eigenvalues 1.567051, .2085342, .000003294318, .01140749 and .000305909 in that order along the diagonal. The matrix Q of the eigenvectors had also converged and appears as Fig. 17-10. Both the $Ax = \lambda x$ and the orthogonality test were passed, with only minor discrepancies in the final place.

$$Q = \begin{array}{ccccc} 1 & 1 & 1 & 1 & 1 \\ .5805669 & -.4584259 & -18.90248 & -3.380281 & -9.174182 \\ .4188010 & -.7059258 & 81.99497 & -.5622993 & 14.15053 \\ .3300610 & -.7375379 & -124.2837 & 1.445165 & 4.940648 \\ .2732582 & -.7127991 & 60.9526 & 2.638451 & -11.82305 \end{array} \qquad \textbf{Fig. 17-10}$$

As for order 6, convergence to these presumed eigenvalues had occurred after four cycles of rotations, but the eigenvectors will be omitted and no followup tests were made:

$$1.6189, \qquad .2423609, \qquad 1.257 \cdot 10^{-5}, \qquad .01632152, \qquad 1.137633 \cdot 10^{-7}, \qquad 6.157708 \cdot 10^{-4}$$

It will have been noticed that the four Hilbert matrices treated have maximum eigenvalues between 1 and 2, and that the least has become minute.

17.127 What are the three main parts of Givens' variation of the Jacobi rotation algorithm for a real symmetric matrix?

▮ In the first part of the algorithm rotations are used to reduce the matrix to triple-diagonal form, only the main diagonal and its two neighbors being different from zero. The first rotation is in the $(2,3)$ plane, involving the elements a_{22}, a_{23}, a_{32} and a_{33}. It is easy to verify that such a rotation, with ϕ determined by $\tan \phi = a_{13}/a_{12}$, will replace the a_{13} (and a_{31}) elements by 0. Succeeding rotations in the $(2, i)$ planes then replace the elements a_{1i} and a_{i1} by zero, for $i = 4, \ldots, n$. The ϕ values are determined by $\tan \phi = a_{1i}/a'_{12}$, where a'_{12} denotes the current occupant of row 1, column 2. Next it is the turn of the elements $a_{24}, \ldots, a_{2n}$, which are replaced by zeros by rotations in the $(3,4), \ldots, (3, n)$ planes. Continuing in this way a matrix of triple-diagonal form will be achieved, since no zero that we have worked to create will be lost in a later rotation. This may be proved by a direct computation and makes the Givens reduction finite whereas the Jacobi diagonalization is an infinite process.

The second step involves forming the sequence

$$f_0(\lambda) = 1 \qquad f_i(\lambda) = (\lambda - \alpha_i) f_{i-1}(\lambda) - \beta_{i-1}^2 f_{i-2}(\lambda)$$

where the αs and βs are the elements of our new matrix

$$B = \begin{bmatrix} \alpha_1 & \beta_1 & 0 & \cdots & 0 \\ \beta_1 & \alpha_2 & \beta_2 & \cdots & 0 \\ 0 & \beta_2 & \alpha_3 & \cdots & 0 \\ & & & \cdots & \beta_{n-1} \\ 0 & 0 & 0 & \beta_{n-1} & \alpha_n \end{bmatrix}$$

and $\beta_0 = 0$. These $f_i(\lambda)$ prove to be the determinants of the principal minors of the matrix $\lambda I - B$, as may be seen from

$$f_i(\lambda) = \begin{vmatrix} \lambda - \alpha_1 & -\beta_1 & 0 & \cdots & 0 \\ -\beta_1 & \lambda - \alpha_2 & -\beta_2 & \cdots & 0 \\ 0 & -\beta_2 & \lambda - \alpha_3 & \cdots & 0 \\ & & & \cdots & -\beta_{i-1} \\ & & \cdots & -\beta_{i-1} & \lambda - \alpha_i \end{vmatrix}$$

by expanding along the last column,

$$f_i(\lambda) = (\lambda - \alpha_i) f_{i-1}(\lambda) + \beta_{i-1} D$$

where D has only the element $-\beta_{i-1}$ in its bottom row and so equals $D = -\beta_{i-1} f_{i-2}(\lambda)$. For $i = n$ we therefore have in $f_n(\lambda)$ the characteristic polynomial of B. Since our rotations do not alter the polynomial, it is also the characteristic polynomial of A.

Now, if some β_i are zero, the determinant splits into smaller determinants that may be treated separately. If no β_i is zero, the sequence of functions $f_i(\lambda)$ proves to be a Sturm sequence (with the numbering reversed from the order given in Problem 16.52). Consequently the number of eigenvalues in a given interval may be determined by counting variations of sign.

Finally, the third step involves finding the eigenvectors. Here the diagonal nature of B makes Gaussian elimination a reasonable process for obtaining its eigenvectors U_j directly (deleting one equation and assigning some component the arbitrary value of 1). The corresponding eigenvectors of A are then $V_j = QU_j$ where Q is once again the product of our rotation matrices.

17.128 Apply the Givens method to the Hilbert matrix of order 3:

$$H = \begin{bmatrix} 1 & \frac{1}{2} & \frac{1}{3} \\ \frac{1}{2} & \frac{1}{3} & \frac{1}{4} \\ \frac{1}{3} & \frac{1}{4} & \frac{1}{5} \end{bmatrix}$$

▮ For this small matrix only one rotation is needed. With $\tan\phi = \frac{2}{3}$, we have $\cos\phi = 3/\sqrt{13}$ and $\sin\phi = 2/\sqrt{13}$. Then

$$Q = \frac{1}{\sqrt{13}} \begin{bmatrix} \sqrt{13} & 0 & 0 \\ 0 & 3 & -2 \\ 0 & 2 & 3 \end{bmatrix} \qquad B = Q^{-1}HQ = \begin{bmatrix} 1 & \dfrac{\sqrt{13}}{6} & 0 \\ \dfrac{\sqrt{13}}{6} & \dfrac{34}{65} & \dfrac{9}{260} \\ 0 & \dfrac{9}{260} & \dfrac{2}{195} \end{bmatrix}$$

and we have our triple diagonal matrix. The Sturm sequence consists of

$$f_0(\lambda) = 1 \qquad f_1(\lambda) = \lambda - 1 \qquad f_2(\lambda) = \left(\lambda - \tfrac{34}{65}\right)(\lambda - 1) - \tfrac{13}{16}$$
$$f_3(\lambda) = \left(\lambda - \tfrac{2}{195}\right)f_2(\lambda) - \tfrac{81}{67,600}(\lambda - 1)$$

which lead to the $\pm$ signs shown in Table 17.5. There are two roots between 0 and 1 and a third between 1 and 1.5. Iterations then locate these more precisely at .002688, .122327 and 1.408319. The eigenvalue so close to zero is another indication of the near singularity of this matrix.

To find the eigenvector for λ_1, we solve $BU_1 = \lambda_1 U_1$ and soon discover $u_1 = 1$, $u_2 = -1.6596$, $u_3 = 7.5906$ to be one possibility. Finally

$$V_1 = QU_1 = (1, -5.591, 5.395)^T$$

which can be normalized as desired. Eigenvectors for the other two eigenvalues respond to the same process.

TABLE 17.5

	f_0	f_1	f_2	f_3	Changes
0	$+$	$-$	$+$	$-$	3
1	$+$	0	$-$	$-$	1
1.5	$+$	$+$	$+$	$+$	0

17.129 A similarity transformation of A is defined by $M^{-1}AM$, for any nonsingular matrix M. Show that such a transformation leaves the eigenvalues unchanged.

▮ Since $Ax = \lambda x$ implies

$$MAM^{-1}(Mx) = \lambda(Mx)$$

we have at once that λ is an eigenvalue of MAM^{-1} with corresponding eigenvector Mx. The orthogonal transformations used in the Jacobi and Givens methods are special cases of similarity transformations.

17.130 Show that a matrix having all distinct eigenvalues and corresponding independent eigenvectors can be reduced to diagonal form by a similarity transformation.

▮ Form the matrix M by using the eigenvectors of A as columns. It follows that

$$AM = MD$$

where D is diagonal and has the eigenvalues along its diagonal. Because the eigenvectors are linearly independent, M^{-1} exists and

$$M^{-1}AM = D$$

as required. This classic theorem on the reduction of matrices to special, or canonical, form has questionable computational value, since to find M appears to presuppose the solution of the entire problem.

17.12 NONSYMMETRIC MATRICES

17.131 What is a Hessenberg matrix?

▮ It is a matrix in which either the upper or the lower triangle is zero except for the elements adjacent to the main diagonal. If the upper triangle has the zeros, the matrix is a lower Hessenberg and vice versa. Here are two small Hessenbergs, the second being also triple diagonal since it is symmetric:

$$\begin{bmatrix} 1 & 1 & 1 & 1 \\ 1 & 1 & 1 & 1 \\ 0 & 1 & 1 & 1 \\ 0 & 0 & 1 & 1 \end{bmatrix} \quad \begin{bmatrix} 1 & 1 & 0 \\ 1 & 1 & 1 \\ 0 & 1 & 1 \end{bmatrix}$$

17.132 Show that a matrix A can be reduced to Hessenberg form by Gaussian elimination and a similarity transformation.

▮ Suppose we take an upper Hessenberg as our goal. The required zeros in the lower triangle can be generated column by column in $n - 2$ stages. Assume $k - 1$ such stages finished and denote the new elements by a'_{ij}. The zeros for column k are then arranged as follows:

1 From the elements $a'_{k+1, k}, \ldots, a'_{nk}$ find the absolutely largest and interchange its row with row $k + 1$. This is the partial pivoting step and can be achieved by premultiplying the current matrix A' by an interchange matrix $I_{r, k+1}$ as introduced in Problem 17.18.

2 Calculate the multipliers

$$c_{jk} = -\frac{a''_{jk}}{a_{k+1, k}} \qquad j = k + 2, \ldots, n$$

(the double prime referring to elements after the interchange). Add c_{jk} times row $k + 1$ to row j. This can be done for all the j simultaneously by premultiplying the current matrix A'' by a matrix G_k similar to the L_i of Problem 17.18:

$$G_k = \begin{bmatrix} 1 & & & & & \\ & \ddots & & & & \\ & & 1 & & & \\ & & -c_{k+2, k} & 1 & & \\ & & \vdots & & \ddots & \\ & & -c_{nk} & & & 1 \end{bmatrix} \quad \text{row } k + 2$$

$$\text{col. } k + 1$$

This is the Gaussian step.

3 Postmultiply the current matrix by the inverses of $I_{r, k+1}$ and G_k. This is the similarity step. Of course, $I_{r, k+1}$ is its own inverse, while that of G_k is found by changing the signs of the c elements. This completes the kth stage of the reduction, which can be summarized by

$$G_k I_{r, k+1} A' I_{r, k+1} G_k^{-1}$$

with A' the input from the preceding stage, or A itself if $k = 1$.

The steps 1, 2 and 3 are carried out for $k = 1, \ldots, n - 2$ and it is easy to discover that the target zeros of any stage are retained.

$$I_{23}A \quad \begin{matrix} 0 & 1 & 2 & 3 \\ 3 & 0 & 1 & 2 \\ 2 & 3 & 0 & 1 \\ 1 & 2 & 3 & 0 \end{matrix} \qquad\qquad I_{34}A' \quad \begin{matrix} 0 & \frac{11}{3} & 1 & 3 \\ 3 & \frac{5}{3} & 0 & 2 \\ 0 & \frac{34}{9} & 2 & \frac{-2}{3} \\ 0 & \frac{11}{9} & 3 & \frac{-1}{3} \end{matrix}$$

$$G_1 I_{23}A \quad \begin{matrix} 0 & 1 & 2 & 3 \\ 3 & 0 & 1 & 2 \\ 0 & 3 & \frac{-2}{3} & \frac{-1}{3} \\ 0 & 2 & \frac{8}{3} & \frac{-2}{3} \end{matrix} \qquad\qquad G_2 I_{34}A' \quad \begin{matrix} 0 & \frac{11}{3} & 1 & 3 \\ 3 & \frac{5}{3} & 0 & 2 \\ 0 & \frac{34}{9} & 2 & \frac{-2}{3} \\ 0 & 0 & \frac{40}{17} & \frac{-2}{17} \end{matrix}$$

$$G_1 I_{23}A I_{23} \quad \begin{matrix} 0 & 2 & 1 & 3 \\ 3 & 1 & 0 & 2 \\ 0 & \frac{-2}{3} & 3 & \frac{-1}{3} \\ 0 & \frac{8}{3} & 2 & \frac{-2}{3} \end{matrix} \qquad\qquad G_2 I_{34}A' I_{34} \quad \begin{matrix} 0 & \frac{11}{3} & 3 & 1 \\ 3 & \frac{5}{3} & 2 & 0 \\ 0 & \frac{34}{9} & \frac{-2}{3} & 2 \\ 0 & 0 & \frac{-2}{17} & \frac{40}{17} \end{matrix}$$

$$\begin{matrix} G_1 I_{23}A I_{23} G_1^{-1} \\ (=A') \end{matrix} \quad \begin{matrix} 0 & \frac{11}{3} & 1 & 3 \\ 3 & \frac{5}{3} & 0 & 2 \\ 0 & \frac{11}{9} & 3 & \frac{-1}{3} \\ 0 & \frac{34}{9} & 2 & \frac{-2}{3} \end{matrix} \qquad G_2 I_{34}A' I_{34} G_2^{-1} \quad \begin{matrix} 0 & \frac{11}{3} & \frac{113}{34} & 1 \\ 3 & \frac{5}{3} & 2 & 0 \\ 0 & \frac{34}{9} & \frac{-1}{51} & 2 \\ 0 & 0 & \frac{186}{289} & \frac{40}{17} \end{matrix}$$

$$I_{23} \quad \begin{matrix} 1 & 0 & 0 & 0 \\ 0 & 0 & 1 & 0 \\ 0 & 1 & 0 & 0 \\ 0 & 0 & 0 & 1 \end{matrix} \qquad\qquad I_{34} \quad \begin{matrix} 1 & 0 & 0 & 0 \\ 0 & 1 & 0 & 0 \\ 0 & 0 & 0 & 1 \\ 0 & 0 & 1 & 0 \end{matrix}$$

$$G_1 \quad \begin{matrix} 1 & 0 & 0 & 0 \\ 0 & 1 & 0 & 0 \\ 0 & \frac{-2}{3} & 1 & 0 \\ 0 & \frac{-1}{3} & 0 & 1 \end{matrix} \qquad\qquad G_2 \quad \begin{matrix} 1 & 0 & 0 & 0 \\ 0 & 1 & 0 & 0 \\ 0 & 0 & 1 & 0 \\ 0 & 0 & \frac{-11}{34} & 1 \end{matrix}$$

Fig. 17-11

17.133 Apply the algorithm of the preceding problem to the matrix

$$\begin{bmatrix} 0 & 1 & 2 & 3 \\ 2 & 3 & 0 & 1 \\ 3 & 0 & 1 & 2 \\ 1 & 2 & 3 & 0 \end{bmatrix}$$

▮ All the essentials appear in Fig. 17-11, the two stages side by side. Remember that as a premultiplier, $I_{r,k+1}$ swaps rows but as its own inverse and postmultiplier it swaps columns. The given matrix A is not symmetric so the result is Hessenberg but not triple diagonal. The matrix M of the similarity transformation MAM^{-1} is $G_2 I_{34} G_1 I_{23}$.

17.134 What is the QR method of finding eigenvalues?

▮ Suppose we have an upper Hessenberg matrix H and can factor it as

$$H = QR$$

with Q orthogonal and R an upper (or right?) triangle. In the algorithm to come what we actually find first is

$$Q^T H = R$$

by reducing H to triangular form through successive rotations. Define

$$H^{(2)} = RQ = Q^T H Q$$

and note that $H^{(2)}$ will have the same eigenvalues as H, because of the theorem in Problem 17.129. (Since Q is orthogonal, $Q^T = Q^{-1}$.) It turns out that $H^{(2)}$ is also Hessenberg, so the process can be repeated to generate $H^{(k+1)}$ from $H^{(k)}$, with H serving as $H^{(1)}$ and $k = 1, \ldots$. The convergence picture is fairly complicated, but under various hypotheses the diagonal elements approach the eigenvalues while the lower triangle approaches zero. (Of course, the R factor at each stage is upper triangular, but in forming the product RQ, to recover the original eigenvalues, subdiagonal elements become nonzero again.) This is the essential idea of the QR method, the eventual annihilation of the lower triangle.

17.135 How can the matrix $Q^{(k)}$, required for the kth stage of the QR method, be found? That is, find $Q^{(k)}$ such that

$$H^{(k+1)} = Q^{(k)T} H^{(k)} Q^{(k)}$$

for $k = 1, \ldots$.

▮ One way of doing this uses rotations, very much as in the Givens method presented in Problem 17.127. Since we are assuming that H is upper Hessenberg, it is only the elements $h_{i+1,i}$ that need our attention, for $i = 1, \ldots, n-1$. But $h_{i+1,i}$ can be replaced by zero using the rotation

$$S_i^T = \begin{bmatrix} 1 & & & & & & \\ & \ddots & & & & & \\ & & \cos\phi & \sin\phi & & & \\ & & -\sin\phi & \cos\phi & & & \\ & & & & \ddots & & \\ & & & & & 1 \end{bmatrix} \begin{matrix} \\ \\ \text{row } i \\ \text{row } i+1 \\ \\ \end{matrix}$$

$$\begin{matrix} \text{col.} & \text{col.} \\ i & i+1 \end{matrix}$$

and calculating $S_i^T H$, provided $\tan\phi = h_{i+1,i}/h_{i,i}$. (It is easier to let $\sin\phi = ch_{i+1,i}$, $\cos\phi = ch_{i,i}$ and choose c to make the sum of squares 1.) Then the product of these rotations

$$Q^T = S_{n-1}^T \cdots S_1^T$$

is what we need. The same argument applies for any stage, so the superscript (k) has been suppressed here.

17.136 How has the idea of eigenvalue shifting, presented in Problem 17.109, been applied to accelerate the convergence of the QR algorithm?

▮ Instead of factoring the matrix H, we try instead the reduction

$$Q^T(H - pI) = R$$

for some suitable value of p. The factorization $H - pI = QR$ is thus implied. Then

$$Q^T(H - pI)Q = RQ = H^{(2)} - pI$$

exhibits the reversed product that is central to the method and also defines $H^{(2)}$. But then

$$H^{(2)} = Q^T(H - pI)Q + pI = Q^THQ$$

so $H^{(2)}$ again has the same eigenvalues as H. With $H^{(2)}$ in hand, we are ready to begin the next iteration. It would be nice to choose p near an eigenvalue, but in the absence of such inside information, the following alternative is recommended. Find the eigenvalues of the 2 by 2 submatrix in the lower right corner of the current H and set p equal to the one closest to h_{nn}, assuming these eigenvalues real. If they are complex, set p to their common real part.

17.137 Given the midget Hessenberg matrix

$$H = \begin{bmatrix} 4 & 2 & 1 \\ 0 & 1 & 0 \\ 0 & 2 & 3 \end{bmatrix}$$

find the eigenvalues by the QR method.

▌ It is easy to discover that the eigenvalues are the diagonal elements $4, 1, 3$. But it is also interesting to watch the QR method perform the triangularization. Choosing a shift of 3, we compute

$$H - 3I = \begin{bmatrix} 1 & 2 & 1 \\ 0 & -2 & 0 \\ 0 & 2 & 0 \end{bmatrix}$$

which will need just one rotation to reach triangular form.

$$S^T = \frac{1}{\sqrt{2}} \begin{bmatrix} \sqrt{2} & 0 & 0 \\ 0 & -1 & 1 \\ 0 & -1 & -1 \end{bmatrix} \qquad S^T(H - 3I) = \frac{1}{\sqrt{2}} \begin{bmatrix} \sqrt{2} & 2\sqrt{2} & \sqrt{2} \\ 0 & 4 & 0 \\ 0 & 0 & 0 \end{bmatrix}$$

Postmultiplication by S then completes the similarity transformation:

$$S^T(H - 3I)S = \tfrac{1}{2} \begin{bmatrix} 2 & -\sqrt{2} & -3\sqrt{2} \\ 0 & -4 & -4 \\ 0 & 0 & 0 \end{bmatrix}$$

Finally we add $3I$ and have

$$H^{(2)} = \begin{bmatrix} 4 & -\dfrac{\sqrt{2}}{2} & -\dfrac{3\sqrt{2}}{2} \\ 0 & 1 & -2 \\ 0 & 0 & 3 \end{bmatrix}$$

the triangular form having been preserved. Ordinarily this will not happen and several stages such as the above will be needed.

17.138 Apply the QR method to the Hessenberg matrix

$$H = \begin{bmatrix} 4 & 1 & 1 & 1 \\ 1 & 4 & 1 & 1 \\ 0 & 1 & 4 & 1 \\ 0 & 0 & 1 & 4 \end{bmatrix}$$

for which the exact eigenvalues are 6, 4, 3 and 3.

▌ A substantial number of rotation cycles eventually reduced this matrix to the triangle

$$\begin{bmatrix} 5.99997 & 1.50750 & -.17830 & .29457 \\ & 3.99997 & -.44270 & .22152 \\ & & 3.00098 & -.60302 \\ & & & 2.99895 \end{bmatrix}$$

in which the eigenvalues are evident along the diagonal. For larger jobs a saving in computing time would be realized by a reduction of the order when one of the subdiagonal elements becomes zero. Here it was entertaining simply to watch the lower triangle slowly vanish. Using the preceding approximate eigenvalues, the corresponding vectors were found directly and matched the correct $(3, 3, 2, 1)$, $(-1, -1, 0, 1)$ and $(0, 0, -1, 1)$ to three decimal places more or less. There is no fourth eigenvector.

17.139 Apply the QR method to the triple diagonal matrix

$$\begin{bmatrix} 4 & 1 & 0 & 0 \\ 1 & 4 & 1 & 0 \\ 0 & 1 & 4 & 1 \\ 0 & 0 & 1 & 4 \end{bmatrix}$$

and then use the results obtained to "guess" the correct eigenvalues.

▮ Once again the rotation cycles were allowed to run their course, with this result. Off-diagonal elements were essentially zero.

$$\begin{bmatrix} 5.618031 & & & \\ & 4.618065 & & \\ & & 3.381945 & \\ & & & 2.381942 \end{bmatrix}$$

Since the given matrix was symmetric, both the lower and upper triangles have become zero, leaving the eigenvalues quite conspicuous. Taking the largest, a direct calculation of the eigenvector managed

$$(1.00002, 1.61806, 1.61806, 1)$$

the fourth component having been fixed in advance. Guessing that this ought to have been $(1, x, x, 1)$ leads quickly to the equations

$$\lambda = x + 4 \qquad x^2 - x - 1 = 0$$

the second of which is familiar by its connection with Fibonacci numbers. The root $x = (1 + \sqrt{5})/2$ is now paired with $\lambda = (9 + \sqrt{5})/2$, while $x = (1 - \sqrt{5})/2$ is paired with $\lambda = (9 - \sqrt{5})/2$ giving us two of the exact solutions. The other two are found similarly.

17.140 What are Householder matrices? Prove them orthogonal.

▮ They are of the form $H = I - 2VV^T$, where I is the identity matrix and V a unit vector in the L_2 norm. They are symmetric by their definition and this allows the computation

$$H^T H = H^2 = (I - 2VV^T)^2 = I - 4VV^T + 4(VV^T)VV^T$$
$$= I - 4VV^T + 4V(V^T V)V^T = I$$

since $V^T V$ is 1 for the unit vector V. So $H^T = H^{-1}$.

17.141 How can Householder matrices be used in the QR method?

▮ They offer an alternative way of finding a QR factorization. First an H_1 is found such that $H_1 A$ has zeros in column one below the diagonal element. Then an H_2 is found such that $H_2 H_1 A$ has zeros in columns one and two below the diagonal elements. This continues until $H_{n-1} \cdots H_1 A = Q^T A = R$, with R an upper triangle. Then $A = QR$ as intended.

The matrix H_k can be formed as follows, assuming H_1 to H_{k-1} have already done their part. Take column k of $H_{k-1} \cdots H_1 A$ and normalize it in the L_2 sense, obtaining vector $d = (d_1, \ldots, d_k)$. Let $D = \pm \sqrt{d_k^2 + \cdots + d_n^2}$, choosing the $+$ sign if $d_k \le 0$. Now form the vector V, starting with $v_1 = \cdots = v_{k-1} = 0$, after which

$$v_k = \sqrt{\tfrac{1}{2}(1 - d_k/D)}$$
$$v_j = -d_j/2Dv_k \qquad \text{for } j = k + 1 \text{ to } n$$

It will be found that $\|V\| = 1$. Finally,

$$H_k = I - 2VV^T$$

which can be verified for its assigned role.

17.142 Use Householder matrices to find a QR factorization of

$$\begin{pmatrix} 2 & -2 & 3 \\ 1 & 1 & 1 \\ 1 & 3 & -1 \end{pmatrix}$$

▌ Dividing first column elements by $\sqrt{6}$, we get

$$d = (.816, .408, .408)^T \qquad D = -1$$

three digits being printed here, though seven were carried. Then

$$V = (.953, .214, .214)^T$$

and we have the Householder matrix H_1 shown in Fig. 17-12. The product $H_1 A$ is also shown and has a bonus zero.

−.816	−.408	−.408	1	0	0	−.816	−.408	−.408
−.408	.908	−.092	0	−.387	−.922	.534	−.267	−.802
−.408	−.092	.908	0	−.922	.387	.218	−.873	.436
	H_1			H_2			$H_2 H_1$	
−2.45	0	−2.45	−2.45	0	−2.45	3	.655	−1.60
0	1.45	−.225	0	−3.74	2.14	.655	−.714	4.20
0	3.45	−2.22	0	0	−.655	.267	.525	−.286
	$H_1 A$			$Q^T A = H_2 H_1 A = R$			$A_1 = Q^T A Q$	

Fig. 17-12

It is now the turn of the second column of $H_1 A$. Normalizing brings $d = (0, .387, .933)^T$ and because of the bonus zero we again have $D = -1$. Then $V = (0, .833, .553)^T$ leading to the H_2 shown in the figure and the triangular $H_2 H_1 A$ below it. The factorization $A = QR$ is now available.

17.143 How can the computation in the preceding problem be continued to find the eigenvalues of A?

▌ The matrix R will almost certainly not have the same eigenvalues as A. To complete a similarity transformation we now find $Q^T A Q$, also shown in the figure, which has the same eigenvalues as A but does not suggest that progress has been made. However, the preceding cycle is now repeated with A_2 replacing $A_1 = A$ as input, and a sequence of output matrices A_j produced in this way. These matrices converge under certain conditions to a triangular limit with the eigenvalues on the diagonal. Here are some results (diagonal elements only) for the matrix A given. The last lower triangle had elements of size 10^{-6} to 10^{-10}. Correct eigenvalues can be guessed.

5 cycles	3.003	−1.905	.902
10 cycles	3.000013	−2.003	1.003
15 cycles	3.000002	−1.999905	.999905

17.144 Use Householder matrices and the QR method to find the eigenvalues of the fourth order matrix

$$\begin{pmatrix} 1 & 1 & 1 & 1 \\ 4 & 3 & 2 & 1 \\ 10 & 6 & 3 & 1 \\ 20 & 10 & 4 & 1 \end{pmatrix}$$

▌ Convergence takes a little time, 30 cycles being needed to produce 11.20065, -3.765505, $.6041101$ and $-.03924772$ along the diagonal, with order 10^{-18} to 10^{-45} in the lower triangle. For comparison, the inverse power method manages 11.20064, -3.765508, $.6041107$ and $-.0392479$, the differences to be blamed on roundoff.

17.145 How well does the QR method perform with the matrix

$$\begin{pmatrix} 2 & 1 & 1 & 1 \\ 1 & 2 & 1 & 1 \\ 0 & 1 & 2 & 1 \\ 0 & 0 & 1 & 2 \end{pmatrix}$$

which has eigenvalues $4, 2, 1, 1$? There is only one eigenvector corresponding to the multiple 1.

▮ The 4 and 2 were produced without enormous difficulty, but after 40 cycles the approximations 1.025 and .975 were in hand for the other pair, with further convergence expected. The presence of a multiple eigenvalue and the absence of an eigenvector did not prevent the method from converging. After a variety of starts, the inverse power method found all the eigenvalues to six figures and, apart from roundoff errors, the eigenvectors

$$(3,3,2,1) \qquad (1,1,0,-1) \qquad (0,0,1,-1)$$

17.146 Try the QR method on the "Fibonacci" matrix

$$\begin{pmatrix} 1 & 1 & 1 & 1 \\ 1 & 0 & 0 & 0 \\ 0 & 1 & 0 & 0 \\ 0 & 0 & 1 & 0 \end{pmatrix}$$

which has two real eigenvalues and a complex pair. (Use this matrix to premultiply some initial vector x several times and discover that the name is not entirely inappropriate.)

▮ After 15 cycles the dominant eigenvalue 1.927562 has settled into the upper left corner of the output matrix and after 100 cycles the figure $-.7759$ more or less has developed in the lower right. The first of these results is accurate but the second eigenvalue is closer to $-.774804$. The lower triangle contains some very small elements, but two others refuse to vanish.

17.147 What success does the QR method have with these two matrices?

$$\begin{pmatrix} 7 & 6 & 5 & 4 \\ 5 & 4 & 3 & 3 \\ 3 & 2 & 2 & 2 \\ 1 & 1 & 1 & 1 \end{pmatrix} \qquad \begin{pmatrix} 1 & 1 & 1 & 1 \\ -1 & 1 & -1 & 1 \\ 1 & 1 & -1 & -1 \\ -1 & 1 & 1 & -1 \end{pmatrix}$$

▮ With the first it rather quickly settles on 13.76126 in the upper left corner, and somewhat later on .2595328 in the lower right. But after 100 cycles the center two diagonal entries still seem secure to only two places, at $-.54$ and .52, though indications of convergence are present. The inverse power method confirms the first pair of eigenvalues and determines the last two as $-.5396428$ and .518852. The slow convergence may be explained by these nearly equal magnitudes.

   The output behavior for the second (Hadamard) matrix, is not productive but is amusing. All entries remain ± 1, with the signs flickering about in an unexplained way.

17.148 Rotating a square a quarter turn clockwise can be simulated by applying the permutation matrix R to the vector $(1,2,3,4)^T$. (See Fig. 17-13.) Reflection in the vertical (dashed) line can be simulated using the matrix V. The eigenvalues of R are easily found to be $1, i, -1, -i$, while those of V are $1, 1, -1, -1$. Both matrices are of Hessenberg type. Will the QR algorithm be convergent in either case?

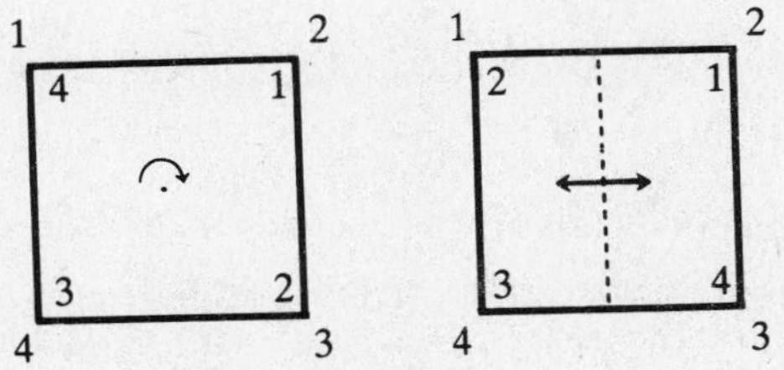

$$R = \begin{bmatrix} 0 & 0 & 0 & 1 \\ 1 & 0 & 0 & 0 \\ 0 & 1 & 0 & 0 \\ 0 & 0 & 1 & 0 \end{bmatrix} \qquad V = \begin{bmatrix} 0 & 1 & 0 & 0 \\ 1 & 0 & 0 & 0 \\ 0 & 0 & 0 & 1 \\ 0 & 0 & 1 & 0 \end{bmatrix}$$

Fig. 17-13

▮ No, apart from some rounding error the zeros remain in place while the signs of the 1 entries undergo periodic fluctuations. The same is true of other permutation matrices, except for the identity, which is never changed at all.

17.13 COMPLEX SYSTEMS

17.149 How can the problem of solving a system of complex equations be replaced by that of solving a real system?

▌ This is almost automatic, since complex numbers are equal precisely when their real and imaginary parts are equal. The equation

$$(A + iB)(x + iy) = a + ib$$

is at once equivalent to

$$Ax - By = a \qquad Ay + Bx = b$$

and this may be written in matrix form as

$$\begin{bmatrix} A & -B \\ B & A \end{bmatrix} \begin{pmatrix} x \\ y \end{pmatrix} = \begin{pmatrix} a \\ b \end{pmatrix}$$

A complex $n \times n$ system has been replaced by a real $2n \times 2n$ system, and any of our methods for real systems may now be used. It is also possible to replace this real system by two systems

$$(B^{-1}A + A^{-1}B)x = B^{-1}a + A^{-1}b$$
$$(B^{-1}A + A^{-1}B)y = B^{-1}b - A^{-1}a$$

of size $n \times n$ with identical coefficient matrices. This follows from

$$(B^{-1}A + A^{-1}B)x = B^{-1}(Ax - By) + A^{-1}(Bx + Ay) = B^{-1}a + A^{-1}b$$
$$(B^{-1}A + A^{-1}B)y = B^{-1}(Ay + Bx) + A^{-1}(By - Ax) = B^{-1}b - A^{-1}a$$

Using these smaller systems slightly shortens the overall computation.

17.150 Reduce the problem of inverting a complex matrix to that of inverting real matrices.

▌ Let the given matrix be $A + iB$ and its inverse $C + iD$. We are to find C and D such that $(A + iB)(C + iD) = I$. Suppose A is nonsingular so that A^{-1} exists. Then

$$C = (A + BA^{-1}B)^{-1} \qquad D = -A^{-1}B(A + BA^{-1}B)^{-1}$$

as may be verified by direct substitution. If B is nonsingular, then

$$C = B^{-1}A(AB^{-1}A + B)^{-1} \qquad D = -(AB^{-1}A + B)^{-1}$$

as may be verified by substitution. If both A and B are nonsingular, the two results are of course identical. In case both A and B are singular, but $(A + iB)$ is not, then a more complicated procedure seems necessary. First a real number t is determined such that the matrix $E = A + tB$ is nonsingular. Then, with $F = B - tA$, we find $E + iF = (1 - it)(A + iB)$ and so

$$(A + iB)^{-1} = (1 - it)(E + iF)^{-1}$$

This can be computed by the first method since E is nonsingular.

17.151 Extend Jacobi's method for finding eigenvalues and vectors to the case of a Hermitian matrix.

▌ We use the fact that a Hermitian matrix H becomes diagonalized under a unitary transformation, that is, $U^{-1}HU$ is a diagonal matrix. The matrices H and U have the properties $\overline{H}^T = H$ and $\overline{U}^T = U^{-1}$. The matrix U is to be obtained as an infinite product of matrices of the form

$$U_1 = \begin{bmatrix} \cos\phi & -\sin\phi e^{-i\theta} \\ \sin\phi e^{i\theta} & \cos\phi \end{bmatrix}$$

all other elements agreeing with those of I. The four elements shown are in positions (i, i), (i, k), (k, i) and (k, k). If the corresponding elements of H are

$$H = \begin{bmatrix} a & b - ic \\ b + ic & d \end{bmatrix}$$

then the (i, k) and (k, i) elements of $U^{-1}HU$ will have real and imaginary parts equal to zero,

$$(d - a)\cos\phi\sin\phi\cos\theta + b\cos^2\phi - b\sin^2\phi\cos 2\theta - c\sin^2\phi\sin 2\theta = 0$$
$$(a - d)\cos\phi\sin\phi\sin\theta - c\cos^2\phi + b\sin^2\phi\sin 2\theta - c\sin^2\phi\cos 2\theta = 0$$

if ϕ and θ are chosen so that

$$\tan\theta = \frac{c}{b} \qquad \tan 2\phi = \frac{2(b\cos\theta + c\sin\theta)}{a - d}$$

This type of rotation is applied iteratively as in Problem 17.121 until all off-diagonal elements have been made satisfactorily small. The (real) eigenvalues are then approximated by the resulting diagonal elements, and the eigenvectors by the columns of $U = U_1 U_2 U_3 \cdots$.

17.152 How may the eigenvalues and vectors of a general complex matrix be found? Assume all eigenvalues are distinct.

▮ As a first step we obtain a unitary matrix U such that $U^{-1}AU = T$ where T is an upper triangular matrix, all elements below the main diagonal being zero. Once again U is to be obtained as an infinite product of rotation matrices of the form U_1 shown in the preceding problem, which we now write as

$$U_1 = \begin{bmatrix} x & -\bar{y} \\ y & x \end{bmatrix}$$

The element in position (k, i) of $U_1^{-1}AU_1$ is then

$$a_{ki}x^2 + (a_{kk} - a_{ii})xy - a_{ik}y^2$$

To make this zero we let $y = Cx$, $x = 1/\sqrt{1 + |C|^2}$, which automatically assures us that U_1 will be unitary and then determine C by the condition $a_{ik}C^2 + (a_{ii} - a_{kk})C - a_{ki} = 0$ which makes

$$C = \frac{1}{2a_{ik}}\left[(a_{kk} - a_{ii}) \pm \sqrt{(a_{kk} - a_{ii})^2 + 4a_{ik}a_{ki}}\right]$$

Either sign may be used, preferably the one that makes $|C|$ smaller. Rotations of this sort are made in succession until all elements below the main diagonal are essentially zero. The resulting matrix is

$$T = U^{-1}AU = \begin{bmatrix} t_{11} & t_{12} & \cdots & t_{1n} \\ 0 & t_{22} & \cdots & t_{2n} \\ & & \cdots & \\ 0 & 0 & \cdots & t_{nn} \end{bmatrix}$$

where $U = U_1 U_2 \cdots U_N$. The eigenvalues of both T and A are the diagonal elements t_{ii}.

We next obtain the eigenvectors of T, as the columns of

$$W = \begin{bmatrix} 1 & w_{12} & w_{13} & \cdots & w_{1n} \\ 0 & 1 & w_{23} & \cdots & w_{2n} \\ 0 & 0 & 1 & \cdots & \\ & & & \cdots & \\ 0 & 0 & 0 & \cdots & w_{nn} \end{bmatrix}$$

The first column is already an eigenvector belonging to t_{11}. To make the second column an eigenvector belonging to t_{22} we require $t_{11}w_{12} + t_{12} = t_{22}w_{12}$ or $w_{12} = t_{12}/(t_{22} - t_{11})$ assuming $t_{11} \neq t_{22}$. Similarly, to make the third column an eigenvector we need

$$w_{23} = \frac{t_{23}}{t_{33} - t_{22}} \qquad w_{13} = \frac{t_{12}w_{23} + t_{13}}{t_{33} - t_{11}}$$

In general the w_{ik} are found from the recursion

$$w_{ik} = \sum_{j=i+1}^{k} \frac{t_{ij}w_{jk}}{t_{kk} - t_{ii}}$$

with $i = k - 1, k - 2, \ldots, 1$ successively. Finally the eigenvectors of A itself are available as the columns of UW.

17.153 Use the following matrix to show that Gaussian elimination can be carried through using complex arithmetic:

$$
\begin{aligned}
(1+i)x + (1-i)y + iz &= 1 \\
(1-i)x + (1+i)y + (1-i)z &= 0 \\
ix + (1-i)y + (1+i)z &= 0
\end{aligned}
$$

I The multipliers $l_{21} = (1-i)/(1+i) = -i$ and $l_{31} = i/(1+i) = (1+i)/2$ arrange zeros as usual in column one, as shown at the top in Fig. 17-14. Then $l_{32} = -i/2(1+i) = -(1+i)/4$ completes the triangularization, as shown at the bottom in the figure. The solution is then found by back-substitution.

The usual factorization is also in hand, the given matrix A being represented as

$$
A = LU = \begin{pmatrix} 1 & 0 & 0 \\ -i & 1 & 0 \\ (1+i)/2 & -(1+i)/4 & 1 \end{pmatrix} \cdot \begin{pmatrix} 1+i & 1-i & i \\ 0 & 2+2i & -i \\ 0 & 0 & (7+i)/4 \end{pmatrix}
$$

$$
\begin{aligned}
(1+i)x + (1-i)y + iz &= 1 \\
2(1+i)y - iz &= i \\
-iy + \tfrac{1}{2}(3+i)z &= -\tfrac{1}{2}(1+i) \\
\hline
(1+i)x + (1-i)y + iz &= 1 \\
2(1+i)y - iz &= i \\
+ \tfrac{1}{4}(7+i)z &= -\tfrac{1}{4}(3+i)
\end{aligned}
$$ **Fig. 17-14**

17.154 Also solve the system of the preceding problem by exchanging it for a system of six real equations.

I Separating the real and imaginary parts of A we have

$$
A = R + iJ = \begin{pmatrix} 1 & 1 & 0 \\ 1 & 1 & 1 \\ 0 & 1 & 1 \end{pmatrix} + i\begin{pmatrix} 1 & -1 & 1 \\ -1 & 1 & -1 \\ 1 & -1 & 1 \end{pmatrix}
$$

and the new system will be

$$
\begin{pmatrix} R & -J \\ J & R \end{pmatrix}\begin{pmatrix} x \\ y \end{pmatrix} = \begin{pmatrix} r \\ j \end{pmatrix}
$$

with $r = (1,0,0)^T$ and $j = (0,0,0)^T$. The solution by Gaussian elimination is $x = (.56,.16,-.44)^T$ and $y = (-.08,.12,-.08)^T$. The solution of the original system is then $x + iy$.

17.155 Invert the matrix

$$
A + iB = \begin{pmatrix} 1 & 0 \\ 1 & 0 \end{pmatrix} + i\begin{pmatrix} 0 & 0 \\ 1 & -1 \end{pmatrix} = \begin{pmatrix} 1 & 0 \\ 1+i & -i \end{pmatrix}
$$

I Improvisation would surely work but suppose we take the opportunity to exercise the routine of Problem 17.150. Since both A and B are singular, the full treatment will be necessary. Since $E = A + B$ is nonsingular, $t = 1$ will do. Pen and paper are then enough to produce these various matrices.

$$
\begin{pmatrix} -1 & 0 \\ 0 & -1 \end{pmatrix} \quad \begin{pmatrix} 1 & 0 \\ 2 & -1 \end{pmatrix} \quad \begin{pmatrix} 2 & 0 \\ 4 & -2 \end{pmatrix} \quad \begin{pmatrix} .5 & 0 \\ 1 & -.5 \end{pmatrix} \quad \begin{pmatrix} -.5 & 0 \\ 0 & -.5 \end{pmatrix} \quad \begin{pmatrix} (1+i)/2 & 0 \\ 1 & (i-1)/2 \end{pmatrix}
$$
$$
F = B - A \qquad E^{-1} \qquad E + FE^{-1}F \qquad (E + FE^{-1}F)^{-1} \qquad E^{-1}F(E + FE^{-1}F)^{-1} \qquad (E + iF)
$$

Multiplication by $(1-i)$ then brings the required inverse:

$$
(A + iB)^{-1} = \begin{pmatrix} 1 & 0 \\ 1-i & i \end{pmatrix}
$$

CHAPTER 18

Optimization

18.1 LINEAR PROGRAMMING

18.1 What is a linear programming problem?

▮ A linear programming problem requires that a linear function

$$H = c_1 x_1 + \cdots + c_n x_n$$

be minimized (or maximized) subject to constraints of the form

$$a_{i1} x_1 + \cdots + a_{in} x_n \le b_i \qquad 0 \le x_j$$

where $i = 1, \ldots, m$ and $j = 1, \ldots, n$. In vector form the problem may be written as

$$H(x) = c^T x = \text{minimum} \qquad Ax \le b, \ 0 \le x$$

18.2 THE SIMPLEX METHOD

18.2 Find x_1 and x_2 satisfying the inequalities

$$0 \le x_1 \qquad 0 \le x_2 \qquad -x_1 + 2x_2 \le 2 \qquad x_1 + x_2 \le 4 \qquad x_1 \le 3$$

and such that the function $F = x_2 - x_1$ is maximized.

▮ Since only two variables are involved it is convenient to interpret the entire problem geometrically. In an x_1, x_2 plane the five inequalities constrain the point (x_1, x_2) to fall within the shaded region of Fig. 18-1. In each case the equality sign corresponds to (x_1, x_2) being on one of the five linear boundary segments. Maximizing F subject to these constraints is equivalent to finding that line of slope 1 having the largest y intercept and still intersecting the shaded region. It seems clear that the required line L_1 is $1 = x_2 - x_1$ and the intersection point $(0, 1)$. Thus, for a maximum, $x_1 = 0$, $x_2 = 1$, $F = 1$.

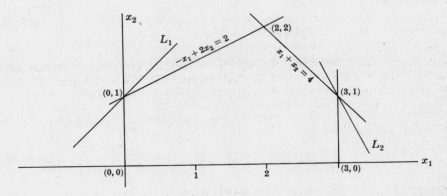

Fig. 18-1

18.3 With the same inequality constraints as in Problem 18.2, find (x_1, x_2) such that $G = 2x_1 + x_2$ is a maximum.

▮ We now seek the line of slope -2 and having the largest y intercept while still intersecting the shaded region. This line L_2 is $7 = 2x_1 + x_2$ and the required point has $x_1 = 3$, $x_2 = 1$ (see Fig. 18-1).

18.4   Find y_1, y_2, y_3 satisfying the constraints

$$0 \le y_1 \qquad 0 \le y_2 \qquad 0 \le y_3 \qquad y_1 - y_2 - y_3 \le 1 \qquad -2y_1 - y_2 \le -1$$

and minimizing $H = 2y_1 + 4y_2 + 3y_3$.

❚ Interpreting the entire problem geometrically, we find that the five inequalities constrain the point (y_1, y_2, y_3) to fall within the region pictured in Fig. 18-2. This region is unbounded in the positive y_1, y_2, y_3 directions, but is otherwise bounded by portions of five planes, shown shaded. These planes correspond to equality holding in our five constraints. Minimizing H subject to these constraints is equivalent to finding a plane with normal vector $(2, 4, 3)$ having smallest intercepts and still intersecting the given region. It is easy to discover that this plane is $1 = 2y_1 + 4y_2 + 3y_3$ and the intersection point is $(\frac{1}{2}, 0, 0)$.

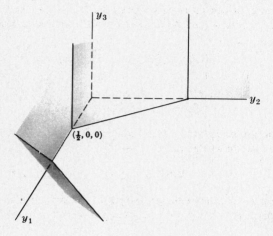

Fig. 18-2

18.5   List three principal features of linear programming problems and their solutions, which are illustrated by the previous problems.

❚ Let the problem be to find a point x with coordinates $(x_1, x_2, \ldots, x_n)$ subject to the constraints $0 \le x$, $Ax \le b$ and minimizing a function $H(x) = c^T x = \Sigma c_i x_i$. Calling a point that meets all the constraints a *feasible point* (if any such exists), then:

1 The set of feasible points is convex, that is, the line segment joining two feasible points consists entirely of feasible points. This is due to the fact that each constraint defines a half space and the set of feasible points is the intersection of these half-spaces.

2 There are certain *extreme feasible points*, the vertices of the convex set, identified by the fact that at least n of the constraints become equalities at these points. In the two-dimensional examples, exactly $n = 2$ boundary segments meet at such vertices. In the three-dimensional example, exactly three boundary planes meet at each such vertex. For $n \ge 3$ it is possible, however, that more planes (or hyperplanes) come together at a vertex.

3 The solution point is always an extreme feasible point. This is due to the linearity of the function $H(x)$ being minimized. (It is possible that two extreme points are solutions, in which case the entire edge joining them consists of solutions, etc.)

These three features of linear programming problems will not be proved here. They are also true if $H(x)$ is to be maximized or if the constraints read $Ax \ge b$.

18.6   What is the general idea behind the *simplex method* for solving linear programs?

❚ Since the solution occurs at an extreme feasible point, we may begin at some such point and compute the value of H. We then exchange this extreme point for its mate at the other end of an edge, in such a way that a smaller (in the case of a minimum problem) H value is obtained. The process of exchange and edge-following continues until H can no longer be decreased. This exchange algorithm is known as the simplex method. The details are provided in the following problem.

18.7 Develop the simplex method.

∎ Let the problem be

$$0 \le x \qquad Ax \le b \qquad H(x) = c^T x = \text{minimum}$$

We introduce *slack variables* $x_{n+1}, \ldots, x_{n+m}$ to make

$$a_{11}x_1 + a_{12}x_2 + \cdots + a_{1n}x_n + x_{n+1} = b_1$$
$$a_{21}x_1 + a_{22}x_2 + \cdots + a_{2n}x_n + x_{n+2} = b_2$$
$$\vdots$$
$$a_{m1}x_1 + a_{m2}x_2 + \cdots + a_{mn}x_n + x_{n+m} = b_m$$

Notice that these slack variables, like the other x_i, must be nonnegative. The use of slack variables allows us to identify an extreme feasible point in another way. Since equality in $Ax \le b$ now corresponds to a slack variable being zero, an extreme point becomes one where at least n of the variables $x_1, \ldots, x_{n+m}$ are zero. Or said differently, at an extreme feasible point at most m of these variables are nonzero. The matrix of coefficients has become

$$\begin{bmatrix} a_{11} & a_{12} & \cdots & a_{1n} & 1 & 0 & \cdots & 0 \\ a_{21} & a_{22} & \cdots & a_{2n} & 0 & 1 & \cdots & 0 \\ & & \cdots & & & & \cdots & \\ a_{m1} & a_{m2} & \cdots & a_{mn} & 0 & 0 & \cdots & 1 \end{bmatrix}.$$

the last m columns corresponding to the slack variables. Let the columns of this matrix be called $v_1, v_2, \ldots, v_{n+m}$. The linear system can then be written as

$$x_1 v_1 + x_2 v_2 + \cdots + x_{n+m} v_{n+m} = b$$

Now suppose that we know an extreme feasible point. For simplicity we will take it that $x_{m+1}, \ldots, x_{m+n}$ are all zero at this point so that $x_1, \ldots, x_m$ are the (at most m) nonzero variables. Then

(1) $$x_1 v_1 + x_2 v_2 + \cdots + x_m v_m = b$$

and the corresponding H value is

(2) $$H_1 = x_1 c_1 + x_2 c_2 + \cdots + x_m c_m$$

Assuming the vectors $v_1, \ldots, v_m$ linearly independent, all $n + m$ vectors may be expressed in terms of this basis.

(3) $$v_j = v_{1j} v_1 + \cdots + v_{mj} v_m \qquad j = 1, \ldots, n + m$$

Also define

(4) $$h_j = v_{1j} c_1 + \cdots + v_{mj} c_m - c_j \qquad j = 1, \ldots, n + m$$

Now, suppose we try to reduce H_1 by including a piece px_k, for $k > m$ and p positive. To preserve the constraint we multiply (3) for $j = k$ by p, which is still to be determined, and subtract from (1) to find

$$(x_1 - pv_{1k}) v_1 + (x_2 - pv_{2k}) v_2 + \cdots + (x_m - pv_{mk}) v_m + pv_k = b$$

Similarly from (2) and (4) the new value of H will be

$$(x_1 - pv_{1k}) c_1 + (x_2 - pv_{2k}) c_2 + \cdots + (x_m - pv_{mk}) c_m + pc_k = H_1 - ph_k$$

The change will be profitable only if $h_k > 0$. In this case it is optimal to make p as large as possible without a coefficient $x_i - pv_{ik}$ becoming negative. This suggests the choice

$$p = \min_i \frac{x_i}{v_{ik}} = \frac{x_l}{v_{lk}}$$

the minimum being taken over terms with positive v_{ik} only. With this choice of p the coefficient of c_l becomes zero, the others are nonnegative and we have a new extreme feasible point with H value

$$H_1' = H_1 - ph_k$$

which is definitely smaller than H_1. We also have a new basis, having exchanged the basis vector v_l for the new v_k. The process is now repeated until all h_j are negative or until for some positive h_k no v_{ik} is positive. In the former case the present extreme point is as good as any adjacent extreme point, and it can further be shown that it is as good as any other adjacent or not. In the latter case p may be arbitrarily large and there is no minimum for H.

Before another exchange can be made all vectors must be represented in terms of the new basis. Such exchanges have already been made in our section on matrix inversion but the details will be repeated. The vector v_l is to be replaced by the vector v_k. From

$$v_k = v_{1k}v_1 + \cdots + v_{mk}v_m$$

we solve for v_l and substitute into (3) to obtain the new representation

$$v_j = v_{1j}'v_1 + \cdots + v_{l-1,j}'v_{l-1} + v_{kj}'v_k + v_{l+1,j}'v_{l+1} + \cdots + v_{mj}'v_m$$

where

$$v_{ij}' = \begin{cases} v_{ij} - \dfrac{v_{lj}}{v_{lk}}v_{ik} & \text{for } i \neq l \\[2mm] \dfrac{v_{ij}}{v_{ik}} & \text{for } i = l \end{cases}$$

Also, substituting for v_l in (1) brings

$$x_1'v_1 + \cdots + x_{l-1}'v_{l-1} + x_k'v_k + x_{l+1}'v_{l+1} + \cdots + x_m'v_m = b$$

where

$$x_i' = \begin{cases} x_i - \dfrac{x_l}{v_{lk}}v_{ik} & \text{for } i \neq l \\[2mm] \dfrac{x_i}{v_{lk}} & \text{for } i = l \end{cases}$$

Furthermore, a short calculation proves

$$h_j' = v_{1j}'c_1 + \cdots + v_{mj}'c_m - c_j = h_j - \dfrac{v_{lj}}{v_{lk}}h_k$$

and we already have

$$H_1' = H_1 - \dfrac{x_l}{v_{lk}}h_k$$

The entire set of equations may be summarized compactly by displaying the various ingredients as

$$\begin{bmatrix} x_1 & v_{11} & v_{12} & \cdots & v_{1,n+m} \\ x_2 & v_{21} & v_{22} & \cdots & v_{2,n+m} \\ & & & \cdots & \\ x_m & v_{m1} & v_{m2} & \cdots & v_{m,n+m} \\ H_1 & h_1 & h_2 & \cdots & h_{n+m} \end{bmatrix}$$

Calling v_{lk} the *pivot*, all entries in the pivot row are divided by the pivot, the pivot column becomes zero except for a 1 in the pivot position, and all the other entries are subject to what was formerly called the *rectangle rule*. This will now be illustrated in a variety of examples.

18.8 Solve Problem 18.2 by the simplex method.

▌ After introducing slack variables, the constraints are

$$\begin{aligned} -x_1 + 2x_2 + x_3 \qquad\qquad &= 2 \\ x_1 + x_2 \qquad + x_4 \qquad &= 4 \\ x_1 \qquad\qquad\qquad + x_5 &= 3 \end{aligned}$$

with all five variables required to be nonnegative. Instead of maximizing $x_2 - x_1$ we will minimize $x_1 - x_2$. Such a switch between minimum and maximum problems is always available to us. Since the origin is an extreme feasible point, we may choose $x_1 = x_2 = 0$, $x_3 = 2$, $x_4 = 4$, $x_5 = 3$ to start. This is very convenient since it amounts to choosing v_3, v_4 and v_5 as our first basis, which makes all $v_{ij} = a_{ij}$. The starting display is therefore

Basis	b	v_1	v_2	v_3	v_4	v_5
v_3	2	-1	②	1	0	0
v_4	4	1	1	0	1	0
v_5	3	1	0	0	0	1
	0	-1	1	0	0	0

Comparing with the format in Problem 18.7, one finds the six vectors b and $v_1, \ldots, v_5$ forming the top three rows and the numbers $H, h_1, \ldots, h_5$ in the bottom row. Only h_2 is positive. This determines the pivot column. In this column there are two positive v_{i2} numbers, but $2/2 < 4/1$ and so the pivot is $v_{12} = 2$. This number has been circled. The formulas of the previous problem now apply to produce a new display. The top row is simply divided by 2 and all other entries are subjected to the rectangle rule:

Basis	b	v_1	v_2	v_3	v_4	v_5
v_2	1	$-\frac{1}{2}$	1	$\frac{1}{2}$	0	0
v_4	3	$\frac{3}{2}$	0	$-\frac{1}{2}$	1	0
v_5	3	1	0	0	0	1
	-1	$-\frac{1}{2}$	0	$-\frac{1}{2}$	0	0

The basis vector v_3 has been exchanged for v_2 and all vectors are now represented in terms of this new basis. But more important for this example, no h_j is now positive so the algorithm stops. The minimum of $x_1 - x_2$ is -1 (making the maximum of $x_2 - x_1$ equal to 1 as before). This minimum is achieved for $x_2 = 1$, $x_4 = 3$, $x_5 = 3$ as the first column shows. The constraints then make $x_1 = 0$, $x_3 = 0$ which we anticipate since the x_j not corresponding to basis vectors should always be zero. The results $x_1 = 0$, $x_2 = 1$ correspond to our earlier geometrical conclusions. Notice that the simplex algorithm has taken us from the extreme point $(0,0)$ of the set of feasible points to the extreme point $(0,1)$, which proves to be the solution point (see Fig. 18-1).

18.9   Solve Problem 18.3 by the simplex method

$\blacksquare$ Slack variables and constraints are the same as in the previous problem. We shall minimize $H = -2x_1 - x_2$. The origin being an extreme point, we may start with the display

Basis	b	v_1	v_2	v_3	v_4	v_5
v_3	2	-1	2	1	0	0
v_4	4	1	1	0	1	0
v_5	3	①	0	0	0	1
	0	2	1	0	0	0

Both h_1 and h_2 are positive, so we have a choice. Selecting $h_1 = 2$ makes v_{13} the pivot, since $3/1 < 4/1$. This pivot has been circled. Exchanging v_5 for v_1 we have a new basis, a new extreme point and a new display:

Basis	b	v_1	v_2	v_3	v_4	v_5
v_3	5	0	2	1	0	1
v_4	1	0	①	0	1	-1
v_1	3	1	0	0	0	1
	-6	0	1	0	0	-2

Now we have no choices. The new pivot has been circled and means that we exchange v_4 for v_2 with the result

Basis	b	v_1	v_2	v_3	v_4	v_5
v_3	3	0	0	1	-2	3
v_2	1	0	1	0	1	-1
v_1	3	1	0	0	0	1
	-7	0	0	0	-1	-1

Now no h_j is positive, so we stop. The minimum is -7, which agrees with the maximum of 7 for $2x_1 + x_2$ found in Problem 18.3. The solution point is at $x_1 = 3$, $x_2 = 1$, which also agrees with the result found in Problem 18.3. The simplex method has led us from $(0,0)$ to $(3,0)$ to $(3,1)$. The other choice available to us at the first exchange would have led us around the feasible set in the other direction.

18.10 Solve Problem 18.4 by the simplex method.

   ▌ With slack variables the constraints become

$$y_1 - y_2 - y_3 + y_4 \qquad = \quad 1$$
$$-2y_1 - y_2 \qquad\qquad + y_5 = -1$$

all five variables being required to be positive or zero. This time, however, the origin ($y_1 = y_2 = y_3 = 0$) is not a feasible point, as Fig. 18-2 shows and as the enforced negative value $y_5 = -1$ corroborates. We cannot therefore follow the starting procedure of the previous two examples based on a display such as

Basis	b	v_1	v_2	v_3	v_4	v_5
v_4	1	1	-1	-1	1	0
v_5	-1	-2	-1	0	0	1

The negative value $y_5 = -1$ in the b column cannot be allowed. Essentially our problem is that we do not have an extreme feasible point to start from. A standard procedure for finding such a point, even for a much larger problem than this, is to introduce an *artificial basis*. Here it will be enough to alter the second constraint, which contains the negative b component, to

$$-2y_1 - y_2 + y_5 - y_6 = -1$$

One new column may now be attached to our earlier display:

Basis	b	v_1	v_2	v_3	v_4	v_5	v_6
v_4	1	1	-1	-1	1	0	0
v_5	-1	-2	-1	0	0	1	-1

But an extreme feasible point now corresponds to $y_4 = y_6 = 1$, all other y_j being zero. This makes it natural to exchange v_5 for v_6 in the basis. Only a few sign changes across the v_6 row are required:

Basis	b	v_1	v_2	v_3	v_4	v_5	v_6
v_4	1	1	-1	-1	1	0	0
v_6	1	②	1	0	0	-1	1
	W	$2W-2$	$W-4$	-3	0	$-W$	0

The last row of this starting display will now be explained.

Introducing the artificial basis has altered our original problem, unless we can be sure that y_6 will eventually turn out to be zero. Fortunately this can be arranged, by changing the function to be minimized from $H = 2y_1 + 4y_2 + 3y_3$ as it was in Problem 18.3 to

$$H^* = 2y_1 + 4y_2 + 3y_3 + Wy_6$$

where W is such a large positive number that for a minimum we will surely have to make $y_6 = 0$. With these alterations we have a starting H value of W. The numbers h_j may also be computed and the last row of the starting display is as shown.

We now proceed in normal simplex style. Since W is large and positive we have a choice of two positive h_j values. Choosing h_1 leads to the circled pivot. Exchanging v_6 for v_1 brings a new display from which the last column has been dropped since v_6 is of no further interest:

v_4	$\frac{1}{2}$	0	$-\frac{3}{2}$	-1	1	$\frac{1}{2}$
v_1	$\frac{1}{2}$	1	$\frac{1}{2}$	0	0	$-\frac{1}{2}$
	1	0	-3	-3	0	-1

Since no h_j is positive we are already at the end. The minimum is 1, which agrees with our geometrical conclusion of Problem 18.4. Moreover, from the first column we find $y_1 = \frac{1}{2}$, $y_4 = \frac{1}{2}$ with all other y_j equal to zero. This yields the minimum point $(\frac{1}{2}, 0, 0)$ also found in Problem 18.4.

18.11 Minimize the function $H = 2y_1 + 4y_2 + 3y_3$ subject to the constraints $y_1 - y_2 - y_3 \leq -2$, $-2y_1 - y_2 \leq -1$, all y_j being positive or zero.

▮ Slack variables and an artificial basis convert the constraints to

$$y_1 - y_2 - y_3 + y_4 \quad - y_6 \quad\quad = -2$$
$$-2y_1 - y_2 \quad\quad + y_5 \quad - y_7 = -1$$

and much as in the preceding problem we soon have the starting display

Basis	b	v_1	v_2	v_3	v_4	v_5	v_6	v_7
v_6	2	-1	1	1	-1	0	1	0
v_7	1	2	①	0	0	-1	0	1
	$3W$	$W-2$	$2W-4$	$W-3$	$-W$	$-W$	0	0

The function to be minimized is

$$H^* = 2y_1 + 4y_2 + 3y_3 + Wy_6 + Wy_7$$

and this determines the last row. There are various choices for pivot and we choose the one circled. This leads to a new display by exchanging v_7 for v_2 and dropping the v_7 column:

v_6	1	-3	0	①	-1	1	1
v_2	1	2	1	0	0	-1	0
	$W+4$	$-3W+6$	0	$W-3$	$-W$	$W-4$	0

A new pivot has been circled and the final display follows:

v_3	1	-3	0	1	-1	1
v_2	1	2	1	0	0	-1
	7	-3	0	0	-3	-1

The minimum of H^* and H is 7, and it occurs at $(0, 1, 1)$.

18.3 THE DUALITY THEOREM

18.12 What is the *duality theorem* of linear programming?

▮ Consider these two linear programming problems:

Problem A	**Problem B**
$c^T x = \text{minimum}$	$y^T b = \text{maximum}$
$x \geq 0$	$y \geq 0$
$Ax \geq b$	$y^T A \leq c^T$

They are called dual problems because of the many relationships between them, such as the following:

1. If either problem has a solution, then the other does also and the minimum of $c^T x$ equals the maximum of $y^T b$.
2. For either problem the solution vector is found in the usual way. The solution vector of the dual problem may then be obtained by taking the slack variables in order, assigning those in the final basis the value zero and giving each of the others the corresponding value of $-h_j$.

These results will not be proved here but will be illustrated using our earlier examples. The duality makes it possible to obtain the solution of both Problems A and B by solving either one.

18.13 Show that Problems 18.2 and 18.4 are dual problems and verify the two relationships claimed in Problem 18.12.

▌ A few minor alterations are involved. To match Problems 18.2 and A we minimize $x_1 - x_2$ instead of maximizing $x_2 - x_1$. The vector c^T is then $(1, -1)$. The constraints are rewritten as

$$x_1 - 2x_2 \geq -2 \qquad -x_1 - x_2 \geq -4 \qquad -x_1 \geq -3$$

which makes

$$A = \begin{bmatrix} 1 & -2 \\ -1 & -1 \\ -1 & 0 \end{bmatrix} \qquad b = \begin{bmatrix} -2 \\ -4 \\ 3 \end{bmatrix}$$

For Problem B we then have

$$y^T A = \begin{bmatrix} y_1 - y_2 - y_3 \\ -2y_1 - y_2 \end{bmatrix} \leq \begin{bmatrix} 1 \\ -1 \end{bmatrix}$$

which are the constraints of Problem 18.4. The condition $y^T b = \text{maximum}$ is also equivalent to

$$y^T(-b) = 2y_1 + 4y_2 + 3y_3 = \text{minimum}$$

so that Problem 18.4 and B have also been matched. The extreme values for both problems proved to be 1, which verifies relationship 1 of Problem 18.12. From the final simplex display in Problem 18.12 we obtain $x^T = (0,1)$ and $y^T = (\frac{1}{2}, 0, 0)$ while from the computations of Problem 18.10 we find $y^T = (\frac{1}{2}, 0, 0)$ and $x^T = (0,1)$, verifying relationship 2.

18.14 Verify that Problems 18.3 and 18.11 are duals.

▌ The matrix A and vector b are the same as in Problem 18.13. However, we now have $c^T = (-2, -1)$. This matches Problem 18.3 with Problem A and Problem 18.11 with Problem B. The final display of Problem 18.9 yields $x^T = (3,1)$ and $y^T = (0,1,1)$ and the same results come from Problem 18.11. The common minimum of $c^T x$ and maximum of $y^T b$ is -7.

18.15 Solve this linear program graphically:

$$F = x_1 - 2x_2 = \text{minimum} \qquad 0 \leq x_1, 0 \leq x_2$$
$$x_1 + 2x_2 \leq 4 \qquad -x_1 + x_2 \leq 1 \qquad x_1 + x_2 \leq 3$$

▌ The graph appears as Fig. 18-3. There are five extreme feasible points, at which the function values range from $-\frac{8}{3}$ to 3. The minimum occurs at $(\frac{2}{3}, \frac{5}{3})$, the maximum at $(3,0)$.

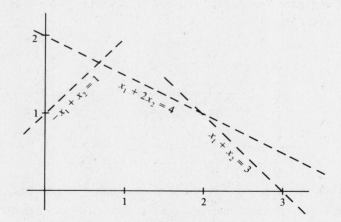

Fig. 18-3

18.16 Solve the preceding problem by the simplex method.

▌ The starting display and the results of two exchanges are given in Fig. 18-4. The minimum value $-\frac{8}{3}$ appears at the lower left corner, with the coordinates of the optimal point $(\frac{2}{3}, \frac{5}{3})$ just above it.

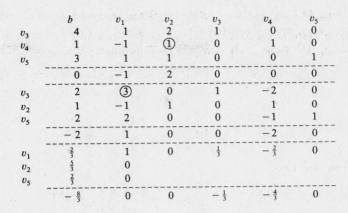

	b	v_1	v_2	v_3	v_4	v_5
v_3	4	1	2	1	0	0
v_4	1	-1	①	0	1	0
v_5	3	1	1	0	0	1
	0	-1	2	0	0	0
v_3	2	③	0	1	-2	0
v_2	1	-1	1	0	1	0
v_5	2	2	0	0	-1	1
	-2	1	0	0	-2	0
v_1	$\frac{2}{3}$	1	0	$\frac{1}{3}$	$-\frac{2}{3}$	0
v_2	$\frac{5}{3}$	0				
v_5	$\frac{2}{3}$	0				
	$-\frac{8}{3}$	0	0	$-\frac{1}{3}$	$-\frac{4}{3}$	0

Fig. 18-4

18.17 What is the dual of Problem 18.16? Using the final display in Fig. 18-4, read off the dual solution.

∎ Problem 18.16 will match the A of Problem 18.12 if all the inequalities are reversed by multiplying through by -1. The B problem, or dual, is then

$$G = -4y_1 - y_2 - 3y_3 = \text{maximum} \qquad 0 \le y_1, 0 \le y_2, 0 \le y_3$$
$$-y_1 + y_2 - y_3 \le 1 \qquad -2y_1 - y_2 - y_3 \le -2$$

which might look better as a minimum problem with inequalities reversed. As it stands, the maximum value will be the same $-\frac{8}{3}$ and the optimal point $(\frac{1}{3}, \frac{4}{3}, 0)$.

18.18 Solve this same dual problem directly by the simplex method.

∎ Since our simplex method seeks minima, we alter our goal to

$$-G = 4y_1 + y_2 + 3y_3 = \text{minimum}$$

and because of the negative $b_2 = -2$, introduce not only the two slack variables v_4 and v_5, but also an artificial basis variable v_6. The constraints then read

$$-y_1 + y_2 - y_3 + y_4 \qquad = 1$$
$$-2y_1 - y_2 - y_3 \qquad + y_5 - y_6 = -2$$

and the exchange of v_5 with v_6 then leads to the initial display at the top of Fig. 18-5. Two exchanges later the optimal value $\frac{8}{3}$ and optimal point $(\frac{1}{3}, \frac{4}{3}, 0)$ appear. The minimum $\frac{8}{3}$ of $-G$ means a maximum $-\frac{8}{3}$ of G.

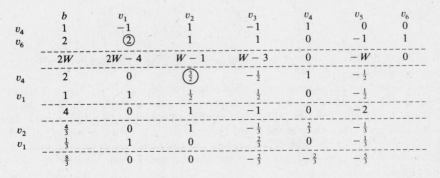

	b	v_1	v_2	v_3	v_4	v_5	v_6
v_4	1	-1	1	-1	1	0	0
v_6	2	②	1	1	0	-1	1
	$2W$	$2W-4$	$W-1$	$W-3$	0	$-W$	0
v_4	2	0	($\frac{3}{2}$)	$-\frac{1}{2}$	1	$-\frac{1}{2}$	
v_1	1	1	$\frac{1}{2}$	$\frac{1}{2}$	0	$-\frac{1}{2}$	
	4	0	1	-1	0	-2	
v_2	$\frac{4}{3}$	0	1	$-\frac{1}{3}$	$\frac{2}{3}$	$-\frac{1}{3}$	
v_1	$\frac{1}{3}$	1	0	$\frac{2}{3}$	0	$-\frac{1}{3}$	
	$\frac{8}{3}$	0	0	$-\frac{2}{3}$	$-\frac{2}{3}$	$-\frac{5}{3}$	

Fig. 18-5

18.19 From the final display in Fig. 18-5, recover the solution of the original problem, the dual of the dual.

∎ The maximum value $-\frac{8}{3}$ of G is still the minimum of F. The optimal point $(\frac{2}{3}, \frac{5}{3})$ appears at the bottom of the slack variable columns, signs changed.

18.20   In Problem 18.15 the maximum of $F = x_1 - 2x_2$, subject to the constraints, was found to be 3, at point $(3, 0)$. Find this same result by the simplex method.

▮ We will minimize $-F$ instead, Fig. 18-6 providing the details. The minimum of $-F$ being -3, the required maximum of F is 3. The optimal point is $(3, 0)$, the zero meaning that v_2 does not appear in the final basis. This agrees with the geometric results.

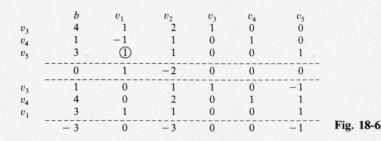

Fig. 18-6

18.21   What is the dual of the preceding problem? Find its solution from the final display in Fig. 18-6.

▮ The preceding problem fits model A of our dual pair if it is written in the form

$$-F = -x_1 + 2x_2 = \text{minimum} \quad \text{all } x_i \geq 0$$
$$-x_1 - 2x_2 \geq -4 \qquad x_1 - x_2 \geq -1 \qquad -x_1 - x_2 \geq -3$$

The dual is then

$$-4y_1 - y_2 - 3y_3 = \text{maximum} \quad \text{all } y_i \geq 0$$
$$-y_1 + y_2 - y_3 \leq -1 \qquad -2y_1 - y_2 - y_3 \leq 2$$

and from the figure we find the optimal value -3 and optimal point $(0, 0, 1)$.

18.22   Use Fig. 18-3 to minimize $F = x_1 - x_2$ subject to the same constraints graphed.

▮ One quickly finds that the minimum value is -1 and that it occurs at each point of the line $-x_1 + x_2 = 1$. Starting at the point $(0, 0)$, the simplex method moves in one exchange to $(0, 1)$, which is optimal. It could also produce the other end of this boundary segment, if allowed to continue in spite of the absence of positive h_i values.

18.23   What is integer programming?

▮ A simple illustration will serve. Suppose the maximum of $F = 4x_1 + 5x_2$ is sought, subject to the constraints

$$2x_1 + x_2 \leq 8 \qquad 3x_1 + 4x_2 \leq 24$$

both x_i nonnegative and integral

Figure 18-7 shows the area in which feasible points are confined. The fact that the solution must be integral is what identifies an integer program.

A common solution technique is called branch and bound. It involves a sequence of steps, in each of which a linear program is solved without the integer requirement. For the problem in hand, the first step finds the optimal point $A = (1.6, 4.8)$ with $F = 30.4$. This nonintegral point is now excluded from the competition by branching to two new linear programs, one with the added constraint $x_1 \leq 1$, the other with $x_1 \geq 2$. This eliminates the part of Fig. 18-7 between the vertical dotted lines. For the triangle remaining at lower right, an integral solution is found at $B = (2, 4)$ with $F = 28$. This now becomes a lower bound for the eventual solution and we try to improve on it.

The optimum for the area at the left occurs at $C = (1, 5.25)$ with $F = 30.25$. Since this is nonintegral, a second branching creates two new linear programs, one with added constraint $x_2 \geq 6$ and the other with $x_2 \leq 5$. However, the first of these has no feasible points and can be abandoned. The second is confined to the rectangle below the dotted horizontal line, and since all its corners are integral we sense the finish. The solution for this rectangle occurs at point $D = (1, 5)$ with $F = 29$. This is higher than the old bound of 28 and so supplants it, but since there are no nonintegral solutions awaiting further branching we can stop.

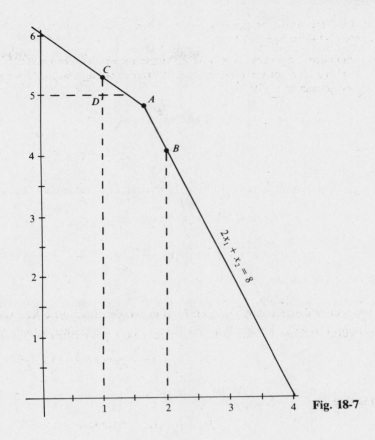

Fig. 18-7

18.24 Solve the integer program $G = x_2 - x_1$ = minimum, with the same constraints used in Problem 18.23.

❙ No branching or bounding is needed, the solution point $(4,0)$ with $G = -4$ being found at once.

18.25 Maximize $F = x_1 - x_2 + 2x_3$ subject to

$$x_1 + x_2 + 3x_3 + x_4 \le 5$$
$$x_1 + x_3 - 4x_4 \le 2$$

and all $x_k \ge 0$.

❙ The full solution process is displayed as Fig. 18-8. The minimum of $-F$ is shown as -4.4, making 4.4 the maximum F. The optimal point is $(4.4, 0, 0, .6)$.

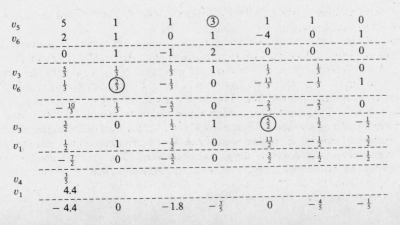

v_5	5	1	1	③	1	1	0
v_6	2	1	0	1	-4	0	1
	0	1	-1	2	0	0	0
v_3	$\frac{5}{3}$	$\frac{1}{3}$	$\frac{1}{3}$	1	$\frac{1}{3}$	$\frac{1}{3}$	0
v_6	$\frac{1}{3}$	$\left(\frac{2}{3}\right)$	$-\frac{1}{3}$	0	$-\frac{13}{3}$	$-\frac{1}{3}$	1
	$-\frac{10}{3}$	$\frac{1}{3}$	$-\frac{5}{3}$	0	$-\frac{2}{3}$	$-\frac{2}{3}$	0
v_3	$\frac{3}{2}$	0	$\frac{1}{2}$	1	$\left(\frac{5}{2}\right)$	$\frac{1}{2}$	$-\frac{1}{2}$
v_1	$\frac{1}{2}$	1	$-\frac{1}{2}$	0	$-\frac{13}{2}$	$-\frac{1}{2}$	$\frac{3}{2}$
	$-\frac{7}{2}$	0	$-\frac{3}{2}$	0	$\frac{3}{2}$	$-\frac{1}{2}$	$-\frac{1}{2}$
v_4	$\frac{3}{5}$						
v_1	4.4						
	-4.4	0	-1.8	$-\frac{3}{5}$	0	$-\frac{4}{5}$	$-\frac{1}{5}$

Fig. 18-8

18.26 Solve the dual of the preceding problem.

▮ Thinking of that problem as $-F$ = minimum, subject to

$$-x_1 - x_2 - 3x_3 - x_4 \geq -5 \qquad -x_1 - x_3 + 4x_4 \geq -2$$

and all $x_k \geq 0$, the dual is $-5y_1 - 2y_2$ = maximum, subject to

$$-y_1 - y_2 \leq -1 \qquad -y_1 \leq 1 \qquad -3y_1 - y_2 \leq -2 \qquad -y_1 + 4y_2 \leq 0$$

with all $y_k \geq 0$. From the final display in Fig. 18-8 we find the maximum to be -4.4 and the optimal point $(\frac{4}{5}, \frac{1}{5})$.

18.27 Maximize $F = 2x + y$ subject to $x_k \geq 0$ and the constraints

$$x - y \leq 2 \qquad x + y \leq 6 \qquad x + 2y \leq A$$

Treat the cases $A = 0$, 6 and 12.

▮ For $A = 0$ there is only one feasible point, at $(0, 0)$. The maximum is therefore 0. For the other values of A the starting display is shown at the left in Fig. 18-9. The two finishing displays appear to its right. For $A = 6$ the maximum value is 8 and the optimal point $(\frac{10}{3}, \frac{4}{3})$. For $A = 12$ the maximum is 10 and the optimal point $(4, 2)$.

v_3	2	1	-1	1	0	0	v_1	$\frac{10}{3}$		v_1	4					
v_4	6	1	1	0	1	0	v_4	$\frac{4}{3}$		v_2	2					
v_5	A	1	2	0	0	1	v_2	$\frac{4}{3}$		v_5	4					
	0	2	1	0	0	0	-8	0	0	-1	0	-1	-10	0	0	$-\frac{1}{2}$	$-\frac{3}{2}$	0

Initial Final, $A = 6$ Final, $A = 12$ **Fig. 18-9**

18.28 Solve the dual of the preceding problem for $A = 6$ and $A = 12$.

▮ The dual can be worked into the form $2y_1 + 6y_2 + Ay_3$ = min subject to

$$y_1 + y_2 + y_3 \geq 2 \qquad -y_1 + y_2 + 2y_3 \geq 1$$

and all $y_k \geq 0$. From the final display in Fig. 18-9 we find the minimum for $A = 6$ to be 8 and the optimal point $(1, 0, 1)$. Similarly, for $A = 12$ we have a minimum of 10 and optimal point $(\frac{1}{2}, \frac{3}{2}, 0)$.

18.4 SOLUTION OF TWO-PERSON GAMES

18.29 Show how a two-person game may be made equivalent to a linear program.

▮ Let the payoff matrix, consisting of positive numbers a_{ij}, be

$$A = \begin{bmatrix} a_{11} & a_{12} & a_{13} \\ a_{21} & a_{22} & a_{23} \\ a_{31} & a_{32} & a_{33} \end{bmatrix}$$

by which we mean that when player R has chosen row i of this matrix and player C has (independently) chosen column j, a payoff of amount a_{ij} is then made from R to C. This constitutes one play of the game. The problem is to determine the best strategy for each player in the selection of rows or columns. To be more specific, let C choose the three columns with probabilities p_1, p_2 and p_3, respectively. Then

$$p_1, p_2, p_3 \geq 0 \quad \text{and} \quad p_1 + p_2 + p_3 = 1$$

Depending on R's choice of row, C now has one of the following three quantities for his expected winnings:

$$P_1 = a_{11}p_1 + a_{12}p_2 + a_{13}p_3$$
$$P_2 = a_{21}p_1 + a_{22}p_2 + a_{23}p_3$$
$$P_3 = a_{31}p_1 + a_{32}p_2 + a_{33}p_3$$

Let P be the least of these three numbers. Then, no matter how R plays, C will have expected winnings of at least P on each play and therefore asks himself how this amount P can be maximized. Since all the numbers involved are positive, so is P, and we obtain an equivalent problem by letting

$$x = \frac{p_1}{P} \qquad x_2 = \frac{p_2}{P} \qquad x_3 = \frac{p_3}{P}$$

and minimizing

$$F = x_1 + x_2 + x_3 = \frac{1}{P}$$

The various constraints may be expressed as $x_1, x_2, x_3 \geq 0$ and

$$a_{11}x_1 + a_{12}x_2 + a_{13}x_3 \geq 1$$
$$a_{21}x_1 + a_{22}x_2 + a_{23}x_3 \geq 1$$
$$a_{31}x_1 + a_{32}x_2 + a_{33}x_3 \geq 1$$

This is the type A problem of our duality theorem with $c^T = b^T = (1, 1, 1)$.

Now look at things from R's point of view. Suppose he chooses the three rows with probabilities q_1, q_2 and q_3, respectively. Depending on C's choice of column he has one of the following quantities as his expected loss:

$$q_1 a_{11} + q_2 a_{21} + q_3 a_{31} \leq Q$$
$$q_1 a_{12} + q_2 a_{22} + q_3 a_{32} \leq Q$$
$$q_1 a_{13} + q_2 a_{23} + q_3 a_{33} \leq Q$$

where Q is the largest of the three. Then, no matter how C plays, R will have expected loss of no more than Q on each play. Accordingly he asks how this amount Q can be minimized. Since $Q > 0$, we let

$$y_1 = \frac{q_1}{Q} \qquad y_2 = \frac{q_2}{Q} \qquad y_3 = \frac{q_3}{Q}$$

and consider the equivalent problem of maximizing

$$G = y_1 + y_2 + y_3 = \frac{1}{Q}$$

The constraints are $y_1, y_2, y_3 \geq 0$ and

$$y_1 a_{11} + y_2 a_{21} + y_3 a_{31} \leq 1$$
$$y_1 a_{12} + y_2 a_{22} + y_3 a_{32} \leq 1$$
$$y_1 a_{13} + y_2 a_{23} + y_3 a_{33} \leq 1$$

This is the type B problem of our duality theorem with $c^T = b^T = (1, 1, 1)$. We have discovered that R's problem and C's problem are duals. This means that the maximum P and minimum Q values will be the same, so that both players will agree on the average payment which is optimal. It also means that the optimal strategies for both players may be found by solving just one of the dual programs. We choose R's problem since it avoids the introduction of an artificial basis.

The same arguments apply for payoff matrices of other sizes. Moreover, the requirement that all a_{ij} be positive can easily be removed since, if all a_{ij} are replaced by $a_{ij} + a$, then P and Q are replaced by $P + a$ and $Q + a$. Thus only the value of the game is changed, not the optimal strategies. Examples will now be offered.

18.30 Find optimal strategies for both players and optimal payoff for the game with matrix

$$A = \begin{bmatrix} 0 & 1 & 2 \\ 1 & 0 & 1 \\ 1 & 2 & 0 \end{bmatrix}$$

▮ Instead we minimize the function $-G = -y_1 - y_2 - y_3$ subject to the constraints

$$
\begin{aligned}
y_2 + y_3 + y_4 &= 1 \\
y_1 \qquad + 2y_3 \qquad + y_5 &= 1 \\
2y_1 + y_2 \qquad\qquad\qquad + y_6 &= 1
\end{aligned}
$$

all y_j including the slack variables y_4, y_5, y_6 being nonnegative. Since the origin is an extreme feasible point, we have this starting display:

Basis	b	v_1	v_2	v_3	v_4	v_5	v_6
v_4	1	0	1	1	1	0	0
v_5	1	1	0	2	0	1	0
v_6	1	②	1	0	0	0	1
	0	1	1	1	0	0	0

Using the indicated pivots we make three exchanges as follows:

v_4	1	0	1	1	1	0	0
v_5	$\frac{1}{2}$	0	$-\frac{1}{2}$	②	0	1	$-\frac{1}{2}$
v_1	$\frac{1}{2}$	1	$\frac{1}{2}$	0	0	0	$\frac{1}{2}$
	$-\frac{1}{2}$	0	$\frac{1}{2}$	1	0	0	$-\frac{1}{2}$
v_4	$\frac{3}{4}$	0	$\frac{5}{4}$	0	1	$-\frac{1}{2}$	$\frac{1}{4}$
v_3	$\frac{1}{4}$	0	$-\frac{1}{4}$	1	0	$\frac{1}{2}$	$-\frac{1}{4}$
v_1	$\frac{1}{2}$	1	$\frac{1}{2}$	0	0	0	$\frac{1}{2}$
	$-\frac{3}{4}$	0	$\frac{3}{4}$	0	0	$-\frac{1}{2}$	$-\frac{1}{4}$
v_2	$\frac{3}{5}$	—	—	—	—	—	—
v_3	$\frac{2}{5}$	—	—	—	—	—	—
v_1	$\frac{1}{5}$	—	—	—	—	—	—
	$-\frac{6}{5}$	0	0	0	$-\frac{3}{5}$	$-\frac{1}{5}$	$-\frac{2}{5}$

From the final display we deduce that the optimum payoff, or value of the game, is $\frac{5}{6}$. The optimal strategy for R can be found directly by normalizing the solution $y_1 = \frac{1}{5}$, $y_2 = \frac{3}{5}$, $y_3 = \frac{2}{5}$. The probabilities q_1, q_2, q_3 must be proportional to these y_j but must sum to 1. Accordingly,

$$q_1 = \frac{1}{6} \qquad q_2 = \frac{3}{6} \qquad q_3 = \frac{2}{6}$$

To obtain the optimal strategy for C we note that there are no slack variables in the final basis so that putting the $-h_j$ in place of the (nonbasis) slack variables,

$$x_1 = \frac{3}{5} \qquad x_2 = \frac{1}{5} \qquad x_3 = \frac{2}{5}$$

Normalizing brings

$$p_1 = \frac{3}{6} \qquad p_2 = \frac{1}{6} \qquad p_3 = \frac{2}{6}$$

If either player uses the optimal strategy for mixing his choices the average payoff will be $\frac{5}{6}$. To make the game fair, all payoffs could be reduced by this amount, or C could be asked to pay this amount before each play is made.

18.31 Find the optimal strategy for each player and the optimal payoff for the game with matrix

$$A = \begin{bmatrix} 0 & 3 & 4 \\ 1 & 2 & 1 \\ 4 & 3 & 0 \end{bmatrix}$$

❚ Notice that the center element is both the maximum in its row and the minimum in its column. It is also the smallest row maximum and the largest column minimum. Such a *saddle point* identifies a game with *pure strategies*. The simplex method leads directly to this result using the saddle point as pivot. The starting display is

Basis	b	v_1	v_2	v_3	v_4	v_5	v_6
v_4	1	0	1	4	1	0	0
v_5	1	3	②	3	0	1	0
v_6	1	4	1	0	0	0	1
	0	1	1	1	0	0	0

One exchange is sufficient:

v_4	$\frac{1}{2}$	—	—	—	—	—	—
v_2	$\frac{1}{2}$	—	—	—	—	—	—
v_6	$\frac{1}{2}$	—	—	—	—	—	—
	$-\frac{1}{2}$	$-\frac{1}{2}$	0	$-\frac{1}{2}$	0	$-\frac{1}{2}$	0

The optimal payoff is the negative reciprocal of $-\frac{1}{2}$, that is, the pivot element 2. The optimal strategy for R is found directly. Since $y_1 = 0$, $y_2 = \frac{1}{2}$, $y_3 = 0$, we normalize to obtain the pure strategy

$$q_1 = 0 \qquad q_2 = 1 \qquad q_3 = 0$$

Only the second row should ever be used. The strategy for C is found through the slack variables. Since v_4 and v_6 are in the final basis we have $x_1 = x_3 = 0$, and finally $x_2 = -h_5 = \frac{1}{2}$. Normalizing, we have another pure strategy

$$p_1 = 0 \qquad p_2 = 1 \qquad p_3 = 0$$

18.32 Find optimal strategies and payoff for the game

$$A = \begin{pmatrix} 1 & 3 \\ 4 & 2 \end{pmatrix}$$

❚ The simplex method is hardly needed but Fig. 18-10 provides the details of its execution. The game has value $\frac{5}{2}$, the optimal strategy for R being found by normalizing $\frac{1}{5}, \frac{1}{5}$ to $\frac{1}{2}, \frac{1}{2}$ and that for C working out to $\frac{1}{4}, \frac{3}{4}$.

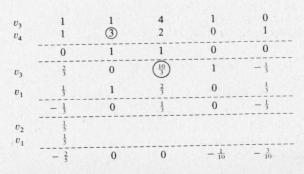

Fig. 18-10

18.33 Use linear programming to find optimal strategies for both players in the game

$$\begin{pmatrix} -6 & 4 \\ 4 & -2 \end{pmatrix}$$

▮ The starting display is at the left and the final display at the right below. The value of the game is $\frac{1}{4}$ and to R's disadvantage, in spite of the zero sum in the payoff matrix. Optimal strategies are $\frac{3}{8}, \frac{5}{8}$ for both players.

v_3	1	−6	4	1	0	v_2		$\frac{5}{2}$
v_4	1	④	−2	0	1	v_1		$\frac{3}{2}$
	0	1	1	0	0	−4	0 0	$-\frac{3}{2}$ $-\frac{5}{2}$

18.34 Solve as a linear program the game with the matrix

$$\begin{pmatrix} 3 & 1 \\ 2 & 3 \end{pmatrix}$$

▮ Pivots at 1, 2 and 2, 3 arrange the two-step computation from the initial to final displays shown below. The game has value $\frac{7}{3}$, with optimal strategies $\frac{1}{3}, \frac{2}{3}$ for R and $\frac{2}{3}, \frac{1}{3}$ for C.

v_3	1	③	2	1	0	v_1		$\frac{1}{7}$
v_4	1	1	3	0	1	v_2		$\frac{2}{7}$
	0	1	1	0	0	$-\frac{3}{7}$	0 0	$-\frac{2}{7}$ $-\frac{1}{7}$

18.35 Solve the game with the matrix

$$\begin{pmatrix} 0 & 3 & -4 \\ 3 & 0 & 5 \\ -4 & 5 & 0 \end{pmatrix}$$

▮ The simplex method takes three steps to progress from the initial to final displays below. The game has value $\frac{10}{7}$. The optimal strategies are $\frac{5}{14}, \frac{8}{14}, \frac{1}{14}$ for both players.

v_4	1	0	3	−4	1	0	0	v_2	$\frac{8}{20}$
v_5	1	3	0	5	0	1	0	v_1	$\frac{5}{20}$
v_6	1	−4	5	0	0	0	1	v_3	$\frac{1}{20}$
	0	1	1	1	0	0	0	$-\frac{7}{10}$ 0 0 0	$-\frac{5}{20}$ $-\frac{8}{20}$ $-\frac{1}{20}$

18.5 NONLINEAR OPTIMIZATION: FUNCTIONS OF ONE VARIABLE

18.36 Find the minimum value of $y(x) = (\tan x)/x^2$ on $(0, \pi/2)$.

▮ The function is positive on this interval and infinite at both ends. It has a derivative at all interior points, so the familiar routine for hunting minima can be used. We have

$$y'(x) = \frac{1}{x^2 \cos^2 x} - \frac{2 \tan x}{x^3}$$

and seek its zero. Newton's method would require the next derivative, and though this is hardly a heavy challenge, the situation was deemed right for *regula falsi*. From the starting interval (.5, 1) it moved in five steps to (.5, .9478233) and then in five more to (.5, .9477473). Two steps later the .5 was finally abandoned and the interval reduced to (.94775, .94775). The value of $y(x)$ at this point was found to be 1.54944 and a brief exploration of the neighborhood found this same figure to hold all the way from $x = .9474$ to .9481.

18.37 What is the *golden section method*?

▮ It is a simple and interesting procedure for finding minimum and maximum points of a function. The function need not be differentiable, but the method assumes that the extreme point has been isolated, that is, there is not more than one in the interval to be investigated. Suppose first that the interval is $(0, 1)$, the target a minimum of $f(z)$ and that two symmetric points $z_1 = t$, $z_2 = 1 - t$ have been chosen, in a manner to be described (see Fig.

18-11). Comparing function values, we find $f(z_1)$ to be greater than $f(z_2)$. This places the minimum to the right of z_1 so we can set z_1 as a new lower bound for the minimum. [If $f(z_2)$ had been the larger, then z_2 would have become a new upper bound on the minimum.] This completes one step of the algorithm. The next begins, if the lower bound was the one replaced, by retaining z_2 in the role just played by z_1, and choosing a new z_3 to replace z_2.

There remains the question of how points are to be chosen. Returning to the initial picture, we require that

$$\frac{t}{1} = \frac{1 - 2t}{1 - t}$$

which assures that if, for example, $z_1 = t$ becomes the new lower bound and $z_2 = 1 - t$ the new z_1, then the interval will remain divided in the same proportions for the next step. (Because of the symmetry this also holds if the upper bound is replaced.) The solution of the quadratic equation is $t = (3 - \sqrt{5})/2$, the other root being greater than 1 and of no use here. This is a familiar figure, encountered earlier in connection with Fibonacci numbers. Its value is .38197 to five places, with $1 - t = .61803$. The ratio $(1 - t)/t = 1.61803$ has long been known as the golden section, and believed to have aesthetic properties.

Finally, for an arbitrary interval (a, b), let $z = (x - a)/(b - a)$ providing a linear mapping of $(0, 1)$ onto (a, b). The points x_1 and x_2 are then to be chosen as follows:

$$x_1 = a + t(b - a) = (1 - t)a + tb$$
$$x_2 = a + (1 - t)(b - a) = ta + (1 - t)b$$

The golden section algorithm is now in order.

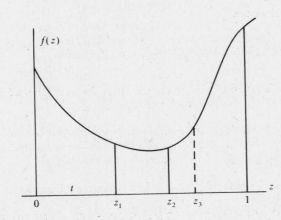

Fig. 18-11

18.38   What is the speed of convergence of the golden section algorithm?

▮ At each step the interval is reduced by $100t$ ($\sim 38\%$) so 62% remains. It follows that after two steps 38% remains, since $(1 - t)^2 = t$. Ten iterations leave about .008 of the original interval, while 20 leave .00007. This is not up to Newtonian speed, but it is often satisfactory and can be counted on for good functions and bad.

18.39   Apply the golden section method to find the minimum of $(\tan x)/x^2$ and compare with Problem 18.36.

▮ The interval of interest is $(0, \pi/2)$ but the endpoints will have to be avoided. If $(.2, 1.3)$ is chosen as the start, then

$$x_1 = (1 - t)(.2) + t(1.3) = .62 \qquad x_2 = t(.2) + (1 - t)(1.3) = .88$$

displaying only two places for simplicity. Next $f(x_1) = 1.86$ and $f(x_2) = 1.56 < f(x_1)$. Replacing a by x_1, we have the new bounds $(.62, 1.3)$. Also replacing x_1 by x_2 and choosing the new

$$x_2 = t(.62) + (1 - t)(1.3) = 1.04$$

we are ready for the next step. This time $f(x_1) = 1.562$ is less than $f(x_2) = 1.575$, so it is the upper bound that is due for replacement. The first 10 steps went as follows, the asterisks indicating no change,

| a | .2 | .62 | * | .78 | .88 | * | .918 | * | .9322 | .9411 | * |
| b | 1.3 | * | 1.04 | * | * | .979 | * | .9556 | * | * | .9501 |

while the thirteenth managed (.94776, .94776). This is probably more accuracy than is needed, since the bottom was found to be flat (to five decimal places) between .9474 and .9481. The value obtained by *regula falsi* was .94775.

18.40 The minimum of $f(x) = \sqrt{|x|}$ is certainly zero, and at $x = 0$, the function having a cusp with infinite derivative at this point. Apply the golden section method through just a few iterations.

▌ The method is undaunted by the infinite derivative and its casual rate of convergence is not further slowed. Here are the first 10 iterations, starting from $(a, b) = (-1, 1)$:

| a | −1 | * | −.53 | −.24 | * | −.125 | −.056 | * | −.029 | −.013 | * |
| b | 1 | .24 | * | * | .056 | * | * | .013 | * | * | .003 |

It appears that the upper bound sets the pace, the lower playing a continual catch-up. Notice that after one step the bounds are 1.24 units apart, or 62% of the original interval. After two steps they are .77 units apart, or 38% of the original interval and so on. The designed reduction is proceeding, with no regard for the bad behavior of the function.

18.41 Suppose the minimum position of $f(x)$ is within the original a, b bounds (as it must be unless a blunder has been made) but not between the first x_1, x_2 pair. This can happen if one bound is too close to the minimum. Does this effect the operation of the golden section algorithm?

▌ Take the function $f(x) = \sqrt{|x|}$ of the preceding problem with $(a, b) = (-.1, 1)$. These initial bounds are quite lopsided with respect to the minimum at zero. The first x_1, x_2 pair is .32, .58 and does not straddle the minimum. It leads to the new set of bounds $(-.1, .58)$, followed by the events

| x_1, x_2 | .16, .32 | .06, .16 | −.0008, .06 | −.04, −.0008 |
| (a, b) | $(-.1, .32)$ | $(-.1, .16)$ | $(-.1, .06)$ | $(-.04, .06)$ |

Notice that the third of these x_1, x_2 pairs does finally straddle the minimum position and then notice that the following pair does not. In fact, as the algorithm proceeds the pairs appear to be open-minded on the issue of straddle, but the bounds steadily close in on zero at the usual speed.

18.42 How can the golden section method be used to find a maximum instead of a minimum of $f(x)$? Or a zero of $f(x)$?

▌ To find the maximum, one looks for the minimum of $-f(x)$. To find a zero, the absolute value of $f(x)$ is minimized. As usual with this method, the minimum should be isolated first. As an example, take $f(x) = (\log x)/x$ over $(1, \infty)$. Its maximum is at $x = e$. Switching to $-f(x)$ and choosing the initial bounds $(1, 4)$, we find these early developments:

| x_1, x_2 | 2.15, 2.85 | 2.85, 3.29 | 2.58, 2.85 |
| (a, b) | $(2.15, 4)$ | $(2.15, 3.29)$ | $(2.58, 3.29)$ |

With all function values negative, the "larger" for each pair will be the smaller in size, and with this in mind the various exchanges can be followed. On the seventeenth step the bounds are (2.7176, 2.7184) and the familiar three place value is in hand.

18.43 What is a constrained minimum?

▌ As an illustration suppose the minimum of $f(x) = -\log(x)/x$ is sought subject to the additional requirement that $g(x) = x^2 - 2x$ be less than or equal to 0. A moment's thought determines that g is negative only in the interval $(0, 2)$, so the absolute minimum of $f(x)$ is ruled out. It is then a short step to the deduction that $x = 2$ is the minimum position, because $f(x)$ is decreasing over the permitted interval. The condition on $g(x)$ is called a constraint and $x = 2$ is a constrained minimum. Constraints are seldom a major hurdle at the one-dimensional level, several being accommodated by reasoning much like the preceding. But with functions of two or more variables things get somewhat more sticky.

18.44   How can the golden section method be adapted to handle one or more constraints?

   ❚   There are at least two principal ways. The direct method simply checks each new x_1, x_2 pair to see if any constraint is violated and sets the new bound accordingly. Take once more the function $f(x) = -\log(x)/x$ with initial bounds $(1, 4)$ and the constraint $g(x) = x^2 - 2x \le 0$. The left bound does satisfy the constraint and the "objective" function $f(x)$ is decreasing, so a minimum toward the right can be expected. An initial left bound with these qualifications is called "feasible" and reduces our constraint checks to the right bound. As each x_1, x_2 pair is generated we have only to check x_2. If $g(x_2)$ is positive, then x_2 becomes the new right bound, with $f(x)$ playing for the moment a passive role. Otherwise things go along as usual. Here are the early stages of the run.

x_1, x_2	2.15, 2.85	1.71, 2.15	1.44, 1.71
(a, b)	$(1, 2.85)$	$(1, 2.15)$	$(1.44, 2.15)$

The first two values of x_2 violate the constraint and these at once become new bounds b. But 1.71 passes the constraint test and it is $f(x)$ that dictates the new value of bound a. After 10 steps the bounds are $(1.979, 2.003)$ and after 20 they reach $(1.9999, 2.0001)$.

18.45   What is the indirect method of handling constraints and what are penalty functions?

   ❚   To use the indirect method one converts the constrained problem to an equivalent unconstrained problem. One way to do this defines a substitute objective function

$$F(x) = f(x) + P \sum_k \delta_k g_k^2(x)$$

in which δ_k is 0 if $g_k(x)$ is satisfied (≤ 0) and 1 otherwise. The underlying idea is that $F(x)$ is penalized for any constraint violations, but this is the only way in which the constraints now influence the problem, which from this point onward is dealt with as an unconstrained problem. P is called the penalty multiplier and is used to adjust the constraint impact. Large values bring the constraints more fully into play, but they may also introduce numerical difficulties into the search for the minimum. A common procedure is to begin with a small multiplier P and gradually increase it, using information gained in earlier stages to assist in the latter. One-dimensional problems hardly require such sequential minimization techniques, but can be used for introductory examples.

18.46   Plot the function $F(x)$ of Problem 18.45 for $f(x) = -\log(x)/x$ on $(1, 3)$ using several values of the penalty multiplier P. The one constraint $g(x) = x^2 - 2x \le 0$ is imposed.

   ❚   With only one constraint, we have

$$F(x) = -\log(x)/x + P(x^2 - 2x)^2$$

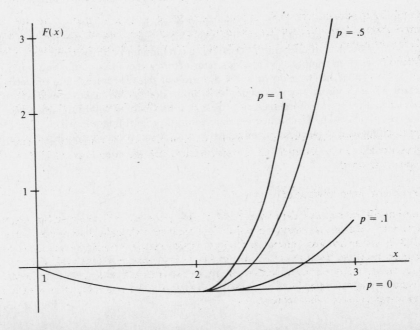

Fig. 18-12

for which four plots are offered in Fig. 18-12, the P multipliers ranging from 0 to 1. $P = 0$ corresponds to Problem 18.42, with no constraint. On this scale the minimum seems extremely flat. As P is increased the minimum retreats fairly quickly toward $x = 2$ and the curves take sharper and sharper turns. The limit for P tending to infinity can be imagined.

18.47 Use the golden section algorithm with the penalty function of the preceding problem to find the constrained minimum at $x = 2$.

I The only difference from our earlier effort is using $F(x)$ in place of $f(x)$. Otherwise the algorithm runs its usual course. For $P = 0$ it rediscovers the minimum at 2.718. Here are a few other discoveries. All runs started with the bounds (1, 4).

P	.01	.1	.25	.5	1	2	5	10
x	2.28	2.07	2.033	2.018	2.009	2.005	2.002	2.001

The last run needed a few more iterations than the others to narrow the gap between bounds enough for three place accuracy.

18.6 FUNCTIONS OF N VARIABLES: NO CONSTRAINTS

18.48 What is the idea of a steepest descent algorithm?

I A variety of *minimization methods* involves a function $S(x, y)$ defined in such a way that its minimum value occurs precisely where $f(x, y) = 0$ and $g(x, y) = 0$. The problem of solving these two equations simultaneously may then be replaced by the problem of minimizing $S(x, y)$. For example,

$$S(x, y) = [f(x, y)]^2 + [g(x, y)]^2$$

surely achieves its minimum of zero wherever $f = g = 0$. This is one popular choice of $S(x, y)$. The question of how to find such a minimum remains. The method of steepest descent begins with initial approximation (x_0, y_0). At this point the function $S(x, y)$ decreases most rapidly in the direction of the vector

$$-\text{gradient } S(x, y)|_{x_0 y_0} = |-S_x, -S_y|_{x_0 y_0}$$

Denoting this by $-\text{grad } S_0 = [-S_{x0}, -S_{y0}]$ for short, a new approximation (x_1, y_1) is now obtained in the form

$$x_1 = x_0 - tS_{x0} \qquad y_1 = y_0 - tS_{y0}$$

with t chosen so that $S(x_1, y_1)$ is a minimum. In other words, we proceed from (x_0, y_0) in the direction $-\text{grad } S_0$ until S starts to increase again. This completes one step and another is begun at (x_1, y_1) in the new direction $-\text{grad } S_1$. The process continues until, hopefully, the minimum point is found.

The process has been compared to a skier's return from a mountain to the bottom of the valley in a heavy fog. Unable to see his goal, he starts down in the direction of steepest descent and proceeds until his path begins to climb again. Then choosing a new direction of steepest descent, he makes a second run of the same sort. In a bowl-shaped valley ringed by mountains it is clear that this method will bring him gradually nearer to home. Figure 18-13 illustrates the action. The dashed lines are contour or level lines, on which $S(x, y)$ is constant. The gradient direction is orthogonal to the contour direction at each point, so we always leave a contour line at right angles.

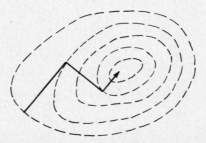

Fig. 18-13

Proceeding to the minimum of $S(x, y)$ along this line means going to a point of tangency with a lower contour line. Actually it requires infinitely many steps of this sort to reach the minimum and a somewhat uneconomical zigzag path is followed.

18.49 Apply a method of steepest descent to solve

$$x = \sin(x + y) \qquad y = \cos(x - y)$$

❚ Here we have

$$S = f^2 + g^2 = [x - \sin(x + y)]^2 + [y - \cos(x - y)]^2$$

making

$$\tfrac{1}{2} S_x = [x - \sin(x + y)][1 - \cos(x + y)] + [y - \cos(x - y)][\sin(x - y)]$$

$$\tfrac{1}{2} S_y = [x - \sin(x + y)][-\cos(x + y)] + [y - \cos(x - y)][1 - \sin(x - y)]$$

Suppose we choose $x_0 = y_0 = .5$. Then $-\operatorname{grad} S_0 \simeq [.3, .6]$. Since a multiplicative constant can be absorbed in the parameter t, we may take

$$x_1 = .5 + t \qquad y_1 = .5 + 2t$$

The minimum of $S(.5 + t, .5 + 2t)$ is now to be found. Either by direct search or by setting $S'(t)$ to zero, we soon discover the minimum near $t = .3$, making $x_1 = .8$ and $y_1 = 1.1$. The value of $S(x_1, y_1)$ is about .04, so we proceed to a second step. Since $-\operatorname{grad} S_1 \simeq [.5, -.25]$, we make our first right angle turn, choose

$$x_2 = .8 + 2t \qquad y_2 = 1.1 - t$$

and seek the minimum of $S(x_2, y_2)$. This proves to be near $t = .07$, making $x_2 = .94$ and $y_2 = 1.03$. Continuing in this way we obtain the successive approximations listed below. The slow convergence toward the result of Problem 16.85 may be noted. Slow convergence is typical of this method, which is often used to provide good starting approximations for the Newton algorithm.

x_n	.5	.8	.94	.928	.936	.934
y_n	.5	1.1	1.03	1.006	1.002	.998
S_n	.36	.04	.0017	.00013	.000025	.000002

The progress of the descent is suggested by path A in Fig. 18-14.

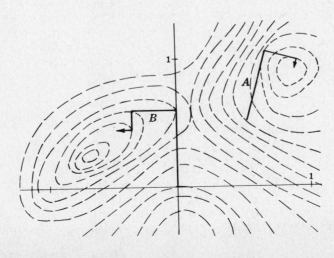

Fig. 18-14

18.50   Show that a steepest descent method may not converge to the required results.

▮ Using the equations of the previous problem, suppose we choose the initial approximations $x_0 = y_0 = 0$. Then $-\text{grad } S_0 = [0, 2]$, so we take $x_1 = 0$ and $y_1 = t$. The minimum of $S(0, t)$ proves to be at $t = .55 = y_1$ with $S(x_1, y_1) = .73$. Computing the new gradient, we find $-\text{grad } S_1 \simeq [-.2, 0]$. this points us *westward*, away from the anticipated solution near $x = y = 1$. Succeeding steps find us traveling the path labeled B in Fig. 18-14. Our difficulty here is typical of minimization methods. There is a secondary valley near $x = -.75$, $y = .25$. Our first step has left us just to the west of the pass or saddle point between these two valleys. The direction of descent at $(0, .55)$ is therefore westward and the descent into the secondary valley continues. Often a considerable amount of experimentation is necessary before a successful trail is found.

18.51   How does the method of random search go about minimizing the function S of Problem 18.49?

▮ The idea is simply to choose points at random throughout the area where the optimum value is expected, evaluate the function and keep track of the current best. Needless to say, a small area is easier to search than a large one, so it pays to confine the optimum as best possible in advance. For the problem in hand we have prior experience and choose the unit square with both x and y between 0 and 1. Since random number generators usually produce values in the $0, 1$ interval, we can use their output raw. Taking two such numbers as x, y the function $S(x, y)$ is computed. If an initial estimate of 2 is assigned as temporary minimum (deliberately too high), then even the first sample will bring improvement. Recording this, we proceed to a new x, y pair. The results of 5000 such samples appear in Table 18.1. Things are moving in the right direction, but random search proceeds at a rate prescribed by sampling theory.

TABLE 18.1

Sample	x	y	S	Sample	x	y	S
1	.71	.99	.08	332	.96	.990	.0008
9	.98	.83	.03	1936	.92	.984	.0007
60	.88	.93	.012	2637	.946	.984	.0003
67	.90	.96	.005	3926	.937	.983	.0003
227	.95	.93	.0045	4762	.931	.994	.00008

18.52   The function S of Problem 18.49 has a second local minimum in the second quadrant. Use a random search to locate it.

▮ Define the search area as the unit square in which both x and y are between 0 and 1 in size. Then x, $y = -r_1, r_2$ will be a suitable random point, where the r_k are standard random numbers. Results of 5000 samples appear in Table 18.2. The minimum does not appear to be headed for zero. Notice that each coordinate falters at times in its effort to converge, when its mate happens to hit on a particularly good value that more than compensates.

TABLE 18.2

Sample	x	y	S	Sample	x	y	S
2	-.85	.35	.138	893	-.96	.27	.108
28	-.83	.33	.117	918	-.93	.25	.107
146	-.89	.21	.116	1024	-.94	.26	.107
209	-.96	.28	.113	1155	-1.00	.24	.104
651	-.92	.25	.109	2606	-.99	.23	.103

18.53   Find by random search the maximum and minimum values of

$$F(x, y) = x(y^2 - x^2)$$

is the rectangle $1 \le x \le 3$, $2 \le y \le 3$.

▮ The function is not very complicated, increasing steadily with y for any positive x and so guaranteeing that the extreme points will be on the boundaries of the rectangle. Confining x, y in this way amounts to imposing four inequality constraints, such as $1 \leq x$. This makes little difference to a random search, which requires only that we choose points without bias throughout the search area. Here the equations

$$x = 1 + 2r_1 \qquad y = 2 + r_2$$

arrange this, with the r_k again standard random numbers. Both the max and the min can be found in one run, by testing each sample point twice. Table 18.3 reports the results of 5000 samples and suggests that we have two decimal place accuracy for the coordinates and two figures for the extreme values. Applying elementary calculus along the top and bottom boundaries will show that the maximum occurs at $x = \sqrt{3}$ and the minimum (constrained) at $x = 3$.

TABLE 18.3

Sample	x_{min}	y_{min}	F_{min}	Sample	x_{max}	y_{max}	F_{max}
2	2.70	2.35	−4.8	1	2.42	2.99	7.5
5	2.87	2.34	−7.9	7	1.47	2.77	8.1
78	2.86	2.11	−10.6	8	1.42	2.95	9.5
102	2.99	2.12	−13.3	30	1.97	2.97	9.75
282	2.96	2.04	−13.6	148	1.79	3.00	10.32
318	2.99	2.05	−14.1	501	1.72	3.00	10.37
1339	2.996	2.05	−14.2				
2057	2.997	2.01	−14.8				

18.54 Why is random search called a zero-order method and steepest descent a first-order method?

▌ Random search uses values of the function being optimized but not values of its derivatives. Steepest descent involves the gradient vector of the function, its first derivatives. The term order is used to indicate the highest level of derivative brought into the algorithm. Zero order usually means simplicity of operation but slow convergence. Higher-order methods converge more rapidly but need good initial approximations. A second-order (Newton) method will be described shortly

18.55 Apply the steepest descent method to the system

$$x = x^2 + y^2 \qquad y = x^2 - y^2$$

to find a solution near $(.8, .4)$.

▌ In Problem 16.92 we found by Newton's method the solution $(.77184, .41964)$. Steepest descent begins by computing the negative gradient vector $(-.256, .288)$ and setting up the search line

$$x = .8 - .256t \qquad y = .4 + .288t$$

which is followed to a minimum value at $t = .09$, $S = .00008$ and $x, y = .777, .426$. A turn in the direction of the new negative gradient $(.00255, -.0279)$ then leads along the line

$$x = .777 + .00255t \qquad y = .426 - .0279t$$

to a minimum at $t = .11$, $S = .00004$ and $x, y = .777, .423$. The next step leads to $x, y = .774, .422$ and $S = .000015$. Many more would be needed to reach the Newton value, but progress is steady and in the right direction.

18.56 What obstacle does steepest descent encounter with the optimization in Problem 18.53?

▌ A direct application is blocked by the constraints. Suppose a start is made at the center of the rectangle $(2, 2.5)$. The negative gradient there is $(5.75, -10)$ and does point more or less in the right direction. But following it to a minimum of the function $F = x(y^2 - x^2)$ leads outside the rectangle. Stopping at the boundary point $(2.2875, 2)$ for a restart does not help, since the indicated direction is outward. The constrained minimum point is not at the bottom of a valley. To the homeward bound skier, the lower boundary of the rectangle is not a natural barrier. Erecting an artificial barrier (even $F = 0$ might do) around this lower corner would allow the method to finish up.

18.57 What is meant by a pattern search algorithm for finding an optimal value of $f(X)$?

   ∎ It is a search algorithm that uses exploratory moves to find a suitable direction for search and acceleration moves, which try to speed things up. Here is one of several possibilities.

1 Choose a first approximation A to the position of the maximum (minimum) of f, and a step size h.

2 Make exploratory moves around A by perturbing its coordinates by amounts $\pm h$. If a perturbation improves the value of f over its current maximum, the perturbed value of that coordinate is retained. After all coordinates have been tested, the new point is B. If $B = A$ go to step 3, otherwise to step 4.

3 Either accept point A as optimal for the grid of mesh h or extend the search around A by testing combinations of coordinate perturbations of sizes 0 or $\pm h$. If an improved value of f is found, let A denote its position and return to step 1; if not, stop.

4 This is the pattern move. Let $T = 2B - A$ be a trial point, reached by passing from A to B and then an equal distance beyond along the same line. If $f(T)$ is smaller than the current maximum, let $A = B$ return to step 1. Otherwise proceed to step 5.

5 Make exploratory moves around T as in step 1. Call the resulting point C. If $C = T$ go to step 6, otherwise to step 7.

6 Set $A = B$ and return to step 1. (The trial point T was too ambitious a move.)

7 Set $A = B$, $B = C$ and return to step 4. (The acceleration was successful and is to be repeated.)

18.58 Apply accelerated search to find the maximum of the function $f(x, y)$ displayed in Table 18.4.

TABLE 18.4 (Circled entries are those considered)

y/x	0	1	2	3	4	5	6	7	8	9	10	11	12	13	14
5	−50	−26	−4	16	34	50	64	76	86	94	100	104	106	106	104
4	−12	10	30	48	64	78	90	100	108	(114)	118	120	120	118	114
3	12	32	50	66	80	(92)	(102)	110	116	120	(122)	122	120	116	(110)
2	22	40	(56)	(70)	82	92	100	106	110	112	112	110	106	100	92
1	18	(34)	48	60	70	78	84	88	90	90	88	84	78	70	60
0	(0)	14	26	36	44	50	54	56	56	54	50	44	36	26	14

   ∎ Take the uninspired $A = (0, 0)$ as initial point, the function value being zero. Also take $h = 1$, as the table presupposes. Then both horizontal and vertical perturbations bring larger values of f, so $B = (1, 1)$. This is step 2, and since $B \neq A$ we move to step 4 with a current maximum of 34. The acceleration move is to $T = (2, 2)$ where the still better value of 56 is found. Coordinate perturbations then discover $C = (3, 2)$ with value 70, so we reset $A = (1, 1)$, $B = (3, 2)$ and return to step 4. Table 18.5 is a summary of all steps, parentheses enclosing function values. The last acceleration move was clearly too ambitious.

At this point there is the option of trying combinations of coordinate perturbations (not just horizontal or vertical moves). Here this leads to no (diagonal) improvements. There is also the option of trying a finer grid. The background function is

$$f(x, y) = 15x + 25y - x^2 + 2xy - 7y^2$$

so the exact solution can be found by calculus to be $\left(\frac{65}{6}, \frac{10}{3}\right)$.

TABLE 18.5

A		B		T		C	
0,0	(0)	1,1	(34)	2,2	(56)	3,2	(70)
1,1	(34)	3,2	(70)	5,3	(92)	6,3	(102)
3,2	(70)	6,3	(102)	9,4	(114)	10,3	(122)
6,3	(102)	10,3	(122)	14,3	(110)		
10,3	(122)		A				

18.59 Also use the accelerated search algorithm on the function shown in Table 18.6, again to find the maximum.

TABLE 18.6 (Circled entries are those considered)

y/x	0	1	2	3	4	5	6	7
11	(209)	197	185	173	161	149	137	125
10	200	190	180	170	160	150	140	130
9	189	181	(173)	165	157	149	141	(133)
8	176	170	164	(158)	152	146	140	134
7	161	157	153	149	(145)	141	137	133
6	144	142	140	138	136	(134)	132	130
5	125	125	125	125	125	(125)	125	125
4	104	106	108	110	112	114	116	118
3	81	85	89	(93)	97	101	105	109
2	56	62	(68)	74	80	86	92	98
1	29	(37)	45	53	61	69	77	85
0	(0)	10	20	30	40	50	60	70

▮ Table 18.7 summarizes developments. One acceleration effort was too ambitious.

TABLE 18.7

A		B		T		C	
0,0	(0)	1,1	(37)	2,2	(68)	3,3	(93)
1,1	(37)	3,3	(93)	5,5	(125)	5,6	(134)
3,3	(93)	5,6	(134)	7,9	(133)		
5,6	(134)	4,7	(145)	3,8	(158)	2,9	(173)
4,7	(145)	2,9	(173)	0,11	(209)		

18.60 In the above method of pattern search, why doesn't the probing at each step include diagonal as well as coordinate directions?

▮ For the two-dimensional problem this might very well be done, the added cost being four extra points to test and a new decision method for selecting the next move. However, when there are n independent variables the total number of test points would be $3^n - 1$, the number of points on the surface of a hypercube. This can be much larger than the current number of $2n$, two tests for each coordinate direction. Experience suggests that it is more efficient to restrict the more exhaustive search to a backup role.

18.61 Define conjugate directions.

▮ The Hessian matrix of a function $f(x, y)$ is defined as

$$H = \begin{pmatrix} f_{xx} & f_{xy} \\ f_{yx} & f_{yy} \end{pmatrix}$$

with the extension to a function of more independent variables apparent. Two vectors (or directions) S_1 and S_2 are conjugate if

$$S_1^T H S_2 = 0$$

18.62 How are conjugate directions used in the Fletcher–Reeves search for an optimal point?

▌ This method modifies steepest descent (or ascent) in such a way that successive directions of search are conjugate instead of orthogonal. This is arranged by choosing each new direction S as a blend of the old S and the new gradient vector. The algorithm can be described in this way, the goal being a maximum point.

1 Choose an initial approximation X.
2 Find $S = \text{grad } f(X)$ and $a = S^T S$.
3 Make a one-dimensional search for t such that $f(X + tS)$ is a maximum. If $t = 0$, stop.
4 $X = X + tS$. (Update X.)
5 Find $G = \text{grad } f(X)$ and $b = G^T G$.
6 $S = G + cS$, where $c = b/a$. (This makes the blend.)
7 $a = b$. (Update a.) Return to step 3.

An additional precaution is often inserted into the algorithm. Sometimes the blending process produces a direction S that points away from the maximum (minimum) instead of toward it. This can be detected by computing $S^T G$. If it is negative (for a maximum search), set $S = G$ and restart.

18.63 Apply the above algorithm to the function $f(x, y)$ of Problem 18.58.

▌ The actual maximum is $f(\frac{65}{6}, \frac{10}{3}) = 122.9167$ to four places. Starting from $(0, 0)$ and using $h = .001$, these two moves were made:

Gradients	S	x	y	f
$(15, 25)$	$(15, 25)$	1.65	2.75	46.9
$(17.2, -10.2)$	$(24.3, 1.56)$	10.843	3.342	122.9163

The third move failed to bring improvement, but the maximum value has already been found to within something like roundoff error. Notice that the gradient for the second move is downward, as orthogonality to the first gradient vector requires. The conjugate vector S, however, aims almost directly at the target.

18.64 Verify that the first two S directions in the preceding problem are conjugate.

▌ The Hessian matrix can be found by inspection,

$$H = \begin{pmatrix} -2 & 2 \\ 2 & -14 \end{pmatrix}$$

and is constant because f is quadratic. Using seven digit values of the vector S we have

$$S_2^T H S_1 = (24.25671, 1.561172) \begin{pmatrix} -2 & 2 \\ 2 & -14 \end{pmatrix} \begin{pmatrix} 15 \\ 25 \end{pmatrix}$$

which proves to be zero to seven places.

18.65 Apply the conjugate direction algorithm to the fourth degree function

$$f(x, y) = 15x + 25y - .03x^4 - .15y^4 + 2xy$$

▌ Starting at $(0, 0)$ and using $h = .001$ for the one-dimensional searches, results for the first few iterations were as recorded in Table 18.8. Five iterations later values $x = 5.71$, $y = 3.97$ and $f = 161.09$ were in hand. Convergence is slower than for the quadratic function of the preceding problem.

TABLE 18.8

Gradient	s	x	y	f
$(15, 25)$	$(15, 25)$	2.49	4.15	116
$(21.4, -12.9)$	$(32.5, 5.5)$	5.51	4.66	152
$(4.22, -24.8)$	$(37.1, -19.2)$	6.22	4.30	158
$(-5.26, -10.2)$	$(2.45, -14.2)$	6.27	4.01	159

18.7 NEWTON'S SECOND-ORDER METHOD

18.66 Develop the Newton second-order method for finding a minimum.

▮ Suppose the descent, or search, has reached the point (x_0, y_0) and level $F(x_0, y_0)$. Truncating Taylor's theorem.

$$F = F_0 + F_{x0}(x - x_0) + F_{y0}(y - y_0)$$
$$+ \tfrac{1}{2}\Big[F_{xx0}(x - x_0)^2 + 2F_{xy0}(x - x_0)(y - y_0) + F_{yy0}(y - y_0)^2 \Big]$$

the zero subscript meaning evaluation at (x_0, y_0). The partial derivatives are

$$F_x = F_{x0} + \tfrac{1}{2}\Big[2F_{xx0}(x - x_0) + 2F_{xy0}(y - y_0) \Big]$$
$$F_y = F_{y0} + \tfrac{1}{2}\Big[2F_{xy0}(x - x_0) + 2F_{yy0}(y - y_0) \Big]$$

and at a smooth minimum must be zero. The system

$$F_{xx0}h + F_{xy0}k = -F_{x0}$$
$$F_{xy0}h + F_{yy0}k = -F_{y0}$$

then determines $h = x - x_0$, $k = y - y_0$, from which the next point (x, y) can be found.

For a truly quadratic function $F(x, y)$ the preceding use of the Taylor theorem involves no truncation and the method finds the minimum point in one step. As a trivial, even mental, example, take $F = x^2 + y^2$ and begin a search for its minimum from position $(1,1)$. The various derivatives being $2x, 2y, 2, 0, 0, 2$, we have $2h = -2$, $2k = -2$ for determining h and k. The new point $(0,0)$ is perfect and would have been found in one step from any other starting point (x_0, y_0).

18.67 Use the Newton method to find the minimum of $F = x^2 + y^4$.

▮ The various derivatives are now $2x, 4y^3, 2, 0, 0, 12y^2$. At the initial point $(1,1)$ they form the system $2h = -2$, $12k = -4$, making $h = -1$, $k = -\tfrac{1}{3}$ and the next point $(0, \tfrac{2}{3})$. Computing new derivative values for this new starting point, the system $2h = 0$, $16k/3 = -\tfrac{32}{27}$ is encountered. It makes $k = -\tfrac{2}{9}$ and we reach $(x, y) = (0, \tfrac{4}{9})$. The y coordinate seems reluctant to close in. A short scrutiny of the equation determining k shows that each y coordinate will be two-thirds its predecessor, which is a disappointing performance for a Newton method. Replacing y^4 by y^{2p} makes things even worse, since then the reduction becomes $(2p - 2)/(2p - 1)$. Of course, $F = x^2 + y^{2p}$ resembles a quadratic function less and less as p grows larger.

18.68 Apply the Newton method to minimize

$$F(x, y) = (x - x^2 - y^2)^2 + (y - x^2 + y^2)^2$$

and compare with the work in Problem 16.92.

▮ Let $u = x - x^2 - y^2$ and $v = y - x^2 + y^2$. Then $F = u^2 + v^2$ and the various derivatives are

$$F_x = 2u(1 - 2x) + 2v(-2x) \qquad F_y = 2u(-2y) + 2v(1 + 2y)$$
$$F_{xx} = -4u + 2(1 - 2x)^2 - 4v + 8x^2$$
$$F_{xy} = -4y(1 - 2x) - 4x(1 + 2y)$$
$$F_{yy} = -4u + 8y^2 + 4v + 2(1 + 2y)^2$$

The recommended start at $(.8, .4)$ now triggers the sequence of points

x	.777	.77196	.7718446	.7718445
y	.424	.41973	.4196434	.4196434

and repeats. The same figures are found as in Problem 16.92 and with the same speed. The earlier Newton method was designed to solve a system of equations, the present one to locate an optimum point. Having arranged equivalent problems, we have a choice.

Restarting at several other points out as far as $(10,10)$ led to convergence in 10 or 15 steps. The function F has only two minima so there is little chance for approximations to run astray. Any reasonable start produces fairly quick convergence either to the desired optimum or to $(0,0)$.

18.8 LAGRANGE MULTIPLIERS

18.69 Illustrate the use of Lagrange multipliers in optimization with equality constraints by maximizing

$$f = 1 - w^2 - x^2 - y^2 - z^2$$

subject to $g_1 = w + x = 1$ and $g_2 = y - z = 1$.

▮ Lagrange multipliers are an alternative to using constraints to eliminate variables. Here, for example, eliminating w and z leads to

$$f = 1 - (1 - x)^2 - x^2 - y^2 - (y - 1)^2$$

which is easy enough to minimize. The Lagrange alternative sets up the function

$$L = f + \lambda_1 g_1 + \lambda_2 g_2$$
$$= 1 - w^2 - x^2 - y^2 - z^2 + \lambda_1(w + x - 1) + \lambda_2(y - z - 1)$$

which it proceeds to minimize with no constraints. The λ_i are the Lagrange multipliers. Taking partial derivatives relative to x, y, z and λ_1, λ_2 and equating them to zero, we have

$$-2w + \lambda_1 = 0 \qquad -2x + \lambda_1 = 0 \qquad -2y + \lambda_2 = 0 \qquad -2z - \lambda_2 = 0$$

plus the two constraints. It is now a short algebraic road to

$$w = x = y = -z = \tfrac{1}{2}$$

with the values of the multipliers also available.

18.70 Use Lagrange multipliers to find the optimal value of $f = xyz$ on the plane $ax + by + cz = 1$.

▮ The function $L = xyz + \lambda(ax + by + cz - 1)$ leads to the attractive set of equations

$$yz + a\lambda = 0 \qquad xz + b\lambda = 0 \qquad xy + c\lambda = 0$$

from which one quickly deduces $ax = by = cz$. The constraint then makes all three of these $\tfrac{1}{3}$. The Lagrange method often preserves natural features of a problem, which can be obscured by elimination of variables.

18.71 Let $V = (x, y, z)^T$ and consider the quadratic form

$$f = V^T A V = a_{11}x^2 + a_{22}y^2 + a_{33}z^2 + 2a_{12}xy + 2a_{13}xz + 2a_{23}yz$$

which will be assumed positive definite. Use Lagrange multipliers to show that finding the extreme values of this form on the sphere $x^2 + y^2 + z^2 = 1$ is equivalent to an eigenvalue problem.

▮ The Lagrange function is taken as

$$L = f - \lambda(x^2 + y^2 + z^2 - 1)$$

and equating its derivatives relative to x, y, z to zero leads at once to $AV = \lambda V$, where the elements of A are a_{ij} and symmetric.

The problem is equivalent to finding the principal axes of an ellipsoid. As the value f increases, say from 0, first contact of the growing ellipsoid with the unit sphere is made at the ends of the longest axis. The vector V out to this point of contact is the unit eigenvector corresponding to this minimum f. As the ellipsoid continues to grow, final contact with the sphere occurs at the ends of the shortest axis. This point is identified with the eigenvector corresponding to the maximum f. The third axis of the ellipsoid is then orthogonal to both the others. Choosing the three eigenvectors for a new coordinate system removes cross-product terms from the form representing the ellipsoid. The problem of reducing the quadratic form to diagonal form is thus another equivalent problem. The Lagrange method of optimization keeps these key features in the forefront of the computation.

18.72 What are the principal axes of the ellipsoid represented by this matrix, used earlier in random walk problems?

$$\begin{pmatrix} 2 & -1 & 0 \\ -1 & 2 & -1 \\ 0 & -1 & 2 \end{pmatrix}$$

∎ The eigenvectors were found in Problem 17.93 to be

$$(1,\sqrt{2},1) \quad (-1,0,1) \quad (1,-\sqrt{2},1)$$

and can easily be reduced to Euclidean length 1. They are orthogonal and are the natural set of coordinate axes for the corresponding ellipsoids.

18.9 PENALTY FUNCTION METHODS

18.73 Find the minimum value of $f = x + y$ subject to the constraint $g = x^2 - 6x - y + 10 \le 0$.

∎ The problem is easy enough to solve. Figure 18.15 shows that we are constrained to the region above a parabola and provides parallel lines as level lines of f. The lowest line still having contact with the feasible area identifies the minimum f. A bit of calculus quickly finds this contact to be at $(\frac{5}{2},\frac{5}{4})$ and a minimum f of 3.75.

Several methods of dealing with nonlinear equalities as constraints use the idea of penalty functions, already introduced in Problems 18.45 and 18.46 for the case of one independent variable. Here we let

$$F = f + p\delta g^2 = x + y + p\delta \left(x^2 - 6x - y + 10 \right)^2$$

and seek the unconstrained minimum of F. As before, the factor δ is to be 0 when the constraint is satisfied and 1 otherwise, providing a penalty for violation of the constraint. The factor p controls the size of the penalty.

For $p = 1, 2, 3, 5, 10$ the minimum values 3.50, 3.63, 3.67, 3.70 and 3.73 were found, all less than the true minimum because with this method we are always on the wrong side of the parabola. The last value corresponded to point $(2.50, 1.21)$ so the x coordinate has been better pinpointed than the y.

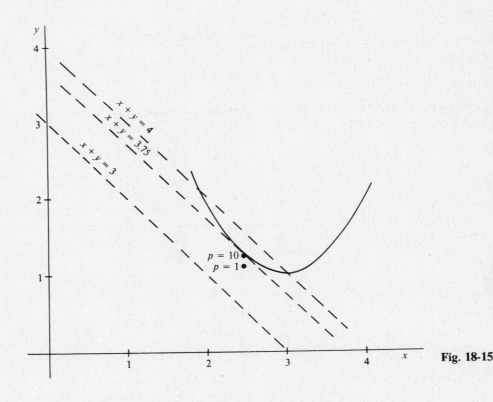

Fig. 18-15

18.74 How can the difficulty of always being in the infeasible region be surmounted?

∎ Several methods have been devised using penalty functions such as $-1/g$ or $-\log(-g)$, for nonpositive g. These suffer to some extent from numerical instability, but can provide feasible minima. Details are available in works on operations research.

CHAPTER 19
Overdetermined Systems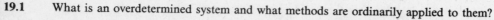

19.1 What is an overdetermined system and what methods are ordinarily applied to them?

   ❚ An overdetermined system of linear equations has the form

$$Ax = b$$

with the matrix A having more rows than columns. Ordinarily no solution vector x will exist and the equation as written is called inconsistent. The problem is to find something as close to a solution as we can, in some sense. Overdetermined systems arise in experimental or computational work whenever more results are generated than would be needed if precision were attainable. In a sense, a mass of inexact, conflicting information becomes a substitute for a few perfect results and one hopes that good approximations to the exact values can somehow be squeezed from the mass.

    The two main methods of attack involve the residual vector

$$R = Ax - b$$

Because R cannot usually be reduced to the zero vector, an effort is made to minimize it, either in the L_2 sense

$$R^T R = r_1^2 + \cdots + r_n^2 = \text{minimum}$$

or in the L_∞ sense

$$\max(|r_1|, \ldots, |r_n|) = \text{minimum}$$

where the r_i are the components of R.

19.1 LEAST-SQUARES SOLUTION

19.2 Derive the *normal equations* for finding the least-squares solution of an overdetermined system of linear equations.

   ❚ Let the given system be

$$a_{11}x_1 + a_{12}x_2 = b_1$$
$$a_{21}x_1 + a_{22}x_2 = b_2$$
$$_2 x_2 = b_3$$

This involves only the two unknowns x_1 and x_2 and is only slightly overdetermined, but the details for larger systems are almost identical. Ordinarily we will not be able to satisfy all three of our equations. The problem as it stands probably has no solution. Accordingly we rewrite it as

$$a_{11}x_1 + a_{12}x_2 - b_1 = r_1$$
$$a_{21}x_1 + a_{22}x_2 - b_2 = r_2$$
$$a_{31}x_1 + a_{32}x_2 - b_3 = r_3$$

the numbers r_1, r_2, r_3 being called residuals, and look for the numbers x_1, x_2 which make $r_1^2 + r_2^2 + r_3^2$ minimal. Since

$$r_1^2 + r_2^2 + r_3^2 = \left(a_{11}^2 + a_{21}^2 + a_{31}^2\right)x_1^2 + \left(a_{12}^2 + a_{22}^2 + a_{32}^2\right)x_2^2 + 2(a_{11}a_{12} + a_{21}a_{22} + a_{31}a_{32})x_1 x_2$$

$$- 2(a_{11}b_1 + a_{21}b_2 + a_{31}b_3)x_1 - 2(a_{12}b_1 + a_{22}b_2 + a_{32}b_3)x_2 + \left(b_1^2 + b_2^2 + b_3^2\right)$$

the result of setting derivatives relative to x_1 and x_2 equal to zero is the pair of normal equations

$$(a_1, a_1)x_1 + (a_1, a_2)x_2 = (a_1, b)$$

$$(a_2, a_1)x_1 + (a_2, a_2)x_2 = (a_2, b)$$

in which the parentheses denote

$$(a_1, a_1) = a_{11}^2 + a_{21}^2 + a_{31}^2 \qquad (a_1, a_2) = a_{11}a_{12} + a_{21}a_{22} + a_{31}a_{32}$$

and so on. These are the *scalar products* of the various columns of coefficients in the original system, so that the normal equations may be written directly. For the general problem of m equations in n unknowns $(m > n)$,

$$a_{11}x_1 + \cdots + a_{1n}x_n = b_1$$
$$a_{21}x_1 + \cdots + a_{2n}x_n = b_2$$
$$\vdots$$
$$a_{m1}x_1 + \cdots + a_{mn}x_n = b_m$$

an almost identical argument leads to the normal equations

$$(a_1, a_1)x_1 + (a_1, a_2)x_2 + \cdots + (a_1, a_n)x_n = (a_1, b)$$
$$(a_2, a_1)x_1 + (a_2, a_2)x_2 + \cdots + (a_2, a_n)x_n = (a_2, b)$$
$$\vdots$$
$$(a_n, a_1)x_1 + (a_n, a_2)x_2 + \cdots + (a_n, a_n)x_n = (a_n, b)$$

This is a symmetric, positive definite system of equations.

It is also worth noticing that the present problem again fits the model of our general least-squares approach in Problems 12.19 and 12.20. The results just obtained follow at once as a special case, with the vector space E consisting of m-dimensional vectors such as, for instance, the column vectors of the matrix A, which we denote by $a_1, a_2, \ldots, a_n$, and the column of numbers b_i, which we denote by b. The subspace S is the range of the matrix A, that is, the set of vectors Ax. We are looking for a vector p in S that minimizes

$$\|p - b\|^2 = \|Ax - b\|^2 = \sum r_i^2$$

and this vector is the orthogonal projection of b onto S, determined by $(p - b, u_k) = 0$, where the u_k are some basis for S. Choosing for this basis $u_k = a_k$, $k = 1, \ldots, n$, we have the usual representation $p = x_1 a_1 + \cdots + x_n a_n$ (the notation being somewhat altered from that of our general model) and substitution leads to the normal equations.

19.3   Find the least-squares solution of the system

$$x_1 - x_2 = 2$$
$$x_1 + x_2 = 4$$
$$2x_1 + x_2 = 8$$

  ❚ Forming the required scalar products, we have

$$6x_1 + 2x_2 = 22 \qquad 2x_1 + 3x_2 = 10$$

for the normal equations. This makes $x_1 = \frac{23}{7}$ and $x_2 = \frac{8}{7}$. The residuals corresponding to this x_1 and x_2 are $r_1 = \frac{1}{7}$, $r_2 = \frac{3}{7}$ and $r_3 = -\frac{2}{7}$, and the sum of their squares is $\frac{2}{7}$. The root-mean-square error is therefore $\rho = \sqrt{\frac{2}{21}}$. This is smaller than for any other choice of x_1 and x_2.

19.4   Suppose three more equations are added to the already overdetermined system of Problem 19.3:

$$x_1 + 2x_2 = 4$$
$$2x_1 - x_2 = 5$$
$$x_1 - 2x_2 = 2$$

Find the least-squares solution of the set of six equations.

  ❚ Again forming scalar products we obtain $12x_1 = 38$, $12x_2 = 9$ for the normal equations, making $x_1 = \frac{19}{6}$, $x_2 = \frac{3}{4}$. The six residuals are 5, -1, -11, 8, 7 and -4, all divided by 12. The rms error is $\rho = \sqrt{\frac{23}{72}}$.

19.5 Two quantities x and y are measured, together with their difference $x - y$ and their sum $x + y$:

$$x = A \qquad y = B \qquad x - y = C \qquad x + y = D$$

Solve this overdetermined system by least squares.

▮ The normal equations are quickly found to be

$$3x = A + C + D \qquad 3y = B - C + D$$

showing that B does not influence x nor does A influence y.

19.6 Four altitudes x_1, x_2, x_3, x_4 are measured, together with the six differences in altitude, as follows. Find the least-squares values.

$$x_1 = 3.47 \qquad x_2 = 2.01 \qquad x_3 = 1.58 \qquad x_4 = .43$$
$$x_1 - x_2 = 1.42 \qquad x_1 - x_3 = 1.92 \qquad x_1 - x_4 = 3.06 \qquad x_2 - x_3 = .44$$
$$x_2 - x_4 = 1.53 \qquad x_3 - x_4 = 1.20$$

▮ To avoid errors one may wish to align these various equations in the standard way, but the normal equations prove to be

$$
\begin{aligned}
4x_1 - x_2 - x_3 - x_4 &= 9.87 \\
-x_1 + 4x_2 - x_3 - x_4 &= 2.56 \\
-x_1 - x_2 + 4x_3 - x_4 &= .42 \\
-x_1 - x_2 - x_3 + 4x_4 &= -5.36
\end{aligned}
$$

with solution $(3.472, 2.01, 1.582, .426)$. Residuals of $.002, 0, .002$ and $-.004$ occur for the first four equations, but range from $-.012$ to $.054$ for the others. Perhaps the differences were harder to measure.

19.7 The three angles of a triangle are measured to be A_1, A_2, A_3. If x_1, x_2, x_3 denote the correct values, we are led to the overdetermined system

$$x_1 = A_1 \qquad x_2 = A_2 \qquad \pi - x_1 - x_2 = A_3$$

Solve by the method of least squares.

▮ The normal equations are

$$2x_1 + x_2 = \pi + A_1 - A_3$$
$$x_1 + 2x_2 = \pi + A_2 - A_3$$

with the solution

$$x_i = A_i + \tfrac{1}{3}(\pi - A_1 - A_2 - A_3) \qquad i = 1, 2, 3$$

19.8 The two legs of a right triangle are measured to be A and B, and the hypotenuse to be C. Let L_1, L_2, and H denote the exact values and let $x_1 = L_1^2$, $x_2 = L_2^2$. Consider the overdetermined system

$$x_1 = A^2 \qquad x_2 = B^2 \qquad x_1 + x_2 = C^2$$

and obtain the least-squares estimates of x_1 and x_2. From these estimate L_1, L_2 and H.

▮ The normal equations are

$$2x_1 + x_2 = A^2 + C^2 \qquad x_1 + 2x_2 = B^2 + C^2$$

from which we find

$$x_1 = A^2 - D \qquad x_2 = B^2 - D$$

with $D = \tfrac{1}{3}(A^2 + B^2 - C^2)$. A bit of algebra then manages

$$H^2 = C^2 + D$$

19.9 Find the least-squares solution of this system.

$$x_1 + x_2 - x_3 = 5 \qquad x_1 + 2x_2 - 2x_3 = 1$$

$$2x_1 - 3x_2 + x_3 = -4 \qquad 4x_1 - x_2 - x_3 = 6$$

▮ The normal equations are

$$22x_1 - 7x_2 - 5x_3 = 22$$

$$-7x_1 + 15x_2 - 7x_3 = 13$$

$$-5x_1 - 7x_2 + 7x_3 = -17$$

with solution $(9, 13, 17)$. The residual vector is $(0, -4, 1, 6)$.

19.10 Find the least-squares solution of $Ax = b$ given the matrix A of Fig. 19-1 and the vector $b^T = (1, 1, 2, 4, 3, 0)$.

▮ The normal equations, their solution and the residual vector also appear in the figure.

$$A = \begin{pmatrix} 2 & -2 & 1 & 2 \\ 1 & 1 & 2 & 4 \\ 1 & -3 & 1 & 2 \\ -2 & -2 & 3 & 3 \\ -1 & -3 & -3 & 1 \\ 2 & 4 & 1 & 5 \end{pmatrix}$$

$$15x_1 + 9x_2 + 4x_3 + 13x_4 = -6$$
$$9x_1 + 43x_2 + 4x_3 + 5x_4 = -24$$
$$4x_1 + 4x_2 + 25x_3 + 23x_4 = 8$$
$$13x_1 + 5x_2 + 23x_3 + 59x_4 = 25$$
$$x^T = (-.64, -.49, -.088, .64)$$
$$r^T = (-.10, .26, .03, -.09, .02, -.13)$$

Fig. 19-1

19.2 CHEBYSHEV SOLUTION

19.11 Show how the Chebyshev solution of an overdetermined system of linear equations may be found by the method of linear programming.

▮ Once again we treat the small system of Problem 19.2, the details for larger systems being almost identical. Let r be the maximum of the absolute values of the residuals, so that $|r_1| \le r$, $|r_2| \le r$, $|r_3| \le r$. This means that $r_1 \le r$ and $-r_1 \le r$, with similar requirements on r_2 and r_3. Recalling the definitions of the residuals we now have six inequalities:

$$a_{11}x_1 + a_{12}x_2 - b_1 \le r \qquad -a_{11}x_1 - a_{12}x_2 + b_1 \le r$$

$$a_{21}x_1 + a_{22}x_2 - b_2 \le r \qquad -a_{21}x_1 - a_{22}x_2 + b_2 \le r$$

$$a_{31}x_1 + a_{32}x_2 - b_3 \le r \qquad -a_{31}x_1 - a_{32}x_2 + b_3 \le r$$

If we also suppose that x_1 and x_2 must be nonnegative, and recall that the Chebyshev solution is defined to be that choice of x_1, x_2 which makes r minimal, then it is evident that we have a linear programming problem. It is convenient to modify it slightly. Dividing through by r and letting $x_1/r = y_1$, $x_2/r = y_2$, $1/r = y_3$, the constraints become

$$a_{11}y_1 + a_{12}y_2 - b_1 y_3 \le 1 \qquad -a_{11}y_1 - a_{12}y_2 + b_1 y_3 \le 1$$

$$a_{21}y_1 + a_{22}y_2 - b_2 y_3 \le 1 \qquad -a_{21}y_1 - a_{22}y_2 + b_2 y_3 \le 1$$

$$a_{31}y_1 + a_{32}y_2 - b_3 y_3 \le 1 \qquad -a_{31}y_1 - a_{32}y_2 + b_3 y_3 \le 1$$

and we must maximize y_3 or, what is the same thing, make $F = -y_3 =$ minimum. This linear program can be formed directly from the original overdetermined system. The generalization for larger systems is almost obvious. The condition that the x_j be positive is often met in practice, these numbers representing lengths or other physical measurements. If it is not met, then a translation $z_j = x_j + c$ may be made, or a modification of the linear programming algorithm may be used.

19.12 Apply the linear programming method to find the Chebyshev solution of the system of Problem 19.3.

❙ Adding one slack variable to each constraint, we have

$$
\begin{aligned}
y_1 - y_2 - 2y_3 + y_4 &= 1 \\
y_1 + y_2 - 4y_3 + y_5 &= 1 \\
2y_1 + y_2 - 8y_3 + y_6 &= 1 \\
-y_1 + y_2 + 2y_3 + y_7 &= 1 \\
-y_1 - y_2 + 4y_3 + y_8 &= 1 \\
-2y_1 - y_2 + 8y_3 + y_9 &= 1
\end{aligned}
$$

with $F = -y_3$ to be minimized and all y_j to be nonnegative. The starting display and three exchanges following the simplex algorithm are shown in Fig. 19-2. The six columns corresponding to the slack variables are omitted since they actually contain no vital information. From the final display we find $y_1 = 10$ and $y_2 = y_3 = 3$. This makes $r = 1/y_3 = \frac{1}{3}$ and then $x_1 = \frac{10}{3}$, $x_2 = 1$. The three residuals are $\frac{1}{3}, \frac{1}{3}, -\frac{1}{3}$ so that the familiar Chebyshev feature of equal error sizes is again present.

Basis	b	v_1	v_2	v_3
v_4	1	1	-1	-2
v_5	1	1	1	-4
v_6	1	2	1	-8
v_7	1	-1	1	2
v_8	1	-1	-1	4
v_9	1	-2	-1	⑧
	0	0	0	1

Basis	b	v_1	v_2	v_3
v_4	$\frac{5}{4}$	($\frac{1}{2}$)	$-\frac{5}{4}$	0
v_5	$\frac{3}{2}$	0	$\frac{1}{2}$	0
v_6	2	0	0	0
v_7	$\frac{3}{4}$	$-\frac{1}{2}$	$\frac{5}{4}$	0
v_8	$\frac{1}{2}$	0	$-\frac{1}{2}$	0
v_3	$\frac{1}{8}$	$-\frac{1}{4}$	$-\frac{1}{8}$	1
	$-\frac{1}{8}$	$\frac{1}{4}$	$\frac{1}{8}$	0

Basis	b	v_1	v_2	v_3
v_1	$\frac{5}{2}$	1	$-\frac{5}{2}$	0
v_5	$\frac{3}{2}$	0	($\frac{1}{2}$)	0
v_6	2	0	0	0
v_7	2	0	0	0
v_8	$\frac{1}{2}$	0	$-\frac{1}{2}$	0
v_3	$\frac{3}{4}$	0	$-\frac{3}{4}$	1
	$-\frac{3}{4}$	0	$\frac{3}{4}$	0

Basis	b	v_1	v_2	v_3
v_1	10	1	0	0
v_2	3	0	1	0
v_6	2	0	0	0
v_7	2	0	0	0
v_8	2	0	0	0
v_3	3	0	0	1
	-3	0	0	0

Fig. 19-2

19.13 Apply the linear programming method to find the Chebyshev solution of the overdetermined system of Problem 19.4.

❙ The six additional constraints bring six more slack variables, $y_{10}, \ldots, y_{15}$. The details are very much as in Problem 19.12. Once again the columns for slack variables are omitted from Fig. 19-3, which summarizes three exchanges of the simplex algorithm. After the last exchange we find $y_1 = \frac{13}{3}$, $y_2 = 1$, $y_3 = \frac{4}{3}$. So $r = \frac{3}{4}$ and $x_1 = \frac{13}{4}$, $x_2 = \frac{3}{4}$. The six residuals are 2, 0, -3, 3, 3 and -1, all divided by 4. Once again three residuals equal the min–max residual r, the others now being smaller. In the general problem $n + 1$ equal residuals, the others being smaller, identify the Chebyshev solution, n being the number of unknowns.

Basis	b	v_1	v_2	v_3
v_4	1	1	-1	-2
v_5	1	1	1	-4
v_6	1	2	1	-8
v_7	1	-1	1	2
v_8	1	-1	-1	4
v_9	1	-2	-1	⑧
v_{10}	1	1	2	-4
v_{11}	1	2	-1	-5
v_{12}	1	1	-2	-2
v_{13}	1	-1	-2	4
v_{14}	1	-2	1	5
v_{15}	1	-1	2	2
	0	0	0	1

Basis	b	v_1	v_2	v_3
v_4	$\frac{5}{4}$	$\frac{1}{2}$	$-\frac{5}{4}$	0
v_5	$\frac{3}{2}$	0	$\frac{1}{2}$	0
v_6	2	0	0	0
v_7	$\frac{3}{4}$	$-\frac{1}{2}$	$\frac{5}{4}$	0
v_8	$\frac{1}{2}$	0	$-\frac{1}{2}$	0
v_3	$\frac{1}{8}$	$-\frac{1}{4}$	$-\frac{1}{8}$	1
v_{10}	$\frac{3}{2}$	0	$\frac{3}{2}$	0
v_{11}	$\frac{13}{8}$	$\left(\frac{3}{4}\right)$	$-\frac{13}{8}$	0
v_{12}	$\frac{5}{4}$	$\frac{1}{2}$	$-\frac{9}{4}$	0
v_{13}	$\frac{1}{2}$	0	$-\frac{3}{2}$	0
v_{14}	$\frac{3}{8}$	$-\frac{3}{4}$	$\frac{13}{8}$	0
v_{15}	$\frac{3}{4}$	$-\frac{1}{2}$	$\frac{9}{4}$	0
	$-\frac{1}{8}$	$\frac{1}{4}$	$\frac{1}{8}$	0

Basis	b	v_1	v_2	v_3
v_4	$\frac{1}{6}$	0	$-\frac{1}{6}$	0
v_5	$\frac{3}{2}$	0	$\frac{1}{2}$	0
v_6	2	0	0	0
v_7	$\frac{11}{6}$	0	$\frac{1}{6}$	0
v_8	$\frac{1}{2}$	0	$-\frac{1}{2}$	0
v_3	$\frac{2}{3}$	0	$-\frac{2}{3}$	1
v_{10}	$\frac{3}{2}$	0	$\left(\frac{3}{2}\right)$	0
v_1	$\frac{13}{6}$	1	$-\frac{13}{6}$	0
v_{12}	$\frac{1}{6}$	0	$-\frac{7}{6}$	0
v_{13}	$\frac{1}{2}$	0	$-\frac{3}{2}$	0
v_{14}	2	0	0	0
v_{15}	$\frac{11}{6}$	0	$\frac{7}{6}$	0
	$-\frac{2}{3}$	0	$\frac{2}{3}$	0

Basis	b	v_1	v_2	v_3
v_4	$\frac{1}{3}$	0	0	0
v_5	1	0	0	0
v_6	2	0	0	0
v_7	$\frac{5}{3}$	0	0	0
v_8	1	0	0	0
v_3	$\frac{4}{3}$	0	0	1
v_2	1	0	1	0
v_1	$\frac{13}{3}$	1	0	0
v_{12}	$\frac{4}{3}$	0	0	0
v_{13}	2	0	0	0
v_{14}	2	0	0	0
v_{15}	$\frac{2}{3}$	0	0	0
	$-\frac{4}{3}$	0	0	0

Fig. 19-3

19.14 Compare the residuals of least-squares and Chebyshev solutions.

▌ For an arbitrary set of numbers $x_1, \ldots, x_n$ let $|r|_{max}$ be the largest residual in absolute value. Then $r_1^2 + \cdots + r_m^2 \le m|r|_{max}^2$ so that the root-mean-square error surely does not exceed $|r|_{max}$. But the least-squares solution has the smallest rms error of all, so that, denoting this error by ρ, $\rho \le |r|_{max}$. In particular this is true when the x_j are the Chebyshev solution, in which case $|r|_{max}$ is what we have been calling r. But the Chebyshev solution also has the property that its maximum error is smallest, so if $|\rho|_{max}$ denotes the absolutely largest residual of the least-squares solution, $|r|_{max} \le |\rho|_{max}$. Putting the two inequalities together, $\rho \le r \le |\rho|_{max}$ and we have the Chebyshev error bounded on both sides. Since the least-squares solution is often easier to find, this last result may be used to decide if it is worth continuing on to obtain the further reduction of maximum residual which the Chebyshev solution brings.

19.15 Apply the previous problem to the systems of Problem 19.3.

▌ We have already found $\rho = \sqrt{\frac{2}{21}}$, $r = \frac{1}{3}$ and $|\rho|_{max} = \frac{3}{7}$, which do steadily increase as Problem 19.14 suggests. The fact that one of the least-squares residuals is three times as large as another already recommends the search for a Chebyshev solution.

19.16 Apply Problem 19.14 to the system of Problem 19.4.

▌ We have found $\rho = \sqrt{\frac{23}{72}}$, $r = \frac{3}{4}$ and $|\rho|_{max} = \frac{11}{12}$. The spread does support a search for the Chebyshev solution.

19.17   Solve the system

$$x = 0 \qquad y = 0 \qquad x + y = 1$$

by least-squares and minmax methods.

▮   The three lines enclose a familiar area of the plane. To discover which point within this triangle will be the solution point we form the normal equations $2x + y = 1$ and $x + 2y = 1$ and find that $x = y = \frac{1}{3}$. The three residuals for this point are all $\pm \frac{1}{3}$, so this is also the rms value, outperforming what might have been a natural guess, that is, $x = y = \frac{1}{2}$.

Equality of absolute residuals is what we have come to associate with the min–max solution and, in fact, any shift from the point $(\frac{1}{3}, \frac{1}{3})$ will increase one residual or another so we do have the min–max.

19.18   Run through the simplex algorithm to further verify the min–max solution just obtained.

▮   The initial display (excluding slack variables) and needed parts of the final display are in Fig. 19-4. We find $y_1 = y_2 = 1$, $y_3 = 3$, making $r = \frac{1}{3}$ and finally $x_1 = x_2 = \frac{1}{3}$.

$$
\begin{array}{rrrrcr}
1 & 1 & 0 & 0 & v_1 & 1 \\
1 & 0 & 1 & 0 & v_2 & 1 \\
1 & 1 & 1 & -1 & & 2 \\
1 & -1 & 0 & 0 & & 2 \\
1 & 0 & -1 & 0 & & 2 \\
1 & -1 & -1 & 1 & v_3 & 3 \\
\hline
0 & 0 & 0 & 1 & & -3
\end{array}
$$

Fig. 19-4

19.19   The solution point for the preceding problem is not equidistant from all three lines. Show how such a point, the center of the inscribed circle of the triangle, can be found from an overdetermined system.

▮   Given three lines $a_i x + b_i y + c_i = 0$, residuals are defined as

$$r_i = \frac{a_i x + b_i y + c_i}{\sqrt{a_i^2 + b_i^2}}$$

which makes them distances from the lines. The min–max solution is now available by the simplex method.

19.20   The angles of a triangle are measured to be 42°, 66° and 78°. Using Problem 19.7, we remove the discrepancy by reducing all three measurements by 2°. This is the least-squares answer. What does the min–max method have to say?

▮   The initial display and needed parts of the final display of the simplex method are in Fig. 19-5. From them we find $y_1 = 20$, $y_2 = 32$ and $y_3 = .5$. Then $r = 2$, leading to $x_1 = 40$, $x_2 = 64$. The min–max and least-squares solutions are the same.

$$
\begin{array}{rrrrcr}
1 & 1 & 0 & -42 & & 2 \\
1 & 0 & 1 & -66 & & 2 \\
1 & 1 & 1 & -102 & & 2 \\
1 & -1 & 0 & 42 & v_2 & 32 \\
1 & 0 & -1 & 66 & v_1 & 20 \\
1 & -1 & -1 & 102 & v_3 & .5 \\
\hline
0 & 0 & 0 & 1 & & -.5
\end{array}
$$

Fig. 19-5

19.21   Find both the min–max and least-squares solutions of the system

$$x = 1 \qquad y = 1 \qquad xy = 5$$

▌   The residuals are $r_1 = x - 1$, $r_2 = y - 1$ and $r_3 = xy - 5$. The intuitive idea of increasing x and y a bit in order to get closer to the hyperbola soon lights on the point $(2, 2)$, which does in fact produce three absolutely equal residuals. So the min–max solution is in hand.

   For the least-squares solution we sum the residual squares, do a bit of calculus and find that x and y must satisfy the cubic equation $x^3 - 4x - 1 = 0$. There is a root between 1 and 2, but the min–max result is certainly the more attractive.

19.22   Solve the overdetermined system

$$x = 1 \qquad y = 1 \qquad x^2 + y^2 = 1$$

▌   The three residuals are

$$r_1 = x - 1 \qquad r_2 = y - 1 \qquad r_3 = x^2 + y^2 - 1$$

and the natural goal seems to be $r_1 = r_2 = -r_3$, reducing x and y equally. This leads quickly to $2r_1^2 + 5r_1 + 1 = 0$ with solution $r_1 = (-5 + \sqrt{17})/4$.

19.23   Solve the overdetermined system

$$x = 1 \qquad y = 2 \qquad y = x^2$$

▌   Of the three residuals

$$r_1 = x - 1 \qquad r_2 = y - 2 \qquad r_3 = y - x^2$$

it seems natural to arrange that $r_2 = -r_1 = -r_3$ since this will bring things closer together. The quadratic $r_1^2 + 4r_1 - 1 = 0$ soon follows and makes $r_1 = -2 + \sqrt{5}$.

CHAPTER 20
Boundary Value Problems

20.1 LINEAR ORDINARY DIFFERENTIAL EQUATIONS

20.1 Find a solution of the second-order equation

$$L(y) = y''(x) - p(x)y'(x) - q(x)y(x) = r(x)$$

satisfying the two boundary conditions

$$c_{11}y(a) + c_{12}y(b) + c_{13}y'(a) + c_{14}y'(b) = A$$

$$c_{21}y(a) + c_{22}y(b) + c_{23}y'(a) + c_{24}y'(b) = B$$

▌ With linear equations, we may rely on the superposition principle which is used in solving elementary examples by analytic methods. Assuming that elementary solutions cannot be found for the preceding equation, the numerical algorithms of an earlier chapter (Runge–Kutta, Adams, etc.) may be used to compute approximate solutions of these three *initial* value problems for $a \le x \le b$:

$$L(y_1) = 0 \qquad L(y_2) = 0 \qquad L(Y) = r(x)$$

$$y_1(a) = 1 \qquad y_2(a) = 0 \qquad Y(a) = 0$$

$$y_1'(a) = 0 \qquad y_2'(a) = 1 \qquad Y'(a) = 0$$

The required solution is then available by superposition,

$$y(x) = C_1 y_1(x) + C_2 y_2(x) + Y(x)$$

where to satisfy the boundary conditions we determine C_1 and C_2 from the equations

$$\left[c_{11} + c_{12}y_1(b) + c_{14}y_1'(b) \right]C_1 + \left[c_{13} + c_{12}y_2(b) + c_{14}y_2'(b) \right]C_2 = A - c_{12}Y(b) - c_{14}Y'(b)$$

$$\left[c_{21} + c_{22}y_1(b) + c_{24}y_1'(b) \right]C_1 + \left[c_{23} + c_{22}y_2(b) + c_{24}y_2'(b) \right]C_2 = B - c_{22}Y(b) - c_{24}Y'(b)$$

In this way the linear boundary value problem is solved by our algorithms for *initial* value problems. The method is easily extended to higher-order equations or to linear systems. We assume that the given problem has a unique solution and that the functions y_1, y_2, etc., can be found with reasonable accuracy. The equations determining C_1, C_2, etc., will then also have a unique solution.

20.2 Show how a linear boundary value problem may be solved approximately by reducing it to a linear algebraic system.

▌ Choose equally spaced arguments $x_j = a + jh$ with $x_0 = a$ and $x_{N+1} = b$. We now seek to determine the corresponding values $y_j = y(x_j)$. Replacing $y''(x_j)$ by the approximation

$$y''(x_j) \simeq \frac{y_{j+1} - 2y_j + y_{j-1}}{h^2}$$

and $y'(x_j)$ by

$$y'(x_j) \simeq \frac{y_{j+1} - y_{j-1}}{2h}$$

the differential equation $L(y) = r(x)$ of Problem 20.1 becomes, after slight rearrangement,

$$\left(1 - \tfrac{1}{2}hp_j\right)y_{j-1} + \left(-2 + h^2 q_j\right)y_j + \left(1 + \tfrac{1}{2}hp_j\right)y_{j+1} = h^2 r_j$$

If we require this to hold at the interior points $j = 1, \ldots, N$, then we have N linear equations in the N unknowns $y_1, \ldots, y_N$, assuming the two boundary values to be specified as $y_0 = y(a) = A$, $y_{N+1} = y(b) = B$. In this case the linear system takes the following form:

$$
\begin{aligned}
\beta_1 y_1 + \gamma_1 y_2 \qquad\qquad\qquad &= h^2 r_1 - \alpha_1 A \\
\alpha_2 y_1 + \beta_2 y_2 + \gamma_2 y_3 \qquad\quad &= h^2 r_2 \\
\alpha_3 y_2 + \beta_3 y_3 + \gamma_3 y_4 &= h^2 r_3 \\
&\;\;\vdots \\
\alpha_N y_{N-1} + \beta_N y_n \qquad\quad &= h^2 r_N - \gamma_N b
\end{aligned}
$$

where

$$
\alpha_j = 1 - \tfrac{1}{2} h p_j \qquad \beta_j = -2 + h^2 q_j \qquad \gamma_j = 1 + \tfrac{1}{2} h p_j
$$

The *band matrix* of this system is typical of linear systems obtained by discretizing differential boundary value problems. Only a few diagonals are nonzero. Such matrices are easier to handle than others which are not so sparse. If Gaussian elimination is used, with the pivots descending the main diagonal, the band nature will not be disturbed. This fact can be used to abbreviate the computation. The iterative Gauss–Seidel algorithm is also effective. If the more general boundary conditions of Problem 20.1 occur these may also be discretized, perhaps using

$$
y'(a) \simeq \frac{y_1 - y_0}{h} \qquad y'(b) \simeq \frac{y_{N+1} - y_N}{h}
$$

This brings a system of $N + 2$ equations in the unknowns $y_0, \ldots, y_{N+1}$.

In this and the previous problem we have alternative approaches to the same goal. In both cases the output is a finite set of numbers y_j. If either method is reapplied with smaller h, then hopefully the larger output will represent the true solution $y(x)$ more accurately. This is the question of *convergence*.

20.3   Show that for the special case

$$
y'' + y = 0 \qquad y(0) = 0 \qquad y(1) = 1
$$

the method of Problem 20.2 is convergent.

$\blacksquare$ The exact solution function is $y(x) = (\sin x)(\sin 1)$. The approximating difference equation is

$$
y_{j-1} + (-2 + h^2) y_j + y_{j+1} = 0
$$

and this has the exact solution

$$
y_j = \frac{\sin(\alpha x_j / h)}{\sin(\alpha / h)}
$$

for the same boundary conditions $y_0 = 0$, $y_{N+1} = 1$. Here $x_j = jh$ and $\cos \alpha = 1 - \tfrac{1}{2} h^2$. These facts may be verified directly or deduced by the methods of our section on difference equations. Since $\lim(\alpha / h)$ is 1 for h tending to zero, we now see that $\lim y_j = y(x_j)$, that is, solutions of the difference problem for decreasing h converge to the solution of the differential problem. In this example both problems may be solved analytically and their solutions compared. The proof of convergence for more general problems must proceed by other methods.

20.4   Solve $y'' + y' + xy = 0$ with $y(0) = 1$ and $y(1) = 0$ by the method of Problem 20.1.

$\blacksquare$ Suppose we take the series approach $y = \sum_{k=0}^{\infty} a_k x^k$. Then

$$
xy = \sum_{k=0}^{\infty} a_k x^{k+1} = \sum_{i=1}^{\infty} a_{i-1} x^i
$$

$$
y' = \sum_{k=1}^{\infty} k a_k x^{k-1} = \sum_{i=0}^{\infty} (i+1) a_{i+1} x^i
$$

$$
y'' = \sum_{k=2}^{\infty} k(k-1) a_k x^{k-2} = \sum_{i=0}^{\infty} (i+2)(i+1) a_{i+2} x^i
$$

and summing brings the identity

$$2a_2 + a_1 + \sum_{i=1}^{\infty} \left[(i+2)(i+1) a_{i+2} + (i+1) a_{i+1} + a_{i-1} \right] x^i = 0$$

This makes $a_2 = -a_1/2$ and leads to the recursion

$$a_{i+2} = -\frac{(i+1) a_{i+1} + a_{i-1}}{(i+2)(i+1)} \quad i = 1, 2, \ldots$$

which can now be exploited to produce the two needed independent solutions. First choose $y_1(0) = 1$, $y_1'(0) = 0$. Then $y_1(x)$ can be computed using the recursion. Only its value at $x = 1$ is needed for the moment and this proves to be $y_1(1) = .87191$. Now choose $y_2(0) = 0$, $y_2'(0) = 1$ leading to $y_2(1) = .58151$. The required solution is now available from the general solution of the differential equation

$$y(x) = c_1 y_1(x) + c_2 y_2(x)$$

with the initial condition forcing $c_1 = 1$ and the terminal condition forcing $c_2 = -y_1(1)/y_2(1) = -1.4994$. For the required solution we now take initial values $y(0) = 1$, $y'(0) = -1.4994$ and appeal to the series and recursion one more time. Here is the output at intervals of .1, the terminal value being zero to six places:

$$1 \quad .8562 \quad .7272 \quad .6081 \quad .4986 \quad .3977 \quad .3045 \quad .2184 \quad .1393 \quad .0664 \quad 0$$

20.5 Also solve the preceding problem by the method of Problem 20.2.

▮ With the line segment discretized as follows,

$$x_0 = 0 \quad\quad x_1 = h \quad\quad\quad x_j = jh \quad\quad\quad\quad x_{N+1} = 1$$

the difference equation takes this shape, for $j = 1, \ldots, N$:

$$\frac{y_{j+1} - 2y_j + y_{j-1}}{h^2} + \frac{y_{j+1} - y_{j-1}}{2h} + x_j y_j = 0$$

Rewriting it as

$$(1 - h/2) y_{j-1} + (-2 + h^2 jh) y_j + (1 + h/2) y_{j+1} = 0$$

and using the boundary values $y_0 = 1$, $y_{N+1} = 0$, allows the system of N equations to be expressed in matrix form:

$$\begin{pmatrix} -2+h^3 & 1+h/2 & & & \\ 1-h/2 & -2+2h^3 & 1+h/2 & & \mathbf{0} \\ & 1-h/2 & -2+3h^3 & 1+h/2 & \\ & & & \ddots & \\ \mathbf{0} & & \cdots & 1-h/2 & -2+Nh^3 \end{pmatrix} \begin{pmatrix} y_1 \\ y_2 \\ \vdots \\ y_N \end{pmatrix} = \begin{pmatrix} -1+h/2 \\ 0 \\ 0 \\ \vdots \\ 0 \end{pmatrix}$$

Choosing $h = .2$ and $h = .1$, Gaussian elimination was used and produced the following values, which are close to the earlier values produced by the series.

x	.1	.2	.3	.4	.5	.6	.7	.8	.9
$h = .2$		.7274		.4990		.3049		.1394	
$h = .1$	.8572	.7272	.6082	.4987	.3978	.3046	.2186	.1393	.0665

20.6 Solve the nonhomogeneous equation $y'' + xy' + y = e^x$ with boundary conditions $y(0) = 0$ and $y(1) = 0$.

▮ Standard procedure first defines three basic functions, two of them solutions of the corresponding homogeneous equation and having initial values

$$y_1(0) = 1 \quad y_1'(0) = 0 \quad y_2(0) = 0 \quad y_2'(0) = 1$$

and the third a solution of the given equation with both $y_3(0)$ and $y_3'(0)$ zero. The general solution is then

$$y(x) = c_1 y_1(x) + c_2 y_2(x) + y_3(x)$$

and the given boundary conditions force $c_1 = 0$ and $c_2 = -y_3(1)/y_2(1)$. This means that it is unnecessary to compute $y_1(x)$ for present purposes. Replacing the given equation by the system

$$y' = p \qquad p' = e^x - \sqrt{x}\, y' - y$$

the Runge–Kutta fourth-order method, with $h = .01$, found $y_2(1) = .66107$ and $y_3(1) = .54810$, making $c_2 = -.82912$. Finally, a run from initial conditions $y(0) = 0$, $p(0) = c_2$ found a terminal value of 10^{-7}. Here are a few other values found along the way

x	.2	.4	.6	.8
y	$-.1400$	$-.2165$	$-.2200$	$-.1475$

20.2 CHARACTERISTIC VALUE PROBLEMS

20.7 Illustrate the reduction of a linear differential eigenvalue problem to an approximating algebraic system.

▮ Consider the problem

$$y'' + \lambda y = 0 \qquad y(0) = y(1) = 0$$

This has the exact solutions $y(x) = C \sin n\pi x$, for $n = 1, 2, \ldots$. The corresponding eigenvalues are $\lambda_n = n^2 \pi^2$. Simply to illustrate a procedure applicable to other problems for which exact solutions are not so easily found, we replace this differential equation by the difference equation

$$y_{j-1} + (-2 + \lambda h^2)\, y_j + y_{j+1} = 0$$

Requiring this to hold at the interior points $j = 1, \ldots, N$, we have an algebraic eigenvalue problem $Ay = \lambda h^2 y$ with the band matrix

$$A = \begin{bmatrix} -2 & 1 & & & \\ 1 & -2 & 1 & & \\ & 1 & -2 & & \\ & \cdots & & \ddots & 1 \\ & & & 1 & -2 \end{bmatrix}$$

all other elements being zero and $y^T = (y_1, \ldots, y_N)$. The exact solution of this problem may be found to be

$$y_j = C \sin n\pi x_j \quad \text{with} \quad \lambda_n = \frac{4}{h^2} \sin^2 \frac{n\pi h}{2}$$

Plainly, as h tends to zero these results converge to those of the target differential problem.

20.8 Use the method of the preceding problem to find eigenvalues and eigenfunctions of $y'' + \lambda y = 0$ with $y(0) = 0$ and $y'(1) = 0$. Prove convergence to the exact $\lambda_n = (n\pi/2)^2$, for odd n, and $y_n(x) = \sin(n\pi x/2)$.

▮ The corresponding difference equation

$$y_{j-1} - 2y_j + y_{j+1} + \lambda h^2 y_j = 0$$

can be written as

$$y_{j-1} - 2a y_j + y_{j+1} = 0$$

with $2a = 2 - \lambda h^2$. Solving the characteristic equation we find $r = a \pm i\sqrt{1 - a^2} = \exp(i\theta)$, with $\cos \theta = 1 - \lambda h^2/2$. Then

$$y_j = c_1 \sin j\theta + c_2 \cos j\theta$$

is the solution family of the difference equation. The initial condition now forces $c_2 = 0$, after which the terminal condition requires that $(N+1)\theta = n\pi/2$ with n odd, since this is where cosines vanish. Recalling that $(N+1)h = 1$, our solution of the discretized boundary value problem is therefore

$$y_j = c_1 \sin jn\pi h/2 = c_1 \sin n\pi x_j/2$$

as expected. As for the eigenvalues,

$$\lambda_n = (2/h^2)(1 - \cos n\pi h/2) = (n\pi/2)^2 + O(h^2)$$

assuring convergence.

20.9 Apply the discretization method to the eigenvalue problem

$$y'' + \lambda xy = 0 \qquad y(0) = 0 \qquad y(1) = 0$$

▮ The variable coefficient makes things more interesting, the difference equation being

$$y_{j-1} - (2 - \lambda jh^3) y_j + y_{j+1} = 0$$

It is convenient to let $L = \lambda h^3$ and multiply through by $-1/j$, converting the equation to

$$(-1/j) y_{j-1} + (2/j - L) y_j - (1/j) y_{j+1} = 0$$

and creating this matrix eigenvalue problem.

$$\begin{pmatrix} 2-L & -1 & & & \\ -\dfrac{1}{2} & 1-L & -\dfrac{1}{2} & & \text{\LARGE 0} \\ & -\dfrac{1}{3} & \dfrac{2}{3}-L & -\dfrac{1}{3} & \\ & \text{\LARGE 0} & & \cdots & \\ & & & -\dfrac{1}{N} & \dfrac{2}{N}-L \end{pmatrix} \begin{pmatrix} y_1 \\ y_2 \\ y_3 \\ \vdots \\ y_N \end{pmatrix} = \begin{pmatrix} 0 \\ 0 \\ 0 \\ \vdots \\ 0 \end{pmatrix}$$

For the case $N = 2$ this is simply

$$\begin{pmatrix} 2-L & -1 \\ -\frac{1}{2} & 1-L \end{pmatrix} \begin{pmatrix} y_1 \\ y_2 \end{pmatrix} = \begin{pmatrix} 0 \\ 0 \end{pmatrix}$$

with eigenvalues $L = \frac{1}{2}(3 \pm \sqrt{3})$ and eigenvectors $(1 + \sqrt{3}, -1)^T$ and $(2, 1 + \sqrt{3})^T$. The elements of these eigenvectors show the behavior of the discrete solutions within the interval. For larger N and smaller h the methods of Chapter 17 are available.

20.3 NONLINEAR ORDINARY DIFFERENTIAL EQUATIONS

20.10 What is the *garden-hose method*?

▮ Given the equation $y'' = f(x, y, y')$, we are to find a solution that satisfies the boundary conditions $y(a) = A$, $y(b) = B$.

One simple procedure is to compute solutions of the initial value problem

$$y'' = f(x, y, y') \qquad y(a) = A \qquad y'(a) = M$$

for various values of M until two solutions, one with $y(b) < B$ and the other with $y(b) > B$, have been found. If these solutions correspond to initial slopes of M_1 and M_2, then interpolation will suggest a new M value between these and a better approximation may then be computed (see Fig. 20-1). Continuing this process leads to successively better approximations and is essentially the *regula falsi* algorithm used for nonlinear algebraic problems. Here our computed terminal value is a function of M, say $F(M)$, and we do have to solve the equation $F(M) = B$. However, for each choice of M the calculation of $F(M)$ is no longer the evaluation of an algebraic expression but involves the solution of an initial value problem of the differential equation.

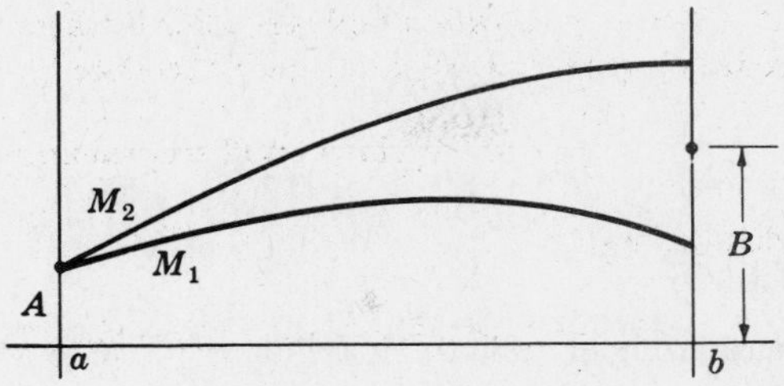

Fig. 20-1

20.11 How may the garden-hose method be refined?

❚ Instead of using the equivalent of *regula falsi*, we may adapt Newton's method to the present problem, presumably obtaining improved convergence to the correct M value. To do this we need to know $F'(M)$. Let $y(x, M)$ denote the solution of

$$y'' = f(x, y, y') \qquad y(a) = A \qquad y'(a) = M$$

and for brevity let $z(x, M)$ be its partial derivative relative to M. Differentiating relative to M brings

(1) $$z'' = f_y(x, y, y') z + f_{y'}(x, y, y') z'$$

if we freely reverse the orders of the various derivatives. Also differentiating the initial conditions, we have

$$z(a, M) = 0 \qquad z'(a, M) = 1$$

Let M_1 be a first approximation to M and solve the original problem for the approximate solution $y(x, M_1)$. This may then be substituted for y in equation (1) and the function $z(x, M_1)$ computed. Then $F'(M) = z(b, M_1)$. With this quantity available the Newton method for solving $F(M) - B = 0$ now offers us the next approximation to M:

$$M_2 = M_1 -- \frac{F(M_1) - B}{F'(M_1)}$$

With this M_2 a new approximation $y(x, M_2)$ may be computed and the process repeated. The method may be extended to higher-order equations or to systems, the central idea being the derivation of an equation similar to (1), which is called the *variational equation*.

20.12 Apply the garden-hose method to $y'' = y^2 + (y')^2$ with $y(0) = 0$ and $y(1) = 1$.

❚ With Runge–Kutta in mind we replace the equation by

$$y' = p \qquad p' = y^2 + p^2$$

and experiment with the initial value of p. Here are results obtained using $h = .1$ and a few rough mental interpolations. A further check using $h = .01$ found no important changes.

Initial p	.5	.7	.6	.605	.6017	.6018
Terminal y	.74	1.35	.995	1.01	.99977	1.00007

20.13 An object climbs from ground level to height 100 ft in 1 s. Assuming an atmospheric drag which makes the equation of motion $y'' = -32 - .1\sqrt{y'}$, what was the initial velocity?

❚ Elementary strategy calls for reduction to a first-order equation in $y' = v$, followed by integration of v. But suppose we apply the Runge–Kutta method to the system

$$y' = v \qquad v' = -32 - .1\sqrt{v}$$

with $y(0) = 0$ and $v(0)$ to be found by the garden-hose method. Here are results of a brief run using ad lib interpolations:

Initial v	100	120	117	116	116.5
Terminal y	83.5	103.5	100.5	99.5	100

20.4 VARIATIONAL PROBLEMS

20.14 Reduce the problem of maximizing or minimizing $\int_a^b F(x, y, y')\, dx$ to a boundary value problem.

▌ This is the classical problem of the calculus of variations. If the solution function $y(x)$ exists and has adequate smoothness, then it is required to satisfy the Euler differential equation $F_y = (d/dx)F_{y'}$. If boundary conditions such as $y(a) = A$, $y(b) = B$ are specified in the original optimization problem, then we already have a second-order boundary value problem. If either of these conditions is omitted, then the variational argument shows that $F_{y'} = 0$ must hold at that end of the interval. This is called the *natural boundary condition*.

20.15 Minimize $\int_0^1 (y^2 + y'^2)\, dx$ subject to $y(0) = 1$.

▌ The Euler equation is $2y = 2y''$ and the natural boundary condition is $y'(1) = 0$. The solution is now easily found to be $y = \cosh x - \tanh 1 \sinh x$ and it makes the integral equal to $\tanh 1$, which is about .76. In general the Euler equation will be nonlinear and the garden-hose method may be used to find $y(x)$.

20.16 Illustrate the Ritz method of solving a boundary value problem.

▌ The idea of the Ritz method is to solve an equivalent minimization problem instead. Consider

$$y'' = -x^2 \qquad y(0) = y(1) = 0$$

sometimes called a Poisson problem in one variable, but in fact requiring only two integrations to discover the solution

$$y(x) = \frac{x(1 - x^3)}{12}$$

Methods are available for finding an equivalent minimization problem for a given boundary problem, but here one is well known:

$$J(y) = \int_0^1 \left[\tfrac{1}{2}(y')^2 - x^2 y \right] dx = \text{minimum}$$

The Euler equation for this integral proves to be our original differential equation.

To approximate the solution by the Ritz method, we need a family of functions satisfying the boundary conditions. Suppose we choose

$$\phi(x) = cx(1 - x)$$

which is probably the simplest such family for this problem. Substituting ϕ for y in the integral, an easy calculation yields

$$J(\phi) = \frac{c^2}{6} - \frac{c}{20} = f(c)$$

which we minimize by setting $f'(c) = 0$. The resulting $c = \frac{3}{20}$ gives us the approximation

$$\phi(x) = \tfrac{3}{20}x(1 - x)$$

which is compared with the true solution in Fig. 20-2. More accurate approximations are available through the use of a broader family of approximating functions, perhaps

$$\phi(x) = x(1 - x)\left(c_0 + c_1 x + c_2 x^2 + \cdots + c_n x^n\right)$$

leading to a linear system for determining the coefficients c_i. The central idea of the Ritz method is the search for the optimum function among members of a restricted family $\phi(x)$, rather than among all $y(x)$ for which the given integral exists.

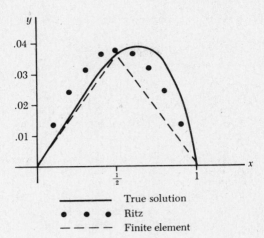

True solution

● ● ● Ritz

– – – – Finite element

Fig. 20-2

20.17 Use the same boundary value problem solved in Problem 20.16 to illustrate a finite element solution method.

▮ The basic idea is the same. It is the nature of the approximating family that identifies a finite element method. Suppose we divide our interval $(0, 1)$ into halves and use two line segments

$$\phi_1(x) = 2Ax \qquad \phi_2(x) = 2A(1 - x)$$

meeting at point $(\frac{1}{2}, A)$ to approximate $y(x)$. In fact, we have a family of such approximations, with parameter A to be selected. The two line segments are called finite elements, and the approximating function is formed by piecing them together. As before we substitute into the integral, and we easily compute

$$J(\phi) = \int_0^{1/2} \phi_1 \, dx + \int_{1/2}^1 \phi_2 \, dx = 2A^2 - \tfrac{7}{48}A = f(A)$$

which we minimize by setting $f'(A) = 0$. This makes $A = \frac{7}{192}$. A quick calculation shows that this is actually the correct value of the solution at $x = \frac{1}{2}$ (see Fig. 20-2). It has been shown that if line segments are used as finite elements (in a one-dimensional problem, of course) correct values are systematically produced at the joins.

20.18 Extend the procedure of the preceding problem to include more finite elements.

▮ Divide the interval $(0, 1)$ into n parts, with endpoints at $0 = x_0, x_1, x_2, \ldots, x_n = 1$. Let $y_1, \ldots, y_{n-1}$ be corresponding and arbitrary ordinates, with $y_0 = y_n = 0$. Define linear finite elements $\phi_1, \ldots, \phi_n$ in the obvious way (see Fig. 20-3). Then

$$\phi_i(x) = y_{i-1} \frac{x_i - x}{x_i - x_{i-1}} + y_i \frac{x - x_{i-1}}{x_i - x_{i-1}}$$

$$= y_{i-1} \frac{x_i - x}{h} + y_i \frac{x - x_{i-1}}{h}$$

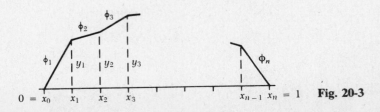

Fig. 20-3

the second equality holding if the x_i are equally spaced. We also have

$$\phi_i'(x) = \frac{y_i - y_{i-1}}{x_i - x_{i-1}} = \frac{y_i - y_{i-1}}{h}$$

Now consider the integral

$$J(\phi) = \sum_{i=1}^{n} \int_{x_{i-1}}^{x_i} \left[\tfrac{1}{2}(\phi_i')^2 - x^2\phi_i \right] dx = \sum_{i=1}^{n} J_i$$

$$= f(y_1, \ldots, y_{n-1})$$

To minimize this, we could obtain f explicitly in terms of the y_i and then compute the partial derivatives, setting them to zero and solving the resulting system of equations. This is what was just done in the simpler case. Here suppose we take derivatives first, integrate second and then form the ultimate system. The dependence of f on a particular ordinate y_k is through only two of the component terms J_k and J_{k+1}. Accordingly, for $k = 1, \ldots, n-1$,

$$\frac{\partial f}{\partial y_k} = \int_{x_{k-1}}^{x_k} \left[\frac{y_k - y_{k-1}}{h}\left(\frac{1}{h}\right) - x^2 \frac{x - x_{k-1}}{h} \right] dx + \int_{x_k}^{x_{k+1}} \left[\frac{y_{k+1} - y_k}{h}\left(\frac{-1}{h}\right) - x^2 \frac{x_{k+1} - x}{h} \right] dx$$

and the integrals being elementary we soon have the system

$$-y_{k-1} + 2y_k - y_{k+1} = \tfrac{1}{12}x_{k-1}^4 + \tfrac{1}{2}x_k^4 + \tfrac{1}{12}x_{k+1}^4 - \tfrac{1}{3}x_{k-1}x_k^3 - \tfrac{1}{3}x_{k+1}x_k^3$$

for $k = 1, \ldots, n-1$.

With $n = 2$ and $k = 1$, this quickly reproduces the $y_1 = \tfrac{7}{192}$ found before. With $n = 3$, the system becomes

$$2y_1 - y_2 = \tfrac{7}{486}$$

$$-y_1 + 2y_2 = \tfrac{25}{486}$$

from which come $y_1 = \tfrac{13}{486}$ and $y_2 = \tfrac{19}{486}$, both of which agree with the true solution values for these positions.

20.19 Find the function $y(x)$ that minimizes $\int_0^1 [xy^2 + (y')^2] \, dx$ and satisfies $y(0) = 0$, $y(1) = 1$.

▮ The Euler equation is the Airy equation $y'' = xy$ which was encountered in Problem 11.128. The series $y = \sum_0^\infty a_k x^k$ was found to be a solution if $a_2 = 0$ and the recursion

$$a_k = a_{k-3}/k(k-1)$$

holds for $k = 3, 4, \ldots$. Here a_0 is also zero by the initial condition and the solution reduces to

$$y(x) = a_1 \left[x + \frac{1}{4 \cdot 3}x^4 + \frac{1}{7 \cdot 6 \cdot 4 \cdot 3}x^7 + \cdots \right]$$

with the terminal condition determining a_1. The value of the remaining series at $x = 1$ proves to be 1.0853 to four places, giving a_1 the value .9214. For comparison with the next problem, the value $y(\tfrac{1}{2})$ was found to be .4655.

20.20 Apply the Ritz method to the preceding problem, using the approximating function $p(x) = ax + bx^2$, with $b = 1 - a$.

▮ The function $p(x)$ does satisfy the boundary conditions so it remains to minimize the integral. Substituting and doing a bit of algebra and calculus, one soon finds

$$I = \tfrac{7}{20}a^2 - \tfrac{3}{5}a + \tfrac{3}{2}$$

which is minimum at $a = \tfrac{6}{7}$. The approximate solution is

$$p(x) = \tfrac{6}{7}x + \tfrac{1}{7}x^2$$

with value $\tfrac{13}{28}$ or .4643 at $x = \tfrac{1}{2}$. This is off by 1 unit in the third place.

20.5 THE DIFFUSION EQUATION

20.21 Replace the diffusion problem involving the equation

$$\frac{\partial T}{\partial t} = a\left(\frac{\partial^2 T}{\partial x^2}\right) + b\left(\frac{\partial T}{\partial x}\right) + cT$$

and the conditions $T(0, t) = f(t)$, $T(l, t) = g(t)$, $T(x, 0) = F(x)$ by a finite difference approximation.

▮ Let $x_m = mh$ and $t_n = nk$, where $x_{M+1} = l$. Denoting the value $T(x, t)$ by the alternate symbol $T_{m, n}$, the approximations

$$\frac{\partial T}{\partial t} \simeq \frac{T_{m, n+1} - T_{m, n}}{k} \qquad \frac{\partial T}{\partial x} \simeq \frac{T_{m+1, n} - T_{m-1, n}}{2h}$$

$$\frac{\partial^2 T}{\partial x^2} \simeq \frac{T_{m+1, n} - 2T_{m, n} + T_{m-1, n}}{h^2}$$

convert the diffusion equation to

$$T_{m, n+1} = \lambda\left(a - \tfrac{1}{2}bh\right)T_{m-1, n} + \left[1 - \lambda(2a + ch^2)\right]T_{m, n} + \lambda\left(a + \tfrac{1}{2}bh\right)T_{m+1, n}$$

where $\lambda = k/h^2$, $m = 1, 2, \ldots, M$ and $n = 1, 2, \ldots$. Using the same initial and boundary conditions above, in the form $T_{0, n} = f(t_n)$, $T_{M+1, n} = g(t_n)$ and $T_{m, 0} = F(x_m)$, this difference equation provides an approximation to each interior $T_{m, n+1}$ value in terms of its three nearest neighbors at the previous time step. The computation therefore begins at the (given) values for $t = 0$ and proceeds first to $t = k$, then to $t = 2k$ and so on.

20.22 Apply the procedure of the preceding problem to the case $a = 1$, $b = c = 0$, $f(t) = g(t) = 0$, $F(x) = 1$ and $l = 1$.

▮ Suppose we choose $h = \frac{1}{4}$ and $k = \frac{1}{32}$. Then $\lambda = \frac{1}{2}$ and the difference equation simplifies to

$$T_{m, n+1} = \tfrac{1}{2}\left(T_{m-1, n} + T_{m+1, n}\right)$$

A few lines of computation are summarized in Table 20.1(a). The bottom line and the side columns are simply the initial and boundary conditions. The interior values are computed from the difference equation line by line, beginning with the circled ① which comes from averaging its two lower neighbors, also looped. A slow trend toward the ultimate "steady state" in which all T values are zero may be noticed, but not too much accuracy need be expected of so brief a calculation.

For a second try we choose $h = \frac{1}{8}$, $k = \frac{1}{128}$, keeping $\lambda = \frac{1}{2}$. The results appear in Table 20.1(b). The top line of this table corresponds to the second line in Table 20.1(a) and is in fact a better approximation to $T(x, \frac{1}{32})$. This amounts to a primitive suggestion that the process is starting to *converge* to the correct $T(x, t)$ values.

In Table 20.1(c) we have the results if $h = \frac{1}{4}$, $k = \frac{1}{16}$ are chosen, making $\lambda = 1$. The difference equation for this choice is

$$T_{m, n+1} = T_{m-1, n} - T_{m, n} + T_{m+1, n}$$

The start of an explosive oscillation can be seen. This does not at all conform to the correct solution, which is known to decay exponentially. Later we shall see that unless $\lambda \le \frac{1}{2}$ such an explosive and unrealistic oscillation may occur. This is a form of numerical *instability*.

TABLE 20.1

0	$\frac{1}{4}$	$\frac{1}{2}$	$\frac{1}{4}$	0		0	$\frac{3}{8}$	$\frac{3}{4}$	$\frac{7}{8}$	1	$\frac{7}{8}$	$\frac{3}{4}$	$\frac{3}{8}$	0		0	−12	17	−12	0
0	$\frac{1}{2}$	$\frac{1}{2}$	$\frac{1}{2}$	0		0	$\frac{1}{2}$	$\frac{3}{4}$	1	1	1	$\frac{3}{4}$	$\frac{1}{2}$	0		0	5	−7	5	0
0	$\frac{1}{2}$	1	$\frac{1}{2}$	0		0	$\frac{1}{2}$	1	1	1	1	1	$\frac{1}{2}$	0		0	−2	3	−2	0
0	①	1	1	0		0	1	1	1	1	1	1	1	0		0	1	−1	1	0
①	1	①	1	1		1	1	1	1	1	1	1	1	1		0	0	1	0	0
																0	1	1	1	0
																1	1	1	1	1
	(a)							(b)									(c)			

20.23   What is the *truncation error* of this method?

▌ As earlier we apply Taylor's theorem to the difference equation and find that our approximation has introduced error terms depending on h and k. These terms are the truncation error

$$\tfrac{1}{2}kT_{tt} - \tfrac{1}{12}ah^2T_{xxxx} + \tfrac{1}{6}bh^2T_{xxx} + O(h^4)$$

subscripts denoting partial derivatives. In the important special case $a = 1$, $b = 0$, we have $T_{tt} = T_{xxxx}$ so that the choice $k = h^2/6$ (or $\lambda = \tfrac{1}{6}$) seems especially desirable from this point of view, the truncation error then being $O(h^4)$.

20.24   Show that the method of Problem 20.21 is convergent in the particular case

$$\frac{\partial T}{\partial t} = \frac{\partial^2 T}{\partial x^2} \qquad T(0,t) = T(\pi,t) = 0 \qquad T(x,0) = \sin px$$

where p is a positive integer.

▌ The exact solution may be verified to be $T(x,t) = e^{-p^2t}\sin px$. The corresponding difference equation is

$$T_{m,n+1} - T_{m,n} = \lambda(T_{m+1,n} - 2T_{m,n} + T_{m-1,n})$$

and the remaining conditions may be written

$$T_{m,0} = \sin\frac{mp\pi}{M+1} \qquad T_{0,n} = T_{M+1,n} = 0$$

This finite difference problem can be solved by "separation of the variables." Let $T_{m,n} = u_m v_n$ to obtain

$$\frac{v_{n+1} - v_n}{v_n} = \lambda\left(\frac{u_{m+1} - 2u_m + u_{m-1}}{u_m}\right) = -\lambda C$$

which defines C. But comparing C with the extreme left member we find it independent of m, and comparing it with the middle member we find it also independent of n. It is therefore a constant and we obtain separate equations for u_m and v_n in the form

$$v_{n+1} = (1 - \lambda C)v_n \qquad u_{m+1} - (2 - C)u_m + u_{m-1} = 0$$

These are easily solved by our difference equation methods. The second has no solution with $u_0 = u_{M+1} = 0$ (except u_m identically zero) unless $0 < C < 4$, in which case

$$u_m = A\cos\alpha m + B\sin\alpha m$$

where A and B are constants and $\cos\alpha = 1 - \tfrac{1}{2}C$. To satisfy the boundary conditions, we must now have $A = 0$ and $\alpha(M+1) = j\pi$, j being an integer. Thus

$$u_m = B\sin\frac{mj\pi}{M+1}$$

Turning toward v_n, we first find that $C = 2(1 - \cos\alpha) = 4\sin^2\{j\pi/[2(M+1)]\}$ after which

$$v_n = \left[1 - 4\lambda\sin^2\frac{j\pi}{2(M+1)}\right]^n v_0$$

It is now easy to see that choosing $B = v_0 = 1$ and $j = p$ we obtain a function

$$T_{m,n} = u_m v_n = \left[1 - 4\lambda\sin^2\frac{p\pi}{2(M+1)}\right]^n \sin\frac{mp\pi}{M+1}$$

which has all the required features. For comparison with the differential solution we return to the symbols $x_m = mh$, $t_n = nk$:

$$T_{m,n} = \left(1 - 4\lambda\sin^2\frac{ph}{2}\right)^{t_n/\lambda h^2}\sin px_m$$

As h now tends to zero, assuming $\lambda = k/h^2$ is kept fixed, the coefficient of $\sin px_m$ has limit $e^{-p^2 t_n}$ so that convergence is proved. Here we must arrange that the point (x_m, t_n) also remain fixed, which involves increasing m and n as h and k diminish, in order that the $T_{m,n}$ values be successive approximations to the same $T(x, t)$.

20.25 Use the previous problem to show that for the special case considered an explosive oscillation may occur unless $\lambda \le \frac{1}{2}$.

▌ The question now is not what happens as h tends to zero, but what happens for fixed h as the computation is continued to larger n arguments. Examining the coefficient of $\sin px_m$ we see that the quantity in brackets may be less than -1 for some values of λ, p and h. This would lead to an explosive oscillation with increasing t_n. The explosion may be avoided by requiring that λ be no greater than $\frac{1}{2}$. Since this makes $k \le h^2/2$ the computation will proceed very slowly, and if results for large t arguments are wanted it may be useful to use a different approach.

20.26 Solve Problem 20.21 by means of a Fourier series.

▌ This is the classical procedure when a is constant and $b = c = 0$. We first look for solutions of the diffusion equation having the product form $U(x)V(t)$. Substitution brings $V'/V = U''/U = -\alpha^2$ where α is constant. (The negative sign will help us satisfy the boundary conditions.) This makes

$$V = Ae^{-\alpha^2 t} \qquad U = B \cos \alpha x + C \sin \alpha x$$

To make $T(0, t) = 0$, we choose $B = 0$. To make $T(1, t) = 0$, we choose $\alpha = n\pi$ where n is a positive integer. Putting $C = 1$ arbitrarily and changing the symbol A to A_n, we have the functions

$$A_n e^{-n^2 \pi^2 t} \sin n\pi x \qquad n = 1, 2, 3, \ldots$$

each of which meets all our requirements except for the initial condition. The series

$$T(x, t) = \sum_{n=1}^{\infty} A_n e^{-n^2 \pi^2 t} \sin n\pi x$$

if it converges properly will also meet these requirements, and the initial condition may also be satisfied by suitable choice of the A_n. For $F(x) = 1$ we need

$$T(x, 0) = F(x) = \sum_{n=1}^{\infty} A_n \sin n\pi x$$

and this is achieved by using the Fourier coefficients for $F(x)$,

$$A_n = 2 \int_0^1 F(x) \sin n\pi x \, dx$$

The partial sums of our series now serve as approximate solutions of the diffusion problem. The exact solution used in Problem 20.24 may be viewed as a one term Fourier series.

20.27 Discretize the problem $\partial T/\partial t = \partial^2 T/\partial x^2 + T$, $0 \le x \le 1$, $0 \le t$, with boundary values $T(0, t) = T(1, t) = 0$ and initial values $T(x, 0) = x(1 - x)$.

▌ The path followed in Problem 20.21 leads to the difference equation

$$T_{m, n+1} = 2T_{m-1, n} + (1 - 2\lambda + k) T_{m, n} + \lambda T_{m+1, n}$$

with $T(0, n) = T(M + 1, n) = 0$ and $T(m, 0) = mh(1 - mh)$. For $\lambda = .5$ this discretization leads to a steady decline in T values as t grows, which is what is to be expected in view of the boundary conditions. At $t = 1$ these values are already zero to four decimal places. But even with $h = .25$ it takes 32 upward steps to reach this level and halving h quadruples the number of steps. The temptation to increase λ and speed things up must, however, be resisted. With $\lambda = 1$ and $h = .25$ these values were found for $T(2, n)$, moving upward on the center of the strip:

$$.14 \qquad .02 \qquad .13 \qquad -.22 \qquad .55 \qquad -1.28 \qquad 3.02 \qquad -7.09$$

The beginning of an error oscillation is apparent.

20.28 Discretize the two-dimensional diffusion equation

$$\frac{\partial T}{\partial t} = \frac{\partial^2 T}{\partial x^2} + \frac{\partial^2 T}{\partial y^2}$$

and comment on truncation error.

$\blacksquare$ The natural extension of ideas just used, with a simplified notation based on Fig. 20-4, is

$$\frac{T_5 - T_0}{k} = \frac{T_1 - 2T_0 + t_2}{h^2} + \frac{T_3 - 2T_0 + t_4}{h^2}$$

or, using $\lambda = k/h^2$

$$T_5 = (1 - 4\lambda)T_0 + \lambda(T_1 + T_2 + T_3 + T_4)$$

With k proportional to h^2, the truncation error of both the time and space discretizations is $O(h^2)$.

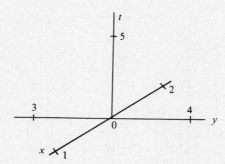

Fig. 20-4

20.29 What is an "implicit" finite difference method?

$\blacksquare$ As a simple example consider the problem

$$\partial T/\partial t = \partial^2 T/\partial x^2 \qquad T(0, t) = T(1, t) = 0 \qquad T(x, 0) = 1$$

already discussed in Problem 20.22. With the same approximations as before except for

$$\partial T/\partial t \simeq (T_{m, n} - T_{m, n-1})/k$$

we obtain

$$\lambda T_{m-1, n} - (1 + 2\lambda)T_{m, n} + \lambda T_{m+1, n} = -T_{m, n-1}$$

Applied first at $n = 1$, the right side of this equation involves known initial values and the left side three unknowns. Using $m = 1, \ldots, M$, we have a linear system of M equations to determine $T_{1,1}$ up to $T_{M,1}$. Solving this system we are ready for a second step with $n = 2$. Each new line is thus obtained as a unit, by solving a linear system. The advantage is that now there proves to be no stability restriction on the size of λ, and the horizontal lines may be more widely separated.

20.30 Use the implicit method together with a variable time step to solve the *free boundary problem*

$$T_t = T_{xx} \quad \text{for } 0 \le t, 0 \le x \le X(t) \text{ where } X(0) = 0$$
$$T_x(0, t) = -1 \qquad T(X, t) = 0 \qquad X'(t) = -T_x(X, t)$$

$\blacksquare$ Problems such as this arise in change of state circumstances, such as the freezing of a lake or the melting of a metal. The position of the boundary between solid and fluid is not known in advance and must be determined by the solution algorithm. One algorithm uses

$$(T_{m-1, n} - 2T_{m, n} + T_{m+1, n})/h^2 = (T_{m, n} - T_{m, n-1})/k_n$$

in place of the diffusion equation and

$$T_{0,n} - T_{1,n} = h \qquad T_{n,n} = 0$$

in place of the first two boundary conditions. The remaining (free) boundary condition is equivalent to

$$X(t) = t - \int_0^{X(t)} T(x,t)\, dx$$

To see this, temporarily let $f(t) = t - \int_0^{X(t)} T(x,t)\, dx$ and differentiate to find

$$f'(t) = 1 - X'(t) T[X(t), t] - \int_0^{X(t)} T_t(x,t)\, dx = 1 - \int_0^{X(t)} T_{xx}(x,t)\, dx$$

$$= 1 - T_x[X(t), t] + T_x(0, t) = X'(t)$$

Since $f(0) = X(0) = 0$, it follows that $f(t) \equiv X(t)$.

We next replace this condition by the discretization

$$nh = t_{n-1} + k_n - \sum_{i=1}^{n-1} T_{i,n-1} h$$

Each step of the computation now consists of determining k_n from this last equation (by the fact that n steps of size h must reach the boundary) and then the $T_{m,n}$ values from a linear system. Since $T_{0,0} = 0$ the first step brings $k_1 = h$, $T_{01} = h$, $T_{11} = 0$ from the boundary conditions alone. But then $k_2 = h$ and the equations

$$T_{0,2} - T_{1,2} = h \qquad T_{0,2} - 2T_{1,2} = hT_{1,2}$$

yield $T_{0,2} = h(2 + h)/(1 + h)$, $T_{1,2} = h/(1 + h)$. Choosing $h = .1$, for example, $T_{0,2} \simeq .191$ and $T_{1,2} \simeq .091$. Of course, $T_{2,2} = 0$. The third step finds $k_3 = .109$, after which the equations

$$T_{0,3} - 2.092T_{1,3} + T_{2,3} = -.0083 \qquad T_{1,3} - 2.092T_{2,3} = 0 \qquad T_{0,3} - T_{1,3} = .1$$

determine $T_{0,3} = .275$, $T_{1,3} = .175$, $T_{2,3} = .084$. Again, $T_{3,3} = 0$. The computation of $k_4 = .144$ now begins step four. Since h is kept fixed, the increasing k_n values suggest an upward curving boundary with $t_n = k_1 + \cdots + k_n$, $X(t_n) = nh$.

20.6   THE LAPLACE EQUATION

20.31   Replace the Laplace equation

$$\frac{\partial^2 T}{\partial x^2} + \frac{\partial^2 T}{\partial y^2} = 0 \qquad 0 \le x \le l, \, 0 \le y \le l$$

by a finite difference approximation. If the boundary values of $T(x, y)$ are assigned on all four sides of the square, show how a linear algebraic system is encountered.

▮   The natural approximations are

$$\frac{\partial^2 T}{\partial x^2} \simeq \frac{T(x - h, y) - 2T(x, y) + T(x + h, y)}{h^2}$$

$$\frac{\partial^2 T}{\partial y^2} \simeq \frac{T(x, y - h) - 2T(x, y) + T(x, y + h)}{h^2}$$

and they lead at once to the difference equation

$$T(x, y) = \tfrac{1}{4}[T(x - h, y) + T(x + h, y) + T(x, y - h) + T(x, y + h)]$$

which requires each T value to be the average of its four nearest neighbors. Here we focus our attention on a square lattice of points with horizontal and vertical separation h. Our difference equation can be abbreviated to

$$T_Z = \tfrac{1}{4}(T_A + T_B + T_C + T_D)$$

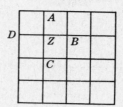

Fig. 20-5

with points labeled as in Fig. 20-5. Writing such an equation for each interior point Z (where T is unknown), we have a linear system in which each equation involves five unknowns, except when a known boundary value reduces this number.

20.32 Apply the method of the previous problem when $T(x,0) = 1$, the other boundary values being 0.

▮ For simplicity we choose h so that there are only nine interior points, as in Fig. 20-5. Numbering these points from left to right, top row first, our nine equations are

$$T_1 = \tfrac{1}{4}(0 + T_2 + T_4 + 0) \qquad T_6 = \tfrac{1}{4}(T_3 + 0 + T_9 + T_5)$$
$$T_2 = \tfrac{1}{4}(0 + T_3 + T_5 + T_1) \qquad T_7 = \tfrac{1}{4}(T_4 + T_8 + 1 + 0)$$
$$T_3 = \tfrac{1}{4}(0 + 0 + T_6 + T_2) \qquad T_8 = \tfrac{1}{4}(T_5 + T_9 + 1 + T_7)$$
$$T_4 = \tfrac{1}{4}(T_1 + T_5 + T_7 + 0) \qquad T_9 = \tfrac{1}{4}(T_6 + 0 + 1 + T_8)$$
$$T_5 = \tfrac{1}{4}(T_2 + T_6 + T_8 + T_4)$$

The system could be rearranged for Gaussian elimination, but as it stands the Gauss–Seidel iteration seems natural. Starting from the very poor initial approximation of zero for each interior T_i the successive results given in Table 20.2 are obtained. Ten iterations bring three place accuracy for this linear system.

TABLE 20.2

Iteration	T_1	T_2	T_3	T_4	T_5	T_6	T_7	T_8	T_9
0	0	0	0	0	0	0	0	0	0
1	0	0	0	0	0	0	.250	.312	.328
2	0	0	0	.062	.078	.082	.328	.394	.328
3	.016	.024	.027	.106	.152	.127	.375	.464	.398
4	.032	.053	.045	.140	.196	.160	.401	.499	.415
5	.048	.072	.058	.161	.223	.174	.415	.513	.422
6	.058	.085	.065	.174	.236	.181	.422	.520	.425
7	.065	.092	.068	.181	.244	.184	.425	.524	.427
8	.068	.095	.070	.184	.247	.186	.427	.525	.428
9	.070	.097	.071	.186	.249	.187	.428	.526	.428
10	.071	.098	.071	.187	.250	.187	.428	.526	.428

20.7 A CONVERGENCE PROOF

20.33 Prove that the linear system encountered in Problem 20.31 will always have a unique solution.

▮ The point is that, since we base our approximation on this system, it is important that it be nonsingular. Denoting the unknown interior values $T_1, \ldots, T_N$, we may rewrite the system in the form

$$(1) \qquad \sum_{k=1}^{N} a_{ik} T_k = b_i$$

where the b_i depend on the boundary values. If all boundary values were zero, then all b_i would be zero also:

$$(2) \qquad \sum_{k=1}^{N} a_{ik} T_k = 0$$

By the fundamental theorem of linear algebra the system (1) will have a unique solution provided that (2) has only the zero solution. Accordingly, we suppose all boundary values are zero. If the maximum T_k value occurred at an interior point Z, then because of $T_Z = \frac{1}{4}(T_A + T_B + T_C + T_D)$ it would also have to occur at A, B, C and D, the neighbors of Z. Similarly this maximum would occur at the neighboring points of A, B, C and D themselves. By continuing this argument we find that the maximum T_k value must also occur at a boundary point and so must be zero. An identical argument proves that the minimum T_k value must occur on the boundary and so must be zero. Thus all T_k in system (2) are zero and the fundamental theorem applies. Notice that our proof includes a bonus theorem. The maximum and minimum T_k values for both (1) and (2) occur at boundary points.

20.34 Prove that the solution of system (1) of Problem 20.33 converges to the corresponding solution of Laplace's equation as h tends to zero.

▌ Denote the solution of (1) by $T(x, y, h)$ and that of Laplace's equation by $T(x, y)$, boundary values of both being identical. We are to prove that at each point (x, y) as h tends to zero,

$$\lim T(x, y, h) = T(x, y)$$

For convenience we introduce the symbol

$$L[F] = F(x+h, y) + F(x-h, y) + F(x, y+h) + F(x, y-h) - 4F(x, y)$$

By applying Taylor's theorem on the right we easily discover that for $F = T(x, y)$, $|L[T(x, y)]| \le Mh^4/6$, where M is an upper bound of $|T_{xxxx}|$ and $|T_{yyyy}|$. Moreover, $L[T(x, y, h)] = 0$ by the definition of $T(x, y, h)$. Now suppose the origin of x, y coordinates to be at the lower left corner of our square. This can always be arranged by a coordinate shift, which does not alter the Laplace equation. Introduce the function

$$S(x, y, h) = T(x, y, h) - T(x, y) - \frac{\Delta}{2D^2}(D^2 - x^2 - y^2) - \frac{\Delta}{2}$$

where Δ is an arbitrary positive number and D is the diagonal length of the square. A direct computation now shows

$$L[S(x, y, h)] = \frac{2h^2\Delta}{D^2} + O\left(\frac{Mh^4}{6}\right)$$

so that for h sufficiently small, $L[S] > 0$. This implies that S cannot take its maximum value at an interior point of the square. Thus the maximum occurs on the boundary. But on the boundary $T(x, y, h) = T(x, y)$ and we see that S is surely negative. This makes S everywhere negative and we easily deduce that $T(x, y, h) - T(x, y) < \Delta$. A similar argument using the function

$$R(x, y, h) = T(x, y) - T(x, y, h) - \frac{\Delta}{2D^2}(D^2 - x^2 - y^2) - \frac{\Delta}{2}$$

proves that $T(x, y) - T(x, y, h) < \Delta$. The two results together imply $|T(x, y, h) - T(x, y)| < \Delta$ for arbitrarily small Δ, when h is sufficiently small. This is what convergence means.

20.35 Prove that the Gauss–Seidel method, as applied in Problem 20.32, converges to the exact solution $T(x, y, h)$ of system (1), Problem 20.33.

▌ This is, of course, an altogether separate matter from the convergence result just obtained. Here we are concerned with the actual computation of $T(x, y, h)$ and have selected a method of successive approximations. Suppose we number the interior points of our square lattice from 1 to N as follows. First we take the points in the top row from left to right, then those in the next row from left to right and so on. Assign arbitrary initial approximations T_i^0 at all interior points, $i = 1, \ldots, N$. Let the succeeding approximations be called T_i^n. We are to prove

$$\lim T_i^n = T_i = T(x, y, h)$$

as n tends to infinity. Let $S_i^n = T_i^n - T_i$. Now it is our aim to prove $\lim S_i^n = 0$. The proof is based on the fact that each S_i is the average of its four neighbors, which is true since both T_i^n and T_i have this property. (At boundary points we put $S = 0$.) Let M be the maximum $|S_i^0|$. Then, since the first point is adjacent to at least one boundary point,

$$|S_1'| \le \frac{1}{4}(M + M + M + 0) = \frac{3}{4}M$$

and since each succeeding point is adjacent to at least one earlier point.

$$|S'_{i+1}| \le \tfrac{1}{4}\left(M + M + M + |S'_i| \right)$$

Assuming for induction purposes that $|S'_i| \le [1 - (\tfrac{1}{4})^i]M$ we have at once

$$|S'_{i+1}| \le \tfrac{3}{4}M + \tfrac{1}{4}\left[1 - \left(\tfrac{1}{4}\right)^i\right]M = \left[1 - \left(\tfrac{1}{4}\right)^{i+1}\right]M$$

The induction is already complete and we have $|S'_N| \le [1 - (\tfrac{1}{4})^N]M = \alpha M$ which further implies

$$|S'_i| \le \alpha M \qquad i = 1, \ldots, N$$

Repetitions of this process then show that $|S^n_i| \le \alpha^n M$ and since $\alpha < 1$ we have $\lim S^n_i = 0$ as required. Though this proves convergence for arbitrary initial T^0_i, surely good approximations T^n_i will be obtained more rapidly if accurate starting values can be found.

20.36   Find an approximate solution to Laplace's equation in the first quadrant region under $y = 1 - x^2$, with $T(0, y) = 1 - y$, $T(x, 0) = 1 - x$ and boundary values on the parabola zero.

▌   The curved boundary must be handled in some way. Suppose we simply transfer the zero boundary values to the nearest points on our grid. For the case $h = \tfrac{1}{4}$ this leads to Fig. 20-6, with six interior points in the grid. Assigning them initial values of zero and taking them in alphabetic order, the averaging formula of Fig. 20-5 produces .062, .141, .035, .410, .236, .122 on the first cycle and converges to

$$.148 \qquad .341 \qquad .170 \qquad .545 \qquad .341 \qquad .148$$

a few cycles later. Reducing h to $\tfrac{1}{8}$ and again transferring boundary values from curve to grid in what seems a reasonable way, we have Fig. 20-7 and 26 more interior points. Focusing our attention on A, B, C, D, E, F we find,

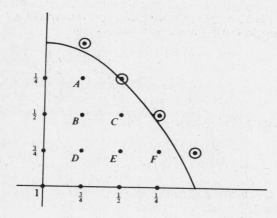

Fig. 20-6

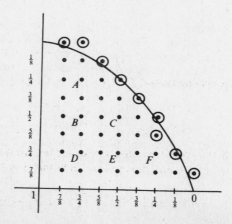

Fig. 20-7

after 40 cycles, convergence to

$$.140 \quad .327 \quad .131 \quad .534 \quad .310 \quad .088$$

which we assume is better than the earlier set. Reducing h to $\frac{1}{16}$ led to the values

$$.132 \quad .330 \quad .150 \quad .538 \quad .324 \quad .113$$

with further improvement clearly desirable. It may be a bit hard to believe that large-scale problems of this type used to be solved with desktop calculators, but only a generation or two ago such efforts were commonplace and crucial. A large team might need a month of coordinated computing to solve a boundary value problem.

20.37 Develop a finite difference formula for the three-dimensional Laplace equation $T_{xx} + T_{yy} + T_{zz} = 0$.

$\blacksquare$ A mental effort is sufficient,

$$T_Z = \tfrac{1}{6}(T_A + T_B + T_C + T_D + T_E + T_F)$$

with the subscript interpretations obvious. Extension to higher dimensions is no harder.

20.38 Develop the basic formulas for a finite element method using triangular elements and the Poisson equation

$$U_{xx} + U_{yy} = K \quad (K \text{ a constant})$$

$\blacksquare$ The region over which this equation is to hold must first be divided up into triangular pieces, making approximations where necessary. Let (x_i, y_i), (x_j, y_j), (x_k, y_k) be the vertices of one such triangle. The solution surface above this triangle is to be approximated by a plane element $\phi^{(e)}(x, y)$, the superscript referring to the element in question. If z_i, z_j, z_k are the distances up to this plane at the triangle corners, or nodes, then

$$\phi^{(e)} = L_i^{(e)} z_i + L_j^{(e)} z_j + L_k^{(e)} z_k$$

where $L_i^{(e)} = 1$ at node i and 0 at the other two nodes, with corresponding properties for $L_j^{(e)}$ and $L_k^{(e)}$. Let Δ_e be the area of the base triangle, formed by the three nodes. Then

$$2\Delta_e = \begin{vmatrix} 1 & x_i & y_i \\ 1 & x_j & y_j \\ 1 & x_k & y_k \end{vmatrix}$$

which leads quickly to the following representations:

$$L_i^{(e)} = \frac{1}{2\Delta_e} \begin{vmatrix} 1 & x & y \\ 1 & x_j & y_j \\ 1 & x_k & y_k \end{vmatrix} \qquad L_j^{(e)} = \frac{1}{2\Delta_e} \begin{vmatrix} 1 & x & y \\ 1 & x_k & y_k \\ 1 & x_i & y_i \end{vmatrix} \qquad L_k^{(e)} = \frac{1}{2\Delta_e} \begin{vmatrix} 1 & x & y \\ 1 & x_i & y_i \\ 1 & x_j & y_j \end{vmatrix}$$

If we also write

$$L_i^{(e)} = \frac{1}{2\Delta_e}(a_i + b_i x + c_i y)$$

then from the determinants

$$a_i = x_j y_k - x_k y_j \qquad b_i = y_j - y_k \qquad c_i = x_k - x_j$$

with these parallel formulas coming from $L_j^{(e)}$ and $L_k^{(e)}$:

$$a_j = x_k y_i - x_i y_k \qquad b_j = y_k - y_i \qquad c_j = x_i - x_k$$
$$a_k = x_i y_j - x_j y_i \qquad b_k = y_i - y_j \qquad c_k = x_j - x_i$$

All these a, b, c coefficients should have the superscript (e) but for simplicity it has been suppressed.

It is now time to consider the minimization problem equivalent to the Poisson equation. It is

$$J(U) = \iint \left[\tfrac{1}{2}\left(U_x^2 + U_y^2\right) + KU\right] dx\, dy = \text{minimum}$$

with the double integral to be evaluated over the given region R of the boundary value problem. We are approximating U by a function ϕ, a composite of plane triangular elements each defined over a triangular portion of R. So we consider the substitute problem of minimizing

$$J(\phi) = \sum J_e(\phi^{(e)})$$

with each term of the sum evaluated over its own base triangle. We want to set the appropriate derivatives of $J(\phi)$ to zero and to this end require the derivatives of the J_e components. Note that

$$\phi_x^{(e)} = \frac{1}{2\Delta_e}(b_i z_i + b_j z_j + b_k z_k)$$

$$\phi_y^{(e)} = \frac{1}{2\Delta_e}(c_i z_i + c_j z_j + c_k z_k)$$

so that, suppressing the superscript,

$$J_e = \iint \left[\tfrac{1}{2}(\phi_x^2 + \phi_y^2) + K\phi \right] dx\, dy = f(z_i, z_j, z_k)$$

The differentiations are straightforward. For example,

$$\frac{\partial f}{z_i} = \iint \left\{ \phi_x \frac{B_i}{2\Delta_e} + \phi_y \frac{c_i}{2\Delta_e} + KL_i \right\} dx\, dy = \frac{1}{\Delta_e}\left(\frac{b_i^2 + c_i^2}{4} z_i + \frac{b_i b_j + c_i c_j}{4} z_j + \frac{b_i b_k + c_i c_k}{4} z_k \right) + \frac{K}{3}\Delta_e$$

with very similar results for $\partial f/z_j$ and $\partial f/z_k$. The three can be grouped neatly in matrix form:

$$\begin{bmatrix} \partial f/z_i \\ \partial f/z_j \\ \partial f/z_k \end{bmatrix} = \frac{1}{4\Delta_e} \begin{bmatrix} b_i^2 + c_i^2 & b_i b_j + c_i c_j & b_i b_k + c_i c_k \\ b_i b_j + c_i c_j & b_j^2 + c_j^2 & b_j b_k + c_j c_k \\ b_i b_k + c_i c_k & b_j b_k + c_j c_k & b_k^2 + c_k^2 \end{bmatrix} \begin{bmatrix} z_i \\ z_j \\ z_k \end{bmatrix} + \frac{K}{3}\Delta_e \begin{bmatrix} 1 \\ 1 \\ 1 \end{bmatrix}$$

The fact that K has been assumed constant makes the integrations needed to achieve this result easy enough. Note also that the integral of each L function is $\frac{1}{3}\Delta_e$, by elementary calculus.

The preceding matrix equation contains the ingredients needed to assemble the partial derivatives of $J(\phi)$. It remains, in a particular application, to do the assembling properly. Specifically, for each element $\phi^{(e)}$ the active nodes i, j, k must be noted and contributions recorded for derivatives relative to the corresponding variables among the $z_1, z_2, z_3, \ldots$.

20.39 Apply the finite element method of the preceding problem given that the region R is the unit square of Fig. 20-8, with the boundary values indicated. The exact solution is easily seen to be $U(x, y) = x^2 + y^2$, since this satisfies $U_{xx} + U_{yy} = 4$.

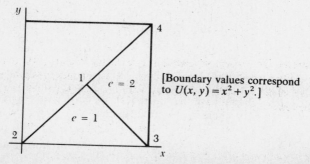

[Boundary values correspond to $U(x, y) = x^2 + y^2$.]

Fig. 20-8

▌ By symmetry only the lower right half of the square needs to be considered, and this has been split into two triangles. The nodes are numbered 1 to 4 and the two triangles are identified by the node numbers involved:

Node	x	y	Elements (by node numbers)
1	$\frac{1}{2}$	$\frac{1}{2}$	1 2 3 $(e=1)$
2	0	0	1 3 4 $(e=2)$
3	1	0	
4	1	1	$\Delta_1 = \Delta_2 = \frac{1}{4}$

From this basic input information we first compute the a, b, c coefficients. Each column below corresponds to a node (i, j, k):

	$e=1$			$e=2$		
a	0	$\frac{1}{2}$	0	1	0	$-\frac{1}{2}$
b	0	$-\frac{1}{2}$	$\frac{1}{2}$	-1	$\frac{1}{2}$	$\frac{1}{2}$
c	1	$-\frac{1}{2}$	$-\frac{1}{2}$	0	$-\frac{1}{2}$	$\frac{1}{2}$

It is useful to verify that columns do provide the desired $L_i^{(e)}$ functions. For instance, the first column gives

$$L_1^{(1)} = 2[0 + (0)\,x + (1)\,y]$$

where the leading 2 is the $1/2\Delta_e$. At node 1 this does produce the value 1, while at nodes 2 and 3 it manages 0. The other columns verify in similar fashion.

For clarity the process of assembling the partial derivatives of $J(\phi) = f(z_1, z_2, z_3, z_4)$ will now be presented in more detail than is probably needed. The matrix equation of the preceding problem contains the contributions to these derivatives from each of our two elements. From element 1 comes

	z_1	z_2	z_3	
$\partial f/z_1$	1	$-\frac{1}{2}$	$-\frac{1}{2}$	$\frac{1}{3}$
$\partial f/z_2$	$-\frac{1}{2}$	$\frac{1}{2}$	0	$\frac{1}{3}$
$\partial f/z_3$	$-\frac{1}{2}$	0	$\frac{1}{2}$	$\frac{1}{3}$

the last column containing constants. Element 2 provides these pieces.

	z_1	z_3	z_4	
$\partial f/z_1$	1	$-\frac{1}{2}$	$-\frac{1}{2}$	$\frac{1}{3}$
$\partial f/z_3$	$-\frac{1}{2}$	$\frac{1}{2}$	0	$\frac{1}{3}$
$\partial f/z_4$	$-\frac{1}{2}$	0	$\frac{1}{2}$	$\frac{1}{3}$

Assembling the two matrices we have this finished product:

	z_1	z_2	z_3	z_4	
$\partial f/z_1$	2	$-\frac{1}{2}$	-1	$-\frac{1}{2}$	$\frac{2}{3}$
$\partial f/z_2$	$-\frac{1}{2}$	$\frac{1}{2}$	0	0	$\frac{1}{3}$
$\partial f/z_3$	-1	0	1	0	$\frac{2}{3}$
$\partial f/z_4$	$-\frac{1}{2}$	0	0	$\frac{1}{2}$	$\frac{1}{3}$

Having thus illustrated the process of assembling elements, it must now be confessed that for the present case only the top row is really needed. The values of z_2, z_3, z_4 are boundary values and are given as $0, 1, 2$. They are not independent variables, and the function f depends only on z_1. Setting this one derivative to zero and inserting the boundary values, we have

$$2z_1 - \tfrac{1}{2}(0) - (1) - \tfrac{1}{2}(2) + \tfrac{2}{3} = 0$$

making $z_1 = \frac{2}{3}$. The correct value is, of course, $\frac{1}{2}$.

20.40 Rework the preceding problem using the finer network of triangles shown in Fig. 20.9.

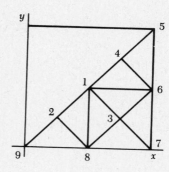

Fig. 20-9

▌ We have these input ingredients: first, the nodes 1 to 4 where the coordinates (x, y) are $(\frac{1}{2}, \frac{1}{2})$, $(\frac{1}{4}, \frac{1}{4})$, $(\frac{3}{4}, \frac{1}{4})$ and $(\frac{3}{4}, \frac{3}{4})$ with corresponding z coordinates to be determined; second, the nodes 5 to 9 at which boundary values are assigned making (x, y, z) coordinates $(1, 1, 2)$, $(1, \frac{1}{2}, \frac{5}{4})$, $(1, 0, 1)$, $(\frac{1}{2}, 0, \frac{1}{4})$ and $(0, 0, 0)$; and third, the eight basic triangles designated by node numbers:

$$2\ 9\ 8 \qquad 2\ 8\ 1 \qquad 1\ 8\ 3 \qquad 3\ 8\ 7 \qquad 3\ 7\ 6 \qquad 1\ 3\ 6 \qquad 1\ 6\ 4 \qquad 4\ 6\ 5$$

A computer program to run the finite element algorithm as described would need this input information.

Suppose we begin a manual run, carrying it through only one of the eight elements, the first. The a, b, c coefficients prove to be

a	0	$\frac{1}{8}$	0
b	0	$-\frac{1}{4}$	$\frac{1}{4}$
c	$\frac{1}{2}$	$-\frac{1}{4}$	$-\frac{1}{4}$

This may be checked as in the preceding problem, the columns representing the three nodes in the given order. The area of each basic triangle is $\frac{1}{16}$. Since partial derivatives will be needed only relative to z_1 to z_4, we can shorten our manual effort by finding only the terms contributing to these. For this element, we have

$$b_i^2 + c_i^2 = 0 + \frac{1}{4} = \frac{1}{4} \qquad b_i b_j + c_i c_j = 0 - \frac{1}{8} = -\frac{1}{8} \qquad b_i b_k + c_i c_k = 0 - \frac{1}{8} = -\frac{1}{8}$$

which, after multiplication by $1/4\Delta_e = 4$, we enter into columns 2, 8 and 9 of the partial derivative matrix. The constant $4\Delta_e/3 = \frac{1}{12}$ is also recorded, all entries in row 2 which pertains to $\partial f/z_2$:

	z_1	z_2	z_3	z_4	z_5	z_6	z_7	z_8	z_9
$\partial f/z_1$	$\frac{1}{2}$	$-\frac{1}{2}$						0	
$\partial f/z_2$		1						$-\frac{1}{2}$	$-\frac{1}{2}$	$\frac{1}{12}$
$\partial f/z_3$									
$\partial f/z_4$									

It remains to find the similar contributions of the other seven elements and to assemble them into the preceding matrix. It is useful to verify that the second element introduces the terms shown in row 1 and to find its further contributions to row 2. The rest of the assembly process will be left to the computer as will the substitution of boundary values and solution of the resulting fourth-order linear system. The following output was obtained:

Node	Computed	True
1	.500000	$\frac{1}{2}$
2	.166667	$\frac{1}{8}$
3	.666667	$\frac{5}{8}$
4	1.166667	$\frac{9}{8}$

The bull's-eye at node 1 is interesting.

20.41 Apply the same finite element method to the problem of a quarter circle, using just a single element as shown in Fig. 20-10. The Poisson equation is again to be used, as are the boundary values $x^2 + y^2 = 1$. The true solution is thus the same $x^2 + y^2$.

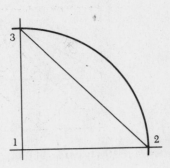

Fig. 20-10

❚ The problem illustrates the approximation of a curved boundary by a straight-line segment. In general, many such segments would be used. The three nodes have the coordinates

Node	x	y	z
1	0	0	—
2	1	0	1
3	0	1	1

The value of z_1 is the independent variable of the optimization. The a, b, c coefficients are

	Node 1	Node 2	Node 3
a	1	0	0
b	-1	1	0
c	-1	0	1

and lead to

$$\frac{\partial f}{z_1} = z_1 - \frac{1}{2}z_2 - \frac{1}{2}z_3 + \frac{2}{3} = 0$$

from which $z_1 = \frac{1}{3}$ follows at once. The true value is, of course, zero. By symmetry the same result would be found for the full circle by using four such triangles.

20.42 Illustrate the concept of convergence, as it applies to finite element methods, by comparing the crude approximation just found with results from two-triangle and four-triangle efforts based on the arrangements shown in Figs. 20.11 and 20.12.

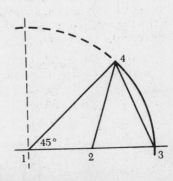

Fig. 20-11

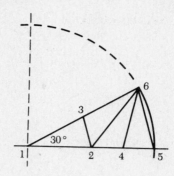

Fig. 20-12

■ Needless to say, all these efforts are relatively crude, but it is interesting to observe the results:

Node	$(0,0)$	$(\frac{1}{2},0)$
Fig. 20-10	.33	—
Fig. 20-11	$-.08$	.35
Fig. 20-12	$-.03$	.26
True	0	.25

Things have begun to move in a good direction. Finite element methods have been shown to be convergent provided the process of element refinement is carried out in a reasonable way.

20.43 Apply the finite element method to the Laplace equation (use $K = 0$ in Problem 20.38) on the triangle with vertices $(0,0)$, $(1,1)$ and $(-1,1)$ with boundary values given by $y^2 - x^2$. Note that this makes $y^2 - x^2$ the true solution.

■ From the symmetry it will be enough to work with the right half of the triangle. Choosing just two interior nodes at $(0,\frac{1}{3})$ and $(0,\frac{2}{3})$, we join these to $(1,1)$ to form three basic triangular elements as in Fig. 20-13. Nodes and elements are thus

Node		Element	Nodes
1	$(0,\frac{2}{3})$	1	$2,3,4$
2	$(0,\frac{1}{3})$	2	$1,2,4$
3	$(0,0)$	3	$1,4,5$
4	$(1,1)$	each $\Delta = \frac{1}{6}$	
5	$(0,1)$		

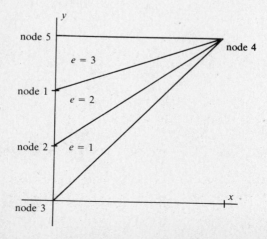

Fig. 20-13

and we are ready for the a, b, c coefficients. These prove to be

	Element 1			Element 2			Element 3		
a	0	$\frac{1}{3}$	0	$-\frac{1}{3}$	$\frac{2}{3}$	0	1	0	$-\frac{2}{3}$
b	-1	$\frac{2}{3}$	$\frac{1}{3}$	$-\frac{2}{3}$	$\frac{1}{3}$	$\frac{1}{3}$	0	$\frac{1}{3}$	$-\frac{1}{3}$
c	1	-1	0	1	-1	0	-1	0	1

and it is time to assemble the partial derivatives. Only those relative to z_1 and z_2 are needed:

	z_1	z_2	z_3	z_4	z_5
$\partial f/z_1$	$\frac{13}{6}+\frac{3}{2}$	$-\frac{11}{6}$		$-\frac{1}{3}$	$-\frac{3}{2}$
$\partial f/z_2$	$-\frac{11}{6}$	$3+\frac{5}{3}$	$-\frac{5}{2}$	$-\frac{1}{2}+\frac{1}{6}$	

There are no constant terms since $K = 0$. Equating these derivatives to zero leads to

$$\tfrac{11}{3}z_1 - \tfrac{11}{6}z_2 = \tfrac{3}{2} \qquad -\tfrac{11}{6}z_1 + \tfrac{14}{3}z_2 = 0$$

from which we find $z_1 = \frac{28}{55}$ and $z_2 = \frac{1}{5}$. The exact values are $\frac{4}{9}$ and $\frac{1}{9}$.

20.8 THE WAVE EQUATION

20.44 Apply finite difference methods to the equation

$$\frac{\partial^2 U}{\partial t^2} - \frac{\partial^2 U}{\partial x^2} = F[t, x, U, U_t, U_x] \qquad -\infty < x < \infty, 0 \le t$$

with initial conditions $U(x,0) = f(x)$, $U_t(x,0) = g(x)$.

▮ Introduce a rectangular lattice of points $x_m = mh$, $t_n = nk$. At $t = n = 0$ the U values are given by the initial conditions. Using

$$\frac{\partial U}{\partial t} \simeq \frac{U(x, t+k) - U(x, t)}{k}$$

at $t = 0$ we have $U(x, k) \simeq f(x) + kg(x)$. To proceed to higher t levels we need the differential equation, perhaps approximated by

$$\frac{U(x, t+k) - 2U(x, t) + U(x, t-k)}{k^2} - \frac{U(x+h, t) - 2U(x, t) + U(x-h, t)}{h^2}$$

$$= F\left[t, x, U, \frac{U(x, t) - U(x, t-k)}{k}, \frac{U(x+h, t) - U(x-h, t)}{2h}\right]$$

which may be solved for $U(x, t+k)$. Applied successively with $t = k, k+1, \ldots$, this generates U values to any t level and for all x_m.

20.45 Illustrate the preceding method in the simple case $F = 0$, $f(x) = x^2$, $g(x) = 1$.

▮ The basic difference equation may be written (see Fig. 20.14)

$$U_A = 2(1 - \lambda^2)U_C + \lambda^2(U_B + U_D) - U_E$$

where $\lambda = k/h$. For $\lambda = 1$ this is especially simple, and results of computation with $h = k = .2$ are given in Table 20.3. Note that the initial values for $x = 0$ to 1 determine the U values in a roughly triangular region. This is also true of the differential equation, the value $U(x, t)$ being determined by initial values between $(x - t, 0)$ and $(x + t, 0)$. (See Problem 20.46.)

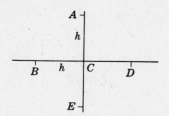

Fig. 20-14

TABLE 20.3

t/x	0	.2	.4	.6	.8	1.0
.6			1.00	1.20		
.4		.52	.64	.84	1.12	
.2	.20	.24	.36	.56	.84	1.20
0	.00	.04	.16	.36	.64	1.00

20.46 Show that the exact solution value $U(x, t)$ of $U_{tt} = U_{xx}$, $U(x, 0) = f(x)$, $U_t(x, 0) = g(x)$ depends on initial values between $(x - t, 0)$ and $(x + t, 0)$.

▌ For this old familiar problem, which is serving us here as a test case, the exact solution is easily verified to be

$$U(x, t) = \frac{f(x + t) + f(x - t)}{2} + \frac{1}{2} \int_{x-t}^{x+t} g(\xi) \, d\xi$$

and the required result follows at once. A similar result holds for more general problems.

20.47 Illustrate the idea of *convergence* for the present example.

▌ Keeping $\lambda = 1$, we reduce h and k in steps. To begin, a few results for $h = k = .1$ appear in Table 20.4. One circled entry is a second approximation to $U(.2, .2)$ so that .26 is presumably more accurate than .24. Using $h = k = .05$ would lead to the value .27 for this position. Since the exact solution of the differential problem may be verified to be

$$U(x, t) = x^2 + t^2 + t$$

we see that $U(.2, .2) = .28$ and that for diminishing h and k our computations seem to be headed toward this exact value. This illustrates, but by no means proves, convergence. Similarly, another looped entry is a second approximation to $U(.4, .4)$ and is better than our earlier .64 because the correct value is .72.

TABLE 20.4

t/x	0	.1	.2	.3	.4	.5	.6	.7
.4				.61	(.68)			
.3			.40	.45	·.52	.61		
.2		.23	(.26)	.31	.38	.47	.58	
.1	.10	.11	.14	.19	.26	.35	.46	.59
0	.00	.01	.04	.09	.16	.25	.36	.49

20.48 Why is a choice of $\lambda = k/h > 1$ not recommended, even though this proceeds more rapidly in the t direction?

▌ The exact value of $U(x, t)$ depends on initial values between $(x - t, 0)$ and $(x + t, 0)$. If $\lambda > 1$ the computed value at (x, t) will depend only on initial values in subset AB of this interval (see Fig. 20-15). Initial values outside AB could be altered, affecting the true solution, but not affecting our computed value at (x, t). This is unrealistic.

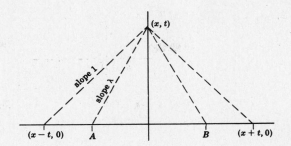

Fig. 20-15

20.49 The boundary value problem

$$U_{tt} + U_{xxxx} = 0 \qquad 0 < x, 0 < t \qquad U(x,0) = U_t(x,0) = U_{xx}(0,t) = 0 \qquad U(0,t) = 1$$

represents the vibration of a beam, initially at rest on the x axis, and given a displacement at $x = 0$. This problem can be solved using Laplace transforms, the result appearing as a Fresnel integral which must then be computed by numerical integration. Proceed, however, by one of our finite difference methods.

▮ A natural discretization of this fourth-order equation, using the simplified notation of Fig. 20-16, is

$$\frac{U_B - 2U_A + U_C}{k^2} = -\frac{U_D - 4U_E + 6U_A - 4U_F + U_G}{h^4}$$

The two initial conditions can be replaced by $U(x,0) = U(x,k) = 0$, which amounts to using a first difference in place of the first time derivative. The boundary condition $U(0,t) = 1$ is indicated and then $U(2h, t) - 2U(h, t) + U(0, t) = 0$ uses the second difference in place of the second derivative relative to x. This equation will be used in the form

$$(*) \qquad U(h,t) = \tfrac{1}{2}[U(0,t) + U(2h,t)]$$

and the basic recursion in the form

$$(**) \qquad U(x, t+k) = 2U(x,t) - U(x, t-k) - \lambda \, \Delta^4 U(x,t)$$

with $\lambda = k^2/h^4$. The problem as posed assumes an "impulsive" action at $x = 0$, which introduces a discontinuity at $(0,0)$, but this is taken in stride by the discretization.

Suppose $k = \tfrac{1}{4}$ and $h = 1$. The algorithm first computes the $-.06$ shown in the figure, using the other circled entries (all of which are available initially) and the difference equation $(**)$. The string of zeros in this third row (upward) in the table now follows in the same way. The boundary condition $(*)$ then fills in the $.47$ in column two, as the average of 1 and $-.06$. It is then the turn of the next row upward, which is handled in the same way. The table shows the beginning of the beam's oscillation up to 1 unit of time. The grid is too wide for accuracy but refinement can bring convergence. As earlier one must not make k too large relative to h, the combination $h = \tfrac{1}{2}$, $k = \tfrac{1}{8}$ for instance leading to instability.

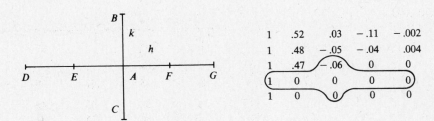

Fig. 20-16

CHAPTER 21
Monte Carlo Methods

2.1 RANDOM NUMBERS

21.1 What are *random numbers* and how may they be produced?

I For a simple but informative first example begin with the number 01. Multiply by 13 to obtain 13. Again multiply by 13, but discard the hundred, to obtain 69. Now continue in this way, multiplying continually by 13 modulo 100, to produce the following sequence of two-digit numbers:

$$01, \quad 13, \quad 69, \quad 97, \quad 61, \quad 93, \quad 09, \quad 17, \quad 21, \quad 73, \quad 49, \quad 37, \quad 81, \quad 53, \quad 89, \quad 57, \quad 41, \quad 33, \quad 29, \quad 77$$

After the 77 the sequence begins again at 01.

There is nothing random about the way these numbers have been generated, and they are typical of what are known as random numbers. If we plot them on a scale from 00 to 99 they show a rather uniform distribution, no obvious preference for any part of the scale. Taking them consecutively from 01 and back again, we find 10 increases and 10 decreases. Taking them in triples, we find double increases (such as 01, 13, 69) together with double decreases occurring about half the time, as probability theory suggests they should. The term *random numbers* is applied to sequences that pass a reasonable number of such probability tests of randomness. Our sequence is, of course, too short to stand up to tests of any sophistication. If we count triple increases (runs such as 01, 13, 69, 97) together with triple decreases, we find them more numerous than they should be. So we must not expect too much. As primitive as it is, the sequence is better than what we would get by using 5 as multiplier $(01, 05, 25, 25, 25, \ldots)$, which are in no sense random numbers). A small multiplier such as 3 leads to $01, 03, 09, 27, 81, \ldots$ and this long upward run is hardly a good omen. It appears that a well-chosen large multiplier may be best.

21.2 Use the random numbers of the preceding problem in a *simulation* of the movement of neutrons through the lead wall of an atomic reactor.

I For simplicity we assume that each neutron entering the wall travels a distance D before colliding with an atom of lead, that the neutron then rebounds in a random direction and travels distance D once again to its next collision and so on. Also suppose the thickness of the wall is $3D$, though this is far too flimsy for adequate shielding. Finally suppose that 10 collisions are all a neutron can stand. What proportion of entering neutrons will be able to escape through this lead wall? If our random numbers are interpreted as directions (Fig. 21-1), then they may serve to predict the random directions of rebound. Starting with 01, for example, the path shown by the broken line in Fig. 21-2 would be followed. This neutron gets through, after four collisions. A second neutron follows the solid path in

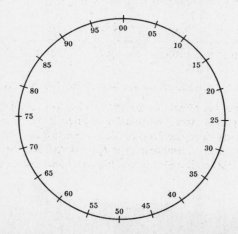

Fig. 21-2

Fig. 21-2, and after 10 collisions stops inside the wall. It is now plain that we do not have enough random numbers for a realistic effort, but see Problem 21.3.

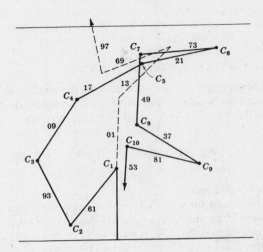

Fig. 21-2

21.3   How may a more extensive supply of random numbers be produced?

▮   There are quite a few methods now available, but most of the best use the modular multiplication idea of Problem 21.1. For example, the recursion

$$x_{n+1} = 7^9 x_n (\mod 10^s) \qquad x_0 = 1$$

generates a sequence of length $5 \cdot 10^{s-3}$ having quite satisfactory statistical behavior. It is suitable for decimal machines. The recursion

$$x_{n+1} = (8t - 3) x_n (\mod 2^s) \qquad x_0 = 1$$

generates a permutation of the sequence $1, 5, 9, \ldots, 2^s - 3$, again with adequate statistical behavior. It is suitable for binary machines. The number t is arbitrary but should be chosen large to avoid long upward runs. In both these method s represents the standard word length of the computer involved, perhaps $s = 10$ in a decimal machine and $s = 32$ in a binary machine.

21.4   Continue Problem 21.2 using a good supply of random numbers.

▮   Using the first sequence of Problem 21.3 on a 10 digit machine ($s = 10$), the results given below were obtained. These results are typical of Monte Carlo methods, convergence toward a precision answer being very slow. It appears that about 28% of the neutrons will get through, so that a much thicker wall is definitely in order.

Number of Trials	5000	10000	15000	20000
Percent Penetration	28.6	28.2	28.3	28.4

21.5   Suppose N points are selected at random on the unit circle. Where may we expect their center of gravity to fall?

▮   By symmetry the angular coordinate of the center of gravity should be uniformly distributed, that is, one angular position is as likely as another. The radial coordinate is more interesting and we approach it by a *sampling* technique. Each random number of the Problem 21.3 sequences may be preceded by a decimal (or binary) point and multiplied by 2π. The result is a random angle θ_i between 0 and 2π, which we use to specify one random point on the unit circle. Taking N such random points together, their center of gravity will be at

$$X = \frac{1}{N} \sum_{i=1}^{N} \cos \theta_i \qquad Y = \frac{1}{N} \sum_{i=1}^{N} \sin \theta_i$$

and the radial coordinate will be $r = \sqrt{X^2 + Y^2}$. Dividing the range $0 \le r \le 1$ into subintervals of length $\frac{1}{32}$, we next discover into which subinterval this particular r value falls. A new sample of N random points is then taken and the process repeated. In this way we obtain a discrete approximation to the distribution of the radial coordinate. Results of over 6000 samples for the cases $N = 2$, 3, and 4 are given in Table 21.1, in abbreviated form. For the case $N = 2$ the exact cumulative result is $(2/\pi)\arcsin(r/2)$ and is also given. We seem to have about three place accuracy. (The choice of $\frac{1}{32}$ was convenient for the SEAC, on which these computations were made in 1958, using input via wire recorder.)

TABLE 21.1

| | $N = 2$ | | $N = 3$ | $N = 4$ |
r	Cum	Exact	Cum	Cum
4/32	.082	.080	.028	.063
8/32	.160	.161	.112	.207
12/32	.246	.245	.335	.394
16/32	.333	.333	.514	.618
20/32	.430	.430	.645	.790
24/32	.540	.540	.775	.892
28/32	.678	.678	.890	.965
32/32	1	1	1	1

21.6   Solve the boundary value problem

$$T_{xx} + T_{yy} = 0 \qquad T(0, y) = T(1, y) = T(x, 1) = 0 \qquad T(x, 0) = 1$$

by a sampling method that uses random walks.

I This is an example of a problem, with no obvious statistical flavor, which can be converted to a form suitable for Monte Carlo methods. The familiar finite difference approximations lead to a discrete set of points (say the nine in Fig. 21-3), and at each of these points an equation such as

$$T_5 = \tfrac{1}{4}(T_2 + T_4 + T_6 + T_8)$$

makes each T value the average of its four neighbors. This same set of nine equations was encountered in Problem 17.38, each unknown standing for the probability that a lost dog will eventually emerge on the south side of our diagram, reinterpreted as a maze of corridors. Though a sampling approach is hardly the most economical here, it is interesting to see what it manages. Starting a fictitious dog at position 1, for example, we generate a random number. Depending on which of the four subintervals $(0, \tfrac{1}{4})$, $(\tfrac{1}{4}, \tfrac{1}{2})$, $(\tfrac{1}{2}, \tfrac{3}{4})$ or $(\tfrac{3}{4}, 1)$ this random number occupies, our dog moves north, east, south or west to the next intersection. We check to see if this brings him outside the maze. If it does not, another random number is generated and a second move follows. When the dog finally emerges somewhere, we record whether it was at the south side or not. Then we start a new fictitious dog at position 1 and repeat the action. The result of 10000 such computer samples was 695 successful appearances at a south exit. This makes the probability of success .0695 and should be compared with the result .071 found by the Gauss–Seidel iteration. The latter is more accurate, but the possibility of solving differential boundary value problems by sampling methods may be useful in more complicated circumstances.

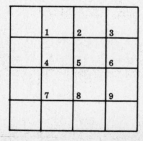

Fig. 21-3

21.2   APPROXIMATE INTEGRATION

21.7   Illustrate approximation integration by Monte Carlo methods.

▌ Perhaps the simplest procedure is the approximation of the integral by an average,

$$\int_a^b f(x)\,dx = \frac{b-a}{N} \sum_{i=1}^N f(x_i)$$

where the x_i are selected at random in (a, b). For example, if we use just the first five random numbers of Problem 21.1, all preceded by a decimal point, then we have

$$\int_0^1 x\,dx \simeq \tfrac{1}{5}(2.41) \simeq .48$$

where the correct result is $\frac{1}{2}$, and we also find $\int_0^1 x^2\,dx \simeq .36$ where the correct result is $\frac{1}{3}$. For the same integrals, with $N = 100$ and using the longer sequences of Problem 21.3, the results .523 and .316 are obtained, the errors being about 5%. This is not great accuracy, but in the case of integration in several dimensions the same accuracy holds and Monte Carlo methods compete well with other integration algorithms.

21.8   Approximate $\int_0^{\pi/2} \sin x\,dx$ using random numbers

▌ We add together the sine values for N random arguments and then divide by N. This estimates the average value of the function, which must be multiplied by the length of the interval to estimate the integral. Here are a few values obtained, the exact integral being 2.

N	10	50	100	200	500
Integral	1.637	1.921	2.118	2.032	1.996

21.9   Use the Monte Carlo method to estimate $\int_1^2 dx/x = \log 2$.

▌ The correct value to three places we know to be .693. Here are the numbers obtained:

N	50	100	200	400	800
First Run	.723	.707	.700	.692	.692
Second Run	.675	.683	.687	.689	.692
Third Run	.732	.708	.702	.701	.700

The third run might better have been omitted, but random processes must be accepted on their own statistical terms. As a rough rule, four times as many points should cut the error in half. A bit of searching will find supporting evidence in these numbers.

21.10   Estimate $\int_0^1 [1/(1 + x^2)]\,dx$ using random numbers.

▌ Evidence to support the above mentioned rule may be sought in the following results. The correct four place value is .7854.

N	50	100	200	400	800	1600	3200
Integral	.7701	.7649	.7853	.7872	.7806	.7802	.7841

21.11   Compute the double integral $\int_0^1 \int_0^1 xy\,dy\,dx$.

▌ The region of integration being a unit square, the average value of the function will equal the integral. Table 21.2 shows the results of two runs using 800 points each.

TABLE 21.2

N	50	100	200	400	800
First Run	.252	.265	.244	.250	.248
Second Run	.274	.243	.245	.250	.256

21.12 Compute the volume cutoff in the first octant by the plane that joins $(1, 0, 0)$, $(0, 1, 0)$ and $(0, 0, 1)$.

▎ The plane is $z = 1 - x - y$ and the volume pyramidal. If we set up as $\int_0^1 \int_0^{1-x} z \, dx \, dy$, then random pairs x, y may be drawn and rejected if $y > 1 - x$. The area of the triangular base in the x, y plane being $\frac{1}{2}$, this factor multiplies the average value of z and is included in these volume estimates. The correct volume is $\frac{1}{6}$.

N	50	100	200	400	800	1600	3200
Volume	.1701	.1716	.1692	.1647	.1680	.1722	.1665

21.13 Find the integral $\int_V x \, dV$ where V is the first octant part of the unit sphere with center $(0, 0, 0)$.

▎ We estimate the average x over this region. It should be multiplied by the volume $\pi/6$ to obtain the integral. Generating three random numbers x, y, z all in $(0, 1)$ we reject the point if $r^2 = x^2 + y^2 + z^2 > 1$. Results for N acceptable points were as follows.

N	50	100	200	400	800	1600
Ave.	.433	.414	.379	.376	.375	.386

21.14 Replace x in the preceding problem by $x^2 + y^2$.

▎ Average values developed as shown in Table 21.3.

TABLE 21.3

N	50	100	200	400	800
First Run	.338	.374	.388	.386	.400
Second Run	.387	.368	.392	.397	.400
Third Run	.374	.388	.400	.382	.396

21.15 Rework the preceding two problems using $r^2 = x^2 + y^2 + z^2$ and $1/r^2$ as integrands.

▎ Either integral is basically one dimensional in r, but here are the average values obtained by the usual three-dimensional effort with random numbers x, y, z. The correct averages are .6 and .3.

N	50	100	200	400	800	1600
r	.575	.604	.586	.582	.603	.600
1/r	2.118	2.661	2.582	2.633	2.651	3.003

21.16 Use random numbers to evaluate the six-dimensional integration

$$\int_0^1 \int_0^1 \int_0^1 \int_0^1 \int_0^1 \int_0^1 \frac{1}{1 + u + v + w + x + y + z} \, du \, dv \, dw \, dx \, dy \, dz$$

▎ With the 0, 1 limits one simply generates random numbers in groups of six, inserts them, totals and divides by N Table 21.4 shows the results of four separate runs.

TABLE 21.4

N	100	200	400	800	1600
First Run	.270	.264	.261	.261	.259
Second Run	.259	.257	.256	.258	.257
Third Run	.263	.258	.258	.259	.260
Fourth Run	.255	.255	.260	.260	.258

21.17 What is to be learned from these various efforts to approximate integrals by Monte Carlo methods?

▌ On the plus side, they produce rough estimates under almost any circumstances. On the minus, they improve those rough estimates at a snail's pace. The last example is typical, with the first hundred sample points suggesting an integral of .270, and later developments estimating about .258. We seem to have come from almost two place accuracy to perhaps three, but the cost has been multiplied by 64. However, no method for multiple integration is cheap. If only 10 points were used for each dimension, a double integral would use 100, a triple integral 1000 and the above six-dimensional integral 1 million.

21.3 SIMULATION

21.18 A baseball player with batting average .300 comes to bat four times in a game. What are his chances of getting 0, 1, 2, 3, 4 hits, respectively? The answers can be found by elementary probability but proceed by simulation.

▌ For each at-bat we generate a standard random number between 0 and 1. If it exceeds .3 the batter is out, otherwise he makes a hit. Four such at-bats make a game. Results over a few seasons of play (100 games a season) appear in Table 21.5. Notice how performance fluctuates even though batting ability is assumed to remain an unchanging .300. The correct values to the nearest whole number are 24, 41, 26, 8 and 1.

TABLE 21.5

Season	1	2	3	4	5	6
Hitless Games	26	22	29	22	17	29
One-Hit Games	44	40	37	45	45	39
Two-Hit Games	22	27	27	24	26	26
Three-Hit Games	7	8	5	8	10	5
Four-Hit Games	1	3	2	1	2	1

21.19 Generate three random numbers and arrange them in increasing order $x < y < z$. Repeat many times and find the average x, y and z.

▌ It is interesting to find the exact averages mentally. For x it is the integral

$$\int_0^1 \int_0^1 \int_0^1 \min(x, y, z) \, dx \, dy \, dz$$

and a little struggle with images of intersecting planes manages the answer $\frac{1}{4}$. For y the result $\frac{1}{2}$ is intuitive, while for z a second effort yields $\frac{3}{4}$.

The simulation amounts to an easy program and outputs the figures in Table 21.6. All's well that ends well seems to be the appropriate comment.

TABLE 21.6

Trials	50	100	200	400	800	1600
Average x	.192	.223	.236	.242	.242	.250
Average y	.476	.515	.514	.501	.506	.504
Average z	.765	.766	.755	.743	.748	.750

21.20 Golfers A and B have these records:

Score	80	81	82	83	84	85	86	87	88	89
A	5	5	60	20	10					
B				5	5	10	40	20	10	10

The numbers in the A and B rows indicate how many times each player has shot the given scores. The stronger player is clearly A. Assume they continue this quality of play and that A allows B to subtract four strokes each time they compete. By simulating matches predict how things will turn out.

▌ Once again probability theory is capable of answering, but a simulation is apt to be more easily understood by the average golfer. The programming is simple, and in a hundred matches we find 40 wins for A and 34 for B, the rest being all square. Repetitions brought the figures (38, 36), (37, 35), (35, 37) and (39, 37), so over 500 matches A appears to have a very slight edge.

Suppose A is not so generous, giving only three strokes per match. Simulation then suggests that the split will be more like (60, 20), and if A were persuaded to offer five strokes per match this imbalance would be approximately reversed. It seems that giving four is the gentlemanly or ladylike thing to do.

21.21 In the "first player back to zero" game A and B take turns moving the same marker back and fourth across this board:

10	9	8	7	6	5	4	3	2	1	0	1	2	3	4	5	6	7	8	9	10

The marker is started at 0. Player A moves to the right and B to the left, the size of the move being determined by the throw of one die. The first player to land on 0 is the winner. If the marker goes off either end of the board the game is restarted at 0 by the alternate player. What are the winning chances for each player?

❚ The answers are no doubt within the reach of probability theory, but simulation seems to offer an easier route. The program is an easy one, and produced the following first few games (minus meaning marker to the left of 0):

```
0  5  3  7   3  4   2   3   0  ///   0  3   2  6  3   7 1 2 0
0  6  1  4  -1  1  -5  -1  -2   4   -1  0  ///  0 1  -5  1 0
```

In 500 games the program found a winning split of (188, 312) with the advantage to B. (B has the first chance to return the marker to the 0 square.) There were 42 restarts, 28 because the marker went off the right end of the board, 13 because it went off the left end and 1 game that exceeded an artificially imposed limit of 100 moves (included to relieve the program of the worry of an endless game) and was terminated.

The thought arises, would it be more fair to start one square to the left of 0, keeping the finish at 0? This would give A the first chance at a win and it would also reduce the amount of time the marker spends on the right half of the board. A shorter simulation found the winning split (161, 139), which does suggest a yes answer to the above question.

21.22 Simulate random walks in the plane, each walk starting at $(0, 0)$ and consisting of 25 steps of length 1. Probability theory says that after N steps the expected distance from the starting point will be $\sqrt{N}$. Does the simulation agree?

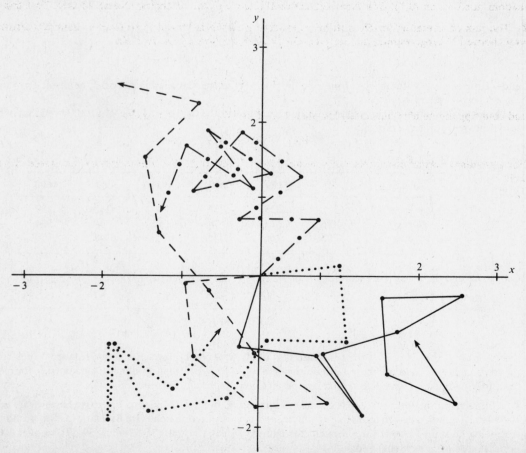

Fig. 21-4

▮ Once again the program is a simple one, but partly as a test and partly for amusement, four short walks of 10 steps each were output and plotted as Fig. 21-4. It is sometimes hard to believe that there is a form of order in such apparent chaos, but average distances from $(0,0)$ after 4, 9, 16 and 25 steps were computed for a sample of 100 walks and were

$$1.9 \qquad 2.8 \qquad 3.8 \qquad 4.7$$

All are a whisker too low.

21.4 NONUNIFORM RANDOM NUMBERS

21.23 Suppose that random numbers y with nonuniform distribution are needed, their density to be $f(y)$. Such numbers can be generated from a uniform distribution by equating the values of the two cumulative distributions at x and y:

$$\int_0^x 1 \cdot dx = \int_0^y f(y)\, dy$$

Treat the particular case in which $f(y) = e^{-y}$.

▮ Apart from constant factors this is the case of the Poisson distribution, often applied in problems of rare events. Integration finds the relationships.

$$x = 1 - e^{-y} \quad \text{or} \quad y = -\log(1 - x)$$

For example, $x = .5$ corresponds to $y = .693$, so half of the y distribution is between 0 and .693. Similarly, one finds that $x = .9$ corresponds to $y = 2.3$, so only one-tenth of the y values will exceed this, though their range extends to infinity. Large values of y are truly rare events.

21.24 For the standard normal distribution $f(y) = e^{-y^2}/\sqrt{2\pi}$ the procedure in the preceding problem can be troublesome. Many systems include a normal generator, but where this is not so a common alternative is to generate 12 uniform random numbers on $(0,1)$, sum then and subtract 6, the last step arranging a mean of zero. Test this process.

▮ The sum of several uniformly distributed random variables is known to be close to normally distributed, but why choose 12? Programming the computation of 1000 numbers by the formula

$$y = \left(\sum_{i=1}^{12} x_i \right) - 6$$

and counting the number absolutely less than 1, 2, 3 and 4, we find the figures

$$830 \qquad 978 \qquad 998 \qquad 1000$$

For a standard normal distribution they would 841, 977, 999 and 1000. The generator has passed this test fairly well.

Index